Tissue Engineering

Edited by

Daniel Eberli MD. PhD.

Tissue Engineering,
Edited by Daniel Eberli MD. PhD.

Republished by InTech
Janeza Trdine 9, 51000 Rijeka, Croatia

Additional hard copies can be obtained from orders@intechopen.com

Tissue Engineering,
Edited by Daniel Eberli MD. PhD.
p. cm.
ISBN 978-953-307-079-7

For EU product safety concerns:
IN TECH d.o.o., Prolaz Marije Krucifikse Kozulić 3, 51000 Rijeka, Croatia,
info@intechopen.com or visit our website at intechopen.com.

Meet the editor

Daniel Eberli MD. Ph.D. is a scientific physician working in the translational field of urologic tissue engineering. He has a medical degree from the Medical School in Zurich, Switzerland, and a Ph.D. in Molecular Medicine from Wake Forest University, Winston Salem, NC. He currently has a faculty position at the Department of Urology at the University Hospital Zurich, where he devotes half of his time to patient care. He is a lecturer at the Medical School of Zurich and the Swiss Federal Institute of Technology. Together with his research team, he is working on novel biomaterials for bladder reconstruction, improving autonomic innervation, cellular treatment of incontinence and tracking of stem cells.

Preface

The term Tissue Engineering was initially defined at the National Science Foundation (NSF) meeting in 1988 as the "application of the principles and methods of engineering and life sciences toward fundamental understanding of structure-function relationships in normal and pathological mammalian tissues and the development of biological substitutes for the repair or regeneration and to restore, maintain or improve tissue or organ function" (Shalak and Fox, 1988). Researchers at that time recognized the potential of Tissue Engineering to facilitate the first medical therapy where engineered tissues become fully integrated within the patient, thus offering a permanent cure for many diseases.

Over the following years Tissue Engineering progressed rapidly and first biological substitutes were developed for several tissues in the body. Today, Tissue Engineering is one of the major approaches of Regenerative Medicine and a growing and exciting field of research. In combination with better understanding of structure, biology, physiology and cell culture techniques Tissue Engineering may offer new treatment options for patients in need for replacement or repair of a deteriorated organ. The concept of Tissue Engineering has been applied clinically for a variety of disorders, for example artificial skin for burn patients, tissue engineered trachea, cartilage for knee-replacement procedures, injectable chondrocytes for the treatment of vesico-ureteric reflux and urinary incontinence to name a few.

The classical approaches of Tissue Engineering have not changed over the last three decades. The principle is to dissociate cells from a tissue biopsy, to expand these cells in culture, and to seed them onto the scaffold material in vitro in order to form a living tissue construct prior to implantation into the recipient's organism. In the appropriate biochemical and biomechanical environment these tissues will unfold their full functional potential and serve as native tissue equivalents. Tissue Engineering products may be fully functional at the time of treatment, or else have potential to integrate and evolve into the expected functional tissue after implantation.

The Tissue Engineering approach has major advantages over traditional organ transplantation and circumvents the problem of organ shortage. Tissues that closely match the patient's needs can be reconstructed from readily available biopsies and subsequently be implanted with minimal or no immunogenicity. This eventually conquers several limitations encountered in tissue transplantation approaches.

This book serves as a good starting point for anyone interested in the application of Tissue Engineering. It offers a colorful mix of topics, which explain the obstacles and possible solutions for TE applications. The first part covers the use of stem cells and adult stem cells and their applications. The following chapters offer an insight into the development of a tailored biomaterial for organ replacement and highlight the importance of cell-biomaterial interaction. As more and more applications move toward clinical application, a reliable preclinical model system to evaluate the developed techniques becomes crucial. Several animal models and Tissue Engineering approaches for a variety of organ systems are presented in the final chapters. In summary, this book offers insights into a wide variety of cells, biomaterials, interfaces and applications of the next generation biotechnology, which is Tissue Engineering.

Finally, I would like to thank all authors who have contributed to this book and hope that the readers will enjoy its reading.

Daniel Eberli MD. PhD.

University Zürich, Switzerland

Contents

Development of human fetal mesenchymal stem cell mediated tissue engineering bone grafts

Zhiyong Zhang[1,2], Swee-Hin Teoh[1,2,3], Mahesh Choolani[4], Jerry Chan[4]
1 Mechanical Engineering, Faculty of Engineering, National University of Singapore, Singapore
2 Centre for Biomedical Materials Applications and Technology (BIOMAT), National University of Singapore, Singapore
3 National University of Singapore Tissue Engineering Programme (NUSTEP), National University of Singapore, Singapore
4 Experimental Fetal Medicine Group, Department of Obstetrics & Gynaecology, Yong Loo Lin School of Medicine, National University of Singapore, Singapore
5 Department of Reproductive Medicine, KK Women's and Children's Hospital, Singapore
6 Cancer and Stem Cell Biology Program, Duke-NUS Graduate Medical School, Singapore

1. Fracture treatment and bone tissue engineering

Although bone is a dynamic and well-vascularized tissue with innate healing and remodeling capacities (Salgado et al., 2004) , up to 10% of the bony fractures are complicated by non-union (Einhorn, 1995; Hayda et al., 1998; Marsh, 1998; Salgado et al., 2004; Bongso, 2005) , which require additional treatment with bone grafts in order to achieve defect union and healing. In fact, bone grafts has become the second most transplanted tissue in the world after blood, with approximately one million cases of bone graft transplantation occurring in United States alone annually (Salgado et al., 2004; Bongso, 2005).

Bone grafts can be categorized into three types: autografts, allografts, and synthetic grafts, with their relative frequency of use illustrated in Figure 1 and a comparison of their advantages and disadvantages in Table 1. *Autografts* are harvested from a secondary site from the patient's own body, commonly the iliac crest. This strategy has been described initially by Chutro and later Phemister in early 20th century (Connolly et al., 1991) , and has been considered the gold standard for many decades. This is because the use of autografts results in the the best clinical outcomes by providing both osteogenic cells and essential osteoinductive factors required for bone healing. However, the use of autografts has been limited by its availability, difficulty in fashioning of grafts to fit defects, an unpredictable efficacy and complexity of two surgeries (Banwart et al., 1995; Fowler et al., 1995; Goulet et al., 1997; Salgado et al., 2004; Hollinger et al., 2005). Futhermore, significant donor site morbidity such as chronic pain, hypersensitivity, infection and paraesthesia occur in up to a third of patients (Prolo and Rodrigo, 1985; Damien and Parsons, 1991; Arrington et al., 1996; Lane et al., 1999). ***Allografts***, bone grafts harvested from another donor (mainly from cadavers), are an alternative with enhanced flexibility of graft size and shapes. However, they introduce the possibility of immune rejection and pathogen transmission. Moreover, processing techniques, such as demineralization, strip the tissue of osteoinductive factors necessary for stimulating bone repair, resulting in impeded healing times, as compared to autografts (Parikh, 2002) . ***Synthetic grafts*** are made from metals or ceramics, which can be fashioned to different shapes and sizes and are non-immunogenic, but thus far have been hampered by their poor speed of healing and inability to remodel in tandem with the natural healing process (Salgado et al., 2004).

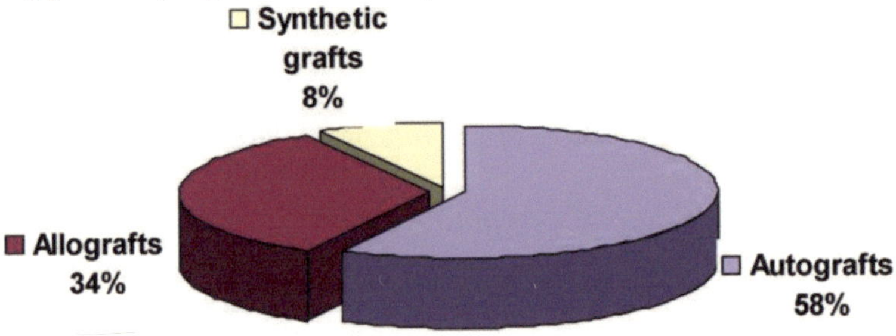

Fig. 1. Breakdown of the current used bone grafts: autograft, allograft and synthetic graft.

	Healing	Remodelling	Immuno-genicity	Availability	Surgery simplicity	Others
Autografts	+++	+++	-	+	+	donor-site morbidities
Allografts	++	++	+	+++	+++	disease transmission
Synthetic grafts	+	0	-	+++	+++	fatigue, fracture, wear
Tissue engineered grafts	+++	+++	-	+++	+++	

Table 1. Comparison of different bone grafts

Thus, in order to fulfill this increasing but unmet clinical need for effective bone grafts, recent research efforts have turned to a tissue engineering approach to develop tissue engineered grafts, which not only have off-the-shelf availability in various shapes and sizes (like allografts and synthetic grafts), but have the ability to stimulate rapid bone healing and undergo remodeling, achieving a better or similar clinical outcome than or as autografts (Table 1).

2. Cell based strategy vs. Growth factor based strategy

Tissue Engineering was first coined by Langer and Vacanti in 1993 as "an interdisciplinary field of research that applies the principles of engineering and the life sciences towards the development of biological substitutes that restore, maintain, or improve tissue function"(Langer and Vacanti, 1993). Since then, several different bone tissue engineering (BTE) approaches have been proposed and explored. In essence, they can be classified into two main categories of a cell or a growth factor based strategy. Both approaches seek to create an optimized bone growing bed with a biological microenvironment suitable for bone regeneration through introducing a critical mass of osteogenic cells onto the scaffold matrix and modulating intercellular communication and cell-matrix interactions. However, the manner in which the introduction of osteogenic cells is different. Cell-based approaches supply exogenous osteogenic cells while relying on the autologous growth factors from the repair site to stimulate bone regeneration, whereas growth factor-based approaches rely on the delivery of growth factors to recruit osteogenic cells from the local environment into the repair site. (Kimelman et al., 2007; Lanza et al., 2007).

The use of growth factor-based BTE approaches has been shown to be efficacious in small animal models. However, the translation from small animal models, through large animal models, to humans has proven to be difficult and the optimal dose of growth factors for human clinical therapy is difficult to determine, because of the large variation of efficacy between small to big animals (up to 100 fold). Another problem relates to the problem of develping an effective delivery and release approach, given the short half-life of the majority of the growth factors are (Hollinger et al., 1996; Bruder and Fox, 1999). In addition, growth factor-based approaches work indirectly by attracting and stimulating the proliferation and differentiation of local osteogenic stem cells to repair bone defect, therefore its effectiveness will be greatly compromised in the absence of sufficient osteogenic stem cell pool (Service, 2000). In contrast, a cell-based approach works independently of the presence of local osteogenic cells, and therefore is an attractive therapy for patients with a diminished pool of osteogenic progenitors such as those with severe trauma, diabetes, a history of tobacco use, irradiation, aging, osteoporosis or other metabolic derangements (Bruder and Fox, 1999). Furthermore, delivering the right quantity of different growth factors in a sequential manner reflecting the different phases of bone growth faces multiple technical challenges (Marsh and Li, 1999; Giannoudis et al., 2007). In contrast, cell-based strategies utilize osteogenic cells which secrete a wide spectrum of growth factors at physiological doses, establishing temporal-spatial microgradients necessary for effective bone regeneration (Rouwkema et al., 2008). Thus, cell based strategies have been pursued in our group.

Two distinct typical cell-based approaches have been proposed to use cultured osteogenic cells in combination with scaffold matrix to engineer bone grafts (Hollinger et al., 2005; Lieberman and Friedlaender, 2005; Lanza et al., 2007). The first approach is to seed cells on the scaffold matrix and the immediate implantation of the resulting construct (Figure 2 A). In the second approach, a period of in vitro culture is performed for cellular expansion and osteogenic pre-differentiation (Figure 2 B), which has been shown to result in significantly higher degree of bone defect healing compared to an immediate implantation approach (Mendes et al., 2002; Sikavitsas et al., 2003; De et al., 2006; Dudas et al., 2006) . In this delay implantation approach, osteogenic cells are seeded into three dimensional highly porous biodegradable scaffolds with the desired shape to fit the bone defect. These cellular scaffolds are then cultured and matured in a suitable *in vitro* environment for cellular expansion and proper pre-differentiation to obtain bone-like constructs with extracellular mineralized matrix and pre-differentiated cells. Finally, the grafts are implanted into the defect to induce and direct the growth of new bone with the controlled degradation of the scaffold matrix (Figure 2B).

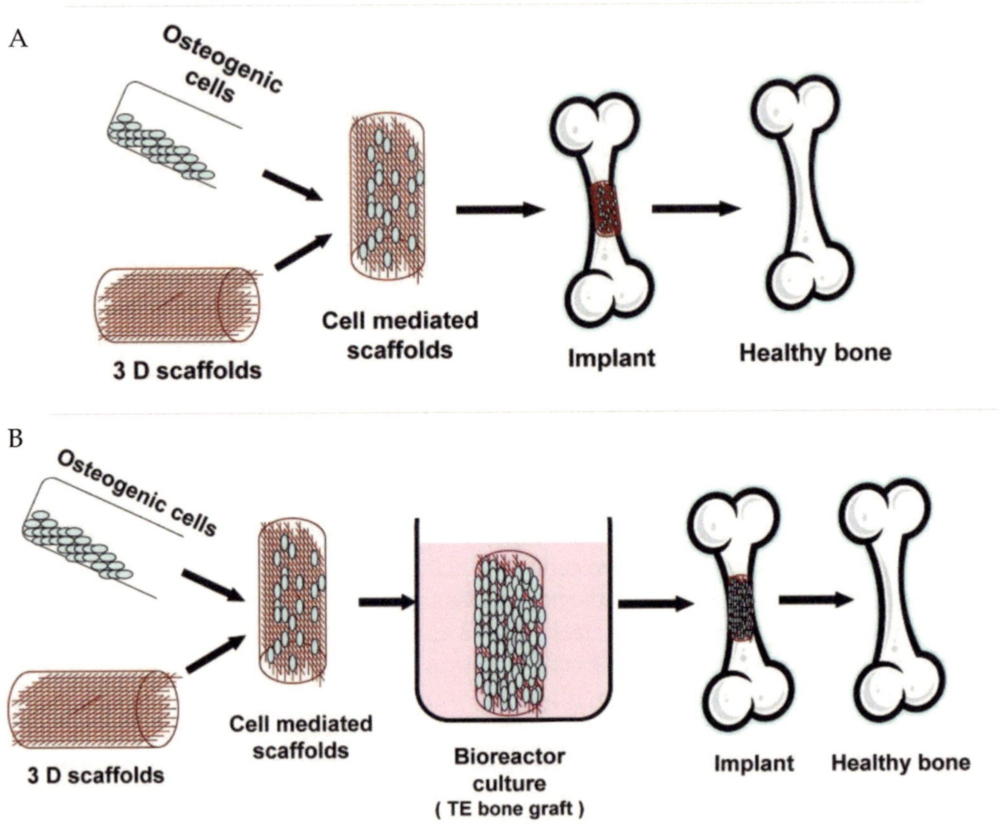

Fig. 2. Two cell based BTE approaches.

3. Polycaprolactone based scaffolds for BTE application

To generate an effective TE bone grafts, three dimensional (3D) scaffolds are required to function as a supportive matrix for cell proliferation, extracellular matrix deposition and consequent bone in-growth and at the same time define the overall shape of the tissue engineered transplant (Laurencin et al., 1999). Moreover, the 3D scaffolds should have rigid mechanical properties to protect the defect area from collapse of surrounding tissue, and proper macroporous structure to prevent the invasion of fibrous tissue but allow the infiltration of blood vessels. Scaffolds should also actively contribute to the regenerative process by enhancing the proliferation and osteogenic differentiation of the cells grew on their surface through the cell-scaffold interaction (Agrawal and Ray, 2001).

Our group has extensively investigated the use of Fused Deposition Modeling (FDM) technique to fabricate the porous three dimensional (3D) scaffolds for BTE applicaiton (Hutmacher et al., 2001; Zein et al., 2002). A polycaprolactone and tricalcium phosphate composite material was selected for the scaffold fabrication. Poly (ε-caprolactone) (PCL) is a semicrystalline, bioresorbable polymer belonging to the aliphatic polyester family. It is highly processable, with low glass-transition temperature (-60°C) and melting point (60°C) but high decomposition temperature of 350°C. It has been previously exploited for drug delivery material (Coombes et al., 2004) and demonstrated great biocompatibility with US

Food and Drug Administration (FDA) approval for its use in several medical and drug delivery devices. It has slow degradation kinetics of only seven percent over a six month in vivo (Lam et al., 2008) , and thus is suitable for utilisation in load bearing bone grafting applications, which requires a prolonged period of mechanical support. In addition, chemical hydrolysis as opposed to enzymatic reactions is responsible for degradation of polymeric chains *in vivo*; thus degradation does not vary from patient to patient.

However, PCL is not without drawbacks, as with all synthetic polymers, PCL implants experience fibrosis and gradually become isolated from the surrounding bone. Consequently, they do not adhere to bone and this has been a critical problem in their use in bone repair (Kokubu et al., 2003). In addition, a common problem to biodegradable polyesters family is that the bulk release of acidic degradation products can cause acidosis (Ciapetti et al., 2003). This led to the search for additional compounds to modify PCL for BTE applications.

β-tricalcium phosphate (TCP), a synthetic ceramic material, is one such compound. Firstly, TCP is bioactive and can generate a carbonated hydroxyapatite layer on its surface that is equivalent chemically and structurally to the biological mineral of nature bone, known as the determining step for the biointegration (Kokubo et al., 2003). Secondly, TCP is osteoconductive and can promote the cellular adhesion, function and expression leading to formation of new bone and a uniquely strong bone-TCP biomaterial interface. Finally, TCP is reported to be osteoinductive, and capable of binding and concentrating endogenous bone morphogenetic proteins from the circulation (LeGeros, 2002). However, pure TCP is a brittle material with very poor mechanical properties, leading to difficulties with its processing for BTE applications. To overcome these shortcomings, a bioactive composite material comprising of a biodegradable polymeric phase (PCL) and a bioactive inorganic phase (TCP) has been developed. This construct takes advantage of the flexibility and processability of the bulk PCL polymer phase, combined with TCP bioactive phase, allowing the scaffold to bond and integrate with bone spontaneously and possess both osteoconductive and osteoinductive properties. Furthermore, the acidic release from the polymer can be compensated by the alkaline calcium phosphate (Blaker et al., 2003; Maquet et al., 2004) . (Figure 3)

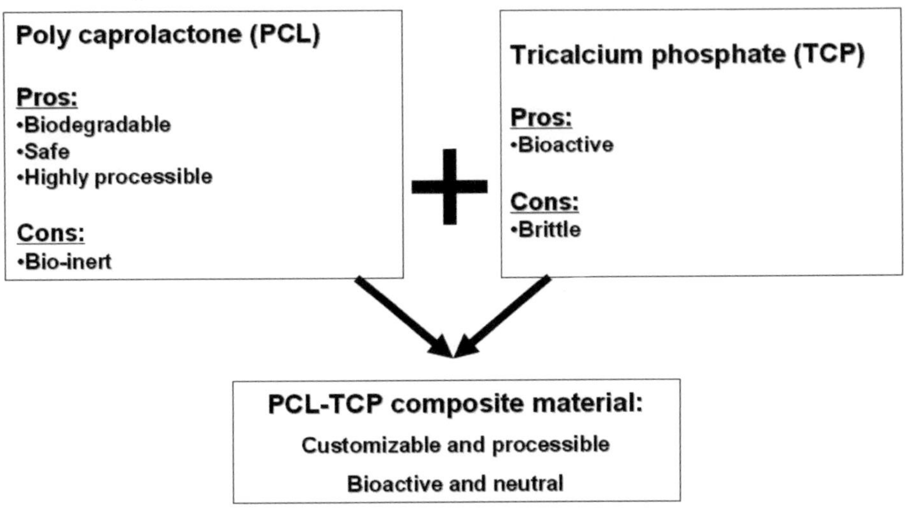

Fig. 3. PCL-TCP composite material, incorporates the strength of PCL and TCP while neutralizes their individual drawbacks

FDM, as a solvent-free rapid prototyping technique, was utilized to fabricate the PCL-TCP composite material into 3D scaffolds (Hutmacher et al., 2001; Zein et al., 2002) in three steps. Firstly, the PCL-TCP filament were fabricated by physically blending the PCL pellets with TCP granules in a desired ratio (e.g. 80% PCL to 20% TCP), melting in a high temperature (190°C) and extruding through spinnerets with a die exit diameter of 1.63 mm. Secondly, the size and shape of the scaffolds were designed through computer assisted design (CAD) systems, for example, patient-specific scaffolds can be created based on the 3D computed tomography(CT) images, then these CAD models were imported into the Stratasys QuickSlice software and sliced into the horizontal layers. The deposition path were created and downloaded to the FDM machine, the 3D Modeler RP system from Stratasys Inc. (Eden Prairie, MN). Lastly, PCL-TCP filament prepared were fed into the FDM head, melted down in the liquefier and extrude out through the nozzle tip and deposit on a platform layer by layer according to the preprogrammed deposition path.

We have fabricated PCL-TCP scaffolds to achieve a honeycomb architecture, with a fully interconnected matrix, and mechanical properties close to cancellous bone. This should allow rapid vascularization to occur and maintain the defect space and the structural integrity of tissue engineered bone grafts in load-bearing applications (Hutmacher et al., 2001; Schantz et al., 2002) (Figure 4). In addition, use of the highly reproducible and computer-controlled FDM technique allows the fabrication of tissue-engineered grafts that have been designed on the basis of computed tomography (CT) scans of individual defect sites(Hutmacher, 2000; Endres et al., 2003) . PCL-TCP scaffolds have a favourable profile for achieving effective cellular adhesion, proliferation and differentiation in vitro (Rai et al., 2004; Zhang et al., 2009; Zhang et al., 2009) , and serve as an osteoconductive matrix for new bone regeneration in both critical-sized rat femoral (Rai et al., 2007) and canine mandibular defects (Rai et al., 2007).

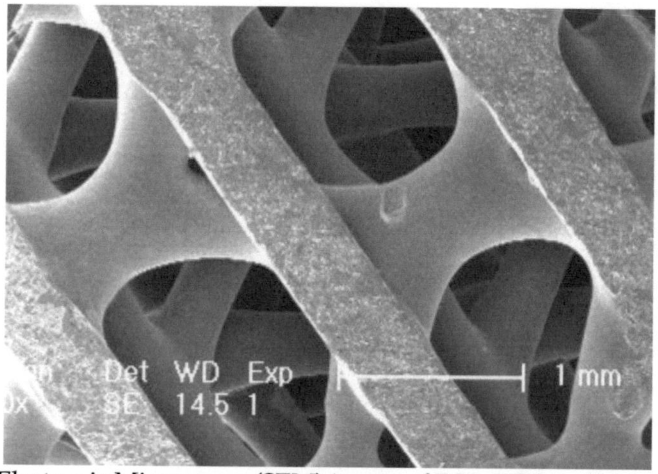

Fig. 4. Scanning Electronic Microscopy (SEM) image of PCL-TCP scaffolds

4. Human fetal mesenchymal stem cells for BTE application

4.1 Available cellular sources for cell based BTE

Cells are the central players for most biological process in human body. As the critical component in the BTE strategy, they play the most essential role and are directly involved in bone regeneration. The selection of cell sources will eventually determine the success of any BTE strategy. An ideal cellular source for cell-based BTE approach should include the following characteristics: (1) no immunogenicity, (2) no tumorigenicity, (3) immediate availability, (4) availability in pertinent quantities, (5) rapid cell proliferation rate, (6) predictable and consistent osteogenic potential as well as; (7) controlled integration into the surrounding tissues (Logeart-Avramoglou et al., 2005) . Different cell sources have been exploited for BTE applications, and can be divided into three categories with the respect to their differentiation status: fresh bone marrow, differentiated osteoblasts and mesenchymal stem cells (Bruder and Fox, 1999; Pioletti et al., 2006) (Table 2).

Fresh bone marrow (BM) has been introduced to heal nonunion bone defects some three decades ago (Connolly et al., 1989; Connolly et al., 1991; Caplan, 2005). BM are readily harvested and contain the osteogenic progenitors, which can enhance bone regeneration. As they are autologous tissues, immunorejection is not an issue, nor are they subject to the regulation by the Food and Drug Administration (FDA) (Bruder and Fox, 1999) . However, their effectiveness is highly dependent on the number of the osteogenic progenitors, which represent approximately 0.001% of the nucleated cell in the healthy bone marrow at best (Bruder et al., 1997) and decreases dramatically with aging or disease process (Pioletti et al., 2006). Thus, this approach may be least applicable in those situations where it is most needed, as fractures tend to occur more frequently in an old age group. Moreover, the use of unmatched allogenic BM is not suitable, as it is beset by problems of either graft rejection or graft-versus-host syndrome (Weissman, 2000).

Osteoblasts participate in bone formation processes via the synthesis and secretion of collagen fibres to form unmineralized bone matrix (osteoid), and have been implicated in osteoid calcification. The use of osteoblasts for the cellular source for BTE applications has demonstrated enhancement in the rate and extent of bone regeneration compared with the use of undifferentiated bone marrow (Okumura et al., 1997; Bruder and Fox, 1999; Montjovent et al., 2004). However, autologous isolation of the osteoblasts is considerably more difficult and complex than fresh bone marrow or mesenchymal stem cells. In addition, they posses a limited capacity for proliferation, and thus do not support in the generation of significant cell numbers for clinical use (Bruder and Fox, 1999; Montjovent et al., 2004; Pioletti et al., 2006).

Mesenchymal stem cells (MSC) are also known as marrow stromal cell or colony forming unit - fibroblast (CFU-F) or more recently named as multipotent mesenchymal stromal cells by International Society for Cellular Therapy (ISCT) (Dominici et al., 2006) . They were first identified and isolated from adult bone marrow (Friedenstein et al., 1966; Friedenstein et al., 1968) and demonstrated a series of favorable advantages for BTE application over other cell sources. Firstly, MSC are easily isolated through plastic adhesion method or simple antibody selection techniques (Simmons and Torok-Storb, 1991) . Secondly, the osteogenic differentiation pathway is well defined and MSC have been shown to generate greater amount of bone tissue than fresh bone marrow in preclinical studies (Kahn et al., 1995; Inoue et al., 1997; Pioletti et al., 2006). Thirdly, both undifferentiated and differentiated MSC have been reported as non-immunogenic, and are thus suitable for allogeneic applications (Le Blanc et al., 2003) . Finaly, cryostorage does not affect the osteogenic potential of MSC, which greatly facilitates their storage (Bruder et al., 1997).

	Collection	**Expansion**	**Storage**	**Allogenic applications**	**Osteogenic potential**
Fresh bone marrow	+++	N.A.	N.A.	N.A.*	+
Osteoblast	+	+	+	N.A.*	+++
MSC	++	++	+++	++	++

N.A.*: allogenic applications can be achieved by proper HLA matching

Table 2. Comparison of different cellular source for BTE.

4.2 MSC biology

Although there are no specific cell markers to distinguish MSC from other cell types, MSC are generally defined by their capabilities to (1) adhere to the plastic surface of tissue culture flask; (2) express D105, CD73 and CD90, with absence of CD45, CD34, CD14 or CD11b, CD79a or CD19 and HLA II; and (3) differentiate into the osteoblasts, adipocytes and chondroblasts under standard *in vitro* differentiating conditions (Dominici et al., 2006). MSC are rare cells, and exist at low frequencies in human adult BM, accounting for 0.00001% to 0.001% of mononuclear cells (MNC) in BM. They are non-haemopoietic and can be separated from the haemopoietic fraction in BM by their adherence to plastic and ability to grow from an initial heterogeneous population into a more homogenous spindle-shaped cell type. MSC have been demonstrated to contain a heterogeneous population of cells. Properties including colony size, proliferation rate, cellular morphology, and multipotency both *in vitro* and *in vivo* have been routinely observed to differ in single colony-derived MSC strains (Muraglia et al., 2000; Mauney et al., 2005) . Other techniques for isolation of MSC include immuno-depletion (Baddoo et al., 2003) or immuno-selection techniques with STRO-1 and CD 271 antibodies (Simmons and Torok-Storb, 1991; Jones and McGonagle, 2008). However, the resulting cultures are still heterogeneous, ranging from spindle shaped cells to broad stromal-like cells (Digirolamo et al., 1999; Colter et al., 2000; Colter et al., 2001)

This is also reflected in intra-sample and inter-species differences in the capacity of MSC to self-renew or form CFU-F in low-density cultures (Digirolamo et al., 1999; Phinney et al., 1999; Banfi et al., 2000; Javazon et al., 2001).

MSC are not inherently immunogenic, making them highly attractive for allogeneic transplantation strategies (Caplan, 2007). MSC are HLA Class I positive and Class II negative and do not express co-stimulatory molecules CD40, CD40L, CD80, or CD86 (Klyushnenkova et al., 2005) . This phenotype is widely regarded as non-immunogenic and suggests that MSC may be effective at inducing tolerance. In the mixed lymphocyte cultures (MLC) assay, MSC do not elicit a proliferation response of allogeneic lymphocytes (Bartholomew et al., 2002; Le Blanc et al., 2003). After interferon γ (INF-γ) induction of HLA II expression, MSC are still capable of escaping recognition by alloreactive T-cells (Le Blanc et al., 2003; Klyushnenkova et al., 2005). Similarly, after cellular differentiation into along the adipogenic, chondrogenic and osteogenic lineages, they remain non-immunogenic (Le Blanc, 2003). MSC have been found to escape lysis by cytotoxic T-cells and alloreactive killer inhibitory receptor mismatched natural killer cells (Le Blanc, 2003; Rasmusson et al., 2003; Gotherstrom, 2007) , making them uniquely suited for allogeneic transplantation applications.

In addition to being non-immunogenic, MSC have been shown to possess immunomodulatory properties both *in vitro* and *in vivo*. *In vitro*, they exhibit immunosuppressive effects and inhibit T-cell alloreactivity (Pittenger et al., 1999; Bartholomew et al., 2002; Krampera et al., 2003; Maitra et al., 2004; Beyth et al., 2005). This suppression effect of MSC is dose dependent and not only influence T cell but also affect the differentiation and maturation of dendritic cells from monocytes (Aggarwal and Pittenger, 2005). *In vivo*, infusion of ex vivo-expanded MSC has been shown to prolong the survival of allogeneic major histocompatibility mismatched skin allografts in immunocompetent out-bred baboons (Bartholomew et al., 2002). Further evidence was provided by Djouad et al, who found that subcutaneously injected melanoma cells in allogeneic immunocompetent mice formed tumors only when co-injected with MSC, suggesting that the presence of MSC prevented immune rejection of the tumor cells (Djouad et al., 2003). More recently MSC have been used to treat graft-versus-host disease (GvHD) after allogeneic hematopoietic stem cell transplantation (Le Blanc et al., 2008). The mechanism governing those effects is not yet understood. However, it is quite likely that the paracrine effect of MSC will play a significant role in it, for example, it was shown that MSC may strongly inhibit T-cell recognition and expansion via inhibiting TNF-α and INF-γ production and, thus, increasing IL-10 levels. (Aggarwal and Pittenger, 2005; Beyth et al., 2005; Caplan, 2007)

MSC are currently defined by their trilineage differentiation capacity (osteogenic, adipogenic and chondrogenic differentiation). In addition, they have been shown to be able to differentiate to muscle, marrow stroma and other connective tissues (Caplan, 1991; Caplan, 2005; Caplan, 2007). Several reports have now expanded repertoire to neuroectodermal lineages such as neurons (Woodbury et al., 2000; Zhao et al., 2002) , hepatocyte (Ong et al., 2006; Banas et al., 2007; Sgodda et al., 2007) and endothelial cell (Oswald et al., 2004; Gang et al., 2006) . A subset of MSC, termed Multipotent adult progenitor cells (MAPC), have been isolated from a number of species. MAPCs have greater differentiation potential than MSC, and have been shown to be pluripotent (Reyes et al., 2001; Jiang et al., 2002). However they are difficult to isolate and have not been reliably reproduced by other laboratories.

The osteogenic differentiation pathway is the default pathway for MSC, which is retained regardless of cell passage number until senescence (Digirolamo et al., 1999). MSC are heterogenous in nature, and consist of subpopulations of cells with tri-potent (osteogenic, adipogenic and chondrogenic), bi-potent (osteogenic and chondrogenic) and uni-potent (osteogenic) differentiation capacities as demonstrated through clonal analysis (Muraglia et al., 2000). Clones have been found to progressively lose their adipogenic and chondrogenic differentiation potential at increasing cell doublings, although osteogenic differentiation ability is retained till senescence (Digirolamo et al., 1999; Muraglia et al., 2000) .

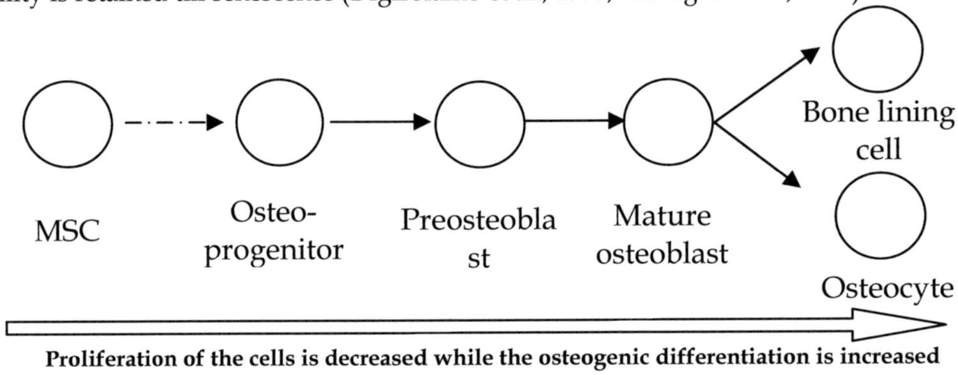

Fig. 5. Osteogenic differentiation pathway of MSC

Under the osteogenic differentiation pathway, MSC can be fully differentiated into osteoblasts, bone lining cells and osteocytes, which is a highly regulated and multi-step process with a sophisticated hierarchy as illustrated in Figure 5. MSC is likely to occupy the top position in this hierarchy, followed by osteoprogenitors, which are osteogenic committed progenitor cells with limited self-renewal and restricted differentiation capacity, but still maintaining substantial proliferation capacity. Further downstream are the preosteoblasts with more limited proliferation capacity. Mature osteoblasts are the terminally differentiated and functional cells, which will finally become the bone lining cells or osteocytes. With the progress of the osteogenic differentiation, proliferation of the cells is decreased while the osteogenic differentiation is increased. (Aubin, 1998; Aubin, 2001; Bilezikian, 2002)

4.3 Limitations of adult BM derived MSC and alternative MSC sources

The adult BM is the first tissue source where MSC are found and isolated, and are thus most investigated. They have been exploited as an autologous cell source for BTE applications, and demonstrated their effectiveness for fracture healing not only in animal but also in a number of reported human clinical trials (Quarto et al., 2001; Schimming and Schmelzeisen, 2004; Ueda et al., 2008; Yamada et al., 2008) . However, several barriers currently exist, limiting their further clinical applications. Firstly, adult BM-MSC are rare cells with a very low frequency, accounting for 0.00001% to 0.001% of mononuclear cells (MNC) in the adult bone marrow (BM). Furthermore, they have limited proliferative capacity, only undergoing an average approximately 38 ± 4 population doublings before reaching cellular senescence (Bruder et al., 1997). As a result, in order to obtain clinically significant cell numbers, a large amount of bone marrow aspiration is required initially. Secondly, for autologous usage, the derivation and expansion of MSC require a long period (4-6 weeks) for *ex vivo* expansion, limiting their clinical use, especially where the cells are required acutely. Secondly, the frequency of MSC within bone marrow and the proliferative capacity decrease dramatically as a function of donor age (it can drop from 0.001-0.0004% before the ages of 30 to 0.00025-0.00005% after the ages of 50) (Caplan, 2007) , with systemic diseases adversely affecting their numbers and function (Suzuki et al., 2001) . Similarly, the osteogenic differentiation ability *in vitro* and *in vivo* has also been found to be inversely correlated with donor age (Mueller and Glowacki, 2001) and reduced in several disease phenotypes (Rodriguez et al., 2000).

More recently, MSC with osteogenic potential have been isolated from a diverse range of other tissue types and ontogeny: (1) **In postnatal tissue**, MSC have been isolated from tissues as diverse as periosteum(Nakahara et al., 1991) , trabecular bone (Sakaguchi et al., 2004) , synovial membrane (De et al., 2001) , adipose tissue (Zuk et al., 2001) and peripheral blood (Eghbali-Fatourechi et al., 2005). (2) **In perinatal tissues,** MSC have been found in the umbilical cord (Sarugaser et al., 2005) , umbilical cord blood (Bieback et al., 2004; Lee et al., 2004) , amniotic fluid (in't Anker, P; Scherjon, S A; Kleijburg-van, Keur C; Noort, W A; Claas, F H; Willemze, R; Fibbe, W E; Kanhai, 2003; De et al., 2007) . (3) **In the prenatal tissue**, human fetal MSC have been isolated and characterized from first-trimester fetal blood, liver, and bone marrow (Campagnoli et al., 2001) , second trimester bone marrow, spleen and lung (in't Anker P ; Noort, W A; Scherjon, S A; Kleijburg-van, Keur C; Kruisselbrink, A B; van, Bezooijen R; Beekhuizen, W; Willemze, R; Kanhai, H H; Fibbe, 2003) , fetal metanephros (Almeida-Porada et al., 2002) , dermis (Zhao et al., 2005) , pancreas (Hu et al., 2003) and thymus (Rzhaninova et al., 2005) . While investigations into their basic biology, immunogenicity and osteogenic potential have been reported, MSC have not been systematically compared for bone tissue engineering applications. Hence, it remains unclear how these novel prenatal, perinatal and postnatal MSC sources compare with their standard adult BM MSC counterparts for osteogenic differentiation and potential for tissue engineering.

4.4 Human fetal BM derived MSC (hfMSC) as a promising cellular source for BTE

Recently, we reported a systematic investigation into four types of MSC from different ontological and anatomical origins in a direct head-to-head manner for BTE applications (Zhang et al., 2009). In this study, MSC were isolated from prenatal (human fetal bone marrow (hfMSC)), perinatal (human umbilical cord (hUCMSC)) and postnatal sources (human adult adipose tissue (hATMSC) and human adult bone marrow (haMSC)). In vitro comparative studies were performed in monolayer cultures and three dimensional bioactive scaffold cultures to investigate their proliferation capacity, osteogenic differentiation and mineralization, and in vivo ectopic bone formation.

hfMSC, haMSC, hUCMSC and hATMSC adopted a similar spindle-shaped morphology when cultured in monolayers, and expressed a consistent MSC immunophenotype which was negative for haemopoietic (CD14, CD34, CD45) and endothelial markers (CD31, and vWF), and positive for mesenchymal markers (CD105 (SH2), CD73 (SH3, SH4)), intracellular markers (Vimentin and Laminin) and cell adhesion molecules (CD29, CD44, CD 106, CD 90). All the four MSC types expressed HLA-I but not HLA-II and possessed the typical trilinage differentition capacities. However, compared with other MSC, hfMSC expressed a lower

level of HLA-I and a higher level of Stro-1 with the implication of lower immunogenicity and higher osteogenic capacity.
In two-dimensional (2D) monolayer cultures, hfMSC proliferated fastest (population doubling time 32.3±2.5 hours), followed by hUCMSC (54.7±4.3 hours), hATMSC (70.4±3.6 hours), with the slowest being haMSC (116.6±22.4 hours, n=3, $p<0.01$) (Figure 6A). In addition, hfMSC had significantly higher self-renewal ability, with 75.1± 5.0% of cells forming colonies, compared to 37.5-47.5% of other three MSC (Figure 6B). When cultured in 3D bioactive scaffold, hfMSC proliferated rapidly and reached confluence within the scaffold, taking up all available spaces by Day 7, while other MSC types achieved confluence only at the end of the 28 day experimental period as shown by a double-stranded DNA (dsDNA) quantification method (Figure 6C). Furthermore, hfMSC mediated scaffold constructs showed more robustly upregulated of key osteogenic genes, deposited significantly higher mineralizaton and demonstrated greater osteogenic capacity compared to other MSC scaffold constructs, as evaluated by a series of assay methods including von-Kossa staining, ALP activity, calcium deposition, calcium visualized on micro-CT and scanning electron-microscopy, and osteogenic gene expression (Figure 6D). Finally, subcutaneous implantation of these 3D constructs in immunodeficient NOD/SCID mice was performed to compare the osteogenic potential of MSC scaffolds in vivo. MSC scaffolds were cultured for two weeks in vitro as osteogenic pre-induction before implantation. Two months after implantation, all MSC scaffolds demonstrated neo-vascularization, with blood vessels infiltrating the scaffolds from the surrounding tissue macroscopically, and shared a similar chimerism rate of human cells in the murine tissue (60 to 67% chimerism) as demonstrated by human specific nuclear stain (lamins A/C), with infiltration of murine cells accounting for a third of the cellular population within the internal spaces of the scaffolds. Compared to other MSC scaffolds, hfMSC-mediated scaffolds generated more ectopic bone formation in vivo (1.8-13.3x, $p<0.01$) through micro-CT quantification (Zhang et al., 2009).

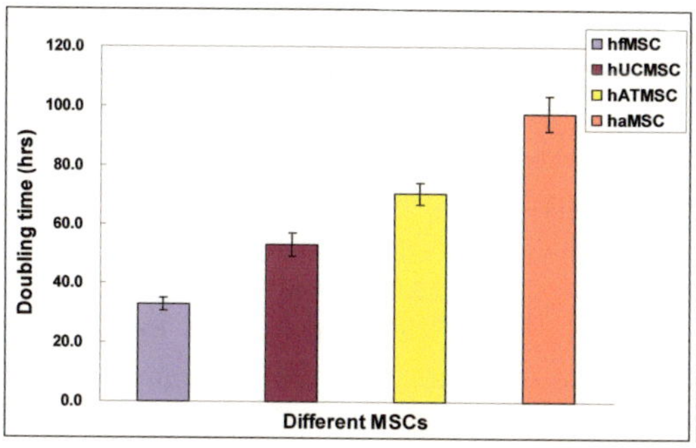

A. Doubling time of different MSCs

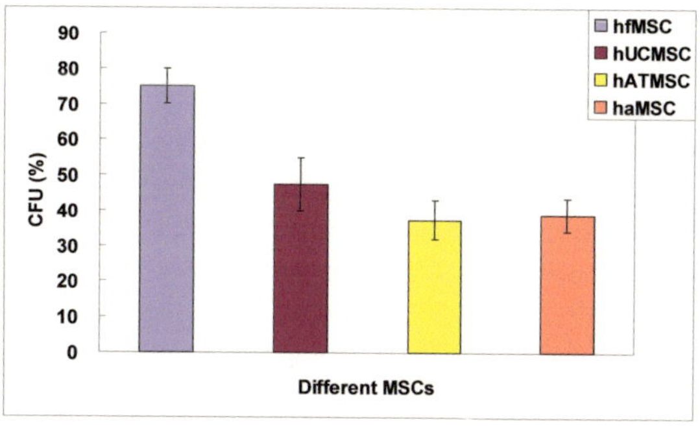

B. CFU of different MSCs

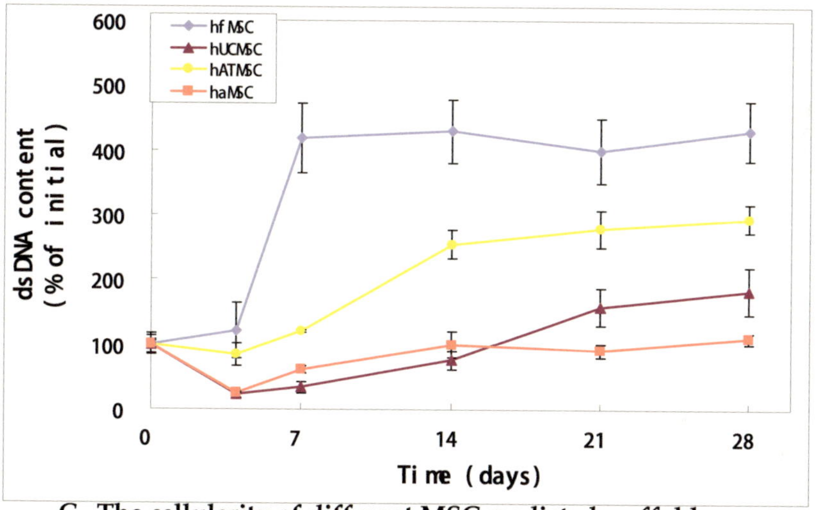

C. The cellularity of different MSC mediated scaffolds

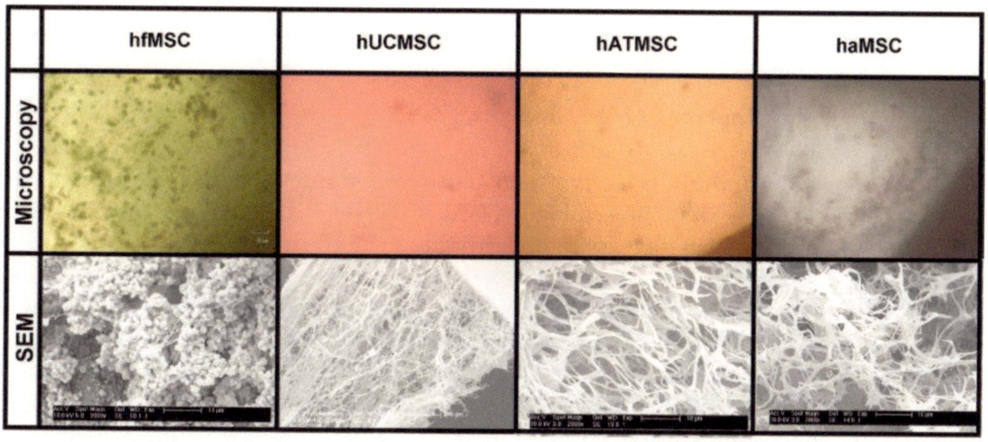

D. Osteogenic differentiation of MSC mediated scaffolds

Fig. 6. Comparison of different MSC for BTE application. (A) hfMSC proliferated faster than other MSCs with the shortest doubling time, and had higher self-renewal capacity on CFU-F

assay (B-C) hfMSC expanded at a faster rate, reaching confluence by day 7 and have a higher final cellularity than other MSC mediated scaffolds. (D) hfMSC mediated scaffolds underwent more robust osteogenic differentiation and mineralization compared to others as seen through light microscopy and scanning electron microscopy (SEM).
Through this study, it was found that the ontological and anatomical origins of MSC have profound influences on the proliferative and osteogenic capacity of MSC. MSC from the ontologically more primitive source have higher proliferative capacities, while MSC from BM related origin showed more osteogenic potential than MSC from other origins. In summary, hfMSC had the most proliferative and osteogenic capacity of MSC sources, as well as being the least immunogenic, suggesting they are superior candidates for BTE applications.

4.5 hfMSC as an off-the-shelf cellular source for BTE application

hfMSC have different immunologic properties to haMSC, as illustrated in Table 3. hfMSC not only express significantly lower levels of HLA Class I than haMSC, they do express any intracellular HLA Class II, which are found haMSC. hfMSC require 7 days of interferon-gamma for the induction of HLA II expression, compared to 2 days for haMSC (Gotherstrom et al., 2003; Le Blanc, 2003) , alluding to their utility for allogeneic transplantation. In fact, hfMSC have been exploited for intrauterine cell therapy (Chan et al., 2007; Guillot et al., 2008; Kennea et al., 2009) , with the first successful clinical applications of hfMSC has been reported through allogeneic hfMSC transplantation in a fetus with osteogenesis imperfecta (Le Blanc et al., 2005).

Immunologic properties	Fetal MSC	Adult MSC
HLA Class I expression	0- (+ +)*	+ + +
Intracellular deposits of HLA Class II	0	+ + +
Induction time to epxress HLA Class II	+ + (7 days)	+ + + (2 days)

"+" means positive, and "0" means no expression. *: no expression or lower

Table 3. Immunologic properties of fetal and adult MSC

Aside from their lower immunogenicity, faster proliferation rate and more robust osteogenic potential, hfMSC possess other favorable advantages over other cellular source, making them become an ideal off-the-shelf cellular candidate for BTE application (Table 4). Firstly, fetal tissues have a lower risk of viral and bacterial contamination compared to adult tissues as the fetus resides in an immunoprivileged environment within the uterus. Secondly, hfMSC have a higher proliferative capacity, and can undergo over 70 population doublings without senescence, compared to haMSC which typically senesces after 38 population doublings (Bruder et al., 1997; Campagnoli et al., 2001; Chan et al., 2005) . Theoretically, an unlimited number of cells can be derived from one hfMSC donor. With a proliferation capacity of more than 70 population doublings, theoretically one single cell isolated at Passage 0 can be proliferated to 1.2×10^{21} (2^{70}) cells at Passage 70 and can be utilised more than 10^{12} patients (utilizing a dose of 1×10^9 cells). "Single-cell banks" can be established using cells derived from a single cell colony with well-investigated biological properties and greatly minimized variation; moreover, all the cells can be frozen down and made available off-the-shelf, eliminating the precious waiting time for patients. Lastly, the ethical conundrum faced with cells derived from embryonic stem cells are less of a problem with the use of fetal stem cells, considering that various human fetal cell types have been explored for therapeutic applications in the clinic already, such as the use of fetal neural progenitors for the treatment of Huntington's (Rosser and Dunnett, 2003) or Parkinson's disease (Clarkson, 2001) , and the use of human fetal liver cells to treat severe immunodeficiencies, haematological disorders and inborn errors of metabolism (Touraine et al., 1993) . More recently human fetal skin cells have been utilized for constructing a tissue engineered skin graft (Hohlfeld et al., 2005; Pioletti et al., 2006).

	Collection	Expansion	Storage	Allogenic applications	Osteogenic potential
Fresh bone marrow	+++	N.A.	N.A.	N.A.*	+
Differentiated osteoblast	+	+	+	N.A.*	+++
MSC from adult BM	++	++	+++	++	++
MSC from fetal BM	++	+++	+++	+++	+++

N.A.*: allogenic applications can be achieved by proper HLA matching

Table 4. Comparison of hfMSC with other cellular source for BTE.

5. Biaxial rotating bioreactors for BTE application

Bioreactors are the biomechanical devices with closely monitored and tightly controlled environmental and operating conditions. Bioreactors have been widely used in industries for fermentation processing, wastewater treatment, food processing and production of pharmaceuticals. Recently bioreactors have been introduced to tissue engineering field in order to mimic the native *in vivo* environment and provide cellular tissue engineered constructs with physiologically relevant stimuli that facilitate and orchestrate the conversion of a "collection of cells" into a specific tissue phenotype (Lanza et al., 2007). Specifically, they can increase the mass transport to mitigate the diffusion limitation of 3D scaffolds, providing adequate nutrient, oxygen and regulatory molecules to the cells while removing metabolites and CO_2 away. Bioreactors can provide proper physiological stimuli, especially the mechanical cues to trigger the mechanotransduction signaling pathway for the differentiation of cellular constructs, and can be used to enable efficient and homogenous cellular seeding in the complex 3D scaffolds. (Martin et al., 2004; Bilodeau and Mantovani, 2006; Chen and Hu, 2006)

Currently, several types of bioreactors such as spinner flasks, perfusion bioreactors and rotating wall vessel (RWV) bioreactors have been investigated for BTE applications. However, various limitations have been reported with the use of these bioreactors. Spinner flasks can improve fluid flow, which can lead to the enhanced cell proliferation, distribution and osteogenic differentiation (Mygind et al., 2007; Song et al., 2008; Stiehler et al., 2008), but the turbulence generated by stirring can be detrimental for seeded cells and newly laid down ECM (Sikavitsas et al., 2002; Chen and Hu, 2006). Perfusion bioreactors have been shown to enhance MSC proliferation and osteogenic differentiation in scaffold constructs (Sikavitsas et al., 2003; van et al., 2003; Gomes et al., 2006), however, they have been beset by non-homogenous cellular distributions, with cells at the frontal zones being washed away by the oncoming perfusion flow (Singh et al., 2007). Rotating wall vessel (RWV) bioreactors (Molnar et al., 1997; Granet et al., 1998; Yu et al., 2004), which generates low shear forces and three dimensional high mass transfer capacity, are prone to problems of non-homogenous cellular growth and ECM deposition (Goldstein et al., 2001; Sikavitsas et al., 2002; Chen and Hu, 2006). Moreover, the free floating culture of RWV bioreactors usually lead to the collision between the scaffolds and the bioreactor walls, which can induce cellular damage and disrupt cellular attachment and matrix deposition on the scaffolds (Goldstein et al., 2001; Sikavitsas et al., 2002; Chen and Hu, 2006).

Consequently, we designed and developed an biaxial rotating bioreactor (Figure 7) to address the deficiencies found in current bioreactor designs. Firstly, a perfusion system was included to allow the media circulation between the vessel and reservoir, allowing maximal mass transfer with consequential low shear stress, and minimizing the washout problem associated with RWV bioreactors. Secondly, the improvement of biaxial rotation movement upon uni-axial rotating, leads to more homogenous cellular and ECM distribution of the scaffold, as previously predicted from in-silico simulation (Singh et al., 2005). Lastly, cellular scaffolds were secured by pins and were not kept in free suspension, avoiding the risk of scaffold collisions with the vessel walls.

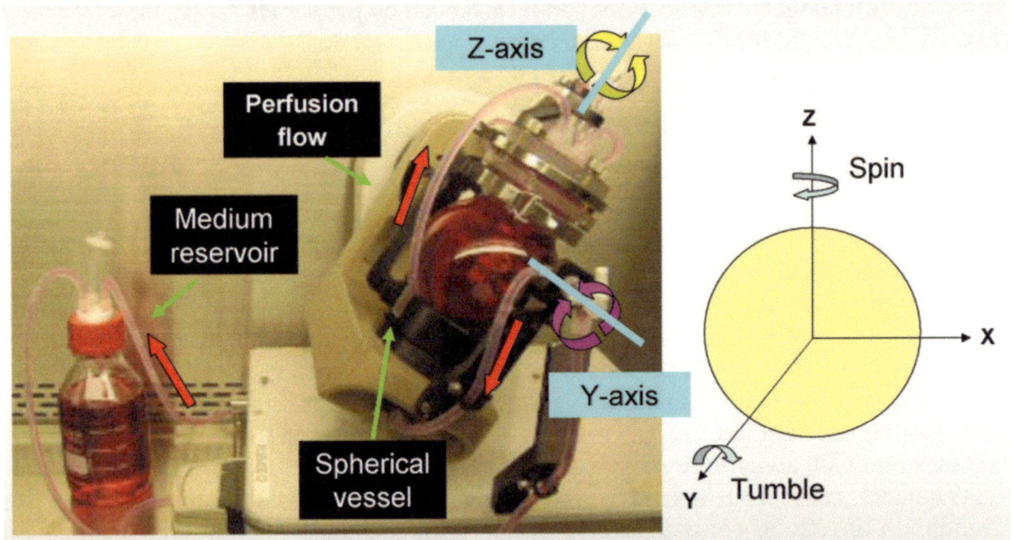

Fig. 7. The working mechanism of biaxial rotating bioreactor. This biaxial bioreactor can rotate in two perpendicular axes simultaneously with an inbuilt perfusion system.

In a recent study, we investigated the performance of this biaxial rotating bioreactor for generation of highly osteogenic bone graft using hfMSC mediated polycaprolactone-tricalcium phosphate (PCL-TCP) scaffolds. hfMSC-scaffolds were cultured and matured in either bioreactor or static cultures, and evaluated for their cellular viability, proliferation and osteogenic differentiation *in vitro*, and after transplantation into immunodeficient mice. Compared with traditional static culture, biaxial bioreactor culture enhanced cellular proliferation, with hfMSC-scaffolds reached cellular confluence earlier (Day 7 vs Day 28) with greater cellularity (2x, $p<0.01$) (Figure 8). Over 28 days of *in vitro* culture, the biaxial bioreactor enabled the maintenance of high cellular viability throughout the scaffolds, including the core, 2 mm from the surface of the scaffold. Thus the use of this bioreactor enabled a ten-fold improvement of mass transfer of nutrients, generally taken to be around 150-200 nm. In contrast, the static culture only allowed the cellular survival at a surface region of the thick scaffolds, with massive cell necrosis within the core of the scaffolds.

Biaxial bioreactor cultured-cellular scaffolds were associated with greater osteogenic induction as indicated by *in vitro* assays such as higher ALP expression (1.5x $p<0.01$), more calcium deposition (5.5x, $p<0.001$) and bony nodule formation observed under scanning electron microscopy (SEM) (Figure 8A), and the *in vivo* ectopic bone formation in immunodeficient mice (3.2x, $p<0.001$) compared with static-cultured scaffolds.

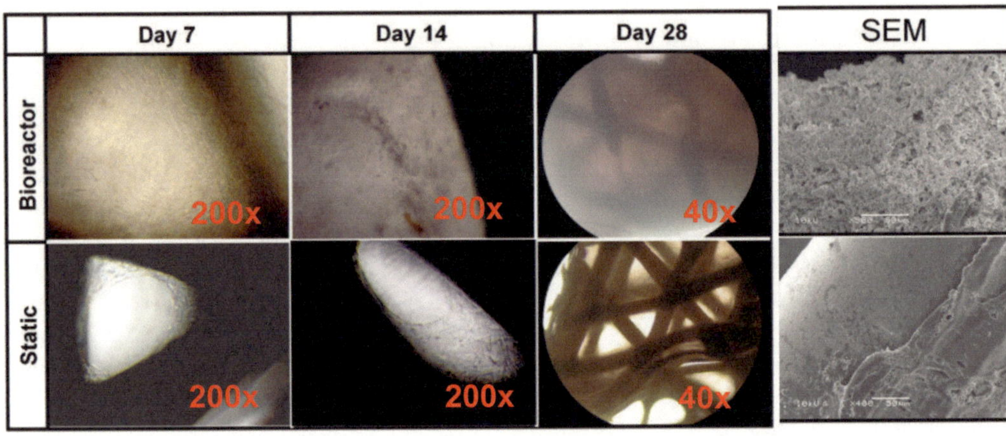

A. Prolfieration and differentiaiton of hfMSC scaffolds under bioreactor culture or static culture

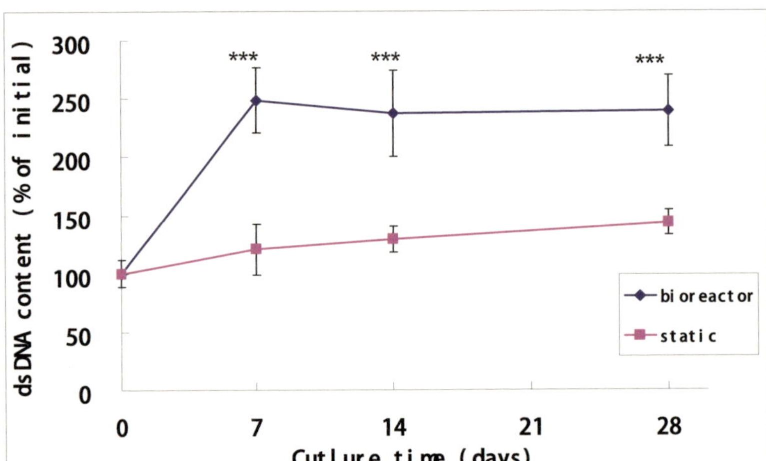

B. Cellularity of hfMSC scaffolds under bioreactor culture or static culture

Fig. 8. Comparison of the biaxial bioreactor culture and static culture. (A) Biaxial bioreactor cultured hfMSC scaffolds proliferated faster and underwent more robust osteogenic differentiation compared to static cultured ones, and (B) resulted in significantly higher cellularity in the cellular scaffolds (*** $p<0.001$).

6. Conclusion –Generating effective TE bone grafts

By combining an inter-disciplinary approach in scaffold technology, bioreactor development and stem cell biology, we have generated an effective bone graft through the seeding of highly proliferative and osteogenic hfMSC onto the osteoconductive PCL-TCP scaffold matrix, and maturing the hfMSC mediated PCL-TCP scaffold under biaxial rotating bioreactor (Figure 9). Our ongoing animal experiment showed that this hfMSC mediated TE bone graft can be used to heal critical sized femoral defect in a rat model (unpublished observations), paving the way for clinical trials to be initiated. This strategy presents a promising solution to fulfill the increasing need for bone grafts and provides effective TE bone grafts, which are available off-the-shelf like allografts and synthetic grafts and can stimulate rapid healing like autografts.

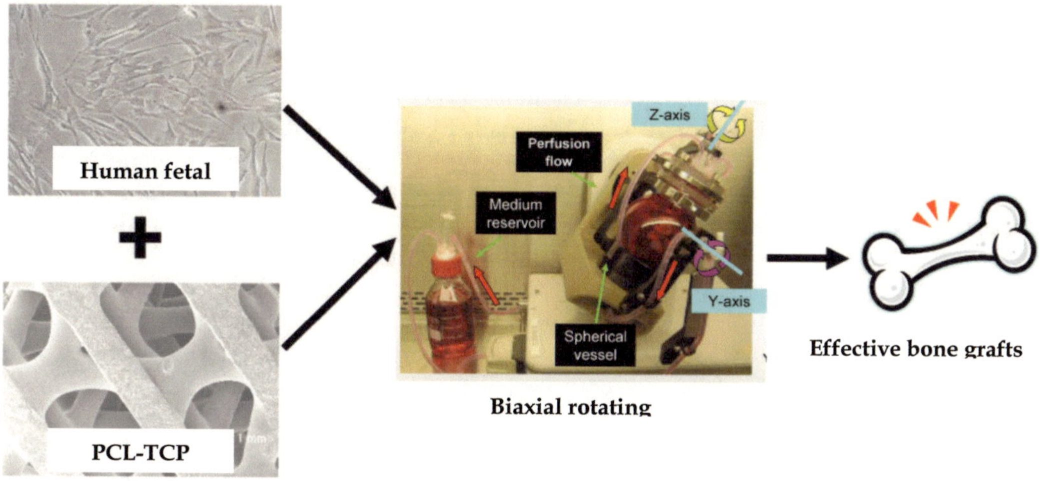

Fig. 9. Generation effective bone grafts with the combinational use of hfMSC, PCL-TCP scaffold and biaxial rotating bioreactor.

7. References

Aggarwal, S. and Pittenger, M. F. (2005). "Human mesenchymal stem cells modulate allogeneic immune cell responses." Blood 105 (4): 1815-22.

Agrawal, C. M. and Ray, R. B. (2001). "Biodegradable polymeric scaffolds for musculoskeletal tissue engineering." J Biomed Mater Res 55 (2): 141-50.

Almeida-Porada, G., El, S. D., Porada, C. and Zanjani, E. D. (2002). "Differentiative potential of human metanephric mesenchymal cells." Exp Hematol 30 (12): 1454-62.

Arrington, E. D., Smith, W. J., Chambers, H. G., Bucknell, A. L. and Davino, N. A. (1996). "Complications of iliac crest bone graft harvesting." Clin Orthop Relat Res(329): 300-9.

Aubin, J. E. (1998). "Bone stem cells." J Cell Biochem Suppl 30-31 : 73-82.

Aubin, J. E. (2001). "Regulation of osteoblast formation and function." Rev Endocr Metab Disord 2 (1): 81-94.

Baddoo, M., Hill, K., Wilkinson, R., Gaupp, D., Hughes, C., Kopen, G. C. and Phinney, D. G. (2003). "Characterization of mesenchymal stem cells isolated from murine bone marrow by negative selection." J Cell Biochem 89 (6): 1235-49.

Banas, A., Teratani, T., Yamamoto, Y., Tokuhara, M., Takeshita, F., Quinn, G., Okochi, H. and Ochiya, T. (2007). "Adipose tissue-derived mesenchymal stem cells as a source of human hepatocytes." Hepatology 46 (1): 219-28.

Banfi, A., Muraglia, A., Dozin, B., Mastrogiacomo, M., Cancedda, R. and Quarto, R. (2000). "Proliferation kinetics and differentiation potential of ex vivo expanded human bone marrow stromal cells: Implications for their use in cell therapy." Exp Hematol 28 (6): 707-15.

Banwart, J. C., Asher, M. A. and Hassanein, R. S. (1995). "Iliac crest bone graft harvest donor site morbidity. A statistical evaluation." Spine 20 (9): 1055-60.

Bartholomew, A., Sturgeon, C., Siatskas, M., Ferrer, K., McIntosh, K., Patil, S., Hardy, W., Devine, S., Ucker, D., Deans, R., Moseley, A. and Hoffman, R. (2002). "Mesenchymal stem cells suppress lymphocyte proliferation in vitro and prolong skin graft survival in vivo." Exp Hematol 30 (1): 42-8.

Beyth, S., Borovsky, Z., Mevorach, D., Liebergall, M., Gazit, Z., Aslan, H., Galun, E. and Rachmilewitz, J. (2005). "Human mesenchymal stem cells alter antigen-presenting cell maturation and induce T-cell unresponsiveness." Blood 105 (5): 2214-9.

Bieback, K., Kern, S., Kluter, H. and Eichler, H. (2004). "Critical parameters for the isolation of mesenchymal stem cells from umbilical cord blood." Stem Cells 22 (4): 625-34.

Bilodeau, K. and Mantovani, D. (2006). "Bioreactors for tissue engineering: focus on mechanical constraints. A comparative review." Tissue Eng 12 (8): 2367-83.

Blaker, J. J., Gough, J. E., Maquet, V., Notingher, I. and Boccaccini, A. R. (2003). "In vitro evaluation of novel bioactive composites based on Bioglass-filled polylactide foams for bone tissue engineering scaffolds." J Biomed Mater Res A 67 (4): 1401-11.

Bongso, A. (2005). Stem cells : from bench to bedside, New Jersey : World Scientific.

Bruder, S. P. and Fox, B. S. (1999). "Tissue engineering of bone. Cell based strategies." Clin Orthop Relat Res(367 Suppl): S68-83.

Bruder, S. P., Jaiswal, N. and Haynesworth, S. E. (1997). "Growth kinetics, self-renewal, and the osteogenic potential of purified human mesenchymal stem cells during extensive subcultivation and following cryopreservation." J Cell Biochem 64 (2): 278-94.

Bruder, S. P., Jaiswal, N. and Haynesworth, S. E. (1997). "Growth kinetics, self-renewal, and the osteogenic potential of purified human mesenchymal stem cells during extensive subcultivation and following cryopreservation." J Cell Biochem 64 (2): 278-94.

Campagnoli, C., Roberts, I. A., Kumar, S., Bennett, P. R., Bellantuono, I. and Fisk, N. M. (2001). "Identification of mesenchymal stem/progenitor cells in human first-trimester fetal blood, liver, and bone marrow." Blood 98 (8): 2396-402.

Caplan, A. I. (1991). "Mesenchymal stem cells." J Orthop Res 9 (5): 641-50.

Caplan, A. I. (2005). "Review: mesenchymal stem cells: cell-based reconstructive therapy in orthopedics." Tissue Eng 11 (7-8): 1198-211.

Caplan, A. I. (2007). "Adult mesenchymal stem cells for tissue engineering versus regenerative medicine." J Cell Physiol 213 (2): 341-7.

Chan, J.; O' Donoghue, K; de, Fuente J; Roberts, I A; Kumar, S; Morgan, J E; Fisk, N. M. (2005). "Human fetal mesenchymal stem cells as vehicles for gene delivery." Stem Cells 23 (1): 93-102.

Chan, J., Waddington, S. N., O, D. K., Kurata, H., Guillot, P. V., Gotherstrom, C., Themis, M., Morgan, J. E. and Fisk, N. M. (2007). "Widespread distribution and muscle differentiation of human fetal mesenchymal stem cells after intrauterine transplantation in dystrophic mdx mouse." Stem Cells 25 (4): 875-84.

Chen, H. C. and Hu, Y. C. (2006). "Bioreactors for tissue engineering." Biotechnol Lett 28 (18): 1415-23.

Ciapetti, G., Ambrosio, L., Savarino, L., Granchi, D., Cenni, E., Baldini, N., Pagani, S., Guizzardi, S., Causa, F. and Giunti, A. (2003). "Osteoblast growth and function in porous poly epsilon -caprolactone matrices for bone repair: a preliminary study." Biomaterials 24 (21): 3815-24.

Clarkson, E. D. (2001). "Fetal tissue transplantation for patients with Parkinson's disease: a database of published clinical results." Drugs Aging 18 (10): 773-85.

Colter, D. C., Class, R., DiGirolamo, C. M. and Prockop, D. J. (2000). "Rapid expansion of recycling stem cells in cultures of plastic-adherent cells from human bone marrow." Proc Natl Acad Sci U S A 97 (7): 3213-8.

Colter, D. C., Sekiya, I. and Prockop, D. J. (2001). "Identification of a subpopulation of rapidly self-renewing and multipotential adult stem cells in colonies of human marrow stromal cells." Proc Natl Acad Sci U S A 98 (14): 7841-5.

Connolly, J. F., Guse, R., Tiedeman, J. and Dehne, R. (1989). "Autologous marrow injection for delayed unions of the tibia: a preliminary report." J Orthop Trauma 3 (4): 276-82.

Connolly, J. F., Guse, R., Tiedeman, J. and Dehne, R. (1991). "Autologous marrow injection as a substitute for operative grafting of tibial nonunions." Clin Orthop Relat Res(266): 259-70.

Coombes, A. G., Rizzi, S. C., Williamson, M., Barralet, J. E., Downes, S. and Wallace, W. A. (2004). "Precipitation casting of polycaprolactone for applications in tissue engineering and drug delivery." Biomaterials 25 (2): 315-25.

Damien, C. J. and Parsons, J. R. (1991). "Bone graft and bone graft substitutes: a review of current technology and applications." J Appl Biomater 2 (3): 187-208.

De, B. C., Dell, A. F., Tylzanowski, P. and Luyten, F. P. (2001). "Multipotent mesenchymal stem cells from adult human synovial membrane." Arthritis Rheum 44 (8): 1928-42.

De, C. P., Bartsch, G. J., Siddiqui, M. M., Xu, T., Santos, C. C., Perin, L., Mostoslavsky, G., Serre, A. C., Snyder, E. Y., Yoo, J. J., Furth, M. E., Soker, S. and Atala, A. (2007). "Isolation of amniotic stem cell lines with potential for therapy." Nat Biotechnol 25 (1): 100-6.

De, K. I., Hicok, K. C., Padilla, R. J., Young, R. G. and Cooper, L. F. (2006). "Effect of vitamin D pretreatment of human mesenchymal stem cells on ectopic bone formation." J Oral Implantol 32 (3): 103-9.

Digirolamo, C. M., Stokes, D., Colter, D., Phinney, D. G., Class, R. and Prockop, D. J. (1999). "Propagation and senescence of human marrow stromal cells in culture: a simple colony-forming assay identifies samples with the greatest potential to propagate and differentiate." Br J Haematol 107 (2): 275-81.

Digirolamo, C. M., Stokes, D., Colter, D., Phinney, D. G., Class, R. and Prockop, D. J. (1999). "Propagation and senescence of human marrow stromal cells in culture: a simple colony-forming assay identifies samples with the greatest potential to propagate and differentiate." Br J Haematol 107 (2): 275-81.

Djouad, F., Plence, P., Bony, C., Tropel, P., Apparailly, F., Sany, J., Noel, D. and Jorgensen, C. (2003). "Immunosuppressive effect of mesenchymal stem cells favors tumor growth in allogeneic animals." Blood 102 (10): 3837-44.

Dominici, M., Le, B. K., Mueller, I., Slaper-Cortenbach, I., Marini, F., Krause, D., Deans, R., Keating, A., Prockop, D. and Horwitz, E. (2006). "Minimal criteria for defining multipotent mesenchymal stromal cells. The International Society for Cellular Therapy position statement." Cytotherapy 8 (4): 315-7.

Dudas, J. R., Marra, K. G., Cooper, G. M., Penascino, V. M., Mooney, M. P., Jiang, S., Rubin, J. P. and Losee, J. E. (2006). "The osteogenic potential of adipose-derived stem cells for the repair of rabbit calvarial defects." Ann Plast Surg 56 (5): 543-8.

Eghbali-Fatourechi, G. Z., Lamsam, J., Fraser, D., Nagel, D., Riggs, B. L. and Khosla, S. (2005). "Circulating osteoblast-lineage cells in humans." N Engl J Med 352 (19): 1959-66.

Einhorn, T. A. (1995). "Enhancement of fracture-healing." J Bone Joint Surg Am 77 (6): 940-56.

Britannica, E. (2007). The New Encyclopaedia Britannica, 2007 Edition, Encyclopedia Britannica, Incorporated.

Endres, M., Hutmacher, D. W., Salgado, A. J., Kaps, C., Ringe, J., Reis, R. L., Sittinger, M., Brandwood, A. and Schantz, J. T. (2003). "Osteogenic induction of human bone marrow-derived mesenchymal progenitor cells in novel synthetic polymer-hydrogel matrices." Tissue Eng 9 (4): 689-702.

Fowler, B. L., Dall, B. E. and Rowe, D. E. (1995). "Complications associated with harvesting autogenous iliac bone graft." Am J Orthop 24 (12): 895-903.

Friedenstein, A. J., Petrakova, K. V., Kurolesova, A. I. and Frolova, G. P. (1968). "Heterotopic of bone marrow.Analysis of precursor cells for osteogenic and hematopoietic tissues." Transplantation 6 (2): 230-47.

Friedenstein, A. J., Piatetzky, S. I. and Petrakova, K. V. (1966). "Osteogenesis in transplants of bone marrow cells." J Embryol Exp Morphol 16 (3): 381-90.

Gang, E. J., Jeong, J. A., Han, S., Yan, Q., Jeon, C. J. and Kim, H. (2006). "In vitro endothelial potential of human UC blood-derived mesenchymal stem cells." Cytotherapy 8 (3): 215-27.

Giannoudis, P. V., Einhorn, T. A. and Marsh, D. (2007). "Fracture healing: a harmony of optimal biology and optimal fixation?" Injury 38 Suppl 4 : S1-2.

Goldstein, A. S., Juarez, T. M., Helmke, C. D., Gustin, M. C. and Mikos, A. G. (2001). "Effect of convection on osteoblastic cell growth and function in biodegradable polymer foam scaffolds." Biomaterials 22 (11): 1279-88.

Gomes, M. E., Bossano, C. M., Johnston, C. M., Reis, R. L. and Mikos, A. G. (2006). "In vitro localization of bone growth factors in constructs of biodegradable scaffolds seeded with marrow stromal cells and cultured in a flow perfusion bioreactor." Tissue Eng 12 (1): 177-88.

Gotherstrom, C. (2007). "Immunomodulation by multipotent mesenchymal stromal cells." Transplantation 84 (1 Suppl): S35-7.

Gotherstrom, C., Ringden, O., Westgren, M., Tammik, C. and Le, B. K. (2003). "Immunomodulatory effects of human foetal liver-derived mesenchymal stem cells." Bone Marrow Transplant 32 (3): 265-72.

Goulet, J. A., Senunas, L. E., DeSilva, G. L. and Greenfield, M. L. (1997). "Autogenous iliac crest bone graft. Complications and functional assessment." Clin Orthop Relat Res(339): 76-81.

Granet, C., Laroche, N., Vico, L., Alexandre, C. and Lafage-Proust, M. H. (1998). "Rotating-wall vessels, promising bioreactors for osteoblastic cell culture: comparison with other 3D conditions." Med Biol Eng Comput 36 (4): 513-9.

Guillot, P. V., Abass, O., Bassett, J. H., Shefelbine, S. J., Bou-Gharios, G., Chan, J., Kurata, H., Williams, G. R., Polak, J. and Fisk, N. M. (2008). "Intrauterine transplantation of

human fetal mesenchymal stem cells from first-trimester blood repairs bone and reduces fractures in osteogenesis imperfecta mice." Blood 111 (3): 1717-25.
Hayda, R. A., Brighton, C. T. and Esterhai, J. L. (1998). "Pathophysiology of delayed healing." Clin Orthop Relat Res(355 Suppl): S31-40.
Hohlfeld, J., de, B. R., Hirt-Burri, N., Chaubert, P., Gerber, S., Scaletta, C., Hohlfeld, P. and Applegate, L. A. (2005). "Tissue engineered fetal skin constructs for paediatric burns." Lancet 366 (9488): 840-2.
Hollinger, J. O., Brekke, J., Gruskin, E. and Lee, D. (1996). "Role of bone substitutes." Clin Orthop Relat Res(324): 55-65.
Hollinger, J. O., Einhorn, T. A., Doll, B. A. and Sfeir, C. (2005). Bone Tissue Engineering, Boca Raton : CRC Press.
Hu, Y., Liao, L., Wang, Q., Ma, L., Ma, G., Jiang, X. and Zhao, R. C. (2003). "Isolation and identification of mesenchymal stem cells from human fetal pancreas." J Lab Clin Med 141 (5): 342-9.
Hutmacher, D. W. (2000). "Scaffolds in tissue engineering bone and cartilage." Biomaterials 21 (24): 2529-43.
Hutmacher, D. W., Schantz, T., Zein, I., Ng, K. W., Teoh, S. H. and Tan, K. C. (2001). "Mechanical properties and cell cultural response of polycaprolactone scaffolds designed and fabricated via fused deposition modeling." J Biomed Mater Res 55 (2): 203-16.
Hutmacher, D. W., Schantz, T., Zein, I., Ng, K. W., Teoh, S. H. and Tan, K. C. (2001). "Mechanical properties and cell cultural response of polycaprolactone scaffolds designed and fabricated via fused deposition modeling." J Biomed Mater Res 55 (2): 203-16.
Inoue, K., Ohgushi, H., Yoshikawa, T., Okumura, M., Sempuku, T., Tamai, S. and Dohi, Y. (1997). "The effect of aging on bone formation in porous hydroxyapatite: biochemical and histological analysis." J Bone Miner Res 12 (6): 989-94.
in't Anker P ; Noort, W A; Scherjon, S A; Kleijburg-van, Keur C; Kruisselbrink, A B; van, Bezooijen R; Beekhuizen, W; Willemze, R; Kanhai, H H; Fibbe, W. E. (2003). "Mesenchymal stem cells in human second-trimester bone marrow, liver, lung, and spleen exhibit a similar immunophenotype but a heterogeneous multilineage differentiation potential." Haematologica 88 (8): 845-52.
in't Anker, P; Scherjon, S A; Kleijburg-van, Keur C; Noort, W A; Claas, F H; Willemze, R; Fibbe, W E; Kanhai, H. H. (2003). "Amniotic fluid as a novel source of mesenchymal stem cells for therapeutic transplantation." Blood 102 (4): 1548-9.
Javazon, E. H., Colter, D. C., Schwarz, E. J. and Prockop, D. J. (2001). "Rat marrow stromal cells are more sensitive to plating density and expand more rapidly from single-cell-derived colonies than human marrow stromal cells." Stem Cells 19 (3): 219-25.
Lieberman, J. R. and Friedlaender, G. E. (2005). Bone Regeneration and Repair: Biology and Clinical Applications, Humana Press.
Jiang, Y., Jahagirdar, B. N., Reinhardt, R. L., Schwartz, R. E., Keene, C. D., Ortiz-Gonzalez, X. R., Reyes, M., Lenvik, T., Lund, T., Blackstad, M., Du, J., Aldrich, S., Lisberg, A., Low, W. C., Largaespada, D. A. and Verfaillie, C. M. (2002). "Pluripotency of mesenchymal stem cells derived from adult marrow." Nature 418 (6893): 41-9.
Bilezikian, J. P. (2002). Principles of Bone Biology, Academic Press.

Jones, E. and McGonagle, D. (2008). "Human bone marrow mesenchymal stem cells in vivo." Rheumatology (Oxford) 47 (2): 126-31.

Kahn, A., Gibbons, R., Perkins, S. and Gazit, D. (1995). "Age-related bone loss. A hypothesis and initial assessment in mice." Clin Orthop Relat Res(313): 69-75.

Kennea, N. L., Waddington, S. N., Chan, J., O, D. K., Yeung, D., Taylor, D. L., Al-Allaf, F. A., Pirianov, G., Themis, M., Edwards, A. D., Fisk, N. M. and Mehmet, H. (2009). "Differentiation of human fetal mesenchymal stem cells into cells with an oligodendrocyte phenotype." Cell Cycle 8 (7): 1069-79.

Kimelman, N., Pelled, G., Helm, G. A., Huard, J., Schwarz, E. M. and Gazit, D. (2007). "Review: gene- and stem cell-based therapeutics for bone regeneration and repair." Tissue Eng 13 (6): 1135-50.

Klyushnenkova, E., Mosca, J. D., Zernetkina, V., Majumdar, M. K., Beggs, K. J., Simonetti, D. W., Deans, R. J. and McIntosh, K. R. (2005). "T cell responses to allogeneic human mesenchymal stem cells: immunogenicity, tolerance, and suppression." J Biomed Sci 12 (1): 47-57.

Kokubo, T., Kim, H. M. and Kawashita, M. (2003). "Novel bioactive materials with different mechanical properties." Biomaterials 24 (13): 2161-75.

Kokubu, T., Hak, D. J., Hazelwood, S. J. and Reddi, A. H. (2003). "Development of an atrophic nonunion model and comparison to a closed healing fracture in rat femur." J Orthop Res 21 (3): 503-10.

Krampera, M., Glennie, S., Dyson, J., Scott, D., Laylor, R., Simpson, E. and Dazzi, F. (2003). "Bone marrow mesenchymal stem cells inhibit the response of naive and memory antigen-specific T cells to their cognate peptide." Blood 101 (9): 3722-9.

Lam, C. X., Hutmacher, D. W., Schantz, J. T., Woodruff, M. A. and Teoh, S. H. (2008). "Evaluation of polycaprolactone scaffold degradation for 6 months in vitro and in vivo." J Biomed Mater Res A.

Lane, J. M., Tomin, E. and Bostrom, M. P. (1999). "Biosynthetic bone grafting." Clin Orthop Relat Res(367 Suppl): S107-17.

Langer, R. and Vacanti, J. P. (1993). "Tissue engineering." Science 260 (5110): 920-6.

Laurencin, C. T., Ambrosio, A. M., Borden, M. D. and Cooper, J. A. (1999). "Tissue engineering: orthopedic applications." Annu Rev Biomed Eng 1 : 19-46.

Le Blanc, K. (2003). "Immunomodulatory effects of fetal and adult mesenchymal stem cells." Cytotherapy 5 (6): 485-9.

Le Blanc, K., Frassoni, F., Ball, L., Locatelli, F., Roelofs, H., Lewis, I., Lanino, E., Sundberg, B., Bernardo, M. E., Remberger, M., Dini, G., Egeler, R. M., Bacigalupo, A., Fibbe, W. and Ringden, O. (2008). "Mesenchymal stem cells for treatment of steroid-resistant, severe, acute graft-versus-host disease: a phase II study." Lancet 371 (9624): 1579-86.

Le Blanc, K., Gotherstrom, C., Ringden, O., Hassan, M., McMahon, R., Horwitz, E., Anneren, G., Axelsson, O., Nunn, J., Ewald, U., Norden-Lindeberg, S., Jansson, M., Dalton, A., Astrom, E. and Westgren, M. (2005). "Fetal mesenchymal stem-cell engraftment in bone after in utero transplantation in a patient with severe osteogenesis imperfecta." Transplantation 79 (11): 1607-14.

Le Blanc, K., Tammik, C., Rosendahl, K., Zetterberg, E. and Ringden, O. (2003). "HLA expression and immunologic properties of differentiated and undifferentiated mesenchymal stem cells." Exp Hematol 31 (10): 890-6.

Le Blanc, K., Tammik, L., Sundberg, B., Haynesworth, S. E. and Ringden, O. (2003). "Mesenchymal stem cells inhibit and stimulate mixed lymphocyte cultures and mitogenic responses independently of the major histocompatibility complex." Scand J Immunol 57 (1): 11-20.

Lee, O. K., Kuo, T. K., Chen, W. M., Lee, K. D., Hsieh, S. L. and Chen, T. H. (2004). "Isolation of multipotent mesenchymal stem cells from umbilical cord blood." Blood 103 (5): 1669-75.

LeGeros, R. Z. (2002). "Properties of osteoconductive biomaterials: calcium phosphates." Clin Orthop Relat Res(395): 81-98.

Logeart-Avramoglou, D., Anagnostou, F., Bizios, R. and Petite, H. (2005). "Engineering bone: challenges and obstacles." J Cell Mol Med 9 (1): 72-84.

Maitra, B., Szekely, E., Gjini, K., Laughlin, M. J., Dennis, J., Haynesworth, S. E. and Koc, O. N. (2004). "Human mesenchymal stem cells support unrelated donor hematopoietic stem cells and suppress T-cell activation." Bone Marrow Transplant 33 (6): 597-604.

Maquet, V., Boccaccini, A. R., Pravata, L., Notingher, I. and Jerome, R. (2004). "Porous poly(alpha-hydroxyacid)/Bioglass composite scaffolds for bone tissue engineering. I: Preparation and in vitro characterisation." Biomaterials 25 (18): 4185-94.

Marsh, D. (1998). "Concepts of fracture union, delayed union, and nonunion." Clin Orthop Relat Res(355 Suppl): S22-30.

Marsh, D. R. and Li, G. (1999). "The biology of fracture healing: optimising outcome." Br Med Bull 55 (4): 856-69.

Martin, I., Wendt, D. and Heberer, M. (2004). "The role of bioreactors in tissue engineering." Trends Biotechnol 22 (2): 80-6.

Mauney, J. R., Jaquiery, C., Volloch, V., Heberer, M., Martin, I. and Kaplan, D. L. (2005). "In vitro and in vivo evaluation of differentially demineralized cancellous bone scaffolds combined with human bone marrow stromal cells for tissue engineering." Biomaterials 26 (16): 3173-85.

Mendes, S. C., Sleijster, M., den Van, M., De, B. J. and Van, B. C. (2002). "A cultured living bone equivalent enhances bone formation when compared to a cell seeding approach." J Mater Sci Mater Med 13 (6): 575-81.

Molnar, G., Schroedl, N. A., Gonda, S. R. and Hartzell, C. R. (1997). "Skeletal muscle satellite cells cultured in simulated microgravity." In Vitro Cell Dev Biol Anim 33 (5): 386-91.

Montjovent, M. O., Burri, N., Mark, S., Federici, E., Scaletta, C., Zambelli, P. Y., Hohlfeld, P., Leyvraz, P. F., Applegate, L. L. and Pioletti, D. P. (2004). "Fetal bone cells for tissue engineering." Bone 35 (6): 1323-33.

Mueller, S. M. and Glowacki, J. (2001). "Age-related decline in the osteogenic potential of human bone marrow cells cultured in three-dimensional collagen sponges." J Cell Biochem 82 (4): 583-90.

Muraglia, A., Cancedda, R. and Quarto, R. (2000). "Clonal mesenchymal progenitors from human bone marrow differentiate in vitro according to a hierarchical model." J Cell Sci 113 (Pt 7) : 1161-6.

Muraglia, A., Cancedda, R. and Quarto, R. (2000). "Clonal mesenchymal progenitors from human bone marrow differentiate in vitro according to a hierarchical model." J Cell Sci 113 (Pt 7) : 1161-6.

Mygind, T., Stiehler, M., Baatrup, A., Li, H., Zou, X., Flyvbjerg, A., Kassem, M. and Bunger, C. (2007). "Mesenchymal stem cell ingrowth and differentiation on coralline hydroxyapatite scaffolds." Biomaterials 28 (6): 1036-47.

Nakahara, H., Dennis, J. E., Bruder, S. P., Haynesworth, S. E., Lennon, D. P. and Caplan, A. I. (1991). "In vitro differentiation of bone and hypertrophic cartilage from periosteal-derived cells." Exp Cell Res 195 (2): 492-503.

Okumura, M., Ohgushi, H., Dohi, Y., Katuda, T., Tamai, S., Koerten, H. K. and Tabata, S. (1997). "Osteoblastic phenotype expression on the surface of hydroxyapatite ceramics." J Biomed Mater Res 37 (1): 122-9.

Ong, S. Y., Dai, H. and Leong, K. W. (2006). "Hepatic differentiation potential of commercially available human mesenchymal stem cells." Tissue Eng 12 (12): 3477-85.

Oswald, J., Boxberger, S., Jorgensen, B., Feldmann, S., Ehninger, G., Bornhauser, M. and Werner, C. (2004). "Mesenchymal stem cells can be differentiated into endothelial cells in vitro." Stem Cells 22 (3): 377-84.

Parikh, S. N. (2002). "Bone graft substitutes: past, present, future." J Postgrad Med 48 (2): 142-8.

Phinney, D. G., Kopen, G., Righter, W., Webster, S., Tremain, N. and Prockop, D. J. (1999). "Donor variation in the growth properties and osteogenic potential of human marrow stromal cells." J Cell Biochem 75 (3): 424-36.

Pioletti, D. P., Montjovent, M. O., Zambelli, P. Y. and Applegate, L. (2006). "Bone tissue engineering using foetal cell therapy." Swiss Med Wkly 136 (35-36): 557-60.

Pittenger, M. F., Mackay, A. M., Beck, S. C., Jaiswal, R. K., Douglas, R., Mosca, J. D., Moorman, M. A., Simonetti, D. W., Craig, S. and Marshak, D. R. (1999). "Multilineage potential of adult human mesenchymal stem cells." Science 284 (5411): 143-7.

Prolo, D. J. and Rodrigo, J. J. (1985). "Contemporary bone graft physiology and surgery." Clin Orthop Relat Res(200): 322-42.

Quarto, R., Mastrogiacomo, M., Cancedda, R., Kutepov, S. M., Mukhachev, V., Lavroukov, A., Kon, E. and Marcacci, M. (2001). "Repair of large bone defects with the use of autologous bone marrow stromal cells." N Engl J Med 344 (5): 385-6.

Rai, B., Ho, K. H., Lei, Y., Si-Hoe, K. M., Jeremy, T. C., Yacob, K. B., Chen, F., Ng, F. C. and Teoh, S. H. (2007). "Polycaprolactone-20% tricalcium phosphate scaffolds in combination with platelet-rich plasma for the treatment of critical-sized defects of the mandible: a pilot study." J Oral Maxillofac Surg 65 (11): 2195-205.

Rai, B., Oest, M. E., Dupont, K. M., Ho, K. H., Teoh, S. H. and Guldberg, R. E. (2007). "Combination of platelet-rich plasma with polycaprolactone-tricalcium phosphate scaffolds for segmental bone defect repair." J Biomed Mater Res A 81 (4): 888-99.

Rai, B., Teoh, S. H., Ho, K. H., Hutmacher, D. W., Cao, T., Chen, F. and Yacob, K. (2004). "The effect of rhBMP-2 on canine osteoblasts seeded onto 3D bioactive polycaprolactone scaffolds." Biomaterials 25 (24): 5499-506.

Rasmusson, I., Ringden, O., Sundberg, B. and Le, B. K. (2003). "Mesenchymal stem cells inhibit the formation of cytotoxic T lymphocytes, but not activated cytotoxic T lymphocytes or natural killer cells." Transplantation 76 (8): 1208-13.

Reyes, M., Lund, T., Lenvik, T., Aguiar, D., Koodie, L. and Verfaillie, C. M. (2001). "Purification and ex vivo expansion of postnatal human marrow mesodermal progenitor cells." Blood 98 (9): 2615-25.

Lanza, R., Langer, R. and Vacanti, J. (2007). Principles of Tissue Engineering, Third Edition, Academic Press.

Rodriguez, J. P., Montecinos, L., Rios, S., Reyes, P. and Martinez, J. (2000). "Mesenchymal stem cells from osteoporotic patients produce a type I collagen-deficient extracellular matrix favoring adipogenic differentiation." J Cell Biochem 79 (4): 557-65.

Rosser, A. E. and Dunnett, S. B. (2003). "Neural transplantation in patients with Huntington's disease." CNS Drugs 17 (12): 853-67.

Rouwkema, J., Rivron, N. C. and van, B. C. (2008). "Vascularization in tissue engineering." Trends Biotechnol.

Rzhaninova, A. A., Gornostaeva, S. N. and Goldshtein, D. V. (2005). "Isolation and phenotypical characterization of mesenchymal stem cells from human fetal thymus." Bull Exp Biol Med 139 (1): 134-40.

Sakaguchi, Y., Sekiya, I., Yagishita, K., Ichinose, S., Shinomiya, K. and Muneta, T. (2004). "Suspended cells from trabecular bone by collagenase digestion become virtually identical to mesenchymal stem cells obtained from marrow aspirates." Blood 104 (9): 2728-35.

Salgado, A. J., Coutinho, O. P. and Reis, R. L. (2004). "Bone tissue engineering: state of the art and future trends." Macromol Biosci 4 (8): 743-65.

Salgado, A. J., Coutinho, O. P. and Reis, R. L. (2004). "Bone tissue engineering: state of the art and future trends." Macromol Biosci 4 (8): 743-65.

Sarugaser, R., Lickorish, D., Baksh, D., Hosseini, M. M. and Davies, J. E. (2005). "Human umbilical cord perivascular (HUCPV) cells: a source of mesenchymal progenitors." Stem Cells 23 (2): 220-9.

Schantz, J. T., Hutmacher, D. W., Chim, H., Ng, K. W., Lim, T. C. and Teoh, S. H. (2002). "Induction of ectopic bone formation by using human periosteal cells in combination with a novel scaffold technology." Cell Transplant 11 (2): 125-38.

Schimming, R. and Schmelzeisen, R. (2004). "Tissue-engineered bone for maxillary sinus augmentation." J Oral Maxillofac Surg 62 (6): 724-9.

Service, R. F. (2000). "Tissue engineers build new bone." Science 289 (5484): 1498-500.

Sgodda, M., Aurich, H., Kleist, S., Aurich, I., Konig, S., Dollinger, M. M., Fleig, W. E. and Christ, B. (2007). "Hepatocyte differentiation of mesenchymal stem cells from rat peritoneal adipose tissue in vitro and in vivo." Exp Cell Res 313 (13): 2875-86.

Sikavitsas, V. I., Bancroft, G. N., Holtorf, H. L., Jansen, J. A. and Mikos, A. G. (2003). "Mineralized matrix deposition by marrow stromal osteoblasts in 3D perfusion culture increases with increasing fluid shear forces." Proc Natl Acad Sci U S A 100 (25): 14683-8.

Sikavitsas, V. I., Bancroft, G. N. and Mikos, A. G. (2002). "Formation of three-dimensional cell/polymer constructs for bone tissue engineering in a spinner flask and a rotating wall vessel bioreactor." J Biomed Mater Res 62 (1): 136-48.

Sikavitsas, V. I., Bancroft, G. N. and Mikos, A. G. (2002). "Formation of three-dimensional cell/polymer constructs for bone tissue engineering in a spinner flask and a rotating wall vessel bioreactor." J Biomed Mater Res 62 (1): 136-48.

Sikavitsas, V. I., van, D. J., Bancroft, G. N., Jansen, J. A. and Mikos, A. G. (2003). "Influence of the in vitro culture period on the in vivo performance of cell/titanium bone tissue-engineered constructs using a rat cranial critical size defect model." J Biomed Mater Res A 67 (3): 944-51.

Simmons, P. J. and Torok-Storb, B. (1991). "Identification of stromal cell precursors in human bone marrow by a novel monoclonal antibody, STRO-1." Blood 78 (1): 55-62.

Singh, H., Ang, E. S., Lim, T. T. and Hutmacher, D. W. (2007). "Flow modeling in a novel non-perfusion conical bioreactor." Biotechnol Bioeng 97 (5): 1291-9.

Singh, H., Teoh, S. H., Low, H. T. and Hutmacher, D. W. (2005). "Flow modelling within a scaffold under the influence of uni-axial and bi-axial bioreactor rotation." J Biotechnol 119 (2): 181-96.

Song, K., Liu, T., Cui, Z., Li, X. and Ma, X. (2008). "Three-dimensional fabrication of engineered bone with human bio-derived bone scaffolds in a rotating wall vessel bioreactor." J Biomed Mater Res A 86 (2): 323-32.

Stiehler, M., Bunger, C., Baatrup, A., Lind, M., Kassem, M. and Mygind, T. (2008). "Effect of dynamic 3-D culture on proliferation, distribution, and osteogenic differentiation of human mesenchymal stem cells." J Biomed Mater Res A.

Suzuki, Y., Kim, K. J., Kotake, S. and Itoh, T. (2001). "Stromal cell activity in bone marrow from the tibia and iliac crest of patients with rheumatoid arthritis." J Bone Miner Metab 19 (1): 56-60.

Touraine, J. L., Roncarolo, M. G., Bacchetta, R., Raudrant, D., Rebaud, A., Laplace, S., Cesbron, P., Gebuhrer, L., Zabot, M. T., Touraine, F. and et (1993). "Fetal liver transplantation: biology and clinical results." Bone Marrow Transplant 11 Suppl 1 : 119-22.

Ueda, M., Yamada, Y., Kagami, H. and Hibi, H. (2008). "Injectable bone applied for ridge augmentation and dental implant placement: human progress study." Implant Dent 17 (1): 82-90.

van, D. J., Bancroft, G. N., Sikavitsas, V. I., Spauwen, P. H., Jansen, J. A. and Mikos, A. G. (2003). "Flow perfusion culture of marrow stromal osteoblasts in titanium fiber mesh." J Biomed Mater Res A 64 (2): 235-41.

Weissman, I. L. (2000). "Translating stem and progenitor cell biology to the clinic: barriers and opportunities." Science 287 (5457): 1442-6.

Woodbury, D., Schwarz, E. J., Prockop, D. J. and Black, I. B. (2000). "Adult rat and human bone marrow stromal cells differentiate into neurons." J Neurosci Res 61 (4): 364-70.

Yamada, Y., Nakamura, S., Ito, K., Kohgo, T., Hibi, H., Nagasaka, T. and Ueda, M. (2008). "Injectable tissue-engineered bone using autogenous bone marrow-derived stromal cells for maxillary sinus augmentation: clinical application report from a 2-6-year follow-up." Tissue Eng Part A 14 (10): 1699-707.

Yu, X., Botchwey, E. A., Levine, E. M., Pollack, S. R. and Laurencin, C. T. (2004). "Bioreactor-based bone tissue engineering: the influence of dynamic flow on osteoblast phenotypic expression and matrix mineralization." Proc Natl Acad Sci U S A 101 (31): 11203-8.

Zein, I., Hutmacher, D. W., Tan, K. C. and Teoh, S. H. (2002). "Fused deposition modeling of novel scaffold architectures for tissue engineering applications." Biomaterials 23 (4): 1169-85.

Zhang, Z. Y., Teoh, S. H., Chong, M. S., Schantz, J. T., Fisk, N. M., Choolani, M. A. and Chan, J. (2009). "Superior Osteogenic Capacity for Bone Tissue Engineering of Fetal Compared To Perinatal and Adult Mesenchymal Stem Cells." Stem Cells 27 (1): 126-37.

Zhang, Z. Y., Teoh, S. H., Chong, W. S., Foo, T. T., Chng, Y. C., Choolani, M. and Chan, J. (2009). "A biaxial rotating bioreactor for the culture of fetal mesenchymal stem cells for bone tissue engineering." Biomaterials 30 (14): 2694-704.

Zhao, L. R., Duan, W. M., Reyes, M., Keene, C. D., Verfaillie, C. M. and Low, W. C. (2002). "Human bone marrow stem cells exhibit neural phenotypes and ameliorate neurological deficits after grafting into the ischemic brain of rats." Exp Neurol 174 (1): 11-20.

Zhao, Z., Liao, L., Cao, Y., Jiang, X. and Zhao, R. C. (2005). "Establishment and properties of fetal dermis-derived mesenchymal stem cell lines: plasticity in vitro and hematopoietic protection in vivo." Bone Marrow Transplant 36 (4): 355-65.

Zuk, P. A., Zhu, M., Mizuno, H., Huang, J., Futrell, J. W., Katz, A. J., Benhaim, P., Lorenz, H. P. and Hedrick, M. H. (2001). "Multilineage cells from human adipose tissue: implications for cell-based therapies." Tissue Eng 7 (2): 211-28.

The Future of Cell Therapy and Tissue Engineering in Cardiovascular Disease: The New Era of Biological Therapeutics

Sepideh Heydarkhan-Hagvall[1], Ali Nsair[1],
Ramin E. Beygui[2] and Richard J. Shemin[1]
[1]*David Geffen School of Medicine at UCLA*, [2]*Stanford University School of Medicine*
USA

1. Introduction

Cardiovascular disease remains a major cause of morbidity and mortality worldwide, resulting 16.7 million deaths each year, accounting for 29% of all deaths globally (WHO, 2006). The major modality of death due to cardiovascular disease remains acute myocardial infarctions (MI) and sudden cardiac death with more than one third of first time MI not making it to the hospital. MIs are mostly caused by sudden occlusion of a coronary artery secondary to plaque rupture and acute thrombosis resulting in loss of oxygen supply to the territory affected; if flow is not restored in a timely fashion, myocardial death ensues. The resultant of that is loss of contractile function, cardiac remodeling and eventual heat failure. The resultant compensatory mechanism however, is cardiac remodeling with fibrous tissue replacing the damaged myocardium early on and overcompensation of the remaining myocardium to maintain cardiac output. This eventually leads to negative remodeling, ventricular dilatation, and eventual loss of pump function and heart failure ensues. Furthermore, fibrous tissue deposition and structural remodeling of the heart in the setting of heart failure leads to electrical uncoupling of myocardium from fibrous tissue and development of re-entrant electrical circuits that lead to the development of significant ventricular dysrrythmias. This increases the risk of sudden cardiac death in patients with heart failure whether this is due to myocyte loss from heart attacks or other causes of heart failure (Leor J 2000; Anversa P 2002; Li SC 2009). Scar formation secondary to cardiac remodeling and fibrosis results in increased risk of fatal dysrrythmias as this tissue becomes a nidus for dyrrthmia origination and propagation. (Strauer BE 2002; Stamm C 2003; Galiñanes M 2004; Pittenger MF 2004; Wollert KC 2004; Dib N 2005; Dimmeler S 2005; Patel AN 2005; Deindl E 2006; Engelmann MG 2006; Schächinger V 2006; Tse HF 2006; Wu KH 2006; Kissel CK 2007; Stamm C 2007; Theiss HD 2007; Yerebakan C 2008).

Although there have been great advances in the knowledge of the mechanisms and treatments of myocardial, therapies for cardiac regeneration and repair are still not available for end stage heart disease. Our current treatments for heart failure are only temporizing whether it is medications such as beta blockers, angiotensin converting enzyme (ACE) inhibitors, diuretics and other classes or devices such as implantable defibrillators and bi-ventricular pacemakers (Moss AJ 2009). Once end stage heart failure ensues, one year survival rates can be as low as 50% for New York Heart Association (NYHA) class III and IV heart failure; the only options then to restore pump function are mechanical ventricular devices (VADs) (Birks EJ 2006) as destination therapy or ultimately open heart transplantation. However, due to the high cost of VADs and the shortage of donor organs these are not viable options as demand for a transplant is ever increasing with a relatively fixed supply of donor hearts.

The concept of the heart as being a post mitotic organ with no cardiomyocyte cell generation and turnover has been challenged over the past decade with culminating evidence that there is slow process of cell turnover in the heart postnatally. This process appears to be slow with a 1% per year turnover rate at the age of 20 decreasing to about 0.45% per year at the age of 75. Fewer than 50% of cardiomyocytes are exchanged during a normal life span (Bergmann O 2009). Although there appears to be postnatal cardiomyocyte turnover and generation, this is insufficient to replenish lost heart cells (myocytes) post MIs and in the setting of non-ischemic cardiomyopathies. Therefore the need for regenerative therapies for heart failure is becoming even more so important with an aging population as more patients are surviving their heart attacks and progressing on to develop heart failure (Akins 2002; Wu KH 2006; Li SC 2009).

2. Tissue Engineering

It is well known that the heart is one of the least regenerative organs in the human body (Wu KH 2006). Nevertheless, in recent years many scientists have envisioned the possibility of regeneration of the myocardium and reproduction of autologous heart tissue *in vitro* or *in vivo* (Zund G 1996; Shimizu T 2002; Leor J 2004; Shin M 2004; Zammaretti P 2004; Leor J 2005; Shaoping Zhong 2005; Ishii O 2006; Sekine H 2006; Zimmerman WH 2006). Heart tissue is composed of myocytes that are arranged in parallel and series so that contraction will generate a coordinated force to empty the ventricle. In brief, myocytes are elongated cells that are dedicated to contraction with most of the cellular cytoplasm occupied by actin/myosin chains, the structures responsible for contraction. They are connected by intercalated discs which conduct electrochemical potentials directly between the cytoplasms of the adjunct cells via gap junctions.

Gap junctions allow action potentials to spread directly between cells by depolarizing the heart via K^+/Na^+ exchange channels. This coordinated wave of electrical discharge coupled with mechanical contraction of the myocardium with each cardiac cycle is essential for normal pump function (Jansen JA 2009). Therefore, one can foresee that any regenerative therapy whether it is cell or tissue implantation will have to ensure appropriate incorporation of the transplanted tissue in a geometric fashion to ensure synchronous contraction with electrical and mechanical coupling. Additionally, the heart also utilizes a unique system for vascularization of the metabolically demanding tissues with a complex vascular bed (Chang AC 2006; Zimmerman WH 2006). This network would be essential for the survival, engraftment and functional incorporation of transplanted tissue.

Tissue engineering in general terms is a process that involves the reconstruction of tissue equivalents by combining biomaterials and living cells, to be used together in repair, maintenance and replacement and augmentation of native tissues or organs (Lanza RP 2000). The complexity of the human heart highlights the current hurdles that need to be overcome before the potential for regenerative therapies and tissue engineering for heart failure are fully realized. These challenges include, but are not limited to, the choice of cells/stem cells, processing, differentiation, engraftment, electrical and mechanical coupling of transplanted and native tissue and appropriate vascularization of the newly incorporated tissue, to name a few. The ideal myocardial construct should mimic the morphological, physiological and functional properties of the native cardiac muscle it intends to replace and needs to remain viable and functional long term after implantation.

Cardiovascular tissue engineering is an emerging field with an enormous potential for revolutionizing the next generation of heart failure therapies; we are witnessing the evolution of therapies from pharmaceuticals and device treatments to the era of biological therapies with cell and engineered tissue. This may eventually become a standard therapy for all heart failure patients post MI and also in the setting of non-ischemic cardiomyopathies to replace damaged and lost myocardium and prevent progression to end stage heart failure. For cell replacement therapy and tissue regeneration to be successful, it is essential to generate sufficient cells in a reliable and reproducible manner with the appropriate functional phenotype to replace the damaged tissue.
Not only do replacement cells or tissues need to be functional (for e.g. contraction in the case of cardiomyocytes, secretory in the case of pancreatic islet cells), these cells have to interact with their environment. This can be secretion of extracellular matrix, or maintenance of tissue and organ homeostasis through regulatory feedback mechanisms (Leor J 2004; Zammaretti P 2004; Leor J 2005; Zhong S 2005; Zimmerman WH 2006). Therefore, selecting the appropriate biomaterials and suitable cell source will be critical for the success of this strategy.

Cardiovascular tissue engineering is a materials-based approach and involves preformed three-dimensional (3D) scaffolds in the form of mesh, patch, or foam. This inert biocompatible material will then implanted with differentiated cells to form appropriate functional tissue to be transplanted in-vivo. This has been the focus of intense research in the cardiovascular tissue regeneration field over the past decade. (Zund G 1996; Shimizu T 2002; Leor J 2004; Shin M 2004; Zammaretti P 2004; Leor J 2005; Shaoping Zhong 2005; Ishii O 2006; Sekine H 2006; Zimmerman WH 2006).

The challenges of producing engineered tissues for *in vivo* use are considerable. The engineered tissue must provide an effective, durable, non-thrombogenic and non-immunogenic substitute that will display functional and morphological properties of the cardiovascular system. Furthermore, this material must possess repair and remodeling capabilities, and retain its viability after implantation (Langer R 1993; Lanza RP 2000; Atala A 2002).

In order for a potential cell source to represent a suitable candidate for the construction of engineered grafts, several criteria must be considered. The ideal cell should be autologous

and available in large quantities. If not autologous, the cells used must be non-immunogenic as to avoid the need for immunosuppression and all the associated difficulties and complications that are encountered in the transplant population. These range from the risk of graft rejection, increased infection and malignancy in patients on immunosuppression. It should also have the capacity to proliferate and differentiate *in vitro*, in a manner that can be reliably reproduced and controlled, and eventually automated for mass production to be practically viable in a clinical setting (Atala A 2002).
There are a number of limitations in the use of differentiated cells in tissue engineering. It is a challenge to generate sufficient numbers of a single cell type, to assemble the needed mixture of multiple cell phenotypes, and to maintain stable phenotypes as needed (Atala A 2002). Thus lies the promise of stem cells as an endless source that can be manipulated and combined with nanostructures and scaffolds to produce engineered tissue grafts to be used in clinical therapies.

Among different pharmaceutical and surgical treatments, the focus of emerging research is to use stem cells in order to heal or replace damaged cardiac tissues (Moffett BS 2006; Ruvinov E 2008; Segers VF 2008). Several kinds of stem cells are potentially useful because of their ability to self renew and differentiate into various types of cells in the body. Embryonic stem cells (ESCs) are capable of indefinite expansion and are pluripotent (able to differentiate into any cell type). However human embryonic stem cells (hESCs) although are immortal and pluripotent, they are burdened with ethical concerns due to their derivation from human embryos. Therefore their use has been very controversial and they have been limited to a very few lines only used in laboratory settings.

The ethical concerns surrounding these cells prevent them from being used in any clinical setting. Furthermore, issues with their immunogenecity, the risk of teratoma formation, the difficulties associated with their differentiation fate, and the ethical challenges arising from their embryonic origin have all excluded them as prime candidates for regenerative medicine (Petersen T 2007).

However, a recent breakthrough has completely revolutionized the field and taken it in a whole new direction. Using four transcription factors, Dr Yamanaka's group was able to reprogram mouse fibroblasts to a stem-cell like state creating induced pluripotent stem cells (iPS). Since then, many groups have been able to reprogram human, porcine, rat and murine somatic cells into these iPS cells using different approaches (Takahashi K 2006; Okita K 2007; Takahashi K 2007; Okita K 2008; Ezashi T 2009; Zhao XY 2009). These iPS cells have been differentiated into different tissue types and promise to revolutionize the regenerative cell and tissue replacement field.

Adult stem cells have been touted as a potential source of stem cells for cell and tissue replacement therapies although these cells present their own unique challenges. They have been found to vary in quality depending on the age and health of the donor/patient and differentiation is often restricted to the origin lineage of the cell source (multipotent) (Smith S 2007). An important consideration is the expansion potential of the adult stem cells to be considered as a viable source for regenerative cell therapy potential. Unlike ESCs that have an almost unlimited expansion capability, adult stem cells may be capable of only a limited

number of doubling cycles. Therefore, candidates that are able to provide an abundant number of cells that would be required for cell and tissue replacement therapies maybe embryonic stem cells, bone marrow derived cells, bone marrow derived mesenchymal stem cells, endothelia progenitor cells, skeletal myoblasts, adipose derived stem cells and induced pluripotent stem cells.
This chapter will focus on adipose derived stem cells (ASCs) and induced pluripotent stem cells (iPS).

3. Adipose Derived Stem Cells

For the production of clinically useful tissue-engineered constructs to replace cardiovascular structures, it is critical to identify a suitable cell source that is autologous and capable of differentiating into the desired cell types. Adipose tissue represents an easily accessible and abundant source of cells. Furthermore, this source represents a potential adult stem cell reservoir for each individual (Aust L 2004). In recent years, interest has rapidly grown in the developmental plasticity and therapeutic potential of stromal cells isolated from human subcutaneous adipose tissue.

Adipose tissue represents an abundant, practical, and appealing source of donor tissue for autologous cell replacement (Planat-Benard V 2004a; Planat-Benard V 2004b; Katz AJ 2005). Adipose tissue is of mesodermic origin, consisting of mature adipocytes and the stromal vascular fraction (SVF). SVF is a heterogeneous cell population, consisting of vascular cells (endothelial cells (ECs) and smooth muscle cells (SMCs)), blood cells, and a fibroblast-like multipotential stem cell population, termed adipose-derived stem cells (ASCs).
These cells can be isolated in large numbers with minimally invasive techniques from liposuctions and grown easily under standard tissue culture conditions (Zuk PA 2001; Zuk PA 2002), and therefore represent an excellent abundant stem cell source with a high therapeutic potential. Since its initial characterization, several groups have demonstrated the ASC population of stem cells within the SVF of subcutaneous adipose tissue displays multilineage developmental plasticity *in vitro* and *in vivo*.

It has been shown that adult stem cells from white adipose tissues can differentiate into multiple cell phenotypes, including the adipocyte, chondrocyte, epithelial, hematopoietic, hepatocyte, neuronal, myogenic, and osteoblast lineages (Halvorsen YC 2000; Halvorsen YD 2001; Zuk PA 2001; Erickson GR 2002; Mizuno H 2002; Safford KM 2002; Zuk PA 2002; Cousin B 2003; Gimble J 2003a; Gimble JM 2003b; Kang SK 2003; Kim DH 2003; Rangappa S 2003; Miranville A 2004; Planat-Benard V 2004a; Planat-Benard V 2004b; Rodriguez LV 2006).
Research efforts towards understanding the nature of how ASCs differentiate into cells of the cardiovascular lineage, including ECs and SMCs, have become more intensive in the past years. Recent studies have shown that ASCs are capable of differentiating into either SMCs (Lee WC 2006; Rodriguez LV 2006) or cells of an EC phenotype (Planat-Benard V 2004; Dimuzio P 2006; Wosnitza M 2007). However, the differentiation protocols employed in those studies are rather diverse, using complex cell culture media that are supplemented with various growth factors. Moreover, the culture time periods that were reported to be necessary for a successful differentiation of ASCs into cardiac cells were rather exhaustive,

making this an unpractical and unfavorable option for potential clinical applications. However, there has yet to be significant research into the relationship between patient's age, gender, comorbidities and the differentiation capabilities of these adipose derived cells.

We have focused on the differentiation of hASCs towards the endothelial and SMC lineage. We have been able to identify specific parameters that are crucial for an efficient and easily reproducible differentiation of human ASCs into ECs and SMCs within 14 days of *in vitro* culture. We found that hASCs, which were cultured in endothelial specific growth medium-2 (EGM-2), exhibited morphological features of mature ECs and expressed EC-specific markers including CD31, CD144 and vWF (Heydarkhan-Hagvall S 2008).
This differentiation pattern appeared to be specific to EGM-2-treated hASCs, cultured in a high-density environment. It has been reported that cell-cell junctions between ECs through CD31 and CD144 are critical for the establishment of a primary vascular network (Dejana E 1995; Baldwin HS 1996; Dejana 1996; Drake CJ 1998).
The failure of hASCs that were cultured in Dulbecco's modified Eagle's medium-20% (DMEM-20%) FBS to form such tubular network may, in part, be due to the absence of these necessary cell-to-cell contacts. Our immunocytochemistry and fluorescence-activated cell sorter (FACS) analyses demonstrated the presence of both immature and mature EC markers, such as CD34, Flk-1/KDR, CD144, CD31 and vWF, in the specifically induced hASCs (Heydarkhan-Hagvall S 2008).

Previous studies have identified three markers, CD133, CD34 and Flk-1/KDR, in early endothelial progenitor cells. In addition, mature ECs are known to express high levels of Flk-1/KDR, CD31, CD144, and vWF, suggesting that our EGM-2-induced hASCs may be assuming a more mature EC phenotype. To date, it is not completely understood as to when an endothelial progenitor cell turns into mature, fully differentiated ECs *in vivo*. One possibility could be the loss of CD133 and a parallel or subsequent expression of vWF in conjunction with the appearance of other endothelial characteristics (Hristov M 2003; Urbich C 2004). An interesting observation was that the differentiation capacity of hASCs towards ECs was significantly diminished when the cells were passaged, whereas the age of the donor was not a critical factor.
In contrast, when hASCs were cultured for 14 days in DMEM-20% FBS they expressed SMC-specific markers including SM myosin, h-caldesmon, basic calponin and SM-α-actin a phenomenon that was independent of passage number, yet affected by the donor age (Heydarkhan-Hagvall S 2008). As the primary function of mature SMC is contraction, the complement of contractile, structural, and regulatory proteins expressed in fully differentiated SMC provides markers of differentiation status (Kuro-o M 1989; Sobue K 1999). These include SM myosin, h-caldesmon, basic calponin, and SM-α-actin. SM-α-actin is an important protein for SMCs structure and contraction, and it is upregulated with differentiation. Indeed, SM-α-actin has been used as a marker of early SMC differentiation, followed by calponin, h-caldesmon and SM myosin (Owens GK 1995; Hungerford JE 1996; Katoh Y 1996; Thyberg J 1996; Hungerford JE 1999).
It is well known that SMCs grown in vitro can undergo a phenotypic cell transition from the normally quiescent 'contractile' status observed *in vivo* to a proliferative-secretory state (Thyberg J 1996; Opitz F 2004). The general feature of cultured SMCs is the downregulation of SM myosin expression, particularly of SM2 concomitant with upregulation of nonmuscle

myosin variants MyHC-A pla1, MyHC-A, and MyHC-B (Somlyo AP 1993). These tendencies depend, however, on cellular density, serum concentration, and substrate for attachment. Myosin gene expression can be regulated during early development (effect on the differentiation phase of the SMC lineage program) or in adulthood (alteration of the stability of the fully differentiated SMC phenotype) (Zanellato AMC 1990; Somlyo AP 1993). The general feature of cultured SMCs is the downregulation of SM myosin expression, particularly of SM2 concomitant with upregulation of non-muscle myosin variants MyHC-A pla1, MyHC-A, and MyHC-B (Somlyo AP 1993). These tendencies depend, however, on cellular density, serum concentration, and substrate for attachment. Myosin gene expression can be regulated during early development (effect on the differentiation phase of the SMC lineage program) or in adulthood (alteration of the stability of the fully differentiated SMC phenotype) (Zanellato AMC 1990; Somlyo AP 1993). High levels of SM-α-actin and basic calponin in particular were observed in the present study, with lower levels of SM myosin and h-caldesmon-expressing cells. Rodriguez et al. (Rodriguez LV 2006) demonstrated an optimal smooth muscle differentiation, phenotypically and functionally, when hASCs were cultured in specific smooth muscle-inductive medium after 6 weeks. However, our data have shown that a prolonged culture time did not increase the number of SMC marker expressing cells, since a plateau was reached after 14 days of culture.

4. Induced Pluripotent Stem Cells

One of the most exciting and revolutionary breakthroughs in stem cell research has been the induction of pluripotent stem cells from mature somatic cells. It has been shown that four factors present in ESC are sufficient to re-induce a pluripotent state in somatic cells. The reprogramming of somatic cells to a pluripotent state can be achieved by simple retroviral overexpression of specific transcription factors, resulting in induced pluripotent stem (iPS) cells that are almost indistinguishable from ESC (Okita K 2007; Mauritz C 2008; Narazaki G 2008; Winkler ME 2008).

Transduction of fibroblasts with only four transcription factors, Oct4, Sox2, cMyc and Klf4, could dedifferentiate the fibroblasts to cells with almost all features of ESC. The key transcription factors needed for reprogramming appear to be Sox2 and Oct4, whereas the other transcription factor used allow for increasing the efficiency of reprogramming likely by opening up the closed chromatin typical of somatic cells ESC (Mitsui K 2003; Silva J 2006; Takahashi K 2006; Takahashi K 2007; Yu J 2007; Park IH 2008).

Initially, reprogramming was accomplished using integrating retroviral or lentiviral vectors, however these vectors carry the risk of insertional mutagenesis, which will make this method of reprogramming unacceptable for use clinically in patients due to increased risk of tumor formation. However, the problem of using retroviral integrating vectors has been overcome as reprogramming has been achieved by many groups using different modalities as a series of very recent studies demonstrate that this can also be achieved using short term overexpression of the transcription factor using plasmids, nonintegrating adenoviral vectors, or using transposons that are subsequently removed. In contrast to reprogramming by somatic cell nuclear transfer (SCNT) or ESC fusion which were the original methods of inducing pluripotency and were very laborious, technically difficult and time consuming, reprogramming via defined transcription factor requires 3 to 4 weeks, suggesting that many more factors may take part in the very quick reprogramming seen by SCNT or fusion (Yu J 2007; Okita K 2008; Shi Y 2008a; Stadtfeld M 2008a; Shi Y 2008b; Woltjen K 2009; Yu J 2009).

The ability of iPS cells to differentiate into all somatic cell typeshas attracted much interest in the field of regenerative medicine and has been a topic of intense research in recent years (Dai W 2006). This interest has become even more relevant from a clinical practical perspective as these reprogrammed fibroblasts exhibit growth and differentiation characteristics comparable to those of murine ES cells. Ease of generation and lack of immunologic and ethical restrictions will likely make iPS cells a highly valuable cell source for applications in regenerative medicine.

Given the appropriate extracellular signals, murine iPS cells differentiate with high efficiency into multipotent mesodermal progenitor cells that possess the potential to differentiate into functional cells of the cardiovascular and hematopoietic lineages (Schenke-Layland K 2008). When exposed to Collagen IV (ColIV), ESC and iPS cells differentiated into cells that showed expression of genes associated with early mesodermal, cardiovascular, and hematopoietic cells.

A cell capable of differentiating into all cardiovascular cell types has a theoretical advantage for more complete tissue regeneration over transplanting cardiomyocytes alone, as has been demonstrated for ESC derivatives (Dai W 2006). Transplanting partially differentiated cardiovascular progenitor cells, such as the Flk1-positive cells which are committed to a cardiovascular lineage will not be associated with the risk of tumor formation as seen with transplanting undifferentiated ES cells into the heart (Behfar A 2007). Flk1-positive progenitor cells have been identified in both iPS cell-derived embryonic bodies (EBs) and ColIV-differentiated iPS cells, most likely representing a population of multipotent mesodermal progenitor cells (Park C 2005; Kattman SJ 2006; Moretti A 2006; Wu SM 2006). To confirm that those Flk1-expressing cells were capable of generating all cardiovascular cell types, ColIV differentiated Flk1-positive cells have been isolated and been exposed to cardiac, smooth muscle, and endothelial cell-specific differentiation conditions. This resulted in the production of spontaneously beating cell clusters and cells expressing cardiac markers as well as hallmark morphological characteristics of mature cardiomyocytes, including the typical cross-striation and generation of Ca^{2+} (Schenke-Layland K 2008).

Gene expression analysis, immunocytochemistry, contractility, and *in vitro* tube formation assays, as well as acLDL uptake tests, further revealed the successful differentiation of iPS cell-derived Flk1-progenitor cells into functional SMC and EC, findings that were similar to those previously reported for murine ES cells (Nishikawa SI 1998; Yamashita J 2000; McCloskey KE 2006). Although iPS cells contributed to mature cardiovascular cells *in vivo* as well, it will be important to determine in future experiments whether transplanted iPS cells can also integrate and differentiate into adult myocardium. In addition, co-culture of the Flk1-positive cells with OP9-GFP stromal cells in hematopoietic cytokine-containing culture medium conferred differentiation into hematopoietic progenitor cells that expressed c-kit, CD41, and the pan-hematopoietic marker CD45 (Mikkola HK 2006).

Furthermore, these ES and iPS cell-derived hematopoietic progenitors demonstrated a multilineage myeloerythroid differentiation potential. Although ColIV-differentiated iPS cell-derived Flk1-positive progenitor cells had properties comparable to those of the ES cell-derived progenitors, differences did exist (Yamashita J 2000; Kattman SJ 2006; McCloskey

KE 2006; Moretti A 2006; Wu SM 2006). Flk1-positive progenitor cells isolated from the ColIV-exposed cultures also possessed hematopoietic differentiation potential when cultured on OP9 stromal cells or in methylcellulose. In addition, Flk1-positive cells were more frequent in ColIV-differentiated iPS cell cultures compared with ES cell cultures, but whether this is a general property of iPS cells or is specific to the 2D4 iPS cell line will require further study and comparison of multiple lines (Schenke-Layland K 2008).

5. Discussion and Conclusion

The use of living cells as a therapeutic option presents several challenges including identification of a suitable source, development of adequate derivation, maintenance and differentiation methods, and very importantly proof of safety and efficacy. One of the major issues for cardiovascular tissue engineering is determining the ideal cell type for use in regenerative therapies.Many clinical trials have used bone marrow derived mononuclear cells (BM-MNC) (Schächinger V 2006). These clinical trials have not shown any significant cardiomyocyte regeneration and the results have been mixed at best with no robust improvement in cardiac function (Coombs 2008). However, this trial and others have provided a proof of concept that intracoronary or intramyocardial transplant of autologous adult stem cells is safe with out any evidence of increased mortality in the treated patients. The next generation of clinical studies will need to demonstrate robust cardiomyocyte regeneration, definite improvement of cardiac pump function, and ultimately improved patient survival as the ultimate goal. The availability of the proposed cell type for regenerative medicine and tissue engineering in sufficient and relatively easily derivable quantities is also critical. To date, engineered tissue constructs containing adult and stem cells have exhibited problems with physical properties, maintenance of cell phenotypes and the host immune response to the engrafted construct. Ideally, the cells used for tissue engineering should have the capacity to proliferate and differentiate in vivo in a manner that can be reproducibly controlled and predicted (Atala A 2002). Due to the high number of cells that is needed for culturing, isolation and expansion require invasive procedures it remains a challenge to generate sufficient numbers of a single cell type, to assemble the needed mixture of multiple cell phenotypes, and to maintain stable phenotypes as needed (Atala A 2002). Furthermore, this process will need to be automated to mass produce sufficient amounts of cells and tissue constructs for clinical therapies on a wide scale.

This discussion also highlights the ushering in of a new era of personalized medicine. One of the possibilities for cell therapies is patient-specific iPS cells generated from his/her own somatic cells to be used to treat that person. These designer iPS cells will be differentiated into the desired tissue types and transplanted in an autologous manner to avoid immune rejection. Although pluripotent hESCs or hiPS cells are non-immunogenic, they loose this characteristic as they become more differentiated with increased risk of immune mediated rejection if transplanted into a non-compatible patient. ASCs derived from patients may be a potential source to use to either differentiate into desired cells and re-transplant into the patient or develop and differentiate iPS cells that would be used in an autologous manner to ensure immunocompatibility as it has been shown that ASCs are much more easily induced into hiPS cells compared to fibroblasts (Sun N 2009).

The isolation of hESCs more than two decades ago was hailed as the beginning of the end of many diseases; but as we can see this potential many years later has not been realized. It was hoped that creating differentiated cells from hESCs might lead to tissue replacement treatments for diseases such as Parkinson's, diabetes, cardiovascular diseases etc. However, the ethical concerns with the use of hESCs due to their origin from human embryos have limited their use to very restricted research with no translational applications to date.
Another obstacle to the use of ESC and iPS derived cells and tissues is the risk of teratoma formation in patients post cell/tissue transplant. The first ever clinical trial to use hESCs was to be done in spinal cord injury patients and was to be launched in 2009 by the biotechnology firm Geron and had FDA approval; before any patients were enrolled, it was put on hold by the FDA due to concerns with results from animal studies regarding higher than acceptable rates of tumor formation in the animal models post ESC transplants.
Moreover, other concerns with ESC based therapies is allogenicity of ESC derived cells which would necessitate life long immunosuppressive therapy so the donor would not reject the transplanted cells or tissue grafts. This is not a favorable option as immunosuppression has many undesirable side effects, such as increased risk of infection and malignancy, and all the side effects of these medications. Thus lies the promise and potential of cells or tissue grafts derived from autologous derived ASCs or patient derived iPS cells, as immunocompatibilty with the host would negate the need for immunosuppression. Although generating patient specific stem cells is, at the current costs to generate good manufacturing processing (GMP) grade cells, not financially tenable, it is not unreasonable to speculate that costs, as with any new technology, will decrease with time making personalized stem cell therapy for at least some diseases possible. One other possibility would be HLA matched banks of hiPs that will be available to use for clinical use that can be matched to the patient of interest.

One other potential for the use of autologous iPS cells lies in the fact that they would encode for the same genetic defects of the patient whether inherited or caused by sporadic genetic mutations. This ability to generate cells with pluripotent characteristics that differentiate into most cell types will also make it possible to generate *in vitro* models of human disease (Dimos JT 2008; Park IH 2008; Wernig M 2008). This presents a great opportunity, for the first time, to study human disease models in vitro and drug discovery directly on diseased or normal human cells derived from iPS cells. Already iPS cells from patient with Parkinson's disease have been generated and differentiated into neurons (Park IH 2008). The other exciting possible application of patient derived iPS cells is the potential for repair of disease causing genetic defects by *in situ* repair of the defect using homologous recombination and re-transplant of these modified cells back into the patient. This has been demonstrated in a mouse model of sickle cell anemia with cure of the mouse from the disease (Hanna J 2007).

Although a number of differentiation protocols induce lineage specification *in vitro*, engraftment of the pluripotent stem cell-derived differentiated cells *in vivo* has been disappointing (Kyba M 2002; Rideout WM 3rd 2002). The availability of pluripotent stem cells will however in the near future influence medicine far and beyond the realm of tissue engineering and replacement. For instance, availability of cells that can generate many differentiated cell types, may lead to the development of protein or small molecule drugs that influence differentiation not only from pluripotent stem cells but also of multipotent stem cells residing in different tissues. The ability to generate cells with pluripotent characteristics that differentiate in most cell types will also make it possible to generate *in vitro* models of human disease (Dimos JT 2008; Park IH 2008; Wernig M 2008). The identification of optimized protocols for the differentiation of ESCs iPS cells into multiple functional cell types *in vitro* and their proper long term engraftment and fate tracking *in vivo* will be the next generation of challenges once the issue of reprogramming and optimal sources of cells is resolved.

Moreover, the risk of oncogenic events caused by the use of potent oncogenes to induce reprogramming to pluripotency and by the random integration of delivery vectors into the genome is a major problem that needs to be overcome before translating iPS cell technology into the clinic. (Marson A 2008; Huangfu D 2008a; Shi Y 2008a; Huangfu D 2008b; Shi Y 2008b) Murine and human iPS cells have recently been derived by using nonintegrative transient expression strategies of the reprogramming factors (Okita K 2008; Okita K 2008; Stadtfeld M 2008a; Stadtfeld M 2008b; Stadtfeld M 2008c).

Regardless of these uncertainties, direct reprogramming of somatic cells to generate patient-matched pluripotent stem cells has the potential to address many of the current limitations and could revolutionize the treatment of many diseases. As with other non-embryonic derived tissue sources, iPS cells escape all of the ethical issues surrounding ESC therapy. The development of efficient, reliable, and easily reproducible differentiation protocols for generating human iPS cell-derived cardiovascular and hematopoietic progenitor cells will facilitate the development of patient-tailored cardiovascular and hematopoietic regenerative therapies as the next generation of approach to these diseases.

It has been shown that hASCs differentiate into different cell types (Halvorsen YC 2000; Zuk PA 2001; Zuk PA 2002; Gimble J 2003a; Gimble J 2003b; Planat-Benard V 2004a; Planat-Benard V 2004b; Zhu Y 2008) and recent work has shown that these hASCs can be induced much more easily into iPS cells than terminally differentiated fibroblasts. This is quite exciting and clinically relevant as hASC-derived hiPS cells can be differentiated into any type of desired cell type (Sun N 2009). ASCs can be a potential alternative cell source for producing iPS cells as they are widely available and easily obtainable from patients.

The expression of the transcription factors that have been used in reprogramming of fibroblasts to iPS is normally very low in adult cells such as fibroblasts. On the other hand these stem cell-related transcription factors, Nanog, Oct-4, Sox-2, and Rex-1 are positively expressed in ASCs (Zhu Y 2008); this makes ASCs easier to reprogram back to an earlier state than fibroblasts since they are not as far along on the differentiation pathway.

The gene expression profiles of the ASCs could be used to identify a subpopulation that is the least differentiated and most easily reprogrammed to pluripoteny (Sun N 2009). The recent advances in iPS cell identification and differentiation and eventual transplant of these iPS derived cells and tissues has ushered in the new era of biological therapeutics to complement the current pharmaceutical modalities. In addition, further development of iPS cell technology may provide a more substantial and dynamic stem cell population available to overcome current stem cell shortages and will very likely be a standard of tissue engineering and patient therapy in the not so distant future.

6. Reference

Akins R (2002). "Can tissue engineering mend broken hearts?" Circ Res 90: 120-122.

Atala A, Lanza RP. (2002). "Methods of Tissue Engineering." Academic Pr.

Aust L, Devlin B, Foster SJ, Halvorsen YD, Hicok K, du LT, Sen A, Willingmyre GD, Gimble JM (2004). "Yield of human adiposederived adult stem cells from liposuction aspirates." Cytotherapy 6: 7-14.

Baldwin HS (1996). "Early embryonic vascular development." Cardiovasc Res 31: E34-E45.

Behfar A, Perez-Terzic C, Faustino RS, Arrell DK, Hodgson DM, Yamada S, Puceat M, Niederländer N, Alekseev AE, Zingman LV, Terzic A (2007). "Cardiopoietic programming of embryonic stem cells for tumor-free heart repair." J Exp Med 204: 405-420.

Bergmann O, Bhardwaj RD, Bernard S, Zdunek S, Barnabé-Heider F, Walsh S, Zupicich J, Alkass K, Buchholz BA, Druid H, Jovinge S, Frisén J (2009). "Evidence for cardiomyocyte renewal in humans." Science 324(5923): 98-102.

Birks EJ, Tansley PD, Hardy J, George RS, Bowles CT, Burke M, Banner NR, Khaghani A, Yacoub MH (2006). "Left ventricular assist device and drug therapy for the reversal of heart failure." N Engl J Med 355(18): 1873-1884.

Chang AC, Towbin JA (2006). "Heart failure in children and young adults: from molecular mechanisms to medical and surgical strategies. Saunders, Elsevier." 828.

Coombs A (2008). "Stem cells for the heart, a new wave of clinical trials Amy Coombs." Nature Reports Stem Cells 10(April).

Cousin B, Andre M, Arnaud E, Penicaud L, Casteilla L (2003). "Reconstitution of lethally irradiated mice by cells isolated from adipose tissue." Biochem. Biophys. Res. Commun 301: 1016-1022.

Dai W, Kloner RA (2006). "Myocardial regeneration by embryonic stem cell transplantation: Present and future trends." Expert Rev Cardiovasc Ther 4: 375-383.

Deindl E, Zaruba MM., Brunner S, Huber B, Mehl U, Assmann G, Hoefer IE, Mueller-Hoecker J, Franz WM (2006). " G-CSF administration after myocardial infarction in mice attenuates late ischemic cardiomyopathy by enhanced arteriogenesis." FASEB J 20: 956-958.

Dejana E (1996). "Endothelial adherense junctions: Implications in the control of vascular permeability and angiogenesis." J Clin Invest 98: 1949-1953.

Dejana E, Corada M, Lampugnani MG. (1995). "Endothelial cell-to-cell junctions." FASEB J 9: 910-918.

Dib N, Michler RE, Pagani FD, Wright S, Kereiakes DJ, Lengerich R, Binkley P, Buchele D, Anand I, Swingen C, Di Carli MF, Thomas JD, Jaber WA, Opie SR, Campbell A, McCarthy P, Yeager M, Dilsizian V, Griffith BP, Korn R, Kreuger SK, Ghazoul M, MacLellan WR, Fonarow G, Eisen HJ, Dinsmore J, Diethrich E (2005). "Safety and feasibility of autologous myoblast transplantation in patients with ischemic cardiomyopathy: four-year follow-up." Circulation 20(112): 1748-1755.

Dimmeler S, Zeiher AM, Schneider MD (2005). "Unchain my heart: the scientific foundations of cardiac repair." J Clin Invest 115: 572-583.

Dimos JT, Rodolfa KT, Niakan KK, Weisenthal LM, Mitsumoto H, Chung W, Croft GF, Saphier G, Leibel R, Goland R, Wichterle H, Henderson CE, Eggan K (2008). "Induced pluripotent stem cells generated from patients with als can be differentiated into motor neurons." Science 321: 1218-1221.

Dimuzio P, Fischer L, McIlhenny S, Dimatteo C, Golesorhki N, Grabo D, Tarola N, Mericli A, Shapiro I, Tulenko T. (2006). "Development of a tissue-engineered bypass graft seeded with stem cells." Vascular Nov-Dec(14(6)): 338-342.

Drake CJ, Hungerford, J. E, Little CD. (1998). "Morphogenesis of the first blood vessels." Ann NY Acad Sci. 857: 155-179.

Engelmann MG, Theiss HD, Hennig-Theiss C, Huber A, Wintersperger BJ, Werle-Ruedinger AE, Schoenberg SO, Steinbeck G, FranzWM (2006). "Autologous bone marrow stem cell mobilization induced by granulocyte colonystimulating factor after subacute ST-segment elevation myocardial infarction undergoing late revascularization: final results from the G-CSF-STEMI (Granulocyte Colony-Stimulating Factor ST-Segment Elevation Myocardial Infarction) trial." J Am Coll Cardiol 48: 1712-1721.

Erickson GR, Gimble JM, Franklin DM, Rice HE, Awad H and Guilak F (2002). "Chondrogenic Potential of Adipose Tissue-Derived Stromal Cells in Vitro and in Vivo " Biochem Biophys Res Commun 290: 763-769.

Ezashi T, Telugu BP, Alexenko AP, Sachdev S, Sinha S, Roberts RM (2009). "Derivation of induced pluripotent stem cells from pig somatic cells." Proc Natl Acad Sci U S A 106(27): 10993-10998.

Galiñanes M, Loubani M, Davies J, Chin D, Pasi J, Bell PR (2004). "Autotransplantation of unmanipulated bone marrow into scarredmyocardium is safe and enhances cardiac function in humans." Cell Transplant 13: 7-13.

Gimble J, Guilak F (2003a). "Adipose-derived adult stem cells: isolation, characterization, and differentiation potential." Cytotherapy 5: 362–369.

Gimble JM, Guilak F (2003b). "Differentiation potential of adipose derived adult stem (ADAS) cells." Curr Top Dev Biol 58: 137-160.

Halvorsen YC, Wilkison WO, Gimble JM (2000). "Adipose-derived stromal cells-their utility and potential in bone formation." Int J Obes Relat Metab Disord 24 Suppl(4): S41-44.

Halvorsen YD, Franklin D, Bond AL et al (2001). "Extracellular matrix mineralization and osteoblast gene expression by human adipose tissue-derived stromal cells." Tissue Eng 7: 729-741.

Hanna J, Wernig M, Markoulaki S, Sun CW, Meissner A, Cassady JP, Beard C, Brambrink T, Wu LC, Townes TM, Jaenisch R (2007). "Treatment of sickle cell anemia mouse model with iPS cells generated from autologous skin." Science 318: 1879-1880.

Heydarkhan-Hagvall S, Schenke-Layland K, Yang JQ, Heydarkhan S, Xu Y, Zuk P, MacLellan WR, Beygui RE (2008). "Human adipose stem cells: a potential cell source for cardiovascular tissue engineering." Cells Tissues Organs 187(4): 263-274.

Hristov M, Erl W, Weber PC (2003). "Endothelial Progenitor Cells, Mobilization, Differentiation, and Homing." Arterioscler Thromb Vasc Biol 23: 1185-1189.

Huangfu D, Maehr R, Guo W, Eijkelenboom A, Snitow M, Chen AE, Melton DA (2008a). "Induction of pluripotent stem cells by defined factors is greatly improved by small-molecule compounds." Nat. Biotechnol. 26: 795-797.

Huangfu D, Osafune K, Maehr R, Guo W, Eijkelenboom A, Chen S, Muhlestein W Melton DA (2008b). "Induction of pluripotent stem cells from primary human fibroblasts with only Oct4 and Sox2." Nat Biotechnol 26: 1269-1275.

Hungerford, JE, Little CD (1999). "Developmental biology of the vascular smooth muscle cell: building a multilayered vessel wall." J Vasc Res 36(1): 2-27.

Hungerford JE, Owens GK, Argraves WS, Little CD (1996). "Development of the aortic vessel wall as defined by vascular smooth muscle and extracellular matrix markers." Dev Biol 178: 375-392.

Ishii O, Shin M, Sueda T, Vacanti JP (2006). "In vitro tissue engineering of a cardiac graft using a degradable scaffold with an extracellular matrix-like topography." J Thorac Cardiovasc Surg 130: 1358-1368.

Jansen JA, van Veen TA, de Bakker JM, van Rijen HV (2009). "Cardiac connexins and impulse propagation." J Mol Cell Cardiol 31.

Kang SK, Lee DH, Bae YC, Kim HK, Baik SY, Jung JS (2003). "Improvement of neurological deficits by intracerebral transplantation of human adipose tissue-derived stromal cells after cerebral ischemia in rats." Exp. Neurol. 183 355-366.

Katoh Y, Periasamy M (1996). "Growth and differentiation of smooth muscle cells during vascular development." Trends Cardiovasc Med 3: 100-106.

Kattman SJ, Huber TL, Keller GM (2006). "Multipotent flk-1 cardiovascular progenitor cells give rise to the cardiomyocyte, endothelial, and vascular smooth muscle lineages." Dev Cell 11: 723-732.

Katz AJ, Tholpady A, Tholpady SS, Shang H, Ogle RC (2005). "Cell Surface and Transcriptional Characterization of Human Adipose-Derived Adherent Stromal (hADAS) Cells." Stem Cells 23(3): 412-423.

Kim DH, Je CM, Sin JY, Jung JS (2003). "Effect of partial hepatectomy on in vivo engraftment after intravenous administration of human adipose tissue stromal cells in mouse." Microsurgery 23: 424-431.

Kissel CK, Lehmann R, Assmus B, Aicher A, Honold J, Fischer-Rasokat U, Heeschen C, Spyridopoulos I, Dimmeler S, Zeiher AM (2007). "Selective functional exhaustion of hematopoietic progenitor cells in the bone marrow of patients with postinfarction heart failure." J Am Coll Cardiol 19(49): 2341-2349.

Kuro-o M, Nagai R, Tsuchimochi H, Katoh H, Yazaki Y, Ohkubo A, and Takaku F (1989). "Developmentally regulated expression of vascular smooth muscle myosin heavy chain isoforms." J Biol Chem 264: 18272-18275.

Kyba M, Perlingeiro RC, Daley GQ (2002). "HoxB4 confers definitive lymphoidmyeloid engraftment potential on embryonic stem cell and yolk sac hematopoietic progenitors." Cell 2002(109): 29-37.

Langer R, Vacanti JP (1993). "Tissue engineering." Science 260(5110): 920-6.

Lanza RP, Langer R, Vacanit J (2000). Principles of Tissue Engineering. San Diego and London, Academic Press.

Lee WC, Rubin JP, Marra KG (2006). "Regulation of alpha-smooth muscle actin protein expression in adipose-derived stem cells." Cells Tissues Organs 183(2): 80-86.

Leor J, Aboulafia-Etzion S, dar A, shapiro L, Barbash IM, Battler A, Granot Y, Cohen S (2000). "Bioengineered cardiac grafts - a new approach to repair the infarcted myocardium?" Circulation 102: 56-61.

Leor J, Amsalem Y, Cohen S (2005). "Cells, scaffolds, and molecules for myocardial tissue engineering." Pharmacology & Therapeutics 105: 151-163.

Leor J, Cohen S(2004). "Myocardial tissue engineering: creating a muscle patch for a wounded heart." Ann. NY. Acad. Sci. May(1015): 312-319.

Li SC, Wang L, Jiang H, Acevedo J, Chang AC, Loudon WG (2009). "Stem cell engineering for treatment of heart diseases: Potentials and challenges." Cell Biol Int 33(3): 255-267.

Marson A, Foreman R, Chevalier B, Bilodeau S, Kahn M, Young RA, Jaenisch R (2008). "Wnt signaling promotes reprogramming of somatic cells to pluripotency." Cell Stem Cell 3: 132-135.

Mauritz C, Schwanke K, Reppel M, Neef S, Katsirntaki K, Maier LS, Nguemo F, Menke S, Haustein M, Hescheler J, Hasenfuss G, Martin U (2008). "Generation of functional murine cardiac myocytes from induced pluripotent stem cells." Circulation 118: 507-517.

McCloskey KE, Stice SL, Nerem RM (2006). "In vitro derivation and expansion of endothelial cells from embryonic stem cells." Methods Mol Biol 330: 287-301.

Mikkola HK, Orkin SH (2006). "The journey of developing hematopoietic stem cells." Development 133: 3733-3744.

Miranville A, Heeschen C, Sengenès C, Curat CA, Busse R, Bouloumié A (2004). "Improvement of Postnatal Neovascularization by Human Adipose Tissue-Derived Stem Cells." Circulation 110: 349-355.

Mitsui K, Tokuzawa Y, Itoh H, Segawa K, Murakami M, Takahashi K, Maruyama M, Maeda M, Yamanaka S (2003). "The homeoprotein Nanog is required for maintenance of pluripotency in mouse epiblast and ES cells." Cell 113: 631-642.

Mizuno H, Zuk P, Zhu M, Lorenz HP, Benhaim P, Hedrick MH (2002). "Myogenic differentiation by human processed lipoaspirate cells." Plast. Reconstr. Surg 199: 199-209.

Moffett BS, Chang AC (2006). "Future pharmacologic agents for treatment of heart failure in children." Pediatr Cardiol 27: 533-551.

Moretti A, Caron L, Nakano A, Lam JT, Bernshausen A, Chen Y, Qyang Y, Bu L, Sasaki M, Martin-Puig S, Sun Y, Evans SM, Laugwitz KL, Chien KR (2006). "Multipotent embryonic Isl1+ progenitor cells lead to cardiac, smooth muscle, and endothelial cell diversification." Cell 127: 1151-1165.

Moss AJ, Hall WJ, Cannom DS, Klein H, Brown MW, Daubert JP, Estes NA 3rd, Foster E, Greenberg H, Higgins SL, Pfeffer MA, Solomon SD, Wilber D, Zareba W; the MADIT-CRT Trial Investigators. (2009). "Cardiac-Resynchronization Therapy for the Prevention of Heart-Failure Events." N Engl J Med Sep 1.

Narazaki G, Uosaki H, Teranishi M, Okita K, Kim B, Matsuoka S, Yamanaka S, Yamashita JK (2008). "Directed and systematic differentiation of cardiovascular cells from mouse induced pluripotent stem cells." Circulation 118: 498-506.

Nishikawa SI, Nishikawa S, Hirashima M, Matsuyoshi N, Kodama H (1998). "Progressive lineage analysis by cell sorting and culture identifies FLK1+VE-cadherin+ cells at a diverging point of endothelial and hemopoietic lineages." Development 125: 1747-1757.

Okita K, Ichisaka T, Yamanaka S (2007). "Generation of germline-competent induced pluripotent stem cells." Nature 448: 313-317.

Okita K, Nakagawa M, Hyenjong H, Ichisaka T, Yamanaka S (2008a). "Induced pluripotent stem cells generated without viral integration." Science 322: 945-949.Okita K, Nakagawa M, Hyenjong H, Ichisaka T, Yamanaka S (2008b). "Generation of mouse induced pluripotent stem cells without viral vectors." Science 322: 949-953.

Opitz F,Schenke Layland K, Richter W, Martin DP, Degenkolbe I, Wahlers T, Stock UA. (2004). "Tissue engineering of ovine aortic blood vessel substitutes using applied shear stress and enzymatically derived vascular smooth muscle cells." Ann Biomed Eng Feb(32(2)): 212-222.

Owens GK (1995). "Regulation of differentiation of vascular smooth muscle cells." Physiol Rev 75(3): 487-517.

Park C, Ma YD, Choi K (2005). "Evidence for the hemangioblast." Exp Hematol 33: 965-970.

Park IH, Arora N, Huo H, Maherali N, Ahfeldt T, Shimamura A, Lensch MW, Cowan C, Hochedlinger K, Daley GQ (2008). "Disease-specific induced pluripotent stem cells." Cell 134: 877-886.

Park IH, Zhao R, West JA, Yabuuchi A, Huo H, Ince TA, Lerou PH, Lensch MW, Daley GQ (2008). "Reprogramming of human somatic cells to pluripotency with defined factors." Nature 451: 141-146.

Patel AN, Geffner L, Vina RF, Saslavsky J, Urschel Jr HC, Kormos R, Benetti F (2005). "Surgical treatment for congestive heart failure with autologous adult stem cell transplantation: a prospective randomized study." J Thorac Cardiovasc Surg 130: 1631-1638.

Petersen T, Niklason L (2007). "Cellular lifespan and regenerative medicine." Biomaterials 28: 3751-3756.

Piero Anversa, Leri A, Kajstura J, Nadal-Ginard B(2002). "Myocyte growth and cardiac repair." J Mol Cell Cardiol 34: 91-105.

Pittenger MF, Martin BJ (2004). "Mesenchymal stem cells and their potential as cardiac therapeutics." Circ Res 95: 9-20.

Planat-Benard V, Menard C, André M, Puceat M, Perez A, Garcia-Verdugo JM, Pénicaud L, Casteilla L (2004a). "Spontaneous Cardiomyocyte Differentiation From Adipose Tissue Stroma Cells." Circ Res 94: 223-229.

Planat-Benard V, Silvestre J, Cousin B, André M, Nibbelink M, Tamarat R, Clergue M, Manneville C, Saillan-Barreau C, Duriez M, Tedgui A, Levy B, Pénicaud L, Casteilla L (2004b). "Plasticity of human adipose lineage cells toward endothelial cells. Physiological and therapeutic perspectives." Circulation 109: 656-663.

Rangappa S, Fen C, Lee EH, Bongso A, Wei ES (2003). "Transformation of adult mesenchymal stem cells isolated from the fatty tissue into cardiomyocytes." Ann. Thorac. Surg 75: 775-779.

Rideout WM 3rd, Hochedlinger K, Kyba M, Daley GQ, Jaenisch R (2002). "Correction of a genetic defect by nuclear transplantation and combined cell and gene therapy." Cell 109: 17-27.

Rodriguez LV, Alfonso Z, Zhang R, Leung J, Wu B, Ignarro LJ (2006). "Clonogenic multipotent stem cells in human adipose tissue differentiate into functional smooth muscle cells." PNAS 103(32): 12167-12172.

Ruvinov E, Dvir T, Leor J, Cohen S (2008). "Myocardial repair:from salvage to tissue reconstruction." Expert Rev Cardiovasc Ther 6: 669-686.

Safford KM, Hicok KC, Safford SD, Halvorsen YDC, Wilkison WO, Gimble JM and Rice HE. (2002). "Neurogenic differentiation of murine and human adipose-derived stromal cells. ." Biochem Biophys Res Commun. 294: 371-379.

Schächinger V, Erbs S, Elsässer A, Haberbosch W, Hambrecht R, Hölschermann H, Yu J, Corti R, Mathey DG, Hamm CW, Süselbeck T, Assmus B, Tonn T, Dimmeler S, Zeiher AM; REPAIR-AMI Investigators (2006). "Intracoronary bone marrow-derived progenitor cells in acute myocardial infarction." N Engl J Med 355(12): 1210-1221.

Schenke-Layland K, Rhodes KE, Angelis E, Butylkova Y, Heydarkhan-Hagvall S, Gekas C, Zhang R, Goldhaber JL, Mikkola HK, Plath K, MacLellan WR (2008). "Reprogrammed Mouse Fibroblasts Differentiate into Cells of the Cardiovascular and Hematopoietic Lineages." STEMCELLS 26: 1537-1546.

Segers VF, Lee RT (2008). "Stem cell therapy for cardiac disease." Nature 451: 937-942.

Sekine H, Shimizu T, Kosaka S, Kobayashi E, Okano T (2006). "Cardiomyocyte bridging between hearts and bioengineered myocardial tissues with mesenchymal transition of mesothelial cells." J Heart Lung Transpl 25: 324-332.

Shi Y, Desponts C, Do JT, Hahm HS, Schöler HR, Ding S (2008a). "Induction of pluripotent stem cells from mouse embryonic fibroblasts by Oct4 and Klf4 with small-molecule compounds." Cell Stem Cell 3: 568-574.

Shi Y, Do JT, Desponts C, Hahm HS, Schöler HR, Ding S (2008b). "A combined chemical and genetic approach for the generation of induced pluripotent stem cells." Cell Stem Cell 5: 525-528.

Shimizu T, Yamato M, Isoi Y, Akutsu T, Setomaru T, Abe K, Kikuchi A, Umezu M, Okano T (2002). "Fabrication of pulsatile cardiac tissue grafts using a novel 3-dimensional cell sheet manipulation technique and temperature-responsive cell culture surfaces." Circ Res 90: 40.

Shin M, Ishii O, Sueda T, Vacanti JP (2004). "Contractile cardiac grafts using a novel nanofibrous mesh." Biomaterials 25: 3717-3723.

Silva J, Chambers I, Pollard S, Smith A (2006). "Nanog promotes transfer of pluripotency after cell fusion." Nature 441: 997-1001.

Smith S, Neaves W, Teitelbaum S, Prentice, DA, Tarne G (2007). "Adult versus embryonic stem cells: treatments." Science 316.

Sobue K, Hayashi K, Nishida W (1999). "Expressional regulation of smooth muscle cell-specific genes in association with phenotypic modulation." Mol Cell Biochem 190: 105-118.

Somlyo AP (1993). "Myosin isoforms in smooth muscle: how may they affect function and structure." J Muscle Res Cell Motil 14: 557-563.

Stadtfeld M, Nagaya M, Utikal J, Weir G, Hochedlinger K (2008a). "Induced pluripotent stem cells generated without viral integration." Science 322: 945-949.

Stadtfeld M, Brennand K, Hochedlinger K (2008b). "Reprogramming of pancreatic beta cells into induced pluripotent stem cells." Curr Biol 18: 890-894.

Stadtfeld M, Maherali N, Breault DT, Hochedlinger K (2008c). "Defining molecular cornerstones during fibroblast to iPS cell reprogramming in mouse." Cell Stem Cell 2(3): 230-240.

Stamm C, Kleine HD, Choi YH, Dunkelmann S, Lauffs JA, Lorenzen B, David A, Liebold A, Nienaber C, Zurakowski D, Freund M, Steinhoff G. I (2007). "ntramyocardial delivery of CD133+ bone marrow cells and coronary artery bypass grafting for chronic ischemic heart disease: safety and efficacy studies." J Thorac Cardiovasc Surg 133: 717-725.

Stamm C, Westphal B, Kleine HD, Petzsch M, Kittner C, Klinge H, Schümichen C, Nienaber CA, Freund M, Steinhoff G (2003). "Autologous bone-marrow stem-cell transplantation for myocardial regeneration." Lancet 361(4): 45-56.

Strauer BE, Brehm M, Zeus T, Köstering M, Hernandez A, Sorg RV, Kögler G, Wernet P (2002). "Repair of infarcted myocardium by autologous intracoronary mononuclear bone marrow cell transplantation in humans." Circulation 106: 1913-1918.

Sun N, Panetta NJ, Gupta DM, Wilson KD, Lee A, Jia F, Hu S, Cherry AM, Robbins RC, Michael T. Longaker MT, Wu JC (2009). "Feeder-free derivation of induced pluripotent stem cells from adult human adipose stem cells." PNAS 106(37): 15720-15725.

Takahashi K, Tanabe K, Ohnuki M, Narita M, Ichisaka T, Tomoda K, Yamanaka S (2007). "Induction of pluripotent stem cells from adult human fibroblasts by defined factors." Cell 131: 861-872.

Takahashi K, Yamanaka S (2006). "Induction of pluripotent stem cells from mouse embryonic and adult fibroblast cultures by defined factors." Cell 126: 663- 676.

Theiss HD, David R, Engelmann MG, Barth A, Schotten K, Naebauer M, Reichart B, Steinbeck G, FranzWM (2007). "Circulation of CD34+ progenitor cell populations in patients with idiopathic dilated and ischaemic cardiomyopathy (DCM and ICM)." Eur Heart J 28: 1258-1264.

Thyberg J (1996). "Differentiated properties and proliferation of arterial smooth muscle cells in culture." Int Rev Cytol 169: 183-265.

Tse HF, Thambar S, Kwong YL, Rowlings P, Bellamy G, McCrohon J, Bastian B, Chan JK, Lo G, Ho CL, Lau CP (2006). "Safety of catheter-based intramyocardial autologous bone marrow cells implantation for therapeutic angiogenesis." Am J Cardiol 1(98): 60-62.

Urbich C, Dimmeler S (2004). "Endothelial Progenitor Cells, Characterization and Role in Vascular Biology " Circ Res 95: 343-353.

Wernig M, Zhao JP, Pruszak J, Hedlund E, Fu D, Soldner F, Broccoli V, Constantine-Paton M, Isacson O, Jaenisch R (2008). "Neurons derived from reprogrammed fibroblasts functionally integrate into the fetal brain and improve symptoms of rats with Parkinson's disease." Proc Natl Acad Sci U S A 105: 5856-5861.

Winkler ME, Mauritz C, Groos S, Kispert A, Menke S, Hoffmann A, Gruh I, Schwanke K, Haverich A, Martin U (2008). "Serum-free differentiation of murine embryonic stem cells into alveolar type II epithelial cells." Cloning Stem Cells 10: 49-64.

Wollert KC, Meyer GP, Lotz J, Ringes-Lichtenberg S, Lippolt P, Breidenbach C, Fichtner S, Korte T, Hornig B, Messinger D, Arseniev L, Hertenstein B, Ganser A, Drexler HI (2004). "Intracoronary autologous bone-marrowcell transfer after myocardial infarction: the BOOST randomised controlled clinical trial." Lancet 364: 141-148.

Woltjen K, Michael IP, Mohseni P, Desai R, Mileikovsky M, Hämäläinen R, Cowling R, Wang W, Liu P, Gertsenstein M, Kaji K, Sung HK, Nagy A (2009). "piggyBac transposition reprograms fibroblasts to induced pluripotent stem cells." Nature 458: 766-770.

Wosnitza M, Hemmrich K, Groger A, Graber S, Pallua N. (2007). "Plasticity of human adipose stem cells to perform adipogenic and endothelial differentiation." Differentiation Jan75(1): 12-23.

Wu KH, Liu YL, Zhou B, Han ZC (2006). "Cellular therapy and myocardial tissue engineering: the role of adult stem and progenitor cells." Eur J Cardiothorac Surg 30: 770-781.

Wu SM, Fujiwara Y, Cibulsky SM, Clapham DE, Lien CL, Schultheiss TM, Orkin SM (2006). "Developmental origin of a bipotential myocardial and smooth muscle cell precursor in the mammalian heart." Cell 127: 1137-1150.

Yamashita J, Itoh H, Hirashima M, Ogawa M, Nishikawa S, Yurugi T, Naito M, Nakao K, Nishikawa S (2000). "Flk1-positive cells derived from embryonic stem cells serve as vascular progenitors." Nature 408: 92-96.

Yerebakan C, Kaminski A, Westphal B, Liebold A, Steinhoff G (2008). "Autologous bone marrow stem cell therapy for the ischemic myocardium during coronary artery bypass grafting. ." Minim InvasiveTher Allied Technol 17: 143-148.

Yu J, Hu K, Smuga-Otto K, Tian S, Stewart R, Slukvin II, Thomson JA (2009). "Human induced pluripotent stem cells free of vector and transgene sequences." Science 324: 797-801.

Yu J, Vodyanik MA, Smuga-Otto K, Antosiewicz-Bourget J, Frane JL, Tian S, Nie J, Jonsdottir GA, Ruotti V, Stewart R, Slukvin II, Thomson JA (2007). "Induced pluripotent stem cell lines derived from human fibroblasts." Science 318: 1917-1920.

Zammaretti P, Jaconi M (2004). "Cardiac tissue engineering: regeneration of the wounded heart." Curr Opin Biotech 15: 430-434.

Zanellato AMC, Borrione AC, Giuriato L, Tonello M, Scannapieco G, Pauletto P, Sartore S (1990). "Myosin isoforms and cell heterogeneity in vascular smooth muscle, I: developing and adult bovine aorta." Dev Biol 141: 431-444.

Zhao XY, Li W, Lv Z, Liu L, Tong M, Hai T, Hao J, Guo CL, Ma QW, Wang L, Zeng F, Zhou Q (2009). "iPS cells produce viable mice through tetraploid complementation." Nature 461(7260): 86-90.

Zhong S, Teo WE, Zhu X, Beuerman R, Ramakrishna S, Yue L, Yung L (2005). "Formation of collagen-glycosaminoglycan blended nanofibrous scaffolds and their biological properties." Biomacromolecules 6: 2998-3004.

Zhu Y, Liu T, Song K, Fan X, Ma X, Cui Z (2008). "Adipose-derived stem cell: a better stem cell than BMSC." Cell Biochem Funct 26(6): 664-675.

Zimmerman WH, Didie M, Doker S, Melnychenko I, Naito H, Rogge C, Tiburcy M, Eschenhagen T (2006). "Heart muscle engineering: an update on cardiac muscle replacement therapy." Cardiovasc Res 71: 419-429.

Zimmerman WH, Melnychenko I, Wasmeier G, Didie M, Naito H, Nixdorff U, Hess A, Budinsky L, Brune K, Michaelis B, Dhein S, Schwoerer A, Ehmke H, Eschenhagen T (2006). "Engineered heart tissue grafts improve systolic and diastolic function in infracted rat hearts." Nat Med 12: 452-458.

Zuk PA, Zhu M, Ashjian P, De Ugarte DA, Huang JI, Mizuno H, Alfonso ZC, Fraser JK, Benhaim P, and Hedrick MH. (2002). "Human adipose tissue is a source of multipotent stem cells." Mol Biol Cell 13: 4279-4295.

Zuk PA, Zhu M, Mizuno H, Huang J, Futrell JW, Katz AJ, Benhaim P, Lorenz HP, Hedrick MH. (2001). "Multilineage cells from human adipose tissue: implications for cell-based therapies." Tissue Eng. 7: 211-228.

Zund G, Breuer CK, Shinoka T, Ma PX, Langer R, Mayer JE, Vacanti JP (1996). "The in vitro construction of a tissue engineered bioprosthetic heart valve." Eur J Cardio Thorac 11: 493-497.

Characteristics of antlerogenic stem cells and their potential application

Marek Cegielski[1,4], Ilona Iżykowska[1], Wojciech Dziewiszek[2,4], Maciej Zatoński[3], Marek Bochnia[3,4] and Olga Kalisiak[5]
[1]Departments of Histology and Embryology, Medical University of Wroclaw
[2] Departments of Pharmacology, Medical University of Wroclaw,
[3]Department of Otolaryngology, Faculty of Dentistry, Medical University of Wroclaw,
[4]Stem Cells spin Ltd. - Wrocław Technology Park, Wroclaw,
[5]Department of Clinical Sciences, Faculty of Veterinary Medicine, Warsaw University of life Sciences SGGW2
Poland
Corresponding author
Marek Cegielski, VD, PhD
Medical University of Wroclaw, Dept. of Histology and Embryology
ul. Chalubinskiego 6a
50-368 Wroclaw, Poland
marcel@hist.am.wroc.pl

1. Introduction – growing antler as a model of tissue regeneration in mammals

Antlers are present in all males in 39 species of the deer family (*Cervidae*) and are a bony product, unlike horns which originate from epidermis in artiodactyls (*Bovidae*). Among reindeer antlers are also present in females, helping them to find fodder. The size and symmetry of the antlers in males are decisive for their hierarchy in a herd, manifest their actual condition and affect the selection of appropriate mating partners. Every year antlers of the deer family are shed and regrown over a short period of time. They develop as elongations of osseous pedicles present in the frontal bone of the skull. The pedicle arises from a specialised region in the frontal bone, the so-called antlerogenic periosteum (Price & Allen, 2004; Price et al., 2005a). Xenogeneic and allogeneic ectopic grafts of the antlerogenic periosteum confirm its antlerogenic potential. At the sites of the implantation a pedicle is formed and in some cases antlers develop. Both organs are built from a specific chondro-osseous tissue which is a mixture of cartilage and bone (Goss & Powel, 1985; Kierdorf & Kierdorf, 2000; Li et al., 2001). During growth of the pedicle, some of the cells of the antlerogenic periosteum migrate to the apical region of the pedicle, where they build its periosteum and participate in creating antler growth centres (Kierdorf et al., 2003; Li & Suttie, 2001). Histological studies of the early stages of the antler regeneration show that these centres create undifferentiated and proliferating clusters of mesenchymal cells (Kierdorf et al., 2003; Li et al., 2005; Price et al., 2005a). The antlerogenic periosteum, the antlerogenic cells of the pedicle and antlers retain their features of embryonic cells for they contain large amounts of glycogen granules (Cegielski et al., 2006; Li & Suttie, 1998; Li & Suttie, 2001). Among the cells of the antlerogenic periosteum, pedicle and mesenchyme of the growing antlers are stem cells, which can differentiate into cartilage, bone and adipose tissue (Berg et al., 2007; Cegielski et al., 2006; Kierdorf et al., 2009; Rolf et. al, 2008).

During the reproduction period, antlers cease to grow and the skin covering them (velvet) is shed. This process coincides with the parallel growth of testosterone concentration, which converts into oestrogen in antler tissues (Bubenik et al., 2005; Muir et al., 1988). Oestrogen inhibits cell proliferation and increases cartilage mineralization via its receptors (Barrell et al., 1999). Until next spring the old antlers remain fixed to the osseous stem and next, they are cast as a result of bone resorption by osteoclasts on the antler-pedicle border. This process is parallel with the decrease in testosterone levels (Goss et al., 1992). The ossified antlers are not a dead structure, because one month before they are cast, inside of them, there are still blood vessels and bone remodelling processes by osteoclasts still take place (Rolf & Enderle, 1999). In red deer, as soon as the old set of antlers is cast, a new one is grown. The antler regenerating process begins before the old set is cast, which can be seen in the ring created around the apical region of the pedicle (Kierdorf et al., 1994). The antler growth is rapid: within three months the daily increments of 2 cm cause that the growing antlers represent one of the most rapidly growing organs in mammals. The most recent studies demonstrate that testosterone is responsible for antler growth in mature stags. IGF-1 is responsible mainly for body mass increase and its role in regulating antler growth is limited to young specimens (Bartos et al., 2009). Other factors affecting the antler growth are factors synthesised by its tissues: BMP2 and BMP4 (bone morphogenetic protein 2, 4) (Feng et al., 1995; Feng et al., 1997), TGFβ1 and TGFβ2 (transforming growth factor β1, β2) (Francis & Suttie, 1998), PTHrP (parathyroid hormone related peptide) (Faucheux et al., 2004), retinoic acid (Allen et al., 2002), FGF-2 (fibroblast growth factor 2) (Lai et al., 2007), VEGF (vascular endothelial growth factor) (Lai et al., 2007), EGF (epidermal growth factor) (Barling et al., 2005), pleiotropin (Clark et al., 2006), PEDF (pigment epithelium-derived factor) (Lord et al., 2007), NGF (nerve growth factor) (Li et al., 2007) as well as IGF-1 (insulin-like growth factor-1) (Gu et al., 2007). Altogether, nearly 800 proteins are present in a growing antler. Among them, there are antler-specific growth factors: MEKK1 (mitogen-activated protein kinase 1), SRP72 (signal recognition particle 72 kDa protein) as well as DRG1 (developmentally regulated GTP-binding protein 1) (Park et al., 2004). The apical region of the antler is the site in which growth and modified endochondral ossification processes take place, while membranous ossification occurs around the antler shaft (Price & Allen, 2004). The apical region of the antler consists of several layers: distal perichondrium and respectively – mesenchyme, transient layer, cartilage and bone (Cegielski et al., 2009; Li et al., 2002). Worth noticing is the mesenchyme, which consists of numerous undifferentiated and actively proliferating cells. They are characterised by low levels of type I collagen and alkaline phosphatase expression and no expression of type II collagen (Price et al., 1994; Price et al., 1996). *In vitro* studies show that IGF-1, IGF-2 and FGF-2 stimulate their proliferation, however BMP-2 and TGFβ-1 inhibit their proliferation (Price et al., 2005b). The mesenchymal cells possess a great potential to proliferate, which has been confirmed by presence of extracellular matrix protein – matrilin-2 – which is a marker characteristic for differentiating cells (Korpos et al., 2005). Moreover, the mesenchymal cells synthesize many growth factors and other substances, which may affect the mesenchyme itself or adjacent layers of cells: PTHrP (Faucheux et al., 2004), retinoic acid (Allen et al., 2002), FGF-2 (Lai et al., 2007), EGF (Barling et al., 2005), pleiotropin (Clark et al., 2006) and PEDF (Lord et al., 2007). The mesenchyme participates in antler growth because constantly provides cells which differentiate into chondroblasts and next, into mature cartilage cells. The newly added cartilage undergoes mineralization and transforms into a bone (Szuwart et al., 1998). The transformation, as well as modelling of the cartilage, take place in the presence of osteoclasts (found in perivascular spaces of the cartilage) which derive from osteoclast progenitor cells influenced by locally secreted PTHrP (Faucheux et al., 2002). Apoptotic processes also play a vital role in antler growth and shaping processes, just as it happens during skeletal development, bone growth and remodelling (Colitti et al., 2005).

2. Histology of growing and mature antler

The terminal fragment of a growing antler (Fig. 1) consists of several layers. Microscopically, from the outside, we can distinguish:

A - hairy skin
B - perichondrium
C - mesenchyme and chondroprogenitor area
D - blood vessels and chondrocyte columns

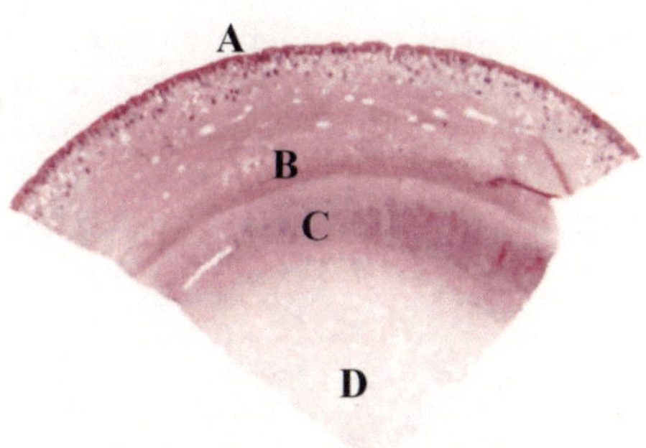

Fig. 1. Cross-section of terminal fragments of the growing antler showing following layers: (A) hairy skin, (B) perichondrium, (C) mesenchyme-chondroprogenitors, (D) central shaft formed of numerous blood vessels and chondrocyte columns. H+E staining, ×40

Antlers are covered by hairy skin, the so-called velvet, consisting of three basic layers: epidermis, dermis and subcutaneous tissue (Fig. 2a). Epidermis is keratinized stratified squamous epithelium built from five layers of cells: basal, squamous, granular, clear and horny (Fig. 2b). Dermis consists of proper connective tissue and has two layers: papillary and reticular. The wave-like organisation of the first layer constitutes the basis of good connection with epidermis (Fig. 2b). The second layer possesses numerous bundles of collagen fibres, hair roots and sebaceous glands (Fig. 2a). The subcutaneous layer has numerous blood vessels, collagen fibres and connective tissue cells. Below the skin region, there is a darker staining zone of tissue constituting the perichondrium (Fig. 2c), the outer part of which is fibrous and the inner part contains cells. Below the perichondrium, there is a wide layer of undifferentiated cells constituting the mesenchyme of the antler. It surrounds the so-called central shaft built from numerous blood vessels stretching from base to apex of the growing antler. Among these vessels, there are chondrocytes organised in characteristic columns (Fig. 2d).

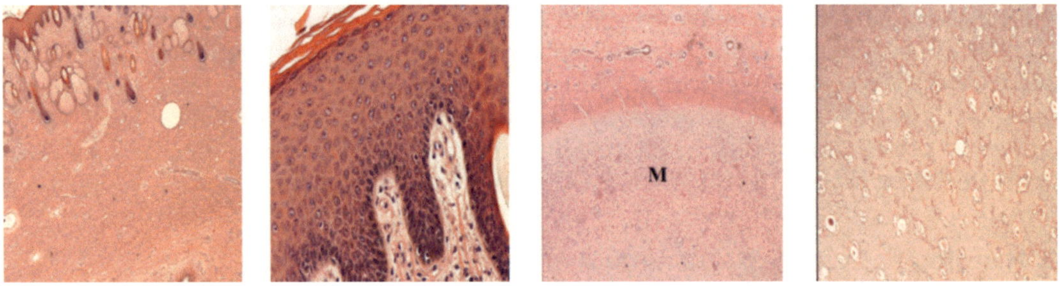

a b c d

Fig. 2. a-d Histology of the apical fragment of the growing antler, longitudinal section. a - hairy skin (velvet) consisting of epidermis, dermis with hair and sebaceous glands, connective tissue, ×100; b - epidermis built from 5 layers of cells and papillary layer of dermis, ×300; c - perichondrium as a darker stained line below which, there is a wide layer of mesenchyme (M), ×100; d - numerous blood vessels and chondrocyte columns lying in the central shaft of the growing antler, ×200. All H+E staining

Immunohistochemical reactions allowed us to identify presence of cells expressing markers characteristic for stem cells (Bcrp1, c-kit) in the zone of intense proliferation - the mesenchyme (Fig. 3a). A large number of Ki-67- and PCNA-positive cells could be observed in the basal layer of epidermis, in skin glands, directly below secretory sections of skin glands, mesenchyme as well as within and in the vicinity of central blood vessels (Fig. 3b-d).

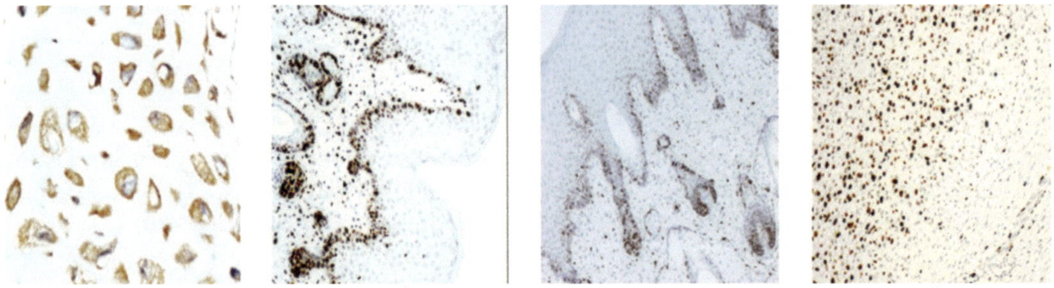

a b c d

Fig. 3. a-d Immunohistochemical reactions for presence of stem cell markers and proliferating cell markers. a - stem cells (c-kit+) within mesenchyme of the growing antler, ×400; b - proliferating cells of the reproductive layer of epidermis and glands of the growing antler (Ki-67+), ×100; c - proliferating cells of the reproductive layer of epidermis as well as hair papillae of the antler (PCNA+), ×100; d - layer of undifferentiated cells constituting antler mesenchyme (Ki-67+), ×100

Unlike growing antlers, mature antlers consist of mineralized lamellar osseous tissue. The architectural unit is the osteon, which is built from the Haversian canal, around which two or three systemic plates are concentrically arranged (Figs. 4a, b). Both in and among these plates, there are osseous lacunae which possess osteocytes. All osseous lacunae within an osteon are connected by a network of bone canaliculi in which osteocyte processes are located.

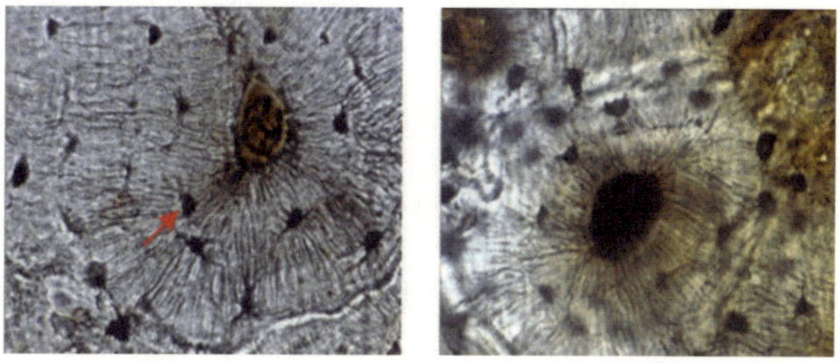

a b

Figs. 4. a, b Cut of an ossified antler shows a typical organisation of lamellar osseous tissue. Around the blood vessel systemic plates, osteocytes (↑) and bone canaliculi can be seen. Stained with silver nitrate, ×400

3. Derivation of MIC-1 cell lineage

To start a culture of antlerogenic cells, we used fragments (100-900 μm) of mechanically disintegrated growing deer antler from which the hairy skin had been removed. By means of cell migration, proliferating cells were isolated and later placed in culture flasks. The culture medium used was DMEM (Dulbecco's Modified Eagle Medium) supplemented with L-glutamine (1 mM/ml), penicillin (100 U/ml) and streptomycin (0.1 mg/ml). The cells were cultured under standard conditions of 5% CO_2 and temperature of 37°C. The culture was maintained over 4 months, with yield of 5×10^6/week per one 75 cm^2 culture flask. Obtained cells were frozen and stored in liquid nitrogen. For study purposes, they were again thawn.

Evaluation of antlerogeneic cells:

- Under an inverted phase contrast microscope. Cells growing in culture are spindle-shaped (fibroblast-like) (Fig. 5a) and their microscopic image is comparable to that of cultured human mesenchymal stem cells obtained from bone marrow (Ryan et al., 2005). The cells possess high proliferative potential (within 72 hours their number in culture doubles). Among cultured cells, there are small, round, undifferentiated, opalescent cells not attached to the surface and possessing no processes or any characteristic morphological features (Fig. 5b).

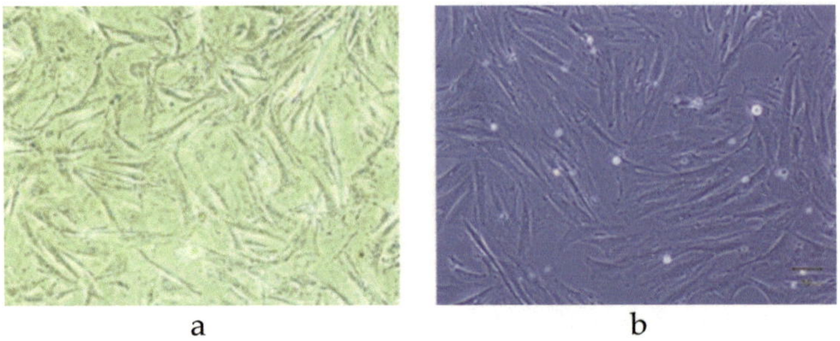

Figs. 5. a, b Primary culture of antlerogenic cells growing in a monolayer. a – differentiating and processes producing antlerogenic stem cells, ×200, b – numerous small, opalescent, non-adherent, undifferentiated cells, ×100. Inverted phase contrast microscopy

- Under light microscopy. Small, oval undifferentiated cells can be observed in fixed 24-h microcultures stained with H+E (Fig. 6).

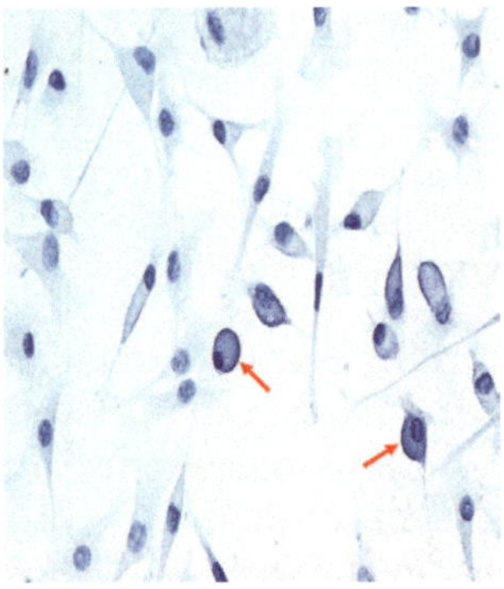

Fig. 6. Microculture of undifferentiated small, oval antlerogenic cells (↑), more mature cells possess processes, H+E staining, ×200

- Under scanning electron microscopy we observed cell morphology (Fig. 7).

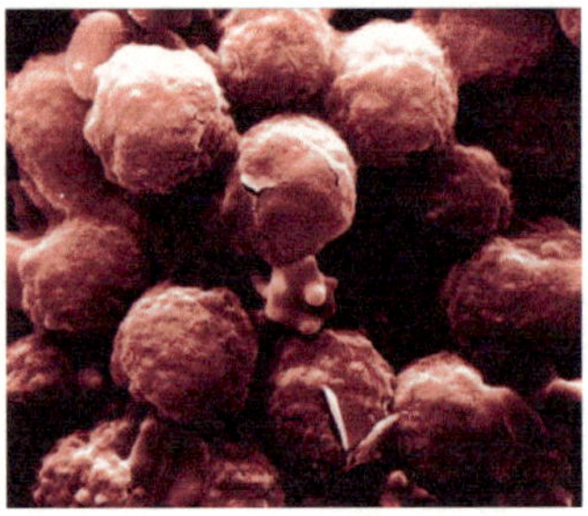

Fig. 7. *In vitro* culture of antlerogenic cells. Scanning electron microscopy, ×6600

- Immunohistochemical reactions for presence of stem cell markers: c-kit, Thy-1. Specific positive membranous reaction with anti-c-kit and anti-Thy1 antibodies was obtained only in small, oval frequently dividing cells (Fig. 8a, b). No expression of the above mentioned antigens was shown in more differentiated cells with processes.

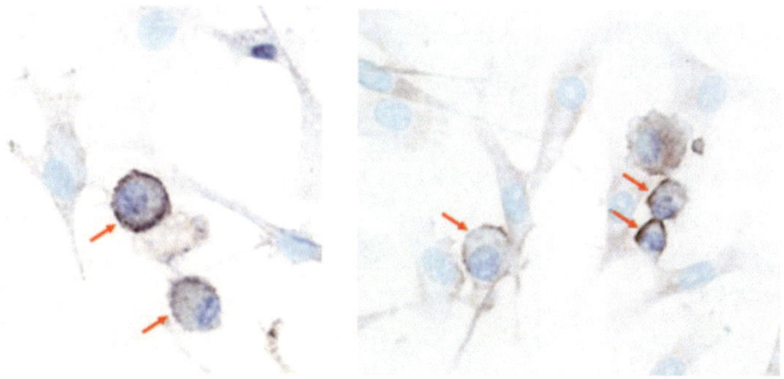

a b

Fig. 8. a, b Microculture: a – immunocytochemical membranous reaction to c-kit (↑) protein presence in small, oval antlerogenic cells, ×400; b – immunocytochemical membranous reaction to Thy1 (↑) protein presence in small, oval antlerogenic cells, ×400

- Using PCR reaction we demonstrated expression of following genes: POU5F1, c-myc, MHC I as well as no expression of MHC II (Fig. 9a, b). From MIC-1 antlerogeneic cells we isolated the total RNA, conducted an reverse transcription and performed the PCR reaction with following primer pairs on the cDNA template (Table 1):

	Primer sequence	PCR product size	References
POU5F1 (OCT4)	5′-ATGACTTGTGTGGAGGGATGG-3′ 5′-GAACACCTTTCCAAAGAGAACC-3′	338 bp	Berg et al., 2007
c-myc	5′-GAGGGTCAAGTTGGACAGTGTCAG-3′ 3′-CTTGGACGGACAGGATGTATGCTG-5′	254 bp	Francis & Suttie, 1998
MHC I	5′-GGATGAAGCATCACTCAG-3′ 3′-CGCTGCTGCGCGCAGACC-5′	530 bp	Holmes et al., 2003
MHC II	5′-GTGTTACTTCACCAACGGGACG-3′ 3′-GTTGTGGTGGTTTAGAGCCTC-3′	207 bp	Swarbrick & Crawford, 1997

Table 1. Sequence of primers used in PCR reactions

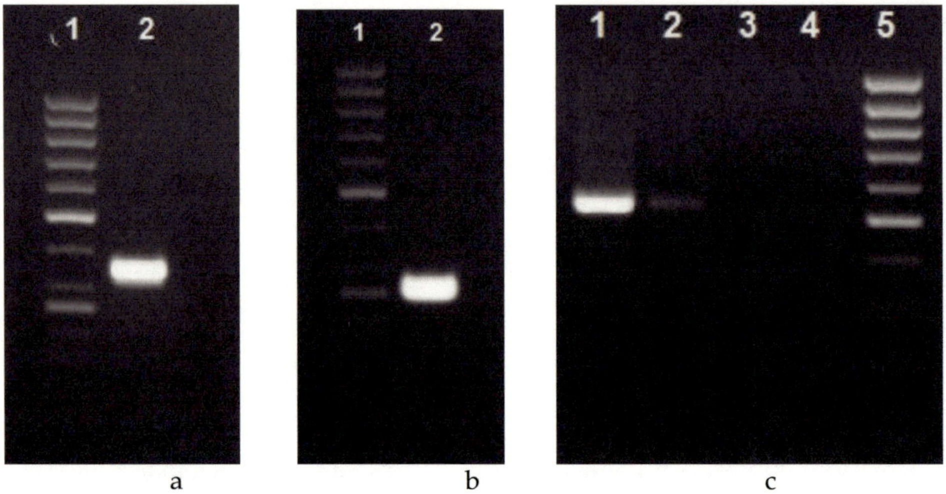

Fig. 9. a, b Expression of various genes in MIC-1 antlerogenic cells. a - lane 1: Molecular weight marker (50-1000 bp, Promega), lane 2 - POU5F1 (338 bp); b - lane 1: Molecular weight marker (50-1000 bp), lane 2: c-myc (254 bp); c – lane 1: MHCI (530 bp), lane 2: MHCI 1:10 cDNA (530 bp), lane 3: MHCII, lane 4: MHCII 1:10 cDNA, lane 5: Molecular weight marker (100-1000 bp, Promega)

- Under electron microscopy. Active cells possess a large nucleus with loose chromatin and nucleoli. The small amount of cytoplasm surrounding the nucleus contained vast amounts of rough endoplasmic reticulum, mitochondria and vacuoles. The presence of glycogen granules makes these cells similar to embryonic cells. The cell surface was covered by numerous microvilli (Fig. 10).

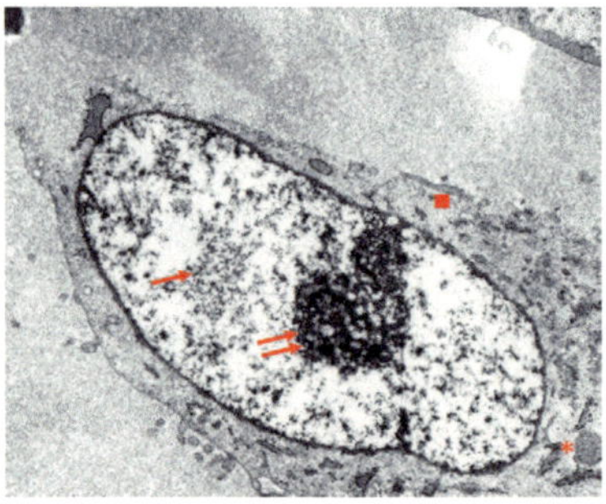

Fig. 10. Undifferentiated antler cell from an *in vitro* culture. Cell nucleus contains loose, transcriptively active chromatin (↑) and a nucleolus (↑↑). The cytoplasm contains rough endoplasmic reticulum (■) and mitochondria (*). Cell membrane forms numerous short processes. EM ×15000

- Marking of antlerogenic cells using a retroviral vector pMINV EGFP. Analysis conducted in a flow cytometer demonstrated 30% of EGFP+ cells (Fig. 11) Transduction of antlerogenic stem cells with an EGFP gene marker enabled us to follow the fates of cells engrafted into auricular cartilage defects.

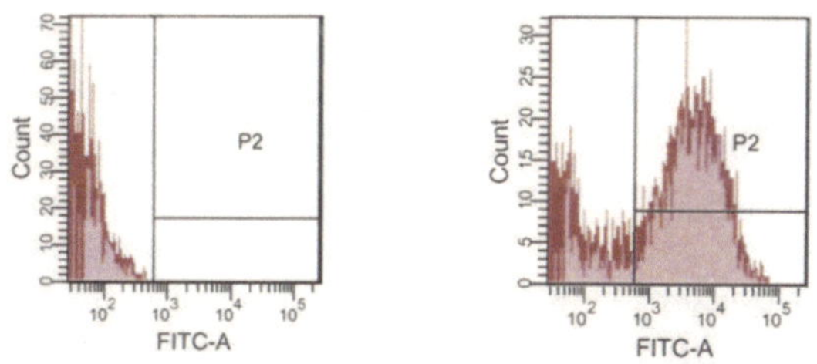

a b

Fig. 11. Evaluation of EGFP expression in antlerogenic cells transducted with a pMINV EGFP by means of flow cytometry. a - no expression of EGFP in non-transduced cells; b - EGFP protein fluorescence level in transduced cells

4. Xenogeneic engraftment of antlerogenic cells and tissue regeneration

Autologous cartilage and bone tissue are commonly used materials for transplants in facial reconstructive surgery. The limited amount of tissue possible to obtain in children, surgical procedures conducted during its collection and, additionally - in case of bone graft - the risk of no graft adhesion with the host's tissue and its resorption, emphasise the imperfection of this method (Rotter et al., 2005). Despite common use of allogeneic cartilage and bone in regeneration of defects, the limiting factor may constitute their availability and possibility of losing them (Komender et al., 2001; Więcko et al., 1980). Chondrocytes amplified *in vitro* and used in tissue engineering may undergo differentiation (Barbero et al., 2003). Studies over the application of *in vitro* amplified osteoblasts show on the other hand, that differentiated cells do not provide proper environment for autogeneic cell regeneration and they require being administered together with growth factors (Sakata et al., 2006; Schliephake et al., 2001).

Another problem we addressed was the aseptic inflammation of the digital flexor muscles and tendons and interosseus muscle inflammation in horses. It affects about 10% of horses from the overall number of orthopaedic patients and constitutes, nowadays, a serious problem for doctors and horse owners. It causes long-term exclusion of horses from their previous activities. The most recent therapies rely on intratendonous injections under ultrasonographic control of: growth factors (platelet-rich plasma, IGF-1), bone marrow and autologous stem cells (Aspenberg, 2007). The last of the mentioned methods is exceptionally difficult to perform. It is necessary to collect the bone marrow, which constitutes an additional procedure. Time required for cell amplification is 3 to 4 weeks, which makes it difficult to administer the cells into the tendon at an optimal moment for after this time, the haematoma formed at the site of injury already undergoes organisation (Richardson et al., 2007).

In recent years, new possibilities for application of various stem cell types in regeneration and reconstruction of all types of tissues, including the connective tissue, appeared (Phinney & Prockop, 2007). Among others, a technology of obtaining chondrocytes from human stem cells and technology of obtaining mesenchymal cell precursors able to trigger cartilage growth and regeneration were devised (Barberi et al., 2005; Vats et al., 2006). More and more often, also mesenchymal stem cells (MSC) are used in combination with natural or synthetic carriers, as well as BMP proteins, for bone reconstruction (Jafarian et al., 2008; Seto et al., 2001). The MSC cells can be obtained, among others, from bone marrow, adipose tissue, periosteum and dental pulp. Their advantages are their great proliferating potential, multipotentiality, as well as little immunogenicity and lack of tendencies to neoplasia (Shanti et al., 2007). Bearing in mind the common difficulties in obtaining the proper material for engraftment and searching for alternatives for human or autologous stem cells, we made an attempt to use a xenogeneic graft of cells obtained from a growing antler of the red deer (*Cervus elaphus*) to reconstruct of auricular cartilage and mandibular bone in rabbits, as well as tendons in horses (Cegielski et al., 2008a; Cegielski & Kalisiak, 2008). A stable MIC-1 antlerogenic cell line (name of the line introduced in patents) - mesenchymal stem cells - was derived from a growing antler of the red deer (Cegielski et al., 2006). The cells cultured by us are characterised by rapid growth and are practically "immortal," require uncomplicated standard procedures, which are generally accepted and used in cell culture laboratories. They can easily be transported, frozen and stored in liquid nitrogen and, if necessary, used for studies. The use of specific antibodies in immunocytochemical reactions allowed us to mark and localise these stem cells both *in vivo* in a growing antler, as well as *in vitro* in cell cultures (Cegielski et al., 2006).

4.1 Cartilage tissue

The aim of this study was application of a xenogeneic graft of stem cells obtained from a growing red deer antler (*Cervus elaphus*) to regenerate defects of ear auricular cartilage in rabbits.

For the experiment, we chose California white rabbits - 8-month-old females weighing each about 4 kg. The Spongostan® fragments matched the size of the defect and constituted scaffolding for antlerogenic cells. The cells in the amount of 2×10^6 cell/ml, rinsed in MEM medium without serum, were placed on Spongostan® through delicate centrifugation (Fig. 12).

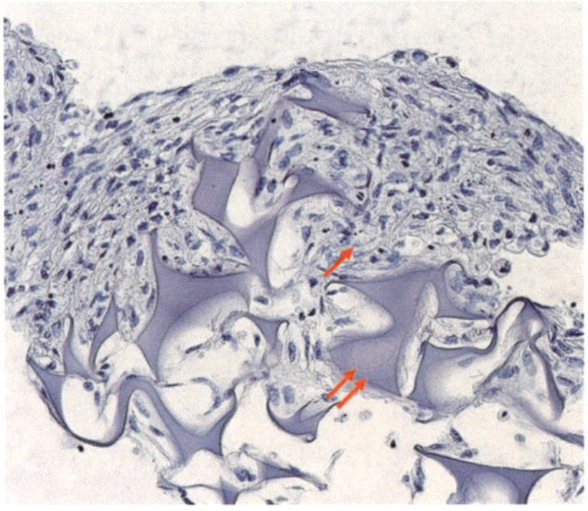

Fig. 12. Cultured antlerogenic cells (↑) placed on Spongostan® (↑↑) and prepared for engraftment. H+E staining, ×200

Surgical procedures included performing a defect half-way up the outer, dorsal part of the right ear auricle. At surgical site a centrally peduncled skin/perichondrium flap measuring 1.5 cm × 1.5 cm was prepared. The exposed cartilage fragment measuring 1 cm × 1 cm was excised and removed. The animals had the cartilage defects implanted with antlerogenic cells suspended on Spongostan®, antlerogenic cells transduced with a retroviral pMINV EGFP vector or Spongostan® alone saturated solely with physiological saline solution (control). The implants were covered with skin/perichondrium flaps. Using clinical observations as guidelines, after a period of 4 and 9 weeks from the surgery, we obtained specimens from the implantation sites, including the implant as well as endogenous cartilage. The collected material was then assessed under optical microscopy (H+E, toluidine blue), electron microscopy and by immunohistochemical examinations (reactions against antigens: Thy1, CXCR4, EGFP, CD3, CD20, CD68).
The implanted cells participated in cartilage reconstruction as a new cartilaginous scaffolding had been created from fibrous cartilage in which sporadic intracartilaginous ossification was observed. Peripherally, in the sites of recovery, there was a visible border between the old and the newly formed cartilage (Fig. 13). Numerous dividing chondrocytes and forming isogeneic groups were present (Fig. 14). After four weeks, the growth in cartilage thickness decreased, however, after nine weeks, the cartilage defect was completely and even excessively reconstructed. Vascularised, mixed chondro-osseous tissue was formed, just like the one participating in formation of deer antlers (Price et al., 2005a). The cartilage reconstruction process (calcification and ossification centres) resembled the remodelling processes taking place after engraftment of auto- and allogeneic cartilage (Lattyak et al., 2003; Więcko et al., 1980). As in application of autogeneic chondrocytes suspended on a polyethylene carrier, after the implantation of antlerogenic cells, we observed no connective-tissue capsule (Wellisz, 1993). In rabbits, at the sites where Spongostan® saturated solely with physiological saline solution was engrafted, we observed no cartilage regeneration, but only a scar formation process.

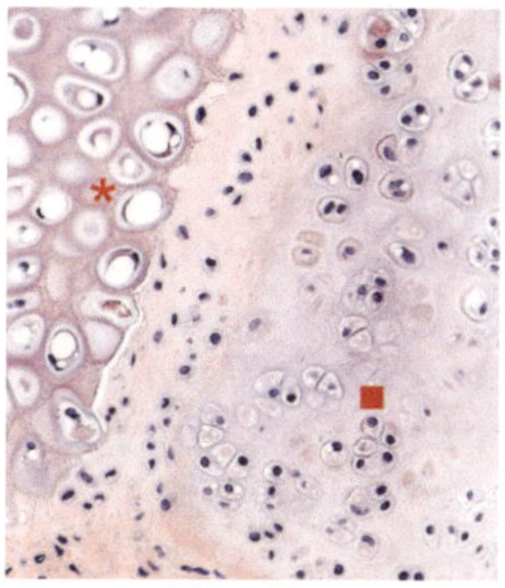
Fig. 13

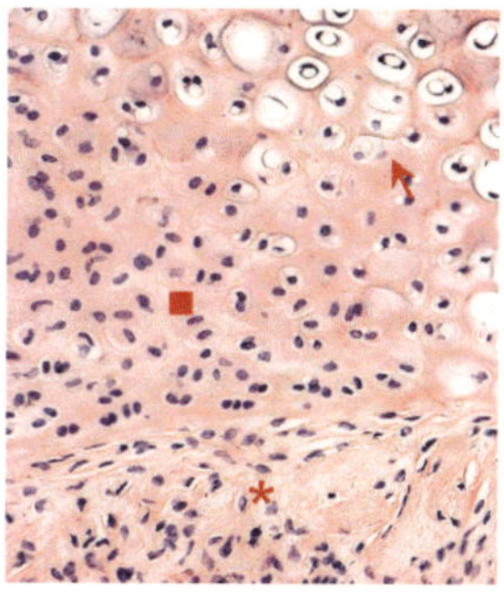
Fig.14

Fig. 13. Engraftment of antlerogenic cells within rabbit's ear auricle. Visible border between the old (*) and the newly formed (■) cartilage, H+E staining, ×200

Fig. 14. Engraft, antlerogenic cells taking part in regeneration of rabbit's cartilage defect. Visible numerous chondroblasts (*), dividing chondrocytes (■) as well as mature cartilage cells (↑), H+E staining, ×200

Conducted immunocytochemical reactions allowed us to localize, at the implantation site, cells showing expression of proteins characteristic of stem cells: Thy1 and CXCR4. The marked cells were located in the vicinity and within blood vessels as well as in the area of proliferating, undifferentiated cells at the implantation/autologous ear auricle cartilage border (Fig. 15a, b). To confirm the participation of antlerogenic stem cells in cartilage regeneration, we engrafted into defect areas cells transduced with a vector containing EGFP marker gene. Stem cells transduction with a retroviral vector containing EGFP marker gene constitutes one of the most commonly used methods to follow the fate of engrafted cells (Szyda et al., 2006). The immunohistochemical reaction conducted four weeks from the engraftment, showed the presence of EGFP-positive cells within the area of cartilage reconstruction (Fig. 15c). The reaction was specific only for small, undifferentiated cells, which proves that antlerogenic stem cells were not destroyed by the cells of the host. Moreover, they did not require any environment (tissue niches) to retain their undifferentiated state and stimulated the rabbit's cartilage to regenerate.

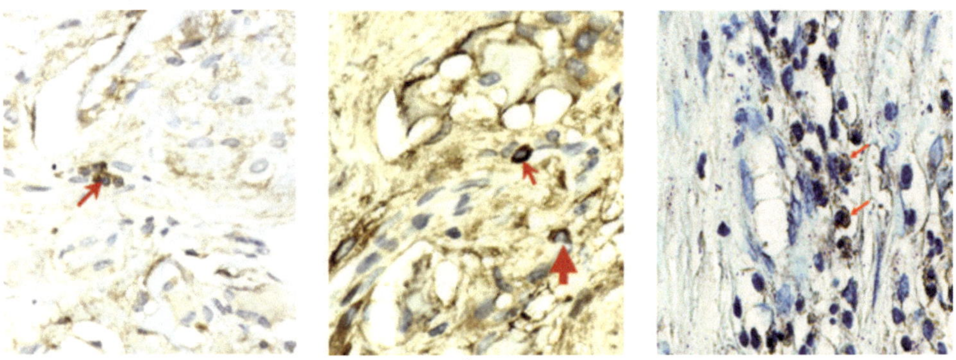

a b c

Figs. 15. a-c Immunohistochemical reactions demonstrating presence of antlerogenic stem cells within the implant. a - immunohistochemical localization of CXCR4+ cells (↑), ×200; b - immunohistochemical localization of Thy1+ cells (↑), ×200; c - immunohistochemical localization of EGFP+ transduced antlerogenic cells (↑), ×400

Participation of antlerogenic stem cells in reconstruction of the damaged tissue may be complex. The reconstruction always occurred from the periphery of the implant outwards and it is possible that stem cells could participate in regeneration of the damaged tissue not by differentiation into cells required for its reconstruction, but by local production of trophic factors modelling the microenvironment at the site of damage as well as inducing survival and proliferation of host cells. To these factors may belong, among others, proteins regulating hematopoesis, angiogenesis, wound healing and immune response (Barbash et al., 2003; Phinney & Prockop, 2007). Nakamura et al. observed regeneration of damaged cardiomyocytes in immunosuppressed mice after implantation of swine MSC. The authors suggest that regeneration is the effect of influence of the trophic MSC-released substances on native tissues (Nakamura et al., 2007). The *in vitro* studies demonstrated that MSC cells secrete cytokines which influence survival and secretory functions of other cells (Le Visage et al., 2004; Parekkadan et al., 2007). While an antler regenerates, certain growth factors are expressed and these are: EGF, FGF-2, VEGF and BMP (Barling et al., 2005; Feng et al., 1995; Lai et al., 2007). In our opinion, engrafted antlerogenic cells may participate in formation of

proper environment for growth of host's cells together with the above mentioned factors. In natural conditions, cartilage is a non-vascularised tissue and it is chondrocytes which probably produce the angiogenesis inhibiting factor (Moses et al., 1990). After implantation of antlerogenic cells into a cartilage defect, we observed numerous small blood vessels on the border of regeneration. This may show that angiogenesis is taking place, creating good conditions for the graft to be accepted. It is worth mentioning that abundant vascularisation of the antler constitutes the most striking difference between its morphology and morphology of other cartilages (Lai et al., 2007). Rich blood supply is necessary here so as to meet high metabolic requirements imposed by rapid cartilage growth. The fact that antlers constitute a valuable model for angiogenesis studies has recently been suggested by Clark and Lai, who demonstrated the presence of VEGF and its receptor VEGFR as well as pleiotropin in tissues of a growing antler (Clark et al., 2006; Lai et al., 2007).
The experiment we conducted did not use immunosupression, and the engrafted xenogenous antlerogenic cells were in direct contact to the perichondrium and remaining cartilage of the host's auricle. Despite this fact, clinically, we observed every time proper wound healing *per primam* and visible increase in cartilage thickness. In an experiment conducted by Ulusal et al. (Ulusal et al., 2005) first, appearing already in the period of 4 to 6 days after the end of immunosupression, symptoms of complete auriculae's allogeneic implant rejection in rats were: edema, localized epidermal desquamation, and formation of erythema. A typical reaction of the host to introduction of an allogeneic implant are massive lymphocyte concentration and the presence of plasma cells (Brown et al., 1980; Romaniuk et al., 1995). On no stage of the conducted experiment, we could observe inflammatory reactions, only in the fourth week we observed minor localized clusters of lymphocytes at the site of cell engraftment. Additionally, immunohistochemical reactions characteristic for T lymphocytes (CD3), B lymphocytes (CD20) and macrophages (CD68) showed none of the mentioned cells at the implantation site, confirming that the host's immune response was minimal (Fig. 16a-c). Because antlerogenic cells are poorly differentiated, they cause minor induction of immune response and good reception of xenogeneic graft. One explanation of the phenomenon responsible for no rejection of the allogeneic grafts with stem cells is the lack of major histocompatibility complex class II antigens (MHC II) and inhibition of T and B lymphocytes activity (Chamberlain et al., 2007; Ryan et al., 2005). Only 6 weeks after Spongostan® with cells or Spongostan® alone were engrafted, we could observe foreign body giant cells (FBGCs) on the surface of the remaining carrier fragments (Fig. 17). In parallel studies, we observed that the FBGCs are produced in the process of macrophage fusion in response to the introduction of a foreign body to an organism (Cegielski et al., 2008b). The carrier we used was completely degraded and resorbed as a result of activity of giant cells. It is probable that Spongostan® had no influence on the survival rate of cells it carried.

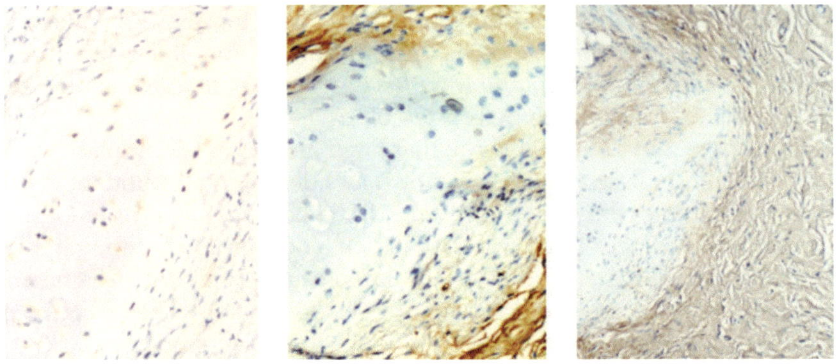

a b c

Figs. 16. a-c Negative immunohistochemical reactions for presence within the implant of: a – T lymphocytes (CD3+), ×200; b – B lymphocytes (CD20+), ×200; c – macrophages (CD68+), ×200

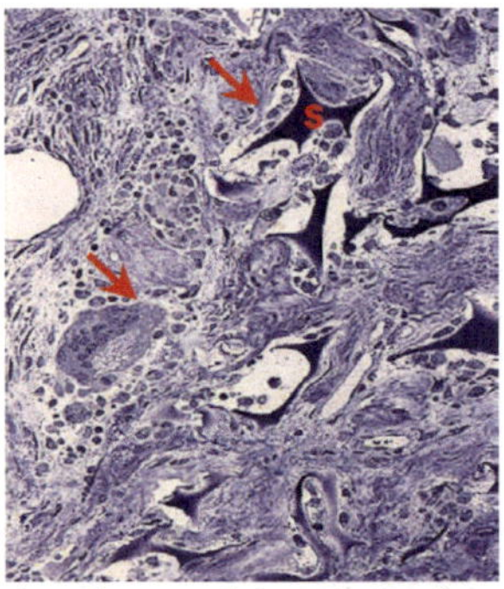

Fig. 17. Spongostan® (S) degradation by multinuclear foreign body giant cells (FBGC) (↑). These cells are visible in the vicinity of the degraded Spongostan®, which takes a typical star-like shape. Toluidine Blue staining, ×200

Electron microscopy analysis of the material collected from implantation sites showed a great number of collagen fibres and relatively numerous chondroblasts and fibroblasts. Active cells contained large amounts of rough endoplasmic reticulum with frequently distended cisterns, vacuoles as well as nuclei with large amounts of loosely packed chromatin (Fig. 18a). Some of the cells underwent apoptosis and membrane-bound cytoplasm fragments separated from them participated in production of extracellular substance (Fig. 18b). Apoptosis plays a vital role during skeletal development as well as bone growth and remodelling. Active apoptotic processes are also responsible for regulation of antler regeneration process, among others – by limiting excessive cell proliferation which may lead to initiation of neoplastic processes (Colitti et al., 2005).

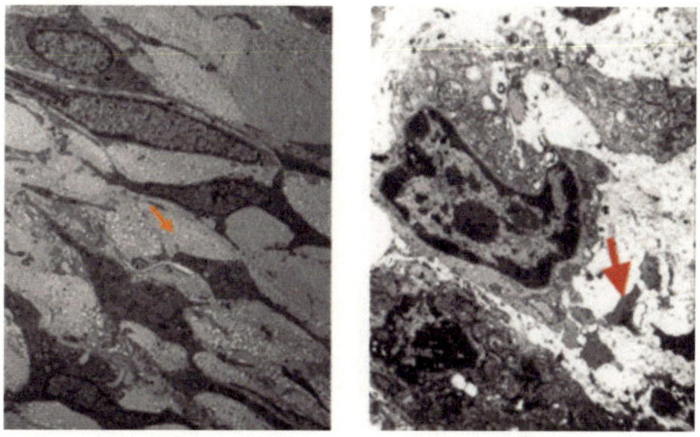

a b

Fig. 18. a, b 9th week after antlerogenic cells implantation: a - numerous collagen fibres (↑), EM ×15000; b - cell containing a nucleus with condensed marginal chromatin, membrane-bound cytoplasm fragments (↑) by separating from the cell become part of extracellular substance, in which numerous collagen fibres are visible, EM ×15000

4.2 Osseous tissue

The aim of the study was the use of a xenogenous stem cell graft from a growing antler of the red deer to regenerate mandibular bone defects in rabbits.

The mandibular bone defect measuring 0.5 cm was performed with a steel surgical drill boring the bone along with the periosteum and reaching the marrow cavity. In the experimental group, inside the bore-hole we placed a fragment of Spongostan® of a matching size saturated with the suspension of antlerogenic cells in the amount of 2×10^6 cell/ml. In the control group, rabbits were implanted with Spongostan® saturated with physiological saline solution. After a period of 1, 2 and 6 months after engraftment, we collected the mandibles for examination.

Preliminary results show, that in the control group autoregeneration of lesion occurred (Fig. 19). In the experimental group antlerogenic cells participated in regeneration of mandibular bone (Fig. 20).

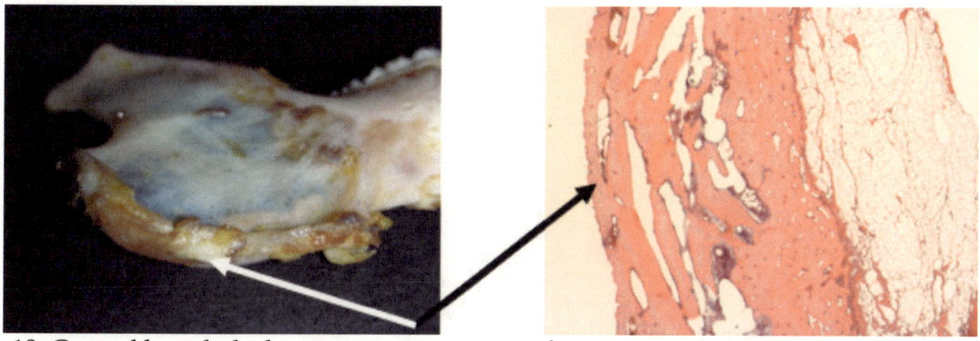

Fig. 19. Control bore-hole, bone autoregeneration after one month, H+E staining, ×40

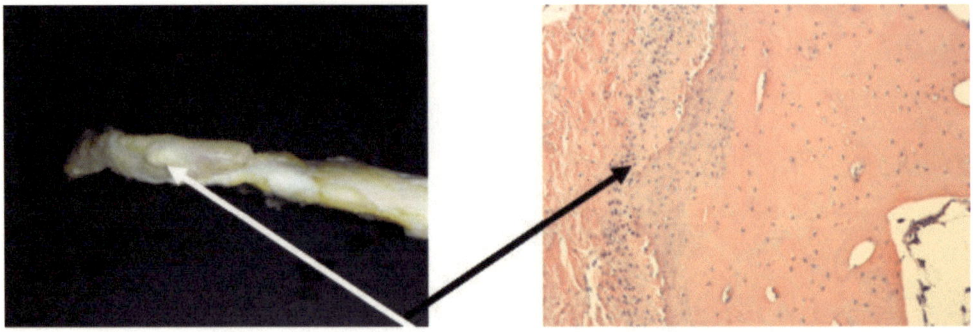

Fig. 20. Bone regeneration with antlerogenic cells at implantation site after one month, H+E staining, ×100

However, between the new periosteum and the muscle layer, we observed oval aggregates of undifferentiated proliferating cells with large amounts of collagen fibres surrounded by a thin connective-tissue capsule (Fig. 21). Isolation of certain areas may be a sign of application of the xenogeneic cells or a carrier. A living organism can react to an xenograft with chronic inflammation, indicating the lack of biocompatibility or can try to isolate the given material by surrounding it with a fibrous capsule of various thickness (usually 0.1-10 µm). The thinner the connective-tissue capsule, the bigger the biocompatibility of the carrier and host cells (Kos et al., 2003). In our experiment, this phenomenon was sporadically observed and probably it did not affect the regenerative processes. On the periphery of the defect, there were well visible ossification areas as well as gradually reconstructed periosteum. In the centre of the regeneration site, there was well vascularized mesenchymal tissue present in a form of a membrane. Angiogenesis constitutes a necessary condition for proper bone formation process. Formation of blood vessels can be a result of the activity of angiogenic factors (VEGF) secreted by antlerogenic cells, just like it takes place in a growing antler (Clark et al., 2006; Lai et al., 2007). The process of formation and mineralisation of the bone occurs from the periphery towards the centre of the regeneration area (Fig. 22).

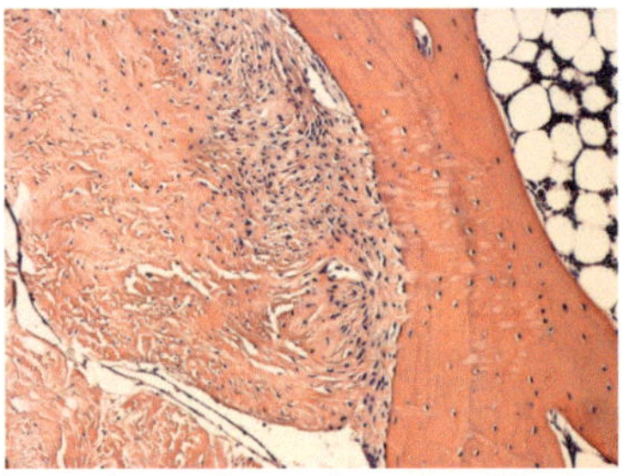

Fig. 21. Two months after implantation, along the mandibular arm we observe the engrafted cells together with a large number of collagen fibres forming oval aggregates surrounded by a thin connective-tissue capsule, H+E staining, ×100

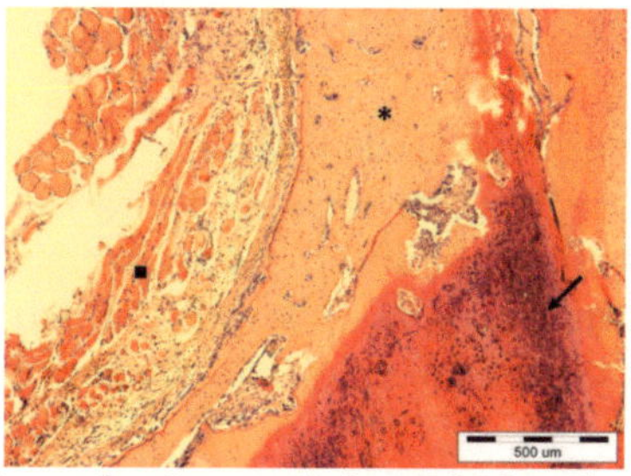

Fig. 22. The area of regeneration of a 1-month bore-hole after implantation of antlerogenic cells. The muscle layer (■) and a visible osseous tissue lying underneath (*). Dark finely flocculating material is the site of early vesicle crystallization and mineralization (↑), H+E staining

After two months, at the implantation site, we observed a reconstructed, two-layered periosteum and a fibrous bone possessing an irregular system of bone plates, containing blood vessels and marrow cavities filled with red marrow. The histological image of the regenerated osseous tissue in our experiment resembles the bone formed after engraftment of autologous MSC of the bone marrow into defects of mandibular bone in dog and monkey. In these experiments, the material collected from implantation sites after respectively 6 and 16 weeks contained trabecular bone with marrow cavity (Jafarian et al., 2008; Seto et al., 2001). 6 months after implantation, we observed the progressing remodelling of the fibrous bone into a more organised lamellar osseous tissue.

The regeneration method in our study proceeded similarly to another long-term xenogeneic model. Human MSC inserted into cranial and mandibular bone defects in mice participated in regeneration of the damaged tissue without inflammatory reaction. MSC cells modulate the function of B lymphocytes and inhibit T lymphocytes proliferation, which induces tolerance for example to allogeneic grafts (Ryan et al., 2005). Both, histological and immunohistochemical evaluations, were performed after 1, 2 and 6 months, and confirmed participation of xenogeneic cells in the regeneration processes, as well as a lack of rejection of the implants. The deficiencies in the bones were replaced by newly formed, thick fibrous bony tissue, that underwent mineralization process and was later remodelled into trabecular bone. Antlerogenic cells participated in regeneration of mandibular bone most probably together with tolerating them host cells. In our opinion, the engrafted cells could have participated in forming of an environment favourable to host's cell growth just like it happened during cartilage tissue regeneration. The results of the experiment with rabbits allow us to believe that the antlerogenic cells could be used in reconstruction of bony tissues in other species as well.

4.3 Tendinous tissue

The aim of the study was the use of a xenogeneic graft of stem cells from a growing antler of red deer to regenerate tendons in horses. Antlerogenic cells synthesize great amounts of collagen (Fig. 23), which is used in a dynamic remodelling of an antler (Price et al., 1996). Tendons and ligaments are built from thick bundles of collagen fibres, among which lie tendinous cells - tenocytes. They are surrounded by loose connective tissue, forming the internal peritendineum. Outside of the tendon, there is the external peritendineum, which participates in formation of a synovial tendon sheath. Guided intratendinous injection of antlerogenic cells with great regenerative potential could rapidly start reconstruction processes.

Considering the results of preliminary studies, we knew that administration of antlerogenic cells to horses does not induce the immune response or lead to development of local inflammatory reaction. Despite the fact that horses are animals greatly susceptible to allergic responses, the test we performed on three horses, which was based on subcutaneous injection of 1.5 ml antlerogenic cells suspension into points of the middle 1/3 of horse's neck (injection sites were shaved and sterilized with 70% alcohol) after three-week-long observation was successful. There was no inflammatory reaction at the site where the cells were injected or hypersensitivity of anaphylactic type. Next, we performed intratendinous injections of antlerogenic cells in two horses suffering from tendon inflammation: in the former peripheral inflammation of superficial digital flexor tendon and muscle occurred, in the latter - interosseous muscle inflammation in the middle of the metacarpus. During six-week-long observation after administration of the cells, we noticed no reaction, despite slight painfulness at the sites of the injection present during 2 first weeks. Follow-up ultrasound examinations conducted at two-week intervals showed intense remodelling of the damaged structures (Fig. 24a-c). In order to assess objectively the effectiveness of this new method, further eighteen-month-long observation of patients lasts and clinical examinations on a bigger number of horses are necessary. However, the preliminary results are promising and encourage further studies.

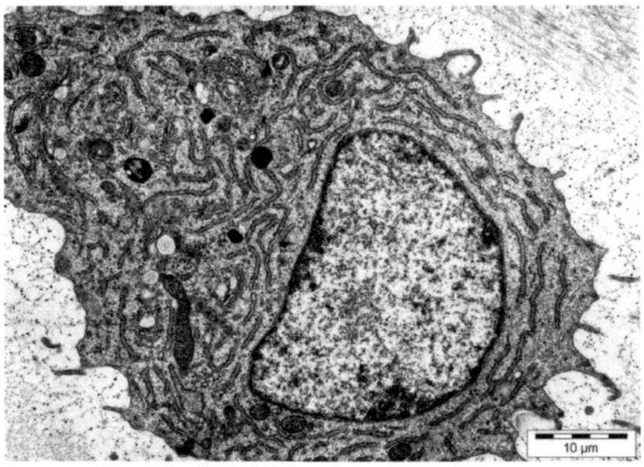

Fig. 23. Growing antler cell (chondroblast) containing numerous canaliculi of the endoplasmic reticulum, mitochondria and collagen fibres

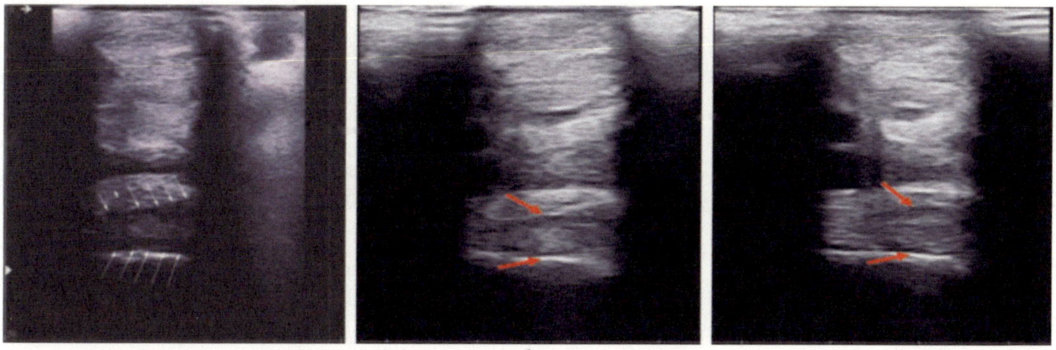

Fig. 24. a - ultrasound image of digital flexor muscle tendon and interosseous muscle in a horse. Arrows indicate the injury of the interosseous muscle; b - ultrasound image of the same horse 4 weeks after the injection; c - ultrasound image of the same horse 8 weeks after the injection. Visible reconstructing processes around the site of antlerogenic cells injection

5. Perspectives of antlerogenic cells application in medicine

More and more often stem cells are used in treatment of injured tissues and organs. In reconstructive surgery of the skull and face, autologous stem cells are successfully used to regenerate bones: mandibular and cranial bones (Lendeckel et al., 2004; Shayesteh et al., 2008; Warnke et al., 2004). A visible improvement has also been gained in the case of using them in treatments of osteogenesis imperfecta and knee joint cartilage defect in human (Horwitz et al., 2002; Kuroda et al., 2007). An alternative for human cells could be properly modified xenogeneic cells. The participation of antlerogenic cell in reconstruction of damaged tissues, poor immunogenicity, good integration of the reconstructed tissues with host's own tissue, rapid *in vitro* amplification as well as low production cost of the engraft constitute its advantages. From the results of the experiments concerning the participation of antlerogenic stem cells in regeneration of cartilaginous and osseous tissue defects in rabbits we know that after the completed reconstruction process, majority of the administered cells undergoes apoptosis and the remaining cells are eliminated by neutrophils. A two-year-long period of observation of experimental animals and lack of any undesirable responses allows us to conclude that there is great probability in using these cells in widely understood regenerative medicine. Our concept is confirmed by studies conducted by other authors, who prove that specificity of the stem cells differentiation process depends on the site of their engraftment. For example, swine mesenchymal stem cells implanted into a heart of a rat do not differentiate into chondroblasts or osteoblasts (as results from their purpose), but stimulate regeneration of injured cardiomyocytes (Nakamura et al., 2007). Full application of stem cells in regenerative medicine will require, however, learning about the differentiation and behaviour of these cells both *in vitro* and *in vivo*, as well as selecting proper carrier for them.

6. References

Allen, S.P.; Maden, M. & Price, J.S. (2002). A role for retinoic acid in regulating the regeneration of deer antlers. *Developmental Biology*, Vol. 251, No. 2, (Nov., 2002) pp. 409-423, ISSN: 0012-1606

Aspenberg, P. (2007). Stimulation of tendon repair: mechanical loading, GDFs and platelets. A mini-review. *International Orthopaedics*, Vol. 31, No. 6, (Dec., 2007) pp. 783-789, ISSN: 0341-2695

Barbash, I.M.; Chouraqui, P.; Baron, J.; Feinberg, M.S.; Etzion, S.; Tessone, A.; Miller, L.; Guetta, E.; Zipori, D.; Kedes, L.H.; Kloner, R.A. & Leor, J. (2003). Systemic delivery of bone marrow-derived mesenchymal stem cells to the infarcted myocardium: Feasibility, cell migration, and body distribution. *Circulation*, 108, (Aug., 2003) pp. 863-868, ISSN: 0009-7322

Barberi, T.; Willis, L.M.; Socci, N.D. & Studer, L. (2005). Derivation of multipotent mesenchymal precursors from human embryonic stem cells. *PLOS Medicine*, Vol. 2, No. 6, (Jun., 2005) pp. 554-560, ISSN: 1549-1277

Barbero, A.; Ploegert, S.; Heberer, M. & Martin, I. (2003). Plasticity of clonal populations of dedifferentiated adult human articular chondrocytes. *Arthritis and Rheumatism*, 48, (May, 2003) pp. 1315-1325, ISSN: 0004-3591

Barling, P.M.; Lai, A.K.W. & Nicholson, L.F.B. (2005). Distribution of EGF and its receptor in growing deer antler. *Cell Biology International*, Vol. 29, No. 3, (Mar., 2005) pp. 229-236, ISSN: 1065-6995

Barrell, G.K.; Davies, R. & Bailey, C.I. (1999). Immunocytochemical localization of oestrogen receptors in perichondrium of antlers in red deer (Cervus elaphus). *Reproduction, Fertility, and Development*, Vol. 11, No. 3, pp. 189-192, ISSN: 1031-3613

Bartos, L.; Schams, D. & Bubenik, G.A. (2007). Testosterone, but not IGF-1, LH, prolactin or cortisol, may serve as antler-stimulating hormone in red deer stags (Cervus elaphus). *Bone*, Vol. 44, No. 4, (Apr., 2009) pp. 691-698, ISSN: 8756-3282

Berg, D.K.; Li, C.; Asher, G.; Wells, D.N. & Oback, B. (2007). Red deer cloned from antler stem cells and their differentiated progeny. *Biology of Reproduction*, Vol. 77, No. 3, (Sep., 2007) pp. 384-394, ISSN: 0006-3363

Brown, B.L.; Kern, E.B. & Neel, H.B. (1980). Transplantation of fresh allografts (homografts) of crushed and uncrushed cartilage and bone: a 1-year analysis in rabbits. *Laryngoscope*, Vol. 90, No. 9, (Sep., 1980) pp. 1521–1533, ISSN: 0023-852X

Bubenik, G.A.; Miller, K.V.; Lister, A.L.; Osborn, D.A.; Bartos, L. & van der Kraak, G.J. (2005). Testosterone and estradiol concentrations in serum, velvet skin, and growing antler bone of male white-tailed deer. *Journal of Experimental Zoology. Part A, Comparative Experimental Biology*, Vol. 303, No. 3, (Mar., 2005) pp. 186-192, ISSN: 1548-8969

Cegielski, M.; Całkosiński, I.; Dzięgiel, P.; Gebarowski, T.; Podhorska-Okolow, M. & Zabel, M. (2006). Search for stem cells in the growing antler stag (Cervus elaphus). *Bulletin of the Veterinary Institute in Pulawy*, Vol. 50, No. 2, pp. 247-251, ISSN: 0042-4870

Cegielski, M.; Dziewiszek, W.; Zabel, M.; Dzięgiel, P.; Iżycki, D.; Zatoński, M. & Bochnia, M. (2008a). Experimental application of xenogenous antlerogenic cells in replacement of auricular cartilage in rabbits. *Xenotransplantation*, Vol. 15, No. 6, (Nov.-Dec., 2008) pp. 374-383, ISSN: 0908-665X

Cegielski, M.; Iżykowska, I.; Podhorska-Okołów, M.; Zabel, M. & Dzięgiel, P. (2008b). Development of foreign body giant cells in response to implantation of Spongostan as a scaffold for cartilage tissue engineering. *In Vivo*, Vol. 22, No. 2, (Apr.-May, 2008) pp. 203-206, ISSN: 0258-851X

Cegielski, M. & Kalisiak, O. (2008). Terapia komórkowa - nadzieja na rozwiązanie problemów związanych z leczeniem ścięgien u koni. *Zycie Weterynaryjne*, Vol. 83, No. 9, pp. 754-757, PL ISSN: 0137-6810

Cegielski, M.; Izykowska, I.; Podhorska-Okolow, M.; Gworys, B.; Zabel, M. & Dziegiel, P. (2009). Histological studies of growing and mature antlers of red deer stags (Cervus elaphus). *Anatomia, Histologia, Embryologia*, Vol. 38, No. 3, (Jun., 2009) pp. 184-188, ISSN: 0340-2096

Chamberlain, G.; Fox, J.; Ashton, B. & Middleton, J. (2007). Concise review: mesenchymal stem cells: their phenotype, differentiation capacity, immunological features, and potential for homing. *Stem Cells*, Vol. 25, No. 11, (Nov., 2007) pp. 2739-2749, ISSN: 0250-6793

Clark, D.E.; Lord, E.A. & Suttie, J.M. (2006). Expression of VEGF and Pleiotrophin in Deer Antler. *The Anatomical Record. Part A, Discoveries in Molecular, Cellular, and Evolutionary Biology*, Vol. 288, No. 12, (Dec., 2006) pp. 1281-1293, ISSN: 1552-4884

Colitti, M.; Allen, S.P. & Price, J.S. (2005). Programmed cell death in the regenerating deer antler. *Journal of Anatomy*, Vol. 207, No. 4, (Oct., 2005) pp. 339-351, ISSN: 0021-8782

Faucheux, C.; Horton, M.A. & Price, J.S. (2002). Nuclear localization of type I parathyroid hormone/parathyroid hormone-related protein receptors in deer antler osteoclasts: evidence for parathyroid hormone-related protein and receptor activator of NF-kappaB-dependent effects on osteoclast formation in regenerating mammalian bone. *Journal of Bone and Mineral Research*, Vol. 17, No. 3, (Mar., 2002) pp. 455-464, ISSN: 0884-0431

Faucheux, C.; Allen, S.; Nicholls, B.M.; Danks, J.A.; Horton, M.A. & Price, J.S. (2004). Recapitulation of the parathyroid hormone-related peptide-Indian hedgehog pathway in the regenerating deer antler. *Developmental Dynamics*, Vol. 231, No. 1, (Sep., 2004) pp. 88-97, ISSN: 1058-8388

Feng, J.Q.; Chen, D.; Esparza, J.; Harris, M.A.; Mundy, G.R. & Harris, S.E. (1995). Deer antler tissue contains two types of bone morphogenetic protein 4 mRNA transcripts. *Biochimica et Biophysica Acta*, Vol. 1263, No. 2, (Aug., 1995) pp. 163-168, ISSN: 0006-3002

Feng, J.Q.; Chen, D.; Ghosh-Choudhury, N.; Esparza, J.; Mundy, G.R. & Harris, S.E. (1997). Bone morphogenetic protein 2 transcripts in rapidly developing deer antler tissue contain an extended 5' non-coding region arising from a distal promoter. *Biochimica et Biophysica Acta*, Vol. 1350, No. 1, (Jan., 1997) pp. 47-52, ISSN: 0006-3002

Francis, S.M. & Suttie, J.M. (1998). Detection of growth factors and proto-oncogene mRNA in the growing tip of red deer (Cervus elaphus) antler using reverse-transcriptase polymerase chain reaction (RT-PCR). *The Journal of Experimental Zoology*, Vol. 281, No. 1, (May, 1998) pp. 36-42, ISSN: 0022-104X

Goss, R.J. & Powel, R.S. (1985). Induction of deer antlers by transplanted periosteum. I. Graft size and shape. *The Journal of Experimental Zoology*, Vol. 235, No. 3, (Sep., 1985) pp. 359-373, ISSN: 0022-104X

Goss, R.J.; Van Praagh, A. & Brewer, P. (1992). The mechanism of antler casting in the fallow deer. *The Journal of Experimental Zoology*, Vol. 264, No. 4, (Dec., 1992) pp. 429-436, ISSN: 0022-104X

Gu, L.; Mo, E.; Yang, Z.; Zhu, X.; Fang, Z.; Sun, B.; Wang, C.; Bao, J. & Sung, C. (2007). Expression and localization of insulin-like growth factor-I in four parts of the red deer antler. *Growth Factors*, Vol. 25, No. 4, (Aug., 2007) pp. 264-279., ISSN: 0897-7194

Holmes, E.C.; Roberts, A.F.; Staines, K.A. & Ellis, S.A. (2003). Evolution of major histocompatibility complex class I genes in Cetartiodactyls. *Immunogenetics*, Vol. 55, No. 4, (Jul., 2003) pp. 193-202, ISSN: 0093-7711

Horwitz, E.M.; Gordon, P.L.; Koo, W.K.; Marx, J.C.; Neel, M.D.; McNall, R.Y.; Muul, L. & Hofmann, T. (2002). Isolated allogeneic bone marrow-derived mesenchymal cells engraft and stimulate growth in children with osteogenesis imperfecta: Implications for cell therapy of bone. *Proceedings of the National Academy of Sciences of the United States of America*, Vol. 99, No. 13, (Jun., 2002) pp. 8932-8937. ISSN: 0027-8424

Jafarian, M.; Eslaminejad, M.B.; Khojasteh, A.; Abbas, F.M.; Dehghan, M.M.; Hassanizadeh, R. & Houshmand, B. (2008). Marrow-derived mesenchymal stem cells-directed bone regeneration in the dog mandible: a comparison between calcium phosphate and natural bone mineral. *Oral Surgery, Oral Medicine, Oral Pathology, Oral Radiology, and Endodontics*, Vol. 105, No. 5, (May, 2008), pp. 14-24, ISSN: 1079-2104

Kierdorf, U.; Kierdorf, H. & Schultz, M. (1994). The macroscopic and microscopic structure of double-head antlers and pedicle bone of cervidae (Mammalia, Artiodactyla). *Annals of Anatomy*. Vol. 176, No. 3, (Jun., 1994) pp. 251-257, ISSN: 0940-9602

Kierdorf, U. & Kierdorf, H. (2000). Delayed ectopic antler growth and formation of a double-head antler in the metacarpal region of a fallow buck (Dama Dama L.) following transplantation of antlerogenic periosteum. *Annals of Anatomy*, Vol. 182, No. 4, (Jul., 2000) pp. 365-370, ISSN: 0940-9602

Kierdorf, U.; Stoffels, E.; Stoffels, D.; Kierdorf, H.; Szuwart, T. & Clemen, G. (2003). Histological studies of bone formation during pedicle restoration and early antler regeneration in roe deer and follow deer. *The Anatomical Record. Part A, Discoveries in Molecular, Cellular, and Evolutionary Biology*, Vol. 273, No. 2, (Aug., 2003) pp. 741-751, ISSN: 1552-4884

Kierdorf, U.; Li, C. & Price, J.S. (2009). Improbable appendages: Deer antler renewal as a unique case of mammalian regeneration. *Seminars in Cell & Developmental Biology*, Vol. 20, No. 5, (Jul., 2009) pp. 535-542, ISSN: 1084-9521

Komender, J.; Marczyński, W.; Tylman, D.; Malczewska, H.; Komender, A. & Sladowski, D. (2001). Preserved tissue allografts in reconstructive surgery. *Cell and Tissue Banking*, Vol. 2, No. 2, pp. 103-112, ISSN: 1389-9333

Korpos, E.; Molnar, A.; Papp, P.; Kiss, I.; Orosz, L. & Deak, F. (2005). Expression pattern of matrilins and other extracellular matrix proteins characterize distinct stages of cell differentiation during antler development. *Matrix Biology*, Vol. 24, No. 2, (Apr., 2005) pp. 124-135, ISSN: 0945-053X

Kos, M.; Staniszewska-Kuś, J.; Matusiewicz, M.; Łuczak, K.; Klempous, R. & Pielka, S. (2003). Tissue reaction after implantation of Spongostan, as the carrier of xenogenic bone morphogenetic protein. Experimental study. *Polimery w Medycynie*, Vol. 33, No. (1-2), pp. 25-33, ISSN: 0370-0747

Kuroda, R.; Ishida, K.; Matsumoto, T.; Akisue, T.; Fujioka, H.; Mizuno, K.; Ohgushi, H.; Wakitani, S. & Kurosaka, M. (2007). Treatment of a full-thickness articular cartilage defect in the femoral condyle of an athlete with autologous bone-marrow stromal cells. *Osteoarthritis and Cartilage*, Vol. 15, No. 2, (Feb., 2007) pp. 226-231, ISSN: 1063-4584

Lai, A.K.W.; Hai, W.L.; Verdon, D.J.; Nicholson, L.F.B. & Barling, P.M. (2007). The distribution of the growth factors FGF-2 and VEGF, and their receptors, in the growing red deer antler. *Tissue & Cell*, Vol. 39, No. 1, (Feb., 2007) pp. 35-46, ISSN: 0040-8166

Lattyak, B.V.; Maas, C.S. & Sykes, J.M. (2003). Dorsal onlay cartilage autografts: comparing resorption in a rabbit model. *Archives of Facial Plastic Surgery*, Vol. 5, No. 3, (May-Jun., 2003) pp. 240–243, ISSN: 1521-2491

Le Visage, C.; Dunham, B.; Flint, P. & Leong, K.W. (2004). Coculture of mesenchymal stem cells and respiratory epithelial cells to engineer a human composite respiratory mucosa. *Tissue Engineering*, Vol. 10, No. (9-10), (Sep.-Oct., 2004) pp. 1426-1435, ISSN: 1076-3279

Lendeckel, S.; Jodicke, A.; Christophis, P.; Heidinger, K.; Wolff, J.; Fraser, J.K.; Hedrick, M.H.; Berthold, L. & Howaldt, H.P. (2004). Autologous stem cells (adipose) and fibrin glue used to treat widesread traumatic calvarial defects: Case report. *Journal of Cranio-Maxillo-Facial Surgery*, Vol. 32, No. 6, (Dec., 2004) pp. 370-373, ISSN: 1010-5182

Li, C. & Suttie, J.M. (1998). Electron microscopic studies of antlerogenic cells from five developmental stages during pedicle and early antler formation in red deer (Cervus elaphus). *The Anatomical Record*, Vol. 252, No. 4, (Dec., 1998) pp. 587-599, ISSN: 0003-276X

Li, C.; Harris, A.J. & Suttie, J.M. (2001). Tissue interactions and antlerogenesis: new findings revealed by a xenograft approach. *The Journal of Experimental Zoology*, Vol. 290, No. 1, (Jun., 2001) pp. 18-30, ISSN: 0022-104X

Li, C. & Suttie, J.M. (2001). Deer antlerogenic periosteum: a piece of postnatally retained embryonic tissue. *Anatomy and Embryology*, Vol. 204, No. 5, (Nov., 2001) pp. 375-388, ISSN: 0340-2061

Li, C.; Clark, D.E.; Lord, E.A.; Stanton, J. & Suttie, J.M. (2002). Sampling technique to discriminate the different tissue layers of growing antler tips for gene discovery. *The Anatomical Record*, Vol. 268, No. 2, (Oct., 2002) pp. 125-130, ISSN: 0003-276X

Li, C.; Suttie, J.M. & Clark, D.E. (2005). Histological examination of antler regeneration in red deer (Cervus elaphus). *The Anatomical Record. Part A, Discoveries in Molecular, Cellular, and Evolutionary Biology*, Vol. 282, No. 2, (Feb., 2005) pp. 163-174, ISSN: 1552-4884

Li, C.; Stanton, J.A.; Robertson, T.M.; Suttie, J.M.; Sheard, P.W.; Harris, A.J. & Clark, D.E. (2007). Nerve growth factor mRNA expression in the regenerating antler tip of red deer (Cervus elaphus). *PLOS One*, Vol. 2, No. 1, (Jan., 2007) e. 148, ISSN: 1932-6203

Lord, E.A.; Martin, S.K.; Gray, J.P.; Li, C. & Clark, D.E. (2007). Cell Cycle Genes PEDF and CDKN1C in Growing Deer Antlers. *The Anatomical Record*, Vol. 290, No. 8, (Aug., 2007) pp. 994-1004, ISSN: 0003-276X

Moses, M.A.; Sudhalter, J. & Langer, R. (1990). Identification of an inhibitor of neovascularization from cartilage. *Science*, Vol. 248, No. 4961, (Jun., 1990) pp. 1408–1410, ISSN: 0193-4511

Muir, P.D.; Sykes, A.R. & Barrell, G.K. (1988). Changes in blood content and histology during growth of antlers in red deer (Cervus elaphus) and their relationship to plasma testosterone levels. *Journal of Anatomy*, Vol. 158, (Jun., 1988) pp. 31-42. ISSN: 0021-8782

Nakamura, Y.; Xiaaohong, W.; Chengsu, X.; Asakura, A.; Yoshiyama, M.; From, A.H.; Zhang, J. (2007). Xenotransplantation of long-term-cultured swine bone marrow-derived mesenchymal stem cells. *Stem Cells*, Vol. 25, No. 3, (Mar., 2007) pp. 612-620, ISSN: 0250-6793

Parekkadan, B.; von Poll, D.; Megeed, Z.; Kobayashi, N.; Tilles, A.W.; Berthiaume, F. & Yarmush, M.L. (2007). Immunomodulation of activated hepatic stellate cells by mesenchymal stem cells. *Biochemical and Biophysical Research Communications*, Vol. 363, No. 2, (Nov., 2007) pp. 247-252, ISSN: 0006-291X

Park, H.J.; Lee, D.H.; Park, S.G.; Lee, S.C.; Cho, S.; Kim, H.K.; Kim, J.J.; Bae, H. & Park, B.C. (2004). Proteome analysis of red deer antlers. *Proteomics*, Vol. 4, No. 11, (Nov., 2004) pp. 3642-3653, ISSN: 1615-9853

Phinney, D.G. & Prockop, D.J. (2007). Concise review: mesenchymal stem/multipotent stromal cells: the state of transdifferentiation and modes of tissue repair-current views. *Stem Cells*, Vol. 25, No. 11, (Nov., 2007) pp. 2896-2902, ISSN: 0250-6793

Price, J.S.; Oyajobi, B.O.; Oreffo, R.O. & Russell, R.G. (1994). Cells cultured from the growing tip of red deer antler express alkaline phosphatase and proliferate in response to insulin-like growth factor-I. *The Journal of Endocrinology*, Vol. 143, No. 2, (Nov., 1994) pp. 9-16, ISSN: 0022-0795

Price, J.S.; Oyajobi, B.O.; Nalin, A.M.; Frazer, A.; Russel, R.G. & Sandell, L.J. (1996). Chondrogenesis in the regenerating antler tip of red deer: expression collagen type I, IIA, IIB and X demonstrated by in situ nucleic acid hybridization and immunocytochemistry. *Developmental Dynamics*, Vol. 203, No. 3, (Mar., 1996) pp. 332-347, ISSN: 1058-8388

Price, J. & Allen, S. (2004). Exploring the mechanisms regulating regeneration of deer antlers. *Philosophical Transactions of the Royal Society of London. Series B, Biological Sciences*, Vol. 359, No. 1445, (May, 2004) pp. 809-822, ISSN: 0962-8436

Price, J.S.; Allen, S.; Faucheux, C.; Althnaian, T. & Mount, J.G. (2005a). Deer antlers: a zoological curiosity or the key to understanding organ regeneration in mammals? *Journal of Anatomy*, Vol. 207, No. 5, (Nov., 2005) pp. 603-618, ISSN: 0021-8782

Price, J.; Faucheux, C. & Allen, S. (2005b). Deer antlers as a model of Mammalian regeneration. *Current Topics in Developmental Biology*, Vol. 67, pp. 1-48, ISSN: 0070-2153

Richardson, L.E.; Dudhia, J.; Clegg, P.D. & Smith, R. (2007). Stem cells in veterinary medicine-attempts at regenerating equine tendon after injury. *Trends in Biotechnology*, Vol. 25, No. 9, (Sep., 2007) pp. 409-416, ISSN: 0167-7799

Rolf, H.J. & Enderle, A. (1999). Hard fallow deer antler: a living bone till antler casting? *The Anatomical Record*, Vol. 255, No. 1, (May, 1999) pp. 69-77, ISSN: 0003-276X

Rolf, H.J.; Kierdorf, U.; Kierdorf, H.; Schulz, J.; Seymour, N.; Schliephake, H.; Napp, J.; Niebert, S.; Wölfel, H. & Wiese, K.G. (2008). Localization and characterization of

STRO-1 cells in the deer pedicle and regenerating antler. *PLOS One*, Vol. 3, No. 4, (Apr., 2008) e2064, ISSN: 1932-6203

Romaniuk, A.; Malejczyk, J. & Kubicka, U. (1995). Rejection of cartilage formed by transplanted allogeneic chondrocytes: evaluation with monoclonal antibodies. *Transplant Immunology*, Vol. 3, No. 3, (Sep., 1995) pp. 251-257, ISSN: 0966-3274

Rotter, N.; Haisch, A. & Bücheler, M. (2005). Cartilge and bone tissue engineering for reconstructive head and neck surgery. *European Archives of Oto-Rhino-Laryngology*, Vol. 262, No. 7, (Jul., 2005) pp. 539-545, ISSN: 0937-4477

Ryan, J.M.; Barry, F.P.; Murphy, J.M. & Mahon, B.P. (2005). Mesenchymal stem cells avoid allogeneic rejection. *Journal of Inflammation*, Vol. 26, (Jul., 2005) pp. 2-8, ISSN: 1078-7852

Sakata, I.; Ueno, T.; Kagawa, T.; Kanou, M.; Fujii, T.; Yamachika, E. & Sugahara, T. (2006). Osteogenic potential of cultured human periosteum-derived cells-A pilot study of human cell transplantation into a rat calvarial defect model. *Journal of Cranio-Maxillo-Facial Surgery*, Vol. 34, No. 8, (Dec., 2006) pp. 461-465, ISSN: 1010-5182

Schliephake, H.; Knebel, J.W.; Aufderheide, M. & Tauscher, M. (2001). Use of cultivated osteoprogenitor cells to increase bone formation in segmental mandibular defects: an experimental pilot study in sheep. *International Journal of Oral and Maxillofacial Surgery*, Vol. 30, No. 6, (Dec., 2001) pp. 531-537, ISSN: 0901-5027

Seto, I.; Asahina, I.; Oda, M. & Enomoto, S. (2001). Reconstruction of the primate mandible with combination graft of recombinant human bone morphogenetic protein-2 and bone marrow. *Journal of Cranio-Maxillo-Facial Surgery*, Vol. 59, No. 1, (Jan., 2001) pp. 53-61, ISSN: 1010-5182

Shanti, R.M.; Li, W.J.; Nesti, L.J.; Wang, X. & Tuan, R.S. (2007). Adult mesenchymal stem cells: biological properties, characteristics, and applications in maxillofacial surgery. *Journal of Oral and Maxillofacial Surgery*, Vol. 65, No. 8, (Aug., 2007) pp. 1640-1647, ISSN: 0278-2391

Shayesteh, Y.S.; Khojasteh, A.; Soleimani, M.; Alikhasi, M.; Khoshzaban, A. & Ahmadbeigi N. (2008). Sinus augmentation using human mesenchymal stem cells loaded into a β-tricalcium phosphate/hydroxyapatite scaffold. *Oral Surgery, Oral Medicine, Oral Pathology, Oral Radiology, and Endodontics*, Vol. 106, No. 2, (Aug., 2008) pp. 203-209, ISSN: 1079-2104

Swarbrick, P.A. & Crawford, A.M. (1997). The red deer (Cervus elaphus) contains two expressed major histocompatibility complex class II DQB genes. *Animal Genetics*, Vol. 28, No. 1, (Feb., 1997) pp. 49-51, ISSN: 0268-9146

Szuwart, T.; Kierdorf, H.; Kierdorf, U. & Clemen, G. (1998). Ultrastructural aspects of cartilage formation, mineralization, and degeneration during primary antler growth in fallow deer (Dama dama). *Annals of Anatomy*, Vol. 180, No. 6, (Dec., 1998) pp. 501-510, ISSN: 0940-9602

Szyda, A.; Paprocka, M.; Krawczenko, A.; Lenart, K.; Heimrath, J.; Grabarczyk, P.; Mackiewicz, A. & Duś, D. (2006). Optimization of a retroviral vector for transduction of human CD 34 positive cells. *Acta Biochimica Polonica*, Vol. 53, No. 4, pp. 815-823, ISSN: 0001-527X

Ulusal, A.E.; Ulusal, B.G. & Hung, L.M. (2005). Establishing a composite auricle allotransplantation model in rats: introduction to transplantation of facial subunits.

Plastic and Reconstructive Surgery, Vol. 116, No. 3, (Sep., 2005) pp. 811–817, ISSN: 0032-1052

Vats, A.; Bielby, R.C.; Tolley, N.; Dickinson, S.C.; Boccaccini, A.R.; Hollander, A.P.; Bishop, A.E. & Polak, J.M. (2006). Chondrogenic differentiation of human embryonic stem cells: the effect of the micro-environment. *Tissue Engineering*, Vol. 12, No. 6, (Jul., 2006) pp. 1687-1697, ISSN: 1076-3279

Warnke, P.H.; Springer, I.N.; Wiltfanf, J.; Acil, Y.; Eufinger, H.; Wehmöller, M.; Russo, P.A.; Bolte, H.; Sherry, E.; Behrens, E. & Terheyden, H. (2004). Growth and tansplantation of a custom vascularised bone graft in man. *Lancet*, Vol. 364, No. 9436, (Aug.-Sep., 2004) pp. 766-770, ISSN: 0140-6736

Wellisz, T. (1993). Reconstruction of the burned external ear using a Medpor porus polyethylene pivoting helix framework. *Plastic and Reconstructive Surgery*, Vol. 91, No. 5, (Apr., 1993) pp. 811–818, ISSN: 0032-1052

Wiecko, J.; Kukwa, A. & Hinek, A. (1980). Auricle reconstruction with allogenous costal cartilage and autogenous perichondrium. *Acta Medica Polona*, Vol. 21, No. 1, pp. 87–92, ISSN: 0001-608X

4

Cell-Protein-Material interaction in tissue engineering

Manuel Salmerón-Sánchez[1,2,3] and George Altankov[4,5]
[1]*Center for Biomaterials & Tissue Engineering, Universidad Politécnica de Valencia, Spain*
[2]*Centro de Investigación Príncipe Felipe, Valencia, Spain*
[3]*Networking Research Center on Bioengineering, Biomaterials and Nanomedicine (CIBER-BBN), Valencia, Spain*
[4]*Institute for Bioengineering of Catalonia, Barcelona, Spain*
[5]*ICREA -Institució Catalana de Recerca i Estudis Avançats, Barcelona, Spain*

1. Introduction

The initial cellular events that take place at the biomaterials interface mimic to a certain extent the natural adhesive interaction of cells with the extracellular matrix (ECM) (Spie, 2002; Griffin & Naughton, 2002; Grinnell, 1986). In fact, the living cells cannot interact directly with foreign materials, but they readily attach to the adsorbed layer of proteins (upon contact with physiological fluids in vivo or culture medium in vitro) such as fibronectin (FN), vitronectin (VN), fibrinogen (FG), representing the so-called soluble matrix proteins in the biological fluids (Grinnell 1986). Upon longer contact with tissues many other ECM proteins, such as collagens and laminins, will also associate with the surfaces, affecting the cellular interaction. The concentration, distribution, and mobility of the adsorbed protein layer on a surface play a fundamental role in the biofunctionality of a synthetic material and are clue factors to understand the biological response of a substrate (Anselme, 2000). Cells recognize these matrix proteins via integrins - a family of cell surface receptors – that provide trans-membrane links between the ECM and the actin cytoskeleton (Hynes, 2002). When integrins are occupied they cluster and develop focal adhesions; discrete supramolecular complexes that contain structural proteins such as vinculin, talin, α-actinin, and signalling molecules, including FAK, Src and paxilin that actually anchorage the cells to the surface and trigger the subsequent cellular response (Geiger et al., 2001). Abnormalities in the cell-ECM integrin mediated interactions are associated with pathologic situations that include tumour formation (Wehrle-Haller & Imhof, 2003). Besides, integrin mediated adhesion involves not only the receptor-ligand but also post-ligation interactions with multiple binding partners (García, 2005). Thus, the initial cell-material interaction is a complex multi-step process consisting of early events, such as adsorption of proteins, followed by cell adhesion and spreading, and late events, related to cell growth, differentiation, matrix deposition and cell functioning. To measure and to quantify some of these parameters comprise the *classical* approach to characterise the cellular biocompatibility of materials (Williams, 1998).

2. Protein-material interaction

Proteins are considered to be clue factors in mediating the cell-material interaction and their status (the amount, conformation and strength of interaction) on a material surface determine the biocompatibility of the system. Even if the design of antifouling surfaces for the repulsion of proteins is an important field of activity (Chen et al, 2008), we are focused in this chapter on the effect of material properties on adsorption of matrix proteins and the influence on cell response.

Protein adsorption on material surfaces is a process driven both by the intensity of the energetic interactions between the molecular groups of the substrate's surface and of the protein (i.e., hydrogen bonding, electrostatic, van der Waals interactions) and by entropic changes as a consequence of the unfolding of the protein as bound water is released from the surface (García, 2006; Werner et al, 2006). Clearly, the amount of protein adsorbed and its conformation depend on the chemical groups of the substrate, which determine the energetic and entropic interactions with the adsorbed proteins. Protein adsorption on different substrates has been extensively investigated in the literature by different techniques. FN, albumin, laminin, collagen, lysozyme, fibrinogen, hemoglobin, and several other proteins were adsorbed on different substrates—mostly, model surfaces such as mica, glass, and self-assembled monolayers—and were investigated by different techniques that include atomic force microscopy, ellipsometry, quartz crystal microbalance, sodium dodecyl sulfate gel electrophoresis, FT-IR, spectroscopic imaging, electron microscopy, and fluorescence probe techniques (Keselowsky et al., 2003; Michael et al., 2003; Tsapikouni & Missirlis, 2007; Benesch et al., 2007; Weber et al., 2007; Steiner et al., 2007; Lord et al., 2006; Noh & Vogler, 2006; Prime & Whitesides, 2006; Sousa et al., 2007). It is convenient to describe the main adhesion proteins -and its behaviour on material surfaces- for the sake of (auto)completeness of the chapter.

fibronectin

FN is a glycoprotein found in blood, extracellular fluids, and connective tissues and attached to the cell surfaces. Both plasma FN and the cell surface forms are dimers, consisting of two subunits of 220 kDa, linked by a single disulfide bond near the carboxyl termini (Erickson & McDonagh, 1981; Erickson & Carell, 1983). Each subunit contains three types of repeating modules (types I, II, and III) that mediate interactions with other FN molecules, other extracellular matrix (ECM) proteins, and cell-surface receptors (Pankov & Yamada, 2002). The importance of FN as a mediator of cell adhesion to a substrate was recognized earlier (Pearlstein, 1980). Since then, many studies have shown the role of FN in promoting cell adhesion and regulating cell survival and phenotype expression on different surfaces (García & Boettiger, 1999; Toworfe et al., 2004; Baugh & Vogel, 2004; Lan et al., 2005; Grinnell & Feld, 1982; Altankov et al., 2000). It has been stressed that, for a fixed surface chemistry, the initial density of integrin–FN bonds is proportional to the surface density of adsorbed FN (García & Boettiger, 1999); moreover, the nature of the surface chemistry is able to modulate FN conformation (Keselowsky et al., 2003). It has been suggested that FN adsorbs preferentially on hydrophobic surfaces (Toworfe et al., 2004) and that it undergoes greater extension of its dimer arms on hydrophilic glass (Baugh & Vogel, 2004) in a conformation that favors the binding of antibodies (Grinnell & Feld, 1982) and strength the cell–material interaction (Kowalczynska et al., 2006). Even the micro/nano surface roughness has been shown to influence FN adsorption (Khang et al., 2007; Costa Martínez et al., 2008). The integrin-FN interaction, governed mainly by the $\alpha_5\beta_1$ dimer, also leads to the formation of extracellular matrix fibrils from the newly secreted FN (Mao & Schwarzbauer, 2005) and even arrangement of those protein molecules adsorbed on the substratum (Altankov & Groth, 1996; Altankov et al., 1996). The thickness of FN matrix fibrils ranges from 10 to 1000 nm in diameter and consists of a few to hundreds of FN molecules across (Singer, 1979). FN binding to integrins induces reorganization of the actin cytoskeleton and activates intracellular signaling complexes. Cell contractility facilitates FN conformational changes, and it allows for the unfolding of the native globular FN structure, thus exposing cryptic domains that were not available in the compact form of soluble FN. Finally, fibrils are formed through FN-FN interactions, usually through binding of I1-5 to either III_{1-2} or III_{12-14} domains (Geiger et al, 2001). Cell-mediated FN reorganization, when adsorbed on a synthetic surface, seems to be also an important factor in determining the biocompatibility of a material, because poor cell adhesion and spreading has been found in cases when integrin-mediated rearrangement of FN did not occur at the material interface (Altankov & Groth, 1994; Altankov & Groth, 1996).

It has been shown that the existence of mechanical tension is necessary for efficient integrin-mediated FN fibrillogenesis (Erickson, 2002; Smith et al., 2007). Although it is generally agreed that FN fibrillogenesis is cell-dependent process, fibrillar networks of FN have been generated also in the absence of cells by means of interactions with the underlying substrate that involves mechanical events at the molecular scale. FN fibrillogenesis upon contact with a lipid monolayer was explained through mechanical tension caused by domain separation in the lipid monolayer that pulls the protein into an extended conformation (Baneyx & Vogel, 1999). The assembly of FN into fibers was obtained also by applying forces to FN molecules via poly(dimethylsiloxane) (PDMS) micropillars at different stages of fibrillogenesis (Ulmer et al, 2008). We have recently shown that FN fibrillogenesis can take place as a consequence of the sole interaction between the protein molecules and a material surface with the appropriate surface chemistry; in concrete, a spontaneous formation of biologically active FN network was found in vitro after its adsorption on poly(ethylacrylate) (PEA) (Gugutkov et al., 2009; Rico et al., 2009).

fibrinogen

FG is a large, complex, fibrous glycoprotein normally present in human blood plasma essential for many biological functions which include haemostasis, wound healing, inflammation, and angiogenesis (Weisel, 2005). It is made up of three pairs of polypeptide chains, designated as Aα, Bβ, and γ, with molecular masses of 66, 52, and 46 kDa, respectively, which are held together by 29 disulfide bonds (Weisel, 2005). These six polypeptides are organized into independently folded units: a central E-domain, which includes the N-terminus of all six polypeptide chains, and two terminal D-domains, which include the Bß and γ chains. The carboxy-terminal of the Aα chain, the aC domain, departs from the D fragment and either associates to the E-domain to constitute a single globular domain close to it or, on the contrary, they form appendages with a certain degree of mobility. In its native form the aC association to the central domain is more common; however, there is equilibrium between these two situations (Veklich, 1993). The cleavage of the small A and B sequences from the Aα and Bβ chains by thrombin in the E-domain yields fibrin, which is able to associate and polymerize. The length of an individual FG molecule is 45–50 nm (Weisel et al., 1985; Gorman et al., 1985).

FG–surface interactions have been investigated on many substrates with different experimental techniques (Brash & Horbett, 1995). Atomic force microscopy (AFM), which is a technique able to provide direct observation of protein conformation on different substrates has been extensively used in recent years, mainly on model surfaces, such as silica, mica, titanium graphite, and self-assembled monolayers (SAMs), flat enough so that the height magnitude is able to reveal the trinodular structure of single-adsorbed FG molecules (Toscano & Santore, 2006; Marchin & Berrie, 2003; Agnihotri & Siedlecki, 2004; Cacciafesta et al., 2000; Tunc et al., 2005; Gettens et al., 2005; Gettens & Gilbert, 2007; Ta & McDermott, 2000; Ishizaki et al, 2007; Ortega-Vinuesa et al., 1998; Mitsakakis et al., 2007; Sit & Marchant, 1999).

The effect of surface wettability, as one of the most important parameters that affects the biological response to a material, on FG adsorption has lead to different, nonconsistent conclusions. Even if there is general agreement in the decrease of FG adsorption with the increase of wettability of the substrate, (Slack & Horbett, 1992; Rodrigues et al., 2006) it is not the case concerning FG conformation. Marchin et al. observed dramatic differences in the conformation of FG adsorbed on hydrophilic mica and hydrophobic graphite: globular conformations were observed on mica, whereas on graphite the trinodular structure of the extended molecule was clearly observed (Marchin & Berrie, 2003). Sit et al. suggested that the spreading of FG increases with the hydrophobicity of the surface (Sit & Marchant, 1999). In addition, Wertz and Santore have shown through total internal reflection fluorescence that the footprint of a FG molecule is larger when adsorbed on a hydrophobic surface (graphite) than on a hydrophilic one (mica) (Wertz & Santore, 2001; Wertz & Santore, 2002). However, other authors have found the trinodular conformation both on graphite and mica (Agnihotri & Siedlecki, 2004). The adhesion force between FG and the substrate has also been investigated by AFM and it has been found to depend on the surface wettability. By functionalizing AFM tips with the protein, Kidoaki et al. found that the strength of adhesion to a hydrophobic SAM was larger than to hydrophilic ones (Kidoaki & Matsuda, 1999). Xu et al. measured adhesion forces to a series of surfaces over a broad wettability range through glow-discharge plasma modification, by using protein modified AFM tips (Xu & Siedlecki); showing a marked transition between protein adherent materials and protein nonadherent materials over the range of water contact angles of 60–65 $^{\circ}$.

laminin

Laminins are trimeric molecules of α, β, and γ chains with molecular masses of 140–400 kDa. Several laminin isomorphs are known, with a large number of genetically distinct chains (α1 to α5, β1 to β3, and γ1 to γ3) (Burgeson et al., 2004). The laminins are important glycoprotein components of basement membranes, where they provide interaction sites for many other constituents, including cell surface receptors (Mercurio, 1995; Beck et al., 1990; Sasaki et al., 2004). Laminin plays an important role in neural cell migration, differentiation, and neurite growth (Kleiman et al., 1987; Heiduschka et al., 2001; Luckenbill-Edds, 1997; He & Bellamkonda, 2005) and it has been used as a coating for improving nerve cell adhesion and growth on different substrates (Rogers et al., 1983; Liesi et al., 1984).

Among the different material properties that influence protein adsorption (and, consequently, cell adhesion and functionality) the hydrophilicity of the system is an important one. The system based on the copolymerisation of ethyl acrylate and hydroxyethyl acrylate provides controlled hydrophilicity maintaining the same chemistry. It consists of a vinyl backbone chain with the side groups -COOCH2CH3 and -COOCH2CH2OH, respectively. Their copolymerization gives rise to a substrate in which the surface density of -OH groups can be varied without modifying any other chemical functionality of the system. The concentration of -OH groups determines both the surface energy and the hydrophilicity of the substrate (Table 1). The interaction of the protein domains with the chemical functionalities of the substrate and with water determines the molecule's adsorbed conformation.

x_{OH}	EWC	WCA (°)
0 (pure PEA)	1.7 ±0.4	89±1
0.3	7.6±0.9	80±2
0.5	18.2±1.7	67±1
0.7	40.6±0.4	55±1
1 (pure PHEA)	134±5	45±2

Table 1. Equilibrium water content (EWC) and water contact angle (WCA) for the different substrates.

The amount of protein adsorbed on the different surfaces was quantified by image analysis of the Western blot bands obtained by analysing the supernatant after adsorption on the material surface (Rodríguez Hernandez et al., 2009; Rico et al., 2009). Two different curves have been observed (included in Figure 1) corresponding to FN and FG adsorption from a solution of concentration 20 μg/mL. The adsorbed FN depends non-monotonically on the -OH density of the substrate, and FN surface density shows a minimum at approximately x_{OH}=0.5. Both higher and lower concentrations of hydroxyl groups in the substrate result in higher amounts of the adsorbed protein. For all concentrations of the original solution, the highest protein adsorption occurs on the most hydrophilic substrate. The situation is completely different when FG is adsorbed on the same family of substrates. The amount of adsorbed protein diminishes monotonically as the OH density increases. The difference of adsorbed FG between pure PEA (OH_0) and OH_{30} is approx. 0.9 μg/cm^2, a huge fall for such a small OH increment, whereas these differences tend to diminish as substrata become more hydrophilic (<0.1 μg/cm).

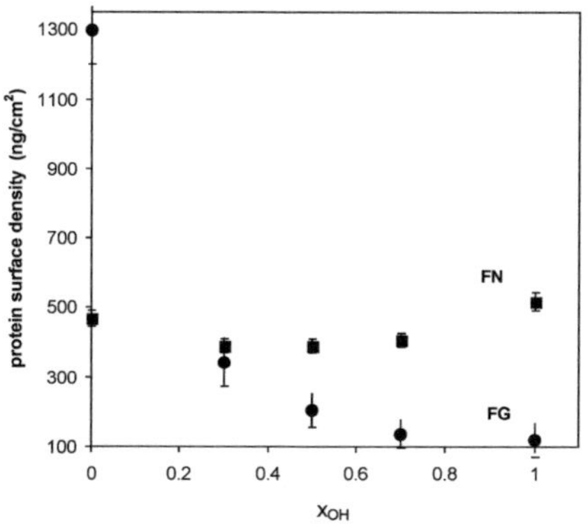

Fig. 1. Surface density of fibronectin (FN, squares) and fibrinogen (FG, circles) adsorbed on substrates with controlled OH density.

The conformation of the protein adsorbed on the substrate can be directly observed by Atomic Force Microscopy. AFM studies of the adsorption of protein on non-model surfaces in cases where the roughness of the substrate is of a size of the order of the protein height cannot be conclusive if monitored with the height signal since the protein features are usually blurred; one would have to polish the neat material to get a surface flat enough (Veklich et al., 1993) Instead of the height, the phase signal of AFM, which is a magnitude sensitive to the different viscoelastic behaviors, (Tamayo & García, 1997; Cleveland et al., 1998; Tamayo & García, 1998; García et al., 1998) can be used to distinctly reveal protein conformation under conditions of usual, non-model polymer surfaces (Holland & Marchant, 2000; Rodríguez-Hernández, 2007). Figure 2 shows protein conformation and distribution after adsorption on the different substrates from a 20 μg/mL protein solution, which is the concentration typically employed when coating a substrate with the protein for cell culture purposes (Keselowsky et al., 2003; Erickson & Carell, 1983; Altankov et al., 2000; Altankov & Groth, 1994; Altankov & Groth, 1996). The more hydrophobic surfaces induce the formation of protein networks, whose density decreases as the fraction of -OH groups increases. FN network is well developed on the PEA ($-OH_0$) and $-OH_{10}$. Protein aggregates with elongated shape are still formed on the $-OH_{30}$ surface, but only weakly connected protein filaments are identified. A higher amount of hydroxyl groups (from x_{OH}=0.5 on) prevents the formation of a protein network on the materials surface, and only disperse (micro) aggregates of the protein are observed on the $-OH_{50}$, $-OH_{70}$, and PHEA ($-OH_{100}$) substrates. The surface density of these globular FN aggregates seems to increase with the fraction of hydroxyl groups from x_{OH}=0.5 to 1.

FG distribution after adsorption on the different substrata at different magnifications reveals rather than single FG molecules, AFM images show protein patterns with different topologies. Nevertheless, some differences between the conformations of FG on the different substrates are worth mentioning. The formation of a FG network takes place on pure PEA (OH_0), but the co-continuity of the protein network is lost when small amounts of OH are introduced in the system (OH_{10} and OH_{20}). However, from this hydroxy content on, FG–FG interactions are somehow enhanced and variable fibril network topologies show up again.

FN conformation on the substrate is not related to the total amount of protein adsorbed on it. This nonmonotonic dependence of adsorbed protein on -OH fraction can be understood if protein adsorption on a substrate's surface is analyzed in terms of the number of available sites on the surface; it is clear that not only the energetic interactions between the substrate and the protein play a role in the adsorption process, but also the conformation of the protein—the configurational entropy—must trigger the amount of molecules directly adsorbed on the substrate: globular conformations of FN on the more hydrophilic substrates must lead to a higher amount of the protein adsorbed. The other way around, minimum adsorption at x_{OH}=0.5 must be a consequence of two opposite processes: energetic and entropic interactions that lead to less efficient FN packing for this substrate composition. In a similar way, FG conformation on the substrate is not directly related to the total amount of protein adsorbed on it. It has been found that the footprint of a FG molecule is larger when adsorbed on a hydrophobic surface than on a hydrophilic one (Wertz & Santore, 2001; Wertz & Santore, 2002), and higher amounts of adsorbed proteins on the most hydrophobic surface result in an ordered FG–FG adsorption, which leads to the formation of a network on the substrate. As hydrophilicity increases, the amount of FG directly in contact with the substrate decreases, as well as the footprint of the molecule, which results in the formation of isolated FG aggregates.

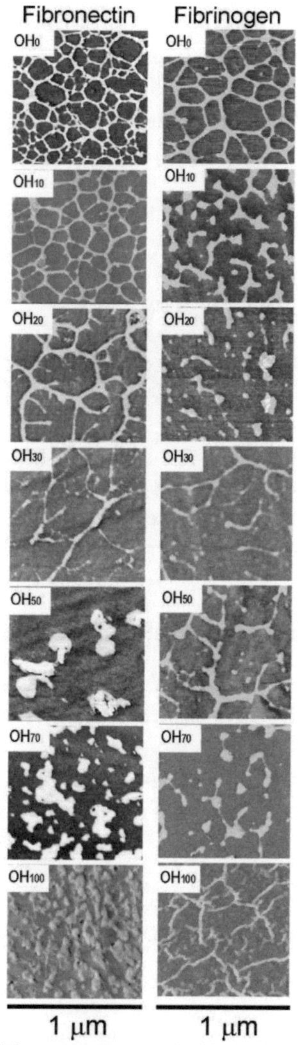

Fig. 2. AFM phase images of fibronectin and fibrinogen adsorbed on substrates with controlled hydroxyl density. Both proteins tend to form networks on the most hydrophobic surface that disaggregates as the hydrophilicity of the substrate increases. (these results are extracted -with permission- from refs Rico et al., 2009 and Rodríguez Hernández et al., 2009).

FN and FG stands for a situation in which similar distribution and conformation of the protein is observed on the substrates (Figure 2) but very different tends in the amount of adsorbed protein are found (Figure 1). When laminin is adsorbed on this set of substrates, protein molecules show globular like morphology on the hydrophilic PHEA and gradually extend as the amount of -OH groups on the surface diminishes, up to a point in which the protein conformation tends again to a more compact, less extended conformation. Additionally, the formation of a laminin network takes place on the 50:50 copolymer in which the N-terminal domain of all three chains of the protein are linked. This polymerized supramolecular aggregation is the typical form of laminin in the basement membrane and its formation depends on time, temperature, and concentration (Luckenbill-Edds, 1997). The formation of protein networks on our surfaces must be conditioned not only by the different conformation of the molecule on the substrate, but also by the surface density of adsorbed protein.

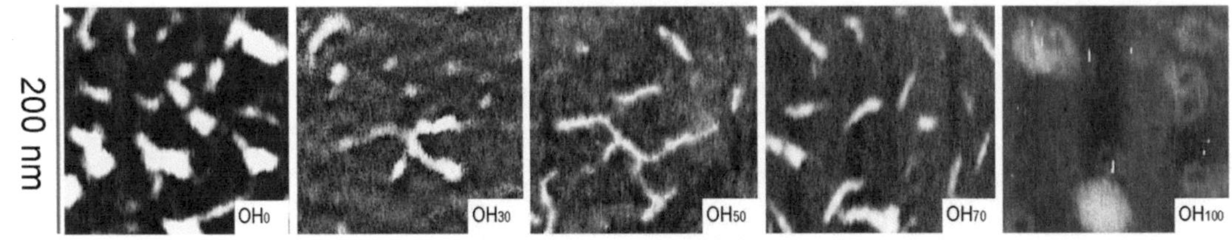

Fig. 3. AFM phase image of laminin molecues adsorbed on substrates with controlled hydroxyl density. Globular protein molecules are observed both on the hydrophilic and hydrophobic polymers while the protein tends to extend their arms for intermediate hydroxyl densities (extracted -with permission- from ref. Rodríguez Hernández et al., 2007).

3. Development of provisional extracellular matrix on biomaterials interface

Cell-matrix interaction in vivo is a complex bi-directional and dynamic process. Cells in the tissues are constantly accepting information on their environment from cues in the ECM (Altankov & Groth, 1994) and, at the same time, cells are producing and frequently remodelling their matrix (Grinnel, 1986; Hynes, 2002; Avnur & Geiger, 1981). Therefore, it is not surprising that many cells cannot adapt and poorly survive in vitro and, conversely, when a foreign material is implanted in the body, the adjacent tissue cells do not interact properly because of lack of their ECM.

A large and growing body of evidence shows that the cells in vivo need to accept distinct physico-chemical signals from the surrounding ECM and because a tight connection between the cytoskeleton and ECM the cells also respond to these properties. Mechanical properties are also important. For instance, as the stiffness of the surrounding ECM is in the same order of magnitude as cells, they are able to reorganize this matrix (Rhees & Grinnel, 2007; Kolakna et al., 2007). However, on stiffer materials cells may fail to do so, and it is an obstacle for their biocompatibility. A line of previous investigations has shown that fibroblasts and endothelial cells tend to rearrange adsorbed matrix proteins, such as FN and FG (Altankov et al., 1997; Tzoneva et al., 2002), as well as collagen (Maneva-Radicheva et al., 2008) in a fibril-like pattern. Using model surfaces -mostly self-assembled monolayers (SAMs) - it has been shown that this cellular activity is abundantly dependent on the surface properties of materials, such as wettability (Altankov et al., 1996), surface chemistry and charge (Pompe et al., 2005; Altankov et al., 2000). It has been shown that for this kind of substrates (SAMs) fibroblast on hydrophilic surfaces may reorganise FN in ECM-like structures whereas on hydrophobic surfaces almost no rearrangement of FN occurs (Fig. 4).

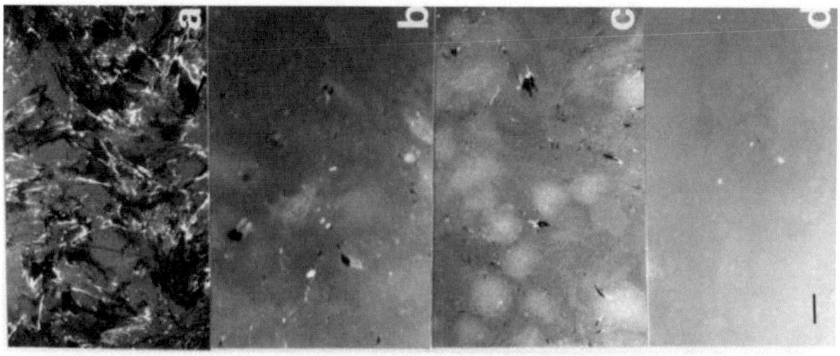

Fig. 4. Reorganization of adsorbed FITC-fibronectin on different wettable model substrates: a) glass (WCA=23^{0}); b) aminopropyl silane (WCA=67^{0}); c) octadecyl silane (WCA=87^{0}); d) silicon (WCA=107^{0}).

These experiments suggests that loosely adsorbed FN on hydrophilic surfaces provide a better substrate for cell growth presumably due to the fact that cells need to modify adsorbed FN for their normal function. Therefore, they remove and organise FN from the substrate into specific fibrillar structures, similar to FN matrix fibrils. Early events of integrin receptor β_1 functioning is also different on hydrophilic and hydrophobic materials. It is known that $\alpha_5\beta_1$ binds FN to the cell surface and induced conformational changes required for FN polymerisation (Smith et al., 2007). It has been shown that clusters of β_1 integrin might organise in a specific linear pattern on the dorsal cell surface of adhering fibroblasts on hydrophilic glass, presumably matching the initial positional organisation of FN matrix. However, on hydrophobic glass, even if the cells formed normal focal adhesions –similar to those on FN-coated glass- they did not develop a linear organisation of the FN receptor (Altankov et al., 1997). This evidence raises the possibility that tissue compatibility of such materials may be connected with the allowance of cells to remodel surface associated proteins presumably as an attempt to form their own matrix, e.g. materials that bind proteins loosely will support the organization of a provisional ECM. This view however does not consider the real molecular architecture of the adsorbed proteins layer and also causes limitations to the materials selection.

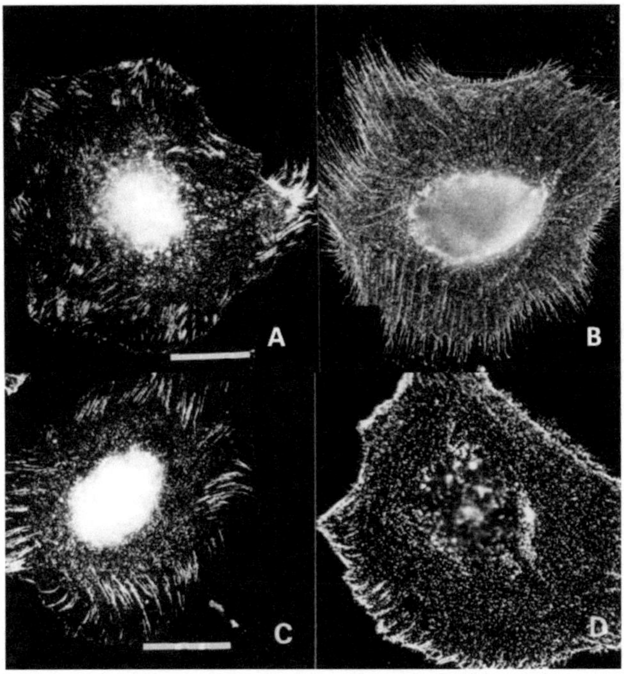

Fig. 5. Organization of β_1 integrins on the dorsal (B,D) and ventral (A,C) cell surface of fibroblasts adhering on hydrophilic glass (A,B) and hydrophobic ODS (C,D) surfaces. Note: Focal adhesions are well developed on both hydrophilic and hydrophobic substrata (A and C) while linear arrangement of ß1 integrins are visible on hydrophilic glass only (B) and absent on hydrophobic ODS (D).

The arrangement of natural FN matrix is also dependent on the ability of cells to reorganize the adsorbed FN layer on the material substrate. When hydrophilic and hydrophobic glass are culture with human fibroblast for longer time (72 h), significant amounts of FN are deposited by cells on the hydrophilic substrate, organised in fibrils and clusters, oriented in the direction of the cell polarization. However, on the hydrophobic glass, less FN fibril formation was observed although cell spreading was almost in the same extent as on the hydrophilic glass (Altankov & Groth, 1996). Figure 6 shows organization of extracellular fibronectin matrix on different wettable model substrates of increasing hydrophobicity as measured by the water contact angle (WCA): glass (WCA=23°) aminopropyl silane (WCA=67°), octadecylsilane (WCA=87°), and silicon (WCA=107°). A clear trend for less FN matrix formation with increasing the hydrophobicity of substratum is demonstrated on Figure 6 below.

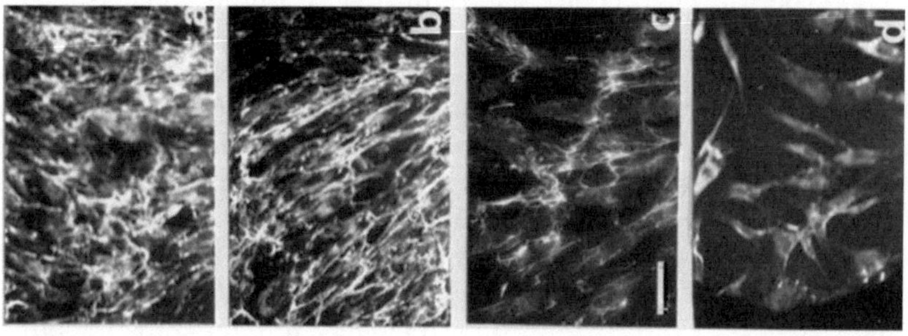

Fig. 6. Organization of extracellular fibronectin matrix on different wettable model substrates: a) glass (WCA=23^0); b) aminopropyl silane (WCA=67^0); c) octadecyl silane (WCA=87^0); d) silicon (WCA=107^0).

Recent investigations however, have shown that even if cells need the adequate environment to synthesize their own matrix at the cell material interface, it is not mandatory that this happens only on hydrophilic substrata. The system based on the copolymerisation of ethyl acrylate and hydroxyethyl acrylate, that we have already commented concerning protein adsorption in the previous section, is an example of how things can happen in a different way. There, it was found that the cells are able to synthesize and deposit FN matrix fibrils on some of the material surfaces. The formation of FN fibrils, the so-called fibrillogenesis, is a process either mediated by integrins or, as it is accounted for previously, induced by the substrate. However, FN fibrils could not be found on the more hydrophilic samples ($-OH_{100}$ and $-OH_{70}$) while on the sample with intermediate composition, $-OH_{50}$ the fibroblasts deposit only small fibrils, located mostly beneath the cells. As the hydroxyl fraction decreases (and the surface becomes more hydrophobic) the FN deposition increases, which moreover is organized into a typical matrix-like structure similar to those on the hydrophilic glass (Altankov et al. 1996). Nevertheless, FN reorganisation does not happen as expected anymore: no reorganization of FN takes place whatever the hydroxyl fraction of groups in the sample, that is, FN reorganization does not depend on the hydrophilicity for this family of substrates. It is noteworthy, however, that the values for the wettability of the PHEA samples (WCA 45°) correspond to values that are optimal for the cellular interaction in other systems. Conversely, surfaces with about 90° WCA, characteristic for pure PEA, and where the best cellular interaction was found, usually abrogate cellular interaction (Garcia, 2006; Grinnell & Feld, 1982; Arima & Iwata, 2007). Collectively this suggests that even unable to organize the preadsorbed FN on the substrate, the fibroblasts respond on the FN network previously formed during protein adsorption on the substrate (Figure 2), presumably because the conformation of the protein provides the adequate signals which stimulate their normal matrix-forming activity.

These results suggest that the distinction between hydrophilic and hydrophobic features of a substrate is insufficient to explain the general trends underlying the cell-material interaction, and more factors must be taken into account. For instance, the conclusion that the ability of fibroblasts to secrete ECM proteins is greatly reduced on hydrophobic substrates (Altankov et al., 1996) even if cell adhesion takes place clearly differs from the results in this recent work. Rather, fibroblast functional behaviour on a synthetic substrate depends in a subtle way on the particular substrate chemistry that triggers the process of protein adsorption. Both protein conformation on the substrate and the intensity of the protein-material interaction play a fundamental role on cell behavior: the adequate protein conformation on the substrate—leading to a substrate induced FN fibrillogenenis—results in excellent cell adhesion and matrix formation (for low -OH contents), even if preadsorbed FN cannot be removed by cells. Alternatively, if protein conformation is good enough so as to support initial cell adhesion, cells will be able to remove the initial FN layer and secrete their own extracellular matrix (as it happens in the control glass). Higher -OH fractions in the substrate lead to inadequate protein conformation on the substrate, which does not support good cell adhesion and consequently leads to diminished functionality.

Despite our knowledge on ECM organization, relatively little is known about the fate of these already arranged matrix proteins. Recent data indicate that polymerized forms of the matrix proteins have properties distinct from protomeric, non-polymerized ones. For example, the state of collagen polymerization has been shown to alter its growth regulatory properties (Schofield, 1978). Emerging evidence also indicates that the ECM form of FN (matrix fibrils) is functionally distinct from the soluble FN (Hynes, 1990; Wagenaar-Mller et al., 2007), having implications on the long-term behaviour of this protein. Thus, depending on the allowance of materials surface to support the development of fibrillar matrix, the biological properties of a material may be altered. Much is known about the interactions between different ECM proteins, but surprisingly less is our knowledge about the ECM composition, organization, and stability at the biomaterials interface. Can it be regulated? A distinct hierarchy of matrix remodelling is already evident from the in vitro studies (Velling et al., 2002). Integrin-mediated assembly of FN into fibrils is well documented (Mosher et al., 1992; Christopher et al., 1997; Sottile & Hocking, 2002), but the fact that FN may tether other matrix proteins to the cell surface, and therefore is required for their organization, is new in the field (Maneva-Radicheva et al., 2008; Dzamba, 2008). It provides also new insights on the mechanisms for other fibrillar matrix proteins assembly, such as collagen and trombospondin (Velling et al., 2002; Sottile & Hocking, 2002). It has been also shown that FN fibrillogenesis is also required for rearrangement of substratum associated fibrinogen (Tzoneva et al., 2002). That is to say, it has been observed that endothelial cells were able to reorganise both adsorbed (on the substrate) and soluble (added to solution after cell adhesion to the substrate) FG in specific fibrillar structures only on hydrophilic glass, while this phenomenon is inhibited on hydrophobic substrata. Thus, endothelial cells mediated fibrinogen fibrillogenesis is altered on hydrophobic substrata, in the same way that FN fibrillogenesis is. Additionally, it was reported that endothelial cell spreading on FG was affected by cell synthesized FN (Dejana et al, 1990). This finding was supported with the observation of a different pattern of integrin organisation during the interaction with substratum-bound soluble FNG (Tzoneva et al., 2002).

It is known that the adhesion on endothelial cells to adsorbed FG is mediated by $\alpha_v\beta_3$ integrin (Cheresh et al, 1989). Indeed, the β_3 integrin clusterizes in structures resembling focal adhesion contacts when endothelial cells adhere to FG-coated substrata. Conversely, on the dorsal cell surface FG fibrils were not co-localized with β_3 integrin, representing a punctuate distribution, in contrast to β_1 integrin, which showed a well-pronounced linear pattern of organisation. The absence of β_1 integrin from the focal adhesion plaques is an indication that the FN receptor does not participate in endothelial cell adhesion to FG. Integrin β_1, however, has clearly shown to be involved in FN fibril formation (Dejana et al, 1990). The co-localisation of FN and FG fibrils was found on the dorsal cell surface of endothelial cells (Tzoneva et al., 2002). Also, the incorporation of FN fibrils into matrix fibrils starts from the distinct place at the cell periphery, near to the focal adhesions, suggesting the leading role of FN in this process. The existence of this joint fibrillogenesis, i.e. coassembly of FG and FN, found for endothelial cells have been reported also for epithelial cells (Guadiz et al., 1989) and fibroblasts (Pereira et al., 2002). Even collagen IV, which is a non-fibrillar protein could undergo fibril-like linear rearrangement along with FN as we recently show (Maneva-Radicheva et al., 2008), a fact that needs to be further elucidated.

In the last decade many studies have been focused on the modifications of biomaterials surfaces with synthetic or natural ECM fragments to provoke an adequate recruitment of cells for in vivo tissue regeneration (Massia & Hubel, 1990; Werner et al., 2006; Ma et al., 2007). However, when a foreign material is implanted in the body, it hampers the local organization of ECM and alters the biocompatibility of the implant, a process further complicated from the non-specific inflammatory response. A possible reason for the lack of adequate tissue response is attributed by many authors to the different dimensionality of the implant. Nowadays tissue engineering strives to replace the damaged tissues with natural or synthetic scaffolds designed to mimic the 3D organization and mechanical properties of ECM. Nevertheless, many of the bioengineered devices such as stents, prosthesis, membranes, metal implants, etc, simply can not avoid the 2D contact with cells. Therefore we argue, if this 3D architecture is always obligatory. When epithelial or endothelial cells reside on the basement membrane, they meet a rather flat environment, which they assess more as topography and as source of positional signals that guide their functionality. The basement membrane is actually a two dimensional structure, common to many types of tissues, providing underlinement for baso lateral cell attachment and functional polarization (Campbell & Teranova, 1988). Therefore, a future prospective for the development of biohybrid organs may require the construction of modules based on 2D permeable membranes colonized with cells that will mimic to a maximal extent the functional arrangement of basal membrane. On the other hand however, dimensionality of cell-material interaction raises a number of obstacles that need to be solved. For example, the blood contacting devices such as small diameter vascular grafts, stents, synthetic heart valves and assist systems, have suffered from a common problem, the lack of significant endothelial cells in-growth and function. While endothelial cells procurement technologies for seeding implants have improved, adhering endothelial cells usually dedifferentiate and act in a counterproductive manner, very often accelerating device failure (Ludwig et al., 2007). Thus, it seems that endothelial cells do not meet an adequate environment, as on the natural basal membrane, but what they are actually missing remains unclear.

4. Remodelling of the extracellular matrix

Except organization, the ECM undergoes proteolytic degradation, which is a mechanism for the removal of the excess ECM usually approximated with remodelling. Remodelling of ECM occurs in various physiological and pathological processes, such as normal development, wound healing and angiogenesis, but also in atherosclerosis, fibrosis, ischemic injury and cancer. Thus, matrix remodelling is a subject of an extensive biomedical research, but how it relates to the biocompatibility of materials remains unclear. Upon

implantation, foreign materials often trigger an uncontrolled deposition of fibrous matrix that, difficult to be predicted, hampers the biocompatibility of the implant.
In fact, ECM remodelling is a dynamic process which consists of two opposite events: assembly and degradation. These processes are mostly active during development and regeneration of tissues but, when miss-regulated, can contribute to diseases. Perturbing matrix remodelling, for example by preventing the turnover of collagen type I or altering the level of matrix-degrading proteases, has been shown to result in fibrosis, arthritis, reduced angiogenesis, and developmental abnormalities (Schofield, 1978; Wagenaar-Mller et al., 2007; Heyman et al., 2006; Holmbeck et al., 1999). The invasive behaviour of cancer cells is also due to up-regulation of matrix remodelling (Reisenawer et al., 2007; Carino et al., 2005). ECM organization in vivo is regulated by the 3D environment and the cellular tension that is transmitted through integrins (Hynes, 2002). It is difficult, however, to create such an environment on the biomaterials surface. Thus, identifying factors that control matrix deposition on the materials interface is an essential step for understanding the mechanisms involved in the pathological host response.
The proteolytic remodelling of matrix proteins such as FN, VN and FG, as well as, collagens and laminins at the biomaterials interface has only recently received attention, although the pericellular proteolysis is extensively studied in various pathological conditions. The proteolytic cleavage of ECM components represents a main mechanism for ECM degradation and removal (Koblinski et al., 2000; Mohamed & Sloane 2006). Several families of proteases operate at the ECM level, including matrix metalloproteinases (MMPs), cysteine proteases and serine proteases. Proteolysis may also regulate the ECM assembly, editing the excess ECM components. During enzymatic remodelling of ECM structures, bioactive fragments and growth factors can be released that will affect cell growth, morphogenesis, tissue repair, and also various pathological processes. The major enzymes that degrade ECM and cell surface associated proteins are MMPs, a family of secreted and membrane bound proteinases. The role of MMPs in both development and diseases has been recently extensively studied and reviewed because is tightly linked with the mechanisms for tumour invasion and metastasis (Page-McCaw et al., 2007). MMPs are family (20 members) of zinc dependent endopeptidases, which together with adamalysin-related membrane proteinases that contain disintegrin and metalloproteinase domains (ADAMs or MDCs), such as thrombin, tissue plasminogen activator (tPA), urokinase (uPA) and plasmin are involved in the degradation of ECM proteins. MMPs are either secreted or anchored to the cell membrane by a transmembrane domain or by their ability to bind directly uPAR and integrin $\alpha_v\beta_3$ (Buck et al., 1982).
In fact, the basement membrane is the most altered structure during remodelling and it depends particularly on the degradation of type IV collagen that is a primary structural component of basement membrane, integrating laminin and nidogen into a microscopically visible two-dimensional network. Assembly of the basement membrane, a strongly specialized form of ECM that underlies epithelial and endothelial cells, but is also in contact with many other cell types, is initiated by laminin. It self-assembles into heterotrimers that bind to the cell surface integrin receptors. The network-forming collagen type IV is the next main component of the basement membrane, however the mechanism of its assembly remains unclear. It is widely accepted that collagen IV also self-assembles into a meshwork by antiparallel interactions and extensive disulfide bounding of four molecules to form 7S domains. The lateral interactions between C-terminal globular domains create an irregular

two-dimensional meshwork that is actually the main constructive element of the basement membrane. Observations have been made however, that collagen IV may be linearly organized during early basement membrane assembly (Fleischmajer et al., 1998) suggesting another type of cell-dependent arrangement. While the molecular mechanisms which endow the spatial distribution and organization of collagen IV in basement membrane are still debatable, our recent results (Maneva-Radicheva et al., 2008) surprisingly suggest that material surface-associated collagen IV also undergo cell dependent rearrangement through reversible association with FN fibrils. It is obvious that ECM remodelling is poorly understood at the biomaterials level.

5. Substrate engineering

Different model substrates have been prepared in the recent years aiming to learn more about cell-material interaction, especially in what cell adhesion is concerned. These works are mainly focused on the effect of material properties on the biological performance of the substrate and, only a few of them, investigate this effect by addressing first protein adsorption and conformation on the material surface and then by correlating this phenomenon with cell behavior. Despite the belief that the issue of cell-protein-material interaction is critical to the engineering of new biomaterials, clear links between the material, the adsorbed protein layer and their influence on the cell remain far from being understood; in particular the behaviour of surface associated matrix proteins is generally missing. Even if the cell material interaction is not a direct one, but it is mediated by ECM proteins previously adsorbed on the substrate's surface, it is said that cells response to three different kinds of surface parameters: chemical, topographical and mechanical. The influence of surface chemistry on protein adsorption and cell adhesion has been addressed mostly on surfaces with well controlled chemistry, in order to investigate the role of concrete chemical groups in self-assembly monolayers SAM (e.g. OH, COOH, NH_2, CH_3 and their mixtures) (Keselowsky et al., 2003; Keselowsky et al., 2004; Faucheux et al.; 2004; Lee et al., 2006; Barrias et al., 2009). Studies on fibronectin (FN) and vitronectin (VN) adsorption under non-competitive and competitive (multi-protein solutions, including FBS) conditions suggested the major role of VN in cell-materials interaction. Different chemistries in SAM substrates were shown to modulate the structure and composition of focal adhesion complexes and fibrillar adhesions, closely linked to the capacity of cells to polymerise FN into fibrils (Keselowsky et al., 2004; Faucheux et al.; 2004). Other attempts to correlate surface chemistry with protein adsorption and cell adhesion were done by preparing substrates based on copolymers, that allow to modulate the material properties (wettability, hydrophilicity, substratum charge, topography, etc.) within the same chemical family (Allen et al., 2006). The response of cells to different material chemistries is a complex process and even minute changes in composition of the substrate produce amplified differences in cell responses (Bae et al., 2006). Plasma surface modifications have allowed to introduce different chemical groups in order to improve cell adhesion and study the effect on the ECM proteins conformation as well as focal adhesion formation (Silva et al., 2008; van Kooten et al., 2004; Pompe et al., 2007) and it is a versatile technique that allows one to produce chemical gradients in the same substrate so that the response to a large range of different chemistries on a single sample can be investigated as well as the effect not only of surface properties but their variations (Zelzer et al., 2008). It has been argued that sequence of

events –contact, attachment, spreading and proliferation- is similar among different surfaces but with very different dynamics, leading in the case of poorly compatible surfaces to long induction periods in which cells are in a life-or-death struggle to improve the pericellular environment by excretion of matrix proteins (Liu et al., 2007).
Surface topography is also a key parameter that is able to modify cell response independently of the chemical composition of the substrate. Even though sometimes topography is only a manifestation of material chemistry, it can be modulated in an independent way. The effect of topography on cell adhesion has been widely studied. Different microtopograhies can promote changes in cell adhesion pattern, cell orientation and cell shape on the substrate (Anselme et al., 2000). Cells cultured on smooth surfaces tend to generate a more organized ECM, with a more homogeneous distribution of focal adhesions. However, on rougher surfaces, focal adhesions are located at cell edges, where the contact with the substrate takes place (Brunette, 1986). Micro and nano patterned surfaces have been prepared for a better understanding of the cell response to topographic features, mainly in what cell adhesion is concerned. Anisotropic surfaces prepared by lithographic and microfabrication techniques can induce cell reorientation following microgrooves, the so-called contact guidance phenomenon (Ohara & Buck, 1979; Lim, 2007); and the scale of anisotropic topography plays an important role in deciding cell alignment (Affrossman et al., 2000). Different techniques have been used to produce controlled isotropic topographies at different scales which include photolithography, electron beam lithography, colloidal lithography, polymer solvent demixing techniques during a high speed spin-casting process (Denis et al., 2002; View et al., 2000; Kriparamanan et al., 2006; Dalby et al., 2004). Polymer solvent demixing techniques make use of phase separation during a high speed spin-casting process. This technique allows obtaining nanotopographic motifs in a broad range (from 9 to 100 nm); however the effect of nanotopography on cell response remains an open question. It seems that the interval 10-30 nm gives rise to better adhesion and higher stimulation of intracellular signaling than going up to 100 nm (Lim et al., 2008; Dalby et al., 2008; Zinger et al., 2005). Cell differentiation and gene expression have also been described to be influenced by surface topography (Dalby et al., 2007; Dalby et al., 2006; Pirouz-Dolatshahi et al., 2008). The effect of surface nanotopography on cell behavior should be a consequence of different protein adsorption patterns. Scarce experimental data exist on the effect of surface nanotopography on protein adsorption but it seems that nanotopography is able to enhance protein adsorption as compared to the same plane chemistry (Khor et al., 2007). Nevertheless, the effect of surface nanotopography on matrix remodelling has not been investigated in the literature and only some qualitative effect has been recently shown (Costa Martínez et al., 2008; Pegueroles et al., 2009). We have developed the technology to prepare different microtopographies with tailored surface chemistry which lead to different cell function as a consequence of the extracellular matrix organisation at the cell-material interface (González-García et al., 2009; Pelham & Wang, 1997).
Mechanical properties of the substrate play also an important role in cell response regardless surface chemistry and topography but it is not completely understood. Cell behaviour depends sensitively on the rigidity of the extracellular matrix. When cells are cultured on classic (rigid) polystyrene dishes, they develop micron-sized focal adhesions connected by actin fibers. However, these structures are gradually lost as cells are cultured on softer substrates, as prepared for example by changing the crosslinking density of gels

and, more recently, polyelectrolyte multilayers (Ryan et al., 2001). Cell spreading and motility are higher in stiff substrates than soft ones, what favours cell-cell interaction compared to the cell-material one and leads to more organised cell aggregates (Wong et al., 2003). Cell proliferation increases on stiff surfaces and, in the case of a rigidity gradient on the substrate, cells migrate to stiffer regions (Lo et al., 2000; Bischofs et al., 2003; Pegueroles et al., 2009). This kind of cell behaviour has been found for different cells types (fibroblasts, muscular VSMC cells, chondrocytes and neurons) independently of the protein coating of the substrate (fibronectin, collagen, etc). It is thought that cells are able to react to substratum rigidity by means of a real tactile exploration, by contractile forces and interpreting the substrate deformation. Because the stiffness of the environment is a passive quantity, it has to be actively sensed by the cell by contracting it and measuring some kind of mechanical response (Schwarz, 2007; Shin, 2008). The relationship between the mechanical properties of the matrix and the activity of cells must lead to the maintenance of a functional mechanical state. The effect of substrate stiffness on the dynamic behaviour of surface associated matrix proteins is generally missing, i.e. is protein conformation determined by substrate stiffness? What is the role of the strength of interaction between the protein layer and the substrate on matrix remodelling as a function of the stiffness of the underlying substrate? Recent studies have indicated that the cells not only detect the roughness, chemistry, or stiffness of the substrates, but can also detect dimensionality (Hollister, 2005). For example, pore sizes of engineered trabecular bone depend on the initial scaffold geometry. Because it is clear that the ECM can affect cell fate and differentiation, tissue engineering is looking toward cells and the developmental biology for guidance in the design of scaffolds (Wnek, 2003).

6. Conclusions and future trends

Understanding the cell-protein-material interaction is fundamental for developing more powerful tools in tissue engineering and regenerative medicine strategies. The design of model substrates including the presence of well defined properties (chemistry, topography, stiffness) and even the gradient of these properties in three dimensional environments must lead in the near future to learn more about the specific roles of protein adsorption and the very dynamic process related to the cell fate of synthetic substrates: cell adhesion, matrix reorganisation, deposition and degradation at the cell-material interface.

7. Acknowledgements

Financial support of the Spanish government through projects MAT2009-14440-C02-01 and MAT2009-14440-C02-02 is kindly acknowledged. CIBER-BBN is an initiative funded by the VI National R&D&i Plan *2008-2011, Iniciativa Ingenio 2010, Consolider Program, CIBER Actions* and financed by the Instituto de Salud Carlos III with assistance from the *European Regional Development Fund.*

8. References

Affrossman, S.; Jerome, R.; O'Neill, S.A.; Schmitt, T. & Stamm, M. (2000). Surface structure of thin films of blends of polystyrene and poly(n-butyl methacrylate). *Colloid Polym Sci*, ,278, 888.

Agnihotri, A.; Siedlecki, C. A. (2004). Time-dependent conformational changes in fibrinogen measured by atomic force microscopy. *Langmuir*, 20, 8846.

Allen, L.T.; Tosetto, M.; Miller, I.S.; O'Connor, D.P.; Penney, S.C.; Lynch, I.; Keenan, A.K.; Pennington; Dawson, K.A.; Gallagher, W.M. (2006). Surface-induced changes in protein adsorption and implications for cellular phenotypic responses to surface interaction. *Biomaterials* 27,3096.

Altankov, G.; Groth, T. (1994). Reorganization of substratum-bound fibronectin on hydrophilic and hydrophobic materials is related to biocompatibility. *J. Mater Sci Mat Med* 5, 732.

Altankov, G.; Groth, T. (1996). Fibronectin matrix formation and the biocompatibility of materials *J Mater Sci Mat Med* 7, 425.

Altankov, G.; Grinnell, F.; Groth, T. (1996). Studies on the biocompatibility of materials: Fibroblast reorganization of sub-stratum-bound fibronectin on surfaces varying in wettability *J Biomed Mater Res* 30, 385.

Altankov, G.; Groth, T.; Krasteva, N.; Albrecht, W.; Paul, D. (1997) Morphological evidence for different fibronectin receptor organization and function during fibroblast adhesion on hydrophilic and hydrophobic glass substrata. *J Biomat Sci Polym E*; 8, 721.

Altankov, G.; Thom, V.; Groth, T.; Jankova, K.; Jonsson, G.; & Ulbricht M. (2000). Modulating the biocompatibility of polymer surfaces with poly(ethylene glycol): effect of fibronectin. *J Biomed Mater Res A* 52, 219.

Anselme, K. (2000). Osteoblast adhesion on biomaterials. *Biomaterials* 21, 667.

Anselme, K.;, Bigerelle, M.; Noel, B.; Dufresne, E.; Judas, D.; Iost, A.; & Hardouin. (2000b). Qualitative and quantitative study of human osteoblasts adhesion on materials with various surfaces roughness. *J Biomed Mater Res* 49,155.

Arima, Y.; & Iwata, H. (2007). Effect of wettability and surface functional groups on protein adsorption and cell adhesion using welldefined mixed self-assembled monolayers. *Biomaterials* 28,3074.

Avnur, Z.; & Geiger, B. (1981). The removal of extracellular fibronectin from areas of cell-substrate contact. *Cell* 25,121

Bae, Y.H.; Johnson, P.A.; Florek, C.A.; Kohn, J.; Moghe, PV. (2006). Minute changes in composition of polymer substrates produce amplified differences in cell adhesion and motility via optimal ligand conditioning. *Acta Biomaterialia* 2:473

Barrias, C.C.; Martins, M.C.L.; Almeida-Porada, G.; & Barbosa, M.A. (2009). The correlation between the adsorption of adhesive proteins and cell behaviour on hydroxyl-methyl mixed self-assembled monolayers. *Biomaterials* 30, 307.

Baugh, L.; & Vogel, L. (2004). Structural changes of fibronectin adsorbed to model surfaces probed by fluorescence resonance energy transfer. *J Biomed Mater Res A* 69, 525.

Baneyx, G.; & Vogel, V. (1999). Self-assembly of fibronectin into fibrillar networks underneath dipalmitoyl phosphatidylcholine monolayers: Role of lipid matrix and tensile forces. *Proc Natl Acad Sci U.S.A* 96, 12518.

Brash JL, T.; & Horbett, A. (1995). Protein at Interfaces II: Fundamentals and Applications, in: ACS Symposium Series No. 602, J. L. Brash, T. A. Horbett, Eds., American Chemical Society, Washington, DC, Chapter 1.

Beck, K., I. Hunter, I.; & Engel, J. (1990). Structure and function of laminin: anatomy of a multidomain protein. *FASEB J* 4, 148.

Benesch, J.; Hungerford, G.; Suhling, K.; Tregidgo, C.; Mano, J.F.; & Reis, R.L. (2007). Fluorescence probe techniques to monitor protein adsorption-induced conformation changes on biodegradablepolymers. *J Coll Interf Sci* 312, 193.

Bischofs, I.B.; Schwarz, U.S. (2003). Cell organisation in soft media due to active mechanosensing. *Proc Natl Acad Sci USA* 100, 9274.

Buck, M.R.; Karustic, D.G.; Day, N.A.; & Honn, K.V.; & Sloane, B.F. (1992). Degradation of extracellular matrix proteins by human cathepsin B from normal and tumour tissues. *Biochem J* 282,273.

Burgeson, R. E., Chiquet, M.; Deutzmann, D.; Ekblom, P.; Engel, J.; Kleinman, H.; Martin, G.R.; Meneguzzi, G.; Paulsson, M.; Sanes, J.; Timple, R.; Tryggvason, K.; Yamada, Y. & Yurchenco, P.D. (1994). A new nomenclature for the laminins. *Matrix Biol* 14, 209.

Brunette, D.M. (1986). Fibroblast on micromachined substrata orient hierarchically to grooves of different dimensions. *Exp Cell Res* 164, 11.

Cacciafesta, P.; Humphris, A. D. L.; Jandt, K. D.; & Miles, M. J. (2000). Human plasma fibrinogen adsorption on ultraflat titanium oxide surfaces studied with AFM. *Langmuir*, 16, 8167.

Campbell, J.H.; & Teranova, V.P. (1988). Laminin. Molecular organization and biological function. *J Oral Pathol* 17, 309

Carino, A.C.; Engelholm, L.H.; Yamada, S.S.; Holmbeck, K.; Lund, L.R.; Molinolo, A.A.; Behrendt, N.; Nielsen, B.S.; & Bugge, T.H. (2005). Intrecellular collagen degradation mediated ny uPAR/Endo 180 is a major pathway of extracellular matrix turnover during malignancy. *J Cell Biol* 169, 977.

Chen, H.; Yaun L.; Song W.; Wu, Z.; & Li, D.(2008). Biocompatible polymer materials: Role of protein-surface interactions. *Progress in Polymer Science*, 33, 1059..

Cheresh, D.; Berliner, S.; Vicente, V.; & Ruggery, Z. (1989). Recognition of distinct adhesive sites on fibrinogen by related integrins on platelets and endothelial cells. *Cell* 58, 945.

Cleveland, J.P.; Anczykowski, B.; Schmid, A.E.; Elings, V.B. (1998). Energy dissipation in tapping-mode atomic force microscopy. *Appl Phys Lett* 72, 2613.

Costa Martínez, E.; Rodríguez Hernandez, J.C.; Machado, M.; Mano, J.F.; Gómez Ribelles, J.L.; Monleón Pradas, M.; & Salmerón Sánchez, M. (2008). Human chondrocyte morphology, its dedifferentiation and fibronectin conformation on different PLLA microtopographies. *Tissue Eng* 14, 1751.

Christopher, R.A.; Kowalchyk, A.; & McKewon-Longo, P.J. (1997). Localization of fibronectin matrix assembly sites on fibroblasts and endothelial cells. *J Cell Sci* 110,569.

Dalby, M.J.; Giannaras, D.; Riehle, M.O.; Gadegaard, N.; Affrossman, S.; & Curtis, A.S.G. (2004). Rapid fibroblast adhesion to 27 nm high polymer demixed nano-topography. Biomaterials 25, 77.

Dalby, M.J.; MCCloy, D.; Robertson, M.; Wilkinson, C.D.W.; & Oreffo, R.O.C. (2006). Osteoprogenitor response to defined topographies with nanoscale depths. Biomaterials 27, 1306

Dalby, M.J.; Gadegaard, N.; Tare, R.; Andar, A.; Riehle, M.O.; Herzyk, P.; Wilkinson C.D.W.; & Oreffo, R.O.C. (2007). The control of human mesenchimal cell differentiation using nanoscale symmetry and sisorder. *Nature Materials* 9,997.

Dalby, M.J.; Hart, A.; & Yarwood, S.J. (2008). The effect of RACK1 signalling protein on the regulation of cell adhesion and cell contact guidance on nanometric grooves. *Biomaterials* 29, 282.

Denis, F.A.; Hanarp, P.; Sutherland, D.S.; & Dufrene, Y.F. (2002) Fabrication of nanostructures polymer surfaces using colloidal litography and spin coating. *Nanoletters* 2, 1419.

Dejama, E; Lampugnani, M.; Giorgi, M.; Gaboli, M.; & Marchisio, P. (1990). *Blood*, 75, 1509.

Dzamba, B.J.; Wu, H.; Jaenisch, R.; & Peters, D.M. (2008). Fibronectin binding site in type I collagen regulates fibronectin fibril formation. *J Cell Biol* 121, 1165.

Erickson, H.P., & McDonagh, J. (1981). Fibronectin molecule visualized in electron microscopy: a long, thin, flexible strand. *J Cell Biol* 91, 673.

Erickson, H.P., & Carell, N.A. (1983). Fibronectin in extended and compact conformations. Electron microscopy and sedimentation analysis. *J Biol Chem* 258, 14539.

Erickson, H. P. J. (2002). Stretching fibronectin. *Muscle Res Cell Motil* 23, 575.

Faucheux, N.; Tzoneva, R.; Nagel, M.D.; & Groth T. (2006). The dependence of fibrillar adhesions in human fibroblasts on substratum chemistry. *Biomaterials;* 27:234.

Fleischmajer, R.; Perlish, J.S.; MacDonald, E.D.; Schechter, A.; Murdoch, A.D.; Iozzo, R.V.; Yamada, Y. (1998). There is binding of collagen IV to beta 1 integrin during early skin basement membrane assembly. *Ann N Y Acad Sci* 857,212.

García, A.J. (2005). Get a grip: integrins in cell-biomaterial interactions. *Biomaterials* 26, 7525.

García, A.J. (2006). Interfaces to control cell-biomaterial adhesive interactions. *Adv Polym Sci* 203, 171.

Geiger, B.; Bershadsky, A.; Pankov, R.; & Yamada, K.M. (2001). Transmembrane crosstalk between the extracellular matrix and the cytoskeleton. *Nat Rev Mol Cell Biol* 2,793.

García, R.; Tamayo, j.; Calleja, M.; & García, F. (1998). *Appl Phys A* 66, S309.

García, A.J.; & Boettiger, D.B. (1999). Integrin–fibronectin interactions at the cell-material interface: initial integrin binding and signaling. *Biomaterials* 20, 2427.

Geiger, B.; Bershadsky, A.; Pankov, R.; & Yamada, K. M. (2001). ransmembrane crosstalk between the extracellular matrix--cytoskeleton crosstalk. *Nat Rev Mol Cell Biol* 2, 793.

Gettens, R.T.T.; Bai, Z.; & Gilbert, J.L. (2005). Quantification of the kinetics and thermodynamics of protein adsorption using atomic force microscopy. *J Biomed Mater Res A* 72, 246.

Gettens, R.T.T.; Gilbert, J.L. (2007). Quantification of the kinetics of protein adsorption onto 316L stainless steel using atomic force microscopy. *J Biomed. Mater Res. A* 81, 465.

González-García, C.; Latorre Ferrus, L.; Moratal, D.; Monleón Pradas, M.; & Salmerón-Sánchez, M. (2009). *Plasma Process Polym* 6, 190.

Gorman, R.R.; Stoner, G. E.; & Catlin, A. (1971). The adsorption of fibrinogen. An electron microscope study.*J Phys Chem* 75, 2103.

Griffin, L.; & Naughton, G. (2002). Tissue engineering–Current challenges and expanding opportunities. *Science* 259,1009.

Grinnell, F. (1986). Focal adhesion sites and the removal of substratum-bound fibronectin. *J Cell Biol* 103, 2697.

Grinnell, F.; & Feld, M.K. (1982). Fibronectin adsorption on hydrophilic and hydrophobic surfaces detected by antibody binding and analyzed during cell adhesion in serum-containing medium. *J Biol Chem* 257, 4888.

Guadiz, G.; Sporn, L.; & Simpson-Haidaris, P. (1997). *Blood*, 90, 2644

Gugutkov, D.; Altankov, G.; Rodríguez Herández, J. C.; Monleón Pradas, M.; & Salmerón-Sánchez, M. (2009). J Biomed. Mater. Res., Part A, DOI: 10.1002/jbm.a.32374.

He, W., & Bellamkonda, R.V. (2005). Nanoscale neuro-integrative coatings for neural implants. *Biomaterials* 26, 2983.

Heiduschka, P.; Romann, I.; Ecken, H.; Schöning, M.; Schuhmann, W.; & Thanos, S. (2001). Defined adhesion and growth of neurons on artificial structured substrates. *Electrochim Acta*. 47:299–.

Heyman, S.; Pauschinger, M.; De Plama, A.; Kollwellis-Opara, A.; Rutschow, S.; Swinnen, M.; Vanhoutte, D.; Gao, F.; Topai, R.; & Baker, A. H. (2006). Inhibition of urokinase type plasminogen activator or matrix metalloproteinases prevents cardiac injury and dysfunction during viral myocarditis. *Circulation* 114, 565.

Holland, N.B.; & Marchant, R. E. (2000). "Imaging individual plasma proteins on rough biomaterials. *J Biomed Mater Res* A 51, 307.

Hollister, S.J. (2005). Porous Scaffold design for tissue engineering. *Nature materials* 4, 518.

Holmbeck, K.; Bianco, P.; & Caterina, S. (1999). MT1 MMP deficient mice develop dwarfism, osteopenia, arthritis and connective tissue disease due to unadequate collagen turnover. *Cell* 99, 8192.

Hynes, R.O. (1990). Fibronectins. Springer –Verlag: New York.

Hynes, R.O. (2002). Integrins: bidirectional, allosteric signaling machines. *Cell* 110, 673.

Ishizaki, T.; Saito, N.; Sato, Y.; & Takai, O. (2007). *Surf Sci* 601, 3861.

Keselowsky, B.G.; Collard, D.M.; & Garcıa, A.J. (2003). Surface chemistry modulates fibronectin conformation and directs integrin binding and specificity to control cell adhesion. *J Biomed Mater Res A* 66, 247.

Keselowsky, B.G.; Collard, D.M.; & García, A.J. (2004). Surface chemistry modulates focal adhesion composition and signalling through changes in integrin binding. *Biomaterials* 25, 5947.

Khang, D.; Yeol, K.; Liu-Snyder, P.; Palmore, T.R.; Durbin, S.M.; & Webster, T.J. (2007). Enhanced fibronectin adsorption on carbon nanotube=poly(carbonate) urethane: independent role of surface nano-roughness and associated surface energy. *Biomaterials* 28, 4759.

Khor, H.L.; Kukula, H.; Tamada, K.; Knoll, W.; Moeller, M.; & Hutmacher, D.W. (2007). Response of cells on surface-induced nanopatterns : fibrolasts and mesenchimal progenitor cells. *Biomacromolecules* 8, 1530.

Kidoaki, S.; & Matsuda, T. (1999). Adhesion Forces of the Blood Plasma Proteins on Self-Assembled Monolayer Surfaces of Alkanethiolates with Different Functional Groups Measured by an Atomic Force Microscope. Langmuir 15, 7639.

Kleiman, H. K.; Cannon, F. B.; Laurie, G. W.; Hasell, J. R.; Aumailley, M.; Terranova, V. P.; Martin, G. R.; & DuBois-Dalcq, M. (1987). Biological activities of laminin. *J Cell Biochem* 27, 317.

Koblinski, J.; Ahram, M.; Sloane, B.F. (2000). Unraveling the role of proteases in cancer. *Clin Chem Acta* 291, 113.

Kolakna, L.; Bakesova, J.; & Varga, F. (2007). Biochemical and biophysical aspects of collagen nanostructure in the extracellular matrix. *Physiol Res* 56, S51.

Kowalczynska, H.M.; Nowak-Wyrzykowska, M.; Kolos, R.; Dobkowski, J.; & Kaminski, J. (2005). Fibronectin adsorption and arrangement on copolymer surfaces and their significance in cell adhesion. *J Biomed Mater Res A* 72, 228.

Kriparamanan, R.; Aswath, P.; Zhou, A.; Tang, L.; Nguyen, K.T. (2006). Nanotopography: cellular responses to nanostructured materials. *J Nanosci Nanotech* 6, 1905.

Lan, M.A.; Gersbach, C.A.; Michael, K.E.; Keselowsky, B.G.; & García, A.J. (2005). Myoblast proliferation and differentiation on fibronectin-coated self assembled monolayers presenting different surface chemistries. *Biomaterials* 26, 4523.

Lee, M.H.; Ducheyne, P.; Lynch, L.; Boettiger, D.; & Composto, R.J. (2006). Effect of biomaterial surface properties on fibronectin-a5b1 integrin interaction and cellular attachment. *Biomaterials* 27, 1907.

Liesi, P., D. Dahl, and A. Vaheri. 1984. Neurons cultured from developing rat brain attach and spread preferentially to laminin. J. Neurosci. Res. 11:241–251.

Liu, X.; Lim, J.Y.; Donahue, H.J.; Dhurjati, R.; Maestro, A.M.; & Vogler, E.A. Influence of substratum surface chemistry/energy and topography on the human fetal osteoblastic cell line hFOB 1.19: Phenotipic and genotypic responses observed in vitro. *Biomaterials* 28, 4535.

Lim, Y.L.; & D, H.J. (2007). Cell sensing and response to micro- and nanostructured surfaces produced by chemical and topographical patterning. *Tissue Engineering* 13, 1897

Lim, J.Y.; Dreiss, D.; Zhou; Hansen, J.C.; Siedlecki, C.A.; Hengstebeck, R.W.; Cheng, J.; Winogard, N.; & Donahue, H.J. (2007). The regulation of integrin-mediated osteoblast focal adhesion and focal adhesion kinase expression by nanoscale topography. *Biomaterials* 28, 1787

Lo, C.M.; Wang, H.B.; & Dembo, M.; & Wang, Y. (2000). Cell movement is guided by the rigidity of the substrate. *Biophys J* 79,144.

Lord, M.S.; Stenzel, M.H.; Simmons, A.; & Milthorpe, B.K. (2006). The effect of charged groups on protein interactions with poly(HEMA) hydrogels. *Biomaterials* 27, 567.

Luckenbill-Edds, L. (1997). Laminin and the mechanism of neuronal outgrowth. *Brain Res Rev* 26, 2983.

Ludwig, N.S.; Yoder, S.; McConey, M.; Vargo, T.G.; Kader, K.N. (2006). Directed type IV collagen self-assembly on hydroxilated PTEF. *J Biomed Mater Res* 78, 615.

Ma, Z.; Mao, Z.; & Gao, C. (2007). Surface modification and property of biomedical polymers used for tissue engineering. *Coll Surf Sci B* 60, 137.

Maneva-Radicheva, L.; Ebert, U.; Dimoudis, N.; & Altankov, G. (2008). Fibroblast remodeling of collagen type IV is altered in contact with cancer cells. *Histol Histopathol* 23, 833.

Marchin, F.L.; & Berrie, C. L. (2003). Langmuir 19, 9883.

Massia, S.P.; & Hubel, J.A. (1990). Covalently attached GRGD on polymer surfaces promotes biospeciffic adhesion of mammalian cells. *Ann NY Acad. Sci* 589,261.

Mao, Y.; & Schwarzbauer, J. E. (2005). Fibronectin fibrillogenesis, a cell-mediated matrix assembly process. *Matrix Biol* 24, 389.

Mercurio, A. M. (1995). Laminin receptors: achieving specificity through cooperation. *Trends Cell Biol* 5, 419.

Michael, K.E.; Vernekar, V.N.; Keselowsky, B.G.; Meredith, J.C.; Latoru, R.A.; & García, A.J. (2003). Adsorption-induced conformational changes in fibronectin due to interactions with well-defined surface chemistries. *Langmuir* 19, 8033.

Mitsakakis, K.; Lousinian, S.; & Logothetidis, S. Early stages of human plasma protein adsorption on biocompatible thin films probed by Atomic Force Microscope. (2007). *Biomol Eng* 24, 119.

Mohamed, M. ; & Sloane, B.F. (2006). Cysteine cathepsins: multifunctional enzymes in cancer. *Nat Rev Cancer* 6, 764.

Mosher, D.F.; Sottile, J.; Wu, C.; McDonald, J.A. (1992). Assembly of extracellular matrix. *Curr Opin Cell Biol* 4, 810.

Noh, H.; & Vogler, E.A. (2006). Volumetric interpretation of protein adsorption: mass and energy balance for albumin adsorption to particulate adsorbents with incrementally increasing hydrophilicity. *Biomaterials* 27, 5801.

Ohara, P.T.; Buck, R.C. (1979). Contact guidance in vitro: A light, transmission, and scanning electron microscopic study. *Exp Cell Res* 121, 235.

Ortega-Vinuesa, J.L.; Tengvall, P.; & Lundström, I. (1998). Aggregation of HSA, IgG, and Fibrinogen on Methylated Silicon Surfaces. *J Colloid Interface Sci* 207, 228.

Page-McCaw, A.; Ewald, A.J.; & Werb Z. (2007). Matrix metaloproteinases and the regulation of tissue remodeling. *Nat Rev Mol Cell Biol* 8, 221.

Pankov, R.; & Yamada, K. M. (2002). Fibronectin at a glance. *J Cell Sci* 2002, 115, 3861.

Pereira, M.; Rybarczyk, B.; Odrljin, T.; Hocking, D.; Sottile, J.; Simpson-Haidaris, P.(2002). *J Cell Sci*, 115, 609.

Pearlstein, E., Gold, L.I., and Garcia-Pardo, A. Fibronectin: a review of its structure and biological activity. Mol Cell Biochem 23, 103, 1980.

Pegueroles, M.; Aparicio, C.; Bosio, M.; Engel, E.; Gil, F.J.; Planell, J.A.; & Altankov, G. (2009). Spatial organization of osteoblast fibronectin matrix on titanium surfaces: Effects of roughness, chemical heterogeneity and surface energy. *Acta Biomaterialia, in press* (doi:10.1016/j.actbio.2009.07.030)

Pelham, J.R.; & Wang, Y.L. (1997). Cell locomotion and focal adhesions are regulated by substrate flexibility. *Proc Natl Acad Sci USA* 94, 13661.

Pompe, T.; Keller, K.; Mitdank, C.; & Werner, C. (2005). Fibronectin fibril pattern displays the force balance of cell-matrix adhesion. *Eur Biophys J* 34, 1049.

Pompe, T.; Keller, K.; Mothes, G.; Nitschke, M.; Teese, M.; Zimmermann, R.; & Werner, C. (2007). Surface modification of poly(hydroxybutyrate) films to control cell-matrix adhesion. Biomaterials 28, 28.

Prime, L.; & Whitesides, G.M. (1991). Self-assembled organic monolayers: model systems for studying adsorption of proteins at surfaces. *Science* 252, 1164.

Reisenawer, A.; Eickelberg, O.; Wille, A.; Heimburg, A.; Reinhold, A.; Sloane, B. F.; Welte, T.; & Buhling F. (2007). Increased carcinogenic potential of myeloid tumor cells induced by aberrant TGF-beta-signaling of cathepsin B. *Biol Chem* 288, 639.

Rhees, S.; & Grinnel, F. (2007). Fibroblast mechanics in 3D collagen gels. *Adv Drug Deliv Rev* 59, 1299.

Rico, P.; Rodríguez Hernández, J.C.; Moratal, D.; Monleón Pradas, M.; Altankov, G.; Salmerón-Sánchez, M. (2009). Substrate-induced assembly of fibronectin into networks. Influence of surface chemistry and effect on osteoblast adhesion. *Tissue Engineering* 15, 3271

Rodrigues, S.N.; Gonçalves, I.C.; Martins, M.C.L.; Barbosa, M.A.; & Ratner, B. (2006). *Biomaterials* 27, 5357.

Rodríguez Hernández, J.C.; Salmerón Sánchez, M.; Soria, J.M.; Gómez Ribelles, J.L.; & Monleon Pradas, M. (2007). *Biophys J* 93, 202.

Rodríguez Hernández, J.C.; Rico, P.; Moratal, D.; Monleón Pradas, M.; & Salmerón Sánchez, M. (2009). Fibrinogen patterns and activity on substrates with tailored hydroxyl density. *Macromolecular Bioscience* 9, 766.

Rogers, S. L.; Letorneau, P.C.; Palm, S.L.; McCarthy, J.; & Furcht, L.T. (1983). Neurite extension by peripheral and central neuron system neurons in response to substratum-bound fibronectin and laminin. *Dev Biol* 98, 212.

Ryan, P.L.; Foty, R.A.; Kohn, J.; & Steinberg, M.S. (2001). Tissue Spreading on implantable substrates is a competitive outcome of cell-cell vs. Cell-substratum sensivity. *Proc Natl Acad Sci* 98, 4323.

Sasaki, T.; Fässler, B.; & Hohenester, E. (2004). Laminin: the crux of membrane assembly. *J Cell Biol* 164, 959.

Schofield, R. (1978). The relationship between the spleen colony-forming cell and the haemopoietic stem cells. *Blood Cells* 4, 7.

Schwarz, U. (2007). Soft matters in cell adhesion : rigidity sensing on soft elastic substrates. *Soft Matter* 3, 263.

Shin, H. (2008). Fabrication methods of an engineered microenvironment for analysis of cell-biomaterial interactions. *Biomaterials* 28, 126.

Silva, S.; Luna, S.M.; Gomes, M.E.; Benesch, J.; Pashkuleva, I.; Mano, J.F.; & Reis, R.L. (2008). Plasma surface modification of chitosan membranes: characterisation and preliminary cell response studies. *Macromol Biosci* 8:568

Singer, I. I. (1979). The fibronexus: a transmembrane association of fibronectin-containing fibers and bundles of 5 nm microfilaments in hamster and human fibroblasts. *Cell* 16, 675.

Sit, P.S.; & Marchant, R.E. (1999). Surface-dependent conformations of human fibrinogen observed by atomic force microscopy under aqueous conditions. *Thromb Haemost* 82, 1053.

Slack S.M.; & Horbett, T. A. (1992). Changes in fibrinogen adsorbed to segmental polyurethanes and hydroxymethacrylate-ethylmethacrylate copolymers. *J Biomed Mater Res* 26, 1633.

Smith, M.L.; Gourdon, D.; Little, W. C.; Kubow, K. E.; Eguiluz, R. A.; Luna-Morris, S.; & Vogel, V. (2007). Force-Induced Unfolding of Fibronectin in the Extracellular Matrix of Living Cells. *PLoS Biol* 5, 2243.

Sottile, J.; & Hocking, D.C. (2002). Fibronectin polymerization regulates the composition and stability of extracellular matrix fibrils and cell-matrix adhesions. *Mol Biol Cell* 13, 3546.

Sousa, S.R.; Manuela Bras, M.; Moradas-Ferreira, O.; & Barbosa, M.A. (2007). Dynamics of fibronectin adsorption on TiO2 surfaces. *Langmuir* 23, 7046.

Spie, J. (2002). Tissue engineering and reparative medicine. *Ann NY Acad Sci* 961, 1.

Steiner, G.; Tunc, S.; Maitz, M.; & Salzer, R. (2007). Conformational changes during protein adsorption. FT-IR spectroscopic imaging of adsorbed fibrinogen layers. *Anal Chem* 79, 1311.

Ta, T.C.; McDermott, M. (2000). Mapping Interfacial Chemistry Induced Variations in Protein Adsorption with Scanning Force Microscopy. *Anal Chem* 72, 2627.

Tamayo, J.; & García, R. (1997). Effects of elastic and inelastic interactions on phase contrast images in tapping-mode scanning force microscopy . *Appl Phys Lett* 71, 2394.

Tamayo, J.; & García, R. (1998). Relationship between phase shift and energy dissipation in tapping-mode scanning force microscopy. *Appl Phys Lett* 73, 2926.

Toscano, A.; & Santore, M. (2006). Fibrinogen Adsorption on Three Silica-Based Surfaces: Conformation and Kinetics. *Langmuir* 22, 2588.

Toworfe, G.K.; Composto, R.J.; Adams, C.S.; Shapiro, I.M.; & Ducheyne, P. (2004). Fibronectin adsorption on surface-activated poly(dimethylsiloxane) and its effect on cellular function. *J Biomed Mater Res A* 71, 449.

Tsapikouni, T.S.; & Missirlis, Y.F. (2007). pH and ionic strength effect on single fibrinogen molecule adsorption on mica studied with AFM. *Coll Surf B* 57, 89.

Tunc, S.; Maitz, M.F.; Steiner, G.; Vázquez, L.; Pham, M.T.; & Salzer, R. (2005). In situ conformational analysis of fibrinogen adsorbed on Si surfaces. *Colloids Surf B* 42, 219.

Tzoneva, R.; Groth, T.; Altankov, G.; & Paul, D. (2002). Remodeling of fibrinogen by endothelial cells in dependence of fibronectin matrix assembly. Effect of substratum wettability. J Mater Sci Mater M 13, 1235.

Ulmer, J.; Geiger, B.; & Spatz, J. P. (2008). Force-induced fibronectin fibrillogenesis in vitro. *Soft Matter* 4, 1998.

Van De Keere, I.; Willaert, R.; Hubin, A.; & Vereecken, J. (2008). Interaction of Human Plasma Fibrinogen with Commercially Pure Titanium as Studied with Atomic Force Microscopy and X-ray Photoelectron Spectroscopy. *Langmuir*, 24, 1844.

Van Kooten, T.G.; Spijker, H.T.; & Busscher, H.J. (2004). Plasma-treated polystyrene surfaces: model surfaces for studying cell-biomaterial interactions. *Biomaterials* 25, 1735.

Veklich, Y.I.; Gorkun, O.V.; Medved, L.V.; Nieuwenhuizen, W.; Weisel, J.W. (1993). Carboxyl-terminal portions of the alpha chains of fibrinogen and fibrin. Localization by electron microscopy and the effects of isolated alpha C fragments on polymerization. *J Biol Chem* 268, 13577.

Velling, T.; & Risteli, J. (2002). Polymerization of type I and III collagens is dependent on fibronectin and enhanced by integrins alpha11-beta1 and alpha2-beta1. *J Biol Chem* 277, 37377.

View, C.; Carcenac, F.; Pepin, A.; Chen, Y.; Mejias, M.; Lebib, A. (2000). Electron bean litography : resolution limits and applications. *Appl Surf Sci* 164, 111.

Wagenaar-Mller, R.; Engelholm, L.; & Gavard, J. (2007). Complementary role of intracellular and pericellular collagen degradation pathways in vitro. *Mol Cell Biol* 27, 6309.

Weisel, J.W.; Stauffacher, C.V.; Bullitt, E.; & Cohen, C. (1985). A model for fibrinogen: domains and sequence. *Science* 230, 1388.

Weisel, J.W. (2005). Fibrinogen and fibrin. *Adv Prot Chem* 70, 2497.

Wehrle-Haller, B.; & Imhof, B.A. (2003). Integrin-dependent pathologies. *J Pathol* 200, 481.

Werner, C. ; Pompe, T.; & Salchert, K. (2006). Modulating extracellular matrix at interfaces of polymeric materials. *Adv Polym Sci* 203, 63.

Wertz, C.F. ; & Santore, M.M. (2001). Effect of Surface Hydrophobicity on Adsorption and Relaxation Kinetics of Albumin and Fibrinogen: Single-Species and Competitive Behavior. *Langmuir* 17, 3006.

Wertz, C.F., & Santore, M.M. (2002). Fibrinogen Adsorption on Hydrophilic and Hydrophobic Surfaces: Geometrical and Energetic Aspects of Interfacial Relaxations. *Langmuir* 18, 706.

Wnek, G.; Carr, M.; Simpson, D.; Bowlin, G. (2003). Electrospinning of nanofiber fibrinogen structure. *Nano Letters* 3, 231.

Wong, J.Y.; Velasco, A.; Rajagopalan, P.; Pham, Q. (2003). Directed movement of vascular smooth muscle cells on gradient-compliant hydrogels. *Langmuir* 19, 1908.

Xu, L-C.; & Siedlecki, C.A. (2007). Effects of surface wettability and contact time on protein adhesion to biomaterial surfaces. *Biomaterials* 28, 3273.

Williams, D.F. (1998). General concepts of biocompatibility, In: Handbook of biomaterial properties, J Black & G Hastings Eds, Chapman & Hill London, p. 481-489.

Zelzer, M.; Majani, R.; Bradley, J.W.; Rose, F.R.A.J.; Davies, M.; & Alexander, M.R. (2008). Investigation of cell-surface interactions using chemical gradientes formed from plasma polymers. *Biomaterials* 29, 172.

Zinger, O.; Zhao, G.; Schwartz, Z.; Simpson, J.; Wieland, M.; Landlolt, D.; & Boyan, B. (2005). Differential regulation of osteoblasts by substrate microstructural features. *Biomaterials* 26, 1837.

5

Learning From Nature: Emulating Macromolecular Crowding To Drive Extracellular Matrix Enhancement For The Creation Of Connective Tissue *in vitro*

Yanxian Peng and Michael Raghunath
National University of Singapore
Singapore

1. Introduction

Tissue engineering, a relatively new field, combines a multitude of separate disciplines: clinical medicine, cell biology, engineering, materials science, biochemistry and also biophysics. With such a diverse pool of research backgrounds, tissue engineers seek a common goal: to replace or repair part of or whole tissues and organs, with functional man-made tissue constructs.

To date, there have been attempts to tissue-engineer bone, cartilage, liver and various other tissues and organs. The more successful tissue engineered constructs have been simpler tissues such as artificial skin, bladder and recently, tissue engineered airway (Macchiarini et. al. 2008). As tissue complexity increases, achieving fully functional tissue constructs becomes exponentially more difficult.

Conventional tissue-engineering strategies can be broadly classified into cell-free and cell-based therapies. Both strategies typically utilize scaffolds, with or without cells, and typically provide biologically active molecules to enhance integration into the body. Scaffolds serve many purposes: mechanical support, cell migration and attachment, cell retention and delivery at the site of repair, and even sequester biochemical factors and/or genes, while allowing for diffusion of nutrients from the host tissue into the construct. Essentially, an ideal scaffold will act as a man-made emulation of the natural extracellular matrix found (ubiquitously) in the body.

Scaffold materials come in many different forms, from synthetic to nature-derived materials, each with a set of properties defining their intrinsic advantages and limitations for application. Polymer scaffold materials have been a mainstay of tissue-engineered constructs, however there has been an increasing awareness of the advantages of biomimicry or nature-derived scaffolds often fabricated from native and denatured "collagen", fibrin, chitosan and glycosaminoglycans. These scaffolds are more biocompatible but remain simplified copies of the complex extracellular milieu in living tissue.

The extracellular matrix is a complex supramolecular assembly of a variety of glycoproteins and proteoglycans. Extracellular matrix provides tissue integrity, acts as a native scaffold for cell attachment and interaction and also serves as a reservoir for growth factors. For most connective tissues, collagen makes up the bulk of the extracellular matrix. Collagen functions in the extracellular matrix as a structural protein as well as a binding partner for ligands that store growth factors. The term collagen encompasses a family of glycoproteins characterised by a particular amino acid signature and triplehelicity. The most abundant collagens are type I, found in skin, tendon and capsules, type IV, found in all basement membranes, and type VII, which is specific for the basement membrane of skin, oral mucosa and cornea. Depending on their tissue location and function, collagen assemblies in the extracellular matrix are often heterotypic. For example, collagen fibers in the dermis represent an alloy of types I, III and V, while cartilage matrix contains types II, IX, and XI (Kielty & Grant, 2002). Depending on the wear and tear of a given tissue, extracellular matrix continuously undergoes remodeling and replacement.

Tissue engineers have long underestimated the capacity of cells to manufacture their own extracellular matrix. We show in the following text, ways to control and augment this cellular capacity to create a well-developed extracellular matrix that can provide cellular cohesion and tissue strength. This cellular ability is an important quality required for downstream applications of tissue engineering. Enhancing the deposition of collagen in tissue-engineered constructs will aid in the maturity of the tissue construct *in vitro* prior to clinical application as well as provide a functional model for the study of fibrogenic processes for therapeutic purposes.

2. Application of Macromolecular Crowding (MMC) in Enhancing Matrix Deposition

2.1 *In vitro* collagen matrix formation (deposition) is tardy

The typical signature of a collagen molecule is its quaternary structure - the triplehelix. It is a trimer of α-chains that is assembled intracellulary (Kielty & Grant, 2002) and secreted by fibrogenic cells into the extracellular space as soluble procollagen. Taking procollagen type I as the most abundant representative, one can envision the polypeptide as a dumbbell-shaped molecule with a central triple helical region of rod-like stiffness and globular propeptide at either end. Once in the extracellular space, both propeptides of procollagen are cleaved off by procollagen N-proteinase and procollagen C-proteinase respectively. The resulting collagen molecules will then self-assemble in *vivo* at pH 7-7.4 into quarter-staggered arrays, thus initiating fibrillogenesis (Kielty & Grant, 2002).

When adequately provided with ascorbic acid (AcA), fibrogenic cells secrete substantial amounts of procollagen (Murad et. al., 1981). However, the enzymatic conversion of procollagen to collagen is extremely tardy under contemporary aqueous cell culture conditions, a fact that is not widely known. Consequently, only small amounts of an insoluble collagen matrix are formed, while most of the (water soluble) procollagen is lost during culture medium change and in bioreactor outflows (Lareu et. al., 2007a, b).

2.2 *In vitro* collagen biosynthesis is enhanced by MMC – effects of negatively charged crowders

We have found that the addition of negatively-charged macromolecules for two days during the culture of fibroblasts reduces procollagen content in the culture medium and enhances the amount of (processed) collagen molecules deposited at the cell layer.

Fibroblasts were first grown in culture medium with or without AcA and macromolecules. The procollagen in the culture medium and the collagen found in the adherent cell layer were isolated by pepsin and quantified by sodium dodecyl sulfate polyacrylamide gel electrophoresis and densitometry as described by Lareu et. al. 2007a, b. Fibroblasts that were not treated with AcA secreted little procollagen into the culture medium and therefore showed no appreciable deposition of collagen deposited into the matrix.[1] Under AcA supplementation the amount of procollagen secreted into the culture medium was substantially increased. However, there was little insoluble collagen at the adherent cell layer. However, with the addition of 100μg/ml of 500kDa dextran sulfate (DxS) for two days, the presence of procollagen in the medium was much reduced, while the amount of collagen associated with the cell layer was dramatically increased (Fig. 1.).

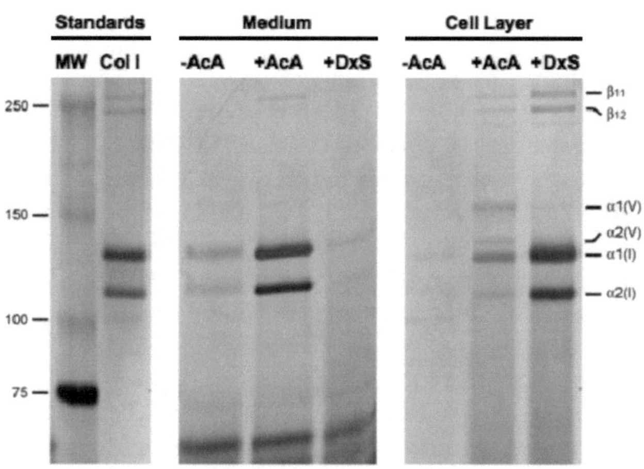

Fig. 1. Addition of 100μg/ml of 500kDa DxS resulted in the reduction of procollagen in the culture medium and the increase of insoluble collagen at the fibroblast cell layer. Note the predominance of collagen V in the non-crowded cell layer. Reprinted from Lareu et. al., 2007a, with permission from the publisher, Mary Ann Liebert

DxS concentrations ranging from 10μg/ml to 500μg/ml were applied to fibroblast cultures. It was observed that collagen bands from cell layer samples became more intense with an

[1] AcA is an essential cofactor for prolyl-4-hydroxylase, an enzyme that catalyses a crucial posttranslational modification to the α-chains resulting in thermal stability (at body temperature) of the folded procollagen triplehelix in the endoplasmic reticulum. In the absence of AcA, prolyl residues of individual α-chains are not sufficiently hydroxylated of form a thermostable triplehelix that can be secreted. (Murad et. al., 1981). Therefore, the omission of ascorbate causes the equivalent of scurvy in culture. Similar to the clinical condition, this can be remedied easily.

increasing concentration of DxS up to 100μg/ml (Fig. 2.). At 500μg/ml, the intensity of collagen bands decreased sharply and, hence, we concluded that the optimum concentration for crowding with DxS was 50-100μg/ml.

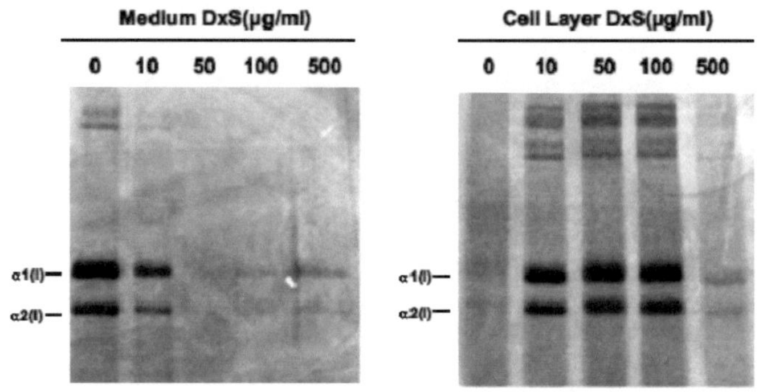

Fig. 2. Collagen deposition was enhanced with increasing concentration of DxS up to 100μg/ml. At 500μg/ml, collagen deposition decreased sharply. The optimum concentration of DxS was determined to be 50-100μg/ml. Reprinted from Lareu et. al., 2007a, with permission from the publisher, Mary Ann Liebert

We also found a marked increase of collagen deposition with 200kDa polysodium-4-styrene sulfonate (PSS), a negatively-charged macromolecule, at a final concentration of 100μg/ml culture medium in fibroblast cultures for two days. Collagen in the adherent cell layer was again isolated by pepsin digestion and analyzed by sodium dodecyl sulfate-polyacrylamide gel electrophoresis. Densitometric analysis of collagen α1(I) and α2(I) bands (Fig. 3A) revealed that the addition of DxS increased collagen deposition by >20-fold, and the addition of PSS increased collagen deposition by >35-fold (Fig. 3B). We have also tested the efficacy of other macromolecules in increasing collagen deposition.. These macromolecules, 50mg/ml of 400kDa Ficoll (Fc 400), 50mg/ml of 70kda Ficoll (Fc 70) and 100μg/ml of 10kDa DxS (DxS 10), did not induce a significant increase in collagen deposition at the cell layer in the time window of 2 days (Fig. 3A.)

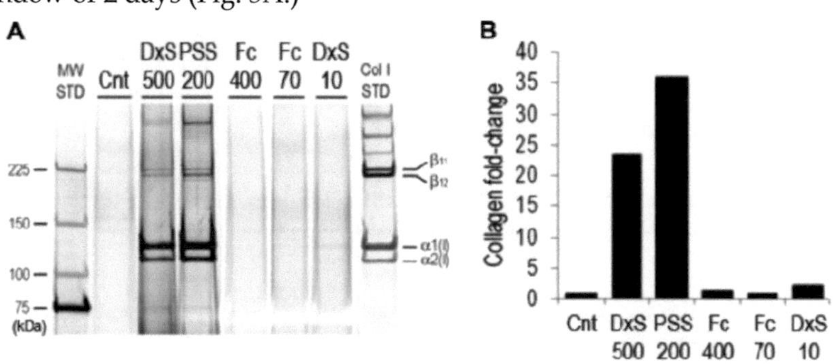

Fig. 3. 100μg/ml of DxS increased collagen deposition by >20-fold. 100μg/ml of PSS was also shown to induce an increase in collagen deposition by >35-fold. Reprinted from Lareu et. al., 2007b, with permission from the publisher, Elsevier

2.3 MMC via negatively charged macromolecules results in a granular collagen deposition

Immunocytochemical analyses of fibroblast cultures treated with DxS or PSS corroborated the SDS-PAGE results. There were significantly stronger signals of collagen immunostaining in these fibroblast cultures compared to the uncrowded control (Fig. 4A.). With regard to the deposition pattern, we noted that the collagen aggregated in a granular form, co-localized with fibronectin staining, whereas the minimal collagen deposited in the uncrowded control was reflected by weakly stained wispy strands.

Transmission electron microscopy revealed that these aggregates produced under DxS crowding contained cross-striated collagen nanofibers surrounded by a three dimensional arrangement of other matrix components (Fig 4B.) We therefore address these aggregates as "microgels".

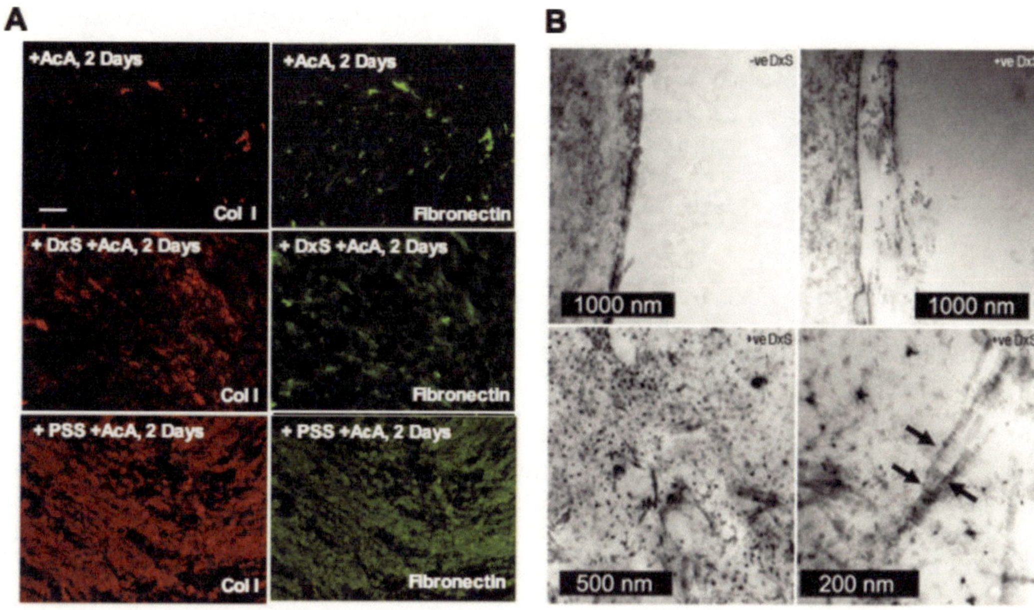

Fig. 4. Morphological analysis of fibroblast cultures in the absence and presence of DxS. (A) Double immunfluorescence staining for collagen and fibronectin reveals granular morphology for DxS-crowded and PSS-crowded fibroblast cell layers. The uncrowded control showed wispy strands of collagen and fibronectin. All magnifications at 10X, scale bar100µm; (B) Transmission electron micrograph of the pericellular matrix in the absence and presence of DxS. Arrows indicate typical collagen fibrillar formation in the presence of DxS. In the presence of DxS, fibroblast cell layer samples showed cross-striated collagen nanofibers, which were absent in non-crowded fibroblast cultures. Fig. 4B was reprinted from Lareu et. al., 2007b, with permission from the publisher, Elsevier

2.4 *In vitro* collagen biosynthesis is enhanced by MMC – effects of neutral crowders

Although we initially found that Fc 400 and Fc 70 as monocrowders were unable to enhance collagen production significantly in two days, we discovered that a combination of Fc 400 and Fc 70 added as a cocktail to the fibroblast cultures increased collagen deposition >10-

fold in six days compared to non-crowded controls (Fig. 5A and 5B). Fc 400 and Fc 70 were used at concentrations of 25mg/ml and 37.5mg/ml, respectively.
However, we noted with interest that under this cocktail, collagen matrices started out in a granular pattern (not shown) but transformed into a reticular pattern reminiscent of the pattern seen with fibronectin staining in the uncrowded controls. This observation is in keeping with the notion that fibronectin serves as a primordial template for collagen deposition (Sottile et. al., 2007). Staining of collagen in Fc-supplemented fibroblast cultures was much stronger in comparison to non-crowded controls. In accordance to earlier observations (see section 2.3), collagen deposits induced by DxS and PSS supplementation showed a coarse granular appearance also after six days (Fig. 5C).

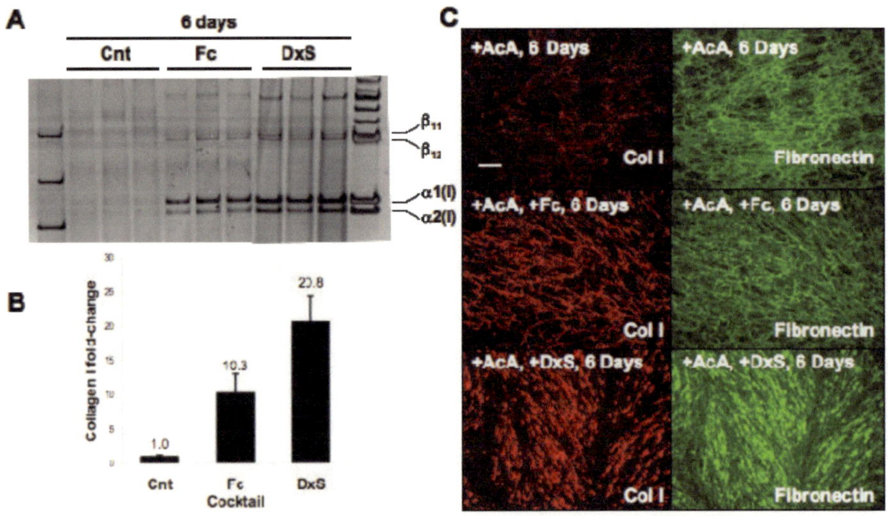

Fig. 5. Application of a mixture of neutral macromolecules Fc 400 and Fc 70 to fibroblast culture medium is also able to increase collagen deposition over the course of 6 days, resulting in a reticular network of collagen deposition rather than the granular collagen deposition observed with negatively charged crowders

2.5 How MMC works - enhancing key enzymatic activities in collagen matrix formation

MMC is a biophysical principle in which inert macromolecules in solution occupy a significant volume of the medium and hence impact on biochemical reactions both *in vivo* and *in vitro* (Minton, 2001). This principle is highly conserved throughout evolution for the interior of cells and the extracellulular environment in multicellular organisms (Zimmerman & Minton, 1993). The density or crowdedness of macromolecules is best determined first by approximation as solute content. To date, the concentrations of macromolecules in the interior of cells have been determined in a range from 50 to 400mg/ml (Minton, 2001). This concentration refers to the cumulative concentration of all species of macromolecules present, rather than the concentration of any single macromolecule. The presence of these macromolecules means that they must significantly occupy space in the given medium, and biophysicists refer to such medium as "crowded" or "confining"(Minton, 2001).
In order to illustrate the principle of MMC, picture a simplified scenario where macromolecules (blue) are confined in a given amount of space (box) (Fig. 6.). When a

reactant molecule (red) of a similar size is introduced to such a system, the presence of these macromolecules limit the space available (white area) to the reactant molecule. It is obvious that the limitation is due to the fact that no two molecules can overlap in the given space. Additionally, the limitation is also due to the dead space (black area) in between macromolecules that the reactant molecule is unable to pass through. Hence the excluded volume is the sum of the occupied space and the dead (or unavailable) space. In such a scenario, there are two consequences on the reactant molecule. Firstly, the reduction in volume will increase the effective concentration of the reactant molecule, and hence the thermodynamic activity of the reactant. Secondly, the diffusion rate and correspondingly the rate of encounter of a reactant molecule with another is also reduced (Chebotareva et. al., 2004). The first consequence will manifest in reactions where the stability of the transition state is limiting, resulting in increased reaction rates. On the other hand, the second consequence will manifest in reactions where the encounter rates are limiting, resulting in decreased reaction rates (Ellis, 2001).

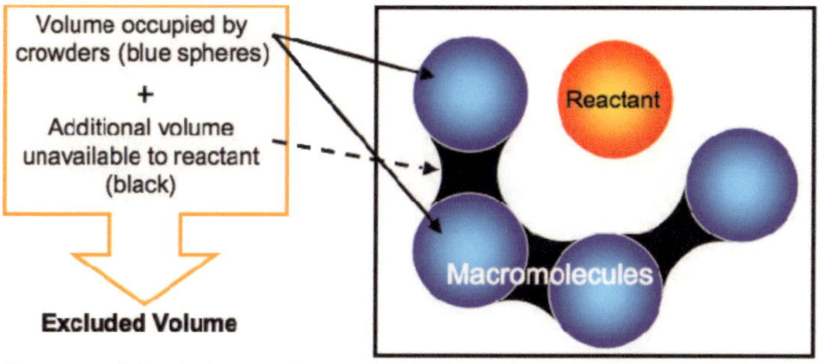

Fig. 6. Macromolecules (blue) limit the space available (white) for the reactant molecule (red) by occupying space and creating dead space (black) in between macromolecules. Both crowder volume and dead space volume contribute to the excluded volume available for the reactant molecule

In addition to the molecular size and shape attributes of a macromolecule, the hydration shells around such crowders also contribute to the total volume of exclusion as they are in an aqueous solvent. In addition, the surface charge on many of the ionic crowders, imparts an electrostatic contribution to this steric exclusion, as a result of like-charge repulsion, thus amplifying the steric exclusion effects.

Finally, the most effective increase in thermodynamic activities is exerted by like-sized molecules on each other (Ellis, 2001). If the reactant molecule is much larger or much smaller than the macromolecule, the effect is less significant.

In contrast to living systems, typical *in vitro* methods involving all kinds of settings (tube, cuvette, ELISA plate, cell culture dish) are done in aqueous solutions with concentrations of macromolecules ranging from 1 to 10 mg/ml (Ellis, 2001). Fibroblasts and other cells are typically cultured on plain or thinly coated tissue culture plastic, bathed in large volumes of non-crowded aqueous medium. This situation is far from physiological. In the body, extracellular space comprises of dense arrays of extracellular matrix macromolecules. Even blood plasma shows solute concentrations of around 80mg/ml (Ellis, 2001). In contrast, typical cell cultures employ supplements of 5%-20% fetal bovine serum resulting in solute concentration of 4- 16 mg/ml. Obviously, such culture conditions - the current norm - do not provide a crowded environment.

In 1986, Bateman et. al. reported that when polyethylene glycol (PEG), dextran T-40 or polyvinylpyrrolidone were added to fibroblast cultures, no collagen or procollagen was found in the medium but instead was associated with the cell layer (Bateman et.al., 1986). In 1994, Hojima et. al. found that either the presence of 500kDa dextran sulfate or PEG increased the respective activities of purified procollagen C-proteinase and purified procollagen N-proteinase *in vitro* in a cell-free system (Hojima et.al., 1994). However, there was no follow-up on these publications until 2007, when we set out to use this system for translational purposes.

Under our control, the addition of inert macromolecules was able to dramatically enhance collagen deposition (Lareu et. al. 2007a,b). The macromolecules employed exerted a crowding effect on the pathways involved in collagen deposition. We have identified at least two enzymatic reactions that are boosted under MMC.

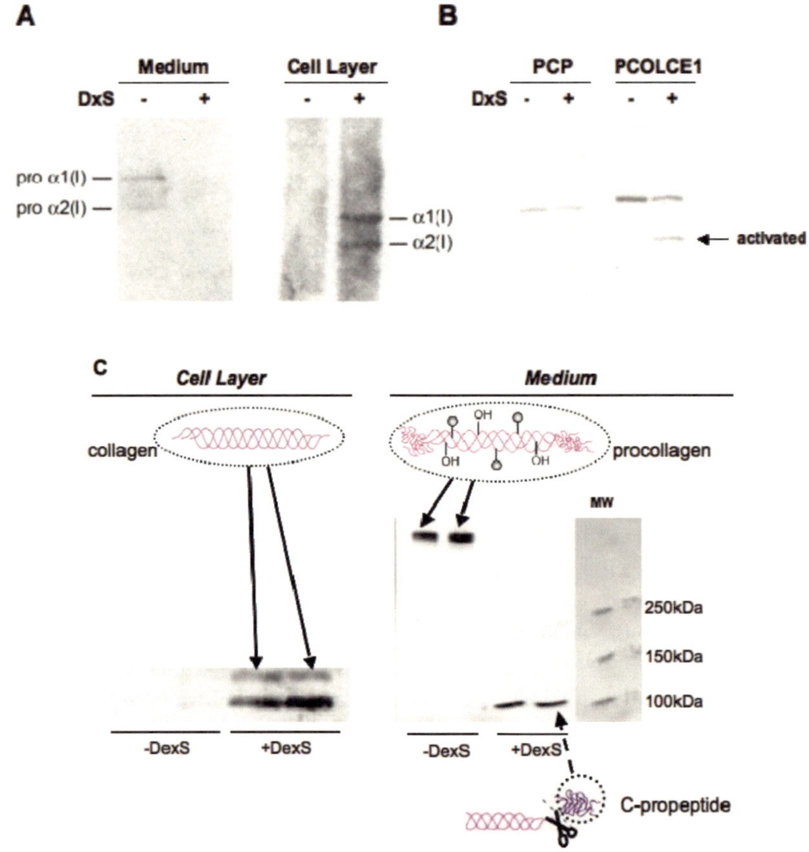

Fig. 7. Proteolytic cleavage of procollagen is enhanced by MMC. (A) Immunoblot for the central portion of collagen showing increased collagen staining from the cell layer fraction under MMC. In the absence of MMC, procollagen remained unprocessed and in the medium fraction. (B) Immunoblot for PCP showed little significant change in quantity of PCP but immunoblot for PCOLCE1, an enhancer of PCP, showed increased activation. (C) Immunoblot for the C-propeptide showed the presence of higher molecular weight

procollagen in the medium of uncrowded controls, and the presence of lower molecular weight C-propeptide in the medium of crowded fibroblast cultures. Fig. 7A was reprinted from Lareu et. al., 2007a, with permission from the publisher, Mary Ann Liebert. Fig. 7B was reprinted from Lareu et. al., 2007b, with permission from the publisher, Elsevier

Firstly, we observed an increase in procollagen conversion in the presence of DxS through Western blot analysis (Fig. 7). An antibody that recognizes only the central portion of the collagen molecule was used. Control cultures contained only procollagen in the medium, with little collagen detected in the cell layer. In the presence of DxS, procollagen in the medium was reduced. On the other hand, procollagen-processed collagen was present in the cell layer, indicating increased procollagen conversion into collagen and subsequent deposition at the cell layer (Fig. 7a). Western blot analysis of procollagen C-proteinase (PCP), the enzyme required for procollagen cleavage into collagen, revealed no significant changes in its amount (Fig. 7B) Therefore, we infer that its *activity* must have changed, which can be confirmed using antibodies against the C-propeptide (Fig 7C). Interestingly, we noted a proteolytic fragmentation under crowding with PCOLCE1, the allosteric enhancer of PCP-procollagen-specific activity (Fig. 7B). Only in the presence of DxS, did we detect a 36kDa active form of PCOLCE along with a C-terminal fragment, which was shown to inhibit matrix metalloproteinase 2 (MMP2) (Mott et.al., 2000).

A second identified enzymatic reaction is the lysyl oxidase-catalyzed reaction which produces large molecular weight dimers, β_{11} and β_{12}, resulting in covalent crosslinking of collagen assemblies. Under crowded conditions, we determined via densitometry an increase of the ratio of β bands versus α bands, suggesting an increased lysyl oxidase-mediated crosslinking of the deposited collagen. (Chen et. al., 2009)
Hence, we hypothesize that MMC influences collagen production and turnover, and thus extracellular matrix production and retention, by at least three ways (Fig. 8). Under the influence of MMC, there is increased proteolytic conversion of a procollagen C-proteinase allosteric enhancer, PCOLCE1, into its 36kDa active form and a 16.5kDa fragment that acts as a MMP2 inhibitor. The former will then enhance PCP enzymatic activity, resulting in increased procollagen cleavage in the extracellular space, thereby producing water-insoluble collagen molecules. These collagen molecules will then self-assemble in a quarter-staggered array at the cell layer. Moreover, the binding of PCOLCE1 to procollagen C-proteinase would be enhanced by MMC. MMC also enhances lysyl oxidase activity, which catalyzes the β crosslinks between the collagen molecules, and so stabilizing these collagen molecules. The other by-product of PCOLCE1 conversion results in a MMP2 inhibitor, which aids in reducing the collagen turnover and thus maintaining the collagen matrix. Finally, crowding might also drive the supramolecular assembly of collagen, as we have observed in gel forming assays (unpublished).

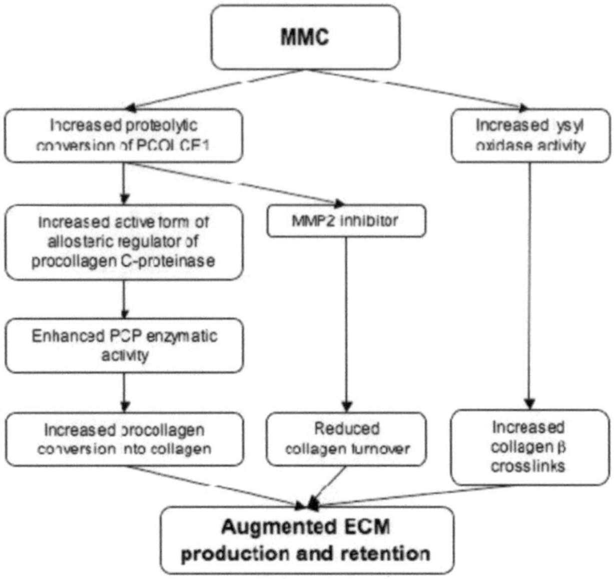

Fig. 8. Mechanisms by which MMC augments extracellular matrix (ECM) production and retention

Other macromolecules that were tried include 100μg/ml of 670kDa neutral dextran (NDx), 50mg/ml of Fc 400, 50mg/ml of Fc 70 and 100μg/ml of 10kDa DxS. These additional macromolecules were unable to significantly increase collagen deposition. As explained earlier in section 2.5, several factors determine the total excluded-volume effect. By definition, all crowder macromolecules are hydrophilic and therefore the hydrodynamic size (R_H), a parameter that captures the effects of size and hydration, *as is*, becomes a critical determinant of the volume of exclusion. In addition, for anionic macromolecules, the surface charge further adds on to the effects of size and hydration. We can illustrate this by a simple comparison of the hydrodynamic radii of charged and neutral macromolecules. Anionic Dextran Sulfate of molecular weight 500kDa with an R_H of 46 nm (Table 1) far exceeds the R_H of a neutral Dextran (21nm) that has a much larger molecular weight at 670kDa. This 2.2-fold greater increase in radius translates to a 10-fold increase in volume per molecule and when calculated to the entire population of macromolecules in solution, this translates into a substantial increase in volume occupancy for the charged crowder. Another similar analogy can be described when we compare PSS (200kDa) with the neutral Dextran (670kDa). Although PSS is three times lower in MW than NDx, both have a similar R_{H}, (21nm). This highlights the significance of charge to determining the hydrodynamic radius. These biophysical findings highlighting the role of charge were corroborated by the cell culture results, which demonstrated the inability of NDx and Fc 70 and Fc 400 as monocrowders to elicit the same degree of collagen deposition when compared with the charged crowders in similar conditions.

Combination of Fc 70 and 400 at 37.5mg/ml and 25mg/ml, respectively, yielded enhanced collagen production due to the synergistic effects of such a mixed crowder cocktail , as described by Harve et. al., 2008.

When the effects of two Dextran Sulfate species, one at 500kDa and another at 10 kDa, were compared, biophysically it was found that the smaller DxS species was undetectable in DLS as the size of this crowder was below the detection threshold. The DxS10 was understandably found to yield minimal amounts of collagen in comparison to its heavier and bigger counterpart. Thus size also does matter when the excluded-volume effects are compared between two charged macromolecules of similar chemical structure.

	DxS 500kDa	PSS 200kDa	NDx 670kDa	Fc 400kDa	Fc 700kDa	DxS 10kDa
Net Charge	Negative	Negative	Neutral	Neutral	Neutral	Negative
Hydro. Radius (nm)	46	22	21	4.5	3	<1

Table 1. Hydrodynamic radii of macromolecules 500kDa DxS, PSS, NDx, Fc 400, Fc 70 and 10kDa DxS

2.6 Why MMC is not equivalent to a simple volume reduction of culture medium

A question often posed by audiences worldwide is since MMC creates a concentration effect via volume exclusion, why not grow the cells in reduced volumes of culture medium? Would this not lead to a similar concentration of secreted factors like enzymes and reactants?

We designed an experiment to address this question. We grew 50,000 fibroblasts in each well of a 24 well culture dish (15 mm diameter) varying the volumes of culture medium from 500 μl to 100 μl. In parallel, we ran one DxS-crowded set of cells in 500 μl. As to be expected, in the smallest volumes, cell cultures dried out, resulting in cell death. Interestingly, there was an increase of collagen deposition in the 200 μl well by 2.2-fold (Fig. 9A). However, this was not substantial compared to the collagen deposition induced in the 500 μl well by DxS, which was increased by an order of magnitude in this experiment (Fig. 9A). We calculated the thermodynamic activity of reactants procollagen and its enzymes procollagen C-proteinase in the case of 500μl of DxS-supplemented culture medium, 200μl of uncrowded medium and 100μl of uncrowded medium and found an order of magnitude difference in the 500μl crowded medium in comparison with the latter two cases (Fig. 9B). Such a large difference in the thermodynamic activities could explain the much-enhanced collagen deposition of the crowded culture.

Thus, MMC and the resulting excluding volume effect are far more powerful than simply reducing the volume of medium, and still allow for ample nutrient-rich culture medium in static or bioreactor settings. We should mention here that MMC in our system did not cause appreciable changes of viscosity (Lareu et. al., 2007a) to the culture medium thus eliminating concerns with diffusion issues.

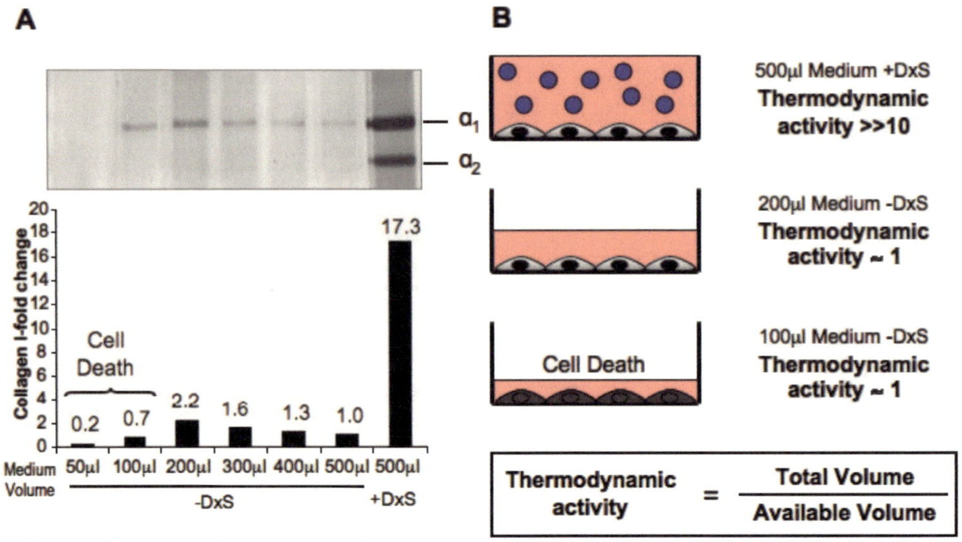

Fig. 9. The reduction of medium volume was unable to enhance collagen deposition to the same degree as the application of MMC. (A) Fibroblast cultures were grown under non-crowded conditions with reduced volumes of medium and showed at best, a 2.2-fold collagen deposition using 200µl of medium. In contrast, DxS-crowded fibroblasts showed a much more enhanced collagen deposition at 17.3-fold. When medium volume was reduced to 100µl and 50µl, the wells dried out, resulting in cell death. (B) Thermodynamic activities in 500µl of DxS-crowded medium far exceeds that of 200µl and 100µl of uncrowded fibroblast cultures

2.7 Increased collagen biosynthesis by MMC for application in connective tissue engineering

We have shown that MMC can be employed to increase collagen biosynthesis in fibroblast cultures. Because collagen comprises the majority of the extracellular matrix components *in vivo*, we are able to produce an extracellular matrix *in vitro* that is more physiological. Our research has brought a technological advancement to the field of connective tissue engineering and other tissue engineering fields as well.

Additionally, we have shown that MMC can be applied in conventional tissue engineering systems comprising scaffolds cultured under static or bioreactor conditions (Lareu et.al. 2007b). We have successfully supplemented DxS into the culture medium of fibroblasts seeded onto Biobrane scaffolds. In comparison to the control, the presence of DxS increased collagen production by >6-fold (Fig. 10A.). Biobrane is a biosynthetic wound dressing comprising a silicone film with a nylon fabric partially embedded into the film. Small collagen peptides are coated onto the nylon fabric, which would be broken down during our pepsin-isolation protocol and hence will not contribute to our collagen detection method. The use of Biobrane as a scaffold is a proof-of-concept of the efficacy of employing MMC in connective tissue engineering.

In another culture system, DxS was supplemented into the culture medium of fibroblasts seeded onto poly-L-lactic acid scaffolds under either static conditions in petri dishes, or under bioreactor conditions. Under both conditions, DxS was able to increase collagen production by >6 fold and 3-fold respectively (Fig. 10B.).

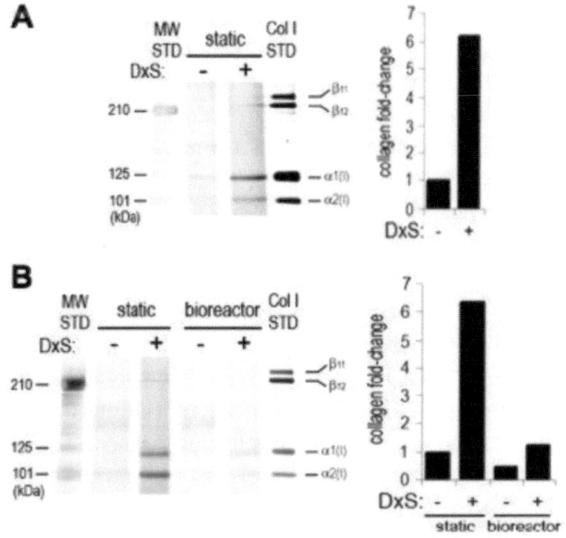

Fig. 10. DxS induced increased collagen production in fibroblast cultures on scaffolds (A) Biobrane and (B) PLLA. PLLA fibroblast cultures were grown under both static and bioreactor conditions. Reprinted from Lareu et. al., 2007b, with permission from the publisher, Elsevier

Thus, we have shown that collagen deposition can be enhanced by the application of MMC in two different scaffold fibroblast culture systems as well as under bioreactor conditions. The application of a biophysical principle, MMC, to solve a tissue engineering problem, is a fine example of how important it is to adopt a multi-disciplinary approach towards solving tissue engineering problems.

The increase in collagen has the potential to benefit most if not all tissue engineering constructs because the protein serves to create cohesion between single cells and cell layers, thereby contributing to mechanical support. Moreover, applying MMC results in an extracellular matrix more similar to that found ***in vivo***, which has the potential to provide biochemical cues such as growth factors, fulfilling a more physiological role in directing the cells.

2.8 Side observations on crowding – growth behaviour

We noted significant differences in cell proliferation depending on the crowders used. In the presence of negatively charged DxS, cell proliferation halted, such that there was 12% less cells in comparison to the uncrowded control in 2 days (unpublished data). In contrast, in the presence of neutral crowders in the Fc cocktail, cell proliferation was increased by 15% in a time window of 6 days (unpublished data). A possible explanation for this phenomenon could be that these observations do not reflect a true stimulation of proliferation by Fc, but rather a longer proliferation slow-down time as it takes longer under these conditions to build up a critical mass of ECM that could move cells into a quiescent state.

2.9 Current immediate application of MMC in cell culture

The ability to rapidly create matrix *in vivo* allowed us to use this technology to build a drug screening tool, the Scar in a Jar (Chen et al, 2009), therefore enabling us to emulate collagen biosynthesis, secretion and deposition in a compressed time window. Thus we can emulate scar formation and test antifibrotic substances in a high content screening setting (Wang et. al., 2009).

3. Future Challenges

One of the current challenges faced by tissue engineers is the size and complexity of tissue constructs. We envision that macromolecular crowding can enhance both static and bioreactor settings and induce cells to coat polymeric biomaterials with their own matrix. Organ printing is an emerging field, where pieces of pelleted tissue are printed onto a hydrogel or substrate and fuse together to form a larger construct. Currently, organ printing has enabled the formation of capillary-like tubes from pellets of endothelial cells (Mironov, 2009). The biggest obstacle to this technique is keeping the printed cell constructs stable. These constructs are most likely unstable due to their lack of structural matrix components after printing. Printing into a scaffold is technically challenging, and hence a less preferred option. However, we propose that exposing printed or extruded cell constructs to macromolecular crowders could enhance collagen deposition by these cell pellets themselves, thereby reducing the period of instability and creating a more cohesive tissue construct. A particular way to create coherent cell sheets was pioneered by Teruo Okano and coworkers (Yang et. al., 2007). This technique is based on thermosensitive polymers that switch hydrophilicity according to temperature. This has led to the discovery that cells grown as a hyperconfluent monolayer can be lifted of as a coherent cell sheet by cooling the culture plates. This elegant technique is dependent on a basic ECM containing of fibronectin and collagen V that lend cohesion to the cell sheet. Unfortunately, it does not appear possible currently to generate matrix-enhanced cell layers and to lift them off the Okano material, as the cells adhere extremely firmly onto the deposited ECM, and the ECM in turn to the polymer. However, we are confident that crowding and cell sheet technology can be combined by designing novel temperature-sensitive cell culture polymers that have a lower critical solution temperature, between 4°C and 10°C and would be predicted to release a matrix-cell- layer completely. This work is currently work in progress.

4. Acknowledgments

We would like to thank Dr Karthik Harve Subramhanya and Clarice Chen, Tissue Modulation Laboratory, for their contributions in this work and critical reading. The authors acknowledge support by the NUS Tissue Engineering Programme and the Bioengineering Programme (Life Science Institute NUS) and the Faculty Research Committee. This work was supported by grants from the National Medical Research Council of Singapore (NMRC grant CPG/003/2004, R-397-000-018-213), the Faculty of Engineering, Office of Research FRC grants (R-397-000-017-112, R-397-000-081-112), Office of Life Science (R-397-000-604-

101, R-397-000-604-712) Y.P. is a recipient of a graduate research scholarship from the Division of Bioengineering, National University of Singapore.

5. References

Bateman, J.F.; Cole, W.G.; Pillow, J.J. & Ramshaw, J.A.M. (1986). Induction of procollagen processing in fibroblast cultures by neutral polymers. *Journal of Biological Chemistry*, Vol. 261, No. 9 (Month 1986) 4198-4203, 0021-9258

Chebotareva, N.A.; Kurganov, B.L. & Livanova, N.B. (2004). Biochemical effects of molecular crowding. *Biochemistry (Moscow)*, Vol. 69, No. 11 (Nov 2004) 1239-1251, 0006-2979

Chen, C.Z.C.; Peng, Y.; Wang, Z.; Fish, P.V.; Kaar, J.L.; Koepsel, R.R.; Russell, A.J.; Lareu, R.R. & Raghunath, M. (epub, ahead of print) The scar-in-a-jar: studying antifibrotic lead compounds from the epigenetic to extracellular level in a single well. *British Journal of Pharmacology*, in press, 0007-1188

Ellis, R.J. (2001). Macromolecular crowding: obvious but underappreciated, *Trends in Biochemical Sciences*, Vol. 26, No. 10 (Month 2001) 597-604, 0968-0004

Harve, K.S.; Vigneshwar, R.; Rajagopalan, R. & Raghunath, M. (2008) Macromolecular crowding in vitro as means of emulating cellular interiors: when less might be more. *Proceedings of the National Academy of Sciences of the United States of America*, Vol 105, No. 51 (Dec 2008), E119, 1091-6490

Hojima, Y.; Behta, B.; Romanix, A.M. & Prockop, D.J. (1994). Cleavage of type I procollagen by C- and N-proteinase is more rapid if the substrate is aggregated with dextran sulfate or polyethylene glycol. *Analytical Biochemistry*, Vol. 223 (Month 1994), 173-180, 0003-2697

Kielty, C.M & Grant, M.E., (2002). The collagen family: Structure, assembly, and organization in the extracellular matrix, In: *Connective Tissue and its Heritable Disorders Molecular, Genetic, and Medical Aspects*, Royce, P.M. & Steinmann, B. (Ed.), 103-147, Wiley-Liss Inc., 0471221929, New York

Lareu, R.R.; Arsianti, I.; Subramhanya, H.K.; Peng, Y. & Raghunath, M. (2007a). *In vitro* Enhancement of Collagen Matrix Formation and Crosslinking for Applications in Tissue Engineering: A Preliminary Study. *Tissue Engineering*, Vol. 13, No. 2, (February 2007) 385-391, 1937-3341

Lareu, R.R.; Subramhanya, H.K.; Peng, Y.; Benny, P; Chen, C; Wang, Z; Rajagopalan, R & Raghunath, M. (2007b). Collagen Matrix Deposition is dramatically enhanced *In vitro* when Crowded with Charged Macromolecules: The Biological Relevance of the Excluded Volume Effect. *FEBS Letters*, Vol. 581, No. 14 (May 2007) 2709-2714, 0014-5793

Macchiarini, P.; Junebluth, P.; Go, T.; Asnaghi, M.A.; Rees, L.E.; Cogan, T.A.; Dodson, A.; Martorell, J.; Bellini, S.; Parnigotto, P.P.; Dickinson, S.C.; Hollander, A.P.; Mantero, S.; Conconi, M.T. & Birchall, M.A. (2008). Clinical Transplantation of a Tissue-Engineered Airway. *The Lancet*, Vol. 372, No. 9655 (December 2008) 2023-2030, 0140-6736

Minton, A.P. (2001). The influence of macromolecular crowding and macromolecular confinement on biochemical reactions in physiological media. *Journal of Biological Chemistry*, Vol 276, No. 14 (February 2001) 10577-10580, 0021-9258

Mironov, V.; Visconti, R.P.; Kasyanov, V.; Forgacs, G.; Drake, C.J. & Markwald, R.R. (2009). Organ printing: Tissue spheroids as building blocks. *Biomaterials* Vol. 30, (Jan 2009) 2164-2174, 0142-9612

Mott, J.D.; Thomas, C.L.; Rosenbach, M.T.; Takahara K.; Greenspan, D.S. & Banda, M.J. (2000). Post-translational proteolystic processing of procollagen C-terminal proteinase enhancer releases a metallo-proteinase inhibitor. *Journal of Biological Chemistry*, Vol. 275, No. 2 (Jan 2000) 1384-1390, 0021-9258

Murad, S.; Grove, D.; Lindberg, K.A.; Reynolds, G.; Sivarajah, A. & Pinnell, S.R. (1981). Regulation of collagen synthesis by ascorbic acid. *Proceedings of the National Academy of Sciences of the United States of America*, Vol. 78, No. 5 (May 1981) 2879-2882, 1091-6490

Sottile, J.; Shi, F.; Rublyevska, I.; Chinag, H.Y.; Lust, J. & Chandler, J. (2007) Fibronectin-dependent collagen I deposition modules cell response to fibronectin. *American Journal of Physiology Cell Physiology*, Vol. 293 (Oct 2007) C1934-C1946, 0363-6143

Wang, Z.; Chen, C.; Finger, S.N.; Kwajah, S.; Jung, M.; Schawz, H.; Swanson, N.; Lareu, R.R. & Raghunath, M. (2009) Suberoylanilide hydroxamic acid: a potential epigenetic therapeutic agent for lung fibrosis? *The European Respiratory Journal*, Vol 34, No. 1 (July 2009) 145-155, 1399-3003

Yang, J.; Yamato, M.; Shimizu, T.; Sekine, H.; Ohashi, K.; Kanzaki, M.; Ohki, T.; Nishida, K. & Okano, T. (2007) Reconstruction of functional tissues with cell sheet engineering. Biomaterials, Vol. 28, No. 34 (Dec 2007) 5033-5043, 0142-9612

Zimmerman, S.B. & Minton, A.P. (1993) Macromolecular crowding: biochemical, biophysical, and physiological consequences. Annual Review of Biophysics and Biomolecular Structure, Vol. 22 (June 1993), 27-65, 1056-8700

6

Scaffolds for the Engineering of Functional Bladder Tissues

Horst Maya M.D.[1], Srinivas Madduri Ph.D.[1], Gobet Rita M.D.[3],Sulser Tullio M.D.[1], Heike Hall Ph.D.[2] and Daniel Eberli M.D. Ph.D.[1]
1 Laboratory for Urologic Tissue Engineering and Stem Cell Therapy Department of Urology, University Hospital, Zurich, Switzerland
2 Cells and Biomaterials, Department of Materials, ETH Zurich, Switzerland
3 Division of Pediatric Urology, Department of Pediatric Surgery, University Children's Hospital, Zurich, Switzerland

1. Introduction

Tissue or organ loss resulting from traumatic or non-traumatic destruction causes major health problems, severely affecting patient's quality and length of life. Traditionally surgical treatment offers use of autologous tissues from a second site to repair or replace the functions of affected tissues or organs, with various outcomes. For some organs, including kidney, liver and pancreas, allogenic transplantation allows for functional restoration. However, the supply of allografts is limited and long waiting lists for tissue and organ transplantations indicate the need for new strategies to overcome the limitations of traditional therapies.

Tissue engineering (TE), one of the major approaches of regenerative medicine, is a rapidly growing and exciting field of research. In combination with better understanding of structure, biology and physiology cell culture techniques or TE may offer new treatment options for patients needing replacement or repair of an organ. The principle is to dissociate cells from a tissue biopsy, to expand these cells in culture, and to seed them onto the scaffold material in vitro in order to form a live tissue construct prior to reimplantation into the recipient's organism. In the appropriate biochemical and biomechanical environment these tissues will achieve their full functional potential and serve as native tissue equivalents. The TE approach has major advantages over traditional organ transplantation. Tissues that closely match the patient's needs can be reconstructed from a generally readily obtainable biopsy. Moreover the new reconstruct can be transplanted into the patient's body without donor site morbidity and with minimal or no immunogenicity. This eventually conquers several limitations, encountered in tissue transplantation approaches.

1.1 Urinary bladder disease

Severe bladder dysfunction can be induced by disease or surgical intervention altering the normal pattern of storage and voiding. This might then lead to an unstable, a non-compliant or a smaller bladder that cannot hold normal volumes of urine. Bladder failure can result in clinical problems ranging from mild to severe chronic urinary incontinence leading to irreversible kidney damage caused by increased upper urinary tract pressure. Severe cases of bladder failure typically do not respond to the most conservative treatment options such as bladder retraining or anticholinergic medications and severely affects the patient's quality of life. Currently, the treatment of choice in these patients is an enterocystoplasty, a surgical enlargement of the bladder using intestinal tissue. The primary aim of the surgical reconstruction is to increase bladder capacity and compliance. This improves continence, reduces intravesical storage pressure and thereby protects the upper urinary tract but fails to restore emptying function. Furthermore, enterocystoplasty is associated with numerous complications such as metabolic disturbance, increased mucus production, urolithiasis, infections and even malignant diseases. Many alternative sources of materials for reconstruction have been proposed to avoid the above mentioned complications however so far with only limited success.

1.2 Application

Tissue engineering, using autologous cells for implantation might offer a solution to this problem. Recently, several studies have confirmed feasibility of bladder reconstruction using engineered segments which were formed using biomaterials seeded with autologous cells in vitro (Yoo, Meng et al. 1998; Atala, Bauer et al. 2006, Oberpenning, Meng et al. 1999). This basic approach was first examined in a canine subtotal cystectomy model, where bladder constructs configured from natural or synthetic scaffolds and seeded with expanded autologous bladder-derived cells were successfully implanted (Yoo, Meng et al. 1998). This approach provided improved findings when compared with earlier attempts. Encouraged by the promising results of several animal studies, a similar approach was applied more recently to a small series of patients with severe neuropathic bladder dysfunction by Antony Atala et al. in 2006 (Atala, Bauer et al. 2006). They reported the results of 7 patients aged 4 to 19 years, with a mean follow up of 4 years. All three patients with the engineered tissues made with composite-scaffolds (collagen-PGA) showed a significant increase in bladder capacity and compliance. Tissue biopsies of the engineered bladder segments were described as showing an adequate structural architecture and phenotype of the different cells. Although some patients benefit was reported, clinical improvement was minimal. So far efficacy of this novel approach does not compare favorable with conventional reconstruction methods using bowel segments as source of material. To date, the clinical application of engineered tissues has been hampered by slow vascularization, poor nutrition leading to cell death and consecutive tissue fibrosis (Oberpenning, Meng et al. 1999, Ko, Milthorpe et al. 2007). For successful bladder reconstruction revascularization of the construct is essential to support short and long term survival. Furthermore, to restore physiological bladder function reinnervation is indispensable.
The concept of tissue engineering has been applied clinically for a variety of disorders, for example artificial skin for burn patients (Metcalfe and Ferguson 2007), cartilage for knee-replacement procedures (Brittberg, Lindahl et al. 1994), injectable chondrocytes for the treatment of vesico-ureteric reflux (Atala, Cima et al. 1993, Caldamone and Diamond 2001)

and urinary incontinence (Bent, Tutrone et al. 2001, Chancellor, Yokoyama et al. 2000). For hollow organs such as bladder, urethra, oesophagus, intestine, vagina, or blood vessels the strategies generally include implantation of either a biomaterial, which subsequently becomes integrated into the host organism through immigration of cells from surrounding tissues, or, with the technical advances in TE preferably an in vitro precombined construct of materials and cells. The basic requirements to achieve functional tissue are proper **scaffolds**, a suitable **environment** and appropriate **cells**.

The ambitious goal to reconstruct a functional, contractile bladder responsive to voluntary control is an ongoing challenge for the future. Our ability to use donor tissue efficiently, to provide the optimal conditions for cultivation, long-term survival, differentiation and growth will lay the foundation for success.

1.3 Scaffold

The purpose of the scaffold is to serve as a temporary supporting structure allowing not only 3-dimensional support of tissue growth and formation but also providing the biological environment needed for cellular growth, differentiation and tissue formation.

In the early days of TE stiff or non-compliant materials have been investigated for their use in tissue engineering. These materials were not suitable to support the formation of healthy tissue due to biomechanical failure and biological incompatibility (Kudish 1957, Bono and De Gresti 1966 Fujita 1978).

The ideal biomaterial for hollow organs should therefore possess adequate mechanical, biomechanical and physical properties while being non-toxic, biocompatible, promoting cellular interactions and tissue development.

Two main classes of scaffold materials have been utilized for the engineering of hollow organs; acellular matrices derived from donor tissues, (e.g., bladder submucosa and small intestinal submucosa), and synthetic polymers such as polyglycolic acid (PGA), polylactic acid (PLA), and poly(lactic-co-glycolic acid) (PLGA). The use of composite scaffolds, composed of acellular matrices together with synthetic polymers for bladder reconstruction is promising, as it combines the advantages of the two types of materials. These different strategies will be discussed in detail later in this chapter.

1.4 Environment

Successful TE approaches depend on meeting a variety of critical experimental conditions. One is to create an environment conductive to cell growth, differentiation and eventually enabling the integration of an implanted TE-construct with the surrounding host tissue. In order to achieve this, the TE-constructs shouldn't induce an immune response i.e. the host cells do not recognize it as a foreign body. Furthermore TE-constructs aim to mimique mechanical and biochemical properties of the native extracellular matrix (ECM). The ECM is the optimized natural milieu that directs tissue development and maintains tissue homeostasis. The ECM refers to a complex network of molecules that provide 2D- or 3D-mechanical support for cells, serves as a barrier between different compartments or cell types and provides guidance cues during development, tissue repair or wound healing. On the individual cell basis ECM induces cell polarity, allows or inhibits cell adhesion, promotes or slows down migration and induces cell and tissue differentiation and might also induce programmed cell death (Cheresh and Stupack 2008). The ECM is composed of chemically very different macromolecules that are assembled into organized structures remaining in close association with the surface of the cells that secreted them. The main components are space filling proteoglycans, containing collagen fibers and non-collagenous glycoproteins such as elastin. Integrated into this hydrogel-like matrix are signaling molecules such as growth factors, cytokines and hormones (Schonherr and Hausser 2000, Uebersax, Merkle et al. 2009). The ECM occurs in many different forms depending on the requirements of the surrounding tissue. In many cases it is a 3D-strucuture of ECM surrounding cells which maintains the tissue specific 3D-architecture. In other cases ECM forms flexible sheet-like structures between 40-120 nm thickness that serve as solid support layers composed of network forming laminin-entactin complexes, type IV collagen and heapran sulphate proteoglycans. This sheet-like ECM called basal lamina is frequently found in hollow organs such as blood vessels or bladder tissue. The ECM is tissue specific and the components self assemble to form spontaneous 2D- or 3D-structures under physiological conditions. Therefore, a great effort has been made to understand the appropriate biological, physical, and chemical cues, with the aim to mimic the ECM to guide morphogenesis in tissue repair (Ghosh and Ingber 2007).

Moreover, the ECM is constantly changing in composition and structure as tissues develop, remodel, repair, and age (Furth, Atala et al. 2007). All body cells, except the blood cells interact directly and in a very specific manner with their surrounding ECM. Specific receptor-ligand contacts are established that enable mutual communication between the ECM and the interior of the cell thus regulating matrix assembly, specific remodeling and local removal or disassembly of the matrix (Ghosh and Ingber 2007). Cell-matrix contacts are mainly formed between different integrins assembled into giant transmembrane protein complexes that regulate and specify their ligand binding affinity as well as matrix assembly. Integrins are transmembrane heterodimeric glycoproteins consisting of one α and one β subunit forming at least 24 integrin-heterodimers known in humans. Many integrins require divalent cations (Ca2+, Mg2+ or Mn2+) for structural integrity and ligand binding as well as activation through cluster formation in order to be fully functional (Hynes 2002, Yamada and Even-Ram 2002, Luo and Springer 2006, Takada, Ye et al. 2007, Banno and Ginsberg 2008, Moser, Legate et al. 2009). Although many cell-matrix contacts are formed and enable the cell to respond to their immediate 2D- or 3D-environment, these contacts are transient and strongly regulated. After implantation of a TE-construct cell-matrix interactions are governed by surface properties of the scaffold material and their imperative interactions found to occur. In addition to these cell-matrix interactions several growth factors and biological molecules are also involved in cell adhesion, cell-cell communication and cell-matrix interaction. This very complex field was covered by recent reviews (e.g. Murugan and Ramakrishna 2007), and will therefore not be further discussed in this chapter

1.5 Cells

The basic bricks of living organisms, the cells, are a predominant factor for successful TE. Tissue renewal requires an adequate number of regeneration-competent cells that do not elicit immune response. Therefore autologous cells are the ideal choice, as their use circumvents many of the inflammatory and rejection issues associated with a nonself donor approach (Atala 2008). With the past decade, major advances in the expansion of a variety of primary human cells have been achieved. The different types of cells commonly used for TE applications can be categorized as differentiated non-stem or adult cells and as

undifferentiated stem cells. Each of these cell types brings along its own advantages and disadvantages.
In the field of bladder reconstruction it has been shown that adult urothelial and smooth muscle cells can be efficiently harvested from biopsy material, expanded extensively in culture and reimplanted into the same host (Oberpenning, Meng et al. 1999, Zhang, Ludwikowski et al. 2001, Atala 2001, Yoo, Park et al. 2000). Their differentiation characteristics, growth requirements and other biological properties have been widely studied (Southgate, Hutton et al. 1994, Turner, Subramaniam et al. 2008, Ludwikowski, Zhang et al. 1999). However, for elderly patients or patients with extensive end stage organ failure, a tissue biopsy may not yield enough normal cells for expansion and transplantation. Therefore adult progenitor cells may not be sufficient or appropriate for tissue engineering and transplantation in all cases (Heath 2000).
Although adult progenitors remain important for TE, the use of pluripotent stem cells has recently been recognized as a promising alternative source of cells from which the desired tissue can be derived. Stem cells, by definition, are immature or undifferentiated cells with the potential to go through numerous cycles of self-renewal, thus replenishing the pool of stem cells as well as to differentiate into more specialized, tissue-/organ-specific cells. Stem cells can be categorized into embryonic, fetal or adult cells according to their source of origination (NIH).
Embryonic stem cells exhibit two remarkable properties: the ability to proliferate in an undifferentiated, but still pluripotent state (self-renewal), and the ability to differentiate into a large number of specified cells (Brivanlou, Gage et al. 2003). As their name implies, embryonic stem cells are derived from the early stage embryo. Although embryonic stem cells research is thought to have much greater potential than adult stem cells, several ethical and legal controversies still exist concerning their use in humans. Furthermore, embryonic stem cells have been shown to transdifferentiate into a malignant phenotype, forming teratomas (Przyborski 2005, Yang, Lin et al. 2008).
Fetal stem cells derived form amniotic fluid and placentas have recently been described and represent a novel source of stem cells (De Coppi, Callegari et al. 2007). The principle stem cell type isolated from amniotic fluid and placentas is mesenchymal, but they have an expansion potential that is superior to that of adult stem cells. They are less immunogenic as they do not express human leukocyte antigen (HLA) and they do not form teratomas in vivo. Fetal stem cells are multipotent and they have been shown do differentiate into myogenic, adipogenic, osteogenic, nephrogenic, neural, and endothelial cells. In addition, the cells have a high replicative potential and could be stored for future use, without the risk of rejection and without ethical concerns (Fauza 2004).
Adult stem cells are basically undifferentiated cells found among differentiated cells in a tissue or organs. They are present in all adult tissues and are critical to tissue health, maintenance, and response to injury or disease throughout life. When compared with embryonic stem cells adult stem cells are more committed but still have the plasticity to differentiate into all three germ layers (Eberli and Atala 2006). However, they demonstrate considerable advantages, including; stable differentiation into specific cell lineages, no transdifferentiation into a malignant phenotype (teratomas), no requirement for the sacrifice of human embryos for their isolation and no or little immune rejection. Furthermore certain ethical and legal issues can also be conquered.

Today, pluripotent stem cells, or differentiable adult stem cells, can be harvested from many different tissues, including bone marrow (Angele, Kujat et al. 1999, Pittenger, Mackay et al. 1999), striated muscle (Bosch, Musgrave et al. 2000, Lee, Qu-Petersen et al. 2000), fat (Zuk, Zhu et al. 2001), skin (Toma, Akhavan et al. 2001), synovial membrane (De Bari, Dell'Accio et al. 2001), and, more recently, testicles (Guan, Nayernia et al. 2006, Kossack, Meneses et al. 2009). These cells can differentiate into committed cells of other tissues, a feature defined as plasticity. This would allow for engineering of composite tissues composed of multiple cell types using one single source of adult stem cells. Therefore, adult stem cells are particularly suitable for cellular therapy and for the engineering of tissues and organs.

Bladder reconstruction using stem cells seeded on a scaffold has recently been shown to be a promising alternative for bladder engineering (Chung, Krivorov et al. 2005),(Zhang, Lin et al. 2005, De Coppi, Callegari et al. 2007, Frimberger, Morales et al. 2006, Oottamasathien, Wang et al. 2007). Recent progress suggests that engineered tissues and cell based therapies using adult stem cells may have an expanded clinical applicability in the future and may represent a viable therapeutic option for those who require tissue replacement or repair.

The most intensively investigated adult stem cells are mesenchymal stem cells (MSCs). This cell type holds significant promise for the engineering of musculoskeletal structures. Bone marrow represents the major source of MSCs. Chung et al. performed studies in a rat model that examined the ability of MSCs to aid in the regeneration of bladder tissue on an acellular matrix scaffold (Chung, Krivorov et al. 2005). The cells were seeded onto an acellular matrix (small intestinal submucosa) and used in bladder augmentation studies. The number of smooth muscle containing bundles was dramatically increased in the seeded grafts versus the unseeded controls, suggesting that the mesenchymal stem cells (MSCs) have differentiated into smooth muscle cells (SMCs) and have contributed to the regeneration of the graft. In a similar study Zhang et al. described the isolation and expansion of bone marrow MSCs from dogs for use as an alternative cellular source for autologous bladder grafts using small intestinal submucosa (SIS) (Zhang, Lin et al. 2005). These authors demonstrated the ability of MSCs to differentiate into SMCs and provide a contractile force on collagen matrices in vitro. MSCs also enhanced the regenerative process when seeded onto SIS for augmentation cystoplasty by enhancing smooth muscle bundle formation (Zhang, Lin et al. 2005). However, in both studies, the cells were not labeled and therefore were indistinguishable from cells that may have migrated in from the surrounding normal tissue.

Another source of adult stem cells is fat tissue. Unlike bone marrow stem cells, which are difficult to isolate and relatively scarce, adipose stem cells (ASCs) are tremendously abundant and easily accessible (Jack, Zhang et al. 2009). Jack et al. demonstrated the feasibility of bladder tissue engineered from adipose stem cells. They showed that ASCs differentiated into bladder smooth muscle cells and showed contractile function in vivo. The vast availability of ASCs combined with their ease of procurement and ability to differentiate into contractile smooth muscle make them a competitive non-embryonic alternative for regeneration of the bladder and other smooth muscle tissues (Jack, Zhang et al. 2009).

Although stem cells are believed to be the key factor for the future of TE, one of the major challenges associated with the use of these cells is to provide appropriate cellular environment cues that regulate cell growth and subsequent tissue formation in a controlled and efficient manner (Murugan and Ramakrishna 2007). A deeper understanding of the complex interplay of stem cells and environment might allow for new strategies where stem cells actively participate in functional tissue and organ formation. Stem cells will differentiate into various cell types in situ depending on the requirements of the regenerating tissue, or they remain stem cells that generate progeny to maintain the tissue. In addition stem cells could be used to secrete factors that enhance cellular ingrowth, neo-vascularisation, and re-innervation.

2. Biomaterials

By providing a temporary supporting structure for growing cells, the scaffold is an important determining factor for the success of TE. Scaffold materials can be of natural or synthetic origin. The underlying principle for the design of scaffolds for TE is similar for different types of tissues. The scaffold preferably mimics the structure and biological functions of native ECM, both in terms of chemical composition and physical properties. Native ECM is a complex and dynamic environment filled with nano-features such as fibers displaying a certain pore size and interconnectivity that should ideally exhibit tissue-specific structures and properties. When naturally derived scaffolds are used they provide specific ligands for cell adhesion and migration, as well as various growth factors inducing specific cell proliferation and functions.

Currently, a variety of materials are available for manufacturing scaffolds for TE, including native and synthetic polymers and their composites. The choice of material depends on the type of tissue to be reconstructed. Most hollow organs are organized in a similar fashion, consisting of epithelium surrounded by a collagen type I-rich connective tissue and a smooth muscle layer. The epithelial or endothelial layer serves as a barrier preventing the content of the lumen from permeating into the body. The collagen type-I rich layer and the muscle layer maintain the structural and functional integrity of the organ. The cells within these layers interact with each other and with structural proteins to regulate cellular differentiation and function (Ziats, Miller et al. 1988, Bacakova, Filova et al. 2004).

The following characteristics are desirable for scaffolds used for TE in general.

The scaffold should

- be biocompatible, meaning that it should not provoke any rejection, inflammation, immune responses or foreign body reactions.
- provide a 3D template for the cells to attach and to guide their growth.
- have a porous architecture with a high surface area for the maximum loading of cells, cell-surface interaction, tissue ingrowth, and transportation of nutrients and oxygen.
- be degradable under physiological conditions and the degradation rate should match the rate of tissue regeneration to sustain tissue functionality.
- be mechanically strong to withstand in vivo biological forces.
- support the cells in synthesizing tissue specific extracellular matrix components and growth factors required for healthy tissue growth.
- be sterilizable to avoid toxic contaminations without compromising any structural and mechanical properties.

Finally, the production process of the scaffold with all the above unique characteristics must be accomplished in a reproducible, economical, and up-scalable manner (Murugan and Ramakrishna 2007).

In respect to bladder TE the ideal scaffold material should

• provide structural support for distinct cell layers, including an adequate surface for stable attachment of urothelial cells.
• give adequate biomechanical support to harbor a high density of smooth muscle cells on the exterior surface without inducing premature collapse of the hollow organ.
• serve as a barrier between luminal contents and the body cavity.
• support the formation of unidirectional muscle tissue in defined layers and allow for rapid innervation and vascularisation.

Since each of the different cell types favors different conditions for optimal growth and differentiation, tissue engineering using multiple cell types must take these factors into account.

Two main classes of biomaterials have been utilized for the engineering of hollow organs; acellular matrices derived from donor tissues, (e.g. bladder submucosa and small intestinal submucosa), and synthetic polymers such as polyglycolic acid (PGA), polylactic acid (PLA), and poly(lactic-co-glycolic acid) (PLGA). These materials have been tested in respect to their biocompatibility in the host tissues (Scriven, Trejdosiewicz et al. 2001; Pariente, Kim et al. 2002). Both types of material were able to support the formation of bladder like tissue. Acellular tissue matrices are extracted form native tissue and therefore contain growth factors, hormones and other signaling factors (Ziats, Miller et al. 1988, Brown, Brook-Allred et al. 2005, Chun, Lim et al. 2007) that promote tissue development and have adhesion domain sequences (e.g. RGD) that may support the phenotype and activity of many types of cells (Dawson, Goberdhan et al. 1996). These matrices are known to slowly degrade upon implantation and are usually replaced and remodeled by ECM proteins synthesized and secreted by transplanted or ingrowing cells (Daniels, Chang et al. 1990; Aharoni, Meiri et al. 1997; Ashammakhi and Rokkanen 1997; Talja, Valimaa et al. 1997; Hodde 2002; Santucci and Barber 2005; Daley, Peters et al. 2008; Mohamed and van der Walle 2008). In contrast, synthetic polymers can be manufactured reproducibly on a large scale with controlled properties of their strength, degradation rate and ultra structure (Hutmacher, Schantz et al. 2007; Ma, Mao et al. 2007). Both classes of biomaterials have been used either with or without cells for the tissue engineering of hollow organs, including bladder (Kropp, Rippy et al. 1996, Yoo, Meng et al. 1998, Oberpenning, Meng et al. 1999), urethra (Olsen 1992, Kropp, Ludlow et al. 1998), oesophagus (Urita, Komuro et al. 2007, Penkala and Kim 2007), intestine (Penkala and Kim 2007), vagina (De Filippo, Yoo et al. 2003), or blood vessels (Amiel, Komura et al. 2006, Lee, Choo et al. 2007).

2.1 Native Acellular Matrices

Native acellular matrices are pioneering materials and offer many potential advantages over synthetic scaffold materials (Southgate, Cross et al. 2003). These collagen-rich matrices are extracted from native tissue by mechanical or chemical decellularization (Chen, Yoo et al. 1999, Dahms, Piechota et al. 1998, Piechota, Dahms et al. 1998). They are either derived from bladder (bladder acellular matrix (BAM)) (Sutherland, Baskin et al. 1996) or from small intestine (SIS) (Kropp, Ludlow et al. 1998, Kropp, Rippy et al. 1996). The tensile backbone of the scaffold consists of fibrillar collagen type I, and the basement membrane, serving as a cyto- and tissue-compatible polymeric scaffold for recellularisation. One important advantage over synthetic materials is the fact that acellular matrices retain their biological activity. They provide specific integrin binding sites and contain endogenous growth factors encouraging the in-growth of tissue (Badylak 2004, Santucci and Barber 2005). Furthermore, given that the composition and structure of the ECM is unique to individual tissues, there may be advantages in orthotopic-derived matrices: BAM may be expected to contain more appropriate growth factors for bladder TE than SIS (Bolland, Korossis et al. 2007). Once implanted into the body, they slowly degrade supporting the ingrowth of host cells which then start to produce new ECM proteins.

In bladder reconstruction acellular matrices have been used either as a graft alone or seeded with urothelial and smooth muscle cells (Kropp, Rippy et al. 1996; Yoo, Meng et al. 1998; Oberpenning, Meng et al. 1999; Sievert, Bakircioglu et al. 2000; Sievert, Amend et al. 2007, Probst, Piechota et al. 2000, Zhang, Frimberger et al. 2006, Reddy, Barrieras et al. 2000).

Analysis of unseeded SIS patches after implantation in dogs showed replacement by normal bladder tissue, vascularisation and re-innervation (Kropp, Rippy et al. 1996). However, successful bladder regeneration using SIS appears to be dependent on the revascularization rate of the graft and the extent of the original bladder damage. SIS was not able to support functional tissue regeneration when used in animals with inflamed and contracted bladder remnants (Zhang, Frimberger et al. 2006).

Implanting BAM into the bladder of rats, rabbits, dogs and pigs resulted in the regeneration of urothelial and muscle layers with innervation and vascularisation of the graft (Probst, Piechota et al. 2000, Sievert, Bakircioglu et al. 2000, Reddy, Barrieras et al. 2000). Moreover, BAM was shown to release exogenous basic fibroblast growth factor (bFGF) in a rat model of bladder augmentation. bFGF is an important growth factor supporting tissue formation and reducing graft shrinkage (Kanematsu, Yamamoto et al. 2003).

However, problems with poor vascularisation, graft shrinkage and incomplete or disorganized smooth muscle development have been associated with the use of decellularised matrices (Zhang, Frimberger et al. 2006, Brown, Farhat et al. 2002, Kropp, Cheng et al. 2004). Graft shrinkage occurs due to the fast ingrowth of fibroblasts (Brown, Farhat et al. 2002). It seems to be a frequent finding of a scaffold when used as a graft for bladder or other hollow structures. Researchers agree that the larger the graft, the more pronounced the shrinkage due to the activity of smooth muscle actin-positive fibroblasts (Brown, Farhat et al. 2002, Kropp, Cheng et al. 2004). Omental coverage, endothelial cell seeding, or application of exogenous angiogenic growth factors were reported to allow ingrowth of capillaries to the graft (Kanematsu, Yamamoto et al. 2003, Baumert, Simon et al. 2007). However, establishment and maintenance of a permanent and sufficiently robust vascular supply to sustain a large graft for the human bladder remains to be demonstrated.

The above mentioned problems led to the concept of ex vivo seeding of autologous cells onto different scaffold materials, with the aim of enhancing tissue integration following implantation. This would minimize the inflammatory response toward the matrix, thus avoiding graft contracture and shrinkage. Yoo et al. showed that there was a major difference between BAM used with autologous cells and matrices used without cells (Yoo, Meng et al. 1998).

A major disadvantage of these systems is the routine variability in protein composition among the batches. There may also be ethical issues regarding their availability, although most naturally derived scaffolds are porcine xenografts.

2.2 Synthetic polymers

Historically the attempt to incorporate of synthetic materials alone into the bladder has mostly failed, primarily as a result of biological and mechanical incompatibilities. Amongst others polyvinyl sponges, silicone, polytetrafluoroethylene (Teflon) and resin-sprayed paper have been used to reconstruct the bladder with variable results, but none of the methods have been pursued to the present day (Kudish 1957, Bono and De Gresti 1966, Fujita 1978). Modern synthetic polymers such as PGA, PLA and PLGA are widely used in tissue engineering and have been applied in bladder reconstruction. These polymers received FDA approval for a variety of applications in human, including suture material. The ester bonds in these polymers are hydrolytically labile, thus allowing degradation by non enzymatic hydrolysis.

The degradation products of PGA, PLA, and PLGA are nontoxic natural metabolites and are eventually eliminated from the body in the form of carbon dioxide and water (Hutmacher 2000). The degradation rate of these polymers can be tailored from several weeks to several years by altering the crystalinity, initial molecular weight, and the copolymer ratio of lactic to glycolic acids. Since these polymers are thermoplastics, they can be easily formed into a 3D scaffold with a desired microstructure, gross shape, and dimension by various techniques, including molding, extrusion (Freed, Vunjak-Novakovic et al. 1994), solvent casting (Mikos, Lyman et al. 1994), phase separation techniques, gas foaming techniques (Harris, Kim et al. 1998) and electrospinning (Bini, Gao et al. 2004, Zong, Bien et al. 2005). Many applications in tissue engineering require a scaffold with high porosity and high ratio of surface area to volume. Other biodegradable synthetic polymers, including poly(anhydrides) and poly(ortho-esters) can also be used to fabricate scaffolds for tissue engineering with controlled properties (Peppas and Langer 1994).

Bladder-derived cells have been propagated on biodegradable synthetic scaffolds (Atala, Bauer et al. 2006, Oberpenning, Meng et al. 1999; Scriven, Trejdosiewicz et al. 2001; Danielsson, Ruault et al. 2006). Compared to natural materials, the advantage of producing a synthetic scaffold material is the full control over processing properties such as strength, biodegradability, microstructure and permeability, however, a fundamental feature of these materials is that they lack the natural signals that regulate cell attachment, growth and differentiation (Danielsson, Ruault et al. 2006, Vacanti and Langer 1999).

Atala and colleagues first demonstrated the feasibility of cells seeding onto a purely synthetic matrix for implantation in vivo (Oberpenning, Meng et al. 1999). PLGA is a well characterized biomaterial with predictable degradation properties, which is widely used as Vicryl® sutures and meshes. It is non-toxic and biocompatible with both urothelial and bladder smooth muscle cells (Pariente, Kim et al. 2002, Scriven, Trejdosiewicz et al. 2001). These qualities make PLGA an attractive candidate for combination with natural materials to form implantable constructs for bladder reconstruction. Oberpenning et al. also used PGA meshes, molded into the shape of a bladder and surface-coated with PLGA. The constructs were seeded with autologous smooth muscle cells on the outer and urothelial cells on the inner surfaces of the scaffold material (Oberpenning, Meng et al. 1999). After subtotal cystectomy in dogs the bladder constructs were then implanted onto the bladder base (trigone) and the neo-bladder was then coated with fibrin glue and surrounded with omentum. The animals were monitored for up to 11 months. There were no complications and at three months, the polymer had degraded. Functionally, the reconstructed bladders provided an adequate capacity with good compliance. Histologically and immunocytochemically, the bladders showed an adequate structural architecture, and phenotypically, the urothelium and muscle retained their program of normal differentiation.

As previously mentioned, synthetic biodegradable polymers lack the presence of extra-cellular matrix components, and therefore of cell adhesion sequences and signaling molecules. However, modification of the polymer scaffolds' chemistry and the method of manufacturing allow improvement of cell adherence potential, growth rate and phenotype regulation (Bisson, Hilborn et al. 2002, Kim, Nikolovski et al. 1999). It is essential that the structural features of the produced scaffolds resemble the natural ECM in order to provide tissue formation and promote rapid clinical translation. Recently, numerous investigations have explored the possibility of producing scaffolds similar to natural ECM (Yang, Murugan et al. 2005, Murugan and Ramakrishna 2006, Xu, Inai et al. 2004). These scaffolds possess a high surface area, high porosity, small pore size, and a low density, all of which are features essential for the improvement of cell adhesion, mandatory for cell migration, proliferation, and differentiation. Polymeric nanofibres matrices are among the most promising ECM-mimetic biomaterials because their physical structure is similar to that of fibrous proteins in native ECM. They are increasingly being used in TE and have advantages over traditional scaffolds due to increased surface-to-volume ratio, which is supposed to be advantageous for cell-scaffold interaction promoting cellular adhesion, proliferation, migration and function.

Additionally, providing a scaffold made of nanofibres may guide the growth of muscle cells in three dimensions. Attitude and orientation of these fibers are considered to be one of the important features of a functional tissue scaffold containing muscle cells. This leads to the concept of nano-fibrous scaffolds for tissue engineering applications. Further, fiber orientation of the scaffolds greatly influences cell orientation and phenotypic expression (Ma, Kotaki et al. 2005, Yang, Murugan et al. 2005). For instance, Xu et al. (Xu, Inai et al. 2004) have evaluated electrospun synthetic biomaterials (poly(l-lactid-co-ε-caprolactone), P(LLA-CL)) using smooth muscle cells. The diameter of the generated fibers was around 500nm with an aligned topography mimicking the circumferential orientation of cells and fibrils found in the medial layer of a native artery. The results show that the cells adhered and migrated along the axis of aligned scaffolds while expressing a spindle-like phenotype. The cytoskeleton organization inside these cells was also parallel to the orientation of the fibrous assembly. A study by Baker et al. showed that smooth muscle cells adapted a more natural organization when grown on electrospun polystyrene scaffolds as compared to collagen fibers in vivo (Baker, Atkin et al. 2006).

Therefore, engineering scaffolds while controlling the fiber orientation is essential for mimicking structural and functional aspects of the native ECM, controlling cell orientation and tissue growth. Currently, there are a number of methods available for manufacturing tissue scaffolds, which include electrospinning, self-assembly, phase separation, solvent-casting and particulate-leaching, freeze drying, melt molding, template synthesis, drawing, gas foaming, and solid-free forming (Murugan and Ramakrishna 2007). Among them, only electrospinning offers the capability to design nanofibrous scaffolds in the form of nonwoven structures that can meet the demands of scaffold-based tissue engineering applications.

2.3 Composite scaffold

Methods to improve cell attachment and proliferation on synthetic materials have already been explored. One approach is to coat a synthetic material with biological substances such as collagen, serum or to use surface modification procedures prior to cell seeding to encourage attachment. In vitro cultured SMCs for example have been shown to attach and proliferate extensively on a biodegradable polyesterurethane foam, which was pre-treated with fetal bovine serum (Danielsson, Ruault et al. 2006) as well as on plasma coated, electrospun polystyrene (Baker, Atkin et al. 2006).

An alternative approach is to combine different scaffold materials with diverse qualities. The so called composite scaffolds can be fabricated with two or more completely different polymer systems for engineering of hollow organs. Scaffolds designed for hollow organs require a special consideration of their barrier function between the cavity and the surrounding tissues while accommodating sufficient amounts of cells that facilitate tissue development. In recent reports a composite scaffold composed of synthetic PGA and a native acellular matrix (collagen) proved to be optimal for the engineering of bladder tissue combining the advantages of the different materials (Eberli, Freitas Filho et al. 2009).

Collagen hybrid matrices have been used with mixed results in vitro. In one study, PLGA mesh was combined with collagen and processed to become either a sponge or a gel (Nakanishi, Chen et al. 2003). Cultured porcine urothelial and bladder smooth muscle cells were seeded onto each of the constructs. Smooth muscle cells were able to proliferate and retain expression of differentiation markers when cultured on the on the gel-based construct, but not on the sponge. The opposite was the case for urothelial cells, which stratified on a sponge but not gel, although unequivocal immunohistochemical markers of differentiation were not tested on the urothelium (Nakanishi, Chen et al. 2003). More promising were the results of Eberli and colleges (Eberli, Freitas Filho et al. 2009), who used a composite scaffolding system of a native acellular collagen matrix bonded to PGA polymer meshes. The acellular matrix served as a barrier that would prevent the luminal content from permeating into the body cavity while providing an optimal surface for epithelial cell adherence and growth. The synthetic polymer layer with large pores was designed to accommodate sufficient numbers of muscle cells and maintained structural integrity of the scaffold at the same time. The study showed that this composite scaffold remained biocompatible, possessed ideal physical and structural characteristics for hollow organ applications and formed bladder tissue in vivo (Eberli, Freitas Filho et al. 2009). Composite scaffolds seem to be an ideal approach for the TE of hollow organs, meeting the demands for a biomaterial addressing the unique needs of the different cells used.

3. Vascularisation and Innervation

Although many studies demonstrated tissue formation similar to native bladder the functionality of these constructs has never been demonstrated. The two main issues limiting the constructs to be both contractile and capable of physiologic voiding are proper innervation and vascularisation of the tissue engineered construct.

Rapid neo-vascularisation is essential for graft survival, and complete restoration of the organ structure and functionality. The two main mechanisms forming new blood vessels are angiogenesis (proliferation and migration of endothelial cells from pre-existing vasculature) and vasculogenesis (formation of new vessels by in situ incorporation, differentiation, migration and/or proliferation of endothelial progenerator cells recruited from peripheral blood). Transplanted matrices relay on vascular ingrowth from the surrounding tissue to support previously seeded cells and promote migration of native cells onto the grafted region (Pope, Davis et al. 1997, Ko, Milthorpe et al. 2007).

The engineering of large organs will require a vascular network of arteries, veins, and capillaries to deliver sufficient nutrients and oxygen to each cell. One possible method to artificially induce re-vascularisation in engineered tissue might be through application of angiogenic agents such as vascular endothelial growth factors (VEGFs) or the implantation of vascular endothelial cells (EC). VEGF is a multifunctional growth factor that functions as an inducer of vascular permeability and endothelial cell specific mitogen (Ferrara and Davis-Smyth 1997). In addition to its angiogenic function, VEGF also functions as anti-apoptotic factor for smooth muscle (Yamanaka, Shirai et al. 2002) and endothelial cells (Gerber, Dixit et al. 1998). Skeletal myoblasts from adult rats were cultured and transduced with an adenovirus encoding $VEGF_{165}$. These cells were injected into a rat with ischemic cardiomyopathy. Neovascularization was assessed histologically four weeks after therapy. A significantly greater increase in vascular density was seen in these animals compared to the control animals treated with adenoviral $VEGF_{165}$ alone. These results indicate that a combination of VEGF and endothelial cells may be useful for inducing neo-vascularisation and volume preservation in engineered tissues (Askari, Unzek et al. 2004).

As graft shrinkage seems to be a natural process of a scaffold material for a hollow organ, enhancement of vascular supply to the graft has been conceived as a measure to sustain the viability of regenerated bladder. In bladder augmentation, omental coverage (Oberpenning, Meng et al. 1999), application of exogenous angiogenic growth factors (Kanematsu, Yamamoto et al. 2003, Kanematsu, Yamamoto et al. 2004, Nomi, Atala et al. 2002) and endothelial cell seeding (Schultheiss, Gabouev et al. 2005) were reported to allow ingrowth of capillaries to the graft, but may still be lacking the ability to provide a permanent and sufficiently robust vascular supply to sustain a large graft for the human bladder.

Innervation of the regenerated bladder tissue is mandatory for long term functional survival of the graft and to avoid secondary degeneration of the smooth muscle. Unfortunately, one of the major obstacles in engineering bladders for clinical use has been the lack of functional innervation.

The influence of different neurotrophic factors in neural development, survival, outgrowth and branching has been investigated by different research groups (McConnell, Dhar et al. 2004, Levenberg, Burdick et al. 2005, Sondell, Sundler et al. 2000). Mitsui et al. transplanted immortalized neural stem cells, neuronal and glial restricted precursors, or fibroblasts expressing neurotrophic factors to contused spinal cord, and reported improved bladder function (Mitsui, Shumsky et al. 2005). NGF is the first and best-characterised member of the neurotrophin family. NGF supports survival, outgrowth, and branching of sensory and autonomic neurons, but does not promote motor neuron regeneration (Kingham and Terenghi 2006). Gene therapy for peripheral nerve regeneration has been used by Sasaki et al. They injected nerve growth factor (NGF) to the bladder wall with a replication-defective adenovirus for the treatment of adult diabetic cystopathy and reported a markedly improved bladder function (Sasaki, Chancellor et al. 2004). This viral vector system has been shown to restore decreased NGF expression in the bladder. However, increased levels of NGF in bladder afferent neurons lead to hyperreflexia, which was significantly reduced when NGF levels were neutralized with anti-NGF antibodies (Seki, Sasaki et al. 2002). Moreover, high doses of NGF delay nerve regeneration by retarding GAP 43 (Hirata, Masaki et al. 2002). Therefore, it is equally important to consider optimal dose and release kinetics for the application of such therapeutic growth factors. Glial cell line-derived neurotrophic factor (GDNF) is a potent survival factor for motor neurons (Henderson,

Phillips et al. 1994). Variations in GDNF release influenced the rate of functional motor nerve recovery in rat primary peripheral nerve repair (Piquilloud, Christen et al. 2007). VEGF a potent angiogenetic factor also serves as neurotrophic factor for nerve regeneration (Sondell, Sundler et al. 2000).

Application of multiple growth factors rather than a single factor may hold great promise to support target organ innervation. The complex neural mechanism regulating bladder function includes various neural subpopulations, which are responsive for different neurotrophic factors. For example, lumbar dorsal root ganglion (DRG) neurons were found to express 65% Ret and 35% TrkA receptors for GDNF and NGF, respectively and 9% of receptors positive for both GDNF and NGF (Kashiba, Hyon et al. 1998). Madduri et al. demonstrated the synergistic effect of GNDF and NGF on axonal elongation and branching form DRG neurons (Madduri, Papaloizos et al. 2009).

NGF combined with VEGF enhanced regeneration of bladder acellular matrix grafts in spinal cord injury induced neurogenic rat bladders and protein gene product 9.5 (PGP) positive nerve fibers were observed most abundantly in the groups treated with combined factors rather than single factor treated groups (Kikuno, Kawamoto et al. 2009). However, the optimal combination of neurotrophic factors supporting bladder regeneration still remains unclear.

Axonal growth direction is well regulated by topographical features. Longitudinally aligned nanofibres guided the axons unidirectionally compared to random fibres, which showed axonal growth distributed in all directions (Corey, Lin et al. 2007).

4. Summery and Perspective

An ideal biomaterial for the engineering of functional bladder should be biocompatible and support tissue formation as well as provide adequate structural support to the neo-organ during tissue development. Many research groups were able to show tissue formation similar to native bladder. However, the functionality of these constructs has never been demonstrated.

The two main issues hampering the tissue engineered constructs to be contractile and allow physiologic voiding are proper innervation and vascularisation.

In near future, tissue engineered scaffolds with controlled topography and multiple neural and angiogenetic factors will provide a potential option to introduce proper biological function to the engineered artificial bladder.

5. Table

Natually derived scaffolds	Synthetic scaffolds
eg. Bladder submucosa	**eg. PGA**
+ Mainly collagen, naturally derived	+ Absorbable
+ Absorbable	+ High porosity (up to 95%)
+ Cell recognition sites	+ Low variability
+ Growth Factors	- Synthetic, no recognition sites
- Low elasticity	
- High variability	

Table 1. Comparison between the 2 main classes of biomaterials utilized for TE of hollow organs

Acknowledgment
The authors would like to thank Damina Balmer Dipl.rer.nat for her support and assistance with the preparation of the manuscript.

6. References

Aharoni, D., I. Meiri, et al. (1997). "Differential effect of components of the extracellular matrix on differentiation and apoptosis." Curr Biol **7**(1): 43-51.
Amiel, G. E., M. Komura, et al. (2006). "Engineering of blood vessels from acellular collagen matrices coated with human endothelial cells." Tissue Eng **12**(8): 2355-65.
Angele, P., R. Kujat, et al. (1999). "Engineering of osteochondral tissue with bone marrow mesenchymal progenitor cells in a derivatized hyaluronan-gelatin composite sponge." Tissue Eng **5**(6): 545-54.
Ashammakhi, N. and P. Rokkanen (1997). "Absorbable polyglycolide devices in trauma and bone surgery." Biomaterials **18**(1): 3-9.
Askari, A., S. Unzek, et al. (2004). "Cellular, but not direct, adenoviral delivery of vascular endothelial growth factor results in improved left ventricular function and neovascularization in dilated ischemic cardiomyopathy." J Am Coll Cardiol **43**(10): 1908-14.
Atala, A. (2001). "Bladder regeneration by tissue engineering." BJU Int **88**(7): 765-70.
Atala, A. (2008). "Bioengineered tissues for urogenital repair in children." Pediatr Res **63**(5): 569-75.
Atala, A., S. B. Bauer, et al. (2006). "Tissue-engineered autologous bladders for patients needing cystoplasty." Lancet **367**(9518): 1241-6.
Atala, A., L. G. Cima, et al. (1993). "Injectable alginate seeded with chondrocytes as a potential treatment for vesicoureteral reflux." J Urol **150**(2 Pt 2): 745-7.
Bacakova, L., E. Filova, et al. (2004). "Cell adhesion on artificial materials for tissue engineering." Physiol Res **53 Suppl 1**: S35-45.
Badylak, S. F. (2004). "Xenogeneic extracellular matrix as a scaffold for tissue reconstruction." Transpl Immunol **12**(3-4): 367-77.
Baker, S. C., N. Atkin, et al. (2006). "Characterisation of electrospun polystyrene scaffolds for three-dimensional in vitro biological studies." Biomaterials **27**(16): 3136-46.

Banno, A. and M. H. Ginsberg (2008). "Integrin activation." Biochem Soc Trans **36**(Pt 2): 229-34.

Baumert, H., P. Simon, et al. (2007). "Development of a seeded scaffold in the great omentum: feasibility of an in vivo bioreactor for bladder tissue engineering." Eur Urol **52**(3): 884-90.

Bent, A. E., R. T. Tutrone, et al. (2001). "Treatment of intrinsic sphincter deficiency using autologous ear chondrocytes as a bulking agent." Neurourol Urodyn **20**(2): 157-65.

Bini, T. B., S. Gao, et al. (2004). "Peripheral nerve regeneration by microbraided poly(L-lactide-co-glycolide) biodegradable polymer fibers." J Biomed Mater Res A **68**(2): 286-95.

Bisson, I., J. Hilborn, et al. (2002). "Human urothelial cells grown on collagen adsorbed to surface-modified polymers." Urology **60**(1): 176-80.

Bolland, F., S. Korossis, et al. (2007). "Development and characterisation of a full-thickness acellular porcine bladder matrix for tissue engineering." Biomaterials **28**(6): 1061-70.

Bono, A. V. and A. De Gresti (1966). "[Partial substitution of the bladder wall with teflon tissue. (Preliminary and experimental note on the impermeability and tolerance of the prosthesis)]." Minerva Urol **18**(2): 43-7.

Bosch, P., D. S. Musgrave, et al. (2000). "Osteoprogenitor cells within skeletal muscle." J Orthop Res **18**(6): 933-44.

Brittberg, M., A. Lindahl, et al. (1994). "Treatment of deep cartilage defects in the knee with autologous chondrocyte transplantation." N Engl J Med **331**(14): 889-95.

Brivanlou, A. H., F. H. Gage, et al. (2003). "Stem cells. Setting standards for human embryonic stem cells." Science **300**(5621): 913-6.

Brown, A. L., T. T. Brook-Allred, et al. (2005). "Bladder acellular matrix as a substrate for studying in vitro bladder smooth muscle-urothelial cell interactions." Biomaterials **26**(5): 529-43.

Brown, A. L., W. Farhat, et al. (2002). "22 week assessment of bladder acellular matrix as a bladder augmentation material in a porcine model." Biomaterials **23**(10): 2179-90.

Caldamone, A. A. and D. A. Diamond (2001). "Long-term results of the endoscopic correction of vesicoureteral reflux in children using autologous chondrocytes." J Urol **165**(6 Pt 2): 2224-7.

Chancellor, M. B., T. Yokoyama, et al. (2000). "Preliminary results of myoblast injection into the urethra and bladder wall: a possible method for the treatment of stress urinary incontinence and impaired detrusor contractility." Neurourol Urodyn **19**(3): 279-87.

Chen, F., J. J. Yoo, et al. (1999). "Acellular collagen matrix as a possible "off the shelf" biomaterial for urethral repair." Urology **54**(3): 407-10.

Cheresh, D. A. and D. G. Stupack (2008). "Regulation of angiogenesis: apoptotic cues from the ECM." Oncogene **27**(48): 6285-98.

Chun, S. Y., G. J. Lim, et al. (2007). "Identification and characterization of bioactive factors in bladder submucosa matrix." Biomaterials **28**(29): 4251-6.

Chung, S. Y., N. P. Krivorov, et al. (2005). "Bladder reconstitution with bone marrow derived stem cells seeded on small intestinal submucosa improves morphological and molecular composition." J Urol **174**(1): 353-9.

Corey, J. M., D. Y. Lin, et al. (2007). "Aligned electrospun nanofibers specify the direction of dorsal root ganglia neurite growth." J Biomed Mater Res A **83A**(3): 636-645.

Dahms, S. E., H. J. Piechota, et al. (1998). "Composition and biomechanical properties of the bladder acellular matrix graft: comparative analysis in rat, pig and human." Br J Urol **82**(3): 411-9.

Daley, W. P., S. B. Peters, et al. (2008). "Extracellular matrix dynamics in development and regenerative medicine." J Cell Sci **121**(Pt 3): 255-64.

Daniels, A. U., M. K. Chang, et al. (1990). "Mechanical properties of biodegradable polymers and composites proposed for internal fixation of bone." J Appl Biomater **1**(1): 57-78.

Danielsson, C., S. Ruault, et al. (2006). "Polyesterurethane foam scaffold for smooth muscle cell tissue engineering." Biomaterials **27**(8): 1410-5.

Dawson, R. A., N. J. Goberdhan, et al. (1996). "Influence of extracellular matrix proteins on human keratinocyte attachment, proliferation and transfer to a dermal wound model." Burns **22**(2): 93-100.

De Bari, C., F. Dell'Accio, et al. (2001). "Multipotent mesenchymal stem cells from adult human synovial membrane." Arthritis Rheum **44**(8): 1928-42.

De Coppi, P., A. Callegari, et al. (2007). "Amniotic fluid and bone marrow derived mesenchymal stem cells can be converted to smooth muscle cells in the cryo-injured rat bladder and prevent compensatory hypertrophy of surviving smooth muscle cells." J Urol **177**(1): 369-76.

De Filippo, R. E., J. J. Yoo, et al. (2003). "Engineering of vaginal tissue in vivo." Tissue Eng **9**(2): 301-6.

Eberli, D. and A. Atala (2006). "Tissue engineering using adult stem cells." Methods Enzymol **420**: 287-302.

Eberli, D., L. Freitas Filho, et al. (2009). "Composite scaffolds for the engineering of hollow organs and tissues." Methods **47**(2): 109-15.

Fauza, D. (2004). "Amniotic fluid and placental stem cells." Best Pract Res Clin Obstet Gynaecol **18**(6): 877-91.

Ferrara, N. and T. Davis-Smyth (1997). "The biology of vascular endothelial growth factor." Endocr Rev **18**(1): 4-25.

Freed, L. E., G. Vunjak-Novakovic, et al. (1994). "Biodegradable polymer scaffolds for tissue engineering." Biotechnology (N Y) **12**(7): 689-93.

Frimberger, D., N. Morales, et al. (2006). "Human embryoid body-derived stem cells in tissue engineering-enhanced migration in co-culture with bladder smooth muscle and urothelium." Urology **67**(6): 1298-303.

Fujita, K. (1978). "The use of resin-sprayed thin paper for urinary bladder regeneration." Invest Urol **15**(5): 355-7.

Furth, M. E., A. Atala, et al. (2007). "Smart biomaterials design for tissue engineering and regenerative medicine." Biomaterials **28**(34): 5068-73.

Gerber, H. P., V. Dixit, et al. (1998). "Vascular endothelial growth factor induces expression of the antiapoptotic proteins Bcl-2 and A1 in vascular endothelial cells." J Biol Chem **273**(21): 13313-6.

Ghosh, K. and D. E. Ingber (2007). "Micromechanical control of cell and tissue development: implications for tissue engineering." Adv Drug Deliv Rev **59**(13): 1306-18.

Guan, K., K. Nayernia, et al. (2006). "Pluripotency of spermatogonial stem cells from adult mouse testis." Nature **440**(7088): 1199-203.

Harris, L. D., B. S. Kim, et al. (1998). "Open pore biodegradable matrices formed with gas foaming." J Biomed Mater Res **42**(3): 396-402.

Heath, C. A. (2000). "Cells for tissue engineering." Trends Biotechnol **18**(1): 17-9.

Henderson, C. E., H. S. Phillips, et al. (1994). "GDNF: a potent survival factor for motoneurons present in peripheral nerve and muscle." Science **266**(5187): 1062-4.

Hirata, A., T. Masaki, et al. (2002). "Intrathecal administration of nerve growth factor delays GAP 43 expression and early phase regeneration of adult rat peripheral nerve." Brain Res **944**(1-2): 146-56.

Hodde, J. (2002). "Naturally occurring scaffolds for soft tissue repair and regeneration." Tissue Eng **8**(2): 295-308.

Hutmacher, D. W. (2000). "Scaffolds in tissue engineering bone and cartilage." Biomaterials **21**(24): 2529-43.

Hutmacher, D. W., J. T. Schantz, et al. (2007). "State of the art and future directions of scaffold-based bone engineering from a biomaterials perspective." J Tissue Eng Regen Med **1**(4): 245-60.

Hynes, R. O. (2002). "A reevaluation of integrins as regulators of angiogenesis." Nat Med **8**(9): 918-21.

Jack, G. S., R. Zhang, et al. (2009). "Urinary bladder smooth muscle engineered from adipose stem cells and a three dimensional synthetic composite." Biomaterials **30**(19): 3259-70.

Kanematsu, A., S. Yamamoto, et al. (2003). "Bladder regeneration by bladder acellular matrix combined with sustained release of exogenous growth factor." J Urol **170**(4 Pt 2): 1633-8.

Kanematsu, A., S. Yamamoto, et al. (2004). "Collagenous matrices as release carriers of exogenous growth factors." Biomaterials **25**(18): 4513-20.

Kashiba, H., B. Hyon, et al. (1998). "Glial cell line-derived neurotrophic factor and nerve growth factor receptor mRNAs are expressed in distinct subgroups of dorsal root ganglion neurons and are differentially regulated by peripheral axotomy in the rat." Neurosci Lett **252**(2): 107-10.

Kikuno, N., K. Kawamoto, et al. (2009). "Nerve growth factor combined with vascular endothelial growth factor enhances regeneration of bladder acellular matrix graft in spinal cord injury-induced neurogenic rat bladder." BJU Int **103**(10): 1424-8.

Kim, B. S., J. Nikolovski, et al. (1999). "Engineered smooth muscle tissues: regulating cell phenotype with the scaffold." Exp Cell Res **251**(2): 318-28.

Kingham, P. J. and G. Terenghi (2006). "Bioengineered nerve regeneration and muscle reinnervation." J Anat **209**(4): 511-26.

Ko, H. C., B. K. Milthorpe, et al. (2007). "Engineering thick tissues--the vascularisation problem." Eur Cell Mater **14**: 1-18; discussion 18-9.

Kossack, N., J. Meneses, et al. (2009). "Isolation and characterization of pluripotent human spermatogonial stem cell-derived cells." Stem Cells **27**(1): 138-49.

Kropp, B. P., E. Y. Cheng, et al. (2004). "Reliable and reproducible bladder regeneration using unseeded distal small intestinal submucosa." J Urol **172**(4 Pt 2): 1710-3.

Kropp, B. P., J. K. Ludlow, et al. (1998). "Rabbit urethral regeneration using small intestinal submucosa onlay grafts." Urology **52**(1): 138-42.

Kropp, B. P., M. K. Rippy, et al. (1996). "Regenerative urinary bladder augmentation using small intestinal submucosa: urodynamic and histopathologic assessment in long-term canine bladder augmentations." J Urol **155**(6): 2098-104.

Kudish, H. G. (1957). "The use of polyvinyl sponge for experimental cystoplasty." J Urol **78**(3): 232-5.

Lee, J. Y., J. E. Choo, et al. (2007). "Characterization of the surface immobilized synthetic heparin binding domain derived from human fibroblast growth factor-2 and its effect on osteoblast differentiation." J Biomed Mater Res A **83**(4): 970-9.

Lee, J. Y., Z. Qu-Petersen, et al. (2000). "Clonal isolation of muscle-derived cells capable of enhancing muscle regeneration and bone healing." J Cell Biol **150**(5): 1085-100.

Levenberg, S., J. A. Burdick, et al. (2005). "Neurotrophin-induced differentiation of human embryonic stem cells on three-dimensional polymeric scaffolds." Tissue Eng **11**(3-4): 506-12.

Ludwikowski, B., Y. Y. Zhang, et al. (1999). "The long-term culture of porcine urothelial cells and induction of urothelial stratification." BJU Int **84**(4): 507-14.

Luo, B. H. and T. A. Springer (2006). "Integrin structures and conformational signaling." Curr Opin Cell Biol **18**(5): 579-86.

Ma, Z., M. Kotaki, et al. (2005). "Potential of nanofiber matrix as tissue-engineering scaffolds." Tissue Eng **11**(1-2): 101-9.

Ma, Z., Z. Mao, et al. (2007). "Surface modification and property analysis of biomedical polymers used for tissue engineering." Colloids Surf B Biointerfaces **60**(2): 137-57.

Madduri, S., M. Papaloizos, et al. (2009). "Synergistic effect of GDNF and NGF on axonal branching and elongation in vitro." Neurosci Res.

McConnell, M. P., S. Dhar, et al. (2004). "In vivo induction and delivery of nerve growth factor, using HEK-293 cells." Tissue Eng **10**(9-10): 1492-501.

Metcalfe, A. D. and M. W. Ferguson (2007). "Bioengineering skin using mechanisms of regeneration and repair." Biomaterials **28**(34): 5100-13.

Mikos, A. G., M. D. Lyman, et al. (1994). "Wetting of poly(L-lactic acid) and poly(DL-lactic-co-glycolic acid) foams for tissue culture." Biomaterials **15**(1): 55-8.

Mitsui, T., J. S. Shumsky, et al. (2005). "Transplantation of neuronal and glial restricted precursors into contused spinal cord improves bladder and motor functions, decreases thermal hypersensitivity, and modifies intraspinal circuitry." J Neurosci **25**(42): 9624-36.

Mohamed, F. and C. F. van der Walle (2008). "Engineering biodegradable polyester particles with specific drug targeting and drug release properties." J Pharm Sci **97**(1): 71-87.

Moser, M., K. R. Legate, et al. (2009). "The tail of integrins, talin, and kindlins." Science **324**(5929): 895-9.

Murugan, R. and S. Ramakrishna (2006). "Nano-featured scaffolds for tissue engineering: a review of spinning methodologies." Tissue Eng **12**(3): 435-47.

Murugan, R. and S. Ramakrishna (2007). "Design strategies of tissue engineering scaffolds with controlled fiber orientation." Tissue Eng **13**(8): 1845-66.

Nakanishi, Y., G. Chen, et al. (2003). "Tissue-engineered urinary bladder wall using PLGA mesh-collagen hybrid scaffolds: a comparison study of collagen sponge and gel as a scaffold." J Pediatr Surg **38**(12): 1781-4.

Nomi, M., A. Atala, et al. (2002). "Principals of neovascularization for tissue engineering." Mol Aspects Med **23**(6): 463-83.

Oberpenning, F., J. Meng, et al. (1999). "De novo reconstitution of a functional mammalian urinary bladder by tissue engineering." Nat Biotechnol **17**(2): 149-55.

Olsen, L. B., S.; Busch, C.; Carlsten, J.; Eriksson, I. (1992). "Urethral reconstruction with a new synthetic absorbable device. An experimental study." Scand J Urol Nephrol **26**(4): 323-6.

Oottamasathien, S., Y. Wang, et al. (2007). "Directed differentiation of embryonic stem cells into bladder tissue." Dev Biol **304**(2): 556-66.

Pariente, J. L., B. S. Kim, et al. (2002). "In vitro biocompatibility evaluation of naturally derived and synthetic biomaterials using normal human bladder smooth muscle cells." J Urol **167**(4): 1867-71.

Penkala, R. A. and S. S. Kim (2007). "Gastrointestinal tissue engineering." Expert Rev Med Devices **4**(1): 65-72.

Peppas, N. A. and R. Langer (1994). "New challenges in biomaterials." Science **263**(5154): 1715-20.

Piechota, H. J., S. E. Dahms, et al. (1998). "In vitro functional properties of the rat bladder regenerated by the bladder acellular matrix graft." J Urol **159**(5): 1717-24.

Piquilloud, G., T. Christen, et al. (2007). "Variations in glial cell line-derived neurotrophic factor release from biodegradable nerve conduits modify the rate of functional motor recovery after rat primary nerve repairs." Eur J Neurosci **26**(5): 1109-17.

Pittenger, M. F., A. M. Mackay, et al. (1999). "Multilineage potential of adult human mesenchymal stem cells." Science **284**(5411): 143-7.

Pope, J. C. t., M. M. Davis, et al. (1997). "The ontogeny of canine small intestinal submucosa regenerated bladder." J Urol **158**(3 Pt 2): 1105-10.

Probst, M., H. J. Piechota, et al. (2000). "Homologous bladder augmentation in dog with the bladder acellular matrix graft." BJU Int **85**(3): 362-71.

Przyborski, S. A. (2005). "Differentiation of human embryonic stem cells after transplantation in immune-deficient mice." Stem Cells **23**(9): 1242-50.

Reddy, P. P., D. J. Barrieras, et al. (2000). "Regeneration of functional bladder substitutes using large segment acellular matrix allografts in a porcine model." J Urol **164**(3 Pt 2): 936-41.

Santucci, R. A. and T. D. Barber (2005). "Resorbable extracellular matrix grafts in urologic reconstruction." Int Braz J Urol **31**(3): 192-203.

Sasaki, K., M. B. Chancellor, et al. (2004). "Gene therapy using replication-defective herpes simplex virus vectors expressing nerve growth factor in a rat model of diabetic cystopathy." Diabetes **53**(10): 2723-30.

Schonherr, E. and H. J. Hausser (2000). "Extracellular matrix and cytokines: a functional unit." Dev Immunol **7**(2-4): 89-101.

Schultheiss, D., A. I. Gabouev, et al. (2005). "Biological vascularized matrix for bladder tissue engineering: matrix preparation, reseeding technique and short-term implantation in a porcine model." J Urol **173**(1): 276-80.

Scriven, S. D., L. K. Trejdosiewicz, et al. (2001). "Urothelial cell transplantation using biodegradable synthetic scaffolds." J Mater Sci Mater Med **12**(10-12): 991-6.

Seki, S., K. Sasaki, et al. (2002). "Immunoneutralization of nerve growth factor in lumbosacral spinal cord reduces bladder hyperreflexia in spinal cord injured rats." J Urol **168**(5): 2269-74.

Sievert, K. D., B. Amend, et al. (2007). "Tissue engineering for the lower urinary tract: a review of a state of the art approach." Eur Urol **52**(6): 1580-9.

Sievert, K. D., M. E. Bakircioglu, et al. (2000). "Homologous acellular matrix graft for urethral reconstruction in the rabbit: histological and functional evaluation." J Urol **163**(6): 1958-65.
Sondell, M., F. Sundler, et al. (2000). "Vascular endothelial growth factor is a neurotrophic factor which stimulates axonal outgrowth through the flk-1 receptor." Eur J Neurosci **12**(12): 4243-54.
Southgate, J., W. Cross, et al. (2003). "Bladder reconstruction--from cells to materials." Proc Inst Mech Eng H **217**(4): 311-6.
Southgate, J., K. A. Hutton, et al. (1994). "Normal human urothelial cells in vitro: proliferation and induction of stratification." Lab Invest **71**(4): 583-94.
Sutherland, R. S., L. S. Baskin, et al. (1996). "Regeneration of bladder urothelium, smooth muscle, blood vessels and nerves into an acellular tissue matrix." J Urol **156**(2 Pt 2): 571-7.
Takada, Y., X. Ye, et al. (2007). "The integrins." Genome Biol **8**(5): 215.
Talja, M., T. Valimaa, et al. (1997). "Bioabsorbable and biodegradable stents in urology." J Endourol **11**(6): 391-7.
Toma, J. G., M. Akhavan, et al. (2001). "Isolation of multipotent adult stem cells from the dermis of mammalian skin." Nat Cell Biol **3**(9): 778-84.
Turner, A. M., R. Subramaniam, et al. (2008). "Generation of a functional, differentiated porcine urothelial tissue in vitro." Eur Urol **54**(6): 1423-32.
Uebersax, L., H. P. Merkle, et al. (2009). "Biopolymer-based growth factor delivery for tissue repair: from natural concepts to engineered systems." Tissue Eng Part B Rev **15**(3): 263-89.
Urita, Y., H. Komuro, et al. (2007). "Regeneration of the esophagus using gastric acellular matrix: an experimental study in a rat model." Pediatr Surg Int **23**(1): 21-6.
Vacanti, J. P. and R. Langer (1999). "Tissue engineering: the design and fabrication of living replacement devices for surgical reconstruction and transplantation." Lancet **354 Suppl 1**: SI32-4.
Xu, C. Y., R. Inai, et al. (2004). "Aligned biodegradable nanofibrous structure: a potential scaffold for blood vessel engineering." Biomaterials **25**(5): 877-86.
Yamada, K. M. and S. Even-Ram (2002). "Integrin regulation of growth factor receptors." Nat Cell Biol **4**(4): E75-6.
Yamanaka, M., M. Shirai, et al. (2002). "Loss of anti-apoptotic genes in aging rat crura." J Urol **168**(5): 2296-300.
Yang, F., R. Murugan, et al. (2005). "Electrospinning of nano/micro scale poly(L-lactic acid) aligned fibers and their potential in neural tissue engineering." Biomaterials **26**(15): 2603-10.
Yang, S., G. Lin, et al. (2008). "Tumor progression of culture-adapted human embryonic stem cells during long-term culture." Genes Chromosomes Cancer **47**(8): 665-79.
Yoo, J. J., J. Meng, et al. (1998). "Bladder augmentation using allogenic bladder submucosa seeded with cells." Urology **51**(2): 221-5.
Yoo, J. J., H. J. Park, et al. (2000). "Tissue-engineering applications for phallic reconstruction." World J Urol **18**(1): 62-6.
Zhang, Y., D. Frimberger, et al. (2006). "Challenges in a larger bladder replacement with cell-seeded and unseeded small intestinal submucosa grafts in a subtotal cystectomy model." BJU Int **98**(5): 1100-5.

Zhang, Y., H. K. Lin, et al. (2005). "Growth of bone marrow stromal cells on small intestinal submucosa: an alternative cell source for tissue engineered bladder." BJU Int **96**(7): 1120-5.

Zhang, Y. Y., B. Ludwikowski, et al. (2001). "Expansion and long-term culture of differentiated normal rat urothelial cells in vitro." In Vitro Cell Dev Biol Anim **37**(7): 419-29.

Ziats, N. P., K. M. Miller, et al. (1988). "In vitro and in vivo interactions of cells with biomaterials." Biomaterials **9**(1): 5-13.

Zong, X., H. Bien, et al. (2005). "Electrospun fine-textured scaffolds for heart tissue constructs." Biomaterials **26**(26): 5330-8.

Zuk, P. A., M. Zhu, et al. (2001). "Multilineage cells from human adipose tissue: implications for cell-based therapies." Tissue Eng **7**(2): 211-28.

7

Biomaterial scaffold fabrication techniques for potential tissue engineering applications

B. Subia, J. Kundu and S. C. Kundu *
** kundu@hijli.iitkgp.ernet.in*
Department of Biotechnology, Indian Institute of Technology,
Kharagpur – 72 13 02,
India

1. Introduction

The dearth of availability of tissues and organs for transplantation as well as inconvenience associated with their transplantation such as donor site morbidity, immune rejection and pathogen transfer led to the emergence of the discipline of tissue engineering. Following its inception in the scientific community since the last two decades, the research and development in this emerging field of tissue engineering and regenerative medicine has progressed in a very rapid rate (Kretlow & Mikos, 2008). Tissue engineering is an interdisciplinary field that applies the principles of engineering and life sciences toward the development of biological substitutes that restore, maintain, or improve function of tissue or whole organ (Chen et al., 1997; Bhatia et al., 1999; Huang & Ingber, 2000; Chiu et al., 2000; 2003). Tissue engineering has emerged as a new method involving the combining of cells, scaffold, and bioactive agents to fabricate functional new tissue to replace damaged tissue (Flanagan et al., 2006 Chen et al., 2002, Solchaga et al., 2001).

A well-designed three-dimensional scaffold is one of the fundamental tools to guide tissue formation *in vitro* and *in vivo.* Frontiers areas in medicine is changing rapidly from utilizing synthetic implants and tissue grafts to a tissue engineering approach that uses degradable porous material scaffolds integrated with biological cells or molecules to regenerate tissues (Hollister et al., 2005). Therefore, the selection of scaffold is important to enable the cells to behave in the desired manner to generate tissues and organs of the desired shape and size.

Characteristics of scaffolds

(1) Biocompatibility, to avoid unwanted host tissue responses to the implant (Ma, & Langer, 1999).

(2) It should have the excellent surface chemistry to allow attachment, migration, proliferation, and differentiation of the cells (Mandal and Kundu, 2008b; 2009a; 2009b).

(3) Interconnected pores with proper pore size to support cell infiltration and vascularization (Hutmacher 2001; Karageorgiou & Kaplan, 2005).

(4) Controlled biodegradability to aid the formation of new tissue (Cima et al., 1991; Kweona et al., 2003).

(5) Adequate mechanical properties to maintain the structure and function immediately after implantation and during remodelling of the implants (Karande & Agrawal, 2008; Kim et al., 2000; Lee et al., 2001).
(6) It should acquire the sufficient mechanical properties to provide the better environment for cells, deliver inductive molecules or cells to the repair site and provide cues to control the structure and function of newly formed tissue (Hutmacher 2000; 2001).
(7) It should support the formation of ECM by promoting cellular functions, and have the ability to provide the bimolecular signals to the cells (Ito et al., 2003).
(8) Scaffolds used for tissue engineering should mimic in part the structure and biological function of the extracellular matrix (Ma, 2008; Li & Shi, 2007).
Many techniques have been developed to fabricate three-dimensional porous architectures to fill this role, such as by particle leaching (Ma & Langer, 1999; Lu et al., 2000), phase separation (Sachlos & Czernuszka, 2003; Smith et al., 2006) and textile technology. However, scaffolds fabricate by these techniques do not adequately mimic the structure of the natural extracellular matrix in terms of architecture, which may be one of the reasons for suboptimal outcome in generating functional tissues (Lavik, & Langer, 2004; Laurencin et al., 1999; Hutmacher 2001). Therefore, new designs and manufacture technologies are required to improve the function and architecture of scaffolds. This chapter discusses the different scaffold fabrication methodologies/techniques utilizing several synthetic and natural polymers, including solvent casting, particulate leaching, gas foaming, phase separation, electrospinning, porogen leaching, fiber mesh, fiber bonding, self assembly, rapid prototyping, melt molding, membrane lamination and freeze drying.

2. Fabrication techniques

In the body, cells and tissue are organized into three-dimensional architecture. To engineer these functional tissue and organs, scaffolds have to be fabricated by different methodology to facilitate the cell distribution and guide their growth into three-dimensional space. The main techniques for scaffolds fabrication are mentioned here.

2.1 Solvent casting

Solvent casting property for the scaffolds preparation is very simple, easy and inexpensive. It does not require any large equipment; it is totally based upon the evaporation of some solvent in order to form scaffolds by one of the two routes. One method is to dip the mold into polymeric solution and allow sufficient time to draw off the solution; as a result a layer of polymeric membrane is created. Other method is to add the polymeric solution into a mold and provide the sufficient time to evaporate the solvent that create a layer of polymeric membrane, which adhere to the mold (Mikos et al., 2004).
One of the main drawbacks of this technique, the toxic solvent denatures the protein and may affect other solvents. There is a possibility that the scaffolds designed by these techniques may also retain some of the toxicity. To overcome these problems scaffolds are fully dried by vacuum process to remove toxic solvent. However, this is very time consuming technique and to overcome these problems some researchers have combined it with particulate leaching techniques (Mikos et al., 1993a; 1993b; 1996) for the fabrication of scaffolds.

2.2 Particulate-leaching techniques

Particulate leaching is one of the popular techniques that are widely used to fabricate scaffolds for tissue engineering applications (Ma & Langer, 1999; Lu et al, 2000). Salt, wax or sugars known as porogens are used to create the pores or channels. Here salt is grounded into small particles and those particles that have desired size are poured into a mold and filled with the porogen. A polymer solution is then cast into the salt-filled mold. After the evaporation of the solvent, the salt crystals are leached away using water to form the pores of the scaffold.

The process is easy to carry out. The pore size can be controlled by controlling the amount of porogen added, the size and shape of the porogen (Plikk et al., 2009). The particulate leached scaffold possesses pore size (~500 μm), percentage porosity (around 94-95%) and desired crystallinity. The advantage of this method is the requirement of very less amount of polymer to fabricate the scaffold. However, certain critical variables such as pore shape and inter-pore openings are not controlled. To conquer these drawback new technologies, are being developed.

2.3 Gas foaming

Many of the fabrication techniques require use of organic solvents and high temperature. The residues that remains after completion of process can damage cells and nearby tissues. This may also denature the biologically active molecules incorporated within the scaffolds. The gas foaming scaffold fabrication techniques does not require the utilization of organic solvents and high temperature.

This technique uses high pressure carbon dioxide gas for the fabrication of highly porous scaffolds. The porosity and porous structure of the scaffolds depend upon the amount of gas dissolved in the polymer. This process involves exposing highly porous polymer with carbon dioxide at high pressure (800 psi) to saturate the polymer with gas (Sachlos & Czernuszka, 2003). Under this condition, dissolved carbon dioxide becomes unstable and will phase separates from the polymer. The carbon dioxide molecule becomes cluster to minimize the free energy; as a result pore nucleation is created. These pores cause the significant expansion of polymeric volume and decrease in polymeric density. A three-dimensional porous structure (scaffolds) is formed after completion of foaming process.

The porosity of the scaffolds is controlled by the use of porogens like sugar, salts and wax (Ikada, 2006). The polymer (eg, PLGA) that expands in foaming process fused together around the porogen to create a continuous polymeric matrix, and also entrap any other molecule which is present in the mixture. The mix polymer and porogen are exposed to high pressure until they have completed its saturation with carbon dioxide, followed by foaming process porogen is removed and a highly interconnected pore structure is formed (Huang & Mooney, 2005).

2.4 Phase separation

Phase separation technique for scaffolds designing requires temperature change that separates the polymeric solution in two phases, one having low polymer concentration (polymer lean phase) and other having the high polymer concentration (polymer rich phase). Polymer is dissolved in phenol or naphthalene, followed by dispersion of biologically active molecule in these solutions. By lowering the temperature liquid-liquid phase is separated and quenched to form a two phase solid and the solvent has removed by extraction, evaporation and sublimation (Mikos et al., 2004) to give porous scaffolds with bioactive molecules integrated in to that structure (Sachlos & Czernuszka, 2003; Hua et al., 2002).

An appropriate liquid-liquid phase separation is critical for the preparation of nanofibers and does not occur in all solvents that's why selection of solvent and phase separation temperature is crucial for the formation of nanofibers. When the condition are favourable, liquid-liquid phase separation produce three dimensional fibrous structure with nano scaled architecture similar to that of collagen type I, and used in various biomedical applications.

Advantage of the phase separation technique is that, it can easily combine with other fabrication technology (Particulate leaching) to design three dimensional structures with control pore morphology. It can also be combined with rapid prototyping to create nano fibrous scaffolds for tissue engineering applications (Smith et al., 2006).

2.5 Electrospinning

The electrospinning technique for the scaffolds designing utilizes the electrostatic force for the production of polymeric fiber ranging from nanoscale to microscale. This process is control by high intensity electric field between two electrodes having electric charges of opposite polarity. One electrode is placed in the polymer solution and other is placed in collector. Generally polymer solution is pumped as result in forming a drop of solution. Afterwards, electric field is generated, which intends to produce a force, due to this the droplets results to overcome the surface tension of the solution. A jet of polymer is ejected, which produces the fibers, same instant the solvent starts evaporating due to jet formation and continues after the nanofibers are deposited to collector.

More than 200 polymers are used for electrospinning like silk fibroin (Zarkoob et al., 2004; Sukigara et al., 2003; Jin et al., 2004), collagen (Mathews et al., 2002), chitosan (Ohkawa et al., 2004), gelatin (Ma et al., 2005) etc. In the field of tissue engineering electrospinning technique is applied for the preparation of nanofiber scaffold design. The process is very versatile in terms of use of polymers, non-invasive and does not require the use of coagulation chemistry or high temperature for fiber generation. Basically, in this process a high voltage is used to create an electrically charged jet of polymer solution or melt, which forms polymer fiber after drying or solidification (Reneker and Chun, 1996; Doshi and Reneker, 1995).

One of the main advantages of this technique is that it can produce the scaffold with main structural feature suitable for growth of the cell and subsequent tissue organization (Li & Tuan, 2009; Liang et al., 2007; Leong et al., 2008). It can produce the ultra fine fibers with special orientation, high aspect ratio, high surface area, and having control over pore geometry. These characteristics are favorable for better cellular growth for *in vitro* and *in vivo* because they directly influence the cell adhesion, cell expression, and transportation of oxygen, nutrients to the cells. This provides spatial environment for the growth of new tissue with appropriate physiological functions. Cell seeding is the main problem of electrospining technology. This is overcome by sacrificial biopolymer or cryospinning, which allows creating the hole of desired size in electrospun matrices (Baker et al., 2008; Leong et al., 2008).

The nano fibrous scaffolds are widely used in biomedical application for scaffolds preparation for tissue engineering (Ma et al., 2005; Yang et al., 2005), wound dressing (Kim et al., 2000), artificial blood vessels (Ma et al., 2005), protective clothing material (Lu and Ding , 2008), drug release membrane (Katti et al., 2004, Chew et al., 2005; Venugopal et al., 2008), nanotube material, chemical catalytic apparatus, bio-transplant material, and hydrogen storage tank for fuel cell (Cho et al., 2003).

The nanofibrous scaffolds are prepared for the biomedical application such as hydrophilicity, mechanical strength, biodegradability, biocompatibility, interaction of cells , which are controlled by chemical composition of the material (Madurantakam et al., 2009). On the basis of that by selecting and adjusting the combination of proper component ratio, property of electrospinning scaffolds can be customized with desired function.

2.6 Porogen leaching

Porogen leaching is one of the most common methods used for preparation of scaffolds with controlled porosity. The particulate leaching method is totally based upon the dispersion of porogen (salt, sugar and wax) either in liquid particulates or powdered materials (Hou et al., 2003; Lee et al., 2004, Nazarov et al., 2004 Vepari & Kaplan, 2007) by the process of evaporation, cross linking or other reaction liquid may be solidified. These porogens act as place holder for pore and interconnection of the pores in the actual scaffolds fabrication technique. Highly porous scaffold with porosity up to 93 % and pore diameter up to 500 micrometers can be prepared by using this technique (Mikos et al., 1993 a; b).

Main objective of this technique is the realization of bigger pore size and increase pore interconnectivity. Main advantage of this technique is its simplicity, versatility and easy to control the pore size and geometry. Pore geometry is control by the selection of the shape for specific porogen agent, where as pore size is control by sieving the porogen particle to the specific dimensional range (Mano et al., 2007). One of the main drawbacks of this technique is that it can only produce thin wafers or membrane up to 3 mm thick and very difficult to design the scaffolds with accurate pore inter-connectivity (Moore et al., 2004).

2.7 Fiber mesh

Fiber mesh technique for scaffold fabrication consists of individual fiber either woven or interweave into three dimensional pattern of variable pore size (Martins et al., 2009). PGA is the first biocompatible and biodegradable polymer to spun into the fiber and used as a synthetic suture thread. It is prepared by the deposition of polymer solution over a nonwoven mesh of another polymer followed by subsequent evaporation (Ikada, 2006).

Main advantage of this technique is to provide the large surface area for cell attachment and rapid diffusion of nutrient that is favourable for cell survival and growth (Chen et al., 2002). However one of the main drawbacks of this technique is lack of structural stability which can partly be overcome by hot drying of PLLA fiber to improve the structure orientation and crystallinity.

2.8 Fiber bonding

Fiber bonding technique for scaffold fabrication is developed by Mikos and his coworkers (Mikos et al., 1993a). Synthetic polymer (PLLA) was dissolved in chloroform followed by non-woven mesh of PGA fiber had added. Subsequently solvent was removed by evaporation as a result a composite material, which consists of non-bonded PGA fiber, embedded in PLLA matrix is formed (Chen et al., 2002). The scaffolds were fabricated by bonding a collagen matrix to PGA polymers with threaded collagen fiber stitches (Eberli et al., 2009). Fiber bonding occurs during post treatments at a temperature above the melting temperature of PGA. As a result PLLA matrix of the composite has removed by dissolving in methylene chloride agent (Sachlos and Czernuszka, 2003) utilizing the fact that PGA is insoluble in this solvent.

This process yields the scaffolds of PGA fiber that is bonded together by heat treatment. PGA mesh provides the high porosity and surface area to polymer mass ratio (Mooney et al., 1996). This provides the mechanical stability and allows the tissue ingrowths. One of the main advantages of using the fiber is its large surface area, which is suitable for scaffolds applications. Therefore, it provides the more surface area for cells attachment and sufficient space for the regeneration of extracellular matrix (Moroni et al., 2008)

2.9 Self assembly

Self assembly is the spontaneous organization of the molecule into well defines into an ordered structure required for specific function (Zhang, 2003). Self assembly of natural or synthetic molecule produced nanoscale fibers known as nanofibers.

Amphiphilic peptide sequence is a common method for the fabrication of 3D nanofibrous structure for tissue engineering. In aqueous solution the hydrophobic and hydrophilic domains within these peptides interact together with the help of weak non covalent bonds (Joshi et al., 2009; Zhang et al., 2006) (eg. Hydrogen bond, Van der Waals interactions, ionic bond and hydrophobic interaction) this produces distinct fast recovering hydrogel, with the hydrophobic interactions as the molecules come together.

Instead of peptides synthetic polymer nanofibers are also prepared by self assembly of diblock polymers (AxBy) when the two blocks separate from one another in bulk due to their incompatibility, the volume formation of A and B can be controlled to obtain B domain of cylindrical shape with nanoscale diameter that is embedded into matrix A. Polymeric dendrimers can also self-assemble into nanofibers (Liu et al., 1996; 1999).

The di- and tri-block peptide ampholites (PAs) are designed that are self-assembled into a rod-like architecture. So a new technique for the self-assembly of PAs into nanofibers by controlling pH and by engineering the peptide head group of the PAs is developed (Hartgerink et al., 2001).

According to Hartgerink et al., (2002) and Tambralli et al., (2009) the salient features for synthesis of the PA involve the following:

1. Phosphoserine residue incorporation to increase hydroxyapatite mineralization.
2. RGD (Arg-Gly-Asp) peptide incorporation to enhance integrin-mediated cell adhesion.
3. Four consecutive cystine residues incorporation forming inter-molecular disulfide bonds that polymerizes to provide improved structural stability.
4. Flexible linker region incorporation consisting of three glycine residues to provide flexibility to the head group.

By virtue of the modifications in the structure of the PA enables a variety of self-assemblies including layered and lamellar structures and by its reversibility properties provides flexibility to the system. Therefore, the self-assembly technique shows designing potential novel scaffolds for tissue engineering applications.

Self assembly shows several advantages over the electrospinning because it produces the much thinner nanofiber with very thin diameter (Ma, 2008). The fabricated nanofibers have amino acid residues that may be chemically modified by the addition of bioactive moieties. Other advantage of this technique is to avoid the use of organic solvent and reduce the cytotoxicity because it is carried out in aqueous salt solution or physiological media (Ma et al., 2005). Main disadvantage of this technique is its complicated and elaborated process.

2.10 Rapid prototyping (RP)

RP is also called as solid free-form technique. This technique is more advanced technique for scaffold fabrication. It is computer controlled fabrication technique. It can rapidly produce 3D object by using layer manufacturing method. RP technique generally comprises the design of scaffold model by using the computer added design (CAD) software, which is then expressed as a series of cross section (Lin et al., 2008, Woodfield et al., 2009). Corresponding to each cross section RP machine lays down a layer of material starting from the bottom and moving up a layer at a time to create the scaffolds. In typical example, image of bone defect in a patient can be taken and develop 3D CAD computer model. The computer then can reduced the model to slice or layers. The 3D objects are constructed layer by layer by using RP techniques such as fused deposition modeling (FDM), selective laser sintering (SLS), 3D printing (3D-P) or stereolithography.

Now a day RP is an efficient way for generating the scaffolds of desired property, other advantage of this technique is to produce the parts with highly reproducible architecture and compositional variations. RP has advantage over other fabrication techniques, it has ability to control matrix architecture (size, shape, inter connectivity, branching, geometry and orientation) yielding biomimetic structure, that varying in design and material composition. It has ability to control the mechanical property, biological effects and degradation kinetics of scaffolds (Kai et al., 2009; Hutmacher et al., 2000; 2001).

RP technique can easily be integrated with the imaging technique to produce the scaffolds that are customized in size and shape allowing the tissue engineered graft to be modified for particular applications or for individual patient. One of the main drawbacks of this technique is achieved low resolution by current systems and types of polymeric materials that are used for this technique

2.11 Melt molding

A large number of techniques are discussed for the fabrication of scaffolds. These scaffolds fabrication techniques are developed to control the pore interconnectivity and geometry, which are important for the exchange of nutrient/waste from pore to pore. Melt molding process involves the filling of teflon mould with PLGA powder and gelatin microspheres of specific diameter followed by heating the mould above the glass transition temperature of PLGA while applying pressure to the mixture (Thompson et al., 1995a, 1995 b). This action causes the PLGA particle to attach together. Once the mould is removed gelatin microspheres is dissolved by immersing the mixture into water and scaffolds are then dried. Scaffolds produced by this technique assume the shape of the mould. Melt molding process was modified to incorporate short fiber of hydroxyapatite (HA). Uniform distribution of HA fiber throughout the PLGA scaffolds could only be achieved by using the solvent casting technique to prepare the composite material of HA fiber, PLGA matrix and gelatin or salt porogen , which are used in melt molding process (Hou et al., 2003).

2.12 Membrane lamination

Membrane lamination is another SFF-like technique used for constructing three-dimensional biodegradable polymeric foam scaffolds with precise anatomical shapes. Membrane lamination is prepared by solvent casting and particle leaching and introducing peptide and proteins layer by layer during the fabrication process. The membranes with appropriate shape are soaked with solvent, and then stacked up in three-dimensional assemblies with continuous pore structure and morphology (Maquet & Jerome, 1997).The bulk properties of the final 3D scaffolds are identical to those of the individual membrane.

This method generates the porous 3D polymer foams with defined anatomical shape, since it is possible to use the computer assisted modeling to design the template with desired implant shape. The disadvantage of this technique is that layering of porous sheets, result in lesser pore interconnectivity (Hutmacher et al., 2000; 2001) and other disadvantage of this technique is that it is a time consuming process since only thin membrane can be used in this process.

2.13 Freeze drying

Freeze drying technique is use for the fabrication of porous scaffolds (Whang et al., 1995; Schoof et al., 2001). This technique is based upon the principle of sublimation. Polymer is first dissolved in a solvent to form a solution of desired concentration. The solution is frozen and solvent is removed by lyophilization under the high vacuum that fabricate the scaffold with high porosity and inter connectivity (Mandal & Kundu, 2009 a, b).This technique are applied to a number of different polymers including silk proteins (Vepari & Kaplan, 2007, Altman et al., 2003) PGA, PLLA, PLGA, PLGA/PPF blends. The pore size can be controlled by the freezing rate and pH; a fast freezing rate produces smaller pores. Controlled solidification in a single direction has been used to create a homogenous 3D-pore structure (Schoof et al., 2001)

Main advantage of this technique is that, it neither requires high temperature nor separate leaching step. The drawback of this technique is smaller pore size and long processing time (Boland et al., 2004). A schematic diagram for scaffold fabrication from silk protein by freeze drying technique is given as an example in figure 1 (Mandal & Kundu, 2008a; 2008b; 2009b; Kundu et al., 2008).

Methods	Merits	Demerits	References
Solvent casting/ particulate leaching	Control over Porosity, pore size and crystallinity	Limited mechanical property, residual solvents and porogen material	Ma, 2007; Xiang et al., 2006
Porogen leaching	Controlled over porosity and pore geometry	Inadequate pore size and pore interconnectivity	Mano et al., 2007
Gas foaming	Free of harsh organic solvents, control over porosity and pore size	Limited mechanical property, inadequate pore interconnectivity	Ikada., 2006
Self assembly	Control over porosity, pore size and fiber diameter	Expensive material, complex design parameters	Zhang et al., 2003; 2006
Electrospinning	Control over porosity, pore size and fiber diameter	Limited mechanical property, pore size decrease with fiber thickness	Liang et al., 2007
Phase separation	No decrease in the activity of the molecule	Difficult to control precisely scaffold morphology	Smith et al., 2006
Rapid prototyping	Excellent control over geometry, porosity, no supporting material required	Limited polymer type, highly expensive equipment	Hutmacher et al., 2000; 2001
Fiber mesh	Large surface area for cell attachment, rapid nutrient diffusion	Lack the structural stability	Chen et al., 2002
Fiber bonding	High surface to volume ratio, high porosity	Poor mechanical property, limited applications to other polymers	Mooney et al., 1996
Melt molding	Independent control over porosity and pore size	Required high temperature for non amorphous polymer	Thompson et al., 1995 a; b
Membrane lamination	Provide 3D matrix	Lack required mechanical strength, inadequate pore interconnectivity	Maquet & Jerome, 1997
Freeze drying	High temperature and separate leaching step not required	Small pore size and long processing time	Boland et al., 2004; Mandal & Kundu, 2008

Table 1. Merits and demerits of different fabrication techniques

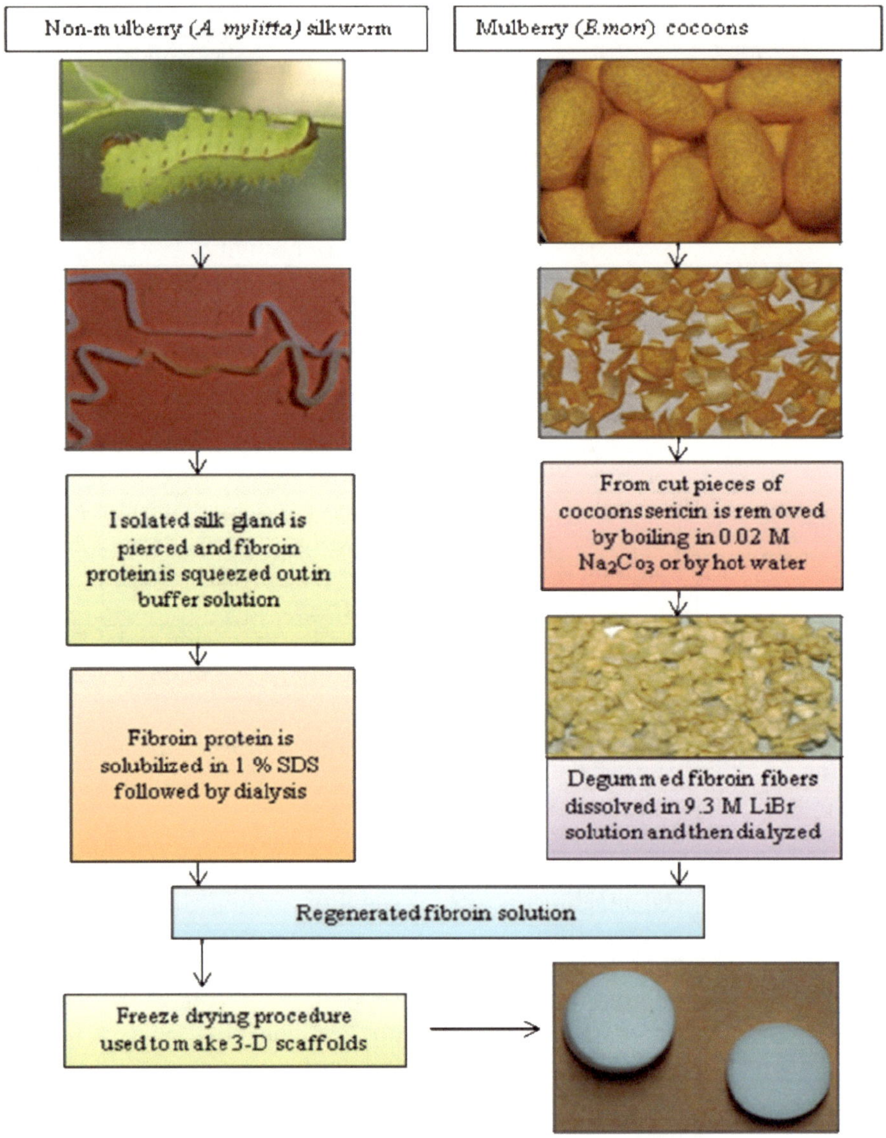

Fig. 1. Schematic diagram of scaffold fabrication from silk fibroin protein by freeze drying technique. The silk fibroin proteins may be obtained from two different silkworm sources either from the silk gland of non mulberry late 5th instar silkworm larva as an example from Indian tropical tasar silkworm (*Antheraea mylitta*). It is difficult to obtain sufficient amount of fibroin from cocoon sources of non mulberry (Mandal & Kundu, 2008a; 2008b; 2009 a) or from cocoons of mulberry silkworm, *Bombyx mori* (Altman et al., 2003; Mandal and Kundu, 2009b; Kundu et al., 2008).

3. Conclusion

Scaffolds are used to provide sites for cells attachment, proliferation, differentiation and migration by up regulating and down regulating the synthesis of protein and growth factors. They provide mechanical support, deliver inductive molecules or cells to the repair site. Scaffolds also provide cues to control the structure and function of newly formed tissue. Recently decellularized extracellular matrix has been suggested as a scaffold for heart valve tissue engineering or direct implantation. However, cell removal damages the physical and biochemical properties of the valve leaflet structure. Matrix /polymer hybrid scaffold with improved biomechanical characteristics may be advantageous for many tissue engineering aspects such as heart valve tissue engineering (Hao et al., 2008). Equilibrium is needed between highly porous scaffolds that allow rapid tissue ingrowths and minimize diffusion limitations and less porous materials that retain both construct shape and the ability to bear mechanical loads in a complex biochemical and mechanical environment.

The incorporation of the bioactive molecules like the growth factor, enzymes, ECM proteins, DNA is another vital consideration while designing the scaffolds. The fabrication methodologies that do not inactivate the biological activity of the bioactive molecule within the scaffolds need to be used judiciously. Scaffolds can be made by different types of fabrication techniques. The fabrication techniques provide the regular architectures; allow mechanical and finite element analyses to be carried out, which permit quantification of suitable mechanical and chemical microenvironments for cells and tissues assisting in the development of the next generation of tissue engineered scaffolds. The fabrication technique for the tissue engineered scaffold is directly related to the bulk and surface properties of the polymer and the proposed function of the scaffold. While each technique has its inherent merits and demerits, the appropriate selection of the methodologies must satisfy the requirements of the specific tissue to be repaired or engineered. Another hindrance in the tissue engineering research and scaffold design is the ignorance of the researchers on the possible interactions that exists between different cell types, such as the communication between the scaffold and the cells seeded within, cells and the different soluble factors, the typical behaviour of cells under the physical forces (shear, compression and vibration forces)

The different techniques are developed to improve the current scaffold design by controlling the regular pore size, pore interconnectivity, porosity and pore characterizations that is suitable for rapid nutrient diffusion and cell attachment. Table 1 shows the most commonly used scaffolds fabrication technologies for potential tissue engineering applications indicating their merits and demerits of each technique. Some references given below in this chapter allow interested readers to go through the detail of fabrication techniques. Recent advances in both computational topology design (CTD) and solid free-form fabrication (SFF) have made it possible to create scaffolds with controlled architecture (Hollister, 2005). Finally, future directions are utilization of modelled scaffolds with *in vivo* experimentation to optimize tissue-engineering treatments, and coupling scaffolds with cells, genes and proteins optimized to create smart scaffolds for tissue regeneration and controlling their delivery through the scaffold with spatial and temporal resolution.

4. Acknowledgements

We wish to thank Ms. Jasdeep Mann for taking interest during the initial stages of this write-up. We fail to accommodate/cite appropriate references of our research colleagues in right places due to limitation of space being book chapter. For this we regret but we are indebted to all of our scientific colleagues. Due to their work the writing of this book chapter has been possible. The authors wish to acknowledge financial support from Bioinformatics SUB-DIC programme, Department of Biotechnology, Govt. of India, and Indo-US Science and Technology Forum, New Delhi.

5. References

Altman, G. H.; Diaz, F.; Jakuba, C.; Calabro, T.; Horan, R. L.; Chen, J. Lu, H.; Richmond, J. & Kaplan, D. L. (2003). Silk based biomaterials. *Biomaterials,* 24, 401–416.

Baker, B. M.; Gee. A. O.; Metter, R. B.; Nathan, A. S.; Marklein, R. A.; Burdick, J. A. & Mauk, R. L., (2008). The potential to improve cell infiltration in composite fiber-aligned electrospun scaffolds by the selective removal of sacrificial fibers. *Biomaterials,* 29, 2348-2358.

Bhatia, S. N.; Balis, U. J.; Yarmush, M. L & Toner, M. (1999). Effect of cell-cell interaction in prevention of cellular phenotype: co-cultivation of hepatocytes and nanoparenchymal cells. *FASEB J.* 13, 1883-1900.

Boland, E. D.; Espy, P. G. & Bowlin, G. L. (2004). Tissue engineering scaffolds. In *Encyclopaedia of Biomaterials and biomedical engineering*. Wenk G. E.; Bowlin, G. L. (Edi). pp 1633-1635. Richmong, Verginia , USA.

Chen, C. S.; Mrksich, M.; Saung, S.; Whitesides, G. M. & Ingber, D. E. (1997). Geometric control of cell life and death. *J. Science* 276, 1425-1528.

Chen, G.; Ushida, T. & Tateishi, T. (2002). Development of biodegradable porous scaffolds for tissue engineering. *J. Mater Sci Eng C.* 17, 63–69.

Chiu, D. T.; Jeon, N. L.; Huang, S.; Kane, R. S.; Wargo, C. J.; Choi, I. S.; Ingber, D. E. & Whitesides, G. M. (2000). Patterned deposition of cells and protein into surface by using three dimensional microfluids system. Proc. *Natl. Acad. Sci.* USA, 97, 2408-2413.

Chiu, J. J.; Chen, L. J.; Lee, P. L.; Lee, C. I.; Lo, L. W.; Usami, S. & Chien, S. (2003). Shear stress inhibits adhesion molecule expression in vascular endothelial cells induced by co-culture with smooth muscles cells. *J. Blood,* 101, 2667-2674.

Chew, S. Y.; Wen, J.; Yim, E. K. & Leong, K. W. (2005). Sustained release of proteins from electrospun biodegradable fibers. *Biomacromolecules,* 6, 2017-2024.

Cho, K. Y.; Moon, J. Y.; Lee ,Y. W.; Lee, K. G.; Yeo, J. H.; Kweon, H. Y.; Kim, K . H. & Cho, S. C. (2003). Preparation of self-assembled silk sericin nanoparticles. *Int. J. Biol . Macromol.* 32, 36-42.

Cima, L. G.; Vacanti, J. P.; Vacanti, C.; Ingber, D.; Mooney, D. & Langer, R. (1991). Tissue engineering by cell transplantation using degradable polymer substrates. *J. Biomechanical Engg.* 113, 143–151.

Doshi, J. & Reneker, D. H. (1995). Electrospinning process and application of electrospun fibers. *J. Electrostat.* 35, 151-160.

Eberli, D.; Freitas, F. L.; Atala, A. & Yoo, J. J. (2009). Composite scaffolds for the engineering of hollow organs and tissues. *Methods,* 47(2), 109-115.

Flanagan, T. C.; Wilkins, B.; Black, A.; Jockenhoevel, S.; Smith, T. J. & Pandit, A. S. (2006). A collagen-glycosaminoglycan co-culture model for heart valve tissue engineering applications, *Biomaterials*, 27, 2233–2246.

Hao, H.; Nian, G.; Jia, S.; Chao, G.; Ping, H (2008) Fabrication of biomatrix/polymer hybrid scaffold for heart valve tissue engineering in vitro. ASAIO J 54(6), 627-632.

Hartgerink, J. D.; Beniash, E. & Stupp, S. I. (2002). Peptide-amphiphile nanofibers: a versatile scaffold for the preparation of self assembling materials. *Proc. Natl. Acad. Sci.* 99, 5133–8.

Hollister, S. J. (2005). Porous scaffold design for tissue engineering. Nature materials, 4, 518-524

Hou, Q. P.; Grijpma, D. W. & Feijen, J. (2003). Porous polymeric structures for tissue engineering prepared by a coagulation, compression moulding and salt leaching technique. *Biomaterials*, 24, 1937–1947.

Hua, F. J.; Kim, G. E.; Lee J. D.; Son, Y. K. & Lee, D. S. (2002). Macroporous poly (l- lactide) scaffold 1. Preparation of a macroporous scaffold by liquid-liquid phase separation of a PLLA- dioxane- water system. *J. Biomed. Mater.* Res. 63 61-167

Hutmacher, D. W. (2000). Scaffold in tissue engineering bone and cartilage. *Biomaterials*, 21, 2529-2543.

Hutmacher, D. W. (2001). Scaffold design and fabrication technologies for engineering tissues- state of the art and future perspectives. *J Biomat Sci. Polym. Ed.* 12, 107-124.

Huang, S. & Ingber D. E. (2000) Shape-dependent control of cell growth, differentiation and apoptosis: switching between attractors in cell regulatory network. Exp. *Cell. Res* 261, 91-103.

Huang, Y. C. & Mooney, D. J. (2005). Gas foaming to fabricate polymer scaffolds in tissue engineering. In*: Scaffoldings in tissue engineering*, Ma X P., Elisseeff J., (Ed.), PP 159,Taylor and Francis group. CRC press, 206, 61-72.

Ikada, Y. (2006). Scope of tissue engineering In: *Tissue engineering: fundamental and applications*, Ikada Y. (Ed.). PP 29, Academic press, USA.

Ito, A.; Mase, A.; Takizawa, Y.; Shinkai, M.; Honda, H.; Hata, K. I.; Ueda, M. & Kobayashi, T. (2003). Transglutaminase-mediated gelatin matrices incorporating cell adhesion factors as a biomaterial for tissue engineering, *J. Ortho* 95, 196–199.

Jin, H. J.; Chen, J.; Karageorgiou, V.; Altman, G.H. & Kaplan, D. L. (2004). Human bone marrow stromal cell responses to electrospun silk fibroin mats. *Biomaterials*, 25, 1039–1047.

Joshi, K. B.; Singh, P. & Verma, S. (2009). Fabrication of platinum nanopillers on peptide-based soft structure using a focused ion beam. *Biofabrication.* doi: 10.1088/1758-5082/1/2/025002.

Kai, H.; Wang, X.; Madhukar, K. S.; Qin, L.; Yan, Y.; Zhang, R.; Wang, X. (2009). Fabrication of two-level tumor bone repair biomaterial based on rapid prototyping technique. *Biofabrication*, doi: 10.1088/1758-5082/1/2/025003.

Karande, S. T. & Agrawal, M. C. (2008). Functions and requirement of synthetic scaffolds in tissue engineering. In: Nanotechnology and Tissue Engineering: The scaffolds. Laurencin, C. T. & Nair, L. S. (Ed.), pp 53, CRC Group, Taylor and Francis group.

Karageorgiou, V. & Kaplan, D. L. (2005). Porosity of 3D biomaterial scaffolds and osteogenesis. *Biomaterials,* 26(27), 5474-91.

Katti, D. S.; Robinson, K. W.; Ko F. K. & Laurencin, C. T. (2004). Bioresorbable nanofiber-based system for wound healing and drug delivery: optimization of the fabrication parameters. *J. Biomed. Mat. Res. Part B, App. Biomaterials*, 70B, 286–296

Kim, B. S.; Baez, C. E & Atala A. (2000). Biomaterials for tissue engineering, World J. Urol. 1, (18), 2–9.

Kretlow, J. D. & Mikos A. G. (2008). From Material to Tissue: Biomaterial Development, Scaffold Fabrication, and Tissue Engineering. *AIChE J.* 54(12), 3048-3067.

Kundu, S. C.; Dash, B. C.; Dash, R. & Kaplan, D. L. (2008). Natural protective glue protein sericin, bioengineered by silkworms: Potential for biomedical and biotechnological applications. *Prog. in Polym. Sci.* 33, 998-1012.

Kweona, H. Y.; Yoo ,M. K.; Park, I. K.; Kim, T. H.; Lee, H. C.; Lee, H-S.; Oh, J-S.; Akaike, T.; Cho, C. S. (2003). A novel degradable polycaprolactone networks for tissue engineering. *Biomaterials*, 24, 801–808.

Laurencin, C. T.; Ambrosio, A. M. A.; Borden, M. D. & Cooper, J. A. (1999). Tissue engineering: orthopedic applications, *Annu. Rev. Biomed. Eng.* 1, 19–46.

Lavik, E. & Langer, R. (2004). Tissue engineering: current state and perspective, *Appl. Microbiol. Biotechnol.* 65 (1), 1–8.

Lee, C. R.; Grodzinsky, A. J. & Spector, M. (2001). The effects of cross-linking of collagen-glycosaminoglycan scaffolds on compressive stiffness, chondrocyte-mediated contraction, proliferation, and biosynthesis. *Biomaterials*, 22, 3145–54.

Lee, S. H.; Kim, B. S.; Kim, S .H.; Kang, S. W. & Kim, Y. H. (2004) Thermally produced biodegradable scaffolds for cartilage tissue engineering. *Macromol. Biosci.* 4, 802–810.

Leong, M. F.; Rasheed, M. Z.; Lim, T. C. & Chian, K. S. (2008). Invitro cell infiltration and invivo cell in filtration and vascularization in fibrous highly porous poly (D,L-Lactic acid) scaffold fabrication by electrospining technique. *J. Biomed. Res.* A, 91, 231-240

Liang, D.; Hsiao, B. S. & Chu, B. (2007). Functional electrospun nanofibrous scaffolds for biomedical applications. *Adv. Drug Deliv. Rev.* 59, 1392–1412.

Lin, L.; Ju, S.; Cen, L.; Zhang, H. & Hu, Q. (2008). Fabrication of porous, TCP scaffolds by combination of rapid prototyping and freeze drying technology. Yi Peng, Xiaohong Weng (Eds.), pp. 88–91APCMBE 2008, IFMBE Proceedings 19, Springer-Verlag, Berlin Heidelberg.

Li, J. & Shi, R. (2007). Fabrication of patterned multi-walled poly-l-lactic acid conduits for nerve regeneration. J. *Neuroscience Methods,* 165, 257-264.

Li, W. J. & Tuan R. S. (2009). Fabrication and application of nanofibrous scaffolds in tissue engineering. *Curr. Protoc. Cell. Biol.* 25 Unit 25.2

Liu, G. J.; Qiao, L. J. & Guo, A. (1996). Diblock copolymer nanofibers. *Macromolecules*, 29, 5508-5510.

Liu, G. J.; Ding, J. F.; Qiao, L. J.; Guo, A.; Dymov, B. P.; Gleeson, J. T.; Hashimoto, T. & Saijo, K.(1999). Polystyrene block- poly (2-cinnamoylethyl methacrylate) nanofibers: Preparation, characterization, and liquid crystalline properties. *Chem. Eur. J.* 5, 2740-2749.

Lu, L.; Peter, S.; Lyman, M.; Lai, H.; Leite, S.; Tamada, J.; Uyama, S.; Vacanti, J.; Langer, R. & Mikos, A. (2000). *In vitro* and *in vivo* degradation of porous poly (DL-lactic-co-glycolic acid) foams. *Biomaterials*, 21, 1595-1605

Lu, P. & Ding, B. (2008) Applications of electrospun fibers. *Recent Pat Nanotechnol.* 2(3),169-82.

Ma, P. X. (2008). Biomimetic materials for tissue engineering. *Adv Drug Deliv Rev.* 60, 184– 98.

Ma, Z.; Kotaki, M.; Inai, R. & Ramakrishna, S. (2005). Potential of nanofiber matrix as tissue-engineering scaffolds, *Tissue Engineering*, 11, 101–109.

Ma, P. X. & Langer, R. (1999). Fabrication of biodegradable polymer foams for cell transplantation and tissue engineering. In Tissue Engineering Methods and Protocols, Morgan, J., and Yarmush, M. (eds.) Humana Press, NJ, 47

Madurantakam, P. A.; Cost, C. P.; Simpson, D.G.; Bowlin, G. L. (2009). Science of nanofibrous scaffold fabrication: strategies for next generation tissue-engineering scaffolds. *Nanomed.* 4(2), 193-206.

Mandal, B. B. & Kundu, S. C. (2008a). Non-bioengineered silk gland fibroin protein: characterization and evaluation of matrices for potential tissue engineering applications. *Biotechnology and Bioengineering*, 100 (6) 1237 - 50.

Mandal, B. B. & Kundu, S. C. (2008b). Non-bioengineered high strength three-dimensional gland fibroin scaffolds from tropical non-mulberry silkworm for potential tissue engineering applications. *Macromolecular Bioscience*, 8, 807-818.

Mandal, B. B. & Kundu, S. C. (2009a). Cell proliferation and migration in silk fibroin 3D scaffolds. Biomaterials, 30, 2956-2965.

Mandal, B. B. & Kundu, S. C. (2009b). Osteogenic and adipogenic differentiation of rat bone marrow cells on non-mulberry and mulberry silk gland fibroin 3D scaffolds. *Biomaterials*, 30, 5019–5030.

Mano, J .F.; Silva, G. A.; Azevedo, H. S.; Malafaya, P .B.; Sousa ,R .A.; Silva, S. S.; Boesel, L. F.; Oliviera, J .M.; Santos, T .C.; Marques, A. P.; Neves, N. M. & Reis. R. L. (2007). Natural origin biodegradable systems in tissue engineering and regenerative medicine: present status and some moving trends. J. R. *Soc. Interface*, 4, 999–1030.

Matthews, J. A.; Wnek, G. E.; Simpson, D. G. & Bowlin G. L. (2002). Electrospinning of collagen nanofibers. *Biomacromolecules* 3, 232-238.

Martins, A. M.; Pham, Q. P.; Malafaya, P. B.; Sousa, R. A,; Gomes, M. E.; Raphael, R. M.; Kasper F. K.; Reis R. L. & Mikos, A. G. (2009). The Role of Lipase and alpha-Amylase in the Degradation of Starch/Poly(varepsilon-Caprolactone) Fiber Meshes and the Osteogenic Differentiation of Cultured Marrow Stromal Cells. *Tissue Eng Part A*. 15(2), 295-305.

Maquet, V. & Jerome, R. (1997). Design of macroporous biodegradable polymer scaffolds for cell transplantations, *Mater Sci Forum*, 250, 15-24.

Mikos, A. G., Sarakinos, G., Leite, S. M., Vacanti, J. P. and Langer, R. (1993a). Laminated three-dimensional biodegradable foams for use in tissue engineering. *Biomaterials*, 14, 323-330.

Mikos, A. G., Bao, Y., Cima, L. G., Ingber, D. E., Vacanti, J. P. & Langer, R. (1993b). Preparation of Poly (glycolic acid) bonded fiber structures for cell attachment and transplantation. Journal of *Biomedical Materials Research*, 27, 183-189.

Mikos, A. G.; Sarakinos, G.; Vacanti, J. P.; Langar R. S. & Cima L. G (1996) Biocompatible polymer membranes and methods of preparation of three dimensional membrane structures. U.S. patent 5, 514, 378.

Mikos, A. G.; Lu, L.; Temenoff, J. S. & Temmser, J. K. (2004). Synthetic Bioresorbable polymer scaffolds. In: *An introduction to material in medicine*, Ratner B D, Hoffman A S, Schoen F J, Lemons J E, (Ed.). pp 743, Elsevier Academic Press. USA.

Mooney, D. J.; McNamara, K.; Hern, D.; Vacanti, J. P. & Langer, R. (1996). Stabilized polyglycolic acid fibre-based tubes for tissue engineering. *Biomaterials*, 17, 115–124.

Moore, M. J.; Jabbari, E.; Ritman, E .L.; Lu, L. C.; Currier, B. L.; Windebank, A. J. & Yaszemski, M. J. (2004). Quantitative analysis of interconnectivity of porous biodegradable scaffolds with micro computed tomography. J. *Biomed. Mater. Res.* Part A, 71, 258–267.

Moroni, L.; Hamann, D.; Paoluzzi, L.; Pieper, J.; de Wijn J. R. & van Blitterswijk C. A. Regenerating articular tissue by converging technologies. *PLoS One* 3(8) e3032.

Nazarov, R.; Jin, H. J. & Kaplan, D. L. (2004). Porous 3-D scaffolds from regenerated silk fibroin. *Biomacromolecules*, 5(3), 718-26.

Ohkawa, K.; Cha, D.; Kim, H.; Nishida, A. & Yamamoto, H. (2004). Electrospinning of Chitosan. *Macromolecular Rapid Communications*, 25, 1600-1605.

Plikk, P.; Målberg, S. & Albertsson A. C. (2009). Design of resorbable porous tubular copolyester scaffolds for use in nerve regeneration. *Biomacromolecules* , 10(5), 1259-64

Reneker, D. H. & Chun, I. (1996). Nanometer diameter fibres of polymer, produced by electrospinning *Nanotechnology*, 7, 216-223

Sachlos, E. & Czernuszka, J. T.(2003). Making tissue engineering scaffolds work. Review on the application of solid free from fabrication technology to the production of tissue engineering scaffolds. *European cells and materials*. 5, 29-40.

Schoof, H.; Apel, J.; Heschel, I. & Rau, G. (2001). Control of pore structure and size in freeze-dried collagen sponges. *J. Biomed Mater Res.* 58, 352–7.

Smith, L. A.; Beck, J. A. & Ma, P. X. (2006). Nano fibrous scaffolds and their biological effects. In: *Tissue Cell and Organ Engineering*, Kumar, C. (Ed.), pp 195, Wiley-VCH.

Solchaga, L. A.; Goldberg, V. M. & Caplan, A. I. (2001) Cartilage regeneration using principles of tissue engineering. *Clin. Orthop. Relat. Res.* 391, S, 161-170.

Sukigara, S.; Gandhi, M.; Ayutsede, J.; Micklus, M. & Ko, F. (2003). Regeneration of Bombyx mori silk by electrospinning—part 1: processing parameters and geometric properties. *Polymer*, 44, 5721–5727.

Tambralli, A.; Blakeney, B.; Anderson, J.; Kushwaha, M.; Andukuri, A.; Dean, D. & Jun, H. W.(2009). A hybrid biomimetic scaffold composed of electrospum Polycaprolactone nanofibers and self-assembled peptide amphiphile nanofibers. *Biofabrication* 1. 025001, 11.

Thompson, R. C.; Wake, M. C.; Yaszemski, & Mikos, A. G. (1995a). Biodegradable polymer scaffolds to regenerate organs. *Adv Polymer Sci* 122: 245-274.aton, FL. pp 173-195.

Thompson, R. C.; Yaszembksi, M. J.; Powers, J. M.; Harrigan, T. P. & Mikos, A. G (1995b). Poly (a-hydroxy ester)/short fiber hydroxyapatite composite foams for orthopaedic applications. In: Polymers in Medicine and Pharmacy, Vol 394. Mikos AG, Leong KW, Yaszemski MJ, (Ed.). pp. 25-30, *Materials Research Society Symposium* Proceedings, Pittsburgh USA.

Vepari, C. & Kaplan, D. L. (2007). Silk as a Biomaterial. *Prog. Polym. Sci.* 32(8-9), 991-1007.

Venugopal, J.; Prabhakaran, M. P.; Low, S.; Choon, A. T.; Zhang, Y. Z.; Deepika, G. & Ramakrishna, S. (2008). Nanotechnology for nanomedicine and delivery of drugs. *Curr Pharm Des.* 14(22), 2184-200.

Woodfield, T. B.; Guggenheim, M.; von Rechenberg, B.; Riesle, J.; van Blitterswijk, C. A. & Wedler, V. (2009). Rapid prototyping of anatomically shaped, tissue engineered implants for restoring congruent articulating surfaces in small joints. *Cell Prolif.* (4), 485-97

Xiang, Z.; Liao, R., Kelly, M. S. & Spector, M. (2006). Collagen–GAG scaffolds grafted onto myocardial infarcts in a rat model: a delivery vehicle for mesenchymal stem cells, *J. Tissue Eng.* 12, 2467–2478.

Yang, F.; Murugan, R.; Wang, S. & Ramakrishna, S. (2005). Electrospinning of nano/micro scale poly (L-lactic acid) aligned fibers and their potential in neural tissue engineering. Biomaterials, 26, 2603–2610,

Zarkoob, S.; Eby, R. K.; Reneker, D. H.; Hudson, S. D.; Ertley D. & Adams, W. W. (2004). Structure and morphology of electrospun silk nanofibers. *Polymer,* 45, 3973–3977.

Zhang, S. (2003). Fabrication of novel biomaterial through molecular self assembly. *J. Nat. Biotechnol.* 21, 1171-1178.

Zhang, S.; Zhao, X. & Spirio, L. (2006). Pura matrix: self assembling peptide nanofiber scaffolds. In "scaffolding in tissue Engineering "(P.X. Ma and J Elisseeff eds.) pp. 217-236. CRC press, Boca raton, FL.

8

Electrospun Functional Nanofibrous Scaffolds for Tissue Engineering

Xiaochuan Yang and Hongjun Wang
Stevens Institute of Technology
United States of America

1. Abstract

Electrospinning, a high-voltage driven spinning technique, has the ability to produce nanofibers with diameter down to nanometer scale from a variety of materials. Although it is predominantly applied to polymeric materials including synthetic and natural polymers, it also allows the functionalization of nanofibers by introducing pores, and incorporating functional elements such as drugs, growth factors, or nanoparticles while electrospinning. Anisotropic nanofibrous scaffolds with spatial orientation of the electrospun nanofibers can be achieved by using the electrode configurations or the rotated collection apparatus. Three dimensional cell-rich nanofibrous constructs can be fabricated while wet electrospinning using the layer-by-layer assembly approach. Taken together, the relatively high production rate and low cost of electrospinning have drawn a great attention in materials and life sciences, especially in tissue engineering research.

2. Introduction

Tissue engineering has proven to be a promising alternative therapy to replace the lost functions of diseased tissues or organs in clinical application (Langer & Vacanti, 1993) and can provide a well defined in vitro model for drug screening or tissue related studies (Griffith & Swartz, 2006). In tissue engineering, scaffolds with high porosity and interconnectivity are normally used to provide the cells with a temporary substrate to adhere, proliferate and form new tissues. It has been recognized that the initial cell-scaffold interaction plays a critical role in regulating the later cellular phenotypic expression. Many factors, such as the surface chemistry and topography of the scaffold are involved in the cell-scaffold interaction. In this regard, more and more efforts in scaffold fabrication are now made to incorporate many instructive external cues in the scaffolds, such as immobilization of cell adhesion domains like Arginine-Glycine-Aspartic acid (RGD) on the scaffold surface or inclusion of growth factors in the scaffolds for a guided tissue formation (Lutolf & Hubbell, 2005).

In order to maintain the proper cell phenotype similar to that in the native tissues, we have proposed the biomimetic design of scaffolds to maximally recapture the major features of the native extracellular matrix (ECM) at a multiscale level, from the composition,

morphology, topography, to spatial organization. In many cases, the cells in native tissues either are embedded in ECM fibres, such as the dermal fibroblasts in skin tissue, or reside on the top of basement membrane consisting of tightly cross-linked ECM fibres with pores, such as the endothelial cells on the luminal surface of blood vessels (Abrams et al., 2000). Clearly, nanofibrous scaffolds will be preferred due to their similarity to ECM fibrils in both dimension and morphology. Indeed, the advantages of nanofibrous scaffold in promoting cell growth and maintaining the proper cell phenotype have been demonstrated in a number of studies (Woo et al., 2007; Min et al., 2004; Ji et al., 2006; Chua et al., 2005), which is considered as the synergistic result of both nanotopography and chemical signalling of the scaffolds (Patel et al., 2007; Wang et al., 2003). Several approaches are available to fabricate nanofibers including self assembly (Zhang 2003), phase separation (Woo et al., 2007; Barnes et al., 2007) and electrospinning (Formhals 1934; Telemeco et al., 2005; Matthews et al., 2002; Michel et al., 1999; Pham et al., 2006). Compared to other techniques, electrospinning represents a simple yet effective approach for preparing non-woven nano/microsized fibres.
Electrospinning, a high voltage driven spinning technique, has been widely used to produce nanofibers due to its easiness for use (Pham et al 2006). The setup of electrospinning is rather simple, composed of a high-voltage power supply, solution reservoir with the spinneret and a grounded collection surface. By far, many polymers including synthetic polymer, natural polymer, and the blend of synthetic and natural polymers have been successfully electrospun into micro/nanosize fibres (Chua et al., 2005; Matthews et al., 2002; Li et al., 2005; Yoshimoto et al., 2003; Moroni et al., 2006; Geng et al 2005). Depending on the materials used and electrospinning conditions, nonwoven fibrous meshes with microscale variation, for instance, in pore size and fibre diameter (Pham et al., 2006), have been successfully produced. However, it remains highly active to fabricate 3D multifunctional scaffolds with electrospun nanofibers to maximally match the native cell growing extracellular matrix. In particular, many in vivo tissues and organs exhibit hierarchical layered structures with distinct ECM composition and arrangement in each layer. These anisotropic properties not only offer the cells with distinct signalling information, but also provide the tissues with a unique mechanical performance to accommodate the physiological requirement. Taking as an example, three distinctive layers, i.e., ventricularis, spongiosa, and fibrosa are clearly recognized in heart valve (Schoen & Levy 1999). Densely packed collagen fibres are found in the fibrosa layer to provide the predominant strength and stiffness, and prevent excessive stretching of valve. Spongiosa, composed of loosely arranged collagen and abundant hydrophilic glycosaminoglycans (GAG), lubricates the relative movement between the ventricularis and fibrosa layers. Ventricularis is rich with elastin and shows radial aligned elastic fibres to enable the recoil and stretch of valve in response to the diastole and systole of heart. Therefore, it is highly desirable to incorporate these hierarchical features into our multiscale design of elaborate scaffolds using the electrospun nanofibers. Additionally, the spatial organization of various types of cells in the hierarchically layered tissues is often recognized, which offers the tissue with specialized function for each layer. For instance, two distinct cell layers are identified in the skin, where the epidermal layer (mainly keratinocytes) stays at the outmost region of the skin and is connected to the underneath dermal layer (mainly fibroblasts) via basement membrane. In this regard, it is also preferred for the scaffolds to help the organization of cells with a distinct and controllable spatial distribution. Electrospinning holds a great promise in

allowing the bottom-up assembly of various types of cells into layered nanofiber/cells constructs.

Synergistic regulation of cell behaviour by growth factors is critical for functional tissue formation (Frenz et al 1994; Wei et al., 2007). Supplement of soluble bioactive molecules in the culture medium can deliver the stimulation to the cells; however, it remains challenging to separately deliver the distinct stimulation to different types of cells in the same culture. In general, the inclusion of bioactive molecules directly into the scaffolds is considered as a practical solution to the aforementioned matter (Corden et al., 2000). Therefore, in the multiscale design of scaffolds to support functional tissue formation, it is critical to combine the hierarchical spatial arrangement nanofibers along with the incorporation of various bioactive molecules in the fibres.

In this chapter, we would like to present the feasibility of fabricating multifunctional nanofibrous scaffolds appropriate for engineering hierarchical tissue construct using electrospun nanofibers, which is a concise summary of the papers recently published from our laboratory (Yang et al., 2008 & 2009a and b) and non-published results. In these papers, polycaprolactone (PCL), a biodegradable and biocompatible synthetic polymer (Kweon et al., 2003; Ball et al., 2004), was used as the base material to explore all the possibilities. Bioactive nanofibers containing either collagen type I, bovine serum albumin (BSA) or fibronectin (FN) were obtained by electrospinning of the PCL mixture. Uniform distribution and bioactivity of these molecules in nanofibers were observed. The release of small molecules from nanofibers was controlled by the fibre diameter and initial loading amount. Nanofiber meshes with various topographical patterns resulted in various cellular responses. Multilayered 3D scaffolds with distinct distribution of bioactive molecules in between were successfully fabricated and demonstrated using PCL/Collagen and PCL/BSA nanofibers. More importantly, the cells of the same type or of different types could be assembled together with nanofibers to form 3D cell-rich constructs with either uniform cell distribution or distinct spatial arrangement.

3. Electrospinning and characterization

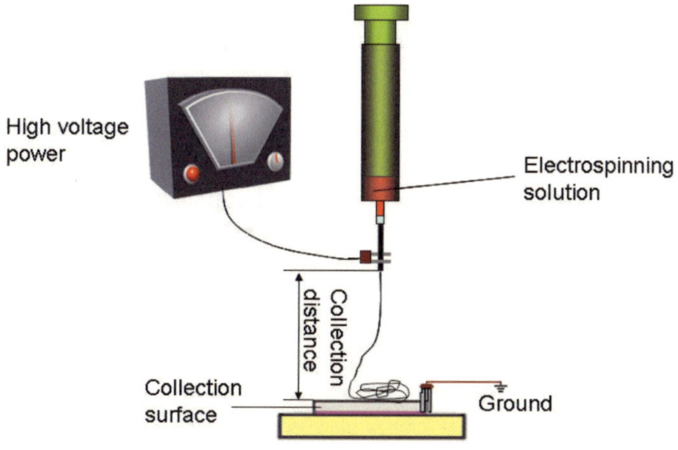

Fig. 1. Illustration of the electrospinning setup with vertical collection of fibers. Copyright 2008 Taylor & Francis Group

The typical electrospinning setup is shown in Figure 1. Using this setup, collagen-containing PCL nanofibers were successfully obtained by electrospinning the collagen-PCL blended solution (1:1 volume ratio) as reported previously (Yang et al., 2008). A common solvent of 1,1,1,3,3,3-hexafluoro isopropanol (HFIP) was used to dissolve both type I collagen and PCL solution for a well-mixed electrospinning solution in this case. The electrospinning solution was transferred to a syringe with a tip-blunt capillary (inner diameter = 0.9mm). The solution pushed out of the capillary tip by syringe pump was continuously pulled and formed the ultrathin filament under a strong electric field as shown in Figure 2A. Due to the small size of electrospun nanofibers, scanning electron microscope (SEM) was used to examine the fibres collected on Si wafer. The average diameter of electrospun fibres was determined using randomly selected SEM images. Upon optimization, the following electrospinning conditions could be used to prepare the PCL-collagen nanofibers, i.e., a flow rate between 5 µL/min and 15 µL/min, electric field strength from 0.8 kV/cm to 2.0 kV/cm and the fibre collection distance between 7 cm and 10 cm. The average diameter of PCL-collagen nanofibers was 454.5 ± 84.9 nm, and meanwhile, the surface of electrospun fibres was smooth by close examination with a SEM (Figure 2B). In addition, the pore size of the PCL/collagen nanofiber meshes that was randomly collected on Si wafer for 1 min, was noticeably below 5 µm (Figure 2A). This small interfiber distance could limit the penetration of cells to the interior of the fibrous meshes and led to the tissue formation on the peripheral surface of the meshes (Powell & Boyce, 2008).

4. Cell response to electrospun nanofibers

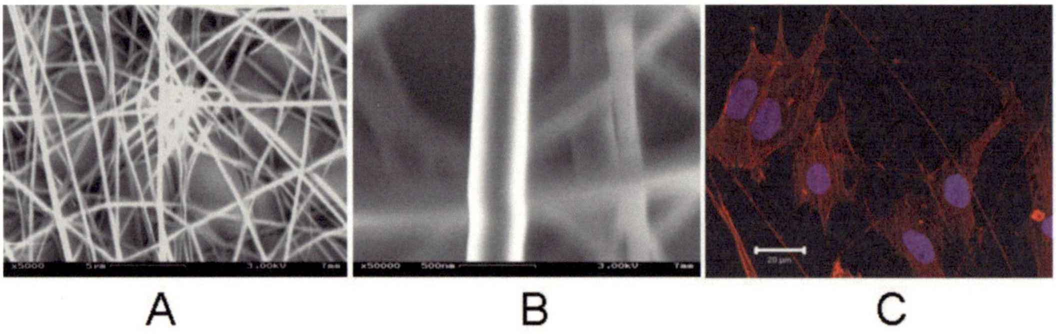

Fig. 2. Growth of neonatal human dermal fibroblasts on electrospun nanofibers. SEM image of 1:1 (w/w) PCL/collagen fibers at a low magnification (A) and high magnification (B). Confocal microscopy images of cells cultured on PCL-collagen fiber meshes (C). Cells were stained with TRITC-phalloidin for F-Actin (red) and with DAPI for cell nuclei (blue). Scale bar: 20 µm. Copyright 2008 Mary Ann Liebert Publishers

The inclusion of collagen into PCL nanofibers was considered favourable to cell attachment. In this regard, human neonatal dermal fibroblasts (HDF, passage 4-6) were used to test the cellular response to the electrospun PCL-collagen nanofibers. In this experiment, 0.5 mL of HDF cell suspension (2×10^4 cells/mL) was seeded onto the electrospun nanofiber meshes collected on glass coverslips. After 24-hour culture at 37 °C in a humidified incubator containing 5% CO_2, the culture was fixed in 4% formaldehyde/PBS fixative for 4 hours at room temperature and the cells were then stained with TRITC conjugated phalloidin (cytoskeleton) and DAPI (cell nuclei). The stained cells were examined at ex360nm/em460nm (DAPI) and ex540nm/em570nm (TRITC) using a confocal microscope. Staining for cytoskeleton protein, F-actin, showed that human dermal fibroblast spread nicely on the surface of PCL/collagen fibrous meshes with a spindle-like morphology (Figure. 2C). In contrast, cells on PCL fibrous meshes remained limited spreading within the investigated time (data not shown), despite that prolonged culture could improve the cell stretch.

The biological superiority of the inclusion of collagen into electrospun nanofibers was demonstrated by the improvement of fibroblast adhesion and rapid spreading with spindle-like cell morphology. The selective attachment of dermal fibroblasts onto PCL-Collagen nanofibers instead of PCL nanofibers demonstrated the advantage of collagen as it can promote the cell-nanofibers interaction via cell membrane integrin receptors such as $\alpha 2\beta 1$ (Tulla et al., 2001; Nykvist et al., 2000; Orr et al,. 2006). Although pure collagen can be electrospun into nanofibers (Nykvist et al., 2000; Matthews et al., 2002) and proved favorable for cell attachment (Shih et al., 2006; Rho et al., 2006), its rapid degradation leads to mechanical instability (Wang et al., 2005). The composite PCL/collagen nanofibers are considered better to mimic the native ECM in both topology and composition and have a good with mechanical stability.

5. Functional electrospun nanofibers with the potential to incorporate drug and growth factor

It is always desirable to locally release drug or growth factors from the scaffold to the attached cells. To demonstrate the possibility of incorporating drugs in electrospun nanofibers, bovine serum albumin (BSA), a model protein for drug or growth factors, was successfully incorporated into the PCL nanofibers by electrospinning the PCL solution containing different amount of BSA (BSA:PCL =1:1, 1:60 to 1:600). The collected nanofiber meshes were cut into discs (1.3 cm in diameter, n=3) and placed in a 24-well plate with 1ml phosphate buffer solution (PBS) for release study. The well plate was incubated at 37°C under gently shaking (80 RPM). Samples (10 μL) were taken from the supernatant at 1 hour intervals for the first 24 hours and 24 hour intervals afterwards in the cumulative release. In continuous release study, supernatant was replaced with fresh PBS at each harvesting time point.

Uniform distribution of BSA in the electrospun fibres was confirmed by using FITC-labelled BSA at a PCL/BSA ratio of 60:1 (Figure. 3A). A rapid release of BSA from PCL nanofibers was observed for those with 1:1 BSA/PCL ratio in the continuous release experiment. Majority of BSA (about 90%) was released within the first 3 hours in 1:1 BSA/PCL nanofibers, and the release of BSA was still detectable even after 24 hours (Figure. 3B). In the cumulative release experiment, BSA release reached its plateau at about 8 hours and remained the same for rest of the experimental time (Figure. 3B). The amount of BSA released into supernatant was proportional to the initial amount loaded into the PCL fibres (Figure. 3C). When the BSA/PCL ratio decreased down to 1:60, the peak release started at approximately 24 hour in the continuous release experiment and lasted more than 6 days (Figure. 3C).

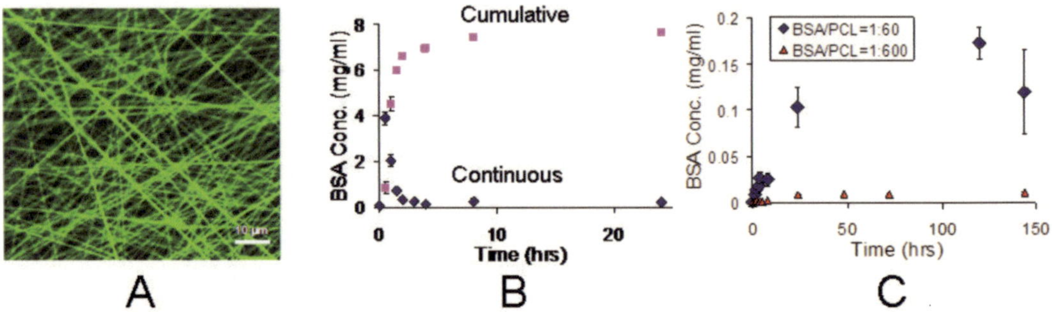

Fig. 3.The release profile of bovine serum albumin (BSA) from electrospun PCL fibers. (A) Fluorescence image of the PCL/FITC-BSA. Scale bar: 10 μm. (B) BSA release from PCL/BSA (1:1, w/w) nanofibers into PBS, measured by Lowry assay. (C) BSA released into PBS from PCL nanofibers with 1:60 or 1:600 FITC-BSA/PCL ratios, determined by fluorescent intensity. Copyright 2008 Taylor & Francis Group

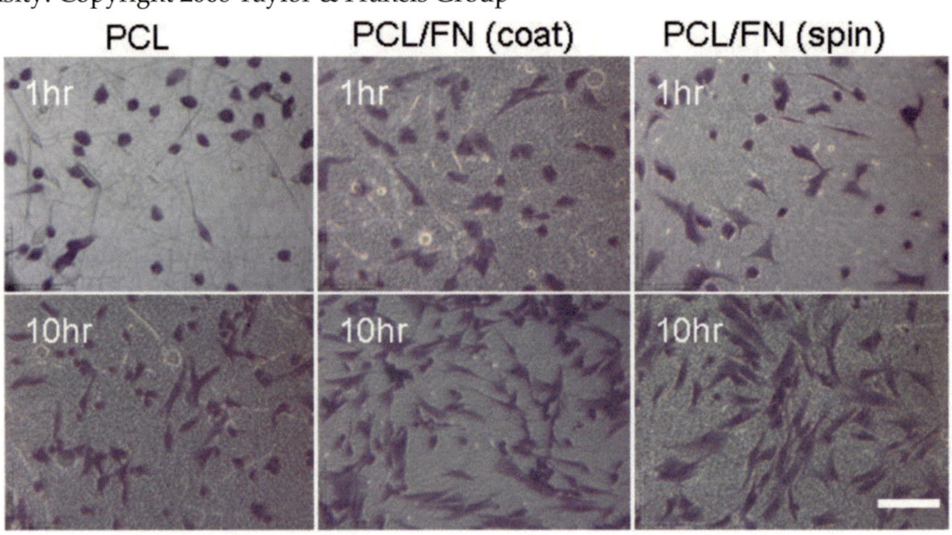

Fig. 4. Cell morphology on different nanofibrous meshes at 1 and 10 hours after cell seeding. Cells were stained with methylene blue (blue). Scale bar: 100 μm. Copyright 2008 Taylor & Francis Group

To investigate if the incorporated biomolecules into PCL fibres remain their biological activity after all the procedures for electrospinning, fibronectin (FN) (0.0125% w/w) was similarly incorporated into the electrospun PCL nanofibers as BSA. Electrospun PCL nanofibers coated with 10 μg/mL fibronectin in PBS for 1 hour were used as the positive controls and pure PCL nanofibers were used as the negative controls. Human dermal fibroblasts were seeded and cultured on all three fibrous meshes for up to 10 hours. The cell morphology on each fibrous mesh at different times was visualized by staining the fixed cells with 0.1% methylene blue. The staining results showed varying cell response to the fibres. Compared to PCL only, both PCL/FN electrospun fibres PCL/FN coated fibres showed a rapid cell spreading as early as 30 min after cell seeding (Figure 4). Majority of the cells showed spindle like morphology 10 hours after the culture on FN containing nanofibers; in contrast, most cells still remained round shape or minimal spreading on the PCL nanofibers. To quantify the difference in cell adhesion onto different fibrous meshes, MTS assay was performed on the cells cultured on the meshes for 1 and 10 hours. Clearly, the cell adhesion onto FN containing nanofibers was significantly improved and this was observed as early as 1 hour after cell seeding (Figure 5). With the culture extended to 10 hours, more cells attached to all the investigated surfaces, but still remained a similar trend in cell adhesion, i.e., more cells were detected on the FN containing nanofibers. Necessary to mention, it seems that PCL/FN electrospun nanofibers are more favourable to cell adhesion than PCL/FN coated fibres although the mechanism is not clearly known

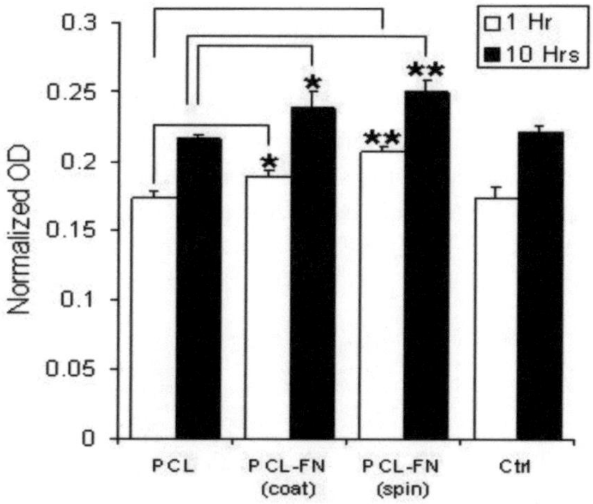

Fig. 5. Cell adhesion to different nanofiber meshes, determined by MTS assay. Representative figure out of 3 separate experiments. * $P<0.05$. ** $P<0.001$. No significant difference was observed between PCL and TCP groups. Copyright 2008 Taylor & Francis Group

It has been shown that bioactive molecules can be easily incorporated into the nanofibers by mixing them into the polymer solution. More importantly, the bioactivity remains after processing as confirmed in both collagen containing nanofibers and FN containing nanofibers. Fibroblasts similarly adhered and spread on both PCL/FN electrospun nanofibers and the PCL nanofibers coated with FN. FN enhances fibroblast adhesion and spreading mainly through the interaction with cell membrane integrins such as $\alpha 5\beta 1$ (Akiyama 1996). Although we did not stain for $\alpha 5\beta 1$, the comparable results between the two FN containing nanofibers suggest the bioactivity still remains intact after the electrospinning. This is a very important evidence for further using electrospun fibers in creation of functional scaffolds, especially for local release bioactive molecules to the attached cells. Thus, the inclusion of bioactive molecules in the nanofibers will allow us to better mimic the native ECM where numerous insoluble or soluble molecules anchor in the ECM fibers to mediate temporal cascades of cell function. Different from collagen, rapid release of BSA from the nanofibers would take place upon contact with medium. The various BSA release profiles indicate that release of bioactive molecules can be tuned to achieve a programmable release by varying the initial drug loading, fiber diameter, spatial arrangement and fibre configuration.

6. Controlled spatial arrangement of nanofibers into various topographical patterns

In native tissues, ECM fibres with preferred orientation are often observed, leading to the anisotropic mechanical performance. The electrospinning technique allows us to control the spatial arrangement of collected nanofibers. Figure 6 (top panel) shows the PCL/collagen/BSA-FITC nanofibers collected on glass coverslips with different electric field modification. Without modifying the electric field, PCL/collagen nanofibers randomly deposited on the collecting surface. By manipulating the intensity distribution of electric field on the collecting surface, a gradient deposition of PCL/collagen nanofibers on the glass coverslip could be obtained. The aligned PCL/collagen nanofibers were achieved by applying parallel grounded metal wires on the collecting surface described previously (Ball & Shuttleworth 2004). The aligned fibres were deposited across the wires. Similarly to the aligned fibres, cross-aligned fibres with two different compositions (PCL/Collagen/FITC-BSA, and PCL/Collagen/TRITC-BSA) were obtained by using parallel wires for the first TRITC fibre and then turning the wire pair 90° for the second FITC-fibre layer.

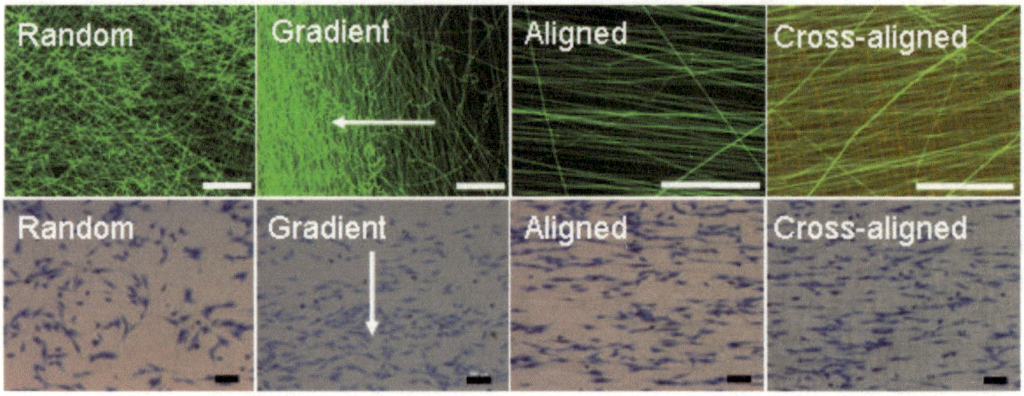

Fig. 6. Cell growth on PCL/Collagen fibrous meshes with distinctive topographical patterns. Fluorescent images of nanofibers collected on glass coverslips (Top). Green fibres labelled with FITC-BSA and red fibres labelled with TRITC-BSA. Cells cultured overnight were stained with methylene blue (blue) (Bottom). Arrow indicates the gradient direction. Scale bar: 100 μm. Copyright 2008 Taylor & Francis Group

It has been shown that the surface properties of substrates play a critical role in regulating the cell attachment. In this regard, human dermal fibroblasts were cultured on these nanofiber meshes for overnight and then stained with methylene blue. The staining results showed that cells had various distribution and spatial orientation in response to the surface configuration of collected nanofibers (Figure 6 bottom panel). On the aligned nanofibers cells oriented themselves along the direction of nanofiber alignment. However, no obvious cell orientation was observed on the randomly collected fibres. When culturing the cells on the meshes with gradient nanofiber deposition, a gradient cell distribution was consequently achieved and higher cell number was observed in the area with high density of PCL/collagen nanofibers. Despite only a few micrometers away between green and red-fluorescent layers, the cells cultured on the cross aligned bi-layer nanofibers could only sense the nanofiber pattern with direct contact, but not the underneath one. This contact guidance inevitably includes the formation of focal adhesion composed of cytoskeletal proteins and integrin receptors (Meyle et al., 1993), and the involvement of integrin receptors can initiate the intracellular cascades to determine further cell function such as anchorage, traction for migration, differentiation, and possibly growth (Dedhar et al., 1987; Zhao et al., 2006). It has been reported that cell alignment could subsequently determine the orientation of newly deposited ECM (Manwaring et al., 2004). This implies that anisotropic new tissue formation can be controlled by designing an anisotropic scaffold. Mechanical anisotropy of aligned fibers was previously reported in several studies (Li et al., 2006; Courtey et al., 2006). A significant increase of aligned fibrous meshes in both Young's modulus (about 33 fold) and tensile modulus (about 5 fold) was measured compared to randomly collected meshes (Li et al., 2006). Ascribed to its anisotropic mechanical performance, biomechanical mimicking of soft tissues or musculoskeletal tissues using aligned fibers has multi-fold advantages and it holds tremendous promises for elaborate tissue formation with multi-functionality (see Figure 7 for the demonstration of spatial control of cell organization (non-published data)). This free control of fibre orientation can accommodate the requirements of native tissues with orientated ECM fibers changing from location to location.

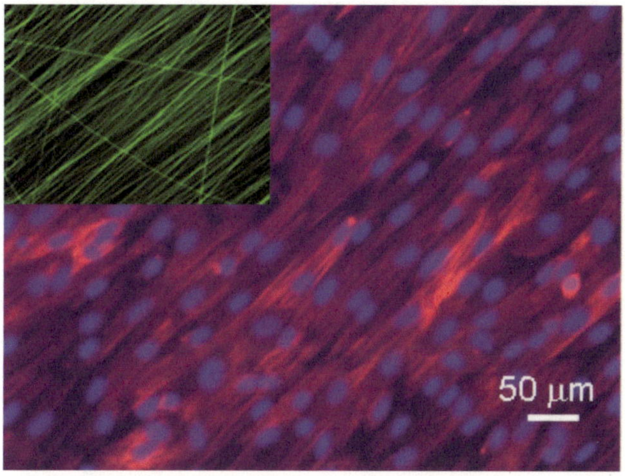

Fig. 7. Fluorescence images of the human cardiomyocytes cultured on the aligned PCL/collagen nanofibrous meshes (inset) for 24 hours. Cells were stained with TRITC-phalloidin for F-Actin (red) and with DAPI for cell nuclei (blue). Green: PCL/collagen-FITC. Scale bar: 50 μm.

7. "Layer-by-layer" 3D tissue formation

7.1 Anisotropic scaffolds with hierarchical structure

To test the capability of manipulating sequential deposition of different layers of fibres into a spatially graded scaffold, a study was performed using PCL only and PCL/Collagen containing FITC-BSA (Yang et al., 2008). The sequence was shown in Figure 8 and sequentially deposited multiple layers were directly collected on aluminium foil (substrate). The cross sections of collected scaffolds were examined under a fluorescence microscope. It was found that different nanofibers were layered into 3D meshes with an identical sequence as designed. By controlling the electrospinning time at a constant flow rate (10μl/min), the thickness of each material layer could be controlled with the designated thickness, from several micro meters to several tens of micro meters. A more confirmative result was obtained by using PCL nanofibers containing either TRITC-labelled BSA (in red color) or FITC-labelled BSA (in green color) for sequential deposition as shown in Figure 8C. The results also indicate that bioactive molecules can be incorporated into nanofibers and thereafter spatially arranged in a high order to form multifunctional scaffolds. In addition, it is intriguing to arrange cells with spatial distribution controlled by materials composition. Collagen containing nanofibers favoured the attachment of fibroblasts; therefore we hypothesized that fibroblasts would primarily attach to PCL/collagen nanofibers instead of PCL nanofibers, and lead to the controllable spatial arrangement of cells. To test this, human dermal fibroblasts were seeded and cultured in between the PCL/collagen and PCL only nanofiber layers for 24 hours. Examination of the cross-sections of cultured constructs revealed that cells only adhered to PCL/collagen nanofibers but not to the PCL ones (data not shown).

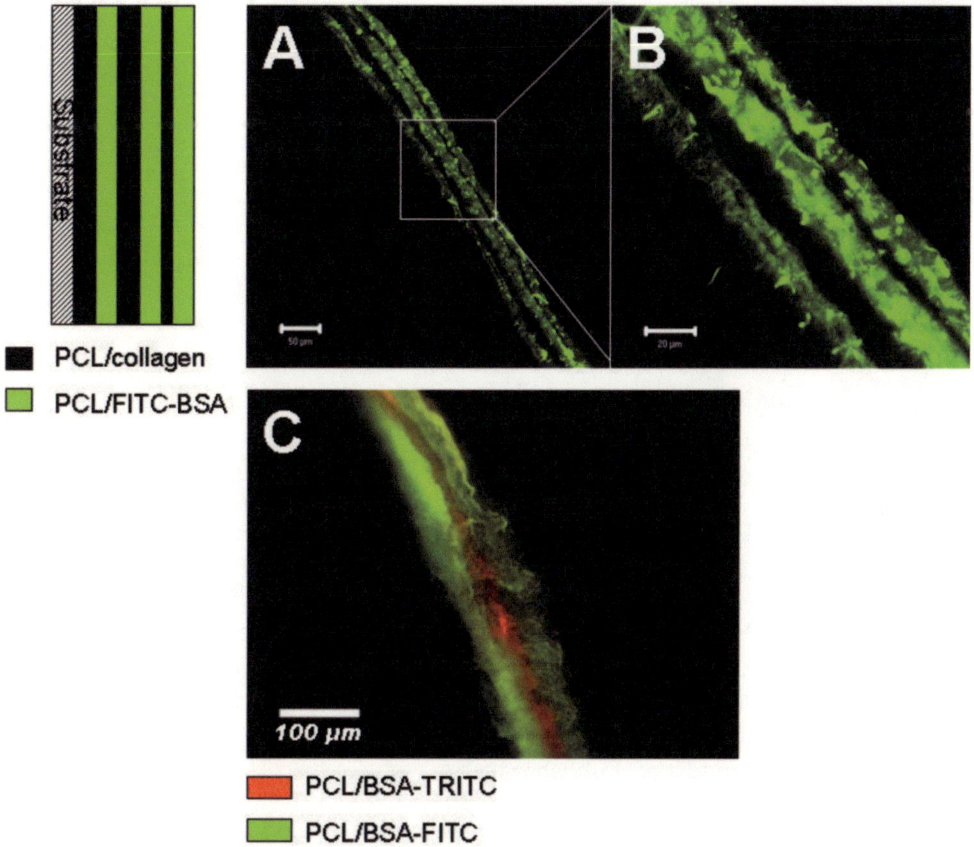

Fig. 8. Fluorescence images of transverse sections of the nanofibrous meshes. (A) and (B), multifunctional scaffolds prepared by sequentially electrospinning of solutions as designated. (B), high magnification highlighting the clear layers. (C), fluorescence images of the transverse section of a three-layer nanofibrous mesh. Red: PCL/BSA-TRITC. Green: PCL/BSA-FITC. Scale bar: (A) 50 μm; (B) 20 μm; (C) 100 μm. Copyright 2008 Mary Ann Liebert Publishers

Cells in native tissue not only experience the anisotropy from ECM fiber arrangement, but also respond to various chemistry changes from the surrounding environment. The chemical stimulation can come from the direct contact or from neighbors, and greatly depends on the spatial concentration distribution. By changing the composition of nanofibers while electrospinning, multilayered scaffolds with specified chemistry and thickness for each layer were made as designed in our study. This demonstrates the potentials to use the layer-by-layer deposition approach towards creation of cell-specific 3D microenvironment by careful selection of the nanofiber composition and the bioactive molecules incorporated into the fibres.

7.2 Layer- by- layer assembly of cell/scaffold construct

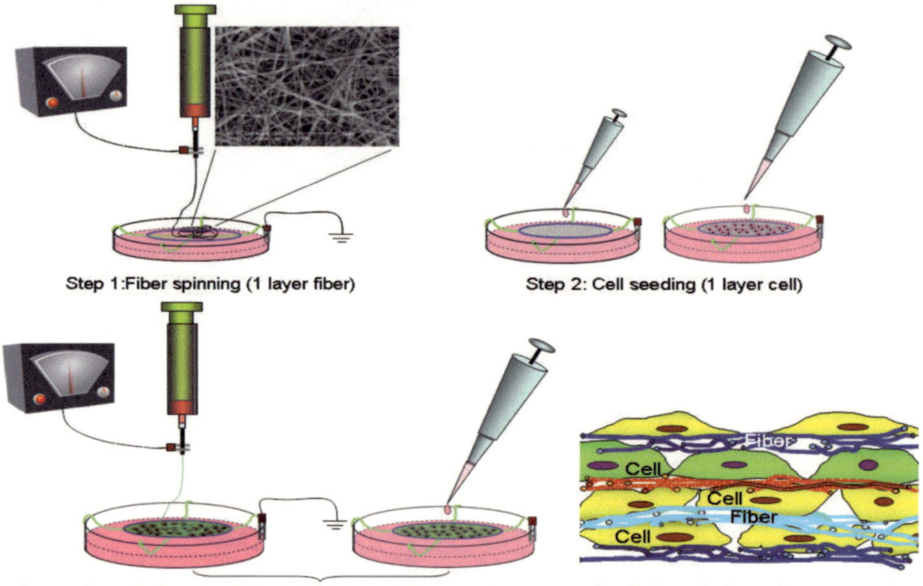

Fig. 9. A schematic illustration of the "layer-by-layer (L-b-L)" cell assembly while electrospinning. As indicated by different colours, both fibre and cell layers can be varied during the cell assembly to create a customized final 3D construct according to the design. Copyright 2008 Mary Ann Liebert Publishers

Layer-by-layer assembly of cell/nanofibers into 3D cell rich construct was recently developed in our lab (Yang et al., 2009b). While assembling the cells together with collected nanofibers to form 3D constructs as shown in Figure 9, a key step is to collect the fibre meshes on the liquid surface that is, keeping the meshes hydrated. Thus, a wet electrospinning was first developed in our lab (Yang et al., 2009b). The setup of wet electrospinning was very similar to that of conventional electrospinning (Figure 1), but with a slight modification on the fibre collection surface. In the wet electrospinning, grounded liquid was used to collect the electrospun fibres instead of using a solid metal surface. It was found the fibres deposited directly onto the liquid surface (we have tested many different liquids including the cell culture medium, phosphate buffered saline (PBS) and distilled water), and the morphology of nanofibrous meshes was very similar to those collected on the metal surface. Although the PCL/collagen fibres rapidly hydrated after depositing on the liquid surface, they remained stretched on the liquid surface without noticeable contraction. So far we have tested several other materials such as chitosan, PLGA and PCL/elastin using our wet electrospinning, and the nanofibrous meshes have been successfully obtained on the grounded water surface.

There is one potential concerns for this on site "L-b-L" cell assembly. That is, the presence of trace amount of solvent (HFIP) in the electrospun nanofibers may cause an adverse effect on the cellular function despite that most of the HFIP would evaporate during the electrospinning process. To determine the adversity to the cells, human dermal fibroblasts were seeded and cultured on the electrospun nanofibers right after they were collected on either the medium surface or on the glass coverslip surface without further vacuum treatment as reported previously (Yoshimoto et al., 2003). The staining of cells with a dead-live kit showed negligible cell death, and most cells remained viable (Fig. 10A). Further quantification of the potential toxicity of trace HFIP was performed, in which nanofibrous meshes collected on the medium surface were left in the same culture medium and incubated for different times. The supernatants were correspondently collected and tested for cytotoxicity using MTT assay. The test result showed that no significant cytotoxicity was observed in all collected supernatants (Fig. 9B).

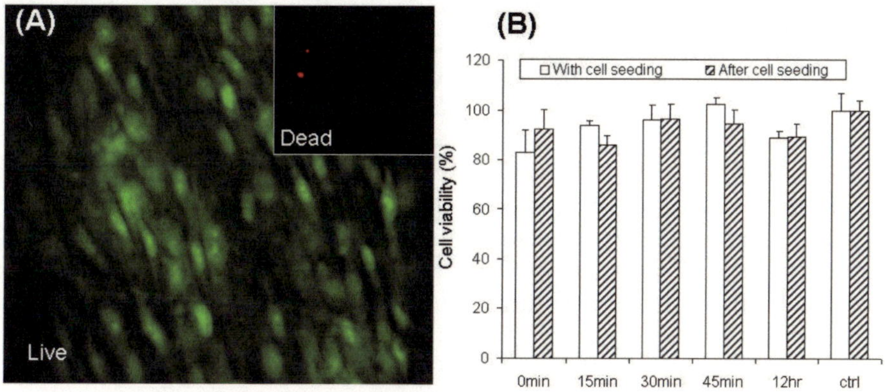

Fig. 10. Cytotoxicity of electrospun nanofibers based on a live/dead cell staining (A) and MTT assay (B). Green = live and red=dead. In MTT assay, medium samples used to extract nanofibers for different times (0 min, 15 min, 30 min, 45 min and 12h) were added to the cells along with cell seeding (white bar) or 12 hours after cell attachment (striped bar). Fresh culture medium served as controls. Data were presented as average ± SD (n = 3). Copyright 2008 Mary Ann Liebert Publishers

The feasibility of on site forming a multilayered cell structure was first evaluated by "L-b-L" assembly of human dermal fibroblasts with PCL/collagen fibres exactly following the steps as shown in Figure 9. The small inter-fibre pore size (less than 5 μm in Figure 2A) could prevent the seeded cells from escaping into the underneath medium. A total of 15 cell/fibre layers were built into a three-dimensional structure. After cultured for 2 days in Dulbecco's Modified Eagle Medium (DMEM) supplemented with 10% fetal bovine serum (FBS) and 1% Penicillin / Streptomycin, the multilayered structure was harvested and cut into thin frozen cross-sections. The staining of sections with DAPI for cell nuclei (blue) allows us to better visualize the cell spatial distribution across the full thickness of the multilayered structure. Figure 11A shows three-dimensional and yet homogeneous distribution of cells throughout the full thickness of the construct and multiple cell layers among electrospun PCL/collagen fibres could be recognized (Figure 11 A). In addition, cells in the multi-layered constructs showed elongated morphology and embedded among fibres. Figure 11B&C are the cross-sections of 7-layer fibroblast/fibre structures but with different electrospinning time for the PCL/collagen/FITC-BSA fibre layers. When the electrospinning time was 30 seconds, then around 5 μm-thick fibre layer was yielded (Figure 11B). When the electrospinning time was increased to 1 min, the thickness of nanofiber layer was correspondently increased to around 10 μm (Figure 11C). The final shape of the built construct was defined by the metal wire loop (circular, and square loops were tested) and remained unchanged during the prolonged culture period (Figure 12).

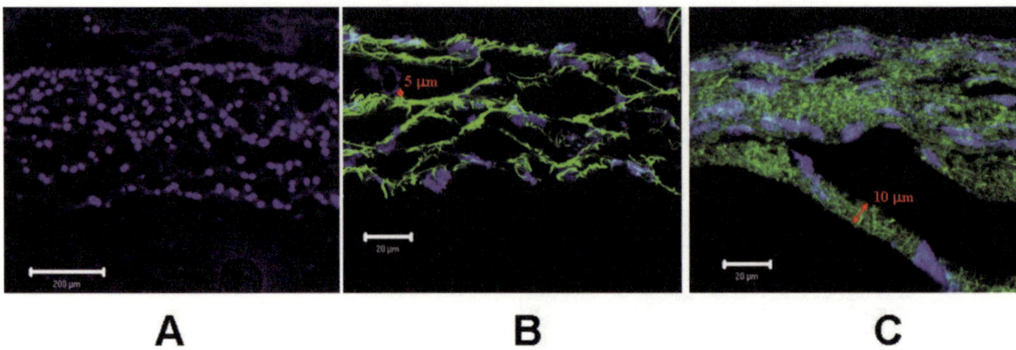

Fig. 11. Microscopic images of multilayered cell-fibre constructs: (A) Fluorescent micrograph of DAPI stained cross-sections of fibre-cell constructs cultured for 2 days. Nuclei = blue. Scale: 200 μm. (B) and (C) Confocal microscopic images of cross-sections of formed cell-fibre constructs with controlled thickness of fibre layer. Scale: 20 μm. Fibres were labelled with FITC (green) and cells were stained blue by DAPI. The images shown are representative of three separate experiments. Copyright 2008 Mary Ann Liebert Publishers

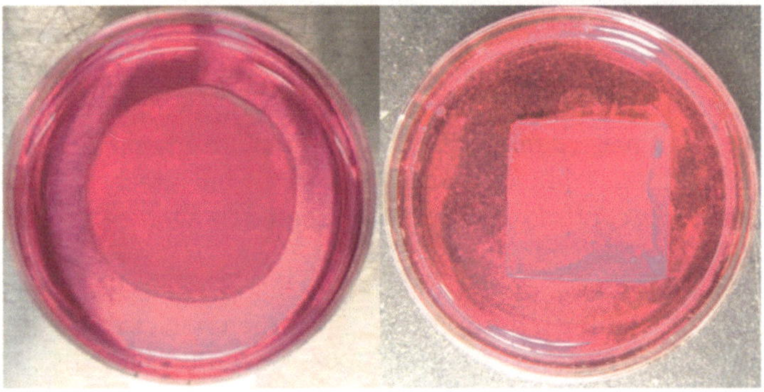

Fig. 12. Images of multilayered cell-fibre constructs cultured for 7 days on the metal wire loop with various shapes. Copyright 2008 Mary Ann Liebert Publishers

Clearly, this "L-b-L" cell assembly represents a bottom-up approach with a great degree to control the spatial distribution of cells. Most importantly, this approach provides a unique solution to the big challenge confronted with the use of electrospun nanofibrous meshes, i.e., the difficulty in penetrating the cells into the interior of the meshes.

7.3 Skin tissue formation from "L-b-L" built cell-fiber construct

As shown above, human dermal fibroblasts could be easily incorporated into the electrospun fibrous meshes along the electrospinning and form a 3D structure with uniform cell distribution. Using the same approach, we have assembled the dermal fibroblasts with PCL/collagen nanofibers into 10-layer fibroblast-PCL/collagen fibre constructs, which can be used as the dermal substitutes. The constructs were further cultured for 3 and 7 days to study the new tissue formation. The cross sections of cultured constructs were examined under a fluorescent microscope. It was found that the cell/fibre construct became packed over a prolonged culture (Figure 13A and B). The staining of the cross-sections of the cultured constructs with hematoxylin and eosin (H & E) showed that fibroblasts evenly distributed in between the fibres without clear recognition of the layers. Meanwhile, the fibroblasts in the construct showed the elongated spindle morphology.

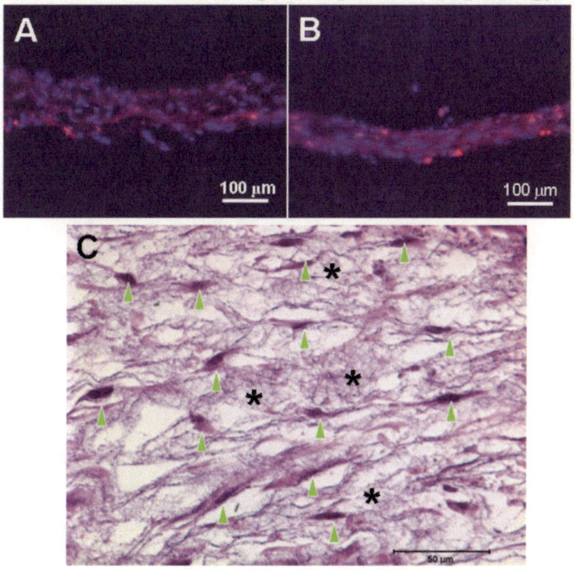

Fig. 13. Formation of dermal tissue from fibroblasts/fibre layered constructs. Fluorescent images of 10-layer fibroblasts/PCL-collagen fibre constructs (see Materials and Methods) cultured for 3 days (A) and 7 days (B). Cell nuclei stained blue, and fibres showed weak red fluorescence, though without labelling. (C) H&E stained cross-section of the construct cultured for 7 days. Arrow head indicates fibroblasts and asterisk shows fibres. Scale: 50μm. The images shown are representative of three separate experiments. Copyright 2008 Mary Ann Liebert Publishers

Using the similar layer-by-layer assembly approach, fibroblasts and keratinocytes were built with the PCL/collagen nanofibers into 3D full-thickness skin substitutes, in which 18 layers of fibroblast/fibre in the lower part were first assembled and then 2 layers of keratinocyte/fibre were assembled on the top. After culturing for three days, a bi-layer skin structure (epidermal and dermal layer) was clearly seen on the H&E stained cross sections (Figure 14A). The seeded keratinocytes remained on the surface and formed a continuous epidermal layer, and fibroblasts retained in the lower part with a uniform distribution. A tight binding between the epidermal layer and dermal layer was observed. The bi-layer structure remained the same even after culture for 7 days. With the successful assembly of two distinct types of skin cells into bi-layer skin substitutes, it is further demonstrated the flexibility of this bottom-up approach in creating multicellular and multifunctional tissues.

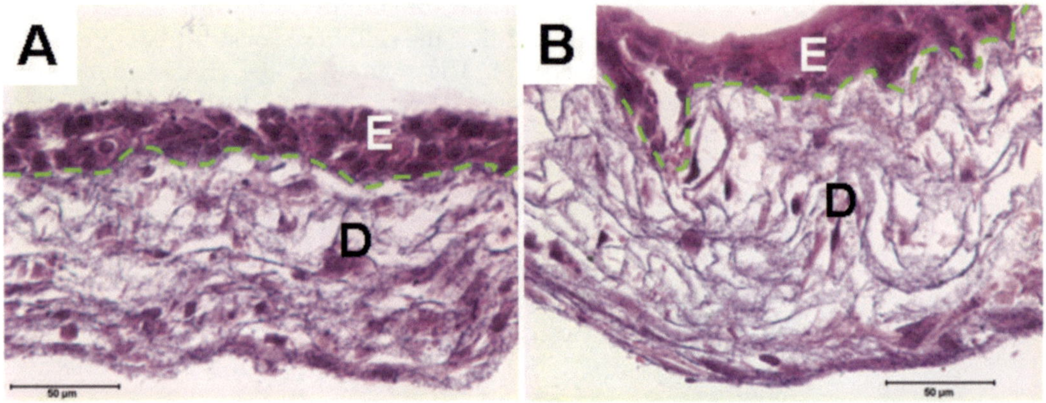

Fig. 14. H & E stained cross-sections of bi-layer skin constructs composed of epidermal (E) and dermal (D) layers and formed by culturing "L-b-L" cell assembled constructs for 3 days (A) and 7 days (B). Green broken line outlines the border between E and D. Scale: 50μm. The images shown are representative of three separate experiments. Copyright 2008 Mary Ann Liebert Publishers

7.4 Mineralized bone-like tissue formation from "L-b-L" assembled cell-fibre construct

Using this "L-b-L" bottom-up assembly approach, we have recently assembled the mouse osteoblastic cells with chitosan-containing PCL nanofibers for bone-like tissue formation (Yang et al., 2009a). The constructs composed of multilayers of chitosan/PCL nanofibers and mouse osteoblasts (MC3T3-E1 cells) were prepared and cultured for up to 28 days to determine whether chitosan-containing nanofibers could support bone-like tissue formation in a 3D setting. SEM examination of the constructs cultured for 28 days showed that the construct surface was fully covered by cells and newly formed ECM (data not shown). Examination of the cross section indicated the formation of an integral tissue across the entire thickness (Figure 15A.) and no individual nanofibers could be identified any more. The cell distribution, tissue formation and mineralization were further evaluated by histological and biochemical analyses. H & E staining of the cross-sections of the cultured constructs showed that cells uniformly distributed through the entire constructs with elongated morphology and formed an integral tissue without recognition of chitosan/PCL nanofibers (Figure 15A). Quantitative measurement of calcium content in the constructs by alizarin red staining indicated that mineralization increased with the incubation time and a significant increase in calcium deposition was observed around 21 days (data not shown). The calcium amount on Day 28 was almost 5.7 folds of that on Day 3 and only 1.6 folds of that on Day 21. Von Kossa staining of the cross-sections further revealed that mineral deposition (brownish precipitates) mainly occurred inside the cells (Figure 15B), indicating the osteogenic differentiation of MC3T3-E1.

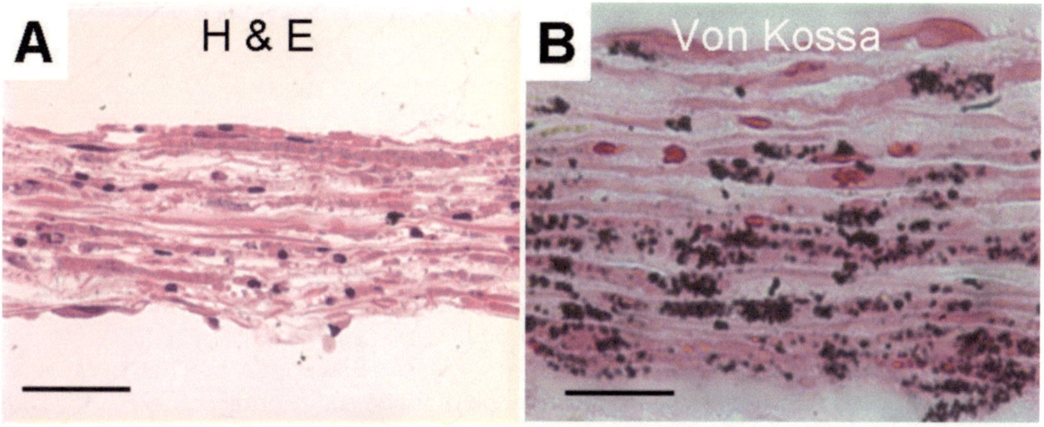

Fig. 15. (A) H&E staining of the cross-sections of 3D constructs with mouse osteoblasts (MC3T3-E1 cells) and chitosan-containing PCL nanofibers cultured for 14 days. (B) von Kossa staining of the cross-sections of 3D constructs with MC3T3-E1 cells and nanofibers cultured for 14 days. Dark brownish dots show the mineral deposition. Scale bar: A = 50μm and B = 25μm. Copyright 2009 American Chemical Society

8. Conclusions

With all the results presented in this chapter, it clearly demonstrates the feasibility of producing multiscale scaffolds with diverse functionality and tuneable distribution of bioactive molecules across the scaffold using electrospun nanofibres. This multifunctional scaffold can offer the cells with specific and customized microenvironment depending on the cell type and therefore achieve various phenotypic expressions for functional tissue formation. With the assistance of electrospun nanofibres, a new layer-by-layer bottom-up approach was developed to assemble the cells with nanofibres into 3D constructs. In this approach, cell/fibre constructs with single or multiple cell types could be assembled in an alternating manner, i.e., alternating cell layer and nanofiber layer on site of electrospinning. This new approach has multi-fold advantages, including 1) providing a uniform cell distribution throughout the nanofiber scaffold, 2) formulating the composition of each fibre layer, 3) controlling cell type for each cell layer, 4) allowing the co-culture of multiple cell types with distinct spatial arrangement. This layering process for tissue formation can provide a useful tool for engineering tissues with hierarchical structure and for in vitro studying the cell-cell or cell-materials interaction. In all, electrospinning has created a new avenue in designing biomimetic scaffolds for supporting the cell and tissue growth. With the continuous improvement of this technology, electrospinning will have many other promising applications in materials and life sciences far beyond the demonstrations presented in this chapter.

9. Acknowledgement

Majority of the results presented here were published. This publication was made possible by Grant Number R21AR056416 from NIAMS.

10. References

Abrams, G. A.; Goodman, S. L.; Nealey, P. F.; Franco, M.; Murphy, C. J., (2000) Nanoscale topography of the basement membrane underlying the corneal epithelium of the rhesus macaque. *Cell Tissue Res* 299, (1), 39-46.

Akiyama, S. K., (1996) Integrins in cell adhesion and signaling. *Hum Cell* 9, (3), 181-186.

Ball, S. G.; Shuttleworth, A. C.; Kielty, C. M., (2004) Direct cell contact influences bone marrow mesenchymal stem cell fate. *Int J Biochem Cell Biol* 36, (4), 714-727.

Barnes, C. P.; Sell, S. A.; Boland, E. D.; Simpson, D. G.; Bowlin, G. L., (2007) Nanofiber technology: Designing the next generation of tissue engineering scaffolds. *Adv Drug Deliv Rev*, 59(14),1413-33.

Chew, S. Y.; Wen, J.; Yim, E. K.; Leong, K. W., (2005) Sustained release of proteins from electrospun biodegradable fibers. *Biomacromolecules* 6, (4), 2017-2024.

Chua, K. N.; Lim, W. S.; Zhang, P.; Lu, H.; Wen, J.; Ramakrishna, S.; Leong, K. W.; Mao, H. Q., (2005) Stable immobilization of rat hepatocyte spheroids on galactosylated nanofiber scaffold. *Biomaterials* 26, (15), 2537-2547.

Corden, T. J.; Jones, I. A.; Rudd, C. D.; Christian, P.; Downes, S.; McDougall, K. E., (2000) Physical and biocompatibility properties of poly-epsilon-caprolactone produced using in situ polymerisation: a novel manufacturing technique for long-fibre composite materials. *Biomaterials* 21, (7), 713-724.

Courtney, T.; Sacks, M. S.; Stankus, J.; Guan, J.; Wagner, W. R., (2006) Design and analysis of tissue engineering scaffolds that mimic soft tissue mechanical anisotropy. *Biomaterials* 27, (19), 3631-3638.

Dedhar, S.; Ruoslahti, E.; Pierschbacher, M. D., (1987) A cell surface receptor complex for collagen type I recognizes the Arg-Gly-Asp sequence. *J Cell Biol* 104, (3), 585-593.

Formhals, A. Process and apparatus for preparing of artificial threads. *U.S. Pat.* 1975504, 1934.

Frenz, D. A.; Liu, W.; Williams, J. D.; Hatcher, V.; Galinovic-Schwartz, V.; Flanders, K. C.; Van de Water, T. R., (1994) Induction of chondrogenesis: requirement for synergistic interaction of basic fibroblast growth factor and transforming growth factor-beta. *Development* 120, (2), 415-424.

Geng, X.; Kwon, O. H.; Jang, J., (2005) Electrospinning of chitosan dissolved in concentrated acetic acid solution. *Biomaterials* 26, (27), 5427-5432.

Griffith, L. G.; Swartz, M. A., (2006) Capturing complex 3D tissue physiology in vitro. *Nat Rev Mol Cell Biol* 7, (3), 211-224.

Ji, Y.; Ghosh, K.; Shu, X. Z.; Li, B.; Sokolov, J. C.; Prestwich, G. D.; Clark, R. A.; Rafailovich, M. H., (2006) Electrospun three-dimensional hyaluronic acid nanofibrous scaffolds. *Biomaterials* 27, (20), 3782-3792.

Kweon, H.; Yoo, M. K.; Park, I. K.; Kim, T. H.; Lee, H. C.; Lee, H. S.; Oh, J. S.; Akaike, T.; Cho, C. S., (2003) A novel degradable polycaprolactone networks for tissue engineering. *Biomaterials* 24, (5), 801-808.

Langer, R.; Vacanti, J. P., (1993) Tissue engineering. *Science 260*, (5110), 920-926.

Li, D.; Ouyang, G.; McCann, J. T.; Xia, Y., (2005) Collecting electrospun nanofibers with patterned electrodes. *Nano Lett* 5, (5), 913-916.

Li, W. J.; Mauck, R. L.; Cooper, J. A.; Yuan, X.; Tuan, R. S., (2006) Engineering controllable anisotropy in electrospun biodegradable nanofibrous scaffolds for musculoskeletal tissue engineering. *J Biomech* 40, (8), 1686-1693.

Luong-Van, E.; Grondahl, L.; Chua, K. N.; Leong, K. W.; Nurcombe, V.; Cool, S. M., (2006) Controlled release of heparin from poly(epsilon-caprolactone) electrospun fibers. *Biomaterials* 27, (9), 2042-2050.

Lutolf, M. P.; Hubbell, J. A., (2005) Synthetic biomaterials as instructive extracellular microenvironments for morphogenesis in tissue engineering. *Nat Biotechnol* 23, (1), 47-55.

Manwaring, M. E.; Walsh, J. F.; Tresco, P. A., (2004) Contact guidance induced organization of extracellular matrix. *Biomaterials* 25, (17), 3631-3638.

Matthews, J. A.; Wnek, G. E.; Simpson, D. G.; Bowlin, G. L., (2002) Electrospinning of collagen nanofibers. *Biomacromolecules* 3, (2), 232-238.

Meyle, J.; Wolburg, H.; von Recum, A. F., (1993) Surface micromorphology and cellular interactions. *J Biomater Appl* 7, (4), 362-374.

Michel, M.; L'Heureux, N.; Pouliot, R.; Xu, W.; Auger, F. A.; Germain, L., (1999) Characterization of a new tissue-engineered human skin equivalent with hair. *In Vitro Cell Dev Biol Anim* 35, (6), 318-326.

Min, B. M.; Lee, G.; Kim, S. H.; Nam, Y. S.; Lee, T. S.; Park, W. H., (2004) Electrospinning of silk fibroin nanofibers and its effect on the adhesion and spreading of normal human keratinocytes and fibroblasts in vitro. *Biomaterials* 25, (7-8), 1289-1297.

Moroni, L.; Licht, R.; de Boer, J.; de Wijn, J. R.; van Blitterswijk, C. A., (2006) Fiber diameter and texture of electrospun PEOT/PBT scaffolds influence human mesenchymal stem cell proliferation and morphology, and the release of incorporated compounds. *Biomaterials* 27, (28), 4911-4922.

Orr, A. W.; Ginsberg, M. H.; Shattil, S. J.; Deckmyn, H.; Schwartz, M. A., (2006) Matrix-specific suppression of integrin activation in shear stress signaling. *Mol Biol Cell* 17, (11), 4686-4697.

Patel, S.; Kurpinski, K.; Quigley, R.; Gao, H.; Hsiao, B. S.; Poo, M. M.; Li, S., (2007) Bioactive nanofibers: synergistic effects of nanotopography and chemical signaling on cell guidance. *Nano Lett* 7, (7), 2122-2128.

Pham, Q. P.; Sharma, U.; Mikos, A. G., (2006) Electrospinning of polymeric nanofibers for tissue engineering applications: a review. *Tissue Eng* 12, (5), 1197-1211.

Pham, Q. P.; Sharma, U.; Mikos, A. G., (2006) Electrospun poly(epsilon-caprolactone) microfiber and multilayer nanofiber/microfiber scaffolds: characterization of scaffolds and measurement of cellular infiltration. *Biomacromolecules* 7, (10), 2796-2805.

Powell, H. M.; Boyce, S. T., (2008) Fiber density of electrospun gelatin scaffolds regulates morphogenesis of dermal-epidermal skin substitutes. *J Biomed Mater Res A*. 84, (4), 1078-1086

Rho, K. S.; Jeong, L.; Lee, G.; Seo, B. M.; Park, Y. J.; Hong, S. D.; Roh, S.; Cho, J. J.; Park, W. H.; Min, B. M., (2006) Electrospinning of collagen nanofibers: effects on the behavior of normal human keratinocytes and early-stage wound healing. *Biomaterials* 27, (8), 1452-1461.

9

Selective Laser Sintering of Poly(L-Lactide)/Carbonated Hydroxyapatite Nanocomposite Porous Scaffolds for Bone Tissue Engineering

Wen You Zhou[1], Min Wang[2], Wai Lam Cheung[2,*] and Wing Yuk Ip[3]
[1]*Biomedical and Tissue Engineering Research Group, Faculty of Dentistry, The University of Hong Kong, 34 Hospital Road, Hong Kong, China*
[2]*Department of Mechanical Engineering, The University of Hong Kong, Pokfulam Road, Hong Kong, China*
[3]*Department of Orthopaedics and Traumatology, The University of Hong Kong, Sassoon Road, Hong Kong, China*

1. Introduction

Bone replacements are frequently required to fix damaged tissue due to trauma, tumor resection, and more generally in many surgeries (Navarro et al. 2004). This requirement for new bone is a major clinical and socioeconomic need. There is a driving need in tissue engineering (TE) field to develop scaffolds that guide bone regeneration across defects that are too large to heal naturally. This may be achieved by a multidisciplinary field: bone tissue engineering (BTE) which seeks to apply the scientific principles to engineering of viable substitutes that restore and maintain the function of human bone tissue. There are three key elements for regenerating bone tissues, namely, osteogenic progenitor cells, osteoinductive growth factors and osteoconductive scaffolds (Schieker et al. 2006). The function of scaffolds is to direct the growth of cells seeded within the porous structure of the scaffold or of cells migrating from surrounding environments. The majority of mammalian cell types are anchorage-dependent resulting in dying if an adhesive surface is not provided. A successful scaffold should balance mechanical strength with biomolecular delivery capacity, providing an environment in which the regenerated tissue assumes function as the scaffold degrades (Hollister 2005). The success of BTE depends, to a large extent, on the performance of the 3D porous biodegradable scaffolds. The BTE scaffolds will fit into the anatomical defect, so they must have some strength and stiffness to keep their integrity at the initial *in vivo* stage. In TE, the selection of appropriate biomaterials for scaffold production is very important because the scaffold properties are largely determined by the intrinsic properties of the materials. Also, the scaffold design and processing method will play an important role in the scaffold properties.

Rapid prototyping (RP) is an excellent alternative compared with conventional methods as it can provide the consistency, flexibility in geometry and reproducibility in scaffold physical properties (Wiria et al. 2007). Among various RP technologies, selective laser sintering (SLS) has been found to be advantageous for TE scaffolds fabrication due to its ability to process a wide range of biocompatible and biodegradable materials (Chua et al. 2004). SLS appears to be useful primarily for bone implants production and it has been used to sinter poly(methyl methacrylate) coated calcium phosphate (Lee & Barlow 1993) and hydroxyapatite reinforced polyethylene composites (known commercially as HAPEX™) for bone implants (Hao et al. 2006).

Disadvantages of SLS include a process more mechanically complex than most other RP techniques and powdery surface may complicate sterilization and cell culture (Venuvinod & Ma 2004). As a layer additive technique, SLS parts also suffer from the stair-stepping problem. The first layers may need a base anchor to reduce thermal effects such as curling. For scaffold production, the resolution of SLS is limited by the laser beam diameter (spot size) and the particle size. Due to heat dissipation, particles outside the laser scan path are sometimes also fused ("growth" effect), causing an irregular boundary and roughness on the part surface (Yang et al. 2002); however, rough surface may attract some types of cells to attach and grow. Furthermore, when building porous scaffolds, the loose, unsintered powder trapped inside of the pores needs to be removed. The "growth" effect can make the cleaning of the loose powder difficult.

So far, only one hydrophobic biodegradable polymer, poly-ε-caprolactone (PCL), has been used to produce TE scaffold by conventional SLS machine because PCL is significantly less expensive and supplied in the fine powder form. Ciardelli et al. (2004) used a CO_2 SLS machine (SYNRAD model J48-5S, Italy) to fabricate scaffolds with the overall size of 15x15x3.3 mm (length x breadth x height) from PCL slurries. Those SLS scaffolds have a pore size of about 300 μm (height) x 700 μm (width). Williams et al. (2005) fabricated sintered PCL cylindrical porous scaffolds (5.0 mm diameter, 4.5 mm height) by Sinterstation® 2000 at University of Michigan. The pore sizes of these scaffolds varied from 1.75 mm to 2 mm. *In vitro* and *in vivo* results showed that laser sintered PCL scaffolds enhance tissue in-growth and their mechanical properties are within the lower range of trabecular bone. From the same group, Partee et al. (2006) optimized the SLS processing parameters for PCL powder using the systematic factorial experimental design method. The scaffolds were in the form of square prism with an overall size of 11.5x14.2x10.8 mm and a pore size of 2 mm. The optimized PCL scaffolds showed designed porous channels and achieved a dimensional accuracy to within 3- 8% of the design specifications and 94% of full density. Using optimal SLS processing parameters, BTE scaffolds were fabricated based on actual minipig and human condyle models.

Recently, Wiria et al. (2007) applied the SLS technique to first fabricate biodegradable composite scaffolds from physically blended PCL with micro-sized hydroxyapatite using Sinterstation® 2500 (3D Systems). The scaffolds were in the form of circular disc with a diameter and thickness of 15 mm and 1 mm respectively. This overall size is good for fitting into a cell culture well. As compared to neat PCL, it was harder to sinter the PCL/HAp biocomposite blends. Similar observation was made by Fan et al. (2001) with other composite materials and they attributed the phenomenon to the increased viscosity of the composite materials. Chua et al. (2004) developed SLS scaffolds on poly(vinyl alcohol) (PVA) blended with micro-sized HAp, intending for craniofacial and joint defects repair.

PVA was chosen due to the cheap price and the approximate particle size for SLS process. PVA is a hydrophilic biodegradable polymer which limits its use in cartilage and corneal tissue engineering (Sinha et al. 2007). Also, other materials, such as polyetheretherketone (PEEK) (Tan et al. 2003) and polyamide (Das et al. 2003), were used for building TE scaffolds. However, these materials are non-biodegradable, which limit their applications as stated earlier.

SLS has been developed primarily for industrial applications and it is still not financially viable to use the technology to produce TE scaffolds from a wide range of biomaterials due to limited availability of biomaterial powders suitable for SLS and extremely high costs of most biomaterial powders. With our success in modifying an existing Sinterstation® 2000 SLS machine to allow small quantities of biopolymer powders to be processed (Lee et al. 2006) and preparation of poly(L-lactide) (PLLA) and PLLA/carbonated hydroxyapatite (CHAp) nanocomposite microspheres (Zhou et al. 2007), we were able to build porous PLLA and PLLA/CHAp TE scaffolds for further investigation.

The effects of sintering conditions; namely, laser power, part bed temperature and scan spacing, on the overall quality of PLLA scaffolds were first studied in this chapter. Then, the results were used to formulate a recommended list of sintering parameters for building good quality PLLA and PLLA/CHAp scaffolds. Finally, the structure and properties, e.g. porosity, compression properties, *in vitro* degradation behaviour and osteoconductivity (such as cell adhesion and growth), of these scaffolds were evaluated.

2. Materials and Processing

2.1 Materials

The PLLA used was 100L 1A with an intrinsic viscosity of 1.9 dL/g (Lakeshore Biomaterials, Birmingham, AL, USA). It was supplied in the form of macro-sized pellets, 1 mm in diameter and 3 mm in length, for conventional extrusion and moulding processes. The emulsifier used in our study was poly(vinyl alcohol) which was cold water soluble and with an average molecular weight of 30,000-70,000 (Sigma-Aldrich). CHAp nanospheres were synthesized by a nanoemulsion method (Zhou et al. 2008). Dichloromethane or methylene dichloride (DCM, Uni-Chem®, Orientalab, China) was used as the organic solvent to dissolve PLLA.

2.2 Preparation of PLLA microspheres

PLLA microspheres were prepared using a normal oil-in-water (O/W) emulsion solvent evaporation technique. In brief, the PLLA was dissolved in DCM to form a solution first and then rapidly added to aqueous PVA solution and stirred for sometime to allow the solvent to evaporate. The hardened microspheres were centrifuged, washed three times with de-ionized water and lyophilized to obtain dry powder.

2.3 Preparation of PLLA/CHAp nanocomposite microspheres

The PLLA/CHAp nanocomposite microspheres were prepared using a solid-in-oil-in-water (S/O/W) emulsion solvent evaporation method (Zhou et al. 2007). Briefly, the CHAp nanospheres were dispersed in the PLLA-dichloromethane solution by ultrasonification (Barnstead Lab-Line ultrasonic cleaner, USA) and homogenization (Ultra-Turrax® T25 basic,

IKA, Germany) to form an S/O nanosuspension. The nanosuspension was mixed with 1 wt% PVA solution to form S/O/W emulsion. The resultant S/O/W emulsion was stirred overnight at 800 rpm and then filtered, washed three times with de-ionized water, and freeze-dried to yield a white PLLA/CHAp biocomposite powder. Nanocomposite microspheres of 10 wt% CHAp in PLLA were used in the SLS process.

2.4 Morphology characterization of microspheres

The surface morphologies of the PLLA microspheres and PLLA/CHAp nanocomposite microspheres were examined using a scanning electron microscope (SEM, Stereoscan 440, Cambridge, UK). The internal structure of the composite microspheres was revealed by the focused ion beam (FIB) milling technique. A focused ion beam (Quanta 200 3D, FEI, USA) was applied to section the PLLA/CHAp composite microspheres for internal examination. The ion beam was tilted 52° with respect to the electron beam and was only used for milling, not for imaging. The FIB milling process was carried out with a Ga^+ ion beam current of 3 nA and the acceleration voltage was 5 KeV. The milled faces were then polished with a low beam current of 0.3 and 0.1 nA prior to SEM imaging.

2.5 Modification of Sinterstation® 2000 system

In order to reduce the consumption of biomaterial powders for TE scaffold construction in the SLS process, modifications were made to an existing Sinterstation® 2000 SLS machine (3D Systems, Valencia, CA, USA). The CO_2 laser of the system has a maximum power output of 50 watts. The CO_2 laser energy intensity across the beam diameter very nearly follows the Gaussian distribution (Nelson et al. 1993). A miniature sintering platform, which consisted primarily of a miniature build cylinder and two powder supply chambers, was designed, fabricated and installed in the build cylinder of the existing SLS machine (Lee 2006; Zhou et al. 2007; Zhou et al. 2008). The miniature build cylinder had a diameter of 50 mm and the movement of its base was directly linked to the base of the existing build cylinder of Sinterstation® 2000. The two powder supply chambers were driven by two additional stepping motors beneath the miniature sintering platform. In the sintering processes, the original powder supply tanks of Sinterstation® 2000 were empty and small amounts of biomaterial powder were fed from the miniature powder supply chambers. The roller positions were sensed and the signals were fed to an additional control panel which controlled the movement of stepping motors and the temperature of the small build cylinder. Other sintering parameters were controlled by the existing Sinterstation® 2000 system and this would ensure good quality of the scaffolds built.

2.6 Design of bone tissue engineering scaffolds

The scaffold models were designed by an extrude-cut patterning method using SolidWorks® (version 2005). They had a tetragonal shape with the ratio of height (H) to width (W) or length (L) equal to 2 (ASTM D695-02a; ASTM D1621-04a). Rectangular channels were cut in all three dimensions to form 3D symmetrical scaffold models. This simple design contains the typical features and feature sizes found in BTE scaffolds (Partee et al. 2006). Type 1 model contained channels of size 0.8×0.8 mm^2 and Type 2 model contained channels of size 0.6×0.6 mm^2. The width of the solid pillar was kept at 0.3 mm in both Type 1 and Type 2 models. The designed porosities for Types 1 and 2 were 78.4% and 70.0% respectively. The detailed design parameters of the scaffold models are listed in Table 1.

In SLS of porous scaffolds, the spot size of the laser beam is very important for building the small features of the scaffolds. The spot size or beam width is defined as the smallest diameter of the focused laser beam. According to the manufacturer of Sinterstation® 2000 system, the spot size of this SLS machine is 0.018" (~457μm). To facilitate handling of the sintered scaffolds, a solid base (9×9×3 mm^3) was incorporated in the scaffold design. If this base was not provided, the first few porous layers would tend to warp significantly, affecting the overall quality of the scaffolds (Partee et al. 2006).

In order to speed up production and to produce scaffolds with similar quality for tests, multiple scaffold models were assembled based on the single scaffold models. Four and nine models were assembled. Then, the data file was exported in an STL format and transferred to the Sinterstation® 2000 system for the sintering process.

Type	Overall Size (mm^3)*	Channel Width (mm)	Pillar Width (mm)	Porosity (%)
1	8×8×16	0.8	0.3	78.4
2	6.6×6.6×13.2	0.6		70.0

Table 1. Design parameters of porous bone tissue engineering scaffolds.

2.7 Production of 3D porous scaffolds by modified Sinterstation® 2000 system

The build orientation (angle between the laser scanning direction and "L" dimension of the scaffold) was 45°. Significant variation in density and anisotropic properties were found in DuraForm™ polyamide (a common SLS material) parts built in 0° and 90° orientations (Caulfield et al. 2007). The 45° build orientation was chosen to reduce the anisotropy, also it was found to be effective in reducing warpage of porous scaffolds (Lee et al. 2004).

In SLS, the laser energy density (ED) level significantly affects the scaffold properties. The ED level is a measure of the amount of energy supplied to the powder particles per unit area of the powder bed surface. The relationship between ED and laser power (P), scan spacing (SS) and beam speed (BS) can be expressed by the following equation (Nelson et al. 1993):

$$ED = \frac{P}{SS \times BS} \tag{1}$$

It is obvious that too low a laser power will render low strength or even incomplete sintering of the scaffold, while too high a laser power will cause the biopolymer to degrade and the scaffold to warp due to high residual stresses. Scan spacing (SS) means the distance between two parallel laser scans during a fill. In this study, the beam speed (BS) was fixed at 1257 mm/s and the effects of laser power, scan spacing and part bed temperature (PBT) were investigated. Prior to this study, DuraForm™ polyamide was used as the reference material for initial settings of the sintering parameters (Lee 2006). After some trials, a workable set of sintering parameters was obtained for PLLA and PLLA/CHAp nanocomposite as shown in Table 2. Fig. 1 shows the SLS process in progress. Four scaffold bases were being sintered and the powder used was a PLLA/CHAp nanocomposite. Also, the powder delivery appeared to be smooth during the sintering process. The production time for one batch of scaffolds was about 2.5 hours.

Sintering Condition	PLLA	PLLA/CHAp
Scan Spacing (mm)	0.15-0.21	0.15-0.21
Part Bed Temperature (°C)	30-40	30-60
Layer Thickness (mm)	0.10	0.10
Roller Speed (mm/s)	127	127
Scan Speed (mm/s)	1257	1257
Stepping Motor Delay (ms)	120	160
Fill Laser Power (W)	11-15	11-19

Table 2. Recommended sintering conditions for PLLA and PLLA/CHAp microspheres.

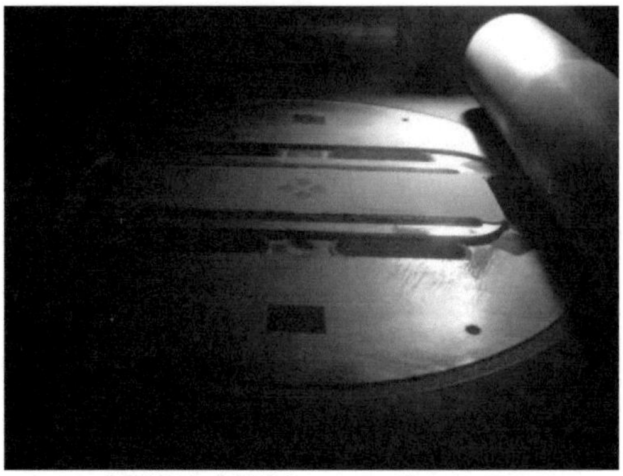

Fig. 1. Selective laser sintering of PLLA/CHAp powder in progress.

2.8 Characterization of 3D porous laser sintered PLLA and PLLA/CHAp nanocomposite scaffolds

2.8.1 Porosity determination

The scaffold porosity was determined using a specific gravity measurement kit and an electronic balance (Shimadzu, AW220, Japan). The detailed procedures and calculation were reported previously (Zhou et al. 2007).

2.8.2 Compression test

Compressive properties of the sintered scaffolds were determined using a MicroTester (Instron 5848, USA) with a load cell of 100 N. The crosshead speed was 0.5 mm/min. The compressive modulus was determined from the initial linear region of compressive stress-strain curve. The compressive yield strength was determined from the first point on the stress-strain curve at which an increase in strain was observed without an increase in stress. The compressive modulus and yield strength for any given set of specimens were obtained by an average of five measurements and expressed as mean ± standard deviation (SD).

2.8.3 *In vitro* degradation test

Specimens containing 4 sintered layers cut from the sintered scaffolds were used in degradation test. The test specimens were accurately weighed, numbered and sterilized by 70% ethanol. The specimens were incubated at 37 °C for 1, 3, 6 and 8 weeks in phosphate buffered saline (PBS) solution (pH 7.4). The PBS solution was changed fresh once a week. At weeks 1, 3, 6, and 8, three samples of each type of scaffold were removed and rinsed with de-ionized water 3 times and then air dried at room temperature. The level of degradation was assessed macroscopically in terms of weight loss and change in compressive properties. The results reported are averages of three measurements. The weight loss (WL) after *in vitro* degradation was calculated by the following equation:

$$WL(\%) = (W_0 - W_t) / W_0 \times 100 \qquad (2)$$

where W_0 and W_t are specimen weights before and after degradation respectively.

2.8.4 *In vitro* cell culture

Human osteoblast-like Saos-2 cells harvested from patients in the Department of Orthopedics and Traumatology (O&T), the University of Hong Kong were used in this research. The cells were cultured and proliferated in a new cell culture plate for one day prior to seeding. The scaffolds were sterilized by immersion in 70% ethanol for 1 day (Wiria et al. 2007). After sterilization, the scaffolds were washed and immersed in PBS for 1 day. Cells were trypsinized out of the plate at 37 °C (1ml trypsin to 1 plate/bottle), 5ml culture medium was added to stop trypsin action, and calculated using haemocytometer. $1\text{x}10^5$ cells/mL were added to each scaffold and allowed to attach to the matrix for 30 min before adding enough medium to submerge the scaffolds (Borden et al. 2003). The medium was changed every 3 days. Cell culture was allowed to continued for 7 and 14 days (Shen et al. 2006). Then, the scaffolds were transferred for SEM examination after fixation with glutaraldehyde solution.

3. Results and Discussion

3.1 Morphology of PLLA and PLLA/CHAp microspheres

Some PLLA microspheres and PLLA/CHAp nanocomposite microspheres are shown in Fig. 2 (a) and (b), respectively. Both types of microsphere exhibited a generally similar range of particle sizes between 5 and 30 μm and which were suitable for SLS process. The pure PLLA microspheres appeared very smooth while some CHAp nanospheres were found partially embedded on the PLLA/CHAp nanocomposite microspheres and this may impart bioactivity (osteoconductivity) for the scaffolds built. Fig. 2 (c) shows the FIB-milled section of a PLLA/CHAp composite microsphere. The CHAp nanospheres were generally well distributed both on and inside the microsphere, forming a nanocomposite structure. The rough surfaces of the nanocomposite microspheres likely facilitate cell attachment. In addition, the nanocomposite microspheres flowed more easily than the pure PLLA microspheres, making powder deposition easier during SLS process.

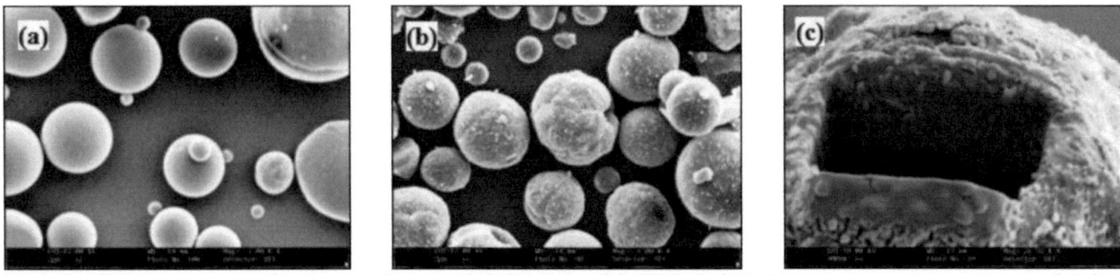

Fig. 2. SEM images of microspheres: (a) PLLA, (b) PLLA/CHAp nanocomposite and (c) an FIB-milled section of a PLLA/CHAp nanocomposite microsphere.

3.2 Effects of sintering parameters on scaffold structure

3.2.1 Effect of laser power

Sintering work was carried out on both PLLA and PLLA/CHAp microspheres covering a range of fill laser power (P) from 7W to 19W. It was found that when P<7W, the scaffolds could not take shape and they fell apart during handling. Fig. 3 (a) is an SEM image showing a surface layer of a Type 1 PLLA scaffold sintered at 19W. It can be seen that the struts of the scaffold were well sintered. The designed dimension of the struts was 300 μm, however, the actual size turned out to be 490±33 μm. The discrepancy was due to "growth", which is a common phenomenon in SLS. When the laser penetrates beyond the design scan range, some polymer particles are either fully or partially melted and adhere onto the designed scaffold structure. Fig. 3 (b) shows a high magnification SEM image of a design pore area. Although the trapped PLLA microspheres still appeared to be in powder form, they were in fact partially fused together and necks were formed between them. It was quite possible when sintering with a high laser power, the temperature of the design pore areas increased to near or even above the melting point of the polymer and caused necking of the microspheres. These trapped microspheres could not be removed by the air-pressure gun and would therefore defeat the function of the porous scaffolds as originally designed. Obviously, a laser power of 19 W was too high and it should be adjusted down so that the powder in the pores could be removed. Finally, a proper set of sintering parameters was obtained and it is shown in Table 2.

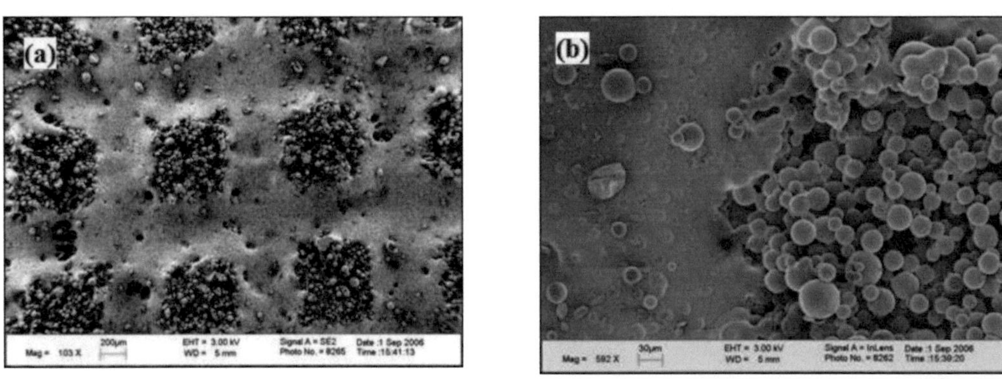

Fig. 3. SEM images of a Type 1 PLLA scaffold sintered at 19W: (a) low magnification showing the overall surface layer structure and (b) high magnification showing necking of PLLA microspheres in the design pore area (on the right side of image).

3.2.2 Effect of part bed temperature

Fig. 4 shows the effect of part bed temperature (PBT) on the structure of scaffolds built from the pure PLLA microspheres. When PBT was within the range as shown in Table 2, the unsintered powder within the pores could be removed easily by an air gun and the resultant scaffolds exhibited a distinctive porous structure (sample a). When PBT was higher than the recommended range however it was difficult to remove all the unsintered powder from the pores (sample b). In normal SLS processes, PBT should be set near the T_g of amorphous polymers and just below the T_m of semi-crystalline polymers, at the time the laser only provides small amounts of extra heat energy for powder fusion. Too high a PBT caused the powder in the pores to partially fuse with the design structure and difficult to be removed. On the other hand, too low a PBT would result in low strength of the scaffolds and they tended to fall apart during handling.

3.2.3 Effect of scan spacing

Fig. 5 shows the effect of scan spacing (SS). Too small an SS would result in excessive amount of laser energy to be deposited on the powder bed surface and cause blocking of the pores (sample a). When SS was too large, the laser energy density became too low for complete sintering of the scaffold structure and hence it could not take shape. In general, extreme values of the processing parameters should be avoided in order to obtain a desirable porous structure (sample b).

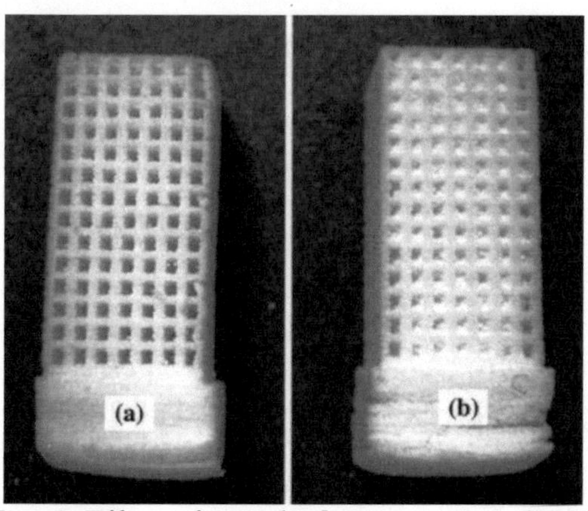

Fig. 4. Effect of part bed temperature (PBT) on structure of PLLA porous scaffolds: (a) PBT = 35 ^{o}C and (b) PBT = 50 ^{o}C.

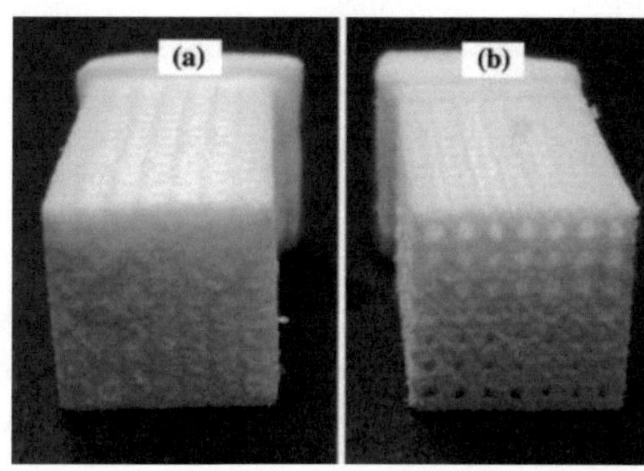

Fig. 5. Effect of scan spacing (SS) on structure of PLLA porous scaffolds: (a) SS = 0.08 mm and (b) SS = 0.18 mm.

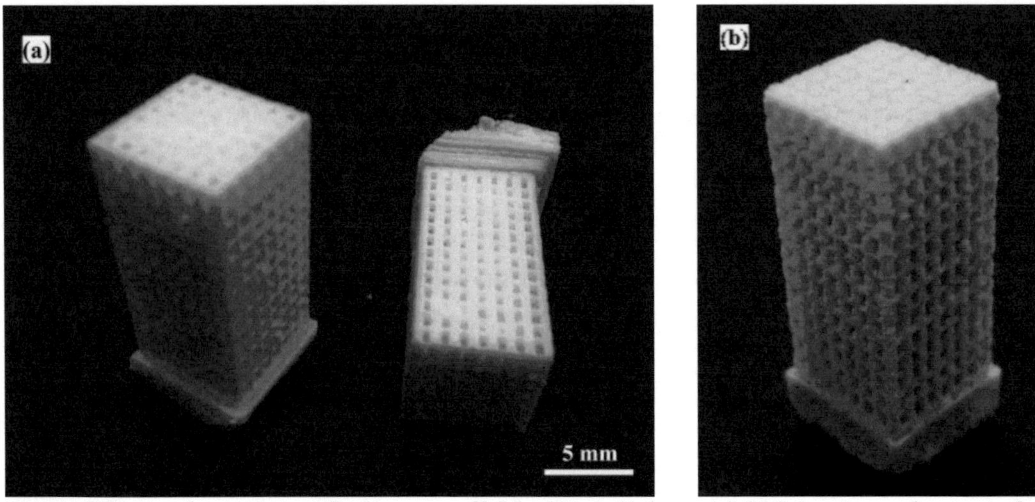

Fig. 6. Selective laser sintered porous scaffolds (Type 1) from (a) PLLA microspheres and (b) PLLA/CHAp nanocomposite microspheres.

3.3 Morphology of scaffolds built under recommended conditions

3.3.1 General appearances

The general appearances of the PLLA and PLLA/CHAp porous scaffolds built under the recommended conditions are shown in Fig. 6 (a) and (b), respectively. After removal of the excessive powder, the pores could be easily recognized and the layers were firmly attached together. Such scaffolds were used for subsequent structural, mechanical, degradation and biological assessments.

3.3.2 Layer morphology

Fig. 7 shows the typical layer morphologies of the PLLA and PLLA/CHAp scaffolds sintered under the recommended conditions. It can be seen that the PLLA microspheres were well fused in the scanned areas, Fig. 7 (a). Nevertheless, the sintered material was not fully dense and some very fine pores were present, resulting in a porous structure possessing a combination of meso-pores (due to design) and micro-pores (due to incomplete fusion of the sintered material). This structure likely facilitates the flow of body fluid and promotes nutrition and metabolism waste exchange, and hence is more beneficial to cell growth than a scaffold with only micro-pores, for example the PCL/HAp disc scaffolds reported by Wiria et al. (2007). Also, the powder preparation methods used in the two studies were different. In this study, the nanocomposite was prepared by emulsion microencapsulation method and the nano-sized CHAp particles were well encapsulated in the PLLA/CHAp microspheres even before sintering, on the other hand the PCL/HAp composites used by Wiria et al. (2007) were physically blended. Furthermore, the biopolymers used in the two studies were different; therefore, the scaffolds produced likely have different *in vitro* and *in vivo* properties. In the laser sintered scaffolds built by Williams et al. (2005) with PCL, the smallest pore size reported was 1.75 mm in diameter. As can be seen from Fig. 7 (b), the pores size of the PLLA scaffold is around 200 μm which is much smaller than the laser spot size of 457 μm. This study showed that it is possible to build scaffolds with pore sizes below the limitation of laser beam width.

When the PLLA and PLLA/CHAp scaffolds were sintered with the respective conditions as shown in Table 2, there was no problem in removing the excessive trapped powder from the pores. For the PLLA scaffolds, a low pressure air gun was needed to remove the excessive powder. The layers were generally well preserved afterward and the pores of the scaffolds could be clearly identified. For the PLLA/CHAp nanocomposite scaffolds, the loose powder could be shaken off by hand. It has been reported in the literature that removing the excessive powder from the pores is a major obstacle for porous scaffold production using the SLS process (Tan et al. 2003). Here, the PLLA/CHAp nanocomposite microspheres seem to offer a solution to the problem provided the sintering conditions are properly set.

There is a difference between the design and actual pore sizes. When the design pore size was 0.8 mm, the actual pore size of sintered scaffolds was only about 438±52 μm as shown in Fig. 7 (a). In Type 2 scaffolds, the pore size was even smaller compared with the design pore size (Fig. 7 (b)). Obviously the reduction in pore size was a result of "growth" which was caused by the penetration of the laser energy beyond the design scan area. Therefore, it is important to take into consideration the effect of "growth" in designing the pore size of the scaffolds built by SLS.

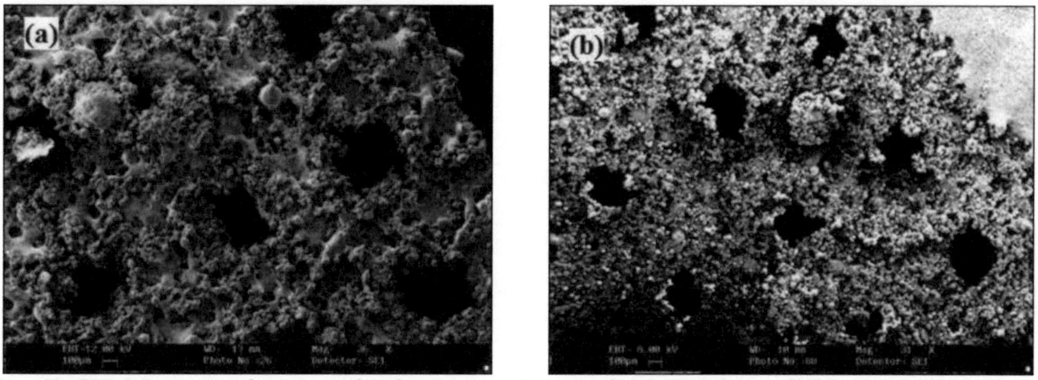

Fig. 7. SEM images showing the layer structures of (a) PLLA scaffold (Type 1) and (b) PLLA/CHAp nanocomposite scaffold (Type 2); sintering condition: P=15W.

3.3.3 Comparison of fusion behavior

It can also be seen from Fig. 8 (a) and (b) that the degree of fusion of the PLLA/CHAp nanocomposite powder was lower than that of the pure PLLA powder. This could be explained by the increased viscosity of the composite material. It has been postulated that a filled polymer will sinter at a different rate as the same polymer without filler, in inverse proportion to the viscosities of the filled and unfilled materials (Childs & Tontowi 2001). Fan et al. studied the material movement and fusion behavior of SiO_2 and glass sphere (GS) filled TrueForm™ composites during selective laser sintering (Fan et al. 2001; Fan 2003). They found that for a given volume fraction of the additive, a smaller particle size of the fillers in dry mixed polymer composites would obstruct fusion of molten polymer particles more significantly and result in weak parts. The embedded fillers increased the material viscosity and adversely affected the densification of the powder bed. Also the CHAp nanoparticles on the powder surface might act as a barrier against fusion. Another possible reason for the lower degree of fusion of the PLLA/CHAp nanocomposite microspheres was the relatively high specific heat capacity of hydroxyapatite. According to Cruz et al. (2005), the molar heat capacity ($C_{p,m}$) of HAp is 694 ± 68 J mol^{-1} K^{-1} in the temperature range 298-1298 K, which is much higher than that of PLLA, from 145.4 to 156.7 J mol^{-1} K^{-1} in the temperature range from 332.5K (T_g) to 480 K (T_m) (Pyda et al. 2004; Malmgren et al. 2006). Therefore, CHAp (a hydroxyapatite based ceramic) can be regarded as a huge energy reservoir to absorb laser energy. So the amount of laser energy input into the PLLA matrix of the composite powder was reduced, resulting in a lower degree of fusion of PLLA/CHAp particles.

Some PLLA microspheres were present in Fig. 8 (a), and most probably bonded to the surface during powder deposition when the sintered material was still hot; as a result they could not be completely removed by the compressed air gun. Such microspheres would increase the surface roughness of the scaffold and this might be advantageous to cell attachment. Underneath the microspheres, the polymer was generally well sintered; nevertheless, some micro-pores were present due to incomplete fusion. This renders a porous structure consisting of both meso-pores (by design) and micro-pores.

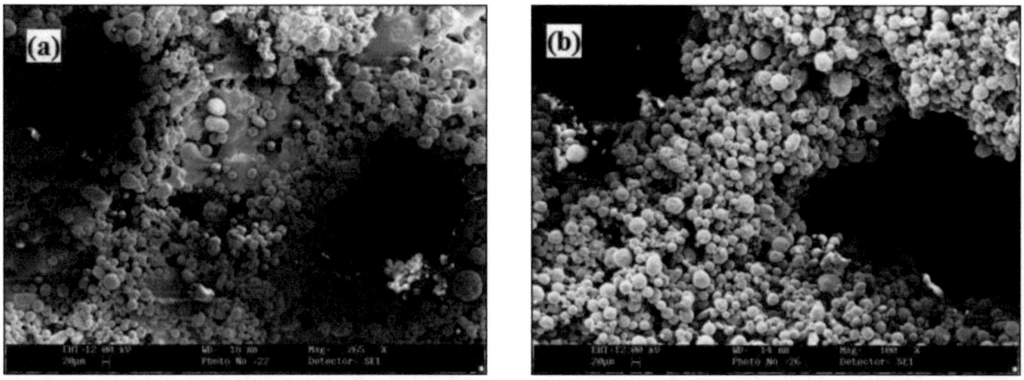

Fig. 8. Comparison of fusion behavior of laser sintered scaffolds (Type 1): (a) PLLA and (b) PLLA/CHAp; sintering condition: P=15W.

3.3.4 Importance of meso-scale feature

As seen above, the scaffolds have similar meso-scale feature compared with some recently developed scaffolds. The 3D channels created into the scaffolds played an important role in cell attachment and proliferation and enhanced nutrient transfer and waste exchange (Rose et al. 2004; Mahoney et al. 2005). A common problem encountered with scaffolds based tissue engineering strategies is the rapid formation of tissue on the outer edge of the scaffold whilst the tissue in the centre becomes necrotic. A way of addressing this problem is to incorporate a specific meso-scale feature design into the scaffold to improve nutrient and cell transfer to the scaffold centre. Recently, Rekow et al. (2006) reviewed that scaffold features at different length scales can have dramatic influences on bone tissue response. Nano-scale features (< 1 μm) of the scaffolds may control interactions at the protein level and thus influence the biocompatibility. Micro-scale scaffold features (1-20 μm) will modulate cell behavior, affecting both the type of cells that adhere to a scaffold surface.

Meso-scale features (200-1000 μm) will influence the structural support of the scaffold and the amount of bone that can develop. The meso-scale has been largely overlooked. Macro-scale features (> 1mm) will define the anatomic shape.
Meso-scale scaffold feature may be achieved by inserting 3D interconnected channels into the body of a scaffold creating a porous structure with different length scales. Frosch and co-workers (2002) assessed the infiltration of human osteoblasts into a range of channel diameters (from 300 to 1000 μm) within titanium implants containing no other form of porosity. They found that cell ingrowth and subsequent matrix formation increased with increasing channel diameter up to 600 μm; however matrix production was sub-optimal in the 1000 μm channel. Xu et al. (2007) pointed out that porous scaffolds should have branched channels to ensure uniform cell feeding and even flow of culture medium through the scaffold to promote uniform cell attachment and growth. Precise control of scaffold architecture (e.g. porosity, pore geometry, size, interconnectivity, orientation, and branching) is needed to optimize the scaffold properties for nutrient diffusion, cell growth and tissue regeneration.

3.4 Properties of scaffolds built under recommended conditions

3.4.1 Effect of laser power on porosity

The porosities of the sintered scaffolds were determined based on the measurement of specific gravity and plotted in Fig. 9 as a function of fill laser power. As expected, the porosities decreased with increasing laser power for both Type 1 (0.8 mm pores) and Type 2 (0.6 mm pores) PLLA and PLLA/CHAp scaffolds. The relationships were almost linear within the recommended range of laser power (see Table 2). All measured porosities of the scaffolds were below the design values. The phenomenon was due to excessive "growth" (especially when sintering was done at high laser power) and the difficulty in removing the unwanted powder from the pores (especially for Type 2 scaffolds). Nevertheless, the measured and design porosities were relatively close for scaffolds sintered at a fill laser power of 11 W, i.e. the lowest value within the recommended range. Apparently, a lower laser power is preferred in terms of controlling the porosity. Furthermore, it is noted that the difference between the measured and design porosities of PLLA scaffolds is much larger than that of the PLLA/CHAp scaffolds, especially in Type 2 scaffolds (i.e. with a smaller design pore size). This indicates that, in order to control the porosity, one not only has to consider the sintering condition, but also the material used and scaffold design. In this respect, PLLA/CHAp seems to be a better material choice than pure PLLA. This can be attributed to its higher viscosity and microsphere surface morphology (containing high content of CHAp nanoparticles) which tend to limit the amount of "growth" and therefore the unsintered powder can be removed more easily. However, it should be pointed out that such material characteristics may affect the fusion behavior (see Fig. 8) and other scaffold properties which will be discussed later.

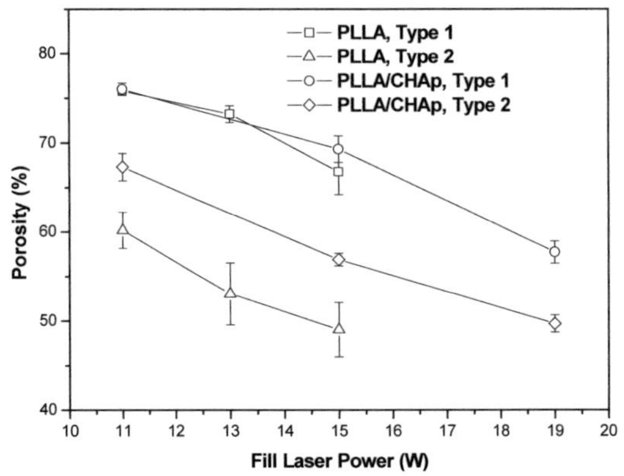

Fig. 9. Effect of fill laser power on porosity of selective laser sintered PLLA and PLLA/CHAp scaffolds, the design porosities of Type 1 and Type 2 scaffolds are 78.4% and 70.0%, respectively.

3.4.2 Effect of laser power on compression properties

Models describing the mechanical behavior of cellular solids such as open-cell foams are well developed (Harley et al. 2007). A typical stress-strain curve for low-density, elastomeric open-cell foam in compression is characterized by three distinct regimes: a linear elastic regime which corresponds to strut bending, a collapse plateau regime which corresponds to strut buckling and pore collapse and a densification regime which corresponds to complete pore collapse throughout the material. Lee (2006) reported two modes of failure of porous selective laser sintered scaffolds built from DuraForm™ PA. Buckling was found to occur when the vertical struts had a higher slenderness ratio or larger pore size, and the scaffold collapsed layer by layer almost vertically. Meanwhile shearing occurred with a lower slenderness ratio of the vertical struts, and the scaffold was compressed into an "S" shape. In this study, shearing was found to be the predominant mode of failure and the scaffolds were compressed into an "S" shape. Fig. 10 shows a typical compressive stress-strain curve of Type 1 PLLA/CHAp scaffolds. It exhibits three regimes which, to a large extent, resemble those commonly observed in cellular structures (Gibson & Ashby 1997). A similar linear elastic regime first appears, from which the compressive modulus can be measured. Beyond this zone, unlike the smooth plateau for low-density, elastomeric open-cell foam, a peaky plateau regime is observed. This peaky plateau can be attributed to sequential collapses of layers due to shearing. Each peak involves a number of layers. Once the shear deformation reaches a critical value, the vertical struts may break causing the compressive load to drop while the heavily deformed layers are collapsing. When all the layers have collapsed, further compressing will give rise to the final regime of densification with further increase of stress.

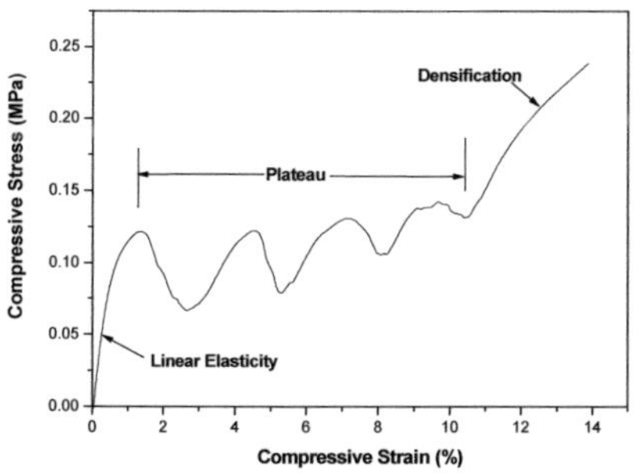

Fig. 10. Typical compressive stress-strain curve of Type 1 PLLA/CHAp scaffold (pore size=0.8mm), sintered at 15W.

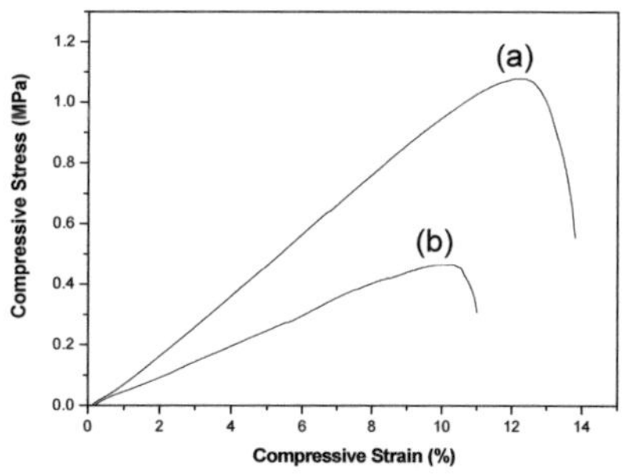

Fig. 11. Typical compressive stress-strain curves of 4-layer PLLA scaffolds (Type 1): (a) sintered at 15W and (b) sintered at 13W.

It was found that sintered scaffolds consisting of 4 layers were sufficient for investigation of the initial linear elastic regime. Therefore, 4-layer specimens were later used to determine the compressive modulus and strength of the different scaffold types in order to save material. Fig. 11 shows the typical compressive stress-strain curves of two such PLLA scaffolds sintered at different laser power. The initial elastic regimes appear relatively linear. It can be seen that the scaffold sintered at 15W has a higher compressive yield strength and strain than that sintered at 13W. Furthermore, the compressive yield strength and strain of the PLLA scaffolds appear to be much higher than those of the Type 1 PLLA/CHAp scaffold shown in Fig. 10.

The overall effects of fill laser power on the compressive yield strength and modulus of the PLLA and PLLA/CHAp scaffolds are shown in Figs. 12 and 13 respectively. It is clear that the compressive yield strength increases with increasing fill laser power. Besides, the compressive yield strength is also affected by both the scaffold type and material. The porosity and fusion behavior are two major factors affecting the mechanical properties of the selective laser sintered scaffolds. For the same material, either PLLA or PLLA/CHAp, Type 2 scaffolds have a lower porosity and hence the compressive yield strength is higher. For the same scaffold type, PLLA gives higher compressive yield strength and modulus than the PLLA/CHAp nanocomposite, and this is mainly due to the better fusion behavior of the pure polymer. Apart from improving the biological property, the ceramic phase (CHAp nanospheres) was added in an attempt also to improve the mechanical property of the scaffolds. On the contrary, the results in Figs. 12 and 13 show a reduction in compressive properties under the current sintering conditions. Perhaps, some sort of post-sintering processing is required to consolidate the structure of the PLLA/CHAp scaffolds before use.

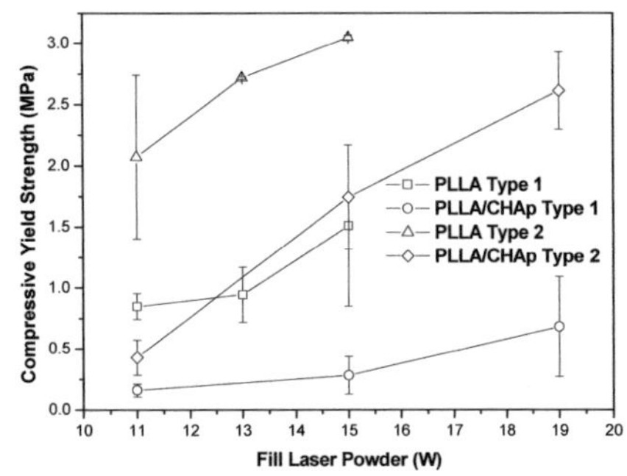

Fig. 12. Effect of fill laser power on the compressive yield strength of laser sintered scaffolds.

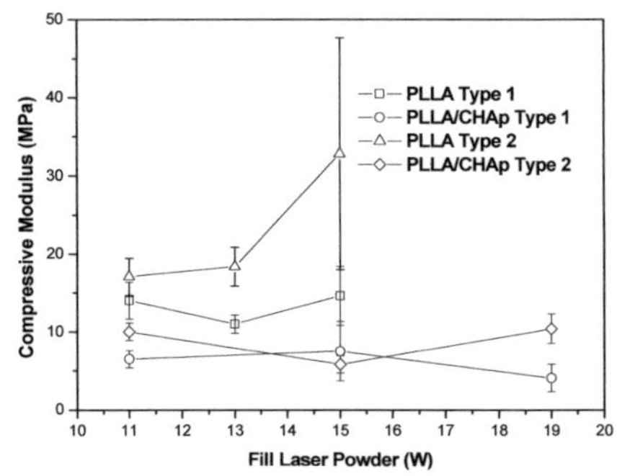

Fig. 13. Effect of fill laser power on the compressive modulus of laser sintered scaffolds.

Mechanical property is an important requirement for scaffold use in bone tissue engineering. As templates to guide tissue regeneration, scaffolds need to have enough mechanical strength in order to retain their shape during *in vitro* and *in vivo* applications. Further requirements for scaffold mechanical properties will differ from applications. For example, for regeneration of trabecular bone, the scaffold will deliver cells and growth factors to the graft site, to mimic cancellous bone grafting where the mass of bony spicules and marrow possess no inherent mechanical strength (Guan & Davies 2004). In such strategy, the scaffold only needs to be strong enough to maintain its dimensional integrity during cell culture. Human cancellous bone (wet condition) has a reported compressive modulus and strength range of 20-900 MPa and 1-15 MPa, respectively. Human cortical bone (wet condition) has a reported compressive modulus and strength range of 5-20 GPa and 100-200 MPa, respectively (Athanasiou et al. 2000; Kofron et al. 2007). In this study, the laser sintered PLLA scaffolds have a compressive modulus and strength range of 15-30 MPa and 1-3 MPa, respectively; the laser sintered PLLA/CHAp scaffolds have a compressive modulus and strength range of 4-10 MPa and 0.2-2.5 MPa, respectively. It indicates that after optimization of the sintering conditions, the SLS scaffolds may achieve mechanical properties suitable for cancellous bone tissue engineering application. Guan & Davies (2004) fabricated a unique composite scaffold for bone tissue engineering by combining poly(lactide-*co*-glycolide) (PLGA) with bioresorbable calcium phosphate cement particles through the process of particle fusion and phase separation/particle leaching. The scaffold is characterized by a highly interconnected macro-porosity, with macro-pores of 0.8-1.8 mm and porosities ranging from 81 to 91%. The addition of calcium phosphate to the matrix prevented scaffold collapse, when placed in media or seeded with cells; however, the compressive strength of the scaffold (0.16 MPa) and modulus (6-8 MPa) are below that of cancellous bone. Williams et al. (2005) fabricated 3D PCL scaffolds by SLS with a pore size 1.75-2.5 mm and porosity 37-55%. The cylindrical scaffolds possessed a compressive modulus of 52-67 MPa and yield strength of 2-3.2 MPa. Polymeric scaffolds fabricated by SLS demonstrated lower mechanical properties when compared with other techniques such as the heat sintered microsphere scaffolds. On the other hand, SLS can fabricate complex scaffolds more easily than microsphere heat sintering method. Kofron et al. (2007) fabricated PLGA/HAp microsphere scaffolds by heat sintering and those scaffolds have a compressive modulus of 100-200 MPa and strength of 20-80 MPa controlled by sintering time and temperature. The difference of mechanical performance for laser sintered microspheres and heat sintered microspheres is due to the energy sources and scaffold design. Heating will generally cause polymer to adhere or melt more easily than laser sintering. In Laurencin's group (Lu et al. 2003), they used 100-200 μm PLGA or PLGA/bioactive glass microspheres to form scaffolds. The average porosity and pore diameter of the PLGA and PLGA/bioactive glass scaffolds were 31 and 43 % and 116 and 89 μm. Thus a lower porosity will give scaffolds higher compression properties. The mechanical properties of the scaffolds are highly dependent on the geometry too. In clinical applications, scaffold geometry will depend on the anatomical location of the injury.

3.4.3 *In vitro* degradation

The weight loss of PLLA and PLLA/CHAp scaffolds over 2 months of *in vitro* degradation test in PBS are shown in Figs. 14 and 15, respectively. In general, the weight loss increases with increasing degradation time and the scattering of data is quite large. The major loss in weight was due to removal of microspheres loosely bonded on the scaffold surface and within the pores, such as those shown in Fig. 8. For Type 1 PLLA scaffolds, Fig. 14 (a), the effect of fill laser power is not clear and the overall weight loss (over 2 months) is around 4-6% for all values of laser power used. Nevertheless, it is quite clear that for Type 1 PLLA/CHAp nanocomposite scaffolds, Fig. 14 (b), the degradation rate decreases with increasing laser power. This is expected as a higher laser power will produce a more compact structure of the scaffolds, reducing the contact between the phosphate buffered saline and the scaffold material. There is a slight trend of abnormal degradation behavior for Type 1 PLLA scaffolds as shown in Fig. 14 (a), i.e. the PLLA scaffolds sintered at a higher fill laser power suffering a higher weight loss than those sintered at a lower laser power. This anomaly was probably due to excessive "growth" under the higher laser power. Although the trapped PLLA microspheres could not be removed by the air-pressure gun, they might be detached from scaffold in PBS during the 2-month degradation test under continuous shaking.

When comparison is drawn between PLLA and PLLA/CHAp scaffolds sintered at 11 W, the nanocomposite scaffolds suffered a much high degradation rate. One possible explanation to the phenomenon is that under this relatively lower laser power (lowest within the recommended range) and due to the poorer fusion behavior of the nanocomposite microspheres, there were a large number of very loosely bonded microspheres within the scaffolds. These microspheres would not survive the test and were detached out relatively quickly. When the laser power was increased to 15 W, however, some of those loosely bonded microspheres became more securely bonded to the solid structure causing the weight loss to drop.

From Fig. 15 (a), Type 2 PLLA scaffolds are found to suffer a much bigger weight loss than and Type 1 PLLA scaffolds. Again, this can be attributed to the larger amount of loosely bonded microspheres being dislodged from the scaffolds structure during the degradation test. Since Type 2 scaffolds possessed smaller pores and therefore more such particles were expected to be trapped within the scaffolds even after the compressed air gun treatment.

The results in Fig. 15 (b) are somewhat unclear. It is suspected that excessive "growth" and different degrees of damage caused during powder removal and degradation test might have contributed to the large variations in the test results.

In general, the degradation time of PLLA is around 2-3 years and there are four steps involved in the degradation process: (1) swelling and hydration of PLLA; (2) breakage of the ester bonds; (3) diffusion of the soluble degradation products; and (4) disappearance of the polymer scaffold struts (Gong et al. 2007). In this study, due to time limit, the degradation tests only lasted for 2 months. Primarily, only the first step was involved and the weight loss was mainly due to detachment of loosely bonded microspheres. There may be some swelling and hydration of PLLA involved too. With regard to the degradation mechanism of PLLA, the commonly held believe is that PLLA is a hydrolyzable polymer that follows a bulk degradation (Gong et al. 2007). The study by von Burkersroda et al. (2002) showed that all degradable polymers can undergo surface degradation or bulk degradation. The way a polymer matrix degrades all depends on the diffusion of water inside the matrix, the degradation rate of the polymer's functional groups and the matrix dimension. The dimension is an important factor. If a matrix is larger than a critical device dimension, it will undergo surface degradation; if not, it will be bulk degradation. Polyanhydrides were found to be surface degradation down to a size of approximately 10^{-4} m while poly(α-hydroxy esters) matrices need to be larger than 10^{-1} m to lose their bulk degradation properties. The strut and microsphere sizes of the sintered PLLA and PLLA/CHAp scaffolds are all less than 0.1 m in this study which would show bulk degradation characteristics. The ester bonds in the backbone chain break randomly to generate carboxyl groups, which in turn autocatalytically accelerate the degradation reaction, resulting in an exponential (Park 1995) or linear (Gong et al. 2007) decrease of the molecular weight of a bulk PLLA material with degradation time. The intermediate degradation products are trapped inside the material before their molecular weights decrease to a critical value of about 1100 to be soluble in water. The loss of the precise shape of the tissue engineering scaffold after implantation is a common problem (Shieh et al. 2004). This study showed that the laser sintered scaffolds could maintain their original shapes after 2 months *in vitro* degradation which may be beneficial for implantation of the scaffolds *in vivo*. However, compared with the degradation *in vitro*, the degradation rate of the PLLA scaffolds *in vivo* was much faster (Dijk et al. 2002; Gong et al. 2007). The *in vivo* environment, including the easy exchange of substances, the enzymes and loading condition, is usually regarded to accelerate the degradation of the polymer scaffold. *In vitro* degradation of biodegradable implants has been commonly used to predict their *in vivo* behavior, but it is difficult to take all the *in vivo* environmental variables into account. In order to optimize the scaffolds for real tissue engineering applications, ideally *in vivo* experiments should be performed. However, such experiments involve animal tests and it is beyond the scope of this study.

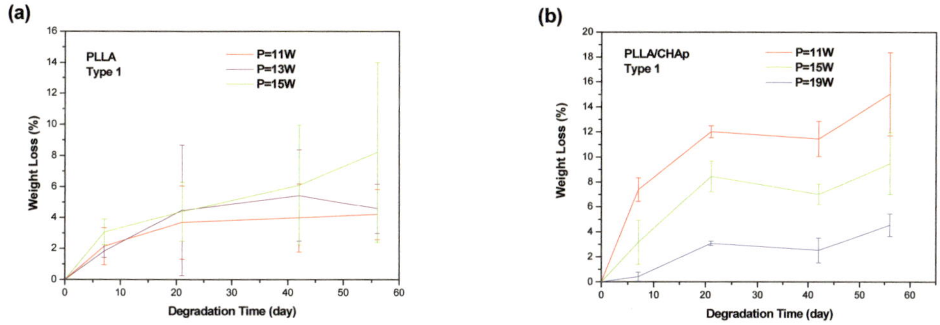

Fig. 14. Weight loss of porous laser sintered Type 1 scaffolds during degradation in PBS: (a) PLLA scaffolds and (b) PLLA/CHAp scaffolds; sampling size n=3.

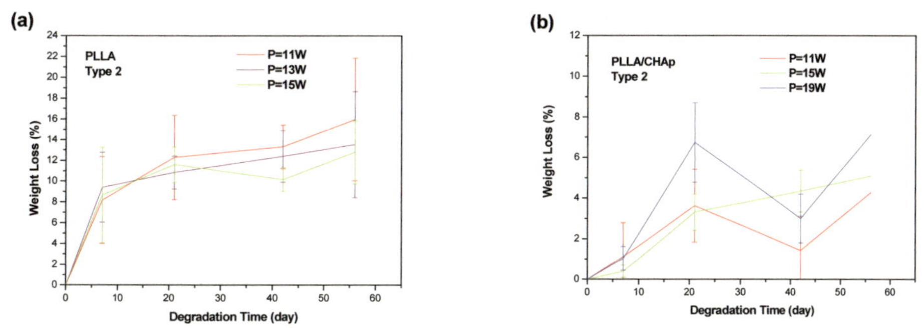

Fig. 15. Weight loss of porous laser sintered Type 2 scaffolds during degradation in PBS: (a) PLLA scaffolds and (b) PLLA/CHAp scaffolds; sampling size n=3.

3.4.4 Compression properties after *in vitro* degradation

The compressive yield strength and modulus of the PLLA and PLLA/CHAp scaffolds were measured through the 2-month *in vitro* degradation test and the results are shown in Figs. 16 and 17. In general, the compression properties decreased with increasing degradation time. The modulus dropped more significantly within the first week of the degradation test and then followed by a more gradual decrease. This trend coincides with the rate of weight loss as shown in Figs. 14 and 15. Furthermore, it can be seen that the change of compressive modulus was much bigger than that of compressive strength. Similar effect was observed by Yang et al. (2006). One possible explanation was that the modulus was more dependent on the amount of loose particles trapped within the scaffold structure. When under compression, some of these particles were squeezed more tightly together giving a support to the scaffold structure, particularly at the early stage of the compression test in which the modulus was determined. Once these particles were removed after the degradation test there was a significant drop in the modulus. In comparison, compressive yielding of the scaffolds occurred at a later stage and under a more complex loading situation. Shearing and bending of the vertical struts would likely involve local tensile stresses and cracking.

Under these circumstances, the loose particles trapped within the scaffold structure would not contribute significantly to the yield strength.

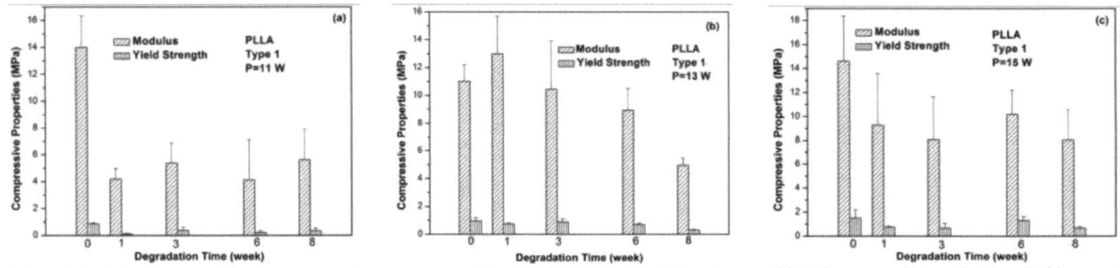

Fig. 16. Change of compression properties of Type 1 PLLA scaffolds produced by different laser powers during *in vitro* degradation.

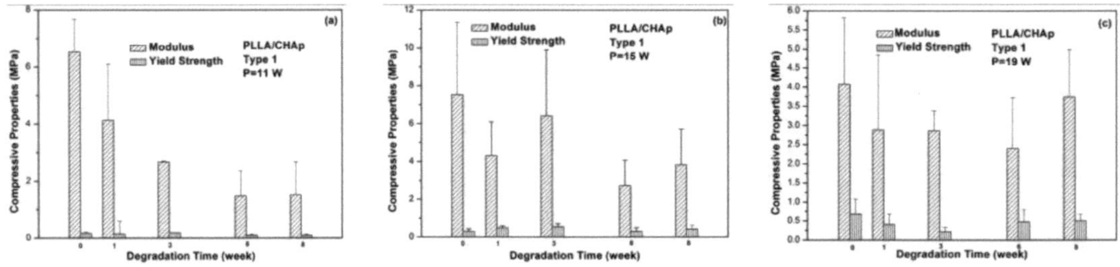

Fig. 17. Change of compression properties of Type 1 PLLA/CHAp scaffolds produced by different laser powers through *in vitro* degradation test.

3.4.5 *In vitro* cell culture

The cells used were Saos-2, the human epithelial-like osteosarcoma cell line. These cells are widely used in studies on bone cell differentiation, proliferation and metabolism and are known to be capable of bone production (Causa et al. 2006; Wiria et al. 2007). The Saos-2 cells possess several osteoblastic features and could be useful as a permanent line of human osteoblast-like cells and as a source of bone-related molecules. *In vitro* studies with these osteoblast-like cells have offered insights into the biological performance of bone implant materials. Figs. 18 and 19 show the SEM images of the cell-seeded Type 1 PLLA and PLLA/CHAp scaffolds respectively after culturing for 7 and 14 days. It can be seen that Saos-2 cells were able to attach to both scaffolds. However, the cells did not spread well on the PLLA scaffold surface and the outlines of many microspheres were still easily observable. In the case of the PLLA/CHAp scaffold, however, the cells were well spread and connected with each other covering most of the scaffold surface. Borden et al. (2003) attached human osteoblast cells on the heat sintered PLGA microsphere scaffolds. Their results, after culturing for 7 and 16 days, showed that the cells had bypassed the top surface of the microspheres and only formed a layer covering the surface of the microspheres within the pores. In another study on laser sintered PCL/HAp scaffolds (Wiria et al. 2007), the Saos-2 cells also did not cover the scaffold surface extensively. Therefore, the result obtained strongly indicates that the PLLA/CHAp nanocomposite microspheres offer a more biomimetic environment for attachment and proliferation of osteoblast-like cells.

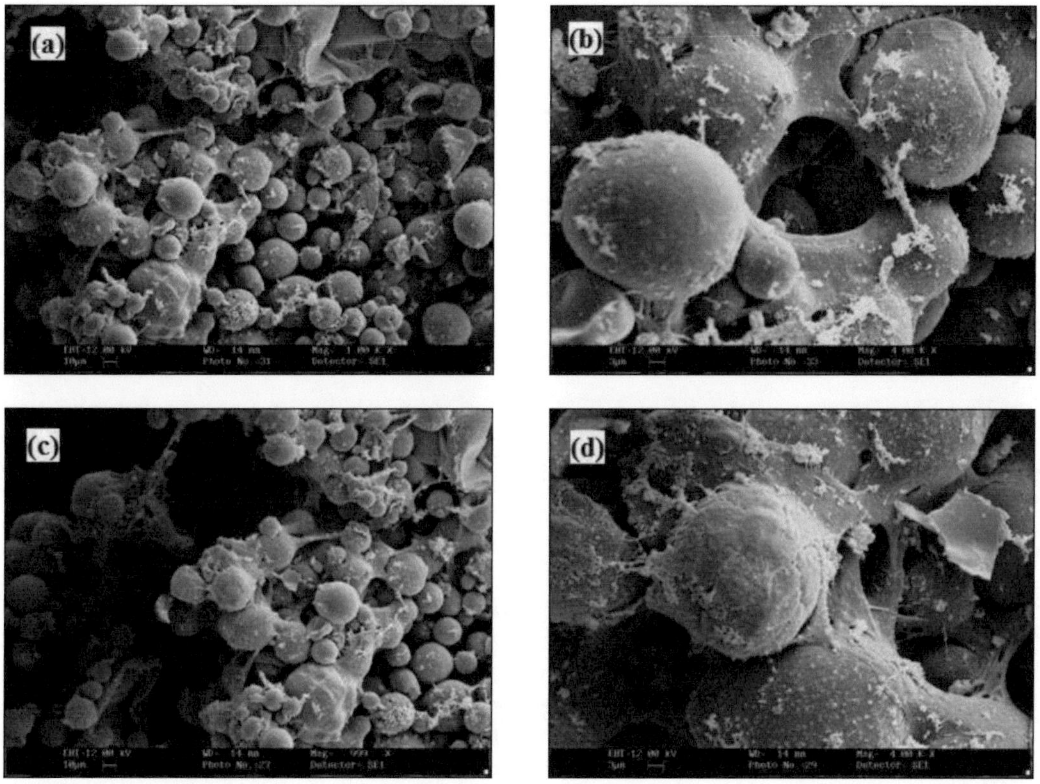

Fig. 18. SEM images of Type 1 PLLA scaffold seeded with Saos-2 cells: (a) 7 days after seeding and (b) close-up view of centre region in (a); (c) 14 days after seeding and (d) close-up view of centre region in (c), scaffold sintered at 15 W.

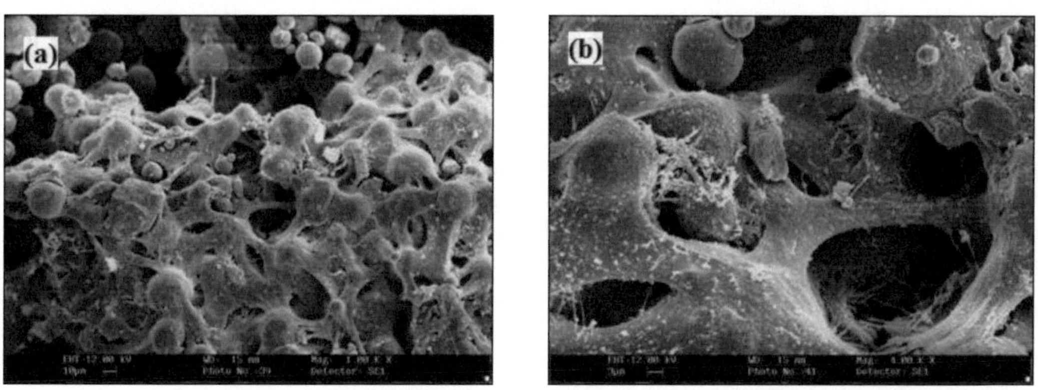

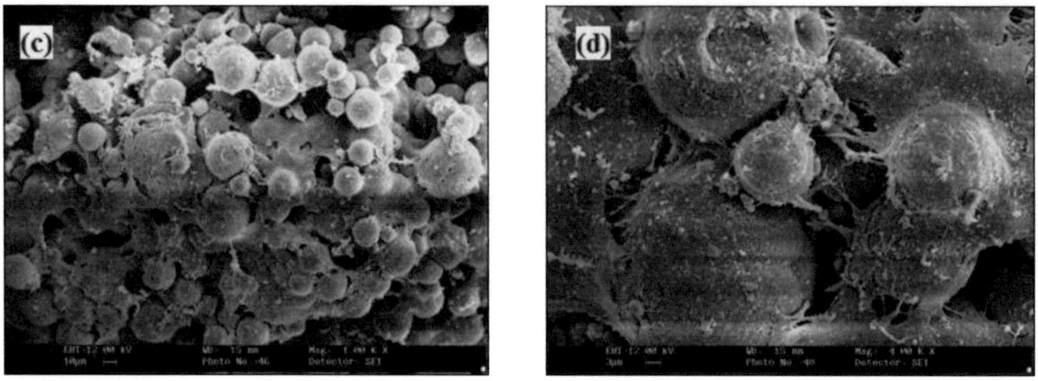

Fig. 19. SEM images of Type 1 PLLA/CHAp scaffold seeded with Saos-2 cells: (a) 7 days after seeding and (b) close-up view of centre region in (a); (c) 14 days after seeding and (d) close-up view of centre region in (c), scaffold sintered at 15 W.

4. Conclusions

From this study, the following conclusions could be drawn:

(1) The custom-made miniature sintering platform allowed small quantities of the biomaterial powders to be processed in a commercial Sinterstation® 2000 SLS machine. With this platform, prototypes of bone tissue engineering scaffolds with meso-scale feature were successfully built from PLLA microspheres and PLLA/CHAp nanocomposite microspheres.

(2) The effects of laser power, scan spacing and part bed temperature on the scaffold structure were studied and discussed. In general, extreme values of the processing parameters should be avoided in order to obtain desirable scaffold properties.

(3) The actual pore sizes obtained were smaller than the design values due to the phenomenon of "growth". Since "growth" cannot be avoided entirely, it should be accounted for in the design stage.

(4) The porosities of laser sintered PLLA and PLLA/CHAp scaffolds decreased almost linearly with increasing fill laser power within the recommended range. Furthermore, the sintered material was not fully dense. The degree of fusion of the PLLA/CHAp nanocomposite microspheres was lower than that of the pure PLLA powder. The phenomenon was caused by an increase in the molar heat capacity and viscosity of the nanocomposite powder after the addition of CHAp.

(5) The stress-strain curves of the scaffolds exhibited three regimes, i.e. linear elasticity, plateau and densification. The PLLA/CHAp scaffolds were found to have lower compressive yield strength and modulus than the PLLA scaffolds of the same type and sintering conditions. This could generally be attributed to the insufficient fusion of the PLLA/CHAp nanocomposite microspheres. Their compressive strength after optimization of the sintering condition would only match those of human cancellous bone.

(6) In the *in vitro* degradation test, most weight loss occurred in the first 2 weeks and it was mostly due to detachment of the microspheres loosely bonded on the scaffold surface and within the pores. This caused a corresponding drop in the compressive properties and the drop was more significant in modulus than in compressive yield strength for both the PLLA and PLLA/CHAp scaffolds.

(7) Saos-2 cells were able to attach to both the PLLA and PLLA/CHAp scaffolds during the *in vitro* cell culture experiments. The PLLA/CHAp nanocomposite scaffolds showed a higher osteoconductivity than the neat PLLA scaffolds. The results strongly suggested that the PLLA/CHAp nanocomposite microspheres offered a more biomimetic environment than the pure PLLA microspheres for attachment and proliferation of osteoblast-like cells.

5. Acknowledgements

This work was fully supported by a General Research Fund (GRF) (HKU 7118/05E) from the Hong Kong Research Grants Council. The authors thank Mr. S.H. Lee (Department of Mechanical Engineering, HKU) for assisting the modification of Sinterstation® 2000 SLS machine and Mr. C.K. Yuen (Department of Orthopaedics and Traumatology, HKU) for his help in the cell culture experiments. Dr. W.Y. Zhou sincerely thanks the University of Hong Kong for providing him the research studentship for the doctorate training.

6. References

ASTM (D695-02a). Standard test method for compressive properties of rigid plastics. PA, US, ASTM International: 1-8.

ASTM (D1621-04a). Standard test method for compressive properties of rigid cellular plastics. PA, US, ASTM International: 1-5.

Athanasiou, K. A., C. Zhu, et al. (2000). Fundamentals of biomechanics in tissue engineering of bone. *Tissue Engineering* 6, 4, 361-381.

Borden, M., S. F. El-Amin, et al. (2003). Structural and human cellular assessment of a novel microsphere-based tissue engineered scaffold for bone repair. *Biomaterials* 24, 4, 597-609.

Caulfield, B., P. E. McHugh, et al. (2007). Dependence of mechanical properties of polyamide components on build parameters in the SLS process. *Journal of Materials Processing Technology* 182, 1-3, 477-488.

Causa, F., P. A. Netti, et al. (2006). Poly-ε-caprolactone/hydroxyapatite composites for bone regeneration: In vitro characterization and human osteoblast response. *Journal of Biomedical Materials Research Part A* 76A, 1, 151-162.

Childs, T. H. C. & A. E. Tontowi (2001). Selective laser sintering of a crystalline and a glass-filled crystalline polymer: Experiments and simulations. *Proceedings of the Institution of Mechanical Engineers Part B: Journal of Engineering Manufacture* 215, 11, 1481-1495.

Chua, C. K., K. F. Leong, et al. (2004). Development of tissue scaffolds using selective laser sintering of polyvinyl alcohol/hydroxyapatite biocomposite for craniofacial and joint defects. *Journal of Materials Science: Materials in Medicine* 15, 10, 1113-1121.

Ciardelli, G., V. Chiono, et al. (2004). Innovative tissue engineering structures through advanced manufacturing technologies. *Journal of Materials Science: Materials in Medicine* 15, 4, 305-310.

Cruz, F., J. N. C. Lopes, et al. (2005). A Molecular dynamics study of the thermodynamic properties of calcium apatites. 1. hexagonal phases. *Journal of Physical Chemistry. B, Condensed Matter, Materials, Surfaces, Interfaces, & Biophysical Chemistry* 109, 51, 24473-24479.

Das, S., S. J. Hollister, et al. (2003). Freeform fabrication of Nylon-6 tissue engineering scaffolds *Rapid Prototyping Journal* 9, 1, 43-49.

Dijk, M. v., D. C. Tunc, et al. (2002). In vitro and in vivo degradation of bioabsorbable PLLA spinal fusion cages. *Journal of Biomedical Materials Research* 63, 6, 752-759.

Fan, K. M. (2003). PhD Thesis: Heat Transfer Propertites and Fusion Behavior of Polymer Based Compoiste Powders in Selective Laser Sintering, The University of Hong Kong.

Fan, K. M., W. L. Cheung, et al. (2001). Material movement and fusion behavior of TrueForm and TrueForm/SiO2 during selective laser sintering. *Proceedings of the 12th Annual Solid Freeform Fabrication Symposium, The University of Texas, Austin, TX*146-154.

Froscha, K.-H., F. Barvencika, et al. (2002). Migration, Matrix Production and Lamellar Bone Formation of Human Osteoblast-Like Cells in Porous Titanium Implants. *Cells Tissues Organs* 170, 214-227.

Gibson, L. J. & M. F. Ashby (1997). *Cellular Solids: Structure and Properties*. Cambridge, UK, Cambridge University Press.

Gong, Y., Q. Zhou, et al. (2007). In vitro and in vivo degradability and cytocompatibility of poly(l-lactic acid) scaffold fabricated by a gelatin particle leaching method. *Acta Biomaterialia* 3, 4, 531-540.

Guan, L. & J. E. Davies (2004). Preparation and characterization of a highly macroporous biodegradable composite tissue engineering scaffold. *Journal of Biomedical Materials Research Part A* 71A, 3, 480-487.

Hao, L., M. M. Savalani, et al. (2006). Selective laser sintering of hydroxyapatite reinforced polyethylene composites for bioactive implants and tissue scaffold development. *Proceedings of the Institution of Mechanical Engineers, Part H: Journal of Engineering in Medicine* 220, 4, 521-531.

Harley, B. A., J. H. Leung, et al. (2007). Mechanical characterization of collagen-glycosaminoglycan scaffolds. *Acta Biomaterialia* 3, 4, 463-474.

Hollister, S. J. (2005). Porous scaffold design for tissue engineering. *Nature Materials* 4, 7, 518-524.

Kofron, M. D., J. A. Cooper Jr, et al. (2007). Novel tubular composite matrix for bone repair. *Journal of Biomedical Materials Research Part A* 82A, 2, 415-425.

Lee, G. & J. W. Barlow (1993). Selective laser sintering of bioceramic materials for implants. *Proceedings of the Solid Freeform Fabrication Symposium, Austin, TX*376-380.

Lee, S. H. (2006). M.Phil. Thesis: Feasibility Study of Selective Laser Sintering of Biopolymer Scaffolds for Tissue Engineering, the University of Hong Kong.

Lee, S. H., W. Y. Zhou, et al. (2004). *Selective laser sintering of porous scaffolds for bone tissue engineering*. Proceedings of the Biomedical Engineering Conference (BME2004), Hong Kong.

Lu, H. H., S. F. El-Amin, et al. (2003). Three-dimensional, bioactive, biodegradable, polymer-bioactive glass composite scaffolds with improved mechanical properties support collagen synthesis and mineralization of human osteoblast-like cellsin vitro. *Journal of Biomedical Materials Research* 64, 3, 465-474.

Mahoney, M. J., R. R. Chen, et al. (2005). The influence of microchannels on neurite growth and architecture. *Biomaterials* 26, 7, 771-778.

Malmgren, T., J. Mays, et al. (2006). Characterization of poly(lactic acid) by size exclusion chromatography, differential refractometry, light scattering and thermal analysis. *Journal of Thermal Analysis & Calorimetry* 83, 1, 35-40.

Navarro, M., S. del Valle, et al. (2004). New macroporous calcium phosphate glass ceramic for guided bone regeneration. *Biomaterials* 25, 18, 4233-4241.

Nelson, J. C., S. Xue, et al. (1993). Model of the selective laser sintering of bisphenol-a polycarbonate. *Industrial & Engineering Chemistry Research* 32, 10, 2305-2317.

Park, T. G. (1995). Degradation of poly(lactic-co-glycolic acid) microspheres: effect of copolymer composition. *Biomaterials* 16, 15, 1123-1130.

Partee, B., S. J. Hollister, et al. (2006). Selective laser sintering process optimization for layered manufacturing of CAPA (R) 6501 polycaprolactone bone tissue engineering scaffolds. *Journal of Manufacturing Science and Engineering* 128, 2, 531-540.

Pyda, M., R. C. Bopp, et al. (2004). Heat capacity of poly(lactic acid). *The Journal of Chemical Thermodynamics* 36, 9, 731-742.

Rekow, D., P. Van Thompson, et al. (2006). Influence of scaffold meso-scale features on bone tissue response. *Journal of Materials Science* 41, 16, 5113-5121.

Rose, F. R., L. A. Cyster, et al. (2004). In vitro assessment of cell penetration into porous hydroxyapatite scaffolds with a central aligned channel. *Biomaterials* 25, 24, 5507-14.

Schieker, M., H. Seitz, et al. (2006). Biomaterials as scaffold for bone tissue engineering. *European Journal of Trauma* 32, 2, 114-124.

Shen, F. H., Q. Zeng, et al. (2006). Osteogenic differentiation of adipose-derived stromal cells treated with GDF-5 cultured on a novel three-dimensional sintered microsphere matrix. *The Spine Journal* 6, 6, 615-623.

Shieh, S.-J., S. Terada, et al. (2004). Tissue engineering auricular reconstruction: in vitro and in vivo studies. *Biomaterials* 25, 9, 1545-1557.

Sinha, A., G. Das, et al. (2007). Poly(vinyl alcohol)-hydroxyapatite biomimetic scaffold for tissue regeneration. *Materials Science and Engineering:* C 27, 1, 70-74.

Tan, K. H., C. K. Chua, et al. (2003). Scaffold development using selective laser sintering of polyetheretherketone-hydroxyapatite biocomposite blends. *Biomaterials* 24, 18, 3115-23.

Venuvinod, P. K. & W. Ma (2004). Selective Laser Sintering (SLS). *Rapid prototyping : laser-based and other technologies*. Boston Kluwer Academic: 248.

von Burkersroda, F., L. Schedl, et al. (2002). Why degradable polymers undergo surface erosion or bulk erosion. *Biomaterials* 23, 21, 4221-4231.

Williams, J. M., A. Adewunmi, et al. (2005). Bone tissue engineering using polycaprolactone scaffolds fabricated via selective laser sintering. *Biomaterials* 26, 23, 4817-4827.

Wiria, F. E., K. F. Leong, et al. (2007). Poly-ε-caprolactone/hydroxyapatite for tissue engineering scaffold fabrication via selective laser sintering. *Acta Biomaterialia* 3, 1, 1-12.

Xu, S., D. Li, et al. (2007). Fabrication of a calcium phosphate scaffold with a three dimensional channel network and its application to perfusion culture of stem cells. *Rapid Prototyping Journal* 13, 2, 99-106.

Yang, F., W. Cui, et al. (2006). Poly(l,l-lactide-co-glycolide)/tricalcium phosphate composite scaffold and its various changes during degradation in vitro. *Polymer Degradation and Stability* 91, 12, 3065-3073.

Yang, S., K. F. Leong, et al. (2002). The design of scaffolds for use in tissue engineering. part II. rapid prototyping techniques. *Tissue Engineering* 8, 1, 1-11.

Zhou, W. Y., S. H. Lee, et al. (2007). Selective laser sintering of tissue engineering scaffolds using poly(L-Lactide) microspheres. *Key Engineering Materials* 334-335, 1225-1228.

Zhou, W. Y., S. H. Lee, et al. (2008). Selective laser sintering of porous tissue engineering scaffolds from poly(L-lactide)/carbonated hydroxyapatite nanocomposite microspheres. *Journal of Materials Science: Materials in Medicine* 19, 7, 2535-2540.

Zhou, W. Y., M. Wang, et al. (2007). Fabrication and characterization of composite microspheres containing carbonated hydroxyapatite nanoparticles. *Key Engineering Materials* 334-335, 1221-1224.

Zhou, W. Y., M. Wang, et al. (2008). Synthesis of carbonted hydroxyapatite nanospheres through nanoemulsion. *Journal of Materials Science: Materials in Medicine* 19, 1, 103-110.

10

Perspectives of Chitin and Chitosan Nanofibrous Scaffolds in Tissue Engineering

R. Jayakumar[1*], S. V. Nair[1], T. Furuike[2] and H. Tamura[2*]
[1]*Amrita Center for Nanosciences, Amrita Institute of Medical Sciences and Research Centre, Amrita Vishwa Vidyapeetham University, Kochi-682 026, India*
[2]*Faculty of Chemistry, Materials and Bioengineering & High Technology Research Centre, Kansai University, Osaka 564-8680, Japan*

Abstract

Chitin and its deacetylated derivative, chitosan, are non-toxic, biodegradable biopolymers currently being developed for use in biomedical applications such as tissue engineering scaffolds, wound dressings, separation membranes, antibacterial coatings, stent coatings, and sensors. Recently, nano fibrous scaffolds based on chitin or chitosan have potential applications in tissue engineering. Tissue engineering is one of the most exciting interdisciplinary and multidisciplinary research areas today, and there has been exponential growth in the number of research publications in this area in recent years. It involves the use of living cells, manipulated through their extracellular environment or genetically to develop biological substitutes for implantation into the body and/or to foster remodeling of tissues in some active manners. Electrospun chitin and chitosan nano fibrous scaffolds would be used to produce tissue engineering scaffolds with improved cytocompatibility, which could mimic the native extracellular matrix (ECM). Electrospinning is truly a feasible means of producing nano fibrous scaffolds that resemble the ECM, however, moreover than this, it is imperative that the effects of an artificial matrix has on cell growth, proliferation, and differentiation. This review summarizes the recent progress in chitin and chitosan based nano fibrous scaffolds with an emphasis in tissue engineering applications.

1. Introduction

Chitin, the second most abundant natural polysaccharide, is synthesized by a number of living organisms. Chitin occurs in nature as ordered microfibrils, and is the major structural component in the exoskeleton of arthropods and cell walls of fungi and yeast. The main commercial sources of chitin are crab and shrimp shells, which are abundantly supplied as

*Corresponding author: Tel: +91-484-2801234; Fax: +91-484-2802020.
E-mail address: rjayakumar@aims.amrita.edu (R. Jayakumar) & tamura@ipcku.kansai-u.ac.jp (H. Tamura)

waste products of the seafood industry. Because chitin is not readily dissolved in common solvents, it is often converted its more deacetylated derivative, chitosan (Kurita, 2001; Rinaudo, 2006; Pillai et al., 2009). Chitosan is often identified by its degree of deacetylation (DD), a percentage measurement of free amine groups along the chitosan backbone. Generally, the material is considered chitosan when it becomes soluble in dilute acidic solutions, which occurs when DD ≥ 60% (Roberts, 1992). Because of its ease of solubility, chitosan is preferred over chitin for a wide range of applications.

Chitin and chitosan are biocompatible, biodegradable, and nontoxic, and are anti-microbial and hydrating agents. Chitin and chitosan are easily processed into gels (Nagahama et al., 2008a; 2008b), membranes (Jayakumar et al., 2005; Jayakumar et al., 2007; Jayakumar et al., 2009; Madhumathi et al., 2009a; Madhumathi et al., 2009b), nanofibers (Schiffman & Schauer, 2007a; Shalumon et al., 2009), beads (Jayakumar et al., 2006), microparticles (Prabaharan & Mano, 2005), nanoparticles (Anitha et al., 2009), scaffolds (Madhumathi et al., 2009c; Maeda et al., 2009) and sponges (Muramatsu et al., 2003; Portero et al., 2007) forms. There are a number of promising applications of nanoscale thin films and fibers of chitin/chitosan (Wang & Hon, 2003; Rinaudo, 2006; Pillai et al., 2009).

Recently, much attention has been paid to electrospinning process as a unique technique because it can produce polymer nanofibers with diameter in the range from several micrometers down to tens of nanometers, depending on the polymer and processing conditions. In electrospinning, a high voltage is applied to create electrically charged jets of a polymer solution. These jets dry to form nanofibers, which are collected on a target as a non-woven fabric. These nanofibers are of considerable interest for various kinds of applications, because they have several useful properties such as high specific surface area and high porosity. Nanofibrous non-woven fibers, containing chitin or chitosan, yield potential applications in areas such as filtrations, recovery of metal ions, drug release, dental, tissue engineering, catalyst and enzyme carriers, wound healing, protective clothing, cosmetics, biosensors, medical implants and energy storage (Zhang et al., 2005; Fang et al., 2008). In this review, we are reporting about the different preparation and tissue engineering applications of electrospun chitin and chitosan based nanofibers in detail.

2. Electrospinning of chitin and chitosan

2. 1. Electrospinning of chitin

Chitin is insoluble in most of the organic solvents. Due to its insolubility, its applications are limited in many applications. Chitin dissolves only in specific solvents such as N, N-dimethylacetamide (DMAC)-LiCl (Cho et al., 2000), hexafluoroacetone, 1,1,1,3,3,3-hexafluoro-2-propanol (HFIP) (Kurita, 2001) and saturated calcium solvent (Jayakumar & Tamura, 2008; Nagahama et al., 2008a; Nagahama et al., 2008b). Min et al., 2004 performed electrospinning of chitin using HFIP solvent. Before electrospinning the chitin was depolymerized by gamma radiation to improve the solubility. Although as-spun chitin nanofibers had the broad fiber diameter distribution, most of the fiber diameters are less than 100 nm. Junkasem et al. 2006 reported that nanocomposite fiber mats composed of electrospun poly(vinyl alcohol) (PVA) containing α-chitin whiskers prepared from shells of Penaeus merguiensis shrimps have been fabricated. This research was conducted since both PVA had previously been electrospun (Ding et al., 2002) in water and chitin/PVA nanocomposite films containing α-chitin whiskers have also been fabricated (Sriupayo et al.,

2005a; Sriupayo et al., 2005b). A maximum tensile strength value of 5.7 ± 0.6 MPa was obtained when the chitin whisker to PVA ratio was approximately 5.1%, after this point, increasing the chitin content decreased the strength of the mats.

Park et al. 2006a developed electrospun chitin/poly(glycolic acid) (PGA) blend nanofibers in HFIP was investigated to fabricate biodegradable and biomimetic nanostructured scaffolds for tissue engineering. PGA was chosen because it was a biocompatible and biodegradable polymer. The PGA/chitin blend fibers have average diameters of around 140 nm, and their diameters have a distribution in the range 50–350 nm. In vitro degradation studies, conducted in phosphate-buffered saline (PBS), pH 7.2 demonstrated that the blend fibers degraded faster than pure PGA fibers. Also, electrospun chitin/silk fibroin (SF) nanofibers in HFIP solvent was reported (Park et al., 2006b). The average diameters of chitin/SF blend fibers decreased from 920 to 340 nm, with the increase of chitin content in blend compositions. Shalumon et al., 2009 developed a novel electrospun water-soluble carboxymethyl chitin (CMC)/PVA blend was successfully prepared by electrospinning technique. The concentration of CMC (7%) with PVA (8%) was optimized, blended in different ratios (0–100%) and electrospun to get nanofibers. Fibers were made water insoluble by cross-linking with glutaraldehyde vapors followed by thermal treatment.

2. 2. Electrospinning of chitosan

The electrospinning process is determining the optimal and suitable solvent system. Since chitosan is soluble in most of the acids. However, the protonation of chitosan changes it into a polyelectrolyte in acidic solutions. Chitosan is the only psuedonatural cationic polymer (Kurita 2001; Rinaudo, 2006; Pillai et al., 2009) and in general there are only a few reports on ionic polymers or polyelectrolytes that have successfully been electrospun (Duan et al., 2004). It has been theorized by Min et al., 2004 that the repulsive force, between ionic groups within polymer backbone that arise due to the application of a high electric field during electrospinning, restrict the formation of continuous fibers and often produce particles. There has been some work (Mckee et al. 2004; Mckee et al. 2006) towards developing an empirical equation for fiber diameter and the effects that the intermolecular associations state of a polyelectrolyte have on electrospinning was reported.

An electrospun nonwoven fabric of chitosan, was successfully prepared (Ohkawa et al., 2004). This study focuses on the effect of the electrospinning solvent and the chitosan concentration on the morphology of the resulting nonwoven fabrics. The solvents tested were diluted hydrochloric acid, acetic acid, formic acid and trifluoroacetic acid (TFA). As the chitosan concentration was increased, the morphology of the deposition on the collector changed from spherical beads to interconnected fibrous networks. The addition of dichloromethane to the chitosan-TFA solution was improved the homogeneity of the electrospun chitosan fiber. Under optimized conditions, homogenous (not interconnected) chitosan fibers with a mean diameter of 330 nm were prepared. Ohkawa et al., 2004 proposed that TFA is the main constituent in the successful solvent system for chitosan because the amino groups of the chitosan can form salts (Hasegawa et al., 1992) with the TFA, which can effectively destroy the rigid interactions between the chitosan molecules thus facilitating electrospinning. Similarly, Haider & Park, 2009 developed electrospun chitosan nanofibers for absorbing metal ions. A follow-up study was conducted by Ohkawa et al. 2006 which focused on idealizing the viscosity of their solutions (Baumgarten et al., 1971; Fridrikh et al., 2003) in order to decrease the average fiber diameter. It was determined that fiber diameter and polymer concentration have an inverse relationship, there was a linear increase of fiber diameter as the concentration of chitosan in solution decreased. In another study, Homayoni et al., 2009 developed electrospinning of chitosan. The problem of chitosan high viscosity, which limits its spinability, is resolved through the application of an alkali treatment which hydrolyzes chitosan chains and so decreases its their molecular weight. Solutions of the treated chitosan in aqueous 70–90% acetic acid produce nanofibers with appropriate quality and processing stability. Decreasing the acetic acid concentration in the solvent increases the mean diameter of the nanofibers. Optimum nanofibers are achieved with chitosan, which is hydrolyzed for 48 h. Such nanofibers result in a moisture regain, which is 74% greater than that of treated and untreated chitosan powder. The diameter of this nanofiber, 140 nm, is strongly affected by the electrospinning conditions as well as by the concentration of the solvent.

The electrospun various grades of chitosan and cross-linked them using glutaraldehyde vapor, utilizing an Schiff base imine functionality by Schiffman & Schauer, 2007a. Chemical, structural, and mechanical analyses have been conducted. Additionally, the solubilities of the as spun and cross-linked chitosan mats have also studied. The solubility was greatly improved after cross-linking. The as-spun medium molecular weight chitosan nanofibers have a Young's modulus of 154.9 ± 40.0 MPa and display a pseudo-yield point that arose due to the transition from the pulling of a fibrous mat with high cohesive strength to the sliding and elongation of fibers. As-spun mats were highly soluble in acidic and aqueous solutions. After cross-linking, the medium molecular weight fibers increased in diameter by an average of 161 nm, have a decreased Young's modulus of 150.8 ± 43.6 MPa, and were insoluble in basic, acidic, and aqueous solutions. Schiffman & Schauer, 2007b demonstrated that Schiff base cross-linked chitosan fibrous mats was produced by a one-step electrospinning process that is 25 times faster and, therefore, more economical than a previously reported two-step vapor-cross-linking method. These fibrous mats were insoluble in acidic, basic, and aqueous solutions for 72 h. This improved production method results in a decreased average fiber diameter, which measures 128 ± 40 nm.

Sangsanoh & Supaphol, 2006 reported an additional method for cross-linking or neutralizing electrospun chitosan nanofiber mats. Further utilization of chitosan nanofibrous membranes that are electrospun from chitosan solutions in TFA with or without dichloromethane (DCM) as the modifying co solvent is limited by the loss of the fibrous structure as soon as the membranes are in contact with neutral or weak basic aqueous solutions due to complete dissolution of the membranes. Dissolution occurs as a result of the high solubility in these aqueous media of $-NH_3^+CF_3COO^-$ salt residues that are formed when chitosan is dissolved in TFA. Traditional neutralization with a NaOH aqueous solution only maintained partial fibrous structure. Much improvement in the neutralization method was achieved with the saturated Na_2CO_3 aqueous solution with an excess amount of Na_2CO_3(s) in the solution. Sangsanoh & Supaphol, 2006 also reported that electrospun chitosan nanofibrous mats, after neutralization in the Na_2CO_3 aqueous solution, could maintain its fibrous structure even after continuous submersion in phosphate buffer saline (pH = 7.4) or distilled water for 12 weeks.

Besides electrospinning chitosan in TFA, the second solvent system that has been demonstrated to effectively produce nanofibers is concentrated acetic acid. Geng et al., 2005 attempted to electrospin three kinds of demineralized and deproteinized chitosan powders. However, uniform fibers were only fabricated from 7% chitosan in 90% aqueous acetic acid solutions. It was noted that the surface tension and charge density were the key factors in determining the spinnability of the system. In preliminary attempts to produce chitosan nanofibers for wound or alternate medical applications, or the removal of metals from solutions for environmental applications, Vrieze et al., 2007 conducted a feasibility study concerning the electrospinning of chitosan in formic, acetic, lactic and hydrochloric acids. It was determined that only the use of concentrated aq acetic acid solutions resulted in fibers. It is notable that the electrospun chitosan in acetic acid has been reported (Li & Hsieh, 2006; Sangsanoh & Supaphol, 2006). Torres-Giner et al., 2008 developed electrospun chitosan nanofibers using TFA and dicholomethane solvent. In addition, Torres-Giner et al., 2008a also developed porous electrospun chitosan nanofibers using pure trichloromethane solvent.

Recently, electrospun chitosan composite nanofibrous mats have been fabricated using synthetic biodegradable polymers such as PVA (Miya et al., 1994; Zheng et al., 2001), poly(ethylene oxide) (PEO) (Yilmaz et al., 2003), poly(vinyl pyrrolidine) (PVP) (Ignatova et al., 2007), poly(lactic acid) (PLA) (Peesan et al., 2006; Torres-Giner et al., 2008b) and poly(ethylene terpthalate (PET) (Jung et al., 2007). These composite fiber mats have more advantageous over the electrospinning of pure chitosan. Because, the mechanical, biocompatible, antibacterial and other properties of the chitosan nanofibers was drastically enhanced by the addition of PVA, PEO, PLA, PVP and PET. Researchers have electrospun composite nanofibers to try to increase their understanding of the role that chitosan plays during electrospinning. Finally, in order for chitosan-containing non-wovens to be applicable for a variety of biomaterial applications, particular mechanical, chemical, biocompatible, and other properties are required. Each of the composite fibrous mats created contain their specific properties and the applications are widen.

PVA is used for a variety of biomedical applications such as bone implants (Allen et al., 2004; Zhang et al., 2005), and artificial organs (Chen et al., 1994). Several researchers have developed nanofibers of PVA with chitosan by electrospinning because PVA has good fiber-forming characteristics (Li & Hsieh, 2006; Zhou et al., 2006; Huang et al., 2007; Jia et al., 2007; Zhou et al., 2007). The electrospun composite chitosan/PVA nanofibers in different ratios for biomedical applications were reported (Ding et al., 2002; Koski et al., 2004; Ohkawa et al. 2004; Zhang et al., 2005; Lin et al. 2006; Zhang et al. 2007; Zhou et al., 2008). Chitosan can be used as a thickener to improve the rheological properties of an aq solution to be electrospun because chitosan is compatible with other biocompatible polymers such as PVA (Miya et al., 1994; Zheng et al., 2001), PEO (Yilmaz et al., 2003) etc. A common defect observed when electrospinning is the formation of beads, to counter this, additives such as salts (Fong et al., 1999) or surfactants (Lin et al., 2004) are added. Alternatively, the addition of cationic and anionic polyelectrolytes (Son et al., 2004) would also increase the conductivity of a solution and thus decrease fiber diameter. Since chitosan is a linear cationic polymer, it was determined that the polyelectrolyte chitosan can act like other ionic additives and does reduce fiber diameter thus producing thinner, uniform, bead free fibers was noted (Lin et al., 2006; Jia et al., 2007). It was also noted by Jia et al., 2007 that the formation of a crystalline microstructure is restricted during electrospinning because the stretched molecular chains solidify rapidly and at high elongation rates. This effect has additionally been noted elsewhere (Deitzel et al., 2001; Zong et al., 2002).

Zhou et al., 2007 developed electrospun fibers from chitosan/PVA in aqueous acrylic acid solutions and in a second article (Zhou et al., 2006) they thermally cross-linked the fibrous mats using triethylene glycol dimethcrylate (TEGDMA) utilizing a higher concentration of an alterative acid than all of the previously mentioned researchers whom electrospun chitosan/PVA. Zhou et al., 2007 was also capable of spinning up to a 90/10 ratio of chitosan/PVA in as high as a 90% concentration of aqueous acrylic acid. To allow for utilization of the chitosan/PVA mats in a variety of applications, the amount of TEGDMA can be chosen to match the desired mechanical properties. It was found that by adding TEGDMA prior to spinning followed by heat-treating the as-spun mats for 2 hr at 80°C cross-linked fibrous mats could be fabricated.

The development of bioinspired or biomimetic materials is essential and has formed one of the most important paradigms in today's tissue engineering research. Zhang et al., 2008a reported a novel biomimetic nanocomposite nanofibers of hydroxyapatite (HAp)/chitosan was prepared by combining an in situ co-precipitation synthesis approach with an electrospinning process. A model HAp/chitosan with the HAp mass ratio of 30 wt% was also synthesized through the co-precipitation method so as to attain homogenous dispersion of the spindle-shaped HAp nanoparticles (ca. 100-30 nm) within the chitosan matrix. Similarly, biocomposite nanofibers were also prepared using of HAp with chitosan/PVA for biomedical applications (Yang et al., 2008).

PEO is also a biocompatible polymer (Griffith, 2000) that has been used as a wound dressing (Yoshii et al., 1999) and cartilage tissue repair (Sims et al., 1996; Subramanian et al., 2005). The PEO had previously been electrospun (Doshi & Renekar, 1995; Fong et al., 1999; Deitzel et al., 2001a; Deitzel et al., 2001b) with chitosan. Duan et al. 2004 electrospun chitosan/PEO based on previous research showing that PEO aided in the electrospinning of silk and collagen (Huang et al., 2001; Jin et al., 2002). The authors stated that the conductivity, surface tension, and moreover the solution viscosity is what allowed for electrospinning to occur when there was a mass ratio of PEO/chitosan of 2/1 or 1/1. Spasova et al. 2004; Duan et al. 2004; Desai et al., 2009, reported the electrospinning of chitosan/PEO system. It was observed that as the proportion of chitosan increased, the fiber diameter was decreased. Bhattarai et al., 2005 also developed electrospun chitosan with PEO. When PEO was added specifically to reduce the viscosity of solution so a higher polymer concentration within a solution would be spinable. It was found that a chitosan/PEO ratio of 9/1 was an appropriate candidate for bone tissue engineering because these mats retained good structural integrity in water and promoted good adhesion of chonrocyte and osteoblast cells. In addition, Vondran, 2007 has also studied the mechanical properties of chitosan/PEO nanofibrous mat as well as mats that were cross-linked by glutaraldehyde vapor. Chitosan/PEO nanofibers with surfactants for filtration applications was reported (Kriegel et al., 2009a; Kriegel et al., 2009b).

Zhang et al., 2008b demonstrated that the preparation of nanofibers by introducing an ultra high-molecular-weight PEO (UHMWPEO) into aqueous chitosan solution. This system leads for certain the chitosan nanofibers good structural stability and handling property during practical applications compared to previous higher PEO loadings. Because of the excellent electrospinnability of the current solution system, it was able to electrospin both the extremely thin nanofibers (<100 nm in diameters) and large microfibers (few tens of micrometers in diameters), which have significant implications in developing biomimetic and bioactive 3-D cell-scaffold complex for engineering tissues. The results suggest that current eco-friendly and easily electrospinnable chitosan formulation could provide great potential for robust and scale-up production of the chitosan nanofibers for efficient practical applications in wound dressings, tissue engineering, drug delivery, and other industrial uses.

PET is used in the textile and plastic industry, its antibacterial properties have been studied (Huh et al., 2001; Yang et al., 2002). Jung et al., 2007 reported that the electrospinning of chitosan with PET was useful for medical applications of the fibrous mats. Chitosan/PET and chitin/PET were electrospun in a TFA/HFIP solution. Antibacterial activity experiments indicated that chitosan/PET nanofibers inhibited the growth much more effectively than both the pure PET and the chitin/PET fibrous mats. Torres-Giner et al., 2008 developed chitosan/PLA blend nanofibers by electrospinning in TFA and tricholoromethane mixture.

Collagen has also been used previously for electrospinning using HFIP solvent (Matthews et al., 2002; Matthews et al., 2003; Rho et al., 2006). To develop a better biomimetic extracellular matrix for the tissue engineering of functional biomaterials, a matrix fibrous mat of chitosan/collagen was electrospun and reported (Chen et al., 2007; Chen et al., 2008). Mo et al., 2007 reported about the electrospun chitosan/collagen in HFIP/TFA. Composite nanofibrous membranes of type I collagen, chitosan, and polyethylene oxide was fabricated by electrospinning, which could be further cross-linked by glutaraldehyde vapor (Chen et al., 2008). Nanofiber diameter was found to be 134±42 nm, which increased to 398±76 nm after cross-linking. The Young's modulus was increased after cross-linking, however, the ultimate tensile strength, tensile strain, and water sorption capability decreased after cross-linking.

Zein is a relatively straightforward biopolymer to electrospin (Miyoshi, Toyohara, & Minematsu, 2005; Torres-Giner et al., 2008b). Additionally, zein has low toxicity and has also being studied in a broad range of areas, such as the food, pharmaceutical, and biodegradable plastics industry (Corradini et al., 2006). Torres-Giner et al., 2009 developed novel antimicrobial ultrathin structures of zein/chitosan blend by electrospinning for biomedical applications.

SF is the fibrous protein that forms the filaments of silkworm silk. Due to its biocompatibility, biodegradability, low inflammatory responses, and good oxygen and water vapor permeability, it has been used for several biomedical applications (Santin et al., 1999; Park et al., 2001). The novel biomimetic nanofibrous scaffolds were prepared using chitosan/SF in formic acid (Park et al., 2004). While a pure chitosan system could not be electrospun, up to a 30/70 chitosan/SF in formic acid could be spun. The influence of a methanol treatment on the secondary structure of as-spun SF versus chitosan/SF was also investigated.

Recently, the chitosan derivatives based nanofibers has also been prepared for biomedical applications. Carboxymethyl (Vondran, 2007), carboxyethyl (Mincheva et al., 2007) and hexanoyl chitosan (Peesan et al., 2006; Neamnark et al., 2006) derivatives were used for the preparation of nanofibers. Hexanoyl chitosan was used for medical applications since it has been proven to be resistant to hydrolysis by lysosome (Lee et al., 1995) and is anti-thrombogenic (Hirano & Noishiki, 1985). Neamnark et al., 2006, prepared hexanoyl chitosan nanofibers in chloroform solvent. The prepared nanofibers displayed ribbon-like morphology with diameters ranging from 0.64 to 3.93 μm. O-Carboxymethyl chitosan (O-CMCS) is a water-soluble derivative of chitosan (Muzzarelli et al., 1994; Chen & Park, 2003). O-CMCS has good moisture retention, gel-forming capability, is antibacterial, and non-cytotoxic, thus making it a good biomaterial (Chen et al., 2002; Chen et al., 2006). Vondran et al., 2007 has prepared electrospun composite fibrous mats using CMCS/PEO using water. The fibers appear to be continuous and cylindrical, while some beading was observed. The average fiber diameter was calculated to be 118.19 ± 40.48 nm. Similarly, Du & Hsieh, 2008 prepared O-CMCS blended PVA, PEO and poly(acrylic acid) (PAA) nanofibers by electrospinning. The optimal fiber formation was observed at equal mass composition of O-CMCS (89 kDa at 0.36 DS) and PVA, producing nanofibers with an average diameter of 130 nm. Heat-induced esterification (at 140 °C for 30 min) produced inter-molecular covalent cross-links within and among fibers, rendering the fibrous membrane water-insoluble. Membranes containing higher O-CMCS carboxyl to PVA hydroxyl ratio retained better fiber morphology upon extended water exposure, indicating more favorable inter-molecular cross-links. The fibrous membranes generated with less substituted O-CMCS were more hydrophilic and retained a greater extent of the desirable amine functionality.

Bicomponent nanofibers of N-carboxyethyl chitosan (N-CECS) and PVA were obtained by electrospinning was studied (Mincheva et al., 2007). The electrospinning of N-CECS containing nanofibers was enabled by the ability of PVA to form an elastically deformable entanglement network based on hydrogen bonds. The average diameters of the bicomponent fibers were in the range 100–420 nm. The average fiber diameter was tuned by varying the applied field strength and the N-CECS content in the spinning solution. The N-CECS/PVA nanofibrous mats were easily cross-linked by thermal treatment resulting in water-resistant materials, which retain their fibrous structure upon one-week contact with water. In addition, a novel biocomposite nanofibers were prepared using of HAp with N-CECS/PVA for biomedical applications (Yang et al., 2008).

Novel quaternized chitosan (QCS)/PVA blend nanofibers were developed for wound-healing applications (Ignatova et al., 2006). The average fiber diameter is in the range of 60–200 nm. UV irradiation of the composite electrospun nano-fibrous mats containing triethylene glycol diacrylate as cross-linking agent had resulted in stabilising of the nano-fibres against disintegration in water or water vapors. Ignatova et al., 2007 also prepared QCS-containing nanofibers by electrospinning with PVP blending. A significant decrease in the fiber diameter of electrospun QCS/PVP mixed fibers and a narrowing of the fiber diameter distribution with increasing the QCS content was observed and explained by the increase in solution conductivity. An increase of the applied field strength led to greater fiber diameters and to broader diameter distribution. The photo-cross-linked QCS/PVP fibers irradiated for 10 h proved to be water insoluble.

The galactosylated chitosan (GC) nanofibrous scaffold was fabricated by electrospinning for tissue engineering applications (Feng et al., 2009) with an average diameter of ~160 nm using formic acid solvent. Jiang et al., 2004 prepared electrospun membranes composed of ibuprofen-loaded poly(lactide-co-glycolide)/poly(ethylene glycol)-g-chitosan (PLGA)/(PEG-g-chitosan) that would be suitable for a trial fibrillation due to the electrospun membranes composed of ultrafine fibers, have high porosity, and can conform to movements (Reneker & Chun, 1996). In this work, ibuprofen was incorporated into the fibrous mats in two different methods: electrostatically conjugated during the electrospinning process and covalently conjugated to the PEG-g-chitosan prior to spinning. This tri-component system had unique properties of being soluble in organic solvents, while being insoluble in neutral pH water (Ouchi et al., 1998). It was also demonstrated that the hydrophilicity, membrane shrinkage, and rate of drug release could be controlled within the system. The fibrous mats in vitro and in vivo biocompatibility and efficacy need to be evaluated. Duan et al., 2006; Duan et al., 2007 simultaneously electrospun from two different syringes have PLGA and chitosan/PVA onto a rotating drum. The PLGA component is expected to enhance the mechanical properties, while the chitosan provides bioactivity and biocompatibility. The fibrous mats were cross-linked with 25% aqueous glutaraldehyde vapor for 4 hr at 37 °C (Chen et al., 2003). It was determined that tri-component systems might have potential in biomedical applications.

Wan et al., 2008 prepared novel poly(chitosan-g-DL-lactic acid) (PCLA) copolymeric nanofibers using an electro-wet-spinning technique. The diameter of fibers in different scaffolds could vary from about 100 nm to around 3 lm. The prepared nanofibrous scaffolds exhibited various pore sizes ranging from about 1 lm to less than 30 lm and different porosities up to 80%. Two main processing parameters, that is, the concentration of PCLA solutions and the composition proportions of coagulation solutions, were optimized for obtaining desired scaffolds with well-controlled structures. The tensile properties of the scaffolds in both dry and hydrated states were examined. Significantly improved tensile strength and modulus for these fibrous scaffolds in their hydrated state were observed.

3. Applications of chitin and chitosan nano-fibers in tissue-engineering

Polymeric nanofibers that mimic the structure and function of the natural extracellular matrix (ECM) are of great interest in tissue engineering as scaffolding materials to restore, maintain or improve the function of human tissues. The natural ECMs in the body are mainly composed of two classes of extracellular macromolecules: proteoglycans and fibrous proteins with fiber diameters ranging from 50 to 150 nm, depending on tissue type (Elsdale & Bard, 1972). Studies showed that the material size feature can substantially influence the morphology and function of cells grown on the ECM, and that cells attach and proliferate well in micro and nanostructured materials (Laurencin et al., 1999;Teixeira et al., 2003).

Noh et al., 2006 studied the cytocompatibility assessment of chitin nanofibers. Chitin nanofibers was found to promote cell attachment and spreading of normal human keratinocytes and fibroblasts compared to chitin microfibers. This may be a consequence of the high surface area available for cell attachment due to their three-dimensional features and high surface area-to volume ratios, which are favorable parameters for cell attachment, growth, and proliferation. Cell studies conducted on chitin/PGA (Park et al., 2006a) and chitin/SF (Park et al., 2006b) fibrous mats proved that a matrix consisting of 25% PGA or SF and 75% chitin had the best results. The chitin/PGA fibers had a bovine serum albumin coating and were considered a good candidate for use as a tissue-engineering scaffold because normal human epidermal fibroblasts (NHEF) attached and spread. The chitin/SF fibrous mats had the highest spreading of NHEF and normal human epidermal keratinocytes (NHEK). Therefore, this scaffold was suggested for wound tissue engineering applications. Shalumon et al., 2009 developed CMC/PVA blend nanofibrous scaffold for tissue engineering applications. The prepared nanofibers were bioactive and biocompatible. Cytotoxicity and cell attachement studies of the nanofibrous scaffold were evaluated using human mesenchymal stem cells (hMSCs) by the MTT assays. The cell attachment studies revealed that cells were able to attach and spread in the nanofibrous scaffolds (**Fig. 1**). These results indicate that the nanofibrous CMC/PVA scaffold supports cell adhesion/attachment and proliferation and hence these scaffolds are useful for tissue engineering applications.

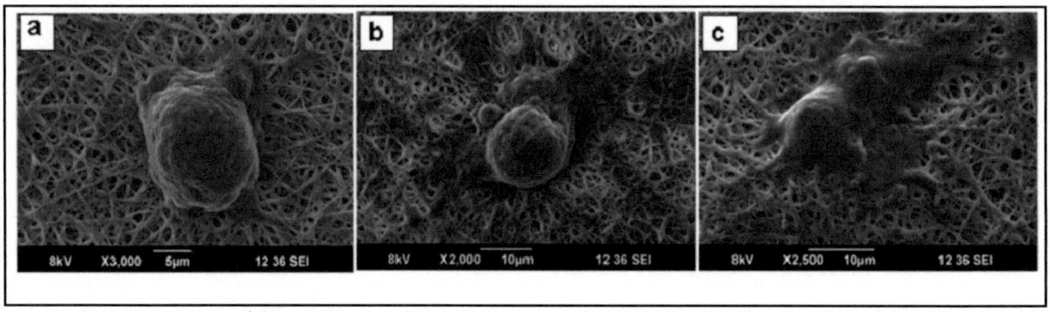

Fig. 1. SEM images of hMSCs attached on the surfaces of CMC/PVA scaffolds after (a) 12 (b) 24 and (c) 48 h of incubation.

Bhattarai et al., 2005 reported that the chitosan/PEO nanofibrous scaffolds were promoted the attachment of human osteoblasts and chondrocytes and maintained characteristic cell morphology and viability throughout the period of study. This nanofibrous matrix is of particular interest in tissue engineering for controlled drug release and tissue remodeling. Similarly, Subramaniyan et al., 2005 prepared chitosan/PEO nanofibers for cartilage tissue engineering. In this study, chitosan/PEO nanofibers was evaluated the biocompatibility with chondrocytes. Cells were attached to the chitosan/PEO nanofiber mats slowly in the fist week. After 10 days, the more cells were attached on the surface of the nanofibers. These results indicated that the electrospun chitosan/PEO mats have been used for cartilage tissue repair. Mo et al., 2007 also reported about that the smooth muscle cells attaching to the electrospun chitosan/collagen nanofibers after 30 days of culture.

The biological evaluations of chitosan/HAp nanfibrous composite scaffolds have been reported (Zhang et al., 2008a). The chitosan/HAp nanofibrous scaffolds have significantly stimulated the bone forming ability as shown by the cell proliferation, mineral deposition, and morphology observation, due to the excellent osteoconductivity of HAp compared to the control chitosan. The results obtained from this study highlight the great potential of using the chitosan/HAp nanocomposite nanofibers for bone tissue engineering applications. A biocomposite of HAp with electrospun nanofibrous scaffolds was prepared by using chitosan/PVA and *N*-CECS/PVA (Fig. 2) for tissue engineering applications (Yang et al., 2008). The cell attachment of the prepared biocomposite nanofibers was studied using mouse fibroblast (L929) cell. The cell studies showed that the L929 cell culture revealed the attachment and growth of mouse fibroblast on the surface of biocomposite scaffolds. Similarly, the potential use of the *N*-CECS/PVA electrospun fiber mats as scaffolding materials for skin regeneration was evaluated *in vitro* was using L929 (Zhou et al., 2008). Indirect cytotoxicity assessment of the fiber mats indicated that the *N*-CECS/PVA electrospun mat was nontoxic to the L929 cell. Cell culture results showed that fibrous mats were good in promoting the cell attachment and proliferation. This novel electrospun matrix would be used as potential wound dressing for skin regeneration.

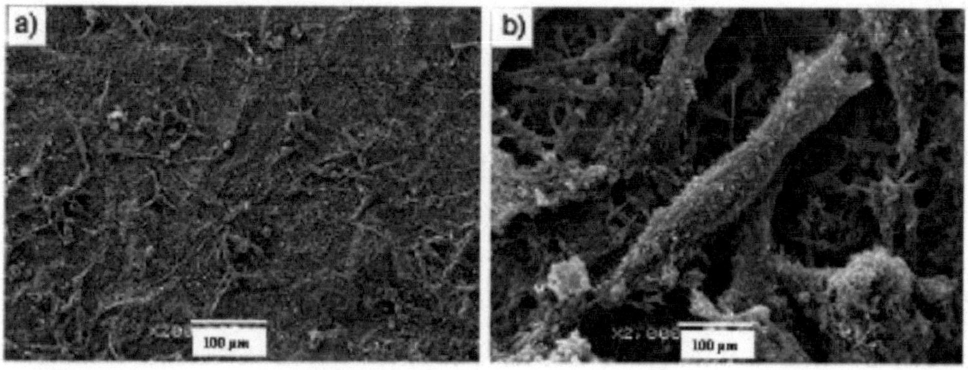

Fig. 2. SEM images of (a) and (b) L929 cells seeded on fibrous membranes of HAp-*N*-CECS/PVA after 48 h culture.

Liver tissue engineering requires a perfect extracellular matrix (ECM) for primary hepatocytes culture to maintain high level of liver-specific functions and desirable mechanical stability. Feng et al., 2009 developed a novel GC nanofibers with surface-galactose ligands to enhance the bioactivity and mechanical stability of primary hepatocytes in culture. The GC nanofibrous scaffolds displayed slow degradation and suitable mechanical properties as an ECM for hepatocytes according to the evaluation of disintegration and Young's modulus testing. The hepatocytes cultured on GC nanofibrous scaffold formed stably immobilized 3D flat aggregates and exhibited superior cell bioactivity with higher levels of liver-specific function maintenance in terms of albumin secretion, urea synthesis and cytochrome P-450 enzyme than 3D spheroid aggregates formed on GC films. These results suggested that the GC-based nanofibrous scaffolds could be useful for various applications such as bioartificial liver-assist devices and tissue engineering for liver regeneration as primary hepatocytes culture substrates.

4. Conclusions

This review summarized the preparation and tissue engineering applications of chitin and chitosan based nanofibers. Additional studies are necessary before clinical applications and for commercialization of the chitin and chitosan based nanofibers. We hope that this review article will bring new innovative types of chitin and chitosan nanofibers for tissue engineering applications in the future.

5. Acknowledgments

The Department of Science and Technology, Government of India supported this work, under a centre grant of the Nanoscience and Nanotechnology Initiative program monitored by Dr. C. N. R. Rao.

6. References

Allen, M. J.; Schoonmaker, J. E.; Bauer, T. W.; Williams, P. F.; Higham, P. A. & Yuan, H. A. (2004). Preclinical evaluation of a poly (vinyl alcohol) hydrogel implant as a replacement for the nucleus pulposus. *Spine, 29*, 515-523.

Anitha, A.; Divya Rani, V. V.; Krishna, R.; Sreeja, V.; Selvamurugan, N.; Nair, S. V.; Tamura, H. & Jayakumar, R. (2009). Synthesis, characterization, cytotoxicity and antibacterial studies of chitosan, *O*-carboxymethyl, *N, O*-carboxymethyl chitosan nanoparticles. *Carbohydrate Polymers*, in press.

Baumgarten, P.K., (1971). Electrostatic spinning of acrylic microfibers. *Journal of Colloid and Interface Science, 36*, 71-79.

Bhattarai, N.; Edmondson, D.; Veiseh, O.; Matsen, F. A. & Zhang, M. (2005). Electrospun chitosan-based nanofibers and their cellular compatibility. *Biomaterials, 26*, 6176-6184.

Chen, D. H.; Leu, J. C. & Huang, T. C. (1994). Transport and hydrolysis of urea in a reactor-separator combining an anion-exchange membrane and immobilized urease. *Journal of Chemical Technology & Biotechnology, 61*, 351-357.

Chen, G.; Sato, T.; Ushida, T.; Hirochika, R.; Shirasaki, Y.; Ochiai, N. & Tateishi, T. (2003). The use of a novel PLGA fiber/collagen composite web as a scaffold for engineering of articular cartilage tissue with adjustable thickness. *Journal of Biomedical Materials Research, 67A*, 1170-1180.

Chen, J. P.; Chang, G. Y. & Chen, J. K. (2008). Electrospun collagen/chitosan nanofibrous membrane as wound dressing. *Colloids and Surfaces A: Physicochemical Engineering Aspects, 313-314*, 183-188.

Chen, R. N.; Wang, G. M.; Chen, C. H.; Ho, H. O. & Sheu, M. T. (2006). Development of *N, O*-carboxymethyl chitosan/collagen matrixes as a wound dressing. *Biomacromolecules, 7*, 1058 -1064.

Chen, X. G. & Park, H. J. (2003). Chemical characteristics of *O*-carboxymethyl chitosans related to the preparation conditions. *Carbohydrate Polymers, 53*, 355-359.

Chen, X. G.; Wang, Z.; Liu, W. S. & Park, H. J. (2002). The effect of carboxymethyl-chitosan on proliferation and collagen secretion of normal and keloid skin fibroblasts. *Biomaterials, 23*, 4609-4614.

Chen, Z.; Mo, X. & Qing, F. (2007). Electrospinning of collagen-chitosan complex. *Materials Letters, 61*, 3490-3494.

Chen, Z.; Mo, X.; He, C. & Wang, H. (2008). Intermolecular interactions in electrospun collagen-chitosan complex nanofibers. *Carbohydrate Polymers, 72*, 410-418.

Cho, Y. W.; Jang, J.; Park, J. R. & Ko, S. W. (2000). Preparation and solubility in acid and water of partially deacetylated chitins. *Biomacromolecules, 1*, 609–614.

Corradini, E. A.; Souto de Medeiros, E.; Carvalho, A. J. F.; Curvelo, A. A. S. & Mattoso, L. H. C. (2006). Mechanical and morphological characterization of starch/zein blends plasticized with glycerol. *Journal of Applied Polymer Science, 101*, 4133–4139.

Deitzel, J. M.; Kleinmeyer, J. D.; Harris, D. & Beck Tan, N. C. (2001a). The effect of processing variables on the morphology of electrospun nanofibers and textiles. *Polymer, 42*, 261-272.

Deitzel, J. M.; Kleinmeyer, J. D.; Hirvonen, J. K. & Beck Tan, N. C. (2001b). Controlled deposition of electrospun poly(ethylene oxide) fibers. *Polymer, 42*, 8163-8170.

Desai, K.; Kit, K.; Li, J.; Davidson, P. M.; Zivanovic, S. & Meyer, H. (2009). Nanofibrous chitosan non-wovens for filteration applications. *Polymer*, in press.

Ding, B.; Kim, H. Y.; Lee, S. C.; Shao, C. L.; Lee, D. R.; Park, S. J.; Kwag, G. B. & Choi, K. J. (2002). Preparation and characterization of a nanoscale poly(vinyl alcohol) fiber aggregate produced by an electrospinning method. *Journal of Polymer Science Part B: Polymer Physics, 40*, 1261-1268.

Doshi, J. & Reneker, D. H. (1995). Electrospinning process and applications of electrospun fibers. *Journal of Electrostatics, 35*, 151-160.

Du, J. & Hsieh, Y. L. (2008). Nanofibrous membranes from aqueous electrospinning of carboxymethyl chitosan. *Nanotechnology, 19*, 125707-125716.

Duan, B.; Dong, C.; Yuan, X. & Yao, K. (2004). Electrospinning of chitosan solutions in acetic acid with poly(ethylene oxide. *Journal of Biomaterials Science, Polymer Edition, 15*, 797-811.

Duan, B.; Wu, L.; Li, X.; Yuan, X.; Li, X.; Zhang, Y. & Yao, K. (2007). Degradation of electrospun PLGA-chitosan/PVA membranes and their cytocompatibility in vitro. *Journal of Biomaterials Science, Polymer Edition, 18*, 95-115.

Duan, B.; Yuan, X.; Zhu, Y.; Zhang, Y.; Li, X.; Zhang, Y. & Yao, K. (2006). A nanofibrous composite membrane of PLGA-chitosan/PVA prepared by electrospinning. *European Polymer Journal, 42*, 2013-2022.

Elsdale, T. & Bard, J. (1972). Collagen substrata for studies on cell behavior. *Journal of Cell Biology, 54*, 626–637.

Fang, J., Niu, H., Lin, T. & Wang, X. (2008). Applications of electrospun nanofibers, *Chinese Science Bulletin, 53*, 2265-2286.

Feng Z. Q.; Chu, X.; Huang, N. P.; Wang, T.; Wang, Y.; Shi, X.; Ding, Y. & Gu, Z. Z. (2009). The effect of nanofibrous galactosylated chitosan scaffolds on the formation of rat primary hepatocyte aggregates and the maintenance of liver function. *Biomaterials, 30*, 2753–2763.

Fong, H.; Chun, I. & Reneker, D. H. (1999). Beaded nanofibers formed during electrospinning. *Polymer, 40*, 4585-4592.

Fridrikh, S. V.; Yu, J. H.; Brenner, M. P. & Rutledge, G. C. (2003). Controlling the Fiber Diameter during Electrospinning. *Physical Review Letters, 90*, 144502.

Geng, X.; Kwon & Jang, J. (2005). Electrospinning of chitosan dissolved in concentrated acetic acid solution. *Biomaterials, 26*, 5427-5432.

Griffith, L. G. (2000). Polymeric biomaterials. *Acta Materialia, 48*, 263-277.

Haider, S. & Park, S. Y. (2009). Preparation of the electrospun chitosan nanofibers and their applications to the adsorption of Cu(II) and Pb(II) ions from an aqueous solution. *Journal of Membrane Science, 328*, 90-96.

Hasegawa, M.; Isogai, A.; Onabe, F.; Usuda, M. & Atalla, R. H. (1992). Characterization of cellulose-chitosan blend films. *Journal of Applied Polymer Science, 45*, 1873-1879.

Hirano, S. & Noishiki, Y. (1985). The blood compatibility of chitosan and N-acylchitosans. *Journal of Biomedical Materials Research, 19*, 413-417.

Homoyoni, H.; Ravandi, S. A. H. & Valizadeh. M. (2009). Electrospinning of chitosan nanofibers: Processing optimization. *Carbohydrate Polymers, 77*, 656-661.

Huang, C.; Chung, Y. C. & Liou, M. R. (1996). Adsorption of Cu(II) and Ni(II) by palletized biopolymer, *Journal of Hazardous Materials, 45*, 265–277.

Huang, L.; Nagapudi, K.; Apkarian, P. R. & Chaikof, E. L. (2001). Engineered collagen PEO nanofibers and fabrics. *Journal of Biomaterials Science, Polymer Edition, 12*, 979-993.

Huang, X. J.; Ge, D. & Xu, Z. K. (2007). Preparation and characterization of stable chitosan nanofibrous membrane for lipase immobilization. *European Polymer Journal, 43*, 3710-3718.

Ignatova, M.; Manolova, N. & Rashkov, I. (2007). Novel antibacterial fibers of quaternized chitosan and poly(vinyl pyrrolidine) prepared by electrospinning. *European Polymer Journal, 43*, 1112-1122.

Ignatova, M.; Starbova, K.; Markova, N.; Manolova, N. & Rashkov, I. (2006). Electrospun nano-fibre mats with antibacterial properties from quaternized chitosan and poly(vinyl alcohol), *Carbohydrate Research, 341*, 2098–2107.

Jayakumar, R. & Tamura, H. (2008). Synthesis, characterization and thermal properties of chitin-g-poly(caprolactone) copolymers using chitin hydrogel. *International Journal of Biological Macromolecules, 43*, 32-36.

Jayakumar, R.; Divya Rani, V. V.; Shalumon, K. T.; Sudhessh Kumar, P. T.; Nair, S. V.; Furuike, T. & Tamura, H. (2009). Bioactive and osteoblast cell attachement studies of novel *a*-, and *β*-chitin membranes for tissue engineering applications, *International Journal of Biological Macromolecules, 45*, 260-264.

Jayakumar, R.; Nwe, N.; Tokura, S. & Tamura, H. (2007). Sulfated chitin and chitosan as novel biomaterials. *International Journal of Biological Macromolecules 40*, 175–181.

Jayakumar, R.; Prabaharan, M.; Reis, R. L. & Mano, J. F. (2005). Graft copolymerized chitosan-Present status and applications. *Carbohydrate Polymers, 62*, 142-158.

Jayakumar, R.; Reis, R. L. & Mano, J. F. (2006). Phosphorous containing chitosan beads for controlled oral drug delivery. *Journal of Bioactive and Compatible Polymers, 21*, 2006, 327-340.

Jia, Y. T.; Gong, J.; Gu, X. H.; Kim, H. Y.; Dong, J. & Shen, X. Y. (2007). Fabrication and characterization of poly (vinyl alcohol)/chitosan blend nanofibers produced by electrospinning method. *Carbohydrate Polymers, 67*, 403-409.

Jiang, H.; Fang, D.; Hsiao, B.; Chu, B. & Chen, W. (2004). Preparation and characterization of ibuprofen-loaded poly(lactide-co-glycolide)/poly(ethylene glycol)-*g*-chitosan electrospun membranes. *Journal of Biomaterials Science, Polymer Edition, 15*, 279-296.

Jin, H. J.; Fridrikh, S. V.; Rutledge, G. C. & Kaplan, D. L. (2002). Electrospinning bombyx mori silk with poly(ethylene oxide). *Biomacromolecules, 3*, 1233-1239.

Jung, K. H.; Huh, M. W.; Meng, W.; Yuan, J.; Hyun, S. H.; Bae, J. S.; Hudson, M. S. & Kang, I. K. (2007). Preparation and antibacterial activity of PET/chitosan Nanofibers. *Jounral of Applied Polymer Science, 105*, 2816-2823.

Junkasem, J.; Rujiravanit, R. & Supaphol, P. (2006). Fabrication of α-chitin whisker-reinforced poly(vinyl alcohol) nanocomposite nanofibres by electrospinning. *Nanotechnology, 17*, 4519–4528.

Koski, A.; Yim, K. & Shivkumar, S. (2004). Effect of molecular weight on fibrous PVA produced by electrospinning. *Materials Letters, 58*, 493-497.

Kriegel, C.; Kit, K. M.; McClements, D. J. & Weiss, J. (2009a). Eelectrospinning of chitosan-poly(ethylene oxide) blend nanofibers in the presence of micellar surfactant solutions. *Polymer, 50*, 189-200.

Kriegel, C.; Kit, K. M.; McClements, D. J. & Weiss, J. (2009b). Influence of surfactant type and concentration of electrospinning of chitosan-poly(ethylene oxide) blend nanofibers. *Food Biophysics*, DOI 10.1007/s11483-009-9119-6

Kurita, K. (2001). Controlled functionalization pf the polysaccharide chitin. *Progress in Polymer Science*, *269*, 1921–1971.

Laurencin, C. T.; Ambrosio, A. M. A.; Borden, M. D. & Cooper, J. A. (1999). Tissue engineering: orthopedic applications. *Annual Review in Biomedical Engineering*, *1*, 19–46.

Lee, K. Y.; Ha, W. S. & Park, W. H. (1995). Blood compatibility and biodegradability of partially *N*-acylated chitosan derivatives. *Biomaterials*, *16*, 1211-1216.

Li, L. & Hsieh, Y. L. (2006). Chitosan bicomponent nanofibers and nanoporous fibers. *Carbohydrate Research*, *341*, 374–381.

Lin, T.; Fang, J.; Wang, H.; Cheng, T. & Wang, X. (2006). Using chitosan as a thickener for electrospinning dilute PVA solutions to improve fiber uniformity. *Nanotechnology*, *17*, 3718-3723.

Lin, T.; Wang, H.; Wang, H. & Wang, X. (2004). The charge effect of cationic surfactants on the elimination of fibre beads in the electrospinning of polystyrene. *Nanotechnology*, *15*, 1375-1381.

Madhumathi, K.; Binulal, N. S.; Nagahama, H.; Tamura, H.; Shalumon, K. T.; Selvamurugan, N.; Nair, S. V. & Jayakumar, R. (2009a). Preparation and charaterization of novel β-chitin-hydroxyapatite composite membranes for tissue engineering applications. *International Journal of Biological Macromolecules*, *44*, 1-5.

Madhumathi, K.; Shalumon, K. T.; Divya Rani, V. V.; Tamura, H.; Furuike, T.; Selvamurugan, N.; Nair, S. V. & Jayakumar, R. (2009b). Wet chemical synthesis of chitosan hydrogel-hydroxyapatite composite membranes for tissue engineering applications. *International Journal of Biological Macromolecules*, *45*, 12-15.

Madhumathi, K.; Sudhessh Kumar, P. T.; Kavya, K. C.; Furuike, T.; Tamura, H.; Nair, S. V. & Jayakumar, R. (2009c). Novel chitin/nanosilica composite scaffolds for bone tissue engineering applications. *International Journal of Biological Macromolecules*, *45*, 289-292.

Maeda, Y.; Jayakumar, R.; Nagahama, H.; Furuike, T. & Tamura, H. (2008). Synthesis, characterization and bioactivity studies of novel β-chitin scaffolds for tissue-engineering applications. *International Journal of Biological Macromolecules*, *42*, 463–467.

Matthews, J. A.; Wnek, G. E.; Simpson, D. G. & Bowlin, G. L. (2002). Electrospinning of collagen nanofibers. *Biomacromolecules*, *3*, 232-238.

Matthews, J. A.; Boland, E. D.; Wneck, G. E.; Simpson, D. G. & Bowlin, G. L. (2003). Electrospinning of collagen type II: A feasibility study. *Journal of Bioactive and Compatible Polymers*, *18*, 125-134.

McKee, M.G.; Elkins, C. L. & Long, T. E. (2004). Influence of self-complementary hydrogen bonding on solution rheology/electrospinning relationships. *Polymer*, *45*, 8705-8715.

McKee, M. G.; Hunley, M. T.; Layman, J. M. & Long, T. E. (2006). Solution rheological behavior and electrospinning of cationic polyelectrolytes. *Macromolecules*, *39*, 575-583.

Min, B. M.; Lee, S. W.; Lim, J. N., You, Y.; Lee, T. S.; Kang, P. H. & Park, W H. (2004). Chitin and chitosan nanofibers: electrospinning of chitin and deacetylation of chitin nanofibers. *Polymer, 45,* 7137-7142.

Mincheva, R.; Manolova, N. & Rashkov, I. (2007). Biocomponent aligned nanofibers of *N*-carboxyethyl chitosan and poly(vinyl alcohol*). European Polymer Journal, 43,* 2809-2818.

Miya, M.; Iwamoto, R. & Mima, S. (1984). FT-IR study of intermolecular interactions in polymer blends. *Journal of Polymer Science. Polymer Physics Edition, 22,* 1149-1151.

Miyoshi, T.; Toyohara, K., & Minematsu, H. (2005). Preparation of ultrafine fibrous zein membranes via electrospinning. *Polymer International, 54,* 1187–1190.

Mo, X.; Chen, Z. & Weber, H. J. (2007). Electrospun nanofibers of collagen-chitosan and P(LLA-CL) for tissue engineering. *Frontier in Material Science, 1,* 20-23.

Muramatsu, K.; Masuda, S.; Yoshihara,; Fujisawa, A. (2003). In vitro degradation behavior of freeze-dried carboxymethyl-chitin sponges processed by vacuum-heating and gamma irradiation. *Polymer Degradation and Stability, 81,* 327-332.

Muzzarelli, R. A. A.; Ilari, P. & Petrarulo, M. (1994). Solubility and structure of *N*-carboxymethylchitosan. *International Journal of Biological Macromolecules, 16,* 177-180.

Nagahama, H.; Kashiki, T.; Nwe, N.; Jayakumar, R.; Furuike, T. & Tamura, H. (2008b). Preparation of biodegradable chitin/gelatin membranes with GlcNAc for tissue engineering applications. *Carbohydrate Polymers, 73,* 456-463.

Nagahama, H.; Nwe, N.; Jayakumar, R.; Koiwa, S.; Furuike, T. & Tamura, H. (2008a). Novel biodegradable chitin membranes for tissue engineering applications. *Carbohydrate Polymers, 73,* 295-302.

Neamnark, A.; Rujiravanit, R. & Supaphol, P. (2006). Electrospinning of hexanoyl chitosan. *Carbohydrate Polymers, 66,* 298-305.

Noh, H. K.; Lee, S. W.; Kim, J. M.; Oh, J. E.; Kim, K. H.; Chung, C. P.; Choi, S. C.; Park, W. H. & Min, B. M. (2006). Electrospinning of chitin nanofibers: Degradation behavior and cellular response to normal human keratinocytes and fibroblasts. *Biomaterials, 27,* 3934-3944.

Ohkawa, K.; Cha, D.; Kim, H.; Nishida, A. & Yamamoto, H. (2004). Electrospinning of chitosan. *Macromolecular Rapid Communications, 25,* 1600-1605.

Ohkawa, K.; Minato, K. I.; Kumagai, G.; Hayashi, S. & Yamamoto, H. (2006). Chitosan nanofiber. *Biomacromolecules, 7,* 3291-3294.

Ouchi, T.; Nishizawa, H. & Ohya, Y. (1998). Aggregation phenomenon of PEG-grafted chitosan in aqueous solution. *Polymer, 39,* 5171-5175.

Park, K. E.; Jung, S. Y.; Lee, S. J.; Min, B. M. & Park, W. H. (2006b). Biomimetic nanofibrous scaffolds: Preparation and characterization of chitin/silk fibroin blend nanofibers. *International Journal of Biological Macromolecules, 38,* 165-173.

Park, K. E.; Kang, H. K.; Lee, S. J.; Min, B. M. & Park, W. H. (2006a). Biomimetic Nanofibrous Scaffolds: Preparation and Characterization of PGA/chitin blend nanofibers. *Biomacromolecules, 7,* 635-643.

Park, W. H.; Ha, W. S.; Ito, H.; Miyamoto, T.; Inagaki, H. & Noishiki, Y. (2001). Relationships between antithrombogenicity and surface free energy of regenerated silk fibrion films. *Fibers and Polymers, 2,* 58-63.

Park, W. H.; Jeong, L.; Yoo, D. & Hudson, S. (2004). Effect of chitosan on morphology and conformation of electrospun silk fibroin nanofibers. *Polymer, 45,* 7151-7157.

Peesan, M.; Rujiravanit, R. & Supaphol, P. (2006). Electrospinning of hexanoyl chitosan/polylactide blends. *Journal of Biomaterials Science, Polymer Edition, 17*, 547-565.

Pillai, C. K. S.; Paul, W. & Sharma, C. P. (2009). Chitin and chitosan polymers: Chemistry, Solubility and fiber formation. *Progress in Polymer Science, 34*, 641-678.

Portero, A.; Teijeiro-Osorio, D.; Alonso, M. J.; Remunan-Lopez, C. (2007). Development of chitosan sponges for buccal administration of insulin. *Carbohydrate Polymers, 68*, 617-625.

Prabaharan, M. & Mano, J. F. (2005). Chitosan-based particles as controlled drug delivery systems, *Drug Delivery, 12*, 41–57.

Reneker, D. H. & Chun, I. (1996). Nanometre diameter fibres of polymer produced by electrospinning. *Nanotechnology, 7*, 216-223.

Rho, K. S.; Jeong, L.; Lee, G.; Seo, B. M.; Park, Y. J.; Hong, S. D.; Roh, S.; Cho, J. J.; Park, W. H. & Min, B. M. (2006). Electrospinning of collagen nanofibers: Effects on the behavior of normal human keratinocytes and early-stage wound healing. *Biomaterials, 27*, 1452-1461.

Rinaudo, M. (2006). Chitin and chitosan: Properties and applications, *Progress in Polymer Science, 31*, 603-632.

Roberts, G. A. F. (1992). Chitin Chemistry, Macmillan, London.

Sangsanoh, P. & Supaphol, P. (2006). Stability improvement of electrospun chitosan nanofibrous membranes in neutral or weak basic aqueous solutions. *Biomacromolecules, 7*, 2710-2714.

Santin, M.; Motta, A.; Freddi, G. & Cannas, M. (1999). In vitro evaluation of the inflammatory potential of the silk fibroin. *Journal of Biomedical Materials Research, 46*, 382-389.

Schiffman, J. D. & Schauer, C. L. (2007a). Cross-linking chitosan nanofibers. *Biomacromolecules, 8*, 594-601.

Schiffman, J. D. & Schauer, C. L. (2007b). One-step electrospinning of cross-linked chitosan fibers. *Biomacromolecules, 8*, 2665-2667.

Shalumon, K. T.; Binulal, N. S.; Selvamurugan, N.; Nair, S. V.; Menon, D.; Furuike, T.; Tamura, H. & Jayakumar, R. (2009). Electrospinning of carboxymethyl chitin/poly(vinyl alcohol) nanofibrous scaffolds for tissue engineering applications. *Carbohydrate Polymers, 77*, 863-869.

Sims, C. D.; Butler, P. E. M.; Casanova, R.; Lee, B. T.; Randolph, M. A.; Lee, W. P. A.; Vacanti, C. A. & Yaremchuk, M. J. (1996). Injectable cartilage using polyethylene oxide polymer substrates. *Plastic and Reconstructive Surgery, 98*, 843-850.

Son, W. K.; Youk, J. H.; Lee, T. S. & Park, W. H. (2004). The effects of solution properties and polyelectrolyte on electrospinning of ultrafine poly(ethylene oxide) fibers. *Polymer, 45*, 2959-2966.

Spasova, M.; Manolova, N.; Paneva, D. & Rashkov, L. (2004). Preperation of chitosan-containing nanofibers by electrospinning of chitosan/poly(ethylene oxide) blend solutions. *E-Polymers, 56*, 1-12.

Sriupayo, J.; Supaphol, P.; Blackwell, J. & Rujiravanit, R. (2005a). Preparation and characterization of *a*-chitin whisker-reinforced chitosan nanocomposite films with or without heat treatment. *Carbohydrate Polymers, 62*, 130-136.

Sriupayo, J.; Supaphol, P.; Blackwell, J. & Rujiravanit, R. (2005b). Preparation and characterization of *a*-chitin whisker-reinforced poly(vinyl alcohol) nanocomposite films with or without heat treatment. *Polymer, 46,* 5637-5644.

Subramaniyan, A.; Vu, D.; Larsen, G. F. & Lin, H. Y. (2005). Preparation and evaluation of the electrospun chitosan/PEO fibers for potential applications in cartilage tissue engineering. *Journal of Biomaterial Science: Polymer Edition, 7,* 861-873.

Tashiro T. (2001). Antibacterial and bacterium adsorbing macromolecules. *Macromolecular Materials Engineering,* 286, 63–87.

Teixeira, A. I.; Abrams, G. A.; Bertics, P. J.; Murphy, C. J. & Nealey, P. F. (2003). Epithelial contact guidance on well-defined micro- and nanostructured substrates. *Journal of Cell Science, 116,* 1881–1892.

Torres-Giner, S.; Gimenez, E. & Lagaron, J. M. (2008b). Characterization of the morphology and thermal properties of Zein Prolamine nanostructures obtained by electrospinning. *Food Hydrocolloids, 22,* 601–614.

Torres-Giner, S.; Ocio, M. J. & Lagaron, J. M. (2009). Novel antimicrobial ultrathin structures of zein/chitosan blends obtained by electrospinning. *Carbohydrate Polymers, 77,* 261-266.

Torres-Giner, S.; Ocio, M. J. & Lagaron, J. M. (2008a). Development of active antimicrobial fiber based chitosan polysaccharide nanostructures by electrospinning. *Engineering in Life Sciences, 8,* 303-314.

Vondran, J. L. (2007). Fabrication, optimization, and characterization of carboxymethylated chitosan nanofiber mats for cartilage regeneration applications, in Materials Science and Engineering. Drexel University: Philadelphia. p. 123.

Vrieze, S. D.; Westbroeak, P.; Camp, T. V. & Langenhove, L. V. (2007). Electrospinning of chitosan nanofibrous structures: feasibility study. *Journal of Materials Science, 42,* 8029-8034.

Wan, Y.; Cao, X.; Zhang, S.; Wang, S. & Wu, Q. (2008). Fibrous poly(chitosan-*g*-DL-lactic acid) scaffolds prepared via electro-wet-spinning. *Acta Biomaterialia, 4,* 876-886.

Wang, J. W. & Hon, M. H. (2003). Preparation and characterization of pH sensitive sugar mediated (polyethylene glycol/chitosan) membrane. *Journal of Materials Science: Materials in Medicine, 14,* 1079-1088.

Yang, D.; Jin, Y.; Zhou, Y.; Ma, G.; Chen, X.; Lu, F. & Nie, J. (2008). In situ mineralization of hydroxyapatite on electrospun chitosan-based nanofibrous scaffolds. *Macromolecular Bioscience, 8,* 239-246.

Yang, M. R.; Chen, K. S.; Tsai, J. C.; Tseng, C. C. & Lin, S. F. (2002). The antibacterial activities of hydrophilic-modified nonwoven PET. *Materials Science and Engineering, C20,* 167-173.

Yilmaz, E.; Erdenizci, N. & Yilmaz, O. (2003). Miscibility of Chitosan and Poly(Ethylene Oxide) in Dilute Solution. *Journal of Polymer Analysis and Characterization, 8,* 327 - 338.

Yoshii, F.; Zhansahan, Y.; Isobe, K.; Shinozaki, K. & Makuuchi, K. (1999). Electron beam crosslinked PEO and PEO/PVA hydrogels for wound dressing. *Radiation Physics and Chemistry, 55,* 133-138.

Zhang, C.; Yuan, X.; Wu, L.; Han, Y. & Sheng, J. (2005). Study on morphology of electrospun poly(vinyl alcohol) mats. *European Polymer Journal, 41,* 423-432.

Zhang, Y. Z.; Venugopal, J. R.; El-Turki, A.; Ramakrishna, S.; Su, B. & Lim, C. T. (2008a). Electrospun biomimetic nanocomposite nanofibers of hydroxyapatite/chitosan for bone tissue engineering, *Biomaterials, 29,* 4314-4322.

Zhang, Y. Z.; Su, B.; Ramakrishna, S. & Lim. C. T. (2008b). Chitosan nanofibers from an easily electrospinnable UHMWPEO-doped chitosan solution system. *Biomacromolecules, 9,* 136-141.

Zhang, Y.; Huang, X.; Duan, B.; Wu, L.; Li, S. & Yuan, X. (2007). Preparation of electrospun chitosan/poly(vinyl alcohol) membranes. *Colloid and Polymer Science, 285,* 855-863.

Zhang, Y.; Lim, C. T.; Ramakrishna, S. & Huang, Z. M. (2005). Recent development of polymer nanofibers for biomedical and biotechnological applications, *Journal of Materials Science: Materials in Medicine, 16,* 933-946.

Zheng, H.; Du, Y., Yu, J.; Huang, R. & Zhang, L. (2001). Preparation and characterization of chitosan/poly(vinyl alcohol) blend fibers. *Journal of Applied Polymer Science, 80,* 2558-2565.

Zhou, Y. S.; Yang, D. & Nie, J. (2006). Electrospinning of chitosan/poly(vinyl alchol)/acrylic acid solutions. *Journal of Applied Polymer Science, 102,* 5692-5697.

Zhou, Y. S.; Yang, D. & Nie, J. (2007). Preparation and characterization of crosslinked chitosan-based nanofibers. *Chinese Chemical Letters, 18,* 118-120.

Zhou, Y. S.; Yang, D.; Chen, X.; Xu, Q.; Lu, F. & Nie, J. (2008). Electrospun water-soluble carboxyethyl chitosan/poly(vinyl alcohol) nanofibrous membrane as potential wound dressing for skin regeneration. *Biomacromolecules, 9,* 349-354.

11

Bioresorbable polymers for tissue engineering

Arnaldo Rodrigues Santos Jr.
Centro de Ciências Naturais e Humanas, Universidade Federal do ABC - UFABC, Avenida dos Estados 5001, CEP 09210-971, Santo André, SP, Brasil,
E-mail: arnaldo.santos@ufabc.edu.br

Key words: bioresorbable polymers, cell culture, cell differentiation, tissue engineering.

Abstract

Polymeric biomaterials are used as substitutes for damaged tissue and for the stimulation of tissue regeneration. One class of polymeric biomaterials are bioresorbable polymers that degrade both *in vitro* and *in vivo* and are used as a temporary support for tissue regeneration. Among the various types of bioresorbable polymers, α-hydroxy acids including the different forms of poly(lactic acid) (PLA), such as poly(L-lactic acid), poly(D-lactic acid) and poly(DL-lactic acid), as well as poly(glycolic acid) and polycaprolactone, have been extensively studied. These polymers are well known for their good biocompatibility, with their degradation products being eliminated from the body by metabolic pathways. Many reports have shown that the different PLA-based substrates do not present toxicity since the cells were found to differentiate over the different polymers, as demonstrated by the production of extracellular matrix components by various cell types. In this chapter, we describe the use of α-hydroxy acids, highlighting the different forms of PLA scaffolds used as cell culture substrates and their applications in clinical practice. The chapter is divided into (1) Introduction; (2) Bioresorbable devices as cell culture substrates; (3) Cell adhesion to polymer substrates; (4) Tissue engineering and bioresorbable polymers; (5) Cell growth and proliferation on bioresorbable polymers; (6) Bioresorbable polymers for cartilage engineering; (7) Bioresorbable polymers for bone tissue engineering; (8) Bioresorbable polymers for skin tissue engineering, and (9) Conclusion.

1. Introduction

For centuries, extensive tissue injuries normally originating from mechanical trauma or degenerative diseases have been a challenge because of the scarcity of therapeutic resources. Removal of the damaged part was the most common practice, which resulted in a series of limitations of the affected patient and in a significant decrease in quality of life. Thus, replacement and/or regeneration of the damaged body regions became the new target. The increase in life expectancy resulting from the discovery of antibiotics and chemotherapy, as well as from improved sanitary and hygiene conditions, has encouraged the search for methods to replace damaged tissues [1].

There are two procedures used for the treatment of the failure or loss of tissues and organs: transplants and implants. In the case of transplants, tissues or organs are obtained from living donors (e.g., heart or kidneys) or from cadavers (e.g., lyophilized and frozen bone). In some cases, immunosuppressive drugs are necessary to prevent rejection of the transplanted organ, or other medications that minimize possible microbial contamination [1]. In addition, transplants have the disadvantage of raising a series of ethical and even religious issues. In contrast, devices developed to serve as implants do not present many of the problems reported above and are designed to act at the recipient tissue interface in the organism, interacting with these tissues [1, 2].

Biomaterials were first developed to remain inert in the organism. Thus, studies were aimed at investigating how to prevent or minimize undesired tissue reactions. At present, new materials are designed to elicit an effective interaction with tissues, provoking physiological responses such as cell growth and/or differentiation at the site of implantation [3]. Significant advances have been made over the past decades in the understanding of the mechanisms underlying the interaction of animal cells with their natural environment, i.e., the extracellular matrix, as well as of the influence of this matrix on cell growth and differentiation [4, 5]. This knowledge is frequently being used for the development of polymers that mimic the characteristics of extracellular matrix, thus playing an active role in tissue restoration.

The biomaterials used can be classified into permanent or temporary materials [6]. Permanent materials are used to replace damaged tissue for an undetermined period of time and are therefore designed to retain their mechanical and physicochemical properties for prolonged periods of time [6]. These types of devices are commonly employed experimentally as prostheses, replacing damaged joints, heart valves and intraocular lenses, among others. On the other hand, in some situations the support only needs to fill the damaged region temporarily until tissue recomposition is completed, or guides the regeneration process. Temporary biomaterials are an alternative in this case.

"Biodegradable" is a term that can be applied to polymers and solid devices that undergo dispersion *in vivo* as a result of macromolecular degradation, but without elimination of products and subproducts by the organism [7]. Biodegradable polymers can be attacked by biological elements in such a way that the integrity of the system is affected, forming fragments and other degradation products that can be removed from the site of action but not necessarily from the organism. "Bioabsorbable" is a term that can be applied to polymeric materials and devices that dissolve in body fluids without breakdown of the macromolecular chain or a reduction in molecular mass [7]. One example is the slow dissolution of soluble implants in organic fluids. "Bioresorbable" materials are polymers and solid devices that degrade through a reduction in size and that are resorbed *in vivo*, i.e., materials that are eliminated by metabolic pathways of the organism. Bioresorption is a concept that reflects the complete elimination of the material and of degradation products (compounds of low molar mass) in the absence of residual side effects [7]. The term "bioresorption" is applied when complete elimination occurs. A polymer can be bioresorbable if its macromolecules are excreted [7,8]. Bioresorbable polymeric materials are preferentially used as temporary devices [6].

2. Bioresorbable devices as cell culture substrates

A wide variety of temporary devices have been employed in biological systems, with the most widely used materials being based on α-hydroxy acid-derived polyesters, such as poly(L-lactic acid) (PLLA), poly(D-lactic acid) (PDLA), poly(DL-lactic acid) (PDLLA), poly(glycolic acid) (PGA), and polycaprolactone (PCL) [6,8]. During degradation, the polymer is broken down into smaller units by simple hydrolysis and its degradation products are eliminated from the body by metabolic pathways, such as the citric acid cycle, or directly by renal excretion (see Figure 1) [9-12].

Although the degradation of bioresorbable polymers mainly occurs by simple hydrolysis, there are reports in the literature indicating that degradation of PGA and PLLA, at least in part, is also mediated by enzymes [12,13]. Figure 1 is a schematic presentation of the degradation of PLLA by hydrolysis [8]. The elimination routes of the degradation products of some polyesters are shown in Figure 2.

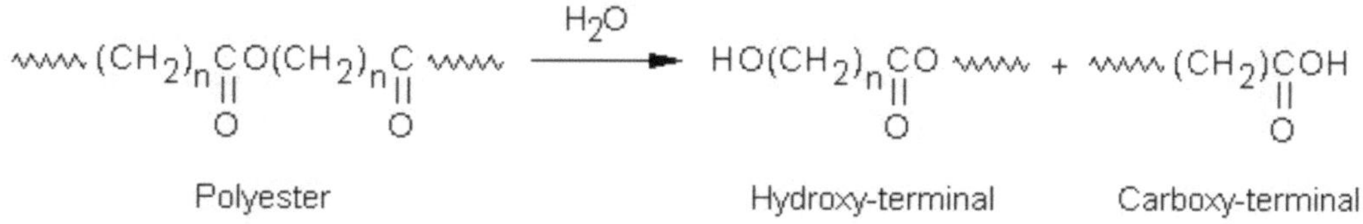

Fig. 1. Degradation of poly(α-hydroxy acids) by hydrolysis (From Barbanti *et al.* [8], with authorization).

Other bioresorbable polymers used are polyhydroxyalcanoates, polyesters produced by microorganisms. These compounds find applications as raw materials of different devices in the areas of biomedicine and tissue engineering [14]. Polyhydroxyalcanoates include poly(3-hydroxybutyrate) (PHB), poly(3-hydroxybutyrate-co-3-hydroxyvalerate) copolymer (PHBV), poly(4-hydroxybutyrate) (P4HB), poly(3-hydroxybutyrate-co-3-hydroxyhexanoate) copolymer (PHBHHx), and poly(3-hydroxyoctanoate) (PHO). These compounds have been used for the development of sutures, devices to guide tissue repair, heart implants, orthopedic pins, stents, tubules for nerve regeneration, and membranes for skin regeneration [14]. The use of PLLA/PHBV blends has appeared as a new proposal in the literature. The biological evaluation of this compound is relevant for tissue engineering [15,16].

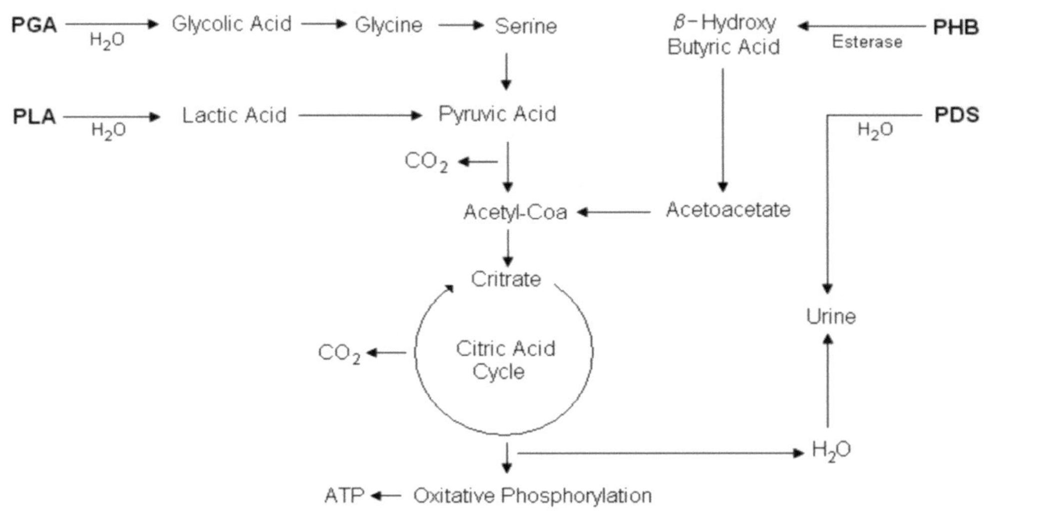

Fig. 2. Routes of degradation and excretion of some polyesters: poly(*p*-dioxane) (PDS), poly(glycolic acid) (PGA), poly(lactic acid) (PLA), and poly(hydroxybutyrate) (PHB).

3. Cell adhesion on polymer substrates

Normally, cell adhesion to the substrate is necessary for a good polymer-cell interaction. Although the substrate does not need to present extracellular matrix-like characteristics for cell adhesion to occur, physicochemical similarity is often desired when the aim is to promote cell differentiation or a more effective interaction of a certain polymer at the implantation site [2-4,17]. Thus, polymers presenting physicochemical and/or mechanical properties as close as possible to those of the tissues into which they will be implanted are currently being developed. These properties include an adequate balance between hydrophilicity/hydrophobicity, electrical charge distribution, hardness, elasticity, and strength [18].

A relationship exists between hydrophilicity and cell adhesion. Among other parameters, more hydrophilic substrates tend to provide a better interaction with cells [19-21]. The relationship between cell adhesion and polar groups on the material surface has been demonstrated for polystyrene. Adhesion of cells was found to increase with increasing polarity of the substrate [22]. In a study investigating hepatic cells cultured on different substrates, cell adhesion increased proportionally to the surface energy of the growth membranes. Adhesion was even higher in the presence of serum proteins adsorbed to the substrates and the metabolic activity of hepatic cells increased on hydrophilic membranes [23]. Other studies demonstrated a relationship between wettability and cell adhesion. Lee *et al.* [24] prepared a wettability gradient on polyethylene surfaces to investigate the interaction of different types of cells (Chinese hamster ovary cells, fibroblasts and endothelial cells) with fetal bovine serum proteins, in terms of surface hydrophilicity/hydrophobicity. Better adhesion, spreading and growth of cells were observed on surfaces with moderate hydrophilicity. Maximum adhesion was found on substrates with a water contact angle of approximately 57°, irrespective of the type of cell used. Serum proteins also adhered better to substrates presenting moderate hydrophilicity [24].

Thus, cell adhesion is an extremely important factor for biomaterials research. Only after adhesion do the cells initiate the process of spreading, division and production of new extracellular matrix [18,19]. Cell spreading is a complex process that involves modifications in cell morphology as a consequence of alterations in the cytoskeleton, thus improving interaction with the substrate. These modifications are shown in Figure 3.

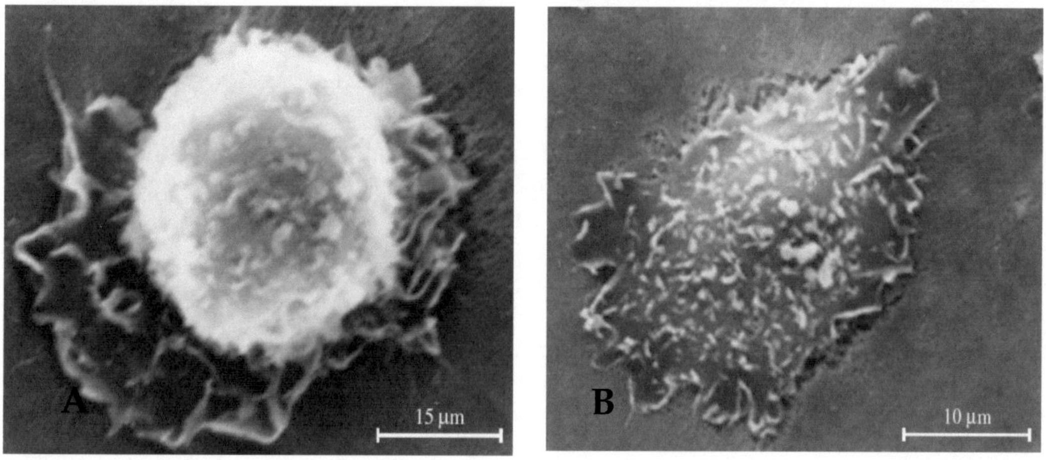

Fig. 3. Spreading of Vero cells on type I collagen gel. Round cells are observed during inoculation, a morphology that is adequate for the state of being in suspension. After adhesion, the cells start their interaction with the substrate. Note the onset of the phenomenon known as spreading in (A), which is characterized by modifications in the cytoskeleton and, consequently, in cell morphology as a result of interactions with the growth surface, with the cell flattening on the growth surface. (B) A cell spread on the substrate. Magnification bar: A = 15 μm; B = 10 μm (From Santos Jr and Wada [5], with authorization).

Good integration of the biomaterial with cells or tissues also depends on the structure of the devices itself. Our experience agrees with literature data indicating that porous materials promote cell growth and induce the production of extracellular matrix components by the cells [21,25]. A uniform distribution and pore interconnections are important to facilitate the formation of tissues in the form of an organized network, with a wide range of applications in tissue reconstruction [3,26,27]. *In vivo*, porosity and pore interconnection are essential for the proliferation of vessels, facilitating tissue nutrition around the implant. In this respect, different types of scaffolds containing PLLA have been developed and tested as substrates for cell growth [3]. Kwon *et al.* [28] evaluated nano- and microfibrous scaffolds with different compositions of PLLA, PCL and poly(L-lactic acid-co-caprolactone) (PLCL) (70/30, 50/50 and 30/70) as cell culture substrate. Human umbilical vein endothelial cells were found to adhere and proliferate well on the small-diameter fiber scaffolds (0.3 to 1.2 mm in diameter), whereas a marked reduction in cell adhesion and spreading, as well as a low proliferative capacity, was observed for large-diameter fiber scaffolds (about 7.0 mm).

Studies from our group have shown an initially slow adhesion to PLLA and PHBV scaffolds [21,25]. This finding does not necessarily indicate that the material tested would not be useful for tissue engineering. Mann *et al.* [29] showed that in the case of slow adhesion the materials may stimulate the early production of extracellular matrix components, permitting cell growth and proliferation. Our observations agree with these authors [21,25]. The extracellular matrix exerts a marked influence on the migration, proliferation and differentiation of cells cultured on biomaterials. Gunawan and coworkers [30] demonstrated the influence of the density of extracellular matrix components on the migration of rat intestinal cells (IEC-6 cells). The cells migrated in the direction of the gradient formed, with the demonstration of the particular role of laminin in this process.

Nevertheless, there is a trend to change the architecture of cell culture substrates used for tissue engineering [31]. The aim of these surface modifications is to improve the interaction with cells both *in vitro* and *in vivo*. With respect to cell cultures, increased cell interaction means to favor the initial interaction of cells with the substrate, i.e., to increase adhesion. In a recent study, nanofibrous PLLA scaffolds were fabricated and tested. The surface modifications introduced significantly increased cell adhesion and spreading. In addition, cell proliferation was maintained for more than 2 weeks and the modifications also increased the production of extracellular matrix components [32].

The surface properties of biomaterials can be altered in such a way as to make them more adequate for biomedical applications. The most commonly used techniques are chemical etching, gas plasma treatment and electron beam radiation [33]. Among these techniques, plasma treatment is particularly versatile because the modification is restricted to the surface without compromising the material properties as a whole. Plasma treatment can be used to modify the polymer surface in a nonspecific manner by introducing a variety of functional groups depending on the type of gas used [34,35]. Plasma is considered to be the fourth state of matter and consists of highly excited states of atoms, molecules, ions and radicals obtained from gases elevated to excited stages by radiofrequency, microwaves or electron discharge [35]. Thus, surface modification by plasma treatment is an economic and effective technique for biomaterials and is gaining space in the area of biomedicine, with its application becoming more frequent. The main advantage of plasma treatment is its capacity to modify the surface of materials, making them more biocompatible or permitting them to more closely mimic tissue without altering their properties. As a consequence, this technique permits a high degree of quality control, offering reliability and reproducing what would be difficult to achieve with conventional techniques. Another advantage of plasma treatment is that it renders the surface sterile by destruction of microorganisms such as bacteria and viruses and is therefore useful for biomedical devices, surgical instruments, tissue engineering and clinical applications [35]. Our experience shows that plasma treatment increases cell adhesion to PHBV matrices, rendering these scaffolds more receptive to cell growth [36].

Nakagawa *et al.* [37] submitted PLLA samples to plasma treatment in a CO_2 atmosphere and observed an increase in the hydrophilicity of the membrane. The cell response was also highly satisfactory for membranes submitted to plasma treatment compared to controls. Plasma treatment provided better cell adhesion and proliferation, although the authors had used short-term cell cultures (up to 3 days only). Ryu and coworkers [38] submitted poly(D,L-lactic acid-co-glycolic acid) (PLGA) membranes to plasma treatment. In that study, the PLGA surface was modified in order to increase the interaction of cells with the material surface. The results showed that the surface modifications increased hydrophilicity and cell adhesion and proliferation.

Several studies have been conducted to develop materials with biomimetic characteristics. Surface modification of biomaterials using bioactive molecules is a relatively simple approach to produce biomimetic materials. Long chains of extracellular matrix proteins such as fibronectin, vitronectin and laminin have been used for this purpose. Biomaterials coated with these proteins normally present increased cell adhesion and proliferation [31,39]. Peptides have also been used for surface modifications, with the most frequently used sequence being Arg-Gly-Asp (RGD), the binding sequence derived from fibronectin and laminin. Other peptide sequences such as Tyr-Ile-Gly-Ser-Arg (YIGSR), Arg-Glu-Asp-Val (REDV) and Ile-Lys-Val-Ala-Val (IKVAV) have also been employed for substrates such as quartz, glass, metal oxide and polymers [31,39].

Porous PDLA and PLLA substrates containing 2-methacryloyloxyethyl phosphorylcholine (MPC) and *n*-butyl methacrylate (BMA) were investigated regarding their capacity to interact with fibroblasts. MPC is an analog of phosphatidylcholine, a typical lipid of the plasma membrane of cells. The results obtained were quite interesting. The number of adhered cells was correlated with PDLA and PLLA content, with the cells presenting a good pattern of adhesion and migrating into the substrate within only 24 h of culture. Furthermore, cell morphology was influenced by the contact with MPC [40], with the addition of MPC rendering the PDLA and PLLA scaffolds more receptive to the initial interaction with cells.

Despite promising results, some investigators have reported discrepancies in the findings regarding the addition of bioactive molecules to growth surfaces. In this respect, an interesting study was conducted by Harnett *et al.* [41] who coated various biomaterials with different adhesion molecules such as fibronectin, polylysine, polyornithine and collagen in order to evaluate changes in surface free energy and the hydrophilicity/hydrophobicity balance. Only fibronectin promoted a homogenous coating for all substrates tested, producing a monopolar acid surface, whereas polylysine, polyornithine and collagen coatings produced hydrophobic or hydrophilic surfaces depending on the underlying substrate they were coated on.

4. Tissue engineering and bioresorbable polymers

Once cell adhesion is established, the material tested is studied as a cell carrier in procedures aimed at the restoration of damaged tissues. Tissue engineering can be understood as the application of the principles of exact sciences to tissue creation and repair. Three general strategies have been adopted to obtain new tissues [17]:

1) Use of autogenous cells i.e., cells isolated from the individual himself; isogeneic cells, i.e., cells obtained from different individuals which, however, are genetically identical or belong to the same species; allogeneic cells, i.e., cells isolated from different individuals of the same species, or xenogeneic cells, i.e., cells obtained from different species [42]. These cells are expanded in culture and implanted into the body by infusion methods. However, limitations of this strategy include the limited capacity of the cells to maintain their differentiated characteristics *in vitro*, the difficulty in sufficiently expanding some cells in culture (for example, hepatic and neural cells cannot be expanded in adequate numbers for clinical use), and immunological rejection when allogeneic and xenogeneic cells are used.

2) Tissue culture for subsequent implantation and replacement of sick or damaged tissues. The most common example used in clinical practice is skin grafts [43]. The main advantage of this strategy is its high biocompatibility and biofunctionality. However, this strategy presents the same disadvantages as described above.
3) Use of substances that induce the regeneration of damaged tissue. The success of this strategy depends on the large-scale purification and production of appropriate signal molecules such as growth factors and adhesion molecules. The proliferation of many cell types (which may induce the formation of a new tissue) depends on a combination of various growth factors that are highly specific proteins. Some growth factors can be released slowly from polymeric capsules and may stimulate the growth of damaged tissue [44]. In contrast, adhesion molecules such as fibronectin, vitronectin and laminin are protein components of biological fluids and/or extracellular matrix that are adsorbed on the material surface and recognized by integrins (cell membrane receptors associated with the cytoskeleton) [4]. Integrins bind to the small domains of adhesion molecules [45], such as the RGD amino acid sequence of fibronectin and vitronectin or the YIGSR sequence of laminin. RGD and other oligopeptides have been incorporated into some biomaterials to stimulate adhesion and consequent cell proliferation.
Autogenous or autograft implants are of special interest. This technique consists of the use of healthy cells derived from the patient himself who will receive the polymer implant. Autogenous implants have some advantages over organ transplants. Since the cell population isolated is expanded *in vitro* by cell culture, only a small number of donor cells is necessary for preparation of the implant. In addition, the use of autogenous cells avoids immunological problems such as rejections or allergic processes [3,42,46].

5. Cell growth and proliferation on bioresorbable polymers

Bioresorbable devices have been used *in vitro* as a support for the growth and proliferation of different cell types. Endothelial cells have been shown to satisfactorily multiply on PLLA scaffolds in the absence of platelet activation [47,48]. Good adhesion to the material and an adequate multiplication rate were reported for NIH/3T3 mouse cells cultured on PHBV membranes. These parameters improved after the introduction of physicochemical modifications (change in polymer hydrophilicity) on the PHBV surface [49]. In another study, cell proliferation on a PHBV substrate continued to be similar to that observed for collagen sponges for up to 35 days of culture [50].
The capacity of a substrate to stimulate cell growth and proliferation is intimately related to its ability to absorb proteins. In a recent study, Zhu and coworkers [48] showed that coating of the surface of PLLA devices with free amine groups increases the spreading and proliferation of endothelial cells. Extracellular matrix proteins such as fibronectin, laminin and collagen have also been shown to stimulate the multiplication of cells on substrates used for tissue engineering [51]. We showed that Vero cells produce extracellular matrix rich in fibronectin and collagen when cultured on dense or porous PLLA membranes, PHBV scaffolds or PLLA/PHBV blends of different proportions [21,25]. This finding may explain the observation of a significant proliferation rate despite the initially slow cell adhesion to these scaffolds [25,52].

6. Bioresorbable devices for cartilage engineering

The use of bioresorbable materials for articular cartilage repair is currently being investigated. Cartilage is an avascular tissue which basically consists of two cell types, chondrocytes and chondroblasts. Chondrocytes produce an extracellular matrix that mainly consists of collagen and glycosaminoglycans. The proportion of these components depends on the type of cartilaginous tissue [53]. Once damaged, cartilage presents little or no regenerative capacity and certain injuries may progress to severe degenerative joint diseases [46,53]. In addition to the fact that the mechanisms underlying the formation of articular cartilage are still unclear, there are few alternative clinical procedures to the replacement of joints with prostheses that can fill small defects resulting from trauma or degenerative diseases. Current treatments for articular cartilage engineering include 1) the creation of a defect by wear or perforation to permit the migration of progenitor cells; 2) arthroscopy; 3) autografts; 4) transplantation of perichondrium and periosteum to introduce undifferentiated cells with a chondrogenic potential; 5) autogenous cell transplants (chondrocytes or undifferentiated cells) previously expanded *in vitro* and reinjected as an autogenous periosteal graft that is able to maintain both chondrogenesis and osteogenesis [54]. Studies on cartilage engineering that combine different materials in the production of scaffolds used as a support for chondrogenic cells are necessary. In this respect, there is an intense search for materials that mimic the biomechanical behavior of articular cartilage and thus can be applied to joint repair [46]. Some polymeric materials including both temporary and permanent polymers are investigated for this application. Bioresorbable materials studied for their use as temporary cartilage matrix include PLLA and PGA polymers and their copolymers and blends.

Chondrocytes cultured on a fibrous PGA matrix and porous PLLA membranes have been studied [55]. The results showed the neoformation of cartilage tissue comparable to that observed for chondrocytes cultured on collagen substrates obtained from articular cartilage. Under these conditions, chondrocytes continued to grow on these polymers for up to 6 months, maintaining the shape of the original device and producing a tissue with characteristics similar to those of cartilage, including the synthesis of glycosaminoglycans and types I and II collagen [55]. In addition, it was observed that cartilage cells cultured on polyesters such as PLLA and PGA tend to show increased synthesis of proteoglycans and collagen when compared to cells cultured on collagen matrix [56]. Puelacher *et al.* [57] studied the *in vitro* and *in vivo* growth of chondrocytes on PGA and PLLA scaffolds that simulated the morphology of human nasal cartilage. The authors observed the formation of hyaline-like cartilage on these substrates. The experimental results indicate that, once improved, these tissue reconstruction techniques have potential applications in orthopedic, plastic and reconstructive and craniomaxillofacial surgery. In addition, the formation of hyaline-like cartilage was observed after 6 weeks when perichondrial cells were cultured on PLLA membranes and implanted into the femoral condyle of rabbits [55,58].

A study of cells obtained from human articular cartilage and maintained in culture on devices consisting of different bioresorbable polyesters showed that the process of adhesion was proportional to the hydrophilicity of the polymers. In addition, no variations in cell spreading were observed between the different biomaterials. Although the cells studied adhered less to the PLLA membranes when compared to PLGA, their proliferative capacity was better when grown on the PLLA membrane [59]. In addition, the production of cartilaginous matrix and type II collagen was found to be lower for human chondrocytes cultured on PLLA membranes compared to those grown on PLGA membranes. On the other hand, cells grown on PLLA scaffolds presented a greater capacity of synthesizing type I collagen [60]. Interesting results were also obtained with PLLA scaffolds used for meniscus reconstruction. Porous implants were found to guide vascular growth into the injury region [61]. Canine meniscus reconstruction using lactic acid/ε-caprolactone copolymers have also been reported [62]. These results demonstrate that the principles of tissue engineering employing bioresorbable materials are a promising study area and certainly will yield significant results in the near future.

In an attempt to better mimic the natural environment of cartilage cells, Takagi and coworkers [63] developed a scaffold consisting of collagen and a copolymer mesh of PLLA and PLGA. Glucuronic acid is one of the components of glycosaminoglycans found in the extracellular matrix of tissues. The authors showed that cartilage cells cultured inside the scaffold consumed glucose from the culture medium and produced typical extracellular matrix components of cartilage tissue. Within this context, another study compared hyaluronic acid scaffolds and polyester scaffolds with different degradation rates [64]. The study showed that the degradation rate of the scaffolds is critical for the cartilage repair process. Cartilage formation was slow with dissolution of the materials, although presenting more cracks and discontinuities. Wang *et al.* [65] tested scaffolds of different origins, such as poly(L-lactide), poly(D,L-lactide) and collagen-hydroxyapatite, for the *in vitro* production of cartilage. Porcine chondrocytes were seeded onto the scaffolds. After 15 weeks of culture, a layer of viable neo-cartilage was produced on each material, with the collagen-hydroxyapatite constructs yielding better results in terms of cell viability and integration. Another approach was the use of a gelatin scaffold obtained from demineralized bone matrix inoculated with rabbit chondrocytes. Neoformation of hyaline-like cartilage was demonstrated [66].

In a well-conceived experiment, chondrocytes were cultured on PLLA microspheres in a bioreactor. In addition, the surface of the microspheres was modified by the addition of RGD peptides, small repetitive sequences of the amino acids arginine (Arg or R), glycine (Gly or G) and aspartate (Asp or D) which are known to stimulate cell adhesion. PLLA degradation was determined after different periods of time (7, 14, 21, 28, 35, 49 and 56 days). The authors observed that the materials continued to be stable to support cell growth after the periods studied. The use of a bioreactor resulted in a significant increase in the number of cells cultured on the biomaterials and the devices studied showed a good capacity to stimulate cell adhesion and proliferation. The authors also observed the formation of microaggregates, a finding that might indicate the production of extracellular matrix [67].

In another study, human mesenchymal stem cells were cultured on a PLDLA scaffold. This construct was maintained in chondrogenic medium and the other end was then loaded with cells of the same origin but previously induced to undergo osteogenesis. This system was then cultured under conditions able to maintain both chondrogenesis and osteogenesis, thus producing *in vitro* a hybrid osteochondral construct [68]. Taken together, these results are promising but the search for a polymeric material that better mimics the function of articular cartilage still continues.

7. Bioresorbable devices for bone tissue engineering

Bone is a natural tissue which consists of an organic component (mainly collagen) and a mineral component composed of hydroxyapatite. Reconstruction of long bone defects is a clinical challenge. Normally, multiple surgeries are required to restore the structure and function of the damage site. The development of tissue engineering techniques has led to the use of new procedures for bone restoration. Polymeric materials may serve as a support for cell growth, permitting the penetration of blood vessels, and even exert morphogenetic activity in some cases. In the case of bioresorbable polymers, the materials are often enriched with hydroxyapatite, growth factors, bone morphogenetic proteins (BMPs) and other bone elements, a fact that renders them highly effective in the stimulation of bone neoformation in the damaged areas [10].

Transplants consisting of different types of isolated cells cultured on PLLA and PGA scaffolds have been investigated as temporary substitutes of damaged tissue portions [17]. Implantation of PLGA copolymers into bone resulted in bioresorption of the material and concomitant bone neoformation at the site of the implant. An additional advantage of PLGA is that its complete degradation is variable and may occur within weeks or years depending on the polyester ratio present in the copolymers [69].

Osteoblastic cells cultured on PLLA, PGA and PLGA films presented a satisfactory pattern of cell adhesion and spreading, in addition to the ability to grow and proliferate on the substrate. Furthermore, cells grown on these polymers presented increased alkaline phosphatase activity, a marker of osteoblast differentiation, and increased synthesis of collagen I [70]. Similar results were obtained when osteoblasts were cultured on PLGA scaffolds. In that case, mineralization of the bone matrix produced was also observed [71,72]. Interestingly, even bone marrow cells cultured on porous PLGA scaffolds and implanted into the rat mesentery were able to initiate ectopic bone formation [71].

It has been postulated that porous materials implanted *in vivo* present a better integration with the recipient tissue. Disagreement still exists regarding the ideal diameter of the pores for tissue growth. *In vitro*, low porosity stimulates osteogenesis, suppressing cell proliferation by forcing cell aggregation. On the other hand, *in vivo*, a higher porosity with larger pores promotes bone growth. However, these factors result in low mechanical properties and, therefore, a functional limit exists for porosity and pore size [73]. On the basis of these studies, a minimum pore size of approximately 100 μm was established considering cell size, migration requirements and transport. *In vitro* and *in vivo* results suggest pore sizes and pore interconnections > 300 μm to facilitate vascularization of the graft. Some investigators recommend diameter variations of 300 to 400 μm, whereas others propose even wider intervals of 200 to 400 μm [74]. Thus, the macrostructure (pre-existing macroporosity), microstructure (microporosity, enhanced surface microroughness or microtexture) and chemical composition of the material play an important role in osteoinduction guided by the biomaterial *in vitro*. In addition, the model used for *in vivo* study may also markedly influence the results. Osteoinduction is only occasionally observed in mice and rats, whereas the same material induces bone formation reproducibly in goats and dogs [75].

Another interesting approach is the adsorption of factors that stimulate cell differentiation on polymer surfaces. In 1965, ectopic bone formation was demonstrated after implantation of demineralized bone matrix into muscles of rabbits, rats, mice and guinea pigs [76]. A protein, called bone morphogenetic protein (BMP), was found to be involved in the cascade of chemotaxis, mitosis, differentiation and bone formation [77]. Since these pioneering reports, the role of BMP in bone formation *in vitro* and *in vivo* has been extensively studied. A recombinant form of human bone morphogenetic protein type 2 (rhBMP-2) was added to bioresorbable PLGA substrates. In those studies, a higher production of bone matrix was observed for osteoblasts cultured on substrates covered with rhBMP-2 when compared to control [78,79]. A similar experiment was conducted by Hollinger *et al.* [80], who adsorbed rhBMP-2 to a collagen I matrix. When this collagen matrix was implanted into fractured bone segments, bone neoformation was observed, as well as integration of the implant into the damaged bone. Subsequently, various groups have shown that BMPs are able to induce endochondral bone formation when implanted at ectopic sites in experimental animals [81]. Although the results are promising, this therapeutic approach might be limited by the size of the fracture produced [78].

The exact mechanism of osteoinduction by biomaterials is still largely unknown. In addition, it is unclear whether the mechanisms of osteoinduction by BMPs and biomaterials are the same. In a recent review [81], it was demonstrated marked differences in osteoinduction mediated by BMPs and biomaterials: 1) biomaterials always induce intramembraneous bone formation, whereas BMP mainly induces endochondral bone formation; 2) in small animals such as rodents, bone formation is rarely induced by biomaterials but is easily mediated by BMPs; 3) newly formed bone is never observed at the periphery of biomaterials but always inside their pores, whereas bone formation induced by BMPs regularly occurs outside the carrier and soft tissue is observed distant from the surface of this carrier.

Gugala *et al.* [82] investigated the adsorption of proteins and activity of osteoblasts cultured for up to 3 weeks on porous and non-porous PDLA membranes. The presence of pores did not influence protein adsorption. The authors observed that the cells maintained their typical phenotype, formed mineralized nodules, i.e., regions where mineralized organic matrix is observed, and produced alkaline phosphatase, an enzyme commonly related to the process of biomineralization both *in vitro* and *in vivo*. The amount of protein, alkaline phosphatase activity and number of cells increased over time and were higher for the porous than for the non-porous membranes. In another study, osteoblasts were cultured on a gradient of PLLA and PDLLA in order to determine whether the gradient alters the pattern of cell interaction with the substrate [83]. Cell adhesion was similar at both ends of the gradient, but proliferation was more significant in the smoother PDLLA-rich region than in the rougher PLLA-rich region of the gradient. These results demonstrate that, in addition to the composition of the substrate, its topography also interferes with the behavior of bone cells. Thus, PLLA might not be the ideal substrate for the culture of bone cells. It was showed that murine osteoblastic cells more intensely adhered to both dense and porous PGA scaffolds than to PLLA scaffolds. Nevertheless, the pattern of cell spreading on the substrates was similar after 24 h of culture [84].

The use of rapidly degrading synthetic or natural polymer matrices with low mechanical properties (high porosity) results in grafts with high biological activity but poor structural properties, such as low strength and rigidity. One approach to correct this problem is the combination of bioactive ceramics, such as calcium phosphate, with bioresorbable polymers in order to improve the mechanical properties of the scaffolds. A composite matrix also increases the osteoconductive properties of the scaffolds [73]. Various of these compounds are investigated for this purpose. Natural polymers combined with hydroxyapatite-collagen, hydroxyapatite compounds (chitosan-hydroxyapatite), and PLA copolymers (PLA-polyethyleneglycol) have been tested.

Porous PLLA and PLLA-hydroxyapatite composite scaffolds were seeded with osteoblasts and cultured. The cells were found to penetrate deep into the PLLA-hydroxyapatite scaffolds and were uniformly distributed. Cell viability and proliferation, as well as the expression of bone differentiation markers, were higher for the composite scaffolds when compared to pure PLLA [85].

Other studies have shown that PCL-calcium phosphate composites confer favorable mechanical and biochemical properties (the ceramic confers strength and the polymer confers hardness and plasticity). In these scaffolds, the ceramic was homogenously distributed in the matrix as being exposed on the surface. The composite was more hydrophilic and degradation was accelerated when compared to the pure PCL scaffold [86]. In addition, the composite was more receptive to cell adhesion [87]. Another study investigating PLA-hydroxyapatite and PCL-hydroxyapatite composites showed that cells tended to adhere and spread among hydroxyapatite particles exposed on the surface. The presence of hydroxyapatite resulted in a higher activity of bone cells [88]. Despite these results, further studies are necessary to identify a material that serves as a substrate for the growth of bone cells for tissue regeneration since there is still no ideal compound that stimulates bone formation.

8. Bioresorbable devices for skin engineering

Several groups have investigated different methods for the creation of dermal equivalents using various substitutes based on biological materials such as collagen, fibrin, culture of epidermal layers, or synthetic materials [89]. However, in patients dermal substitutes have shown a slow growth of vascular structures into the dermal components. As a consequence, a second surgical procedure is necessary to transplant epidermal components into the regenerating wound [90].

Much of the knowledge about the mechanisms of skin regeneration by tissue engineering stems from three-dimensional scaffold-based fibroblast cultures. The mechanism of action of these grafts is based on 1) fibroblast colonization, 2) production of growth factors, 3) a substrate for keratinocyte migration, and 4) the wound immune response. Another important factor is the migration of keratinocytes to the site of injury. A large number of keratinocytes is observed around chronic wounds. However, these cells do not migrate onto the wound surface. One of the possible explanations is the degradation of extracellular matrix by proteases present at high concentrations in many types of injuries. The control of the synthesis and/or degradation of extracellular matrix seem to have marked implications in the outcomes of wound healing. Genes encoding extracellular matrix components are expressed during wound reconstruction. These compounds include tenascin, decorin and some types of collagen (I, III, V and VI). A provisional extracellular matrix is expected to provide a good substrate for the migration of cells such as keratinocytes and leukocytes. The maintenance of this matrix is therefore important [91].

Natural polymers, especially collagen, are extensively studied as a substrate for skin regeneration. Collagen has been obtained from different xenogenic sources such as cattle, swine and horses. Some types of human collagen are used in the United States with approval by the Food and Drug Administration (FDA). Collagen is extremely receptive to the culture of fibroblasts. We showed that fibroblasts grown on collagen gels produce extracellular matrix components such as glycosaminoglycans and fibronectin, forming a tissue that resembles reconstituted connective tissue [92]. However, variations in culture conditions change the behavior of fibroblasts into that of epithelial cells, with a reduction in their migration behavior on the collagen matrix and the production of molecules such as collagen IV and laminin [92,93]. Thus, the use of collagen is an interesting model not only for the area of tissue reconstruction but also for the study of the differentiation itself of cells cultured on it.

Collagen presents a series of advantages such as availability, biodegradation, bioresorption, and resistance to distension. In addition, the properties of collagen can be altered by modifying its functional groups. However, there are disadvantages such as rapid degradation, high hydrophilicity that may cause significant swelling after implantation, low resistance to mechanical compression forces, and high cost of purification, factors limiting the use of collagen [94]. The dermal replacement layer of collagen serves as a matrix for the growth of macrophages, lymphocytes and endothelial cells derived from the wound. During the tissue repair process, an endogenous collagen matrix is deposited, whereas the dermal substitute is degraded within approximately 30 days. Furthermore, the skin graft is found to be flexible at the site of injury and does not adhere to deeper layers, thus permitting free and functional joint movement [95].

In view of the positive experiences obtained with collagen, development of an artificial dermis was the next target. This type of structure initially consisted of collagen I and chondroitin sulfate [96-98]. The so-called artificial dermis is composed of a bilayer membrane, in which a matrix of collagen I and chondroitin-6-sulfate is bonded to a layer of polysiloxane polymer (silicone). The collagen matrix serves to attract macrophages, lymphocytes, fibroblasts and capillaries. In addition to being bioresorbable, the collagen matrix is gradually eliminated, permitting the formation of a new dermal matrix. In the presence of adequate vascularization, the silicone layer is removed and an epidermal autograft is added to the newly formed dermis. The silicone layer represents an additional barrier that prevents or minimizes the risk of microbial contamination. The formation of this neodermis occurs within 14-21 days. After the completion of regeneration, the neodermis is histologically and functionally similar to normal dermis [98,99]. This dermal substitute is considered to be an effective treatment for burns and tissue reconstruction in patients with cancer [100,101].

Since these first results, various research groups have worked to improve the device in an attempt to optimize outcomes and to reduce production costs. Among the modifications introduced are the removal of chondroitin-6-sulfate and alterations in the storage format of the substrate for implantation [102]. Long-term studies have subsequently evaluated the efficiency of the implants fabricated, tissue receptivity, degree of regeneration, and strength of the skin formed. The data reported were highly satisfactory. The advantages of these implants include 1) that the inner layer of the implant spontaneously becomes a structure resembling connective tissue, 2) a good long-term postoperative appearance even in the case of thin implants, and 3) minimal damage to the donor tissue [43].

Intensive research into dermal substitutes and tissue engineering devices has inevitably led to bioresorbable polymers. The latter present some advantages over biological substrates: 1) since bioresorbable polymers are synthetic, their production can be standardized and variations between different production lots are therefore small; 2) these polymers can be modified to better fulfill clinical requirements without the loss of their mechanical properties; 3) since the possibility of harboring viruses or prions is practically zero, these polymers are safer for patients [5,89]. Among the bioresorbable polymers available, the most widely used are variations of PLA, such as PLLA, PDLA, PDLLA, PGA, and PLGA. Fibroblasts cultured on three-dimensional dense and porous (different pore diameters) PLLA membranes were found to adhere to the polymers, to proliferate on them and to produce extracellular matrix molecules such as collagen IV and fibronectin [21]. This behavior suggests that PLLA can be used as a substrate for skin injuries. However, PLA presents low mechanical strength. One alternative would be the formation of copolymers or blends to modify the mechanical properties of the material. It was also evaluated the efficacy of PLGA membranes with and without plastifier as a dressing for skin injuries. The *in vitro* results showed that the addition of the plastifier reduced the vitreous transition temperature (Tg) of the membranes and increased their flexibility. *In vivo* analysis demonstrated that the polymer degraded rapidly when in contact with the skin without causing serious inflammation and protected the ulcerated area from the action of external agents. Wound healing was faster in the presence of the membranes, a fact indicating their potential use as skin dressings [103].

Another polymer that is currently gaining interest is poly(hydroxybutyrate-co-hydroxyvalerate) (PHBV) copolymer, a biodegradable and bioresorbable polymer derived from microbial activity. PLLA/PHBV blends are more resistant and, at the same time, more malleable than pure PLLA membranes. Since PHBV is more stable, its degradation is slow. Thus, the desired strength and degradability are obtained by varying the proportions of PLLA and PHBV in the blend. Fibroblasts cultured on different proportions of PLLA/PHBV blends (100:0, 60:40, 50:50, 40:60 and 0:100) were found to adhere to the substrates and to multiply on them. Again, these cells produced collagen IV and fibronectin on the supports [25].

New substrates such as poly(ethylene terephthalate)/poly(butylene terephthalate) (PEGT/PBT) are also being tested. A fragment of skin equivalent was reconstructed using PEGT/PBT scaffolds and cultures of fibroblasts and keratinocytes. The structure formed presented various characteristics of differentiated skin, such as an epidermal layer expressing several types of keratins on which grew layers of fibroblasts. In addition, basement membrane components such as collagen VII, laminin and nidogen were observed between these two layers [89].

Other alternatives also aimed at dermal regeneration and/or tissue engineering are being developed. Some investigators are carefully observing the use of chitosan. Chitosan, or *N*-carboxybutylchitosan, is derived from the deacetylation of chitin. Chitosan is a non-immunogenic compound that presents slow degradation in aqueous medium, even in the presence of lysozymes, and its degradation product, glycosamine, is not toxic [104]. Chitosan sponges should present, at least in theory, a stable shape and size during the period of cell culture. Porous chitosan membranes showed good integrity and favored cell spreading [104]. The use of collagen/chitosan blends was proposed taking advantage of the characteristics of chitosan and combining them with the properties of collagen. The blends were found to present a great water retention capacity, i.e., they were highly hydrophilic, and also markedly stimulated the growth of fibroblasts on their surface. Furthermore, the addition of chitosan did not reduce the interactivity of cells with collagen. These blends support fibroblast infiltration *in vivo*, a fact facilitating the formation of a neodermis [94].

Finally, our experience and the literature data available indicate that bioresorbable polymers are a suitable alternative as long as a growth substrate is chosen whose degradation coincides with the regeneration of the dermis-epidermis and that permits neovascularization of the skin formed at the site of the implant. Dense or porous PLGA membranes may show these characteristics. The search for materials with good properties might be a promising alternative for the reconstruction of damaged skin.

9. Conclusion

The advantages of the use of bioresorbable compounds for tissue engineering are numerous when compared to other more traditional surgical procedures. Internal fixation devices used in orthopedic surgeries lose their function of maintaining tissues together when structural recomposition is completed. The use of bioresorbable implants for internal fixation has the advantage of eliminating the second surgical intervention necessary for their removal. In addition, the risks of metal implants such as corrosion or friction with bone are avoided. Although still under study, the use of bioresorbable polymers is highly promising. New compounds are being developed and tested every year and their application perspectives are immense. The search for materials that are ideal for each tissue and clinical approach continues to represent a challenge.

10. References

[1] Hench, L.L. Biomaterials: a forecast for the future. *Biomaterials*, 19: 1419-1423, 1998.

[2] Hubbell, J.A. Biomaterials in tissue engineering. *Biotechnology*, 13: 565-576, 1995.

[3] Place, E.S.; Evans, N.D.; Stevens, M.M. Complexity in biomaterials for tissue engineering. *Nature Materials*, 8: 457-470, 2009.

[4] Alberts, B.; Johnson, A.; Lewis, J.; Raff, M.; Roberts, K. & Walter, P. *Molecular Biology of the Cell*, 5th ed., Garland Science, New York, 2008.

[5] Santos Jr., A.R; Wada, M.L.F. Polímeros biorreabsorvíveis como arcabouços para cultura de células e engenharia tecidual. *Polímeros: Ciência e Tecnologia*, 17: 308-317, 2007.

[6] Törmälä P, Pohjonen T & Rokkanen P. Bioabsorbable polymers: materials technology and surgical applications. *Proc. Instn. Mech. Eng. Part H - J. Eng.Med.*, 212: 101-111, 1998.

[7] Vert, M.; Li, M. S.; Spenlehauer, G. & Guerin, P. Bioresorbability and biocompatibility of aliphatic polyesters. *J. Mater. Sci. Mater. Med.*, 3: 432-446 1992.

[8] Barbanti, S.H.; Zavaglia, C.A.C. & Duek, E.A.R. Polímeros bioreabsorvíveis na engenharia de tecidos. *Polímeros: Ciência e Tecnologia*, 15: 13-21, 2005.

[9] Hollinger, J.O.; Battistone, G.C. Biodegradable bone repair materials: synthetic polymers and ceramics. *Clin. Orthop. Rel. Res.* 207: 290-305, 1986.

[10] An, Y.H.; Woolf, S.K.; Friedman, R.J. Pre-clinical in vivo evaluation of orthopedic bioabsorbable devices. *Biomaterials*, 21(24): 2635-2652, 2000.

[11] Barbanti, S.H.; Zavaglia, C.A.C.; Duek, E.A.R. Porous and dense poly(L-lactic acid) membranes: in vitro degradation. *Acta Microscopica*, 11: 85-89, 2002.

[12] Willians, D.F.; Mort, E. Enzyme-accelerated hydrolysis of polyglycolic acid. *J. Bioeng.*, 1: 231-238, 1977.

[13] Li, S.M.; Garreau, H.; Vert, M. Structure property relationship in the case of the degradation of massive poly(α-hydroxy acids) in aqueous media, Part 1: poly (DL-lactic acid). *J. Mater Sci. Mater. Res.* 1: 123-130, 1990

[14] Chen, G.-Q. ; Wu, Q. The application of polyhydroxyalkanoates as tissue engineering materials. *Biomaterials*, 26: 6565-6578, 2005.

[15] Ferreira, B.M.P.; Zavaglia, C.A.C.; Duek, E.A.R. Films of poly(L-lactic acid)/poly(hydroxybutyrate-co-hydroxyvalerate) blends: in vitro degradation. *Mater. Res.*, 4: 34-42, 2001.

[16] Ferreira, B.M.P; Zavaglia, C.A.C.; Duek, E.A.R. Films of PLLA/PHBV: thermal, morphological and mechanical characterization. *J. Applied. Polymer Sci.*, 86: 2898-2906, 2002.

[17] Langer, R.; Vacanti, J.P. Tissue engineering. *Science*, 260: 920-926, 1993.

[18] Davies, J.E.; Causton, B.; Bovell, Y.; et al. The migration of osteoblasts over substrata of discrete surface charge. *Biomaterials* 7: 231-233, 1986.

[19] Dewez, J-L; Lhoest, J.-B.; Detrait, E.; Berger, V.; Dupont-Gillain, C.C.; Vincent, L.-M.; Schneider, Y.-J.; Bertrand, P.; Rouxhet, P.G. Adhesion of mammalian cells to polymer surfaces: from physical chemistry of surfaces to selective adhesion of defined patterns. *Biomaterials*, 19: 1441-1445, 1998.

[20] Neff, J.A.; Caldwell, K.D. & Tesco, P.A . A novel method for surface modification to promote cell attachment to hydrophobic substrates *J. Biomed. Mater. Res.*, 40: 511-519, 1998.

[21] Santos Jr, A.R.; Barbanti, S.H.; Duek, E.A.R.; Dolder, H.; Wada, R.S.; Wada, M.L.F. Growth and differentiation of Vero cells on poly(L-lactic acid) membranes of different pore diameters. *Artif. Organs*, 25: 7-13. 2001.

[22] Birdi, K. Cell adhesion on solids and the role of surface forces, *J. Theor. Biol.*, 93: 1-5, 1981.

[23] De Bartolo, L.; Morelli, S.; Bader, A.; Drioli, E. Evaluation of cell behaviour related to physico-chemical properties of polmeric membranes to be used in bioartifical organs, *Biomaterials*, 23: 2485-2497, 2002.

[24] Lee, J.H.; Khang, G.; Lee, J.W.; Lee, H.B. Interaction of different types of cells of polymer surfaces with wettability gradient. *J. Colloid Interface Sci.*, 205: 323-330, 1998.

[25] Santos Jr, A.R.; Ferreira, B.M.P; Duek, E.A.R.; Dolder, H.; Wada, R.S.; Wada, M.L.F. Differentiation pattern of Vero Cells Cultured on Poly(L-Lactic Acid)/Poly(Hydroxybutyrate-co-Hydroxyvalerate) Blends. *Artif. Organs*, 28: 381-89, 2004.

[26] Wald, H.L.; Sarakinos, G.; Lyman, M.D.; Mikos, A.G.; Vacanti, J.P.; Langer, R. Cell seeding in porous transplantation devices. *Biomaterials*, 14: 270-278, 1993.

[27] van Sliedregt, A.; Van Loon, J.A.; Van Der Brink, C.; De Groot, K.; Van Blitterswijk, C.A. Evaluation of polylactice monomers in a in vitro biocompatibility assay. *Biomaterials*, 15: 251-256, 1994.

[28] Kwon, I.K.; Kidoaki, S. & Matsuda, T. Electrospun nano- to microfiber fabrics made of biodegradable copolyesters: structural characteristics, mechanical properties and cell adhesion potential. *Biomaterials*, 26: 3929-3939, 2005.

[29] Mann, B.K.; Tsai, A.T.; Scott-Burden, T.; West, J.L. Modification of surfaces with cell adhesion peptides alters extracellular matrix deposition. *Biomaterials*, 20: 2281-2286, 1999.

[30] Gunawan, R.C.; Silvestre, J.; Gaskins, H.R.; Kenis, P.J.A.; Leckband, D.E. Cell migration and polarity on microfabricated gradients of extracellular matrix proteins. *Langmuir*, 22: 4250-4258, 2006.

[31] Shin, H.; Jo, S.; Mikos, A.G. Biomimetic materials for tissue engineering. *Biomaterials*, 24: 4353-4364, 2003.

[32] Liu, X.H.; Won, Y.; Ma, P.X. Porogen-induced surface modification of nano-fibrous poly(L-lactic acid) scaffolds for tissue engineering. *Biomaterials*, 27: 3980-3987, 2006.

[33] Keen, I.; Broota, O.; Rintoul, L.; Fredericks, P.; Trau, M.; Grondahl, L. Introducing amine functionalities on a poly(3-hydroxybutirate-co-3-hydroxivalerate) surface: compring the use of ammonia plasma treatment and ethylenediamine aminolysis. *Biomacromolecules*, 7: 427-434, 2006.

[34] Yang, J.; Bei, J.; Wang, S. Enhanced cell affinity of poly(D,L-Lactide) by combining plasma treatment with collagen anchorage. *Biomaterials*, 23: 2607-2614, 2002.

[35] Chu, P.K.; Chen, J.Y.; Wang, L.P.; Huang, N. Plasma-surface modification of biomaterials. *Mater Sci Eng Ver.*,36:143-206, 2002.

[36] Lucchesi, C.; Ferreira, B.M.P.; Duek, E.Ar.; Santos Jr, A.R.; Joazeiro, P.P. Increased response of Vero cells to PHBV matrices treated by plasma. *J. Mater. Sci. Mater. Med.*, 19: 635-643, 2008.

[37] Nakagawa, T.; Tagawa, T. Experimental Pathology - Ultrastructural study of direct bone formation induced by BMPs-collagen complex implanted into an ectopic site. *Oral Dis.*, 6, 172-179, 2000.

[38] Ryu, G.H.; Yang, W.-S.; Roh, H.-W.; Lee, I.-S.; Kim, J.K.; Lee, G.H.; Lee, D.H.; Park, B.J.; Lee, M.S. & Park, J.C. Plasma surface modification of poly(D,L-lactic-co-glycolic acid)(65/35) film for tissue engineering. *Surf. Coat. Technol.*, 193: 60-64, 2005.

[39] Hersel, U.; Dahmen, C.; Kessler, H. RGD modified polymers: biomaterials for stimulated cell adhesion and beyond. *Biomaterials*, 24: 4385-4415, 2003.

[40] Watanabe, J.; Eriguchi, T.; Ishihara, K. Cell adhesion and morphology in porous scaffold based on enantiomeric poly(lactic acid) graft-type phospholipid polymers. *Biomacromolecules*, 3: 1375-1383, 2002.

[41] Harnett, E.M.; Alderman, J.; Wood, T. The surface energy of various biomaterials coated with adhesion molecules used in cell culture. *Colloids and Surfaces B: Biointerfaces*, 55: 90-97, 2007.

[42] Atala, A.; Lanza, R. Methods of Tissue Engineering, Academic Press, Florida, 2002.

[43] Suzuki, S.; Kawai, K.; Ashoori, F.; Morimoto, N.; Nishimura, Y.; Ikada, Y. Long-term follow-up study of artificial dermis composed of outer silicone layer and inner collagen sponge. *Br. J. Plastic Surg.*, 53: 659-666, 2000.

[44] Lieberman, J.R.; Daluiski, A.; Einhorn, T.A. The role of growth factors in the repair of bone - Biology and clinical applications. *J. Bone Joint. Surg. Am.*, 84-A: 1032-1044, 2002.

[45] Yamada, K.M. Adhesive recognition sequences. *J. Biol. Chem.*, 266: 12809-12812, 1991.

[46] Temenoff, J.S.; Mikos, A.G. Review: tissue engineering for regeneration of articular cartilage. *Biomaterials*, 21: 431-440, 2000.

[47] Hsu, S.H.; Tseng, H.J.; Fang, Z.H. Polyurethane blended with polylactides for improved cell adhesion and reduced platelet activation. *Artif. Organs*, 23: 958-61, 1999.

[48] Zhu YB, Gao CY, Liu XY, He T, Shen JC. Immobilization of biomacromolecules onto aminolyzes poly(L-lactic acid) toward acceleration of endothelium regeneration. *Tis. Engineering*, 10: 53-61, 2004.

[49] Lee, S.J.; Lee, Y.M.; Khang, G.; Kim, I.Y.; Lee, B.; Lee, H.B. Effect of poly(3-hydroxybutyrate-co-hydroxyvalerate) surface with different wettability on fibroblast behavior. *Macromolecular Res.*, 10: 150-157, 2002.

[50] Rivard, C.H.; Chaput, C.; Rhalmi, S.; Selmani, A. Bioabsorbable synthetic polyesters and tissue regeneration: A study on the three-dimensional proliferation of ovine chondrocytes and osteoblasts. *Ann. Chirurgie*, 50, 651-658, 1996.

[51] Aframian, D.J.; Cukierman, E.; Nikolovski, J.; Mooney, D.J.; Yamada, K.M.; Baum, B.J. The growth and morphological behavior of salivary epithelial cells on matrix protein-coated biodegradable substrata. *Tis. Engineering*, 6: 209-16, 2000.

[52] Santos Jr, A.R.; Ferreira, B.M.P.; Duek, E.A.R.; Dolder, H.; Wada, M.L.F. Use of blend of bioabsorbable poly(L-lactic acid)/poly(hydroxybutyrate-co-hydroxyvalerate) as surface for Vero cell cultured. *Braz. J. Med. Biol. Res.*, 38: 1623-1632, 2005.

[53] Junqueira, L.C.; Carneiro, J. "*Histologia Básica*", 11ª ed, Guanabara Koogan, Rio de Janeiro, 2008.

[54] Mano, J.F.; Reis, R.L. Osteochondral defects: present situation and tissue engineering approaches. *J. Tissue Eng. Regen. Med.*, 1: 261-273, 2007.

[55] Freed, L.E.; Marquis, J.C.; Nohria, L.E.; Emmanual, J.; Mikos, A.G.; Langer, R. Neocartilage formation in vitro and in vivo using cell cultures on synthetic biodegradable polymers. *J. Biomed. Mater. Res.*, 27: 11-23, 1993.

[56] Grande, D.A.; Halbertadt, C.; Naughton, G.; Schwartz, R.; Manji, R. Evaluation of matrix scaffolds for tissue engineering of articular cartilage grafts. *J. Biomed. Mat. Res.* 34: 211-220, 1997.

[57] Puelacher, W.C.; Mooney, D.; Langer, R.; Upton, J.; Vacanti, J.P.; Vacanti, C.A. Design of nasoseptal cartilage replacements synthesized from biodegradable polymers and chondrocytes. *Biomaterials*, 15: 774, 1994.

[58] Chu, C.R.; Coutts, R.D.; Yoshioka, M.; Harwood, F.L.; Monosov, A.Z. & Amiel, D. Articular-cartilage repair using allogeneic perichondrocyte-seeded biodegradable porous polylactic acid (PLA) - a tissue-engineering study. *J. Biomed. Mater. Res.*, 29: 1147-1154, 1995.

[59] Ishaug-Riley, S.L.; Okun, L.E.; Prado, G.; Applegate, M.A.; Ratcliffe, A. Human articular chondrocytes adhesion and proliferation on synthetic biodegradable polymer films. *Biomaterials*, 20: 2245-2256, 1999.

[60] Rotter, N.; Aigner, J.; Naumann, A.; Planck, H.; Hammer, C.; Burmester G.; Sittinger, M. Cartilage reconstruction in head and neck surgery: comparison of resorbable polymer scaffolds for tissue engineering of human septal cartilage. *J. Biomed. Mater. Res.*, 42: 347-356, 1998.

[61] Klompmaker, J.; Jansen, H.W.; Verth, R.P. De Groot, J.H.; Nijenhuis, A.J.; Pennings, A.J. Porous polymer implant for repair of meniscal lesions: a preliminary study in dogs. *Biomaterials*, 12: 810-816, 1991.

[62] de Groot, J.H.; Zijlstra, F.M.; Kuipers, H.W. Meniscal tissue regeneration in porous 50/50 copoly(L-lactide/epsilon-caprolactone) implantes. *Biomaterials*, 18: 613-622, 1997.

[63] Takagi, M.; Fukui, Y.; Wakitani, S. & Yoshida, T. Effect of Poly DL-Lactic-Co-Glycolic Acid Mesh on a Three-Dimensional Culture of Chondrocytes. *J. Biosci. Bioeng.*, 98: 477-481, 2004.

[64] Solchaga, L.A.; Temenoff, J.S.; Gao, J.Z.; Mikos, A.G.; Caplan, A.I.; Goldberg, V.M. Repair of osteochondral defects with hyaluronan- and polyester-based scaffolds. *Osteoarthr. Cartil.*, 13: 297-309, 2005.

[65] Wang, X.H.; Grogan, S.P.; Rieser, F.; Winkelmann, V.; Maquet, V.; La Berge, M.; Mainil-Varlet, P. Tissue engineering of biphasic cartilage constructs using various biodegradable scaffolds: an in vitro study. *Biomaterials*, 25: 3681-3688, 2004.

[66] Li, X.D.; Jin, L.; Balian, G.; Laurencin, C.T.; Anderson, D.G. Demineralized bonematrix gelatin as scaffold for osteochondral tissue engineering. *Biomaterials*, 27: 2426-2433, 2006.

[67] Chen, R.; Curran, S.J.; Curran, J.M. & John A. Hunt, J.A. The use of poly(L-lactide) and RGD modified microspheres as cell carriers in a flow intermittency bioreactor for tissue engineering cartilage. *Biomaterials*, 27: 4453-4460, 2006.

[68] Tuli, R.; Nandi, S.; Li, W.J.; Tuli, S.; Huang, X.X.; Manner, P.A.; Laquerriere, P.; Noth, U.; Hall, D.J.; Tuan, R.S. Human mesenchymal progenitor cell-based tissue engineering of a single-unit osteochondral construct. *Tissue Eng.*, 10: 1169-1179, 2004.

[69] Reed, A.M.; Gilding, D.K. Biodegradable polymers for use in surgery – poly(glycolic)/poly(lactic acid) homo and copolymers 2: in vitro degradation. *Polymer* 22: 342-346, 1981.

[70] Ishaug, S.L.; Yaszemski, M.J.; Brizios, R.; Mikos, A.G. Osteoblast function on synthetic biodegradable polymers. *J. Biomed. Mater. Research.*, 28: 1445-1453, 1994.

[71] Ishaug-Riley, S.L.; Crane, G.M.; Gurlek, A.; Miller, M.J.; Yasko, A.W.; Yaszemski, M.J.; Mikos, A.G. Ectopic bone formation by marrow stromal osteoblast transplantation using poly(L-lactic-co-glycolic acid) foams implanted into the rat mesentery. *J. Biomed. Mater. Res*, 36: 1-8, 1997.

[72] Ishaug-Riley, .S.L.; Crane-Kruger, G.M.; Yaszemski, M.J.; Mikos, A.G. Three-dimensional culture of rat calvarial osteoblasts in porous biodegradable polymers. *Biomaterials*, 19: 1405-1412, 1998.

[73] Hutmacher, D.W.; Schantz, J.T.; Lam, C.X.F.; Tan, K.C.; Lim, T.C. State of the art and future directions of scaffold-based bone engineering from a biomaterials perspective. *J. Tissue Eng. Regen. Med.*, 1: 245-260, 2007.

[74] Burg, K.J.L.; Porter, S.; Kellam, J.F. Biomaterial development for bone tissue engineering. *Biomaterials* 21: 2347-2359, 2000.

[75] Yuan, H.; van Blitterswijk, C.A.; de Groot, K.; de Bruijn, J.D. Crossspecies comparison of ectopic bone formation in biphasic calcium phosphate (BCP) and hydroxyapatite (HA) scaffolds. *Tissue Eng.*, 12: 1607-1615, 2006.

[76] Urist, M.R. Bone: formation by autoinduction. *Science*, 150: 893-899, 1965.

[77] Urist, M.R.; Grant, T.T.; Lindholm, T.S.; Mirra, J.M.; Hirano, H.; Finerman, G.A. Induction of new-bone formation in the host bed by human bone-tumour transplants in athymic nude mice. *J. Bone Joint Surg. Am.*, 61: 1207-1216, 1979.

[78] Lee, S.C.; Shea, M.; Battle, M.A.; Kozitza, K.; Ron, E.; Turek, T.; Schaub, R.G.; Hayes, W.C. Healing of lange segmental defects in rat femurs in aided by RhBMP-2 in PLGA matrix. *J. Biomed. Mater. Research.*, 28: 1149-1156, 1994.

[79] Whang, D.C.; Nam, E.K.; Aitken, M.; Sprague, S.M.; Patel, P.K.; Healy, K.E. Ectopic bone formation via rhBMP-2 delivery from porous bioabsorbable polymer scaffolds. *J. Biomed. Mater. Res.*, 42: 491-499, 1998.

[80] Hollinger, J.O.; Schmitt, J.M.; Buck, D.L.; Shannon, R.; Joh, S.-P.; Zegzula, H.D. Wozney, J. Recombinant human bone morphogenetic protein-2 and collagen for bone regeneration. *J. Biomed. Mater. Res.* 43: 356-364, 1998.

[81] Habibovic, P.; de Groot, K. Osteoinductive biomaterials - properties and relevance in bone repair. *J. Tissue Eng. Regen. Med.*, 1: 25-32, 2007.

[82] Gugala, Z.; Gogolewski, S. Protein adsorption, attachment, growth and activity of primary rat osteoblasts on polylactide membranes with defined surface characteristics *Biomaterials*, 25: 2341-2351, 2004.

[83] Simon Jr, C.G.; Eidelmanb, N.; Kennedya, S.B.; Sehgala, A.; Khatria, C.A. & Washburna, N.R. Combinatorial screening of cell proliferation on poly(L-lactic acid)/poly(D,L-lactic acid) blends. *Biomaterials*, 26: 6906-6915, 2005.

[84] Barbanti, S.H.; Santos Jr, A.R.; Zavaglia, C.A.C.; Duek, E.A.R. Porous and dense poly(L-lactic acid) and poly(D,L-lactic acid-co-glycolic acid) scaffolds: in vitro degradation in culture medium and osteoblasts culture. *J. Mater. Sci. Mater. Med.*, 15: 1315-1321, 2004.

[85] Ma, P.X.; Zhang, R.; Xiao, G.; Franceschi R. Engineering new bone tissue *in vitro* on highly porous poly(*a*-hydroxyl acids)/hydroxyapatite composite scaffolds. *Biomed. Mater. Res.*, 54: 284-293, 2001

[86] Lam, C.X.; Savalani, M.M.; Teoh, S.H.; Hutmacher, D.W. Dynamics of in vitro polymer degradation of polycaprolactone-based scaffolds: accelerated versus simulated physiological conditions. *Biomed. Mater.*, 3: 034108, 2008.

[87] Rai, B.; Teoh, S.H.; Hutmacher, D.W.; Cao, T.; Ho, K.H. Novel PCL based honeycomb scaffolds as drug delivery systems for rhBMP-2. *Biomaterials,* 26: 3739-3748, 2005.

[88] Rizzi, S.C.; Heath, D.J.; Coombes, A.G.; Bock, N.; Textor, M.; Downes, S. Biodegradable polymer/hydroxyapatite composites: surface analysis and initial attachment of human osteoblasts. *J. Biomed. Mater. Res.*, 55: 475-486, 2001.

[89] El Ghalbzouri, A.; Lamme, E.N.; van Blitterswijk, C.; Koopman, J.; Ponec, M. The use of PEGT/PBT as a dermal scaffold for skin tissue engineering. *Biomaterials*, 25: 2987-2996, 2004.

[90] Dantzer, E.; Braye F.M. Reconstructive surgery using an artificial dermis (Integra): results with 39 grafts. *Br. J. Plast. Surg.*, 54: 659-664, 2001.

[91] Mansbridge, J. Skin tissue engineering. *J. Biomater. Sci. Polymer Edn.*, 19: 955-968, 2008.

[92] Santos Jr, A.R.; Dolder, H.; Wada, M.L.F. Effects of fetal calf serum and dexamethasone in the differentiation of fibroblastic cells cultured on collagen I gel. *J. Submicrosc. Cytol. Pathol.*, 35: 35-42, 2003.

[93] Haas, V.R.; Santos Jr, A.R.; Wada M.L.F. Behaviour of fibroblastic cells cultured in collagen I using the sandwich technique. *Cytobios*, 106: 255-267, 2001.

[94] Ma, L.; Gao, C.; Mao, Z.; Zhou, J.; Shen, J.; Hu, X.; Han, C. Collagen/chitosan porous scaffold with improved biostability of skin tissue engineering. *Biomaterials*, 24: 4833-4841, 2003.

[95] Stern, R.; McPherson, M.; Longaker, M.T. Histologic study of artificial skin used in the treatment of full-thickness thermal injury. *J. Burn. Care Rehabil.*, 11:7-13, 1990.

[96] Yannas, I.V.; Burke, J.F. Design of an artificial skin 1: Basic design principles. *J. Biomed. Mater. Res. (Appl. Biomat.)*, 14: 65-81, 1980.
[97] Yannas, I.V.; Burke. J.F.; Gordon, P.L.; Huang, C.; Rubenstein, R.H. Design of an artificial skin 2: Control of chemical composition. *J. Biomed. Mater. Res (Appl. Biomat.)*, 14: 107-131, 1980.
[98] Burke, J.F.; Yannas, I.V.; Quinby Jr, W.C.; Bondoe, C.C.; Jung, W.K. Successful use of physiologicaly acceptable artificial skin in the treatment of extensive burn injury. *Ann. Surg.*, 194: 413-428, 1981.
[99] Burke, J.F.; Naughton, G.; Cassai, N. A histological, immunological and electron microscopy study of bovine collagen implanted in the human. *Ann. Plastic Surg.*, 14: 515-522, 1985.
[100] Frame, J.D.; Still, J.; Lakhel-LeCoadou, A.; Carstens, M.H.; Lorenz, C.; Orlet, H.; Spence, R.; Berger, A.C.; Dantzer, E.; Burd, A. Use of dermal regeneration template in contracture release procedures: a multicenter evaluation. *Plast. Reconstr. Surg.*, 113: 1330-1338, 2004.
[101] Tufaro, A.P.; Buck II, D.W.; Fischer, A.C. The use of artificial dermis in the reconstruction of oncologic surgical defects. *Plast. Reconstr. Surg.*, 120: 638-646, 2007.
[102] Matsuda, K.; Suzuki, S.; Isshiki, N.; Ikada, Y. Re-freeze dried bilayer artificial skin. *Biomaterials*, 14: 1030-1035, 1993.
[103] Rezende, C.A.; Luchesi, C.; Barbo, M.L.P.; Duek, E.A.R. Membranas de poli (ácido lático-co-ácido glicólico) como curativos para pele: degradação *in vitro* e *in vivo*. *Polímeros: Ciência e Tecnologia*, 15: 232-238, 2005.
[104] Ma, J.; Wang. H.; He, B.; Chen, J. A prelimilary *in vitro* study on the fabrication and tissue engineering applications of a novel chitosan bilayer material as a scaffold of human neofetal dermal fibroblasts. *Biomaterials*, 22: 331-336, 2001.

Physical Limitations to Tissue Engineering of Intervertabral Disc Cells

Shigeru Kobayashi,[1,2] Hisatoshi Baba,[1,2] Kenichi Takeno,[1]
Tsuyoshi Miyazaki,[1] Adam Meir[3] and Jill Urban[3]
[1]*Department of Orthopaedics and Rehabilitation Medicine, Faculty of Medical Sciences, The University of Fukui, Fukui, Japan.*
[2] *Research and Education Program for Life Science.The University of Fukui, Fukui Japan.*
[3] *Dept of Physiology, Anatomy and Genetics, Oxford University, Parks Road, Oxford, UK*

1. Introduction

In humans, the intervertebral disc becomes avascular by the fourth year of life. It has been the major target of basic and clinical research on low back pain. Disc degeneration is a common cause of chronic low back pain and neurological dysfunction. So far, there have been numerous advances in our basic knowledge of the anatomy, biochemistry, physiology, and biomechanical features of the disc cells. However, disc cell research and fundamental knowledge are certainly lacking in comparison to research on cells of other organs.

It is generally considered that the notochord gives rise to the nucleus pulposus during development of the intervertebral discs. However, it remains unclear why the cells composing the nucleus pulposus differ among animal species and stages of maturation and what role notochordal cells play in the growth process. Development of the human spinal column begins in about embryonic week of 3 (Keys & Compere, 1932). In about week 8 of gestation, the cartilaginous vertebral bodies are completed, and the notochord is gradually pushed into the gaps between the vertebral bodies, where it later becomes the intervertebral discs. In week of 10, the notochord disappears from the vertebral bodies and the nucleus pulposus is formed. Embryonic notochordal cells remain active producers of matrix components until the end of the first decade of life, at which time a ''notochordal'' nucleus pulposus can no longer be defined because of its increased collagen content and the loss or metaplasia of notochordal cells (Keys & Compere, 1932, Smith, 1930, Coventry, et al., 1945a, Peacock, 1952, Taylor & Twomey, 1988). Thus, the notochordal cells disappear from the nucleus pulposus during adolescence and are replaced by nonnotochordal (chondrocyte-like) cells. Although the role of notochordal cells in the nucleus pulposus is unknown, it is interesting to note that the loss of these cells from the nucleus pulposus occurs before early signs of degeneration appear, and that the combination of notochordal cells and cartilage-like cells actually determines the properties of this tissue. During weeks 12 to 14 of gestation, blood vessels invade the cartilaginous vertebral bodies and primary ossification centers begin to appear. Primary cartilaginous ossification remains active until about puberty, and a cartilaginous region including the growth plate is observed at the vertebral epiphysis. From the early stage of puberty, small secondary ossification centers appear and gradually extend around the periphery of the vertebral body to form a ring apophysis. At this stage, the cartilage canal plays an important role in nourishment and growth of the intervertebral disc and osteogenesis at the epiphyseal nucleus. The ring apophysis fuses completely with the vertebral body to form the vertebral bone at 18 to 25 years of age, if growth is normal, and the cartilage canal becomes buried in the bone marrow of vertebral body during the process of ossification (Peacock, 1951, Walmsley, 1953, Verbout, 1985).

The largely avascular intervertebral disc relies for nutrition on the outer one-third of the annulus fibrosus, which contains clearly defined blood vessels, and vascular channels in the cartilaginous end plate (Crock & Yoshizawa, 1977). In maintaining the nucleus pulposus, the central region of the end plate is probably more important than the outer region since it has a relatively higher density of vascular channels. Nutrient supply to the disc is affected by the architecture of the vascular buds and the porosity of the subchondral plate. nucleus pulposus cells may be up to 8 mm away from the closest blood supply in the adult lumbar disc (Bibby, et al., 2001). Steep gradients in metabolites develop with oxygen and glucose concentrations low in the disc center and lactic acid concentrations high. The pH in the disc center is thus acidic. The endplate does not normally provide a barrier to diffusion. In aging and degeneration, however, this end plate tends to calcify by unknown mechanisms and the apparent permeability of this plate decreases with age, as the end plate becomes more sclelotic (Robert et al., 1993). The consequent fall in supply of nutrients to the disc cells inhibits their ability to synthesise and maintain the matrix and even leads to cell death. Thus, it is noteworthy that the first signs of the disc degeneration are seen in the disc center.

There is now an increasing interest in developing biological repair methods for the treatment of disc degeneration that achieves the correct biomechanical properties for a successful outcome. Many researchers have performed basic studies on regeneration and transplantation of intervertebral discs that have applied tissue engineering using cell culture methods (Alini et al., 2002, Ganey &b Meisel, 2002, Masuda et al., 2004, Shimer et al., 2004, Brisby et al., 2004, Richardson et al. 2007, O'Halloran & Pandit, 2007, Sebastine & Williams, 2007, Kalson et al., 2008). So far, rat (Walsh et al., 2002, Yoon et al., 2003a, Risbud et al., 2003), rabbit (Nishida et al., 1999, Okuma et al., 2000, Kroeber et al., 2002, Yoon et al., 2003b, Masuda et al., 2003, An H, et al., 2005, Sakai et al., 2006, Iwashina et al., 2006)., porcine (Baer et al., 2001, Chen et al., 2004), canine (Thompson et al., 1991, Ganey et al., 2003, Hohaus et al., 2008), ovine (Sun et al., 2001, Mizuno et al., 2004), and bovine discs (Osada et al., 1996, Matsumoto et al., 2001, Alini et al., 2003, Hamilton et al., 2005) have been used in previous studies as tissue sources. However, the effects of aging and degeneration on the metabolism of human discs have been less frequently studied because suitable human discs for in vitro experiments are difficult to obtain (Alini et al.,2008, Hunter et al., 2003b). Therefore, comparison of metabolic activity between notochordal and nonnotochordal cells is considered to be important for determining the type of cell to use for transplantation when attempting to regenerate intervertebral discs. And also, tissue engineering and biological approach for disc degeneration is a potential approach for disc repair. However, these attempts to develop tissue with properties similar to those of native disc have been a challenge largely because of the difficulty of providing optimum physiological environments for the cells and lack of good control of extracellular conditions. The balance cellular demand and transport limits the number of cells that can be supported in any tissue and governs the inverse relationship between thickness and cell density. Such limitations apply to all avascular tissues, including tissue-engineered constructs.

2. Nutritional supply to the disc by cartilage canal

It is well known that a cartilage canal is present in the epiphyseal cartilage of the long bones and spine from the fetal to juvenile stages, and this canal plays an important role in osteogenesis at the epiphysis as well as in nourishment and growth of the epiphyseal cartilage (Strayer, 1943, Brookes, 1971, Gardner & Gray, 1970). It has been variously reported that the cartilage canal was derived from differentiation during angiogenesis (Haines, 1933, Hurrell, 1934), progression of angiogenesis (Hurrell, 1934, Anderson & Matthiessen, 1966), transformation of mesenchymal cells to chondrocytes (Levene, 1964), and extension of the perichondrium (Moss-Salentijn, 1975, Wilsman, 1970), but no definitive conclusion has been reached. Even at the early stage of vertebral formation, a cartilage canal nourishing the unossified cartilaginous part of the vertebrae and intervertebral discs is present in the unossified cartilaginous region, and is subsequently buried within the vertebral marrow during the progression of ossification (Fig.1A) (Crock & Yoshizawa, 1977, Kobayashi et al., 2008). During the formation of vertebrae, the cartilage canal enlarges and undergoes anastomosis to form the bone marrow of the secondary epiphyseal center (Fig.2A,B). It also persists in the unossified cartilaginous region at a suitable location to provide nourishment for the proliferating cartilage until the completion of ossification (Haines, 1933, Yoshizawa et al., 1986). The cartilage canal made up of arteries, veins, capillaries and mesenchymal elements surrounded by chondral tissue (Fig.2C). The blood vessels in the cartilage canal have an important role in providing nutrition for intervertebral discs and enchondral ossification. True capillaries are classified into 3 types, which are discontinuous, fenestrated, and continuous: the capillary permeability increases in this order and the morphological features of the endothelium indicate organ specificity (Bennett et al. 1959). Fenestrated capillaries are observed in metabolically active tissues, such as endocrine glands, the dorsal root ganglion, the kidney, and the intestinal mucosa (Kobayashi et al., 1993, Kobayashi & Yoshizawa, 2002). In the cartilage canal, fenestrated capillaries were observed. This suggests that there was enhanced permeability and active metabolism in the canal, with the capillaries providing nutritional support for immature cells surrounding the vessels and chondrocytes outside the cartilage canal.

A perichondrium exists between the cartilage canal and surrounding cartilage matrix, and this is composed of fibroblast-like cells (Fig.2C). Electron microscopy suggested that immature chondrocytes observed near the cartilage canal might have originated from the perichondrium. A large number of fibroblastic-like cells as well as blood vessels were observed in the cartilage canal, and these cells were connected by processes to form an intercellular network in the cartilage canal (Fig.2D). Therefore, these cells were assumed to be the source of fibroblasts forming the perichondrium and chondrocytes outside the cartilage canal (Fig.3). This network was also considered to be involved in intercellular signaling that controlled the direction of cell differentiation. Chondroclasts also appeared in the cartilage canal to make the bone marrow of the secondary epiphysial center (Fig.2E). Concerning the origin of macrophages and chondroclasts, some hold that they arise from osteoprogenitors (Crelin & Koch, 1967), others that they are derived from monocytes (Fischman & Hay, 1962).

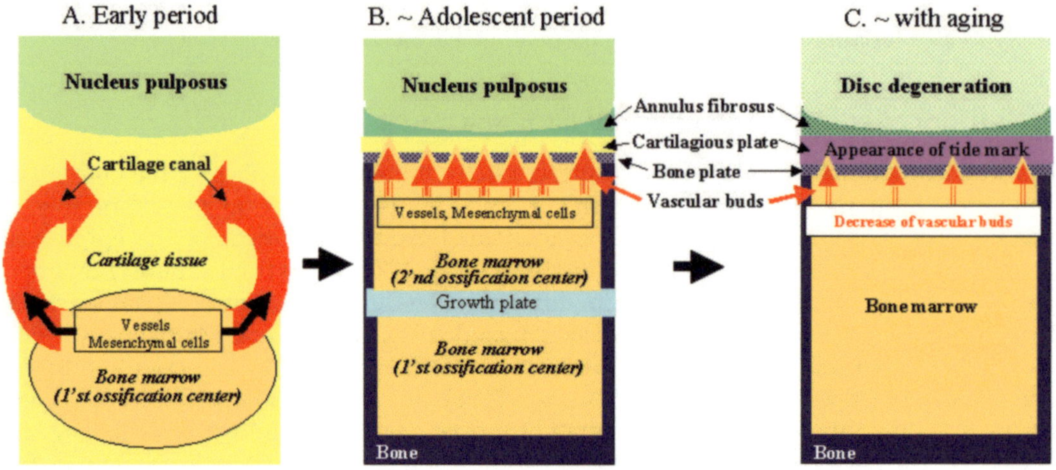

Fig. 1. A schematic drawing to show the developing process of cartilage canal and vascular bud in the vertebral end-plate. (Reproduced with permission from Kobayashi S, et al: Fine structure of cartilage canal and vascular buds in the vertebral end-plate in rabbit. J Neurosurg Spine 2008; 9: 96-103.)

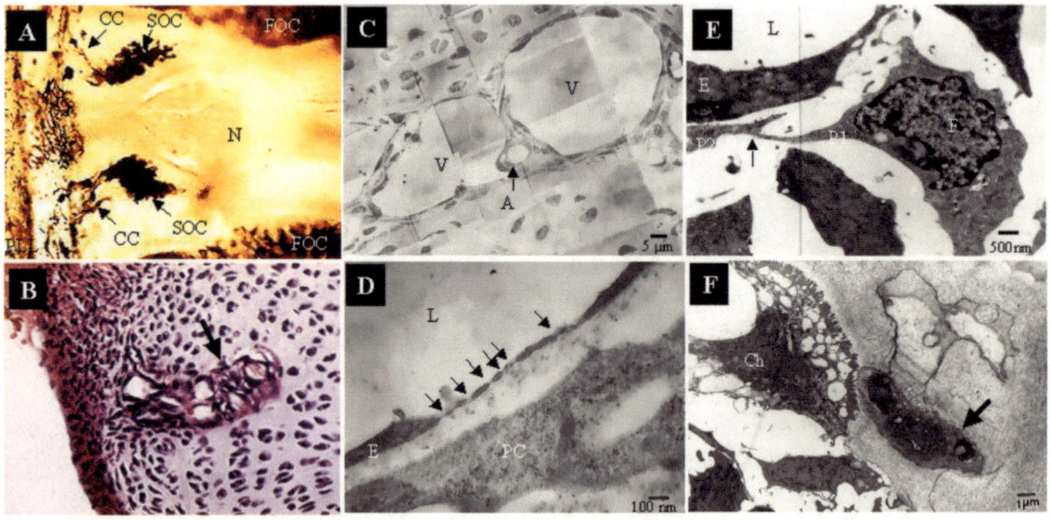

Fig. 2. The cartilage canal in the chondral cap of 5th lumbar vertebral body after its birth.

(A) A transparent specimen showing cartilage canal of a thin median sagittal section at 1 week after its birth. This specimen showing cartilage canal and second ossification center. Cartilage canals and second ossification center stained with India ink. CC: cartilage canal, FOC: first ossification center, N: nucleus pulposus, PLL: posterior longitudinal ligament, SOC: secound (epiphysial) ossification center.
(B) A hematoxylin and eosin stained medial saggital section showing immature chondrocytes around the cartilage canal (arrow).
(C) A transverse electron photograph of the cartilage canal. The cartilage canal is made

up of vessels of various size and fibroblastic-like cells around vessels. A: arteriole, V: venule.

(D) A high magnification of endothelium. This capillary has many fenestrations with a diaphragm in an endothelial cell (arrows). E: vascular endothelium, L: vascular lumen, PC: perichondral cell (fibroblast-like cell).

(E) A fibroblastic-like cell in the cartilage canal. These cells were connected each other by gap junction (arrow) and thus forming an intercellular network. Ch: chondroclast, E: vascular endothelium, L: vascular lumen, M: macrophage, P1 and P2: process of fibroblastic-like cells.

(F) A chondroclast with ruffled border was presented adjacent to the uncalcified matrix in canal. Chondrocytes undergoing apoptosis were seen around cartilage canal, the cells and nuclei were reduced in size and chromatin condensation was visible in the nuclei (arrow). Ch: chondroclast. (Reproduced with permission from Kobayashi S, et al: Fine structure of cartilage canal and vascular buds in the vertebral end-plate in rabbit. J Neurosurg Spine 2008; 9: 96-103.)

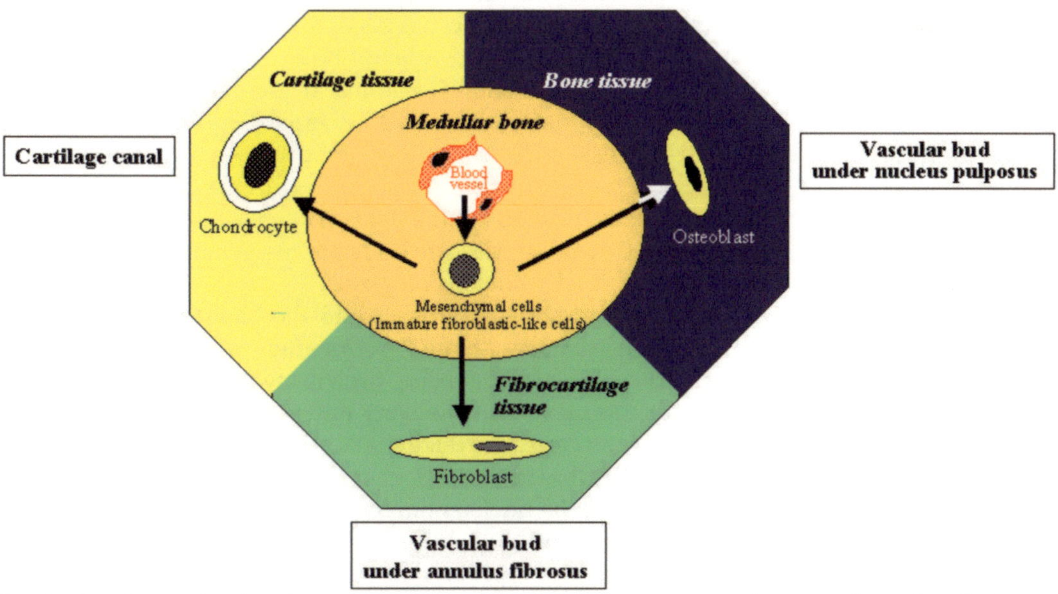

Fig. 3. A schematic drawing to show the developing process of mesenchymal cells in cartilage canal and vascular bud. The mesenchymal cells seen surrounding the cartilage canal and vascular bud represent a common precursor for the three main types connective tissue cells seen during early vertebral development. (Reproduced with permission from Kobayashi S, et al: Fine structure of cartilage canal and vascular buds in the vertebral end-plate in rabbit. J Neurosurg Spine 2008;9:96-103.)

3. Nutritional supply to the disc by vascular buds and disc degeneration

Previous histological studies of the bone-disc junction have pointed out that the bone-cartilage interface is perforated by small holes corresponding to the bone marrow the vascular terminations projecting into the cartilage, which are called as capillary beds or vascular buds (Crock & Yoshizawa, 1977, Nachemson et al., 1970, Ogata & Whiteside, 1981, Crock et al., 1973, Maroudas et al., 1975). These vascular buds in the bone-disc junction area are structurally very similar to the cartilage canal made up of vessels and mesenchymal elements (Crock & Goldwasser,1984, Yoshizawa et al., 1986). A segmental arrangement of small vessels communicating with the bone marrow cavity is known to exist in the adult human vertebral body end-plate, and Crock et al. found a large number of small blood vessels that were systematically arranged around the end-plate in clear specimens of adult human or canine vertebral bodies, which they termed capillary buds (Crock & Yoshizawa, 1977, Crock & Goldwasser, 1984). Ogata et al. assigned the term vascular buds to these small blood vessels (Ogata & Whiteside, 1981). Thus, capillary buds are identical to vascular buds, and these are suggested to be an important nutritional pathway for the intervertebral disc. When ossification of vertebral body has progressed and the vertebral end-plate has formed, vascular buds appear in small foramina. These vascular buds consist of capillaries and mesenchymal cells, with a structural resemblance to the cartilage canal, although smaller in size, and their tips are in contact with the intervertebral disc. That is, special tissue consisting of blood vessels and mesenchymal cells is present in the undifferentiated and unossified part of the vertebra, while the cartilage canal persists in the bone marrow after it is buried during the process of ossification of the vertebral body. It then reappears in the form of vascular buds occupying numerous small foramina at the vertebral end-plate in association with formation of the end-plate, and plays an important role in supplying nutrition to the intervertebral discs. These vascular buds are morphologically similar among humans, monkeys, and rabbits, and numerous buds are observed in the central part of the end-plate corresponding to the nucleus pulposus of the intervertebral disc (Yoshizawa et al., 1986). This gives us the impression that these vascular buds are important structure for the nutrition of the intervertebral disc, especially nucleus pulposus.

In the case of rabbits, this vascular bud starts to project into the bone cartilage interface through the bone plate from the bone marrow about 4 weeks after its birth (Fig. 4A). This corresponds to the adolescent period of the human being (Yoshizawa et al., 1986, Kobayashi et al., 2008). Mature vascular buds are present in the bony end-plate and their ends are in contact with the cartilage of the intervertebral disc, so nutrients are considered to be transferred across this junction and distributed to the deeper areas of the intervertebral discs by diffusion (Fig.1B, 4B). In maintaining the nucleus pulposus, the central region of the end plate is probably more important than the outer region since it has a relatively higher density of vascular channels. There are reports that these channels disappear with increasing age, and eventually become occluded by calcification (tide mark) (Fig.1C, 4C). Schmorl and Junghanns noted "ossification gaps" representing degenerative foci in the mature cartilage end plate (Schmorl & Junghanns, 1971). These were seen to appear at the same time as the blood vessels disappeared and with the concomitant degeneration of the disc, it was assumed that the vessels had a major role in disc nutrition. Histological studies of intervertebral disc degeneration conducted by Conventry et al. (Coventry, 1945b, Coventry et al., 1945c) and Roberts et al. (Robert et al., 1993). Roberts et al., 1996) have demonstrated that degeneration begins near the vertebral end-plate with aging, while a study by Ogata et al. demonstrated that blockade of the vertebral route interfered greatly with intervertebral disc nutrition (Ogata & Whiteside, 1981). Thus, the tide mark which is the certified zone in the cartilage, covers the bone-disc junction by the end of growing period and looks to close the nutritional route through these vascular buds. On the other hand, the tide mark may need to tolerance the body weight with age. These is still controversy concerning the role of the vascular buds after the appearance of this tide mark, and further studies may be required.

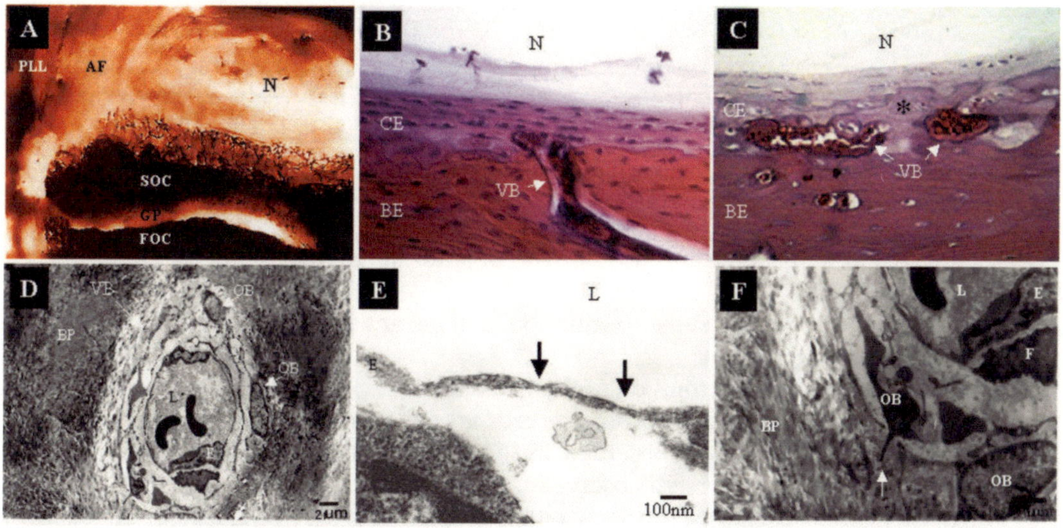

Fig. 4. The vascular buds in the vertebral end-plate of adult rabbit.

(A) A transparent specimen showing vascular buds at 6 month after its birth. The vascular buds were numerous in number in the central potion of the end-plate.

(B) A vascular bud in the central portion under the light microscope at 6 month after its birth. A vascular bud is contacting directly to the cartilage end-plate (Hematoxylin and eosin–stained section).

(C) A vascular bud in the central portion under the light microscope at 2 year after its birth. The tide mark (*) which is the calcified zone in the cartilage plate, covers the bone-disc junction by the end of growing period and looks to close the nutritional route through the vascular buds.

(D-F) Transverse sections of transmission electron microscope at 6 month after its birth.

(D) A photograph of the vascular bud in the center portion. A diameter of a capillary is around 10 μm and some fibloblastic-like cells surround it. These cells surrounding the capillary were connected each other by gap junction and thus forming an intercellular network.

(E) High magnification shows fenestrae in the endothelium (arrows). The capillary is fenestrated type.

(F) A osteoblastic-like cell that had differentiated from fibroblastic-like cells were observed and this cell were bound to the surrounding bone through processes (arrow). AF: annulus fibrosus, BP: bone plate, CP: cartilaginous end-plate, E: vascular endothelium, F: fibroblastic-like cell, FOC: first ossification center, GP: growth plate, L: vascular lumen, N: nucleus pulposus, OB: osteoblastic like cell, PLL: posterior longitudinal ligament, SOC: secound (epiphysial) ossification center, VB: vascular bud. (Reproduced with permission from Kobayashi S, et al: Fine structure of cartilage canal and vascular buds in the vertebral end-plate in rabbit. J Neurosurg Spine 2008; 9: 96-103.)

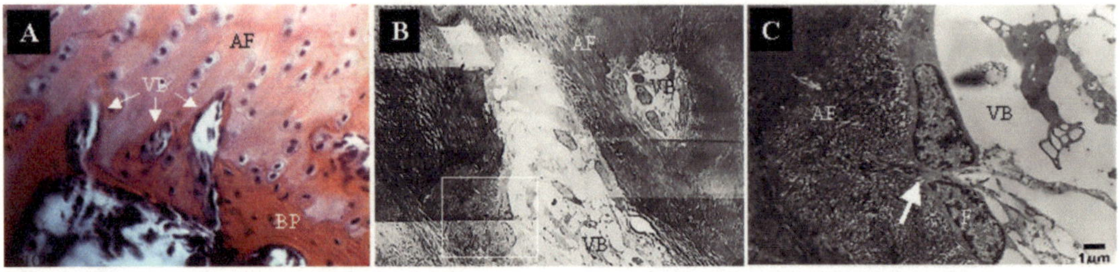

Fig. 5. The vascular buds in the annulus fibrosus of adult rabbit (at 6 month of age).

(A) A sagittal section showed some vascular buds in the annulus fibrosis (Hematoxylin and eosin–stained section).
(B) A sagittal constructed photograph of the vascular buds. The vascular buds are made up of vessels of various size and fibroblastic cells around vessels.
(C) Some fibroblastic cells were observed in the wall of vascular buds. Fibroblastic cells producing collagen fibers were observed (arrow). AF: annulus fibrosus, BP: bone plate, FB: fibroblastic like cell, N: nucleus pulposus, VB: vascular bud. (Reproduced with permission from Kobayashi S, et al: Fine structure of cartilage canal and vascular buds in the vertebral end-plate in rabbit. J Neurosurg Spine 2008; 9: 96-103.)

Vascular buds are resemble glomerulus-like capillary vessels (Oki et al., 1996). These are fenestrated capillaries and show increased vascular permeability like blood vessels in the cartilage canal, suggesting active metabolism of tissues around the vertebral end-plate (Fig.4D,E). A large number of fibroblastic-like cells were observed around capillaries in the vascular buds, and these cells was connected with each other by gap junctions to form an intercellular network as seen the cartilage canal. Osteoblast-like cells that had probably differentiated from fibroblastic-like cells were also observed, and some of the processes of these cells were buried in the bone tissue (Fig.3, 4F). In the vascular buds of the annulus fibrosus (Fig.5A,B), fibroblast-like cells produced collagen fibers in the region adjacent to the fibrocartilage, a finding which suggests that vascular buds in the annulus fibrosus have a marked influence on the composition of the fibrocartilagious matrix (Fig.3, 5C). Bessett et al. cultured mesenchymal cells under various conditions, and reported that bone tissue was generated when compression was applied under aerobic conditions and cartilaginous tissue was generated when compression was applied under anaerobic conditions (Bassett, 1962). When tension was applied under aerobic conditions, connective tissue was formed. Although cell differentiation is not only determined by the local oxygen tension and mechanical force, this research is interesting because blood vessels are not present in the cartilage although abundant vessels are observed in the bone.

4. A phenotypic comparison of morphological features of nucleus pulposus cells

It is generally considered that the notochord gives rise to the nucleus pulposus during development of the intervertebral discs. However, it remains unclear why the cells composing the nucleus pulposus differ among animal species and stages of maturation and what role notochordal cells play in the growth process. Morphological changes with aging have been actively studied since a report by Luschka in 1858 (von Luschka, 1858), and comparative studies of humans and animals have also been performed (Hansen, 1959, Bulter, 1988). As a result, it has been demonstrated that the nucleus pulposus of cattle (Hansen, 1959), horses (Bulter, 1988), sheep (Carlier, 1890), and chondrodystrophoid dogs (e.g., beagles) (Hansen, 1959, Bulter, 1988, Hansen,1952, Braund et al.,1975) shows similar changes to that of humans, while the nucleus pulposus of mice (Berry, 1961), rats (Williams,1908), rabbits (Souter & Taylar,1970, Scott et al.,1980), pigs (Hansen, 1959, Williams,1908), cats (Hansen, 1959, Bulter, 1988, Butler & Smith, 1967) and nonchondrodystrophoid dogs (e.g., mongrels) (Hansen, 1959, Bulter, 1988, Hansen,1952, Hunter et al.,2003a) retains notochordal cells. These studies have also demonstrated that notochordal and cartilaginous nucleus pulposus cells display considerable morphological differences. Although the role of notochordal cells in the nucleus pulposus is unknown, it is interesting to note that the loss of these cells from the nucleus pulposus occurs before early signs of degeneration appear, and that the combination of notochordal cells and cartilage-like cells actually determines the properties of this tissue.

Here, comparison of morphological features between notochordal and nonnotochordal cells is considered to be important for determining the type of cell to use for transplantation when attempting to regenerate intervertebral discs. In this study, we used notochordal cells from the nucleus pulposus of rats (Fig.6A) and rabbits (Fig.6B), as well as chondrocyte-like bovine nucleus pulposus cells (Fig.6C). At first, we determined the age of the animals based on their skeletal maturity so that nucleus pulposus tissues from different species could be compared under similar conditions. The age when skeletal maturity is reached is 2, 4-6, and 24-30 months after birth for rat (Adler et al., 1983, Moskowitz et al.,1990), rabbits (Kobayashi et al., 2008, Scott et al.,1980, Smith & Serafini-Fracassini,1968), and cattle (Lawrence et al.,2001), respectively. Loss of notochordal cells occurs at 12 and 6 months after birth in rats (Adler et al., 1983, Moskowitz et al.,1990) and rabbits (Scott et al.,1980, Smith & Serafini-Fracassini,1968), respectively, while notochordal cells are completely replaced by cartilage-like nucleus pulposus at birth in cattle (Hansen,1956). Accordingly, the discs from 2-month-old rats, 6-month-old rabbits, and 18 to 24-month-old cattle were considered to correspond approximately to the age of adolescence in humans. The cell density of the nucleus pulposus was noticeably higher in notochordal discs and was low (4,264 cells/mm^3 on average) in bovine discs (Fig.6G), which was close to the value (about 6,000 cells/mm^3) reported for the human nucleus pulposus by Maroudas et al (Maroudas et al., 1975). Butler and Smith et al. classified the nucleus pulposus cells into 3 stages: in stage 1, the NP consists entirely of notochordal cells; in stage 2, notochordal cells form clusters and the extracellular matrix increases; and in stage 3, notochordal cells show a marked decrease and extracellular matrix and cartilage-like cells become predominant (Butler & Smith, 1967). According to this classification, humans are already in stage 2 at birth and progress to stage 3 in their teens.

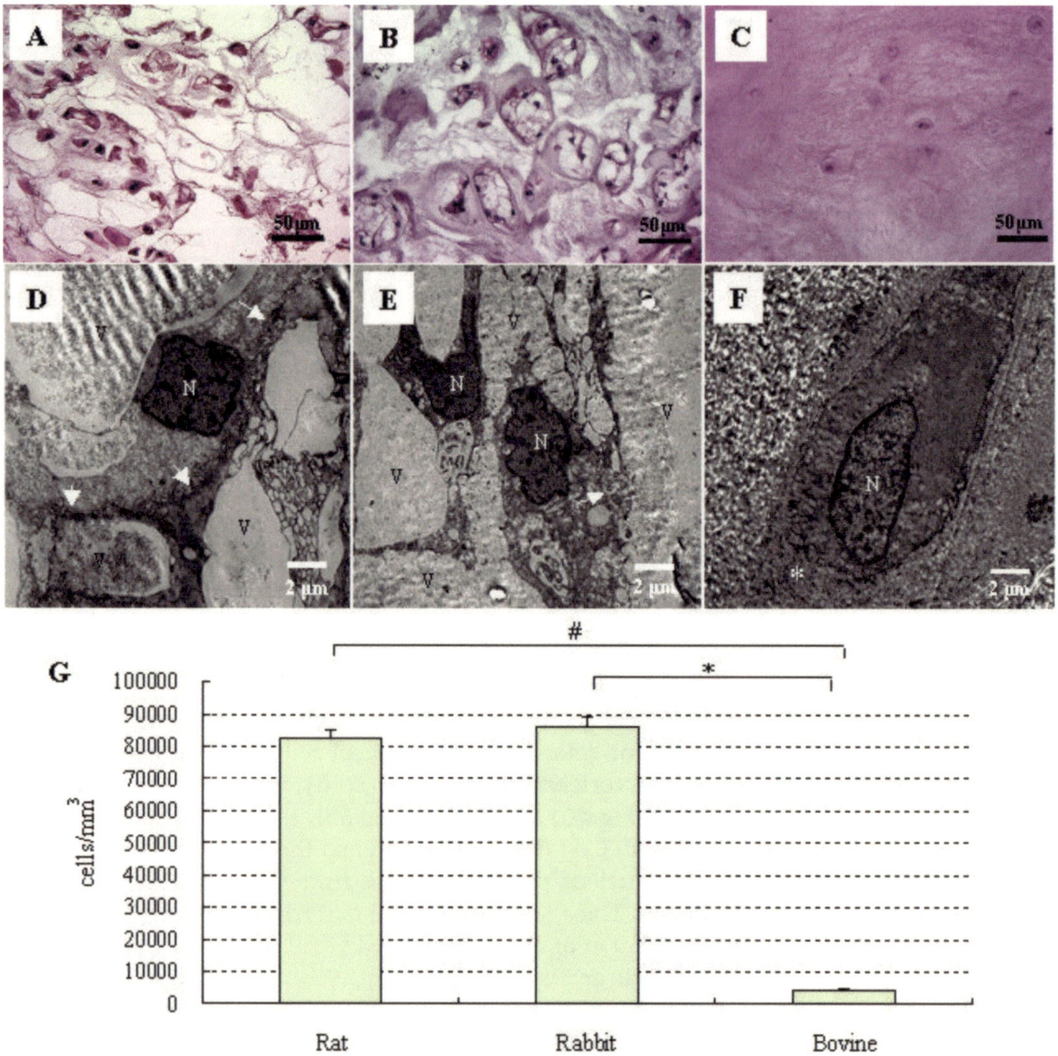

Fig. 6. Photomicrographs (A–C) and electron micrographs (D-F) of transverse sections of rat, rabbit, and bovine lumbar nucleus pulposus.

(A–C) Photomicrographs: Rat and rabbit discs have a gelatinous and well-hydrated nucleus pulposus. In contrast, the bovine nucleus pulposus is much more fibrous and cartilage like. The rat nucleus pulposus only contains notochordal cells with a number of vacuoles (A). In the rabbit nucleus pulposus, a few notochordal cells form clusters surrounded by mucoid material (B). In the bovine nucleus pulposus, no notochordal cells are observed, and chondrocyte-like cells with lacunae are scattered in a hyaline-like matrix (C) (X40, scale bar.50 mm, hematoxylin and eosin–stained sections).
(D-F) Transmission electron micrographs of the rat nucleus pulposus (D), rabbit nucleus pulposus (E), and bovine nucleus pulposus (F). In the rat nucleus pulposus, a number of vacuoles of various sizes, which are empty or contain an amorphous substance, can be

observed around the cells. Cells are connected to adjacent cells by tight junctions (white arrow) (D). In the rabbit nucleus pulposus , as in rats, notochordal cells with a number of vacuoles of varying sizes are observed. The cells are connected with adjacent cells to form an intercellular network as in rats (E). Bovine nucleus pulposus cells have larger nuclei than notochordal cells. The cytoplasm does not contain vacuoles and the cells are found individually in the matrix. Lacunae (*) are observed around the cells (F). The cells are completely different from notochordal cells, closely resembling chondrocytes from articular cartilage. N: nucleus; V: vacuoles.
(G) The number of nucleus pulposus cells per unit volume was counted in 10 discs from each animal, and the mean values were compared. As a result, the number of cells in the bovine nucleus pulposus (chondrocyte-like cells) was significantly lower than that of notochordal cells in the nucleus pulposus of rats and rabbits (*, #: $p<0.05$). (Reproduced with permission from Miyazaki T, Kobayashi S, et al. A phenotypic comparison of proteoglycan production of intervertebral disc cells isolated from rats, rabbits, and bovine tails; Which animal model is most suitable to study tissue engineering and biological repair of human disc disorders? Tissue Eng Part A. 2009 Aug 15. [Epub ahead of print] PMID: 19681728 [PubMed - as supplied by publisher].)

The cells of the adult nucleus pulposus more closely resemble articular chondrocytes. The rat, rabbit, and bovine nucleus pulposus used in this study were considered to correspond to stages 1, 2, and human stage 3, respectively. Overall, the structure of bovine and human discs is similar, but some differences exist with regard to the diameter, height, and thickness of the wall of the annulus (Race et al., 2000). Although bovine tails support a lower load than human lumbar discs, the swelling pressure of bovine coccygeal discs has been shown to be similar to that of the discs in a person resting in the prone position (Ohshima et al., 1993, Ishihara H, et al. 1996). Biochemically, similar rates of *in vitro* proteoglycan synthesis (Ohshima et al., 1993). and matrix synthesis in response to hydrostatic pressure (Ishihara H, et al. 1996, Ohshima et al., 1995) have also been found in human and bovine coccygeal discs. Both bovine and human discs are also similar with regard to the types and distribution of aggrecan and collagen (Ohshima et al., 1993, Roberts et al., 1991). The bovine disc closely resembles the human disc because adult human and adult bovine nucleus pulposus have almost no notochordal cells and contain chondrocyte-like nucleus pulposus cells that produce a hyaline cartilage-like matrix. It has been demonstrated that notochordal cells persist throughout life in rats and rabbits, but decrease with growth, and are replaced by hyaline cartilage-like matrix and chondrocyte-like nucleus pulposus cells infiltrating from the annulus fibrosus (Keys, & Compere, 1932, Coventry, et al., 1945, Peacock, 1951, Kim et al., 2003). It is unknown, however whether this decrease is caused by differentiation of notochordal cells into non-notochordal cells, apoptosis, or lack of nutrients diffusing from the end plates. Thus, in each animal species, notochordal cells decrease and chondrocyte-like cells become predominant in the matrix of the nucleus pulposus with aging, but the reason for the decrease of notochordal cells and the origin of the chondrocyte-like nucleus pulposus cells remain to be elucidated. Therefore, investigation of the relationship between the decrease of matrix formation by notochordal cells and the onset of intervertebral disc degeneration is an important subject for the future.
When compared with a report by Trout et al., who observed the human nucleus pulposus at various ages from fetal to 91 years by electron microscopy (Trout et al., 1982a, Trout et al.,

1982b), the morphological characteristics of notochordal cells observed in rats and rabbits in this study were similar to those of nucleus pulposus cells observed in human discs up to adolescence, while the features of cartilage-like cells observed in bovine discs were similar to human nucleus pulposus cells from adolescence onward. That is, the notochordal discs contained physaliphorous cells aggregated in close contact with each other (Fig.6D,E), while non-notochordal discs contained cartilage-like cells scattered through the matrix and characteristically surrounded by lacunae (Fig.6F), which were not observed around notochordal cells. The mechanism leading to formation of lacunae and their significance remains unclear, but lacunae may reflect the influence of the cell over its immediate pericellular matrix, either through accumulation of excreted products or some other process. The formation of lacunae may also be related to changes of the nucleus pulposus that occur with age, including the apparent increase of some nutrients and substrates, making the extracellular matrix less accessible to cell products (Happy & Bradford,1964, Happy, et al. 1969). Changes in the ability of substances to diffuse in or out of the cell and a reduction in the number of viable cells might gradually alter the extracellular matrix and thereby lead to age-related changes in the mechanical properties of the disc.

5. Metabolic activities of notochordal cells and non-notochordal cells

Proteoglycans, particularly the large proteoglycan known as aggrecan, play a major role in load-bearing by the intervertebral disc. Because of the high osmotic pressure due to the sulphated glycosaminoglycans in aggrecan, it absorbs water that expands the collagen network and maintains tissue turgor (Grodzinsky, 1983, Maroudas & Bannon, 1981). The stiffness of intervertebral disc tissues is thus strongly dependent on the aggrecan content (Kempson et al., 1970, Treppo et al., 2000). Thus, it is necessary to have an adequate content of proteoglycans (glycosaminoglycans) when attempting to produce intervertebral disc tissue with the mechanical strength to withstand stress of about 2.8-13.0 kN by tissue engineering (Adams & Hutton, 1982). The structure and composition of the intervertebral discs are intimately related to their function. By studying species differences and similarities, many researchers have attempted to define the functions of the discs. However, caution is necessary when interpreting and extrapolating to humans from such data, because even though all vertebrates have discs, not all discs are the same. For example, comparison of proteoglycan synthesis in the discs of chondrodystrophoid dogs (beagles) and nonchondrodystrophoid dogs (greyhounds) has revealed significantly lower proteoglycan synthesis in chondrodystrophoid dogs (Cole et al., 1985). The relative risk of disc herniation is about 10 to 12 times higher in nonchondrodystrophoid dogs than in chondrodystrophoid dogs (Gage, 1975, Priester, 1976). Autoradiography with ^{35}S-sulfate, a radio label that is largely incorporated by the glycosaminoglycan chains of Proteoglycans, has established that notochordal cells also synthesize proteoglycans (Souter & Taylar, 1970). A similar finding has been reported for human fetal and pediatric discs in which notochordal cells are more abundant compared with mature discs (Bayliss, et al., 1988, Johnstone & Bayliss, 1995). Aguiar et al. have suggested that the notochordal cells found in the human nucleus pulposus up to approximately 10 years of age play an active role in nucleus pulposus development and in the maintenance of disc integrity through the production of soluble factors that induce nucleus pulposus cells to increase proteoglycan synthesis (Aguiar et al., 1999). Interestingly, it has been found that co-culture of immature cells such as mesenchymal stem cells with notochordal cells and adult nucleus pulposus cells (chondrocyte-like cells) stimulates proteoglycan synthesis more rapidly, and this method has been extensively used in studies on disc regeneration (Sakai et al., 2006, Iwashina et al., 2006, Cappello et al., 2006, Erwin & Inman, 2006). However, previous studies on proteoglycan production by notochordal cells and chondrocyte-like nucleus pulposus cells have only compared proteoglycan synthesis on the basis of sulphate synthesis, and no studies have compared glycosaminoglycan accumulated by these cells.

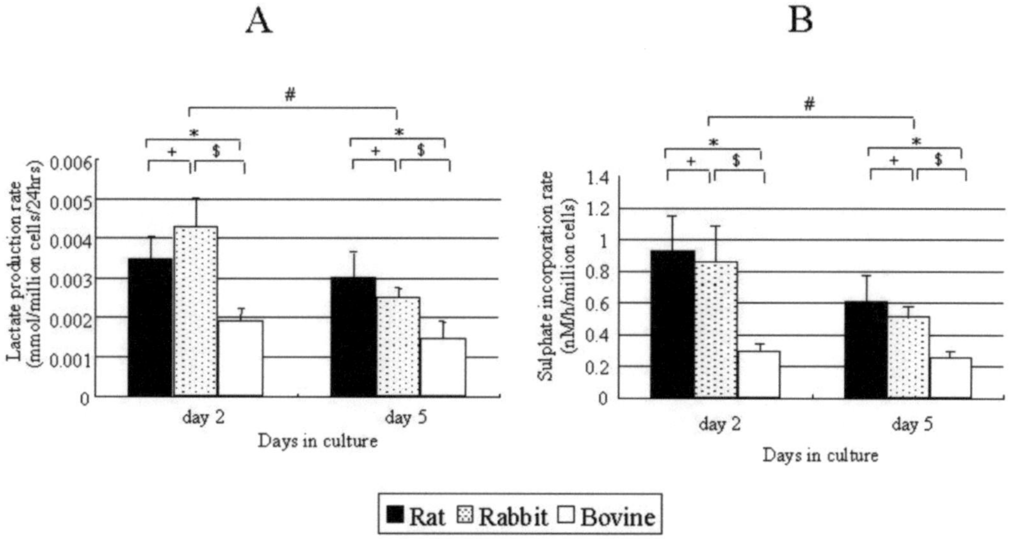

Fig. 7. Metabolic activity of rat, rabbit, and bovine nucleus pulposus cells.

(A) Lactate production rate. Cells were cultured under standard conditions in beads containing 4 million cells/ml (1.0ml medium and five beads per well) for up to 5 days, with a complete medium change everyday. Representative beads were dissociated for cell counting, and the viable cell density per bead was determined. Lactate in the medium was measured after each 24-h culture period and production per million cells/24 hours is shown. Notochodal nucleus pulposus cells from rats and rabbits produced about twice as much lactate as bovine nucleus pulposus cells (*: $p<0.05$ by the paired t-test for rat vs. bovine cells; $: $p<0.05$ by the paired t-test for rabbit vs. bovine cells; +: p=n.s., by the paired t-test for rat vs. rabbit cells.) Lactate production decreased over time during culture (#: $p<0.05$ by two-way ANOVA with repeated measures for comparison among days 2 and 5).

(B) Sulfate incorporation rate. On days 2 and 5, radiolabeled sulfate was added to fresh medium in three wells, and the beads were cultured for 4 hours. Then the beads were dissociated and cell density and sulfate incorporation were measured. Results are given as the mean_standard error of six independent experiments. The sulfate incorporation rate decreased over time (#: $p<0.05$ by two-way ANOVA with repeated measures for 2 vs. 5 days). The rate was highest for rat cells and lowest for bovine cells up to 5 days of culture (*: $p<0.05$ by the paired t-test for rat vs. bovine cells; $: $p<0.05$ by the paired t-test for rabbit vs. bovine cells; +: p=n.s., by the paired t-test for rat vs. rabbit cells.). ANOVA, analysis of variance. (Reproduced with permission from Miyazaki T,

Kobayashi S, et al. A phenotypic comparison of proteoglycan production of intervertebral disc cells isolated from rats, rabbits, and bovine tails; Which animal model is most suitable to study tissue engineering and biological repair of human disc disorders? Tissue Eng Part A. 2009 Aug 15. [Epub ahead of print] PMID: 19681728 [PubMed - as supplied by publisher].)

We examined that a 3-dementional culture system was employed to compare the metabolism of notochordal cells from rat and rabbit nucleus pulposus, as well as chondrocyte-like cells from the bovine nucleus pulposus (Miyazaki et al., 2009). As a result, notochordal nucleus pulposus cells showed levels of lactate production (Fig.7A) and proteoglycan synthesis (Fig.7B) about twice as high as those of chondrocyte-like nucleus pulposus cells, indicating the high metabolic activity of notochordal cells.

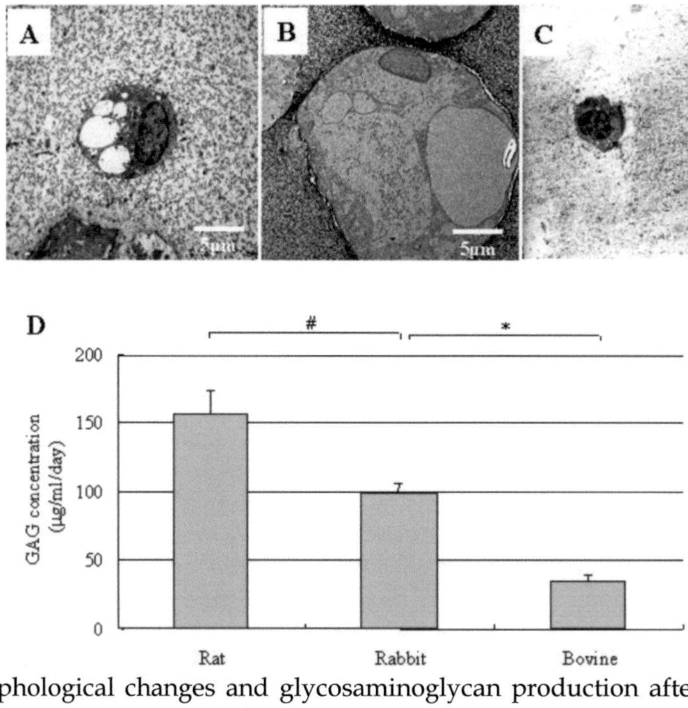

Fig. 8. Morphological changes and glycosaminoglycan production after three-dimensional culture. Cells were encapsulated in alginate beads and cultured in Dulbecco's modified Eagle's medium with 6% serum under air for 5 days.

(A-C) Electron micrographs of rat (A), rabbit (B), and bovine (C) nucleus pulposus cells cultured in alginate beads for 5 days. Notochordal (A,B) and non-notochordal (cartilaginous)(C) nucleus pulposus cells showed essentially similar findings to those observed *in vivo*. (D) Glycosaminoglycan accumulated/day in rat, rabbit, and bovine nucleus pulposus cells. The GAG concentration was measured after 5 days. Pooled data from six separate experiments are shown. After culture for 5 days, accumulative glycosaminoglycan production was higher in notochordal nucleus pulposus cells than in cartilaginous nucleus pulposus cells, being about 156.5, 99.5, and 35.4 µg/ml/day for rat, rabbit, and bovine nucleus pulposus cells, respectively. The glycosaminoglycan accumulated/day was lowest in the bovine cells. Values are the mean ± standard error. (*, #: $p<0.05$). (Reproduced with permission from Miyazaki T, Kobayashi S, et al. A phenotypic comparison of proteoglycan production of intervertebral disc cells isolated from rats, rabbits, and bovine tails; Which animal model is most suitable to study tissue engineering and biological repair of human disc disorders? Tissue Eng Part A. 2009 Aug 15. [Epub ahead of print] PMID: 19681728)

When nucleus pulposus cells were encapsulated in alginate beads, cultured for 5 days, and examined under a electron microscope, notochordal nucleus pulposus cells of rats (Fig.8A) and rabbits (Fig.8B) were larger than chondrocyte-like bovine nucleus pulposus cells (Fig.8C), and both cell types were surrounded by gaps. However, electron microscopy only revealed lacunae around the bovine nucleus pulposus cells (Fig.8C) and not around rat or rabbit nucleus pulposus cells, which contained a number of vacuoles. Both types of cell showed essentially similar findings to those observed in vivo and contained few intracellular organelles. After culture for 5 days, accumulative glycosaminoglycan production was higher in notochordal nucleus pulposus cells than in cartilaginous nucleus pulposus cells, being about 156.5, 99.5, and 35.4 μg/ml/day for rat, rabbit, and bovine nucleus pulposus cells, respectively (Fig.8D). Thus, the notocordal cells were more active metabolically than were chondrocyte-like nucleus pulposus cells. One index that in used for disc repair and regeneration is the glycosaminoglycan content of the disc. It has been reported that a 7-10 % glycosaminoglycan content is required to obtain disc tissue with enough strength for clinical application (Urban & Roberts, 1995). Based on the results of the present study, it was calculated that it would respectively take at least 1 year, 2 years, and 5 years for rat, rabbit, and bovine nucleus pulposus cells to produce a 7%-glycosaminoglycan content (i.e., 70 mg/ml). This provides important information about prospective autologous transplantation of cells with a high proteoglycan production capacity such as notochordal cells and mesenchymal stem cells.

6. Which animal model is most suitable to study tissue engineering and biological repair of human disc disorders?

Accordingly, glycosaminoglycan accumulation by chondrocyte-like bovine and adult human nucleus pulposus cells is extremely slow compared with that by notochordal cells, and in vitro production of a construct with a 7%-glycosaminoglycan content by using chondrocyte-like nucleus pulposus cells alone would be assumed to take an extremely long time and thus would be impractical. In fact, transformation of the matrix by cells in human discs is very slow, and studies have shown that proteoglycan turnover takes about 20 years (Roughley, 2004) and collagen turnover takes more than 100 years (Verzijl et al., 2000). However, autologous transplantation of cells with a high proteoglycan production capacity such as notochordal cells and mesenchymal stem cells may lead to successful repair of degenerated discs (Sakai et al., 2006, Iwashina et al., 2006, Aguiar et al.,1999, Cappello et al., 2006, Erwin & Inman, 2006). Even if the transplantation of cells with a high proteoglycan production capacity into the nucleus pulposus is feasible, it is also essential for preventing the progression of disc degeneration that the transplanted cells survive in the degenerated disc and continue to produce appropriate macromolecules for maintenance of disc mechanical strength throughout life.

When cells with a high metabolic activity such as notochordal cells are implanted at such a location, it is unclear that the cells will be able to obtain enough nutrients to achieve adequate regeneration of the extracellular matrix. In fact, notochordal nucleus pulposus cells have a higher metabolic activity than chondrocyte-like nucleus pulposus cells, and thus should require a larger amount of nutrients to maintain their metabolism than the chondrocyte-like cells. It can be assumed from the results of this study that the glycosaminoglycan content of the nucleus pulposus decreases when chondrocyte-like cells take over the major role of extracellular matrix production from notochordal cells during the growth process in humans. The immature disc shows a much higher cell density in vivo than the mature nucleus pulposus with chondrocyte-like cells, and lactate and glycosaminoglycan production were elevated when culture was done at the same cell density (4 x 10^6 cells/mL). This indicates that notochordal nucleus pulposus cells have a higher metabolic activity than chondrocyte-like nucleus pulposus cells and require more nutrients. Based on these results, it was considered preferable to study disc regeneration using discs from animals with chondrocyte-like cells resembling those in the adult human disc.

One of the initial signs of disc degeneration is said to be a decrease of proteoglycan, particularly glycosaminoglycans, in the nucleus pulposus, and one possible cause may be a difference of metabolic activity between nucleus pulposus cells with different phenotypes. Although such changes are often considered to be signs of degeneration, these changes may merely be stages in the normal evolution of discs subjected to mechanical stress. Also, given that many degenerated discs are asymptomatic in humans (Boden et al., 1990, Jensen et al., 1994) identification of the features that distinguish physiological from pathological degeneration is critical. That is, the decrease of the proteoglycan content in the nucleus pulposus that is observed in human discs during the growth process does not necessarily indicate pathological degeneration. Its major cause is considered to be a decline in the function of chondrocyte-like nucleus pulposus cells or the death of cells due to impairment of nutrient supply from the end plates, which occurs after notochordal cells are completely replaced by chondrocytelike cells in the nucleus pulposus from adolescence onward.

7. Changes of physico-chemical environments of the disc cells with aging.

Nutritional supply to the disc is affected by the architecture of the vascular buds and the porosity of the end plate (Fig.9A). Nutrients move from the vascular buds that supply the disc, through the end plates and the dense matrix of the disc, to the cells. Its limits transport of large molecules into and out of the disc (Roberts et al., 1996). For small solutes such as glucose, lactate acid, and oxygen, both experimental and modeling studies have shown that solute transport is accomplished mainly by diffusion (Urban et al.,1982, Katz et al., 1986, Ferguson et al., 2004, Holm et al.,1981), hence, the movement of fluid in and out of the disc as a result of the diurnal loading pattern has little direct influence on transport. Gradients in the concentration arise depending on the balance between the rate of supply of glucose or oxygen from the blood supply to the cells and the rate of cellular consumption (Holm et al.,1981, Sélard, et al., 2003). The intervertebral disc is avascular, and the metabolic activity of its cells is regulated by various factors in the extracellular matrix, such as oxygen (Holm et al.,1981, Bibby et al., 2005), osmolality (Urban & Maroudas, 1979, Maroudas, 1981, Ishihara et al, 1997, Takeno et al., 2007, Negoro et al., 2008), and pH (Ohshima & Urban,

1992, Razaq et al., 2004). The cell density of the normal human nucleus pulposus is 2- 4 x 10^6 cells/mL, and the extracellular environment differs markedly from that of other tissues, with an oxygen saturation of 1-5%, pH of 7.2-7.4, and extracellular osmolality of 370-400mOsm. Disc cells activity is regulated by extracellular oxygen concentrations and extracellular pH. Synthesis rates appear to be highest at around 5% oxygen, where they are greater than the rates found in air (Ishihara & Urban, 1999). However, once oxygen tension falls below 5%, synthesis is inhibited appreciably in an oxygen-tension dependent manner. Extracellular pH also has a marked effect on the synthesis rates of matrix components. The synthesis rates are 40% higher at pH 7.0 than pH 7.4 but fall steeply once the environment become acidic (Ohshima & Urban, 1992). Extracellular matrix degradation, however, appears less sensitive to pH. In fact, the production of active matrix metalloproteinases is similar at pH 7.0 and pH 6.4. Thus, acidic pH levels, by inhibiting synthesis but not degradation, may increase the rate of matrix breakdown.

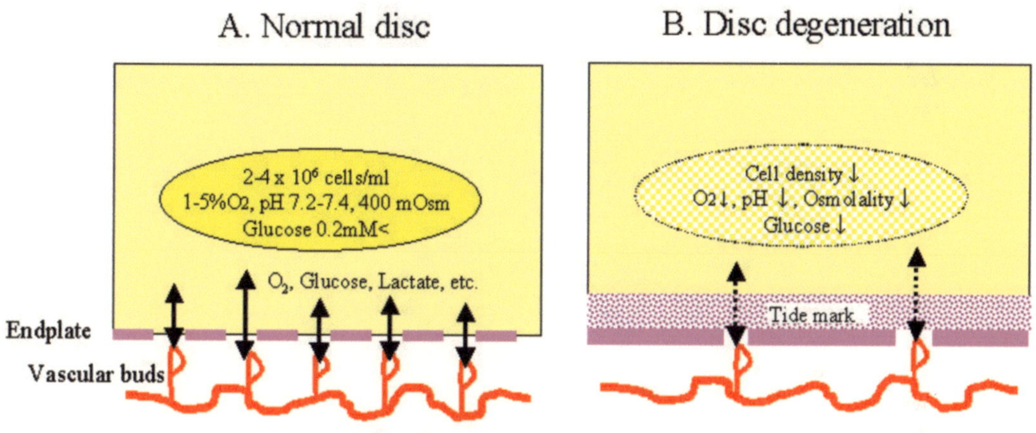

Fig. 9. Changes of nutrition and the extracellular environment in nucleus pulposus with aging. (A) Normal disc, (B) Degenerated disc.

The number of disc cells decreases with aging, but it is unknown whether this decline is caused by differentiation of notochordal cells into nonnotochordal cells, apoptosis, or insufficient supply of nutrients from the end plate. Ossification of the end plate (tide mark) occurs with aging and is one of the major causes of disc degeneration. This leads to deterioration of the extracellular environment in the nucleus pulposus and causes cellular impairment that is followed by a decline of matrix metabolism, resulting in progression to disc degeneration. The nutrient supply for cells and the extracellular environment of the disc have a considerable influence on the outcome of treating disc degeneration by bioengineering techniques. (Reproduced with permission from Miyazaki T, Kobayashi S, et al. A phenotypic comparison of proteoglycan production of intervertebral disc cells isolated from rats, rabbits, and bovine tails; Which animal model is most suitable to study tissue engineering and biological repair of human disc disorders? Tissue Eng Part A. 2009 Aug 15. [Epub ahead of print] PMID: 19681728)

In aging, the end plate tends to calcify by unknown mechanisms (Fig.9B). This tide mark (calcification) acts as a barrier to nutrients transport and is thought to be a major factor in the development of disc degeneration (Urban & Roberts, 1995, Nachemson et al., 1970, Rajasekaran et al., 2004). Cellular parameters are very important in regulating nutrient levels, with levels of oxygen or pH falling with increases in rates of cell metabolism or cell density (Sélard, et al., 2003, Mokhbi-Soukane et al., 2005). For the disc cells to remain viable, the levels of extracellular nutrients and pH must remain above critical values.

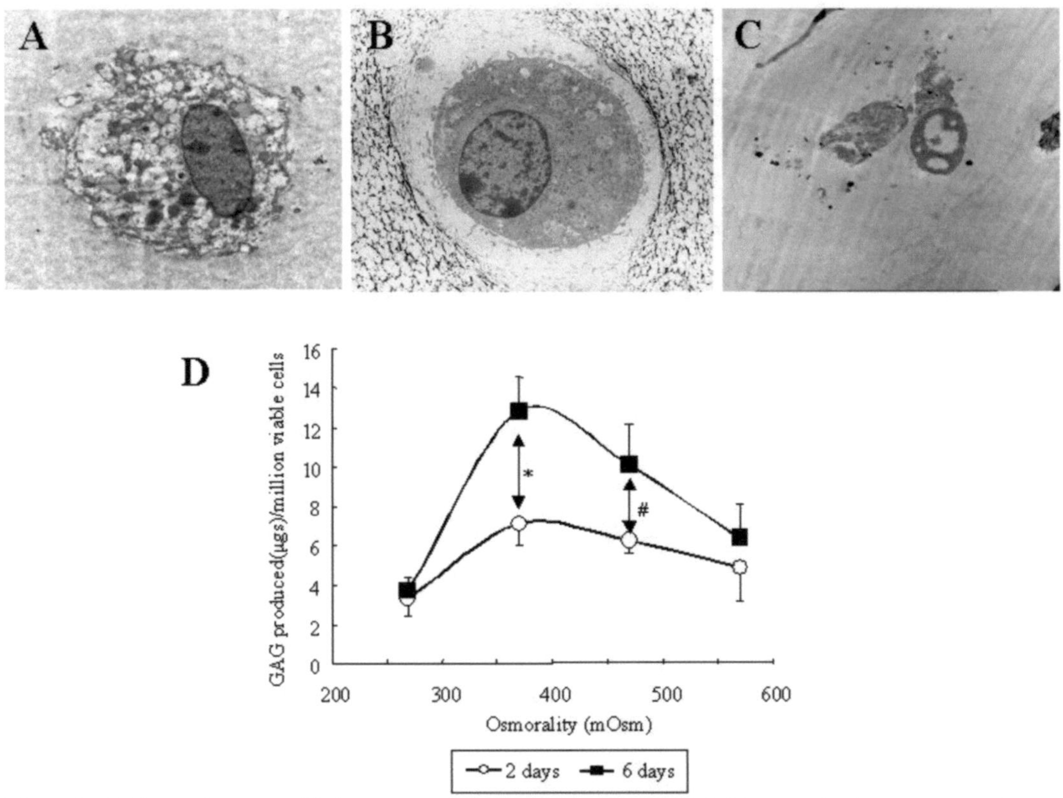

Fig. 10. Effect of extracelluler osmotic change on cell viability and glycosaminoglycan production. Cells were isolated from the nucleus pulposus of 18-24 month bovine caudal discs. They were cultured for 6 days in alginate beads at 4 million cells/ml in Dulbecco's modified Eagle's medium containing 6% serum under 21% O_2, Medium osmolality was altered by NaCl addition over the range 270-570 mOsm and monitored using a freezing point osmometer.

(A-C) Electron micrograms of nucleus pulposus cells in the centre of the beads. At 270mOsm, the cell was swelling with numerous vacuoles and cytoplasmic organelles destroyed were visible (A). This cell undergoing oncosis were seen. At 370 mOsm, all cells appeared viable (B). At 570mOsm, the cells were reduced in size and budding was visible pinching off from the cytoplasm (C). (D) Effect of extracelluler osmotic change on glycosaminoglycan produced per million cells. Total glycosaminoglycans produced per million cells was the highest in the 370 mOsm group and the lowest in the hypo-osmolality

(270 mOsm) group. Glycosaminoglycan produced per million cells was significantly increased at 370 and 470 mOsm with time in culture (Paired t test between day 2 and day 6 [*,#: P<0.05]. However, there was no significant values of glycosaminoglycan produced per million cells at 270 and 570 mOsm with time in culture. Values are mean ± standard error. (Reproduced with permission from Takeno K, Kobayashi S, et al.: Physical limitations to tissue engineering of intervertabral disc cells. Effect of extracellular osmotic change on glycosaminoglycan production and cell metabolism. J Neurosurg Spine 7: 637-644, 2007).

Because disc cells obtain ATP primarily by glycolysis, glucose is a critical nutrient. The cells start to die within twenty-four hours if glucose concentration falls below 0.2 mM and the efficiency of glucose transport into the cell is likely reduced at this glucose concentration (Windhaber et al., 2003). The rate of cell death increases when pH levels are acidic. The cell viability is reduced even with adequate glucose at pH 6.0. The osmotic environment of nucleus pulposus cells in the discs changes with loading and pathologic states. The osmolality of the extracellular matrix is regulated by negatively charging the glycosaminoglycan chains of proteoglycans which adjust ionic composition. Particularly, extracellular osmolarity is controlled by negatively charged proteoglycans. It is now evident that an increase in the concentration of proteoglycans which control ionic composition causes an increase in the osmolality, and conversely, a decrease in proteoglycans reduces osmolality (Ishihara et al., 1997, Maroudas, 1981). Urban and Maroudas et al. assessed the osmotic pressure across the sagittal section of the discs and noted that the osmolality in the nucleus pulposus was about 370-400 mOsm (Maroudas, 1975, Urban & Maroudas, 1979), and the osmolality was decreased in degenerated disc. glycosaminoglycan production was largest in the 370mOsm, and the capacity for glycosaminoglycan production and cell metabolism (lactate production) was low under hypo-osmolality and hyper-osmolality, and cell deaths were observed on electron microscopy (Fig. 10) (Takeno et al., 2007). Thus, it may be said that osmotic pressure gradient disturbance associated with reduced proteoglycans is an important factor contributing to the development of disc degeneration. The results also suggest that standard culture mediums do not provide an appropriate ionic and osmotic environment for nucleus pulposus cells.
The physico-chemical environment created and maintained by disc cells in turn has a powerful effect on disc cell metabolism. However, the supply of nutrients from vascular buds at the end plates to the nucleus pulposus of a degenerative disc is likely to be affected, causing the extracellular environment to deteriorate (Urban & Roberts, 1995, Nachemson et al.,1970, Rajasekaran et al., 2004). Roberts et al. demonstrated that the cells in degenerate discs are senescent (Roberts et al., 2006). This environment is often neglected by it can strongly influence matrix turnover or the responses of disc cells to growth factors or other external stimuli. Such limitations apply to all avascular tissues including tissue engineered constructs.

8. Effect of cell density on the rate of glycosaminoglycan accumulation by nucleus pulposus cells

Glycosaminoglycan accumulation in constructs is dependent on the rate of glycosaminoglycan production per cell and on the cell density. It seems intuitive, therefore, that increasing cell density should increase rate of glycosaminoglycan deposition, as indeed has been shown in several studies (Almarza & Athanasiou, 2005, Mauck et al., 2002, Mauck et al., 2003, Saini & Wick, 2003, Williams et al., 2005, Kobayashi et al., 2008). However, it is apparent from these studies that glycosaminoglycan accumulation in the construct does not increase in proportion to cell density and, indeed, glycosaminoglycan production per cell appears to fall at high cell densities. When bovine nucleus pulposus cells were cultured in three-dimensional constructs such as alginate beads, the amount of glycosaminoglycan accumulated increased with time in culture and also with increase in cell density as expected (Fig. 11A) (Kobayashi et al., 2008). However, the rise in cell density did not lead to a proportional rise in the amount of glycosaminoglycan accumulated; rather the amount of glycosaminoglycan produced per live cell fell with cell density and was significantly lower in beads cultured at a density of 22.7 million cells per ml than in those cultured at 2.02 million cells per ml for nucleus pulposus cells (Fig. 11B). This difference arose at least in part from a fall in metabolic activity of the cells rather than because of increased loss of glycosaminoglycan; the rate of energy production (Fig.12A) and of sulfated glycosaminoglycan production/viable cell (Fig.12B) was lower in cells cultured at high cell densities.

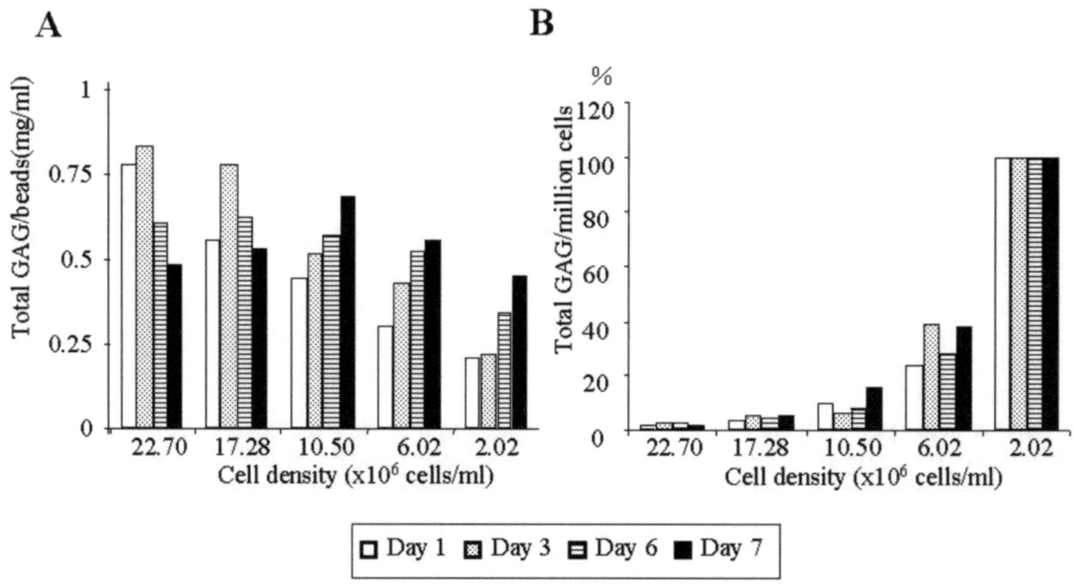

Fig. 11. Effect of cell density on glycosaminoglycan production.

Typical results showing effect of cell density on glycosaminoglycan deposition (A) and on glycosaminoglycan accumulation per million cells (B) by nucleus pulposus cells. Cells were isolated, encapsulated in alginate beads at cell densities ranging from 2.02 to 22.7 million cells/ml. Beads were cultured for 7 days at 5 wells/bead in 2 ml medium, 2 wells for each cell density and cultured for 7 days in Dulbecco's modified Eagle's medium containing 6% serum. Beads were then dissociated for cell counting and assay of total glycosaminoglycans. (A). Effect of cell density on glycosaminoglycan accumulation per beads (mg/ml). More glycosaminoglycan at high cell density than low cell density. Amount of

glycosaminoglycan/bead increased with time in culture at low cell density, but not at high cell density.
(B). Effect of cell density in beads on glycosaminoglycan production per million cells. glycosaminoglycan per cell higher at low cell density. Data normalized to results at 4 million cells/ml.

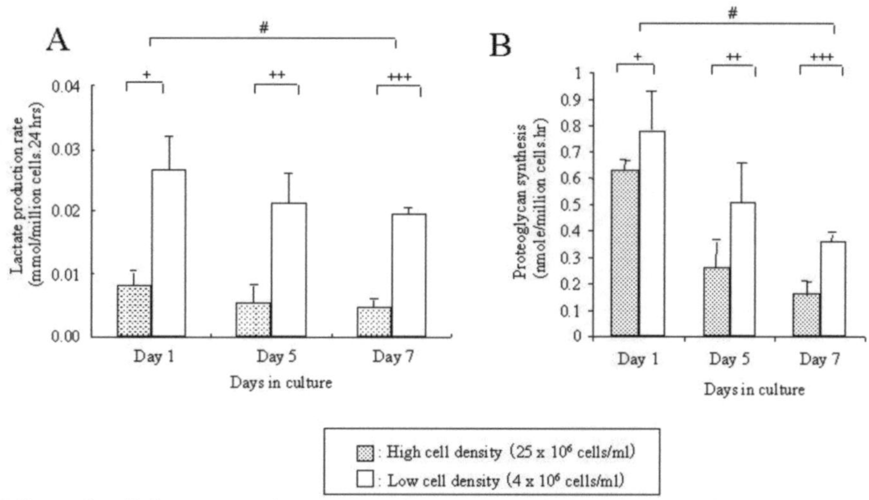

Fig. 12. Effect of cell density on lactate production rate (A) and ^{35}S-sulfate incorporation rate (B). (A) Cells were cultured under standard conditions in beads containing 4 and 25 million cells/ ml (1.0ml medium, 5 beads/well) for up to 7 days, with complete medium change daily. Representative beads were dissociated for cell counting and viable cell density/bead recorded. Lactate in the medium was measured at days 1,5 and 7, after 24 hours culture and rates per million cells/24 hrs reported. High cell density lead to a fall in cellular metabolism (+,++,+++: $P<0.05$, Paired t test between high [25 million cells/ml] and low cell density [4 million cells/ml]). Lactate production rate fall with time in culture (#: $P<0.05$, 2 way ANOVA with repeated measures among 1, 5 and 7 days). (B) At days 1,5 and 7, tracer sulphate was added to the fresh medium of 3 wells, the beads were cultured in the radioactive solution for 4 hours, the beads dissociated and cell density and sulphate incorporation measured (Fig 3B). Results are given as means standard error of 3 independent experiments. Sulfate incorporation rates fall with increase in cell density (+,++,+++: $P<0.05$, Paired t test between high [25 million cells/ml] and low cell density [4 million cells/ml]) and with time in culture (#:$P<0.05$, 2 way ANOVA with repeated measures among 1, 5 and 7 days). (Reproduced with permission from Kobayashi S, Meir A, Urban J. Effect of cell density on the rate of glycosaminoglycan accumulation by disc and cartilage cells in vitro. J Orthop Res 26:493-503,2008.)

The change in percentage of live and dead cells with time in culture at the periphery and centre of beads is shown in Fig.13 for cells cultured at low (4 million cells/ml) and high cell densities (25 million cells/ml), respectively. For cells cultured at 4 million cells/ml, 100% of the cells were viable at the both the periphery (Fig 13A) and in the centre (Fig 13B). It can be seen that by day 2 of culture at high cell densities, while almost all the cells at the periphery were alive (Fig.13C,E), 30 percent of the cells in the bead centre were dead (Fig.13D,E).

Similar percentages were dead at day 5 of culture, suggesting the profile of viable cells across the bead was established early in culture (Fig.13E). At low cell density, transmission electron micrographs indicated that all cells appeared viable and active.

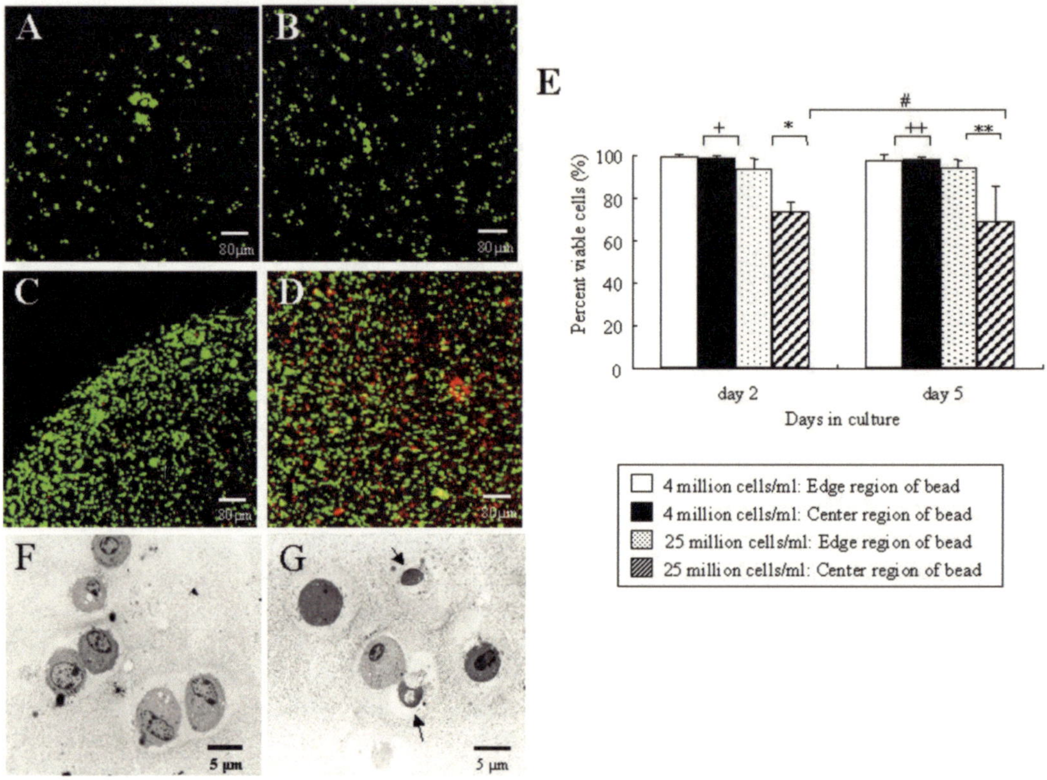

Fig. 13. Effect of cell density on cell viability under conforcal (A-E) and electron (F,G) microscope.

(A-E) A and B shows the periphery and central region respectively of a typical bead cultured at 4 million cells/ml after 5 days. C and D shows the periphery and central region of a bead cultured at 25 million cells/ml after 5 days. Cell viability was determined using a live/dead assay kit; live cells (green) and dead cells (red) were counted manually. E shows the variation of cell viability at the edge and centre of beads with time and cell density. Results are means and s.e.ms of percentage of viable cell from 4 representative beads. At high cell density (25 million cells/ml), cell viability is lower in the centre than at the edge (+: P=0.961, ++: P=0.932, *, **: P<0.05, 2 way ANOVA with repeated measures between edge and centre).

(F,G) Electron micrographs of peripheral (F) and central (G) nucleus pulposus cells (A-D) cultured at high cell density. The figures show representative cells from the central and peripheral regions of beads cultured at high (25 million cells/ml) for 5 days. In the bead periphery, some cells appear normal. However, the central region showing cells undergoing apoptosis (arrows). The cells and nuclei were reduced in size and chromatin condensation was seen in the nuclei in comparison of the cells in the periphery. (Reproduced with permission from Kobayashi S, Meir A, Urban J. Effect of cell density on the rate of glycosaminoglycan accumulation by disc and cartilage cells in vitro. J Orthop Res 26:493-503,2008.)

Nucleus pulposus cells cultured at high cell density appeared viable at the bead periphery (Fig.6F). However cells undergoing apoptosis were seen in the centre; the cells and nuclei were reduced in size and chromatin condensation was visible in the nuclei (Fig.6G). These results are in agreement with those of others who have found regions of cell death in the center of constructs or even of microsphere aggregates (Martin et al., 1999, Mercier et al., 2004, Obradovic et al., 1999), and that glycosaminoglycan accumulation may highest at the construct peripheries. In addition, others have also found that increasing cell density or cell number does not necessarily increase matrix accumulation (Mercier et al., 2004).

These avascular constructs, unless experimentally perfused, rely on diffusion for supply of nutrients to the cells (Grunhagen et al., 2006, Obradovic et al., 2000) simulating the condition seen in intervertebral disc. In avascular tissues and in constructs, there are steep gradients of oxygen and other nutrients between the surface and center of the tissue or constructs (Kellner et al., 2002, Malda et al., 2004). The steepness of these gradients, and hence the nutrient concentrations in the center of the construct, depend not only on the geometry and properties of the tissue or construct but also on the cell density and the cellular activity. (Haselgrove et al., 1993, Zhou et al., 2004, Soukane et al., 2005). Thus, in any particular construct or tissue, an increase in cell density will lead to a corresponding fall in the concentration of nutrients such as oxygen and glucose, and an increase of metabolic by-products such as lactic acid (Zhou et al., 2004), leading, once cell density has risen sufficiently, to a fall in rates of cell metabolism and glycosaminoglycan synthesis (Gray et al., 1988, Ysart & Mason, 1994). If cell density is sufficiently great, oxygen and glucose concentrations and pH levels can fall to levels which can no longer sustain viable cells (Horner & Urban, 2001) leading to the necrotic region in the construct center. Diffusional nutrient transport is thus a limitation on the number of viable and active cells which can be maintained in any construct or tissue; indeed, viable cell density is inversely related to diffusion distance both in disc and in constructs (Horner & Urban, 2001, Stockwell, 1971).

9. Physical limitations to biological repair and tissue engineering

The interrelationships between cell density, cell viability and activity, and diffusion distance resulting from nutrient supply constraints, limit the rate at which glycosaminoglycan can be accumulated in three-dimensional constructs. Glycosaminoglycan accumulation depends on glycosaminoglycan production per cell and on cell density. At low cell densities, cells may be functioning optimally but the low cell density limits the rate of glycosaminoglycan accumulation. At high cell densities, more glycosaminoglycan is deposited at least initially, but nutrient gradients particularly in the center of constructs, reduce the rate of glycosaminoglycan deposition per cell and may even lead to a fall in cell number if cells die. Glycosaminoglycan accumulation thus appears necessarily slow, and the general finding that cultures of >7 months are required to achieve concentrations of glycosaminoglycan similar to those seen in vivo may not be easily overcome (Kellner et al., 2002, Roughley, 2004). The different maneuvers which have been tried to increase glycosaminoglycan production all have limitations. An increase in glycosaminoglycan production rate per cell can be induced by addition of growth factors, by providing mechanical or ultrasound stimulation or through alterations to scaffold properties (Blunk et al., 2002, Richmon et al.,

2005, van der Kraan et al., 2002, Kuo & Lin, 2006), but the relative increase which can be achieved is limited (usually two–threefold under optimal conditions) and the consequent increase in metabolic demand can lead to a fall in pH in the construct center (Razaq et al., 2004) and thus severely limit growth factor efficacy. Indeed, addition of growth factors to constructs was found to have little effect on the concentration of accumulated glycosaminoglycan although it increased construct size (Walsh et al., 2002, Yoon et al., 2003a, Yoon et al., 2003b, Masuda et al., 2003, An et al., 2005, Malda et al., 2004). In addition, glycosaminoglycan production rates appear to fall with time in culture in many different systems also limiting glycosaminoglycan accumulation (Mercier et al., 2004). Increasing cell density potentially should increase glycosaminoglycan deposition, but leads to a lower activity per cell, and also, in general, has not been found to increase glycosaminoglycan deposition rates (Panossian et al., 2001). It should also be noted that tissue *in vivo* cannot support too high a cell density, so *in vitro* culture of constructs at high cell density could lead to cell death after implantation.
Culture conditions such as stirring or perfusion (Freyria et al., 2000, Seidel et al., 2004) appear able to overcome diffusive transport initially, but as glycosaminoglycan concentrations rise and the hydraulic permeability of the construct falls, convective transport also is reduced and rates of glycosaminoglycan deposition slow. Glycosaminoglycan concentrations were reported to reach 5%by wet weight within 2 months but took a further 5 months to increase to 7% glycosaminoglycan. In view of the long culture times which appear necessary to achieve the required glycosaminoglycan composition in vitro, achievement of in vivo concentration before implantation of a construct may be an unrealistic and possibly unnecessary goal for tissue engineered disc.
Cellular repair using autologous chondrocyte transplantation appears successful even though chondrocytes are implanted with no matrix at all. Under these conditions, remodeling in vivo appears to produce a cartilage-type matrix under some conditions. Tissue engineered composites implanted with low glycosaminoglycan appeared to accumulate glycosaminoglycan in vivo, withstand physiological loading, and remodel towards a hyaline-type matrix. Perhaps optimization of such processes is a more useful goal.

10. Conclusions

There is increasing interest in the using biological methods to repair degenerate discs. Biological repair depends on the disc maintaining a population of viable and active cells. Adequate nutrition of the disc influences the outcome of such therapies and, hence, must be considered to be a crucial parameter. Therefore, it is very important to maintain an appropriate physicochemical environment to achieve successful disc repair by biological methods and tissue engineering procedures.

11. References

Adams, M.A. & Hutton, W.C. (1982). Prolapsed intervertebral disc. A hyperflexion injury. *Spine*, 7, 184-191.

Adler, J.H.; Schoenbaum, M. & Silberberg, R. (1983). Early onset of disk degeneration and spondylosis in sand rats (Psammomys cbesus). *Vet Pathol*, 20, 13-22.

Aguiar, D.J.; Johnson, S.L. & Oegema, T.R. (1999). Notochordal cells interact with nucleus pulposus cells: Regulation of proteoglycan synthesis. *Exp Cell Res,* 246, 129-137.

Alini, M.; Roughley, P.J. & Antoniou, J.; et al. (2002). A biological approach to treating disc degeneration: not for today, but maybe for tomorrow. *Eur Spine J,* 11, S215-S220.

Alini, M.; Li, W.; Markovic, P.; et al. (2003). The potential and limitations of a cell-seeded collagen/hyaluronan scaffold to engineer an intervertebral disc-like matrix. *Spine,* 28, 446-453.

Alini, M.; Eisenstein, S.M.; Ito, K.; et al. (2008). Are animal models useful for studing human disc disorders/degeneration? *Eur Spine J* , 17, 2-19.

Almarza, A.J. & Athanasiou, K.A. (2005). Effects of initial cell seeding density for the tissue engineering of the temporomandibular joint disc. *Ann Biomed Eng,* 33, 943-950.

An, H.; Takegami, K.; Kamada, H.; et al. (2005). Intradiscal administration of osteogenic protein-1 increases intervertebral disc height and proteoglycan content in the nucleus pulposus in normal adolescent rabbits. *Spine,* 30, 25-31.

Anderson, H. & Matthiessen, M.E. (1966). The histiocyte in human oetal tissues. *Z Zellforsch Mikrock Anat,* 72, 193-211.

Baer, A.E.; Wang, J.Y.; Kraus, V.B.; et al. (2001). Collagen gene expression and mechanical properties of intervertebral disc cell-alginate cultures. *J Orthop Res,* 19, 2-10.

Bassett, C.A.L. (1962). Current concepts of bone formation. *J Bone Joint Surg,* 44A, 1217-1244.

Bayliss, M.T.; Johnstone, B.; O'Brien. J.P.; et al. (1988). Proteoglycan synthesis in the human intervertebral disc. Variation with age, region and pathology. *Spine,* 13, 972-981.

Bennett, H.S.; Luft, J.H. & Hampton, J.C. (1959). Morphological classification of vertebrate blood capillaries. *Am J Physiol,* 196, 381-390.

Berry, R.J. (1961). Genetically controlled degeneration of the nucleus pulposus in the mouse. *J Bone Joint Surg,* 43B, 387-393.

Bibby, S.R.; Jones, D.A.; Lee, R.B.; et al. (2001). The pathophysiology of the intervertebral disc. *Joint Bone Spine,* 68, 537-542.

Bibby, S.R.; Jones, D.A.; Ripley, R.M.; et al. (2005). Metabolism of the intervertebral disc: effects of low levels of oxygen, glucose, and pH on rates of energy metabolism of bovine nucleus pulposus cells. *Spine,* 30, 487-496.

Blunk, T.; Sieminski, A.L.; Gooch, K.J.; et al. (2002). Differential effects of growth actors on tissue-engineered cartilage. *Tissue Eng,*8,73-84.

Boden, S.D.; McCowin, P.R.; Davis, D.O.; et al. (1990). Abnormal magnetic-resonance scans of the cervical spine in asymptomatic subjects. A prospective investigation. *J Bone Joint Surg,* 72A, 1178-1184.

Braund, K.G.; Ghosh, P.; Taylor, T.K.F.; et al. (1975). Morphological studies of the canine intervertebral disc: the assignment of the beagle to the achondrplastic classification. *Res Vet Sci,* 19, 167-172.

Brisby, H.; Tao, H.; Ma, D.D; et al. (2004). Cell therapy for disc degeneration. Potentials and pitfalls. *Orthop Clin North Am,* 35, 85-93.

Brookes, M. (1971). The blood supply of bone. *An Approach to Bone Biology.* Butterworth, London.

Butler, W.F. & Smith, R.W. (1967). Age change in the nucleus pulposus of the nonruptured intervertebral disc of the cat. *Res Vet Sci,* 8, 151-156.

Bulter, W.F. (1988). Comparative anatomy and development of the mammalian disc. In: *The Biology of the Intervertebral Disc.* Ghosh, P., (ed.), pp.84-108, CRC Press, Boca Raton, FL.

Cappello, R.; Bird, J.L.E.; Pfeiffer, D.; et al. (2006). Notochordal cell produce and assemble extracellular matrix in a distinct manner, which may be responsible for the maintenance of healthy nucleus pulposus. *Spine,* 31, 873-882.

Carlier, E.W. (1890). The fate of the notochord and development of the intervertebral disc in the sheep, with observations on the structure of the adult disc in these animals. *J Anat Phys,* 24, 573-585.

Chen, J.; Yan, W. & Setton, L.A. (2004). Static compression induces zonal-specific changes in gene expression for extracellular matrix and cytoskeletal proteins in intervertebral disc cells in vitro. *Matrix Biology,* 22, 573-583.

Cole, T.C.; Buckhardt, D.; Frost, L.; et al. (1985). The proteoglycans of the canine intervertebral disc. *Biochim Biophys Acta,* 829, 127-138.

Coventry, M.B.; Ghormley, P.K. & Kernohan, J.W. (1945a). The intervertebral disc: its microscopic anatomy and pathology. Part I. anatomy, development and physiology. *J Bone Joint Surg,* 27, 105-112.

Coventry, M.B.; Ghormley, R.K. & Kernohan, J.W. (1945b). The intervertebral disc: Its microscopic anatomy and pathology. Part II. Changes in the intervertebral disc concomitant with age. *J Bone Joint Surg,* 27, 233-247.

Coventry, M.B.; Ghormley, R.K. & Kernohan, J.W. (1945c). The intervertebral disc: Its microscopic anatomy and pathology. Part III. Pathologic changes in the intervertebral disc concomitant with age. *J Bone Joint Surg,* 27, 460-474.

Crelin, E.D. & Koch, W.E. (1967). An autoradiographic study of chondrocyte transformation into chondroclasts and osteocytes during bone formation in vitro. *Anatomical Record,* 158, 473-483.

Crock, H.V.; Yoshizawa, H. & Kame, S.K. (1973). Observations on the venous drainage of the human vertebral body. *J Bone Joint Surg,* 55B, 528-533.

Crock, H.V. & Yoshizawa, H. (1977). *The Blood Supply of the Vertebral Column and Spinal Cord in Man.* Springer, Wien, New York.

Crock, H.V. & Goldwasser, M. (1984). Anatomic studies of the circulation in the region of the vertebral end-plate in adult greyhound dogs. *Spine,* 9, 702-706.

Erwin, W.M. & Inman, R.D. (2006). Notochord cells regulate intervertebral disc chondrocyte proteoglycan production and cell proliferation. *Spine,* 31, 1094-1099.

Ferguson, S.J.; Ito, K. & Nolte, L.P. (2004). Fluid flow and connective transport of solutes within the intervertebral disc. *J Biomech,* 37, 213-221.

Fischman, D.A. & Hay, E.D. (1962). Origin of osteoclasts from mononuclear leucocytes in regenerating newt limbs. *Anatomical Record,* 143, 329-334.

Freyria, A.M.; Ronzi, S.; Roche, C.F.; et al. (2000). Regulation of growth, protein synthesis, and maturation of fetal bovine epiphyseal chondrocytes growth in high-density culture in the presence of ascorbic acid, retinoic acid and dihydrocytochalasin B. *J Cell Biochem,* 76, 84-98.

Gage, E.D. (1975). Incidence of clinical disc disease in the dog. *J Am Anim Hosp Assoc,* 11, 135-138.

Ganey, T.M. & Meisel, H.J. (2002). A potential role for cell-based therapeutics in the treatmant of intervertebral disc herniation. *Eur Spine J,* 11, S206-214.

Ganey, T.; Libera, J.; Moos, V.; et al. (2003). Disc chondrocyte transplantation in a canine model: a treatment for degenerated or damaged intervertebral disc. *Spine,* 28, 2609-2620.

Gardner, E. & Gray, D.J. (1970). The Prenatal development of the human femur. *Am J Anat,* 129, 163-182.

Gray, M.; Pizzanelli, A.; Grodzinsky, A.; et al. (1988). Mechanical and physiochemical determinants of the chondrocyte biosynthetic response. *J Orthop Res,* 6, 777-792.

Grodzinsky, A.J. (1983). Electromechanical and physicochemical properties of connective tissue. *Crit Rev Biomed Eng,* 9, 133-199..

Grunhagen, T.; Wilde, G.; Soukane, D.M.; et al. (2006). Nutrient supply and intervertebral disc metabolism. *J Bone Joint Surg,* 88A, 30-35.

Haines, R.W. (1933). Cartilage canals. *J Anat,* 68, 45-64.

Hamilton, D.J.; Séguin, C.A.; Wang, J.; et al. (2005). Formation of a nucleus pulposus-cartilage endplate construct in vitro. *Biomaterials,* 27, 397-405.

Hansen, H.J. (1952). A pathologic-anatomical study on disc degeneration in the dogs: with special reference to the so-called enchondrosis intervertebralis. *Acta Orthop Scand,* 11(Suppl), 1-117.

Hansen, H.J. (1956). Studies on the pathology of the lumbosacral disc in female cattle. *Acta Othop Scand,* 25, 161-182.

Hansen, H.J. (1959). Comparative views on the pathology of disc degeneration in animals. *Lab Invest,* 8, 1242-1265.

Happy, F. & Bradford, A.G.J. (1964). Preliminary observations concerning the fine structure of the intervertebral disc. *J Bone Joint Surg,* 46B, 563-567.

Happy, F.; Pearson, C.H.; Naylor, A.; et al. (1969). The aging of the human intervertebral disc. *Gerontologia,* 15, 174-188.

Haselgrove, J.C.; Shapiro, I.M. & Silverton, S.F. (1993). Computer modeling of the oxygen supply and demand of cells of the avian growth cartilage. *Am J Physiol,* 265, C497-C506.

Hohaus, C.; Ganey, T.M.; Minkus, Y.; et al. (2008). Cell transplantation in lumbar spine disc degeneration disease. *Eur Spine J,* 17, 492-503.

Holm, S.; Maroudas, A,; Urban, J.P.; et al. (1981). Nutrition of the intervertebral disc: Solute transport and metabolism. *Connect Tissue Res,* 8, 101-119.

Horner, H.A. & Urban, J.P. (2001). Volvo Award Winner in Basic Science Studies: Effect of nutrient supply on the viability of cells from the nucleus pulposus of the intervertebral disc. *Spine,* 26, 2543-2549.

Hunter, C.J.; Matyas, J.R. & Duncan, N.A. (2003a). The three-dimensional architecture of the notochordal nucleus pulposus: novel observations on cell structures in the canine intervertebral disc. *J Anat,* 202, 279-291.

Hunter, C.J.; Matyas, J.R. &, Duncan, N.A. (2003b). The notochordal cell in the nucleus pulposus: A review in the content of tissue engineering. *Tissue Eng,* 9, 667-677.

Hurrell, D.J. (1934). The vascularization of cartilage. *J Anat,* 69, 47-61.

Ishihara, H.; McNally, D.S.; Urban, J.P.; et al. (1996). Effects of hydrostatic pressure on matrix synthesis in different regions of the intervertebral disk. *J Appl Physiol,* 80, 839-846.

Ishihara, H.; Warensjo, K.; Roberts, S.; et al. (1997). Proteoglycan synthesis in the intervertebral disk nucleus: the role of extracellular osmolality. *Am J Physiol,* 272, C1499-C1506.

Ishihara, H. & Urban, J.P. (1999). Effects of low oxygen concentrations and metabolic inhibitors on proteoglycan and protein synthesis rates in the intervertebral disc. *J Orthop Res,* 17, 829-835.

Iwashina, T.; Mochida, J.; Sakai, D.; et al. (2006). Feasibility of using a human nucleus pulposus cell line as a cell source in cell transplantation therapy for intervertebral disc degeneration. *Spine,* 31, 1177-1186.

Jensen, M.C.; Brant-Zawadzki, M.N.; Obuchowski, N.; et al. (1994). Magnetic resonance imaging of the lumbar spine in people without back pain. *N Engl J Med,* 331, 69-73.

Johnstone, B. & Bayliss, M.T. (1995). The large proteoglycan of the human intervertebral disc. Changes in their biosynthesis and structure with age, topography, and pathology. *Spine,* 20, 674-684.

Kalson, N.S.; Richardson, S.M. & Hoyland, J.A. (2008). Strategies for regeneration of the intervertebral disc. *Regen Med,* 3, 717-729.

Katz, M.M.; Hargen, A.R. & Garfin, S.R. (1986). Intervertebral disc nutrition. Diffusion versus convection. *Clin Orthop Relat Res,* 210, 243-245.

Kellner, K.; Liebsch, G.; Klimant, I.; et al. (2002). Determination of oxygen gradients in engineered t issue using a fluorescent sensor. *Biotechnol Bioeng,* 80, 73-83.

Kempson, G.E.; Muir, H.; Swanson, S.A.; et al. (1970). Correlations between stiffness and the chemical constituents of cartilage on the human femoral head. *Biochim Biophys Acta,* 215, 70-77.

Keys, D.C, & Compere, E.L. (1932). The normal and pathological physiology of the nucleus pulposus of the intervertebral disc. An anatomical, clinical and experimental study. *J Bone Joint Surg* , 14, 897-938.

Kim, K.W.; Lim, T.H.; Lim, J.G.; et al. (2003). The origin of chondrocytes in the nucleus pulposus and histologic findings associated with the transition of a notochordal nucleus pulposus to a fibrocartolagious nucleus pulposus in intact rabbit intervertebral discs. *Spine,* 28, 982-990.

Kobayashi, S.; Yoshizawa, H.; Hachiya, Y.; et al. (1993). Volvo awards in basic science. Vasogenic edema induced by compression injury to the spinal nerve root. Distribution of intravenously injected protein tracers and gadolinium enhanced magnetic resonance imaging. *Spine,* 18, 1410-1424.

Kobayashi, S. & Yoshizawa, H. (2002). Effect of mechanical compression on the vascular permeability of the dorsal root ganglion. *J Orthop Res,* 20, 730-739.

Kobayashi, S.; Baba, H.; Takeno, K.; et al. (2008). Fine structure of catilage canal and vascular buds in the vertebral end-plate. *J Neurosurg Spine,* 9, 96-103.

Kobayashi, S.; Meir, A. & Urban, J. (2008). Effect of cell density on the rate of glycosaminoglycan accumulation by disc and cartilage cells in vitro. *J Orthop Res,* 26, 493-503.

Kroeber, M.W.; Unglaub, F.; Wang, H.; et al. (2002). New in vivo animal model to create ntervertebral disc degeneration and to investigate the effects of therapeutic strategies to stimulate disc regeneration. *Spine,* 27, 2684-2690.

Kuo, Y.C. & Lin, CY. (2006). Effect of genipin-crosslinked chitin-chitosan scaffolds with hydroxyapatite modifications on the cultivation of bovine knee chondrocytes. *Biotechnol Bioeng,* 95, 132-144.

Lawrence, T.E.; Whatley, J.D.; Montgomery, T.H.; et al. (2001). A comparison of the USDA ossification-based maturity system to a system based on dentition. *J Anim Sci,* 79, 1683-1690.

Levene, C. (1964). The patterns of cartilage canals. *J Anat,* 98:515-538.

Malda, J.; Rouwkema, J.; Martens, D.E.; et al. (2004). Oxygen gradients in tissue-engineered PEGT/PBT cartilaginous constructs: measurement and modeling. *Biotechnol Bioeng,* 86, 9-18.

Maroudas, A. (1975). Biophysical chemistry of cartilaginous tissues with special reference to solute and fluid transport. *Biorheology,* 12, 233-248.

Maroudas, A.; Stockwell R.A.; Nachemson, A.; et al. (1975). Factors involved in the nutrition of the n human lumbar intervertebral disc: cellularity and diffusion of glucose in vitro. *J Anat,* 120, 113-130.

Maroudas, A. & Bannon, C. (1981). Measurement of swelling pressure in cartilage and comparison with the osmotic pressure of constituent proteoglycans. *Biorheology,* 18, 619-632.

Maroudas, A. (1981). Metabolism of cartilaginous tissue, a quantitative approach. In: *Studies in Joint Disease 1,* Maroudas, A.; Holborow, J. (Ed.), pp 59-86, Pitman Medical, London.

Martin, I.; Obradovic, B.; Freed, L.E.; et al. (1999). Method for quantitative analysis of glycosaminoglycan distribution in cultured natural and engineered cartilage. *Ann Biomed Eng,* 27, 656-662.

Masuda, K.; Takegami, K.; An, H.; et al. (2003). Recombinant osteogenic protein-1 upregulates extracellular matrix metabolism by rabbit annulus fibrosus and nucleus pulposus cells cultured in alginate beads. *J Orthop Res,* 21, 922-930.

Masuda, K.; Oegema, T.R.Jr. & An, H.S. (2004). Growth factors and treatment of intervertebral disc degeneration. *Spine,* 29, 2757-2769.

Matsumoto, T.; Masuda, K.; An, H.; et al. (2001). Formation of transplantable disc-shaped tissues by nucleus pulposus and annulus fibrosus cells: biochemical and biomechanical properties. *Ortho Res Soc Trans,* 26, 897.

Mauck, R.L.; Seyhan, S.L.; Ateshian, G.A.; et al. (2002). Influence of seeding density and dynamic deformational loading on the developing structure/function relationships of chondrocyte-seeded agarose hydrogels. *Ann of Biomed Eng,* 30, 1046-1056.

Mauck, R.L.; Wang, C.C.; Oswald, E.S.; et al. (2003). The role of cell seeding density and nutrient supply for articular cartilage tissue engineering with deformational loading. *Osteoarthritis Cartilage,* 11, 879-890.

Mercier, N.R.; Costantino, H.R.; Tracy, M.A.; et al. (2004). A novel injectable approach for cartilage formation in vivo using PLG microspheres. *Ann Biomed Eng,* 32, 418-429.

Miyazaki, T.; Kobayashi, S.; Takeno, K.; et al. (2009). A phenotypic comparison of proteoglycan production of intervertebral disc cells isolated from rats, rabbits, and bovine tails; Which animal model is most suitable to study tissue engineering and biological repair of human disc disorders? *Tissue Eng Part A,* 2009 Aug 15. [Epub ahead of print] PMID: 19681728 [PubMed - as supplied by publisher]

Mizuno, H.; Roy, A.K.; Vacanti, C.A.; et al. (2004). Tissue-engineered composites of annulus fibrosus and nucleus pulposus for intervertebral disc replacement. *Spine*, 29, 1290-1297.

Mokhbi-Soukane, D.; Shirazi-Adl, A. & Urban, J.P.G. (2005). Analysis of nonlinear coupled diffusion of oxygen and lactic acid in intervertebral discs. *J Biomech Eng*, 127, 1121-1126.

Moskowitz, R.W.; Ziv, I.; Denko, C.W.; et al. (1990). Spondylosis in sand rats: A model of intervertebral disc degeneration and hyperostosis. *J Orthop Res*, 8, 401-411.

Moss-Salentijn, L. (1975). Cartilage canals in the human spheno-occipital synchondrosis during fetal life. *Acta Anat*, 92, 595-606.

Nachemson, A.; Lewin, T.; Maroudas, A.; et al. (1970). In vitro diffusion of dye through the end-plates and annulus fibrosus of human lumbar intervertebral discs. *Acta Orthop Scand*, 41, 589-607.

Negoro, K.; Kobayashi, S.; Taneno, K.; et al. (2008). Effect of osmolarity on glycosaminoglycan production and cell metabolism of articular chondrocyte under three dimensional culture system. *Clin Exp Rheumatol*, 26, 527-533.

Nishida, K.; Kang, J.D.; Gilbertson, L.G.; et al. (1999). Modulation of the biologic activity of the rabbit intervertebral disc by gene therapy: an in vivo study of adenovirus-mediated transfer of the human transforming growth factor beta 1 encoding gene. *Spine*, 24, 2419-2425.

Obradovic, B.; Carrier, R.L.; Vunjak-Novakovic, G.; et al. (1999). Gas exchange is essential for bioreactor cultivation of tissue engineered cartilage. *Biotechnol Bioeng* 63, 197-205.

Obradovic, B.; Meldon, J.H.; Freed, L.E.; et al. (2000). Glycosaminoglycan deposition in engineered cartilage: Experiments and mathematical model. *Aiche Journal*, 46, 1860-1871.

Ogata, K. & Whiteside, A. (1981). Nutritional pathways of the intervertebral disc: an experimental study using hydrogen washout technique. *Spine*, 6, 211-216.

Ohshima, H. & Urban, J.P.G. (1992). The effect of lactate and pH on proteoglycan and protein synthesis rates in the intervertebral disc. *Spine*, 17, 1079-1082.

Ohshima, H.; Ishihara, H.; Urban, J.P.; et al. (1993). The use of coccygeal discs to study intervertebral disc metabolism. *J Orthop Res*, 11, 332-338.

Ohshima, H.; Urban, J.P.G. & Bergel, D.H. (1995). Effect of static load on matrix synthesis rates in the intervertebral disc measured in vitro by a new perfusion technique. *J Orthop Res*, 13, 22-29.

Oki, S.; Matsuda, Y.; Shibata, T.; et al. (1996). Morphological differences of the vascular buds in the vertebral endplate. *Spine*, 21, 174-177.

Okuma, M.; Mochida, J.; Nishimura, K.; et al. (2000). Reinsertion of stimulated nucleus pulposus cells retards intervertebral disc degeneration: An in vitro and in vivo experimental study. *J Orthop Res*, 18, 988-997.

Osada, R.; Ohshima, H.; Ishihara, H.: et al. (1996). Autocrine/paracrine mechanism of insulin-like growth factor-1 secretion, and the effect of insulin-like growth factor-1 on proteoglycan synthesis in bovine intervertebral discs. *J Orthop Res*, 14, 690-699.

O'Halloran, D.M. & Pandit, A.S. (2007). Tissue-engineering approach to regenerating the intervertebral disc. *Tissue Eng*, 13, 1927-1954.

Panossian, A.; Ashiku, S.; Kirchhoff, C.H.; et al. (2001). Effects of cell concentration and growth period on articular and ear chondrocyte transplants for tissue engineering. *Plast Reconstr Surg*, 108, 392-402.

Peacock, A. (1951). Observations on the pre-natal sturucture of the intervertebral disc in man. *J Anat*, 85, 260-274.

Peacock, A. (1952). Obsevations on the postnatal development of the intervertebral disc in man. *J Anat, 86*, 162-179.

Priester, W.A. (1976). Canine intervertebral disc disease: Occurrence by age, breed, and sex among 8117 cases. *Theriogenology*, 6, 293-301.

Race, A.; Broom, N.D.; Robertson, P.; et al. (2000). Effect of loading rate and hydration on the mechanical properties of the disc. *Spine*, 25, 662-669.

Rajasekaran, S.; Babu, J.N.; Arun, R.; et al. (2004). A study of diffusion in human lumbar discs: a serial magnetic resonance imaging study documenting the influence of the endplate on diffusion in normal and degenerate discs. *Spine*, 29, 2654-2667.

Razaq, S.; Wilkins, R.J. & Urban, J.P.G. (2004). The effect of extracellular pH on matrix turnover by cells of the bovine nucleus pulposus. *Eur Spine J*, 13, 685-701.

Richardson, S.M.; Mobasheri, A.; Freemont A.J.; et al. (2007). Intervertebral disc biology, degeneration and novel tissue engineering and regenerative medicine therapies. *Histol Histopathol*, 22, 1033-1041.

Richmon, J.D.; Sage, A.B.; Shelton, E.; et al. (2005). Effect of growth factors on cell proliferation, matrix deposition, and morphology of human nasal septal chondrocytes cultured in monolayer. *Laryngoscope*, 115, 1553-1560.

Risbud, M.V.; Izzo, M.W.; Adams, C.S.; et al. (2003). An organ culture system for the study of the nucleus pulposus: description of the system and evaluation of the cells. *Spine*, 28, 2652-2658.

Roberts, S.; Menage, J.; Duance, V.; et al. (1991). Collagen types around the cells of the intervertbral disc and cartilage end plate: an imminolocalization study. *Spine*, 16, 1030-1038.

Robert, S.; Menage, J.; Eisenstein SM.; et al. (1993). The cartilage end-plate and intervertebral disc in scoliosis: calcification and other sequelae. *J Orthop Res*, 11, 747-757.

Roberts, S.; Urban, J.P.; Evans, H.; et al. (1996). Transport properties of the human cartilage endplate in relation to its composition and calcification. *Spine*, 21, 415-420.

Roberts, S.; Evans, E.H.; Kletsas, D.; et al. (2006). Senescence in human intervertebral discs. *Eur Spine J*, 15, 312-316.

Roughley, P.J. (2004). Biology of intervertebral disc aging and degeneration: Involvement of the extracellular matrix. *Spine*, 29, 2691-2699.

Saini, S. & Wick, T.M. (2003). Concentric cylinder bioreactor for production of tissue engineered cartilage: effect of seeding density and hydrodynamic loading on construct development. *Biotechnol Prog*, 19, 510-521.

Sakai, D.; Mochida, J.; Iwashina, T.; et al. (2006). Regenerative effects of transplanting mesenchymal stem cells embedded in atelocollagen to the degenerated intervertebral disc. *Biomaterials*, 27, 335-345.

Schmorl ,G. & Junghanns, H. (1971). *The Human Spine in Health and Disease.* Grune & Stratton, New York, London.

Scott ,N.A.; Harris, P.F. & Bagnal,l K.M. (1980). A morphological and histological study of the postnatal development of the intervertebral discs in the lumbar spine of the rabbit. *J Anat,* 130, 75-81.

Sebastine, I.M. & Williams, D.J. (2007). Current developments in tissue engineering of nucleus pulposus for the treatment of intervertebral disc degeneration. *Conf Proc IEEE Eng Med Biol Soc 2007,* 6400-6405.

Seidel, J.O.; Pei, M.; Gray, M.L.; et al. (2004). Long-term culture of tissue engineered cartilage in a perfused chamber with mechanical stimulation. *Biorheology,* 41, 445-458.

Shimer, A.L.; Chadderdon, R.C.; Gilbertson, L.G., et al. (2004). Gene therapy approaches for intervertebral disc degeneration. *Spine,* 29, 2770-2778.

Smith, N.R. (1931). The intervertbral discs. *Br J Surg,* 18, 358-375.

Smith, J.W. & Serafini-Fracassini A. (1968). The distribution of the protein- polysaccaride complex in the nucleus pulposus matrix in young rabbits. *J Cell Sci,* 3, 33-40.

Soukane, D.M.; Shirazi-Adl, A. & Urban, J.P. (2005). Analysis of nonlinear coupled diffusion of oxygen and lactic acid in intervertebral discs. *J Biomech Eng,* 127, 1121-1126.

Souter, W.A & Taylar, T.K.F. (1970). Salphated acid mucopolysaccharide metabolism in the rabbit intervertebral disc. *J Bone Joint Surg,* 52B, 371-384.

Stockwell, R. (1971). The inter-relationship of cell density and cartilage thickness in mammalian articular cartilage. *J Anat,* 109, 411-422.

Strayer, L.M. (1943). The embryology of the human hip joint. *Yale J Biol Med,* 11, 13-26.

Sun, Y.; Hurtig, M.; Pilliar, R.M.; et al. (2001). Characterization of nucleus pulposus-like tissue formed in vitro. *J Orthop Res,* 19, 1078-1084.

Sélard, E.; Shirazi-Adl, A. & Urban, J.P.G. (2003). Finite element study of nutrient diffusion in the human intervertebral disc. *J Biomech,* 28, 1954-1953.

Takeno, K.; Kobayashi, S.; Negoro, K.; et al. (2007). Physical limitations to tissue engineering of intervertabral disc cells: Effect of extracellular osmotic change on glycosaminoglycan production and cell metabolism. *J Neurosurg Spine,* 7, 637-644.

Taylor, J.R, & Twomey, L.T. (1988). The development of human intervertebral disc. In: *The Biology of the Intervertebral Disc.* Ghosh, P. (ed.) pp.39-82. CRC Press, Boca Raton, FL.

Thompson, J.P.; Oegema, T.J. & Bradford, D.S. (1991). Stimulation of mature canine intervertebral disc by growth factors. *Spine,* 16, 253-260.

Treppo, S.; Koepp, H.; Quan, EC.; et al. (2000). Comparison of biomechanical and biochemical properties of cartilage from human knee and ankle pairs. *J Orthop Res,* 18, 739-748..

Trout, J.J.; Buckwalter, J.A.; Moore, K.C.; et al. (1982a). Ultrastructure of the human intervertebral disc. I. Changes in notochordal cells with age. *Tissue Cell,* 14, 359-369.

Trout, J.J.; Buckwalter, J.A. & Moore, K.C. (1982b). Ultrastructure of the human intervertebral disc. II. Cells of the nucleus pulposus. *Anat Rec,* 204, 307-314..

Urban, J.P.G. & Maroudas, A. (1979). The measurement of fixed charge density in the intervertebral disc. *Biochim Biophys Acta,* 586, 166-178.

Urban, J.P.; Holm, S.; Maroudas, A.; et al. (1982). Nutrition of the intervertebral disc: Effect of fluid flow on solute transport. *Clin Orthop Relat Res,* 170, 296-302.

Urban, J.P.G & Roberts, S. (1995). Development and degeneration of the intervertebral discs. *Mol Med Today,* 1, 329-335.

Verbout, A.J. (1985). The development of the vertebral column. *Adv Anat Embryol Cell Biol,* 90, 1-122.

Verzijl, N.; DeGoot, J.; Thorpe, S.R.; et al. (2000). Effect of collagen turnover on the accumulation of advanced glycation end products. *J Biol Chem,* 275, 39027-39031.

Walmsley, R. (1953). The development and growth of the intervertebral disc. *Edinb Med J,* 60, 341-363.

Walsh, A.L.; Lotz, J. & Bradford, D. (2002). Single and multiple injections of GDF-5, IGF-1, or TGF beta into degenerated intervertebral discs. *Ortho Res Soc Trans,* 27, 820.

Williams, E.W. (1908). The latter development of the notochord in mammals. *Am J Anat ,* 8, 251-284.

Williams, G.M.; Klein, T.J. & Sah, R.L. (2005). Cell density alters matrix accumulation in two distinct fractions and the mechanical integrity of alginate-chondrocyte constructs. *Acta Biomater,* 1, 625-633.

Wilsman, N.J. (1970). The relationship of cartilage canals to the initial osteogenesis of secondary centers of ossification. *Anat Rec,* 168, 381-392.

Windhaber, R.A.; Wilkins, R.J. & Meredith, D. (2003). Functional characterisation of glucose transport in bovine articular chondrocytes. *Pflugers Arch,* 446: 572-577.

Yoon, S.T.; Kim, K.S.; Li, J.; et al. (2003a). The effect of bone morphogenetic protein-2 on rat intervertebral disc cells in vitro. *Spine,* 28, 1773-1780.

Yoon, S.T.; Li, J.; Kim, K.S.; et al. (2003b). BMP-2 stimulates disc cell matrix production and other BMPs. *Spine J,* 3, 144-145.

Yoshizawa, H.; Ohiwa, T.; Kubota, K.; (1986). Morphological study on the vertebral route for the nutrition of the intervertebral disc. *Neuro-Orthop,* 1, 17-32.

Ysart, G.E. & Mason, R.M. (1994). Responses of articular cartilage explant cultures to different oxygen tensions. *Biochim. Biophys Acta,*1221,15-20.

Zhou, S.; Cui, Z. & Urban, J.P. (2004). Factors influencing the oxygen concentration gradient from the synovial surface of articular cartilage to the cartilage-bone interface: a modeling study. *Arthritis Rheum,* 50, 3915-3924.

van der Kraan, P.M.; Buma, P.; van Kuppevelt, T.; et al. (2002). Interaction of chondrocytes, extracellular matrix and growth factors: relevance for articular cartilage tissue engineering. *Osteoarthritis Cartilage,* 10, 631-637.

von Luschka, H. (1858). *Die Halbgelenke des Menschlichen Körpers.* In: Reimer, G. (ed.), pp 1-144, Berlin.

13

Enamel Tissue Engineering

Masaki J. Honda[1] and Ken-ichiro Hata[2]
[1]*Department of Anatomy, Nihon University School of Dentistry*
[2]*Japan Tissue Engineering, Co. Ltd.*

1. Short description of the book chapter

Dental scientists are eagerly embracing regenerative medicine using tissue engineering technology. Enamel, one component of teeth, is the hardest tissue in the body. It cannot regenerate by itself, because it is formed by a layer of ameloblasts that is lost by the time the tooth crown has completely formed. Although the enamel is spent during lifetime, it is vulnerable to wear, damage, and decay. For these reasons, the development of a technique that produces artificially-grown enamel using culture and transplantation techniques is strongly desired. In this chapter we report a new technique for culturing enamel organ epithelial cells that have the capacity to produce enamel.

After completion of the formation of the tooth crown, which is composed of an enamel-dentin complex, tooth root formation begins. Generally, no new ameloblasts or enamel is generated from the enamel organ epithelial cells after the tooth root is formed. Epithelial cell rests of Malassez, derived from Hertwig's epithelial root sheath fragments, are located in the periodontal ligament, and recently we reported that these cells are also capable of enamel production using the same technique.

Dental epithelial cells including enamel organ epithelial cells and epithelial cell rests of Malassez can continue to proliferate when they are subcultured on top of feeder layer cells. These subcultured cells were placed onto collagen sponge scaffolds along with dental pulp cells and the constructs were transferred into the abdominal cavities of rats. When removed after 4 weeks, enamel production was observed in the scaffolds. The key finding of this study is that even after multiple divisions, the cells retained the capacity to produce enamel. Now that dental epithelial cells can be generated by a culture and transplantation system, the next step will be to combine the tissue-engineered enamel with the injured enamel in original teeth.

2. Abbreviations

EOE, enamel organ epithelial; ERM, epithelial cell rests of Malassez; HERS, Hertwig's epithelial root sheath; KLK4, kallikrein-4; MMP-20, enamelysin; PDL, periodontal ligament

3. Introduction

One purpose of studying the development and structure of enamel is to understand how it responds to enamel loss caused by caries or fractures. One of frequent dental disease is dental caries which is a specific infectious disease of the teeth that results in localized dissolution and destruction of the calcified enamel and dentin (1, 2). Enamel is the hardest tissue in the body and it does not regenerate because the cells that synthesize the enamel matrix are lost during tooth eruption. To replace enamel, dentistry has formulated artificial materials that mimic the hardness of enamel but replacing enamel with artificial substitutes may not be the most appropriate therapy. The establishment of a novel therapy to engineer natural enamel would provide an important approach to repair enamel loss.

Recently, tissue engineering therapy for tooth replacement has been addressed using various strategies and all such attempts share a common feature, which is that odontogenic capacity is necessary to generate a tooth structure. The first successful report described tooth structures such as the enamel-dentin complex generated in *vivo* by seeding dissociated odontogenic cells from a 6-month porcine third molar onto polyglycolic acid fiber mesh (3). The data suggested the presence of epithelial and mesenchymal progenitor/stem cells in the early stages of crown formation in the developing tooth. Furthermore, these results indicated a high likelihood of developing a technique to generate enamel. Importantly, this successful attempt provided one novel strategy to create natural enamel. In this chapter we describe our strategy to generate enamel using tissue engineering technology. However, many issues still remain to be resolved before we can apply the method to clinical practice.

So why is it difficult to generate enamel? There are two big obstacles. Firstly, no one has yet achieved successful tooth regeneration using subcultured odontogenic cells. One of the reasons is that it is difficult to expand the ameloblast-lineage cells that are capable of secreting amelogenin which is a constituent of enamel. Recently, dental tissues such as the periodontal ligament (PDL), dental papilla and dental follicle have been identified as easily accessible sources of undifferentiated cells. Dental mesenchymal stem cells could be feasible tools for dental tissue engineering. However, dental epithelial stem cells have not been discovered in teeth. Secondly, how can ameloblasts be harvested if they disappear when tooth-crown formation is completed? Practically this approach is obviously limited by the very small number of cells that can be obtained.

This article mainly reviews the novel therapy developed to generate natural enamel using a subculture and transplantation system, and the new cell sources for enamel regeneration in light of the two main difficulties mentioned above.

4. Tooth development

Tooth development is regulated by inductive signals between ectoderm-derived epithelium and neural crest-derived ectomesenchyme (4, 5). Teeth are initiated with an invagination from the oral epithelium. The oral epithelium at this stage has been shown to have the potential to instruct tooth development. Thereafter, the horseshoe-shaped dental lamina generates the future dental arches as the cap stage. During the cap and bell stages, transient signaling centers are generated in the epithelium as primary and secondary enamel knots. Secondary enamel knots appear at the site of the future cusp tips and contribute to the shaping of the tooth crown at the bell stage. At the crown formation stage, the enamel-secreting ameloblasts and dentin-secreting odontoblasts differentiate from the inner enamel epithelium and dental pulp, respectivelyand dental pulp which being derived from the dental papilla, respectively.. Both ameloblasts and odontoblasts are subsequently involved in the formation of tooth crowns, which consist of the enamel-dentin complex.

5. Enamel

Enamel, the most highly mineralized dental tissue, is composed of crystalline calcium phosphate and is approximately 96% mineral with the remaining 4% consisting of organic components and water. The basic micro-structural unit of enamel is an enamel rod, which is tightly packed and adherent to other enamel rods. The interwoven architecture of enamel rods contributes to both its strength and resistance to fracture.

Enamel is generated by ameloblasts which are epithelial cells derived from the enamel organ in the developing tooth, which covers the dentin-pulp complex. Differentiation of epithelial cells towards ameloblasts is regulated by an interaction between the epithelial and mesenchymal cells. The major difference between enamel and mucosa is the absence of any epithelial response because the cells that form enamel are lost at the time of tooth eruption, which means that enamel is neither replaceable nor repairable after eruption.

Amelogenin is produced by ameloblasts and is the most abundant protein within the enamel, accounting for more than 90% of the total enamel-related proteins. Studies in amelogenin knockout mice showed that amelogenin plays an important role in enamel biomineralization (6, 7), and while ameloblast differentiation is relatively normal in these mice, an abnormally thin enamel layer is formed (8). Amelogenins are essential for the formation of hydroxyapatite in enamel (9), and they are also recognized as a key factor in controlling the orientation and elongated growth of enamel rods during the mineralization process (10). In addition to amelogenin, the developing enamel is composed of several structural proteins such as ameloblastin (sheathlin/amelin), enamelin, and tuftelin. They are eventually degraded by the action of proteinases such as enamelysin (MMP-20) and kallikrein 4 (KLK4) (formerly known as enamel matrix serine proteinase 1). Ameloblastin is the most abundant non-amelogenin enamel-specific glycoprotein (11-13), and functions as a cell adhesion molecule for ameloblasts, but not for dental epithelial cells (14). Ameloblastin expression in ameloblasts peaks at the secretory stage and diminishes at the maturation stage, and maintenance of normal levels of ameloblastin in enamel formation is important because overexpression of ameloblastin in ameloblasts impairs the enamel structures (15). Furthermore, although dental epithelial cells can differentiate into ameloblasts in ameloblastin knockout mice, the ameloblasts detach from the matrix surface at the secretory stage and lose polarity (14), suggesting that ameloblastin is essential for maintaining normal ameloblast differentiation and attachment to the enamel matrix. In a recent study however , amelogenin and ameloblastin double knockout mice still showed a very thin layer of enamel, which was due to the presence of enamelin which was still expressed in the teeth of these double knockout mice. The enamelin gene has also been implicated in human amelogenesis imperfecta. Enamelin knockout mice do not form normal enamel, because of the lack of mineralization at the secretory surface of the ameloblasts (16), which suggests that enamelin is also necessary for enamel formation.

6. Enamel organ epithelial cell culture

Tissue engineering protocols generally involve the proliferation of tissue-specific stem cells in vitro that support development of a particular cellular phenotype (17). These cells are grown under specific culture conditions and can be transplanted into the body, in order to generate new tissue or organs (18). Recently, dental pulp stem cells isolated from post-natal pulp (19) and from deciduous tooth pulp (20) have been described. Dental mesenchymal stem cells have been highlighted for their ability to form dentin or periodontium (21), however, in contrast, the characterization of dental epithelial stem cells has proven to be more elusive because they are hard to grow in vitro. To date, it has proved difficult to maintain primary enamel organ epithelial (EOE) cells in cultures for prolonged periods with the primary phenotype (22). In fact, although the establishment of a primary culture system for porcine ameloblasts has been previously reported, the approach involved the transformation and immortalization of EOE cells with the transforming genes of simian virus 40 (23). Recently, a system for multiplication without transformation has been reported, using mouse ameloblasts as the cell source (24). We have tried to develop a strategy using only culture techniques to grow EOE cells to generate enamel however there are two obstacles that must be overcome in order to successfully grow EOE cells. The first obstacle is that porcine epithelial cell growth is disrupted if contaminated with dental mesenchymal cells, due to the high proliferation capacity of the mesenchymal cells (22, 25). Secondly, expansion of porcine dental epithelial cells in vitro has been problematic due to their tendency to terminally differentiate after only a few passages (25). In order to overcome these obstacles, we combined two techniques: a non-serum culture system and a 3T3-J2 feeder layer system, derived from a mouse teratoma (Fig. 1). Den Besten and colleagues had previously reported that a non-serum selective medium (LHC-9) enables the growth of only EOE cells while the accompanying mesenchymal cells are killed (22), however one disadvantage of this non-serum selective medium is the low growth potential of the EOE cells. We were encouraged by reports of the adaptation of a feeder layer technique for the culture of keratinocytes which enabled the long-term culture of these cells during which time they retained an epithelial phenotype (26, 27) and we considered that the disadvantages of a non-serum medium could be overcome using the feeder layer technique. Then, to determine whether the feeder layer system would be effective for EOE cells, enamel organs were harvested from the third molar teeth of a 6-month-old-pig (Fig. 2). Since enamel organs are connected to dental follicles, primary cultures of the EOE cells appeared to be contaminated with dental follicle cells, as we expected. The EOE cells were isolated from the mesenchymal cells using the non-serum medium, and then seeded onto a 3T3-J2 feeder layer for sequential subcultures (Fig. 3a). The EOE cell colonies grew with time in culture (Fig. 3b) and were able to be maintained for at least 4 passages. The subcultured EOE cells from both the primary culture and after 4 passages expressed amelogenin, ameloblastin, MMP-20 (28), KLK4 (29), and enamelin (30), but not alkaline phosphatase (Fig. 4a). No obvious changes in the expression of ameloblast-related genes were detected during long-term culture. Furthermore, expression of amelogenin in subcultured EOE cells was shown by fluorescence immunocytochemistry (Fig. 4b), and Western blotting revealed secretion of amelogenin by the EOE cells into the culture medium.

By using 3T3-J2 cells as a feeder layer, we have successfully overcome the limitations of odontogenic epithelial cell culture. The passaged cells express the ameloblast-specific markers amelogenin, ameloblastin and enamelysin, but not enamelin. The culturing of epithelial cells on a feeder layer of 3T3 cells has proven to be a major advance, because this provides an environment that considerably improves cell proliferation (26, 27). The development and characterization of these cultured epithelial cells is also useful for further studies of enamel tissue engineering.

7. Enamel tissue engineering

Developing a technique to manipulate EOE cells is a significant advance towards enamel replacement and therefore we attempted to develop a strategy to generate enamel based on subcultured EOE cells using tissue engineering technology. To examine the enamel-forming capability of subcultured EOE cells, we made use of techniques we had developed previously by transplanting cells onto a biodegradable scaffold in vivo (31-33). Newly generated tooth tissue was formed by tissue engineering techniques requiring an interaction between dental mesenchymal cells and EOE cells (31). As previously described, the constructs were prepared by isolating fresh dental pulp cells from the third molars of pigs during the early stage of crown formation (32). Dental pulp cells were first plated on top of a scaffold and then subcultured EOE cells were seeded directly on top of the pulp cells. In the primary control group, the scaffolds were seeded with subcultured EOE cells alone. In the secondary control group, oral epithelial cells obtained from the oral mucosa were subcultured (34) and seeded in combination with fresh dental pulp cells onto the scaffolds. We had previously determined that the use of a collagen sponge was superior to that of polyglycolic acid fiber mesh (Nipro Corporation, Japan), and so collagen sponges were used as scaffolds for the dental cells (35). Interestingly, histological analysis showed subcultured EOE cells gave rise to ectopic cartilage tissue under the primary control group conditions (36), while the implants grown under the conditions of the secondary control group not only generated enamel but also cartilage. Amelogenin gene splice products have the ability to enhance the transcription factor sox9 (37, 38), therefore, amelogenin secreted from the cultured EOE cells may induce chondrocytes from the mesenchymal stem cells in the surrounding omentum of the rat host. Four weeks after transplantation of EOE cells combined with dental pulp cells in scaffolds, several phenomena related to amelogenesis were distinguished in the implants. In the most mature structures, enamel was readily found in the implants (Fig. 5a). Furthermore, amelogenin immunoreactivity was detected in tall columnar epithelial cells on the surface of the dentin or enamel, indicating that the tissue-engineered enamel contains well-developed ameloblasts (Fig. 5b). Together, these results indicate that the subcultured EOE cells have the potential to generate enamel. Interestingly, enamel formation was always observed after dentin was formed in the implants. On the other hand, when subcultured EOE cells were combined with subcultured dental pulp cells, enamel-dentin complexes were not observed in any of the implants. In addition, the production of enamel by EOE cells was observed within 4 weeks of transplantation of the constructs (33), which was markedly earlier than that observed with constructs established from non-cultured EOE cells which generated enamel at 15-25 weeks after transplantation (3, 31, 39). Enamel production may have been facilitated in our culture model because EOE cells were maintained at an undifferentiated stage in the ameloblast-lineage cell phenotype by the 3T3 feeder layer. Our culture model provides a promising step towards a new therapy for reforming enamel but further study is needed on how to combine the newly-generated enamel with the original dental dentin or enamel in the tooth. As EOE cells disappear in adult teeth after tooth eruption we also need to discover new cell sources to advance this technology in the clinical setting.

8. New cell sources for enamel tissue engineering

8.1 Epithelial cell rest of Malassez

The epithelium surrounding the teeth after eruption is anatomically classified into 4 types of epithelium; the oral gingival epithelium, the oral sulcular epithelium, the junctional epithelium, and the epithelial cell rests of Malassez (ERM). The epithelium, except for ERM, covers the surface, while ERM, developmental residues of Hertwig's epithelial root sheath (HERS), are embedded within the PDL tissue. Since ERM are a direct lineage from HERS, derived from the enamel organ through the cervical loop structures, we hypothesized that ERM retain their original ability to secrete a matrix conducive to generating enamel in a fashion similar to the enamel organ in the crown formation stage. Previously, there was a report supporting our hypothesis showing that removal of stem cells from their natural milieu may change their differentiation properties. A well-studied example of this phenomenon involves hematopoietic stem cells. Furthermore, we recently reported that the epithelial cells from HERS can differentiate into ameloblasts and produce enamel-dentin complexes when combined with non-cultured dental pulp cells in the core of the dental pulp (40).

The epithelial cell rests of Malassez in the periodontium were thought to be completely quiescent (41, 42), however it is now known that ERM can be induced to differentiate into cementoblasts (43, 44) with disruption of Patched, the hedgehog receptor, leading to shorter roots (45). Moreover, it has been suggested that ERM might prevent resorption of the root surface (46) and although ERM are generally in the G0 phase of the cell cycle, ERM can be stimulated to proliferate in response to injury (47). ERM also have the potential to differentiate in two different directions: squamous metaplasia and formation of cystic lesions, and ameloblastic differentiation during the formation of odontogenic tumors (48, 49), suggesting that ERM have a mitotic activity (50, 51). In fact, when the ERM cells are cultured in vitro, they actively proliferate but there have been no reports of the in vivo behavior of subcultured ERM cells. To investigate the potential of ERM cells for generating enamel, we established a cell culture system for ERM cells in order to preserve as many of the original cellular characteristics as possible, and then designed a study to examine the tissue-forming capability of ERM cells(53).

Firstly, we examined the presence of ERM cells in deciduous incisor teeth. In histological sections, clusters of aggregated cells were easily recognized in the PDL of porcine deciduous incisor teeth. The clustered cells expressed cytokeratin-14, an epithelial cell marker. Interestingly, immunolabeling for amelogenin was not observed in any of the epithelial cell clusters.

We then prepared three types of epithelial cells (ERM, EOE cells, and oral gingival epithelial cells) and PDL cells from the same mandible of a 6-month-old pig. The deciduous incisor teeth were removed from the mandible and then the PDLs were removed from the middle one-third of the incisor tooth roots. The PDLs were placed in culture dishes and then ERM cells and PDL cells migrated from the explants. Before the cells reached confluence, the serum containing growth medium was replaced with a non-serum epithelial cell-selective medium. After replacement with the non-serum medium, the PDL cells were detached from the culture dish and the ERM cells subsequently survived. The growth of the ERM cells in non-serum medium was quite slow, similar to the behavior of EOE cells. ERM cells were then subcultured on 3T3-J2 feeder cell layers with complete-MEM medium, the same reagents used in the culture of EOE cells (36) and oral gingival epithelial cells (52). The ERM cells colonies increased with time and became confluent after 2 weeks. Subcultured ERM cells were passaged and this was repeated ten times with feeder cell layers. All passaged cells showed the typical polygonal-shaped epithelial morphology. Interestingly, when ERM were subcultured on collagen type I coated without a feeder cell layer, the ERM cells displayed an extended morphology.

We compared the cell characteristics of the subcultured ERM, EOE cells, and oral gingival epithelial cells, produced using the same feeder cell layer technique. No appreciable differences were found in any of the cell morphologies (Fig. 6a-c), and all cell populations were able to grow on the feeder cell layer at similar growth rates. Gene expression patterns of the subcultured ERM were then compared to EOE cells using ameloblast-related markers. Both ERM and EOE cells expressed ameloblastin and tuftelin, while EOE cells alone expressed amelogenin, MMP20, and KLK4. None of the cell populations expressed enamelin. The expression patterns of the ameloblast-related genes in ERM were inconsistent with those in EOE cells. In particular, subcultured ERM did not show expression of amelogenin as well as the histological observation of the root surface of deciduous teeth. These results suggest ERM would not have the potential to grow enamel.

The possibility that ERM cells can differentiate into ameloblasts was examined in vitro using a co-culture model with dental pulp cells. Presumably, dental pulp has the potential to induce dental epithelial cells into ameloblasts (40). Surprisingly, although no immunoreactivity for amelogenin was observed in the ERM cells alone over a 4-week culture period (Fig. 7a-c), when ERM cells were co-cultured with dental pulp cells, both cytokeratin-14 and amelogenin were clearly positive (Fig. 7d-f). Since there was now a possibility of ERM differentiating into ameloblasts, the capacity of subcultured ERM cells to produce enamel was examined by transplanting subcultured ERM cells, seeded onto scaffolds, into the omentum of athymic rats. Histological analysis showed that all implants displayed the appropriate stages of amelogenesis from initiation to maturation. At 8 weeks post-transplantation, enamel-dentin complex structures were recognized in the implants (Fig. 8a). At high magnification, the width of the enamel was approximately 100 μm (Fig. 8b). Amelogenin expression was also detected in the region of the tall columnar cells. These results demonstrate that subcultured ERM have the potential to differentiate into ameloblasts and generate enamel in vivo, suggesting that ERM could be a new cell source for tissue engineering techniques investigating enamel replacement in the future (53).

Generally, tissue-specific stem cells are thought to be activated by a repair and regeneration process and they are maintained in a quiescent state until they are required to act on behalf of the tissue. The normal behavior of stem cells is strictly governed by the cellular microenvironment. ERM are able to grow in vitro and become free of the influence of the PDL cells. It could be that ERM have the characteristics of epithelial stem/progenitor cells.

8.2 Bone marrow stroma cells

It has been known that bone marrow contains stem cells that can give rise to various types of epithelial cells. One experimental study was designed to determine whether bone marrow cells could differentiate into ameloblasts (55) and the success of this study provides a new cell source for enamel tissue engineering. Basically, bone marrow contains two types of stem cell - hematopoietic and mesenchymal. The ability of hematopoietic stem cells to give rise to epithelial cells has already been shown for tissues such as skin, lung, and liver. However, although bone marrow cells have the potential to differentiate into dental mesenchymal cells (56), their ability to differentiate into ameloblasts has never been examined. Bone marrow cells were collected from tibias and femurs of CD1 mice and isolated using c-kit, a cytokine tyrosine kinase receptor expressed by hematopoietic progenitor cells. Dental epithelium was dissociated into single cells and the single cell suspension was mixed with the bone marrow cell suspension to form a pellet. The pellet was re-associated with dental mesenchyme obtained from the tooth germ at embryonic day 14. After 20 days of culture, a tooth crown was generated from the constructs. The c-kit-positive bone marrow cells were present in the inner dental epithelium and expressed both amelogenin and ameloblastin, as shown by in situ hybridization (53). This was the first report to show that bone marrow cells can differentiate into ameloblasts. However, so far, no studies have shown that bone marrow cells alone, without dental epithelial cells, can differentiate into ameloblasts.

8.3 Oral keratinocytes

At an early stage the embryonic oral epithelium provides the instructive signals for tooth initiation and development. These signals are received by the mesenchymal cells, which are then primed to become odontogenic and in turn act as a source of reciprocal signals back to the epithelium. In mice, dental epithelium prior to embryonic day 12.5 has these instructions, and then at this stage, the ability shifts to the dental mesenchyme from the epithelium. Of note, non-dental embryonic epithelium can differentiate into dental epithelium when there is interaction with odontogenic mesenchyme (55). Thus it is possible that non-dental epithelial cells may be a new cell source for enamel tissue engineering technology.

Of particular interest is whether post-natal non-dental oral epithelium can differentiate into ameloblasts to generate enamel. A recent study (57) designed to investigate this question obtained oral epithelium from the palatal epithelium of ICR mice at post-natal day 1 and the palatal epithelial cells were separated from the mesenchyme. Mandibular molar teeth from embryonic day 12.5-16.5 were isolated and dental mesenchyme was separated from the dental epithelium. The dissociated epithelial cells were combined with the mandibular molar mesenchyme for re-association. The epithelium covered the entire surface of the molar mesenchyme. After 2 days of culture, the re-associations were transplanted under the kidney capsule of an adult mouse as the host. Samples were obtained 14 days after transplantation and analyzed by histology. The re-association of palatal epithelium with dental mesenchyme formed tooth germs associated with enamel-dentin complexes (58). Therefore palatal mucosal epithelium, not only from embryos but also from newborn mice, has the potential to differentiate into ameloblasts and then generate enamel. However, the palatal epithelia of mice more than 2 days of age could not differentiate into ameloblasts. As we prefer to use older cells for enamel tissue engineering, clearly further studies are needed to resolve this problem for the clinical setting.

9. Conclusions and future directions

For a long time metal and resin have been investigated as potential sources to replace enamel. In our laboratory, we are focusing on producing enamel by tissue engineering methods as these techniques are believed to be an extremely powerful approach to replacing enamel. This goal has remained elusive until recently when we finally identified a new tissue engineering method to generate enamel using a specialized enamel organ culture technique and transplantation system. However, two problems still remain to be solved before this technique can be used for clinical applications. Firstly, there is still no method to combine the tissue-engineered enamel with the injured enamel of the original teeth. Secondly, we still cannot control the shape and size of the tissue-engineered enamel. If these two above issues could be addressed then a method to reform the injured enamel on teeth could be developed. The authors thank Dr. Haward Green (Harvard Medical School) for critical reading of the manuscript.

10. Acknowledgments

This work was supported by Grants-in-Aid for Scientific Research from the Ministry of Education, Science, Sports and Culture, Japan. We thank all laboratory staff who helped with this work. The authors thank Dr. Haward Green (Harvard Medical School) for critical reading of the manuscript.

11. References

1. Bergenholtz G. Inflammatory response of the dental pulp to bacterial irritation. *J Endod.* 1981 Mar;7(3):100-104.
2. Smith AJ. Pulpal responses to caries and dental repair. *Caries Res.* 2002 Jul-Aug;36(4):223-232.
3. Young CS, Terada S, Vacanti JP, Honda M, Bartlett JD, Yelick PC. Tissue engineering of complex tooth structures on biodegradable polymer scaffolds. *J Dent Res.* 2002 Oct;81(10):695-700.
4. Miletich I, Sharpe PT. Normal and abnormal dental development. *Hum Mol Genet.* 2003 Apr 1;12 Spec No 1:R69-73.
5. Thesleff I, Sharpe P. Signalling networks regulating dental development. *Mech Dev.* 1997 Oct;67(2):111-123.
6. Gibson CW, Yuan ZA, Hall B, Longenecker G, Chen E, Thyagarajan T, et al. Amelogenin-deficient mice display an amelogenesis imperfecta phenotype. *J Biol Chem.* 2001 Aug 24;276(34):31871-31875.
7. Hatakeyama J, Sreenath T, Hatakeyama Y, Thyagarajan T, Shum L, Gibson CW, et al. The receptor activator of nuclear factor-kappa B ligand-mediated osteoclastogenic pathway is elevated in amelogenin-null mice. *J Biol Chem.* 2003 Sep 12;278(37):35743-35748.
8. Gibson CW, Kulkarni AB, Wright JT. The use of animal models to explore amelogenin variants in amelogenesis imperfecta. *Cells Tissues Organs.* 2005;181(3-4):196-201.

9. Li Y, Suggs C, Wright JT, Yuan ZA, Aragon M, Fong H, et al. Partial rescue of the amelogenin null dental enamel phenotype. *J Biol Chem*. 2008 May 30;283(22):15056-15062.
10. Du C, Falini G, Fermani S, Abbott C, Moradian-Oldak J. Supramolecular assembly of amelogenin nanospheres into birefringent microribbons. *Science*. 2005 Mar 4;307(5714):1450-1454.
11. Cerny R, Slaby I, Hammarstrom L, Wurtz T. A novel gene expressed in rat ameloblasts codes for proteins with cell binding domains. *J Bone Miner Res*. 1996 Jul;11(7):883-891.
12. Krebsbach PH, Lee SK, Matsuki Y, Kozak CA, Yamada KM, Yamada Y. Full-length sequence, localization, and chromosomal mapping of ameloblastin. A novel tooth-specific gene. *J Biol Chem*. 1996 Feb 23;271(8):4431-4435.
13. Fong CD, Cerny R, Hammarstrom L, Slaby I. Sequential expression of an amelin gene in mesenchymal and epithelial cells during odontogenesis in rats. *Eur J Oral Sci*. 1998 Jan;106 Suppl 1:324-330.
14. Fukumoto S, Kiba T, Hall B, Iehara N, Nakamura T, Longenecker G, et al. Ameloblastin is a cell adhesion molecule required for maintaining the differentiation state of ameloblasts. *J Cell Biol*. 2004 Dec 6;167(5):973-983.
15. Paine ML, Wang HJ, Luo W, Krebsbach PH, Snead ML. A transgenic animal model resembling amelogenesis imperfecta related to ameloblastin overexpression. *J Biol Chem*. 2003 May 23;278(21):19447-19452.
16. Hu JC, Hu Y, Smith CE, McKee MD, Wright JT, Yamakoshi Y, et al. Enamel defects and ameloblast-specific expression in Enam knock-out/lacz knock-in mice. *J Biol Chem*. 2008 Apr 18;283(16):10858-10871.
17. Honda M, Yada T, Ueda M, Kimata K. Cartilage formation by cultured chondrocytes in a new scaffold made of poly(L-lactide-epsilon-caprolactone) sponge. *J Oral Maxillofac Surg*. 2000 Jul;58(7):767-775.
18. Honda MJ, Yada T, Ueda M, Kimata K. Cartilage formation by serial passaged cultured chondrocytes in a new scaffold: hybrid 75:25 poly(L-lactide-epsilon-caprolactone) sponge. *J Oral Maxillofac Surg*. 2004 Dec;62(12):1510-1516.
19. Gronthos S, Mankani M, Brahim J, Robey PG, Shi S. Postnatal human dental pulp stem cells (DPSCs) in vitro and in vivo. *Proc Natl Acad Sci U S A*. 2000 Dec 5;97(25):13625-13630.
20. Miura M, Gronthos S, Zhao M, Lu B, Fisher LW, Robey PG, et al. SHED: stem cells from human exfoliated deciduous teeth. *Proc Natl Acad Sci U S A*. 2003 May 13;100(10):5807-5812.
21. Seo BM, Miura M, Gronthos S, Bartold PM, Batouli S, Brahim J, et al. Investigation of multipotent postnatal stem cells from human periodontal ligament. *Lancet*. 2004 Jul 10-16;364(9429):149-155.
22. Den Besten PK, Mathews CH, Gao C, Li W. Primary culture and characterization of enamel organ epithelial cells. *Connect Tissue Res*. 1998;38(1-4):3-8; discussion 35-41.
23. DenBesten PK, Gao C, Li W, Mathews CH, Gruenert DC. Development and characterization of an SV40 immortalized porcine ameloblast-like cell line. *Eur J Oral Sci*. 1999 Aug;107(4):276-281.
24. Nakata A, Kameda T, Nagai H, Ikegami K, Duan Y, Terada K, et al. Establishment and characterization of a spontaneously immortalized mouse ameloblast-lineage cell line. *Biochem Biophys Res Commun*. 2003 Sep 5;308(4):834-839.

25. Kukita A, Harada H, Kukita T, Inai T, Matsuhashi S, Kurisu K. Primary and secondary culture of rat ameloblasts in serum-free medium. *Calcif Tissue Int.* 1992 Nov;51(5):393-398.
26. Rheinwald JG, Green H. Serial cultivation of strains of human epidermal keratinocytes: the formation of keratinizing colonies from single cells. *Cell.* 1975 Nov;6(3):331-343.
27. Rheinwald JG, Green H. Formation of a keratinizing epithelium in culture by a cloned cell line derived from a teratoma. *Cell.* 1975 Nov;6(3):317-330.
28. Fukae M, Tanabe T, Uchida T, Lee SK, Ryu OH, Murakami C, et al. Enamelysin (matrix metalloproteinase-20): localization in the developing tooth and effects of pH and calcium on amelogenin hydrolysis. *J Dent Res.* 1998 Aug;77(8):1580-1588.
29. Hu JC, Ryu OH, Chen JJ, Uchida T, Wakida K, Murakami C, et al. Localization of EMSP1 expression during tooth formation and cloning of mouse cDNA. *J Dent Res.* 2000 Jan;79(1):70-76.
30. Hu JC, Zhang CH, Yang Y, Karrman-Mardh C, Forsman-Semb K, Simmer JP. Cloning and characterization of the mouse and human enamelin genes. *J Dent Res.* 2001 Mar;80(3):898-902.
31. Honda MJ, Sumita Y, Kagami H, Ueda M. Histological and immunohistochemical studies of tissue engineered odontogenesis. *Arch Histol Cytol.* 2005 Jun;68(2):89-101.
32. Honda MJ, Tsuchiya S, Sumita Y, Sagara H, Ueda M. The sequential seeding of epithelial and mesenchymal cells for tissue-engineered tooth regeneration. *Biomaterials.* 2007 Feb;28(4):680-689.
33. Honda MJ, Shinohara Y, Hata KI, Ueda M. Subcultured odontogenic epithelial cells in combination with dental mesenchymal cells produce enamel-dentin-like complex structures. *Cell Transplant.* 2007;16(8):833-847.
34. Hata K, Kagami H, Ueda M, Torii S, Matsuyama M. The characteristics of cultured mucosal cell sheet as a material for grafting; comparison with cultured epidermal cell sheet. *Ann Plast Surg.* 1995 May;34(5):530-538.
35. Sumita Y, Honda MJ, Ohara T, Tsuchiya S, Sagara H, Kagami H, et al. Performance of collagen sponge as a 3-D scaffold for tooth-tissue engineering. *Biomaterials.* 2006 Jun;27(17):3238-3248.
36. Honda MJ, Shimodaira T, Ogaeri T, Shinohara Y, Hata K, Ueda M. A novel culture system for porcine odontogenic epithelial cells using a feeder layer. *Arch Oral Biol.* 2006 Apr;51(4):282-290.
37. Veis A, Tompkins K, Alvares K, Wei K, Wang L, Wang XS, et al. Specific amelogenin gene splice products have signaling effects on cells in culture and in implants in vivo. *J Biol Chem.* 2000 Dec 29;275(52):41263-41272.
38. Nebgen DR, Inoue H, Sabsay B, Wei K, Ho CS, Veis A. Identification of the chondrogenic-inducing activity from bovine dentin (bCIA) as a low-molecular-mass amelogenin polypeptide. *J Dent Res.* 1999 Sep;78(9):1484-1494.
39. Honda MJ, Shinohara Y, Sumita Y, Tonomura A, Kagami H, Ueda M. Shear stress facilitates tissue-engineered odontogenesis. *Bone.* 2006 Jul;39(1):125-133.
40. Sumita Y, Tsuchiya S, Asahina I, Kagami H, Honda MJ. The location and characteristics of two populations of dental pulp cells affect tooth development. *Eur J Oral Sci.* 2009 Apr;117(2):113-121.

41. Grupe HE, Jr., ten Cate AR, Zander HA. A histochemical and radiobiological study of in vitro and in vivo human epithelial cell rest proliferation. *Arch Oral Biol.* 1967 Dec;12(12):1321-1329.
42. Valderhaug JP, Nylen MU. Function of epithelial rests as suggested by their ultrastructure. *J Periodontal Res.* 1966;1:69-78.
43. Thomas HF. Root formation. *Int J Dev Biol.* 1995 Feb;39(1):231-237.
44. Bosshardt DD, Schroeder HE. Cementogenesis reviewed: a comparison between human premolars and rodent molars. *Anat Rec.* 1996 Jun;245(2):267-292.
45. Nakatomi M, Morita I, Eto K, Ota MS. Sonic hedgehog signaling is important in tooth root development. *J Dent Res.* 2006 May;85(5):427-431.
46. Loe H, Waerhaug J. Experimental replantation of teeth in dogs and monkeys. *Arch Oral Biol.* 1961 Apr;3:176-184.
47. Talic NF, Evans CA, Daniel JC, Zaki AE. Proliferation of epithelial rests of Malassez during experimental tooth movement. *Am J Orthod Dentofacial Orthop.* 2003 May;123(5):527-533.
48. Buchner A, Sciubba JJ. Peripheral epithelial odontogenic tumors: a review. *Oral Surg Oral Med Oral Pathol.* 1987 Jun;63(6):688-697.
49. Hamamoto Y, Hamamoto N, Nakajima T, Ozawa H. Morphological changes of epithelial rests of Malassez in rat molars induced by local administration of N-methylnitrosourea. *Arch Oral Biol.* 1998 Nov;43(11):899-906.
50. McCulloch CA, Melcher AH. Continuous labelling of the periodontal ligament of mice. *J Periodontal Res.* 1983 May;18(3):231-241.
51. Trowbridge HO, Shibata F. Mitotic activity in epithelial rests of Malassez. *Periodontics.* 1967 May-Jun;5(3):109-112.
52. Hata K, Ueda M. Fabrication of cultured epithelium using oral mucosal cells and its clinical applications. *Hum Cell.* 1996 Mar;9(1):91-96.
53. Hu B, Unda F, Bopp-Kuchler S, Jimenez L, Wang XJ, Haikel Y, et al. Bone marrow cells can give rise to ameloblast-like cells. *J Dent Res.* 2006 May;85(5):416-421.
54. Shinmura Y, Tsuchiya S, Hata KI, Honda MJ. Quiescent epithelial cell rests of Malassez can differentiate into ameloblast-like cells. J Cell Physiol. 2008; 217: 728-738.
55. Ohazama A, Modino SA, Miletich I, Sharpe PT. Stem-cell-based tissue engineering of murine teeth. *J Dent Res.* 2004 Jul;83(7):518-522.
56. Mina M, Kollar EJ. The induction of odontogenesis in non-dental mesenchyme combined with early murine mandibular arch epithelium. *Arch Oral Biol.* 1987;32(2):123-127.
57. Nakagawa E, Itoh T, Yoshie H, Satokata I. Odontogenic potential of post-natal oral mucosal epithelium. *J Dent Res.* 2009 Mar;88(3):219-223.

Figure legends

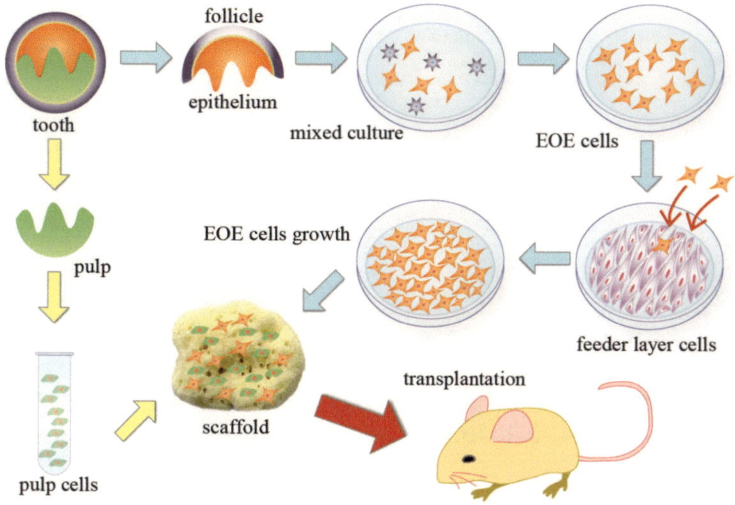

Fig. 1. A strategy for tissue-engineered enamel.

First, enamel organ epithelium was isolated from dental pulp and dental follicles. Enamel organ epithelial (EOE) cells were isolated from mesenchymal cells in culture using a non-serum medium. EOE cells were subcultured on the 3T3-J2 feeder layer cells. Thereafter, subcultured EOE cells were combined with primary dental pulp cells and the constructs were transplanted into the omentum of immunodeficient rats.

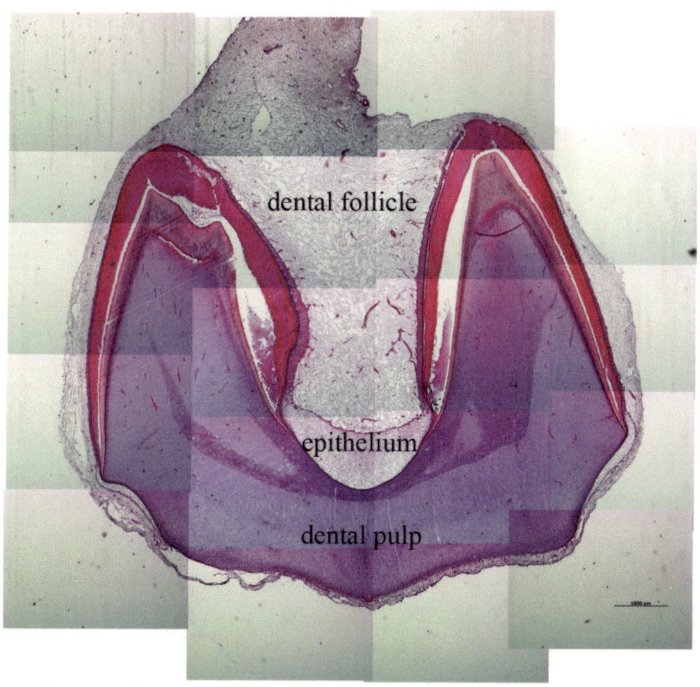

Fig. 2. A third molar tooth stained with hematoxylin and eosin (H & E). Tooth crown

formation is partly achieved. Enamel organ epithelium is connected with the dental follicle.

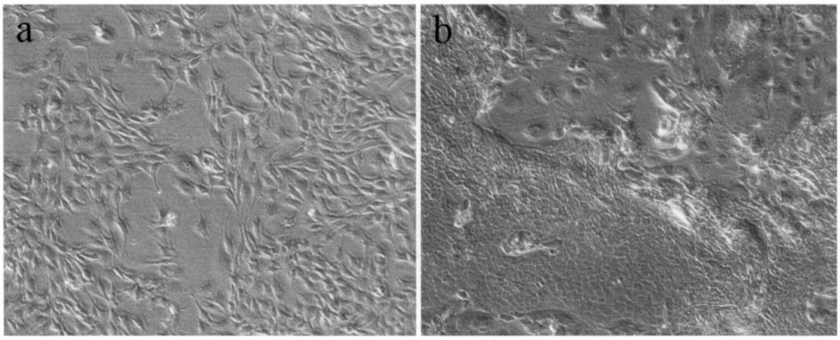

Fig. 3. a: Phase contrast micrograph of 3T3-J2 feeder layer cells.
b: The isolated enamel organ epithelial (EOE) cells were seeded onto the 3T3-J2 feeder layer and the EOE cell colonies grew with time in culture.

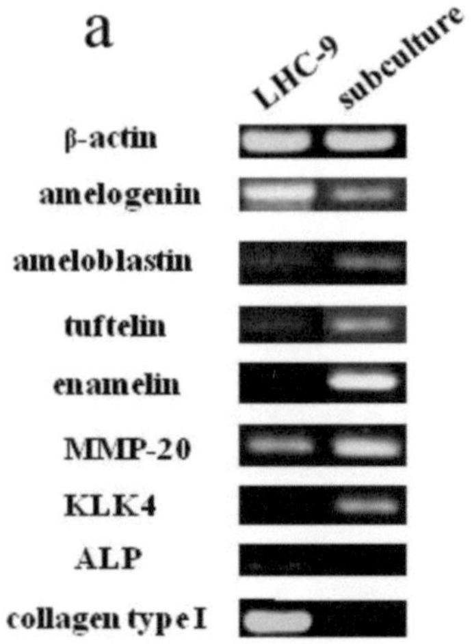

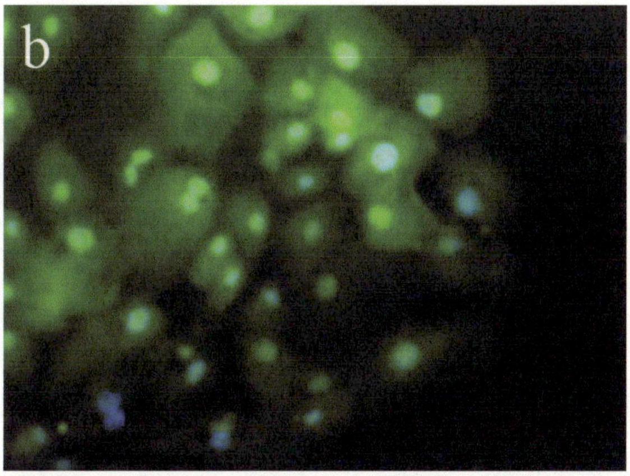

Fig. 4. a: The subcultured enamel organ epithelial (EOE) cells from both the primary culture and after 4 passages expressed amelogenin, ameloblastin, MMP-20, kallikrein-4 (KLK4), and enamelin, but not alkaline phosphatase.
b: Fluorescence immunocytochemistry showing the expression of amelogenin in subcultured EOE cells at passage 2.

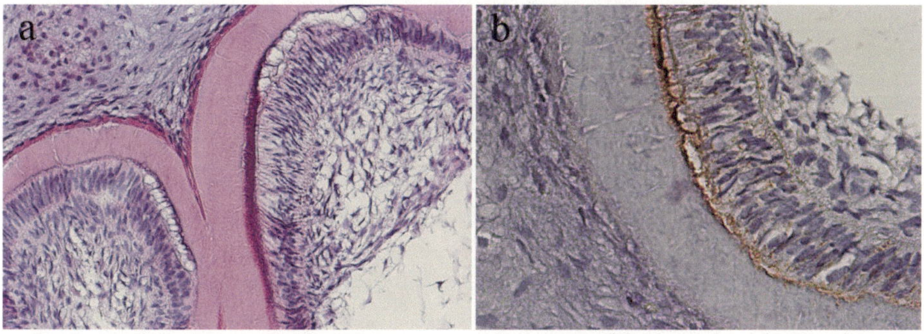

Fig. 5. a: Four weeks after transplantation, enamel was readily found in the implants.
b: Amelogenin immunoreactivity was detected in tall columnar epithelial cells on the surface of the dentin or enamel.

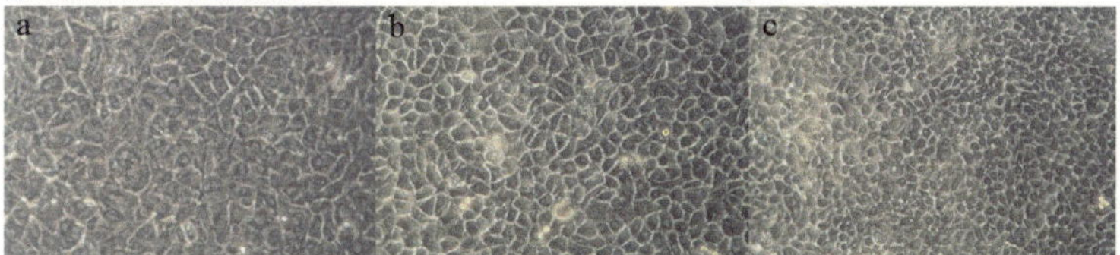

Fig. 6. a-c: There was no appreciable difference in cell morphology among epithelial cell rests of Malassez (ERM) (a), enamel organ epithelial (EOE) cells (b), and oral gingival epithelial cells (c).

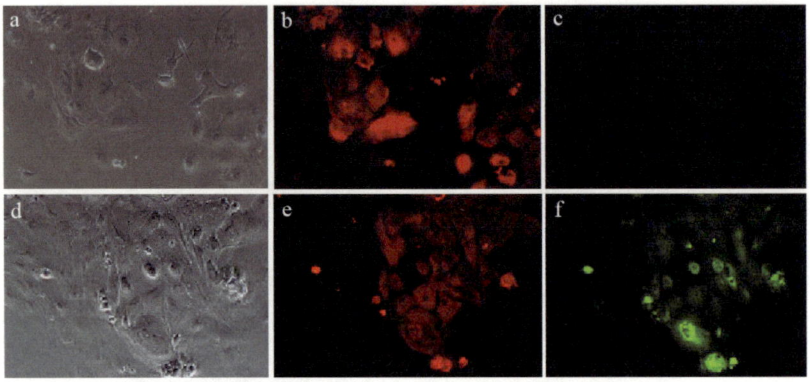

Fig. 7. Cytokeratin14 (CK14) and amelogenin expressions in vitro
a-c: When ERM were co-cultured with 3T3-J2 cells (a), immunofluorescence showed that ERM were positive for cytokeratin-14 (b) and were negative for amelogenin (c).
d-f: When ERM were co-cultured with primary dental pulp cells (d), both cytokeratin-14 (e) and amelogenin (f) were clearly positive.

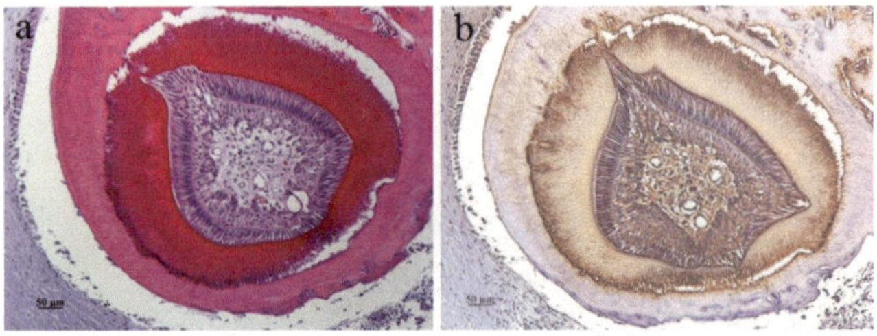

Fig. 8. a: At 8 weeks post-transplantation, enamel-dentin complex structures were recognized in the implants.
b: At high magnification view, the width of the enamel was approximately 100 μm.

14

Design of injectable bone tissue engineering scaffold consists of ß-tricalcium phosphate beads and alginate

Tomonori Matsuno
Department of Oral and Maxillofacial Surgery, School of Life Dentistry at Tokyo, The Nippon Dental University
Japan

1. Introduction

Autologous bone is generally considered the gold-standard graft material. The advantage of an autograft is that it contains viable osteoblasts and osteogenic precursor cells that can contribute to the formation of new bone (Arrington et al, 1996). In addition, the autograft possesses the three essential elements of bone regeneration -osteogenesis, osteoinduction, and osteoconduction- that are required for bone regeneration. However, only a minimal amount of bone tissue can be harvested for autografts, the harvesting procedure may lead to donor site discomfort and morbidity, and it may be difficult to form this tissue into the desired shape (Goldberg & Stevenson, 1987; Damien & Parsons, 1991), a problem that is particularly important in the craniofacial region.
Therefore, bone graft substitutes have been used to reconstruct bone defects. A bone graft substitute should be osteoconductive, osteoinductive, biocompatible, biodegradable, structurally similar to bone, easy to use, and cost-effective (Giannoudis et al, 2005). Hydroxyapatite (HA) and β-tricalcium phosphate (β-TCP) are both well-known ceramics that possess high tissue compatibility and osteoconductivity. However, neither HA nor β-TCP has osteoinductive or osteogenic abilities, and HA usually shows minimal biodegradation (Spector, 1994; Schmitz et al, 1999).
To overcome these limitations, bone tissue engineering has been promoted as a new way to regenerate bone tissue. This approach combines cells capable of osteogenic activity and osteoinductive signal molecules with an appropriate material (Livingston et al, 2005). For bone tissue engineering to succeed, osteoconductive scaffolding biomaterials must provide a suitable environment for the cells. Furthermore, it is desirable that the scaffolds can control the release of growth factors. Accordingly, functional composite scaffolds for bone tissue engineering have been developed in combination with synthetic ceramics and natural polymers.
Recently, minimally invasive treatments have been developed using an injectable system for bone tissue engineering. Several injectable gels have been used to carry cells in order to engineer bone. Amongst which are collagen (Tsuchida et al, 2003), alginate (Shang et al,

2001; Wang et al 2003), and fibrin (Perka et al 2001) gels, but these substances cannot be molded to the shapes of the bone defects when injected *in situ* or that they cannot follow the shapes of 3D cell culture molds *in vitro*.
In this chapter, I introduce the design of injectable composite scaffold fabricated from β-TCP beads and alginate for bone tissue engineering (Matsuno, 2008). The β-TCP bead/alginate composite is not only injectable, but also instantaneously formed from β-TCP beads and alginate. Therefore, the composite is sufficiently adaptable to the irregularities of bone defects and which possessed adequate mechanical strength. Additionally, the composite act as an appropriate 3D scaffold for osteogenic cells, and function as drug delivery carrier of growth factors.

2. Preparation of injectable composite scaffold

2.1 β-TCP beads

β-tricalcium phosphateis a synthetic calcium phosphate ceramic used as an alternative autologous bone graft. It has been reported that β-TCP has good biodegradability and osteoconductivity as a scaffold material for bone tissue engineering (Le geros, 2002; Matsuno et al, 2006). With regard to shape availability, both block and granular forms are available. However, both shapes pose an injectability problem and are difficult to be molded into a 3D structure.
Thus, we used the β-TCP beads for injectable bone substitute. β-TCP beads were prepared in a manner similar to a previously described method (Ushida et al, 2001; Furukawa et al , 2004; Miyauchi et al , 2004). An aqueous slurry of β-TCP powder (Advance Ltd., Saitama, Japan) and polyvinyl alcohol (Shin-Etsu Chemical Co. Ltd., Tokyo, Japan) aqueous solution as a binder reagent were mixed using an ultrasonic homogenizer. The resultant slurry was added drop-by-drop to liquid nitrogen, whereby slurry drops were frozen, and the frozen beads were freeze-dried. The beads were then sintered at 1100°C for 10 hours. Diameter of the beads was controlled at 100−1,000 μm by regulating the opening and shutting times of an electromagnetic valve.
Figure 1 shows the SEM photographs of β-TCP bead and the surface of the bead. Diameter of the bead was about 500 μm, and micropores were observed on the surface.

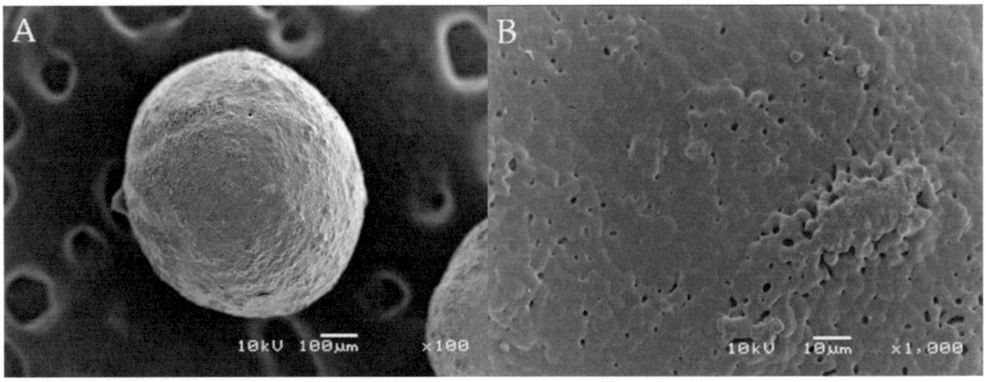

Fig. 1. SEM photographs of a β-TCP bead (A) and the surface of the bead (B).

Figure 2 shows the XRD patterns of the synthesized β-TCP beads and commercially available β-TCP (OSferion®; Olympus, Tokyo, Japan). There were no significant differences in the intensity of β-TCP peaks between β-TCP beads and commercially available β-TCP.

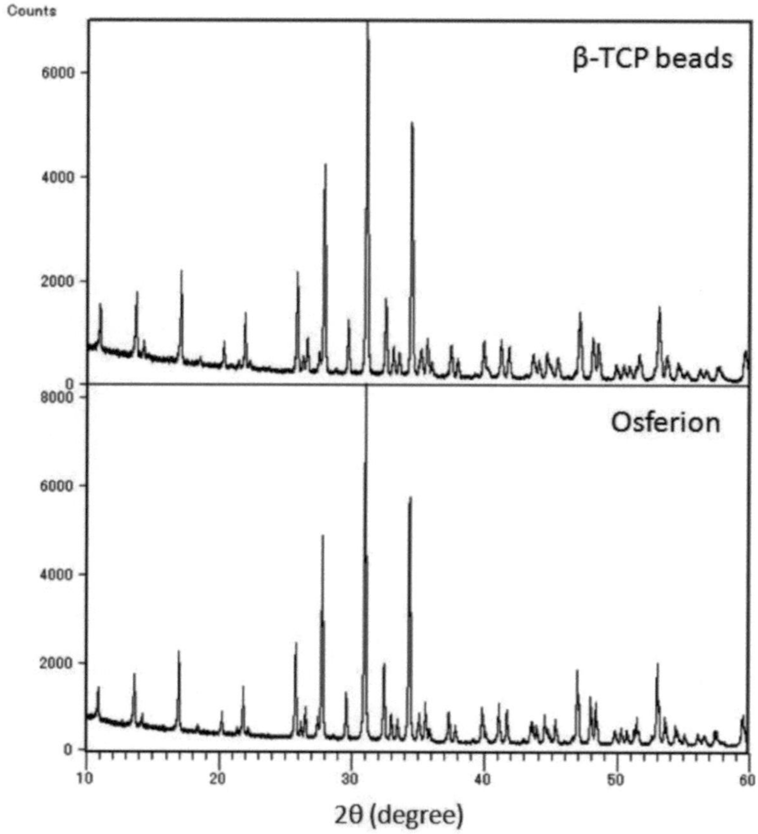

Fig. 2. X-ray diffraction patterns of β-TCP beads and commercially available β-TCP

2.2 Alginate hydrogel

Alginate has many uses in bioengineering, such as in polymer films, cell encapsulation, wound dressings, and surgical sponges (De Vos et al, 1997; Matthew et al, 1997). Alginic acid is soluble in water and can be ionically crosslinked with a non-toxic divalent cation solution, such as calcium chloride (De Vos et al, 1997). The calcium ions bind the guluronic acid sites of alginate strands together to form a stable alginate gel (Becker et al, 2001). However, alginate lacks the initial mechanical strength needed for bone tissue engineering. Alginate hydrogel was made by dissolving alginic acid sodium (Protanal LF 10/60, FMC BioPolymer, PA, USA) in phosphate-buffer saline (PBS) to give two final concentrations of 1.0 and 2.0 wt%.

2.3 Injectable 3D-formed composite scaffold

The β-TCP beads, with a diameter of 710−850 μm, were dipped in 1.0% $CaCl_2$ solution (Sigma-Aldrich Co.) and then dried. β-TCP beads in the syringe were sterilized with ethylenoxide gas and alginate hydrogel with Stericup® (0.22 μm; Millipore Co., MA, USA). As alginate hydrogel was pushed out from the tip of the syringe, it passed through the $CaCl_2$- treated β-TCP beads. As soon as a 3D-formed composite of β-TCP beads and alginate was obtained due to the instantaneous crosslinking that occurred in the syringe.
Figure 3 shows a photograph of injectable 3Dformed composite of β-TCP beads and alginate and a light microscope photograph of the composite. Each bead was wrapped in the anastomosing network of the alginate gel, and the beads were bonded tightly to form a 3D structure. Minute crystal projections were observed on the cross-sectional surface of the β-TCP beads treated with $CaCl_2$ (Fig. 4).

Fig. 3. Injectable 3D-formed composite composite.

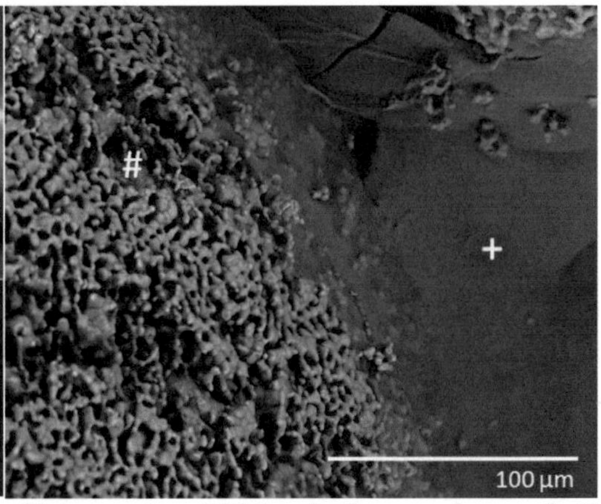

Figure. 4 SEM of the surface of the β-TCP beads (#), alginate (+)

3. Evaluation of β-TCP bead/alginate composite

3.1 Mechanical strength of the composite

The compression tests were performed in the two types of composites (1.0 and 2.0 wt% alginate hydrogels; 5.0 mm in diameter and 10.0 mm in length). The compressive strength increased with increased alginate concentration. The composite of 2.0 wt% alginate concentration exhibited a compressive strength of 69.0±4.6 kPa, which was significantly greater ($P<0.05$) than the strength seen with 1.0 wt% (6.11±2.28 kPa).

3.2 *In vitro* evaluation of 3D mesenchymal stem cell culture

Human mesenchymal stem cells (MSCs) were dispersed into the alginate hydrogel. Following which, MSC loaded alginate hydrogel was added into the syringe which was packed with $CaCl_2$-treated beads. Consequently, MSC-loaded alginate was crosslinked and MSCs were loaded into the 3D scaffold at about 2×10^5 cells/scaffold. They were cultured in an osteogenic differentiation medium (10mM β-glycerophosphate, 100 nM dexamethasone, and 50 μg/ml ascorbic acid) for 1, 7, and 14 days.

Figure 5 shows the SEM photographs of MSC-loaded composite after 1 day and 14 days in the osteogenic differentiation culture. MSCs were observed in the composite at day 1 (Fig. 5A). After 14 days, many calcified nodules appeared on the surface of β-TCP beads and alginate (Fig. 5B).

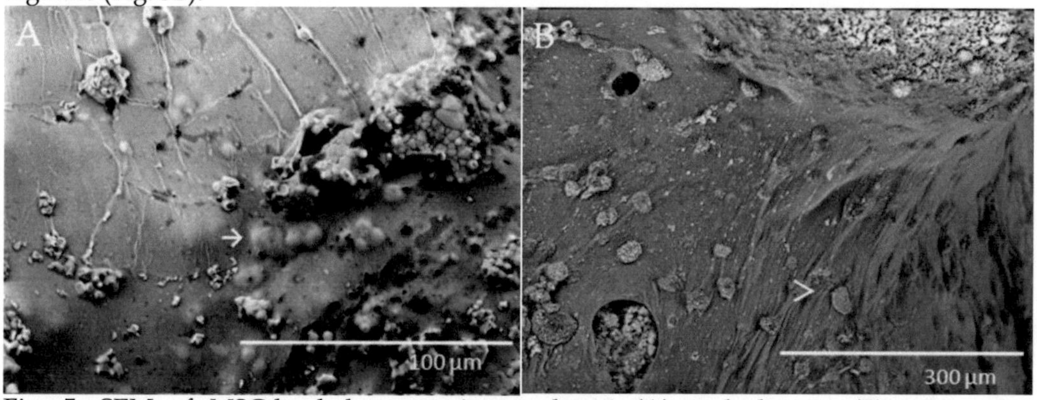

Fig. 5. SEM of MSC-loaded composite at day 1 (A) and day 14 (B) in osteogenic differentiation culture. MSCs (arrow), calcified nodules (>)

Figure 6 shows that the ALP activity of MSCs grown within the composite of osteogenic differentiation culture increased from day 1 to day 7, but decreased from day 7 to day 14. There was a significant increase between day 1 and day 7(*P<0.05), and similarly a significant decrease between day 7 and day 14 (**P<0.05).

No OCN mRNA expression was detected at day 1. However, OCN mRNA expression increased from day 1 to day 7, and increase between day 7 and day 14 was statistically significant (*P<0.05) (Fig. 7).

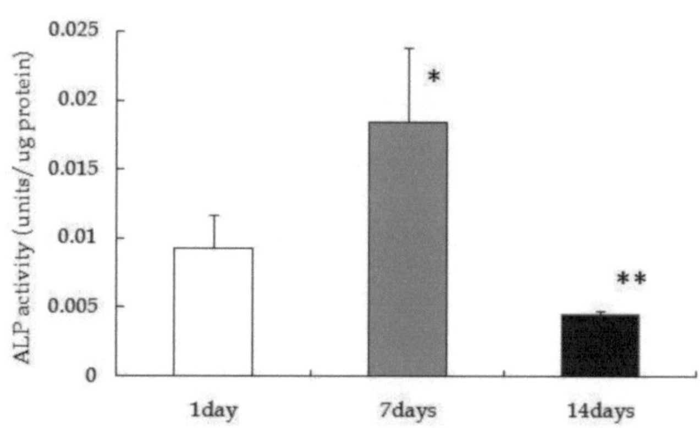

Fig. 6. ALP activity of MSCs grown within the composite in osteogenic differentiation culture.

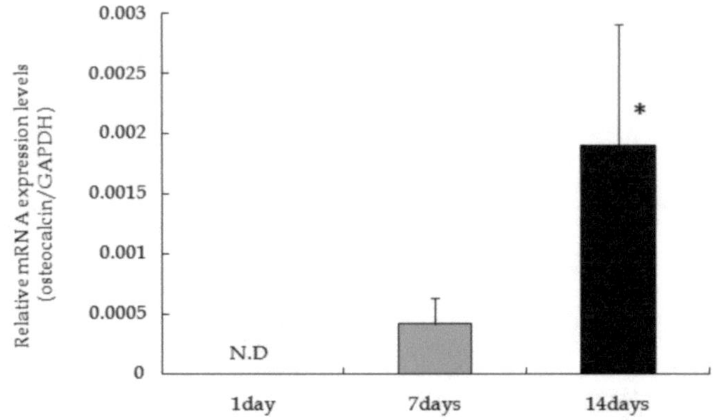

Fig. 7. OCN mRNA expression levels of MSCs grown within the composite in osteogenetic differentiation culture.

3.3 *In vivo* evaluation of MSCs loaded composite scaffold

Non-osteogenic-differentiated MSCs loaded into the 3D composite scaffold (5×10^6 cells/scaffold) and MSC-free composite as control were subcutaneously implanted into the backs of 6-week-old KSN nude mice. The mice were killed at 8 weeks after implantation.

Figure 8 shows the histological sections of the 3D composite scaffold loaded with non-osteogenicdifferentiated MSCs at 8 weeks after implantation. Newly formed, bone-like calcified tissue was directly deposited on the surface of the β-TCP beads. Interconnected alginate was completely degraded, and bone-like calcified tissue was observed in the connective tissue. The surface of the β-TCP beads was slightly degraded and vacuolization was observed in the beads. On the other hand, cell-free composite contained only fibrovascular tissue around nondegraded β-TCP beads (data was not shown).

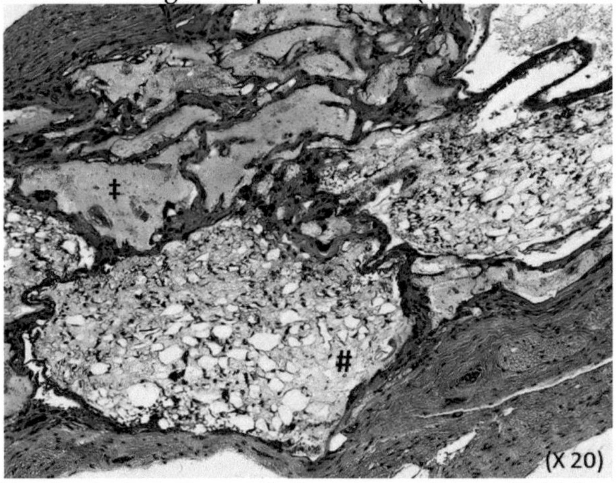

Fig. 8. HE-stained histological sections of the composite scaffold loaded with MSCs at 8 weeks after implantation.

Newly formed, bone-like calcified tissue (‡), β-TCP beads (#)

3.4 *In vitro* release of growth factor from composite

A solution of basic fibroblast growth factor (bFGF) was dispersed into the alginate hydrogel. Following which, bFGF incorporated alginate hydrogel was added into the syringe which was packed with $CaCl_2$-treated beads. Consequently, bFGF-incorporated composite scaffold was obtained. In vitro release tests were performed as follows. The composite scaffolds completely sank in the Hank's balanced solution, and incubated at 37°C . bFGF concentration was determined using ELISA.
Figure 9 shows that the release of bFGF from the composite. There was a significant increase between day 1 and day 4 (*$P<0.05$).

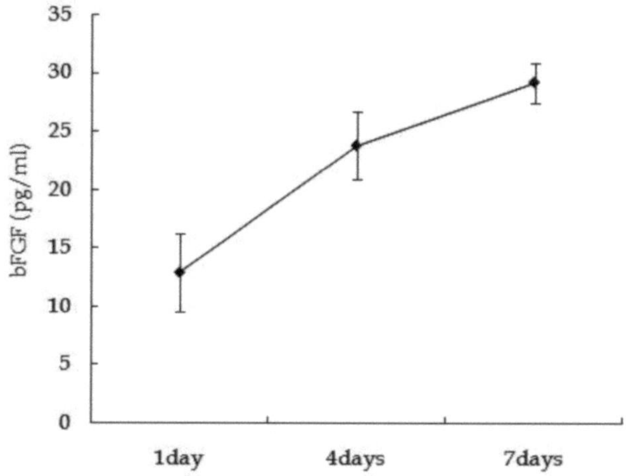

Fig. 9. The release of bFGF from the composite.

4. Conclusion

Both ceramics and polymers have their own merits and drawbacks, and a better solution may be to synergize the advantageous properties of both materials for composite scaffolds. Therefore, many different composite forms such as sponges, gels, films, and blocks have been developed using different methods (Wahl & Czernuszka, 2006). However, multi functional composite such as our developed composite are little or no.
Recently, the increasing popularity of minimally invasive techniques has led to the development of an injectable system that can be molded to the shape of the bone defect and which polymerizes when injected into the site (Niemeyer et al, 2004). Such devices should be easy to implant, thereby shortening the surgical time and minimizing the damage to healthy tissue. β-TCP has been used for bone tissue engineering. However, it is difficult to fit into the surgical sites around implants or to be shaped to the desired forms using β-TCP blocks or granules. The composites consisting of β-TCP beads and alginate can be injects into bone defect site. On the other hand, the composite can simply be cut with scissors or a sharp knife, and can therefore be easily molded for use for various tissue disorders.
Bone tissue engineering approach is to combine osteogenetic cells with an appropriate composite scaffold. In keeping with the tissue engineering concept, cells are cultivated on an appropriate 3D scaffolds to replace 3D tissue defects (Liu et al, 2006). However, elaborate

seeding and culture conditions are needed to achieve uniform cellular distribution, sustain cell viability, and provide nutrients for tissue formation in scaffolds (Weinand et al, 2006).
In this study, alginate hydrogel was used to facilitate the delivery and distribution of MSCs. It has been previously demonstrated that alginate hydrogels induce bone formation *in vitro* (Weinand et al, 2006; Alsberg et al, 2001). However, the lack of adequate initial strength of hydrogel/cell constructs requires additional mechanical support for implantation (Niemeyer et al, 2004). Therefore, a uniform size of $CaCl_2$-coated β-TCP beads was selected to reinforce the initial strength. Calcium ions can be crosslinked with the guluronic acid sites of alginate strands to form a stable alginate gel (Becker et al, 2001). Leveraging on all these properties in this study, an instantaneously formed, injectable 3D scaffold was obtained after alginate hydrogel reacted with $CaCl_2$-coated β-TCP beads and then loaded with MSCs.
In vitro evaluation, ALP activity at 7 days was significantly increased as compared to day 1, but decreased from day 7 to day 14. As a marker of early osteogenic differentiation, the ALP activity of the construct was measured on days 1, 7, and 14 of the culture period. The typical rise and fall of ALP activity was characteristic of osteogenic differentiation and common to all culture systems (Shin et al, 2004). Furthermore, OCN mRNA expression, another marker of osteogenic differentiation, was significantly increased at day 14 as compared to day 7. Therefore, it is conceivable that calcification progressed in the composite.
In vivo experiment demonstrated that the β-TCP bead/alginate composite had an osteoinductive effect in soft tissue — without an osteogenic differentiation medium for MSCs. Similarly, Livingston *et al.* (2005) demonstrated that hydroxyapatite/β-TCP ceramic scaffold promoted bone formation in soft tissue, which was induced by non-osteogenic-differentiated MSCs. Furthermore, this study demonstrated that β-TCP bead/alginate composite induced the formation of bone-like calcified tissue. Taken together, these results suggested that a 3D structure consisting of β-TCP beads and alginate allowed an appropriate spatial arrangement of osteogenic cells as well as vascular invasion. These findings also indicated that injectable β-TCP bead/alginate composites supported bone regeneration with a minimally invasive technique.
In addition, the β-TCP bead/alginate composite can locally release growth factors from the composite, used as drug delivery carriers. It will be enhance bone formation to treat bone defects.
In light of the encouraging results obtained for osteogenesis, the novel composite developed in this study may be useful as an injectable scaffold for bone tissue engineering.

5. References

Alsberg, E.; Anderson, KW.; Albeiruti, A.; Franceschi, RT. & Moony, DJ.(2001). Cell-interactive alginate hydrogels for bone tissue engineering. *J Dent Res*, 80, 2025-2029, 0022-0345

Arrington, ED.; Smith, WJ.; Chambers, HG.; Bucknell, AL. & Davino, NA. (1996). Complications of iliac crest bone graft harvesting. *Clin Orthop Relat Res*, 329., 300-309, 0009-921X

Becker, TA.; Kipke, DR. & Brandon, T. (2001). Calcium alginate gel: A biocompatible and mechanically stable polymer for endovascular embolization. *J Biomed Mater Res,* 54, 76-86, 0021-9304

Damien, J. & Parsons, R. (1991). Bone graft and bone graft substitutes: a review of current technology and applications. *J Appl Biomater*, 2., 187-208, 1045-4861

De Vos, P.; De Haan, B. & Van Schilfgaarde, R. (1997). Effect of the alginate composition on the biocompatibility of alginate-polylysine microcapsules. *Biomaterials* , 18, 273-278, 0142-9612

Furukawa, KS.; Miyauchi, S.; Suzuki, D.; Umezu, Y, Shinjo, T.; Ushida, T.; Eguchi, & Tateishi, T. (2004). Bone tissue engineering based on bead-cell sheets composed of calcium phosphate beads and bone marrow cells. *Mater Sci Eng C*, 24, 437-440, 0928-4931

Giannoudis, PV.; Dinopoulos, H. & Tsiridis E. (2005). Bone substitutes: An update. *Injury*, 36., 20-27, 0020-1383

Goldberg, M. & Stevenson, S. (1987). Natural history of autografts and allografts. *Clin Orthop Relat Res*, 225., 7-16, 0009-921X

Le Geros, RZ. (2002). Properties of osteoconductive biomaterials: calcium phosphates. *Clin Orthop Relat Res* , 395, 81-98, 0009-921X

Liu, H.; Li, H.; Cheng, W.; Yang, Y.; Zhu, M. & Zhou, C. (2006). Novel injectable calcium phosphate/chitosan composites for bone substitute materials. *Acta Biomater*, 2, 557-565, 1742-7061

Livingston, AT.; Tran, T.; Mcalary, J. & Daculsi, G. (2005). A comparative study of biphasic calcium phosphate ceramics for human mesenchymal stem cell-induced bone formation. *Biomaterials*, 26: 3631-3638, 0142-9612

Matsuno, T.; Nakamura, T.; Kuremoto, K.; Notazawa, S.; Nakahara, T.; Hashimoto, Y.; Satoh, T. & Shimizu, Y. (2006). Development of β-tricalcium phosphate/collagen sponge composite for bone tissue regeneration. *Dent Mater J*. 25, 138-144, 0287-4547

Matsuno, T.; Hashimoto, Y.; Adachi, S.; Omata, K.; Yoshitaka, Y.; Ozeki, Y.; Umezu, Y.; Tabata, Y.; Nakamura, M. & Satoh, T. (2008). Preparation of injectable 3D-formed β-tricalcium phosphate bead/ alginate composite for bone tissueengineering. *Dent Mater J*. 27, 827-834, 0287-4547

Matthew, IR.; Browne, RM.; Frame, JW. & Millar, BG. (1997). Subperiosteal behavior of alginate and cellulose wound dressing materials. *Biomaterials*, 18, 707-713, 0142-9612

Miyauchi, S.; Furukawa, KS.; Umezu, Y.; Ozeki, Y.; Ushida, T. & Tateishi, T. (2004). Novel bone graft model using bead-cell sheets composed of tricalcium phosphate beads and bone marrow cells. *Mater Sci Eng C*, 24, 875-879, 0928-4931

Niemeyer, P.; Krause, U.; Fellenberg, J.; Kasten, P.; Seckinger, A.; Ho, AD. & Simank, HG. (2004). Evaluation of mineralized collagen and α-tricalcium phosphate as scaffolds for tissue engineering of bone using human mesenchymal stem cells. *Cells Tissues Organs*, 177, 68-78, 1422-6405

Perka, C.; Arnold, U.; Spitzer, RS. & Lindenhayn, K. (2001). The use of fibrin beads for tissue engineering and subsequential transplantation. *Tissue Eng*, 7: 359-361, 1076-3279

Schmitz, JP.; Hollinger, JO. & Milam, SB. (1999). Reconstruction of bone using calcium phosphate bone cements: A critical review. *J Oral Maxillofac Surg*, 57: 1122–1126, 0278-2391

Shang, Q.; Wang, Z.; Liu, W.; Shi, Y.; Cui, L. & Cao Y. (2001). Tissue-engineered bone repair of sheep cranial defects with autologous bone marrow stromal cells. *J Craniofac Surg*, 12: 586-593, 1049-2275

Shin, H.; Zygourakis, K.; Farach-Carson, MC.; Yaszemski, MJ. & Mikos, AG. (2004). Modulation of differentiation and mineralization of marrow stromal cells cultured on biomimetic hydrogels modified with Arg-Gly-Asp containing peptides. J Biomed Mater Res A, 69: 535-543. 1549-3296

Spector, M. (1994). Anorganic bovine bone and ceramic analogs of bone mineral as implants to facilitate bone regeneration. *Clin Plast Surg*, 21: 437-444, 0094-1298

Tsuchida, H.; Hashimoto, J.; Crawford, E.; Manske, P. & Lou, J. (2003). Engineered allogeneic mesenchymal stem cells repair femoral segmental defect in rats. *J Orthop Res*, 21: 44-53, 0736-0266

Ushida, T.; Chen, G.; Tamaki, T.; Umezu, Y. & Tateishi, T. (2001). A three-dimensional porous scaffold of biodegradable synthetic polymers and porous hydroxyapatite beads for bone tissue engineering. *Bioceramics.* 13, 519-522

Wahl, DA. & Czernuszka, JT. (2006). Collagen-hydroxyapatite composites for hard tissue repair. *Eur Cell Mater*, 11: 43-56, 1473-2262

Wang, L.; Shelton, RM.; Cooper, PR.; Lawson, M.; Triffitt, JT. & Barralet JE. (2003). Evaluation of sodium alginate for bone marrow cell tissue engineering. *Biomaterials*, 24: 3475-3481, 0142-9612

Weinand, C.; Pomerantseva, I.; Neville, CM.; Gupta, R.; Weinberg, E.; Madisch, I.; Shapiro, F.; Abukawa, H.; Troulis, MJ. & Vacanti JP. (2006). Hydrogel-β-TCP scaffolds and stem cells for tissue engineering bone. *Bone*, 38, 555-563, 8756-3282

15

Low Intensity Pulsed Ultrasound: A Laboratory and Clinical Promoter in Tissue Engineering Abstract

Tarek H. El-Bialy

This chapter will focus on the current and future applications of low intensity pulsed ultrasound (LIPUS) in tissue engineering at laboratory and clinical levels. Also, this chapter will explain the molecular basis of LIPUS stimulatory effects on cell expansion, differentiation and matrix production of different tissues. In addition, this chapter will provide information about the differences between LIPUS application in the laboratory settings and in the clinical applications with emphasis on tissue engineering applications. Moreover, this chapter will discuss why variability in outcome might occur when applying LIPUS in laboratory or clinical settings. Also, we will provide some tips on how to prevent potential inconsistency in treatment results with emphasis on the technique sensitivity and how a scientist or a clinician should be aware of this sensitivity of the application technique in order to optimize the utilization of LIPUS in different settings. This chapter will set a few hypotheses and potential new applications of utilizing LIPUS in other tissue engineering applications. The outcome of this chapter will be an understanding of the advantages and limitations of using LIPUS in tissue engineering applications either at the laboratory or in clinical settings.

1. Introduction

Tissue engineering utilizes cells, scaffold or housing and growth factors to maintain the specific characteristics of the tissue to be engineered. Tissue engineering efforts are faced with challenges including the availability of cells to be used for engineering specific tissue (Chu et al., 2002). Another challenge in tissue engineering is the availability of biocompatible, biodegradable scaffolds that enhance cell adhesion as well as cell proliferation and matrix production. The degradation rate of the scaffolds is optimum when it occurs in a timely pattern with matrix production and maturation of the cell seeded within the scaffold. For some tissues, like bone, cartilage, skin and nerve tissue engineering, differentiation growth factors are readily available. However the key growth factors that are important in differentiating specialized tissues, like dentine in teeth tissue engineering, for example, are still not available. This again puts another roadblock in engineering these specialized tissues. Ultrasound is a form of mechanical acoustic pressure wave at frequencies above the limit of human hearing that when transmitted through biological tissues can produce different biological effects. The most widely used applications in medicine are operative (usually has a frequency that ranges between 2 - 8 K Hz), therapeutic (usually has a of frequency that ranges between 20 K Hz 3M Hz either in continuous or pulsed modes), and diagnostic (usually has a frequency that ranges between 1.6-12 M Hz) (Ziskin, 1987; Maylia & Nokes, 1999; Harris, 2005). The most commonly used type of therapeutic ultrasound is US that utilizes intensities between 30 mW to as high as 1-3 W/cm^2 and hence it may cause considerable heating in living tissues. The low intensity pulsed ultrasound (LIPUS) is widely accepted in cell and tissue repair due to the nonthermal effects of this type of ultrasound.

2. LIPUS utilization in cell expansion

The availability of optimized cell density (CD) is a major challenge in tissue engineering, either due to donor site morbidity (if to be harvested from another organ within the same individual) or precursor cells, sometimes also called undifferentiated mesenchymal stromal cells or (mesenchymal stem cells) (Puelacher et al., 1994; Carrier et al., 1999; Holy et al., 2000; Goldstein, 2001; Panossian et al., 2001). It has been reported that one of the important parameters impacting tissue engineering is the cell seeding density. Seeding densities more than 8×10^6 cells cm^3 of tissue engineering scaffold promotes extracellular matrix production, cell distribution and/or cell alignment when compared to low cell seeding densities ($<4 \times 10^6$ cells cm^3 of scaffold). Low cell seeding densities is reported to be associated with decreased cell proliferation and minimal or no mechanical integrity (Freed et al., 1998; Carrier et al., 1999; Li et al., 2001; Saini & Wick, 2003; Leddy et al., 2004, Hsieh-Bonassera et al., 2009).

In this context, several studies have attempted to expand mesenchymal stem cells (MSCs) before differentiating them for engineering different tissues. These reports include, but are not limited to using different growth factors and /or mechanical stresses to enhance stem cell expansion. It has been reported that chondrocytes increased in number by 100-fold within 17 days when a special growth factor cocktail was used in the culture medium (Chen, et al., 2009). Also, ultra-nanocrystalline diamond films have been shown to enhance neural stem cell expansion for tissue engineering (Wang et al., 2006).

Moreover other growth factors that have been used to enhance stem cell expansion include ascorbic acid-functionalized poly (methyl methacrylate) (AA-PMMA) (Tamama et al., 2005); epidermal growth factor receptor (EGFR) (Hsieh-Bonassera et al., 2009) ; and Flt3 ligand (FL) was used in addition to IL-11 or G-SCF (Granulocyte-Colony Stimulating factor) to stimulate the growth of rat MSCs (Jacobsen et al., 1995). These (conjugated FL) factors expanded the progenitors more than 40 times after 2 weeks incubation. On the other hand, prolactin-like protein E (PLP-E) + human thrombopoietin (TPO) showed no effect on the expansion of human MSCs (Lefebvre et al., 2001). Glucocorticoids were found to be effective in recruiting cells rather than stimulating their proliferation (Purpura et al., 2004). It was reported that Dkk-1 stimulates MSCs expansion (Horwitz, 2004). Culture flasks coated with heparin and N-(O-beta-(6-O-sulfogalactopyranosyl)-6-oxyhexyl)-3,5-bis(dodecyloxy)-benzamide have been shown to expand MSCs (Takagi, 2005). A similar study has shown that Angiopoietin-like proteins expanded HSCs 24-30 fold in ten days (Zhang et al., 2006). However, more studies are needed to evaluate the effect of these factors on MSCs behavior (especially differentiation potential) after expansion. Other attempts using pulsed electromagnetic fields (PEMF) showed that chondrocytes- and osteoblast-like cell proliferation increased (De Mattei et al., 2001; Hartig et al., 2000). However, no studies tested the effect of PEMF on MSCs expansion. Others used LIPUS to enhance periosteal cell expansion for 5, 10, and 20 minutes for 2 and 4 days. No cell proliferation was noted (Leung et al., 2004). This may be due to the small doses of ultrasound that were used. It has been shown that ultrasound effect on tissue is evident after three weeks of application (Tsai et al., 1992). Also, LIPUS was used for 20 minutes for one day to stimulate MSCs expansion and differentiation into chondrocytes. Matrix production was enhanced; however, cell proliferation was not enhanced (Ebisawa et al., 2004). This may also be due to using LIPUS in pellet culture. MSCs are known to have cell-cell growth inhibition when they are in close proximity to each other.

On the other hand, another report using LIPUS showed that LIPUS increases matrix production and proliferation of the intervertebral disc cell culture (Iwashina et al., 2006). The effect of LIPUS was different depending on the type of cells from different origins (nucleus pulposus [NP] or from annulus fibrosus [AF]). More recently, LIPUS has been shown to enhance chondrocyte proliferation in three-dimensional collagen scaffolds (Kobayashi et al., 2009).

Another precursor cell proliferation and differentiation that has been recently reported to be stimulated by LIPUS are human periodontal ligament cells that have been differentiated into cementoblasts (teeth root outer layer cells) (Inubushi et al., 2008). This is an important report on the potential use of LIPUS in dental tissue engineering. The scantiness of cementoblasts for teeth tissue engineering made it almost impossible in the past to tissue engineer teeth. However, with this recent report, it could be possible to tissue engineer teeth in the near future using LIPUS as a cell expansion and differentiation enhancer. Also, it has been reported that LIPUS enhanced cell proliferation in human fibroblasts, osteoblasts, and monocytes (Doan et al., 1999).

In a pilot study in our lab, we have shown that LIPUS enhances the expansion of bone marrow stem cells in vitro (Figure 1). This stimulatory effect is dose dependent.

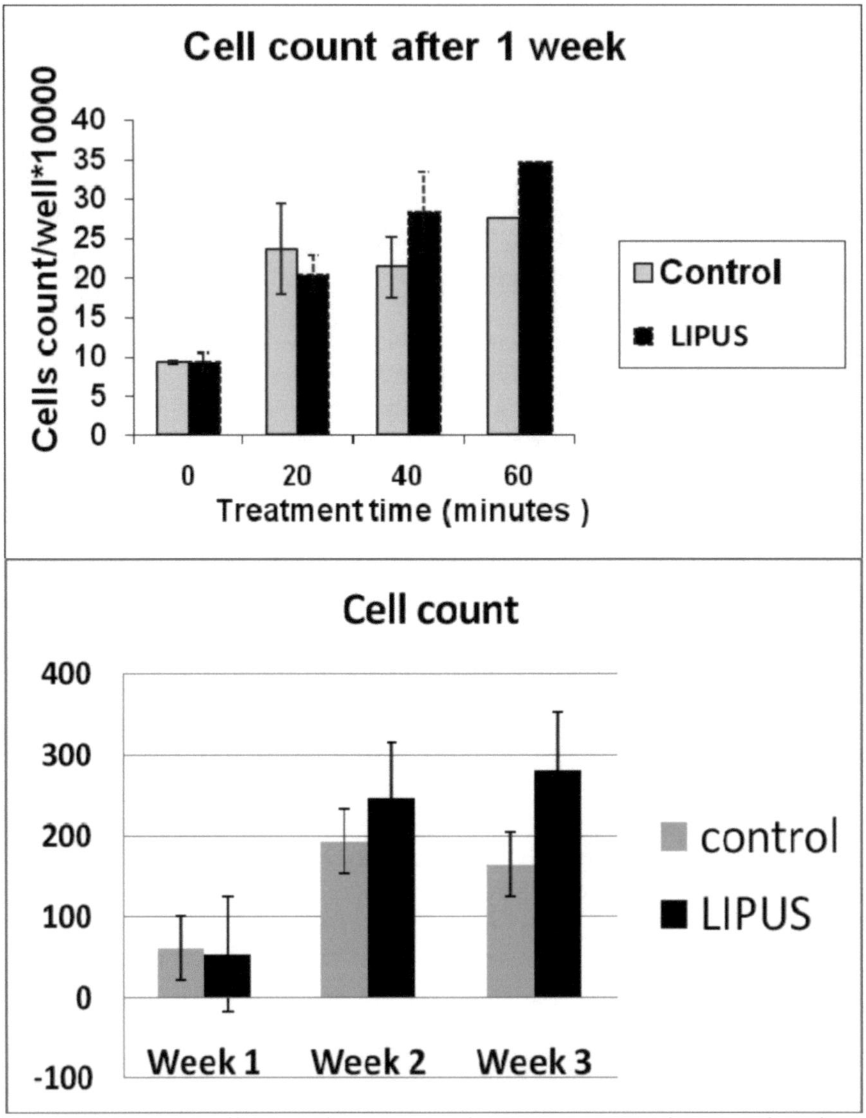

Fig. 1. Bone marrow stem cell (BMSC) expansion by LIPUS. IT can be seen that the increased LIPUS treatment time (20, 40, 60 minutes per day) or 2 and 3 weeks application led to increased BMSC expansion as measured by cell count.

3. Effect of LIPUS on in-vitro matrix formation for tissue engineering

Recent studies showed consistent agreement that LIPUS enhances different tissue matrix production by different cells in vitro. These cells include chondrocytes from different tissue origins (Wu et al., 1996; Doan et al., 1999; Iwabuchi et al., 2005; Mukai et al., 2005; Hsu et al., 2006; Iwashina, et al., 2006; Schumann et al., 2006a; Inubushi et al., 2008; Korstjens et al., 2008; Takeuchi et al., 2008; Tien et al., 2008; Kobayashi et al., 2009). It is to be noted that the stimulatory effect on chondrocyte matrix production has been recently shown to be dose dependent (Schumann et al., 2006a; Tien et al., 2008). A more interesting in-vitro study showed that herniated disc resorption (HDR) treatment is accelerated by LIPUS. This effect was mediated by activation of matrix metalloproteinase-3 (MMP-3) maturation through tumor necrosis factor-alpha (TNF-alpha) and macrophage chemoattractant protein-1 (MCP-1) pathways (Iwabuchi et al; 2005). When compared to Bioreactors, LIPUS produced more matrix at 28 days in tissue engineered cartilage, while bioreactors continued matrix production until day 42 in-vitro (Hsu et al., 2006). It is to be noted that bioreactors cannot be applied clinically to engineered tissues while LIPUS can be applied. Moreover, the longer tissue culture periods could be questionable regarding possible errors or potential culture infection. The in-vitro conditioning of chondrogenic differentiated mesenchymal stem cells by LIPUS for cartilage tissue engineering in vivo has been recently tested. It has been concluded that type I and type X collagens and matrix metalloproteinase-13 are stimulated by LIPUS which indicate that LIPUS preconditioning in-vitro can be an effective treatment to upregulate chondrogenic differentiation of MSCs in vivo (Cui et al., 2007).

Studies also reported that LIPUS stimulated matrix formation and maturation by osteoblasts from different bone origins (Tsai et al., 1992; Saito et al., 2004; Sena et al., 2005; Tam et al., 2008; Suzuki et al., 2009). The foundation for the applications of LIPUS to enhance bone fracture healing in clinical situations was based on the detailed in-vitro study by Tsai et al. (Tsai et al., 1992) Since then, many in-vitro and in-vivo studies were conducted and reported on the molecular basis of LIPUS stimulation of bone cell matrix production and hence its clinical applications. The most recent studies showed that LIPUS enhances different early and late osteoblastic differentiation gene expressions. In most of these reports, alkaline phosphates were markedly increased by LIPUS in different bone cell lines. The enhanced osteogenic differentiation by LIPUS is suggested to be due to activation of early differentiation genes (c-jun, c-myc, COX-2, Egr-1, TSC-22) as well as the bone differentiation marker genes, osteonectin and osteopontin, phosphorylation (Tam et al., 2008) as well as activation of ERK1/2 and p38 MAPK pathways (Sena et al., 2005). Most importantly, LIPUS enhances bone morphogenetic proteins expression in bone cells (Suzuki et al., 2009). The importance of this report that LIPUS enhances BMPs expression is that LIPUS might be a future substitute of many BMPs-based therapies.

Another cell type that has been reported to be stimulated by LIPUS to enhance matrix production is macrophages. It has been shown that LIPUS up-regulated the phagocytosis of human primary macrophages through the activation of Src and ERKs and promoted the protein expression of CD147 and MMPs, as well as increased the level of protein tyrosine phosphorylation (Ikeda et al., 2006; Li et al., 2007a). Also, it has been reported that the effect of LIPUS on macrophages depends on cell adhesion, and relates to the integrin-MEK-ERK pathway (Li et al., 2007b). The stimulatory effect of LIPUS on the phagocytosis of macrophages may be important in tissue engineering, especially during tissue replacement and scaffold degradation for optimum integration between the engineered and original tissues. Another effect of LIPUS on macrophages is the ability of LIPUS to liberate fibroblast mitogenic factors from macrophages which are known to be a source of many important growth factors which can act as wound mediators during tissue repair. This stimulatory effect of ultrasound appeared to be mediated by producing permeability changes in these cells as well as to stimulate the cell's ability to synthesize and secrete fibroblast mitogenic factors (Young & Dyson, 1990a).

Also, LIPUS has been shown to enhance epithelial cell repair (Hill et al, 2005) as well as to accelerate tendon healing via regulation of vascular endothelial growth factor (VEGF) expression (Hill et al., 2005; Lu et al., 2008). Other reports showed that LIPUS enhances VEGF expression by fibroblasts, osteoblasts and monocytes (Doan et al., 1999). Moreover, other angiogenesis-related cytokines, IL-8 and bFGF, were significantly stimulated in osteoblasts by LIPUS (Reher et al., 1999). Earlier reports also supported the importance of LIPUS in supporting angiogenesis (Young & Dyson, 1990b; Young & Dyson, 1990c). The importance of VEGF and angiogenic factors expression by LIPUS in different tissues is very important in tissue engineering, especially bone tissue engineering. Angiogenesis is reported to be a key factor in bone maturation and bone fracture healing (Rabie et al., 2002; Rabie et al., 2001; El-Bialy et al., 2003).

Very recently, it has been reported that dental tissue matrices can be formed by LIPUS in vitro and in vivo. The first report about new dental tissue formation was reported while LIPUS was applied to osteodistracted rabbit mandibles (El-Bialy, 2003). A follow up pilot clinical study in human patients showed for the first time that human dental tissue matrices (dentine and cementum) are formed by LIPUS in 28 days (El-Bialy et al., 2004). These two studies triggered several in-vitro studies and potential tissue engineering teeth using LIPUS (Dalla-Bona et al., 2006; Scheven et al., 2007; Dalla-Bona et al., 2008; Inubushi et al., 2008; Scheven et al., 2009a; Scheven et al., 2009b). Moreover, recently, we have shown that LIPUS has an anabolic effect on human gingival fibroblasts, which in turn may help to use these cells in craniofacial tissue engineering (Mostafa et al., 2009). The preliminary conclusion of these studies may suggest that LIPUS can play an important role in dental and craniofacial tissue engineering.

Another potential application of LIPUS in enhancing tissue matrix formation for tissue engineering is via enhancing gene delivery to promote dentine, bone and other tissue engineering. Recent appreciable number of studies reported that LIPUS can provide a synergistic tool to enhance gene delivery to different cell types for tissue engineering (Inagaki et al., 2006; Kaigler et al., 2006; Kimelman et al., 2006; Nozaki et al., 2006; Tachibana et al., 2006; Saito et al., 2007; Shen et al., 2008; Sheyn et al., 2008; Kamimura & Liu, 2009; Osawa et al., 2009; Sakai et al., 2009;). The sonoporation technique, which utilizes ultrasound to facilitate drug delivery or gene transfection into cells has been accepted both in vitro and in vivo, however clinical application of this technique for tissue engineering still needs further optimization in-vitro experiments before it might be applicable clinically. The ultrasound used in sonoporation is quite different from LIPUS, however an optimization experiment for utilizing LIPUS in sonoporation might be justified.

4. Therapeutic ultrasound and its role in enhancing tissue engineering in vivo

It has been widely accepted that LIPUS can enhance cartilage repair based on the previously mentioned in-vitro studies. Moreover, it has been suggested that LIPUS can precondition chondrogenic derived stem cells for in-vivo cartilage tissue engineering (Schumann et al., 2006b; Cui et al., 2007). In an in-vivo pilot study, we have shown that LIPUS can enhance tissue engineered articular condyles in rabbits (El-Bialy et al., 2009). The optimum LIPUS treatment for maximum tissue engineered cartilage/or articular joints is still yet to be established.

On the other hand, there are controversial reports about the clinical efficacy of LIPUS in enhancing repair of fractured bone in human patients. While earlier reports showed that LIPUS enhances and accelerates repair of fractured bones (Heckman et al., 1994; Cook et al., 1997; Kristiansen et al., 1997; Hadjiargyrou et al., 1998; Mayr et al., 2000; Warden et al., 2000; Nolte et al., 2001; Jingushi et al., 2007; Walker et al., 2007; Griffin et al., 2008; Busse et al., 2009; El-Bialy et al., 2009), recent reports showed no difference in the healing of the LIPUS treated and nontreated fractured bones (Emami et al., 1999; Rue et al., 2004; Handolin et al., 2005; Lubbert et al., 2008).

The controversial reports about the clinical success of using LIPUS in repairing bone fractures could be due to the difference in the proximity of the fractured bone to the LIPUS transducer. It can be noted that most published reports with successful LIPUS treatment were performed on clinically accessible bones for LIPUS application, while deep bones or bones with thick muscle covering showed less success in accelerating bone fracture healing by LIPUS. A simple explanation of this is that the LIPUS power attenuates exponentially as a function of distance. In a rough estimate of LIPUS power attenuation, we have presented a simplified mathematical calculation as follows based on our experience in LIPUS attenuation of rabbit mandibles. An order-of magnitude estimate of the degree of attenuation and relative dose strengths between the direct and indirect sides can be made based on the following assumptions: 1) plane wave propagation through one side of the jaw and mandible, through the tongue and into the other side of the jaw and mandible (in reality, there will be additional loss due to scattering, especially with the presence of air regions); and 2) the pressure attenuation coefficient for teeth/mandible is α bone= 1.5 Np/cm and for the tongue is α tongue = 0.15 Np/cm at 1.5 MHz (Goss et al., 1978; Goss et al., 1980) assuming the bone thickness is 0.5 cm and the tongue thickness is 2 cm. With these assumptions the fraction of the intensity reaching the other side of the jaw would be:

$$I/I_0 = \text{Exp}\,[-2\,\alpha\text{ bone d bone}]\ \text{Exp}\,[-2\,\alpha\text{ tongue d tongue}] \quad (1)$$

$$= \text{Exp}\,[-2(1.5)(0.5)]\ \text{Exp}\,[-2(0.15)(2)]$$

$$= 0.122.$$

Where I is the intensity at the other side and I_0 is the intensity incident on the first (treated) side. This is a very rough estimate that may be taken as an upper bound; but, it predicts that 12.2% of the energy gets to the far (indirect) side relative to the near (direct) side. It has to be noted as previously mentioned that the stimulatory effect of LIPUS in enhancing tissue matrix production and bone fracture healing is dose dependent (El-Bialy et al., 2002; Schumann et al., 2006a). This could explain in part the differences in clinical outcomes between studies when considering differences in bone sizes.

Another consideration that needs to be taken into account while using LIPUS for tissue repair and/or tissue engineering is that the differences in LIPUS power could be different with different applications. In another words, when using LIPUS to enhance cell proliferation and/or matrix production in vitro, lower power is needed as there is no intervening tissues between the transducer and the cells. While in animal or clinical applications, many tissues can be intervening between the point of LIPUS application and the target bone or tooth, which needs further experimentation to reach the optimum LIPUS needed to reach each target for optimum results.

It is to be noted that different cell types might respond differently to LIPUS treatment time. While stem cells are stimulated by LIPUS in a dose dependent response (Schumann et al., 2006a); the optimum LIPUS treatment for skin fibroblasts is 10 minutes per day (Zhou et al., 2004). Moreover, clinical applications using LIPUS has been accepted to be 20 minutes per day for 3-4 weeks (Heckman et al., 1994; Cook et al., 1997; Kristiansen et al., 1997; Hadjiargyrou et al., 1998; Mayr et al., 2000; Warden et al., 2000; Nolte et al., 2001; Jingushi et al., 2007; Walker et al., 2007; Griffin et al., 2008; Busse et al., 2009), in-vitro responses vary from one day to 21 days. It is to be remembered that in-vitro experiments allow direct effect of LIPUS to the cells, while in-vivo application suffers from LIPUS power attenuation as explained above. In this context, in-vitro results might not be completely applicable to in-vivo applications.

Conclusion and future directions:

We can conclude that there is ample literature that supports that LIPUS can be utilized to enhance cell proliferation and differentiation, matrix production with the differentiated cells, gene transfection for cell differentiation as well as tissue repair both in vitro and in vivo. The optimum LIPUS treatment time and dose is still yet to be studied. The potential applications of LIPUS in tissue engineering could be used for bone, cartilage, skin, nerve, and possibly teeth tissue engineering. A final remark is that the daily use of LIPUS for an extended period of time should not be of concern as the safety of this type of low level power ultrasound has been addressed before and proven to be safe (Mende et al., 1996; Hata et al., 1997; Blaas & Eik-Nes, 1998; Turnbull & Foster, 2002).

5. References

Blaas, H.G. & Eik-Nes, S.H. (1998). Advances in the imaging of the embryonic brain. *Croatian Medical Journal,* Jun;39(2):128-131, ISSN 0353-9504 (Print) 1332-8166 (Electronic).

Busse, J.W.; Kaur, J.; Mollon, B.; Bhandari, M.; Tornetta, P. 3rd; Schünemann, H.J. & Guyatt, G.H. (2009). Low intensity pulsed ultrasonography for fractures: systematic review of randomised controlled trials. *BMJ (Clinical research ed),* Feb 27;338:b351, ISSN 0959-8138 (Print) 1468-5833 (Electronic).

Carrier, R.L.; Papadaki, M.; Rupnick, M.; Schoen, F.J.; Bursac, N.; Langer, R.; Freed, L.E. & Vunjak-Novakovic, G. (1999). Cardiac tissue engineering: cell seeding, cultivation parameters, and tissue construct characterization, *Biotechnology and Bioengineering,* Sep 5; 64(5):580–589, ISSN 0006-3592 (Print) 1097-0290 (Electronic).

Chen, Y.C.; Lee, D.C.; Hsiao, C.Y.; Chung, Y.F.; Chen, H.C.; Thomas, J.P.; Pong, W.F.; Tai, N.H.; Lin, I.N. & Chiu, I.M. (2009). The effect of ultra-nanocrystalline diamond films on the proliferation and differentiation of neural stem cells. *Biomaterials,* Jul;30(20):3428-3435, ISSN 0142-9612 (Print) 1878-5905 (Electronic).

Chu, T.M.; Orton, D.G.; Hollister, S.J.; Feinberg, S.E. & Halloran, J.W. (2002). Mechanical and in vivo performance of hydroxyapatite implants with controlled architectures. *Biomaterials,* Mar;23(5):1283-1293, ISSN 0142-9612 (Print) 1878-5905 (Electronic).

Cook, S.D.; Ryaby, J.P.; McCabe, J.; Frey, J.J.; Heckman, J.D. & Kristiansen, T.K. (1997). Acceleration of tibia and distal radius fracture healing in patients who smoke. *Clinical Orthopaedics and Related Research,* Apr;(337):198-207, ISSN 0009-921X (Print) 1528-1132 (Electronic).

Cui, J.H.; Park, S.R.; Park, K.; Choi, B.H. & Min, B.H. (2007). Preconditioning of mesenchymal stem cells with low-intensity ultrasound for cartilage formation in vivo. *Tissue Engineering,* Feb;13(2):351-360, ISSN 1076-3279 (Print) 1557-8690 (Electronic).

Dalla-Bona, D.A.; Tanaka, E.; Oka, H.; Yamano, E.; Kawai, N.; Miyauchi, M.; Takata, T. & Tanne, K. (2006). Effects of ultrasound on cementoblast metabolism in vitro: *Ultrasound in Medicine & Biology,* Jun;32(6):943-948, ISSN 0301-5629 (Print) 1879-291X (Electronic).

Dalla-Bona, D.A.; Tanaka, E.; Inubushi, T.; Oka, H.; Ohta, A.; Okada, H.; Miyauchi, M.; Takata, T. & Tanne, K. (2008). Cementoblast response to low- and high-intensity ultrasound. *Archives of Oral Biology,* Apr;53(4): p. 318-323, ISSN 0003-9969 (Print) 1879-1506 (Electronic).

De Mattei, M.; Caruso, A.; Pezzetti, F.; Pellati, A.; Stabellini, G.; Sollazzo, V. & Traina G.C. (2001). Effects of pulsed electromagnetic fields on human articular chondrocyte proliferation. *Connective Tissue Research,* 42(4):269-279, ISSN 0300-8207 (Print) 1607-8438 (Electronic).

Doan, N.; Reher, P.; Meghji, S. & Harris, M. (1999). In vitro effects of therapeutic ultrasound on cell proliferation, protein synthesis, and cytokine production by human fibroblasts, osteoblasts, and monocytes. *Journal of Oral and Maxillofacial Surgery,* Apr;57(4):409-419, ISSN 0278-2391 (Print) 1531-5053 (Electronic).

Ebisawa, K.; Hata, K.; Okada, K.; Kimata, K.; Ueda, M.; Torii, S. & Watanabe, H. (2004). Ultrasound enhances transforming growth factor beta-mediated chondrocyte differentiation of human mesenchymal stem cells. *Tissue Engineering,* May-June;10(5-6):921-929, ISSN 1076-3279 (Print) 1557-8690 (Electronic).

El-Bialy, T.; Uludag, H.; Doschak, M.; Jomha, N. & Badylak, S. (2009). Functionally integrated tissue engineered mandibular condyle. Canadian Biomedical Engineering Conference, Calgary, May, 2009. (Proceeding of the Canadian Acoustics Association meeting, October, 2008).

El-Bialy, T.H.; el-Moneim Zaki, A. & Evans, C.A. (2003). Effect of ultrasound on rabbit mandibular incisor formation and eruption after mandibular osteodistraction. *American Journal of Orthodontics and Dentofacial Orthopedics,* Oct;124(4):427-434, ISSN 0889-5406 (Print) 1097-6752 (Electronic).

El-Bialy, T.; El-Shamy I. & Graber, T.M. (2004). Repair of orthodontically induced root resorption by ultrasound in humans. *American Journal of Orthodontics and Dentofacial Orthopedics* Aug;126(2):186-193, ISSN 0889-5406 (Print) 1097-6752 (Electronic).

El-Bialy, T.H.; Royston, T.J.; Magin, R.L.; Evans, C.A.; Zaki, A.E. & Frizzell, L.A. (2002). The effect of pulsed ultrasound on mandibular distraction. *Annals of Biomedical Engineering,* Nov-Dec;30(10):1251-1261, ISSN 0090-6964 (Print) 1521-6047 (Electronic).

Emami, A.; Petrén-Mallmin, M. & Larsson, S. (1999). No effect of low-intensity ultrasound on healing time of intramedullary fixed tibial fractures. *Journal of OrthopaedicTrauma,* May;13(4):252-257, ISSN 0890-5339 (Print) 1531-2291 (Electronic).

Freed, L.E.; Hollander, A.P.; Martin, I.; Barry, J.R.; Langer, R. & Vunjak-Novakovic, G. (1998). Chondrogenesis in a cell-polymer-bioreactor system. *Experimental Cell Research,* Apr 10;240(1):58-65, ISSN 0014-4827 (Print) 1090-2422 (Electronic).

Goldstein, A.S. (2001). Effect of seeding osteoprogenitor cells as dense clusters on cell growth and differentiation. *Tissue Engineering,* Dec;7(6):817-27, ISSN 1076-3279 (Print) 1557-8690 (Electronic).
Goss, S.A.; Johnston, R.L. & Dunn, F. (1980). Compilation of empirical ultrasonic properties of mammalian tissues, II. *The Journal of the Acoustical Society of America,* Jul;68(1):93-108, ISSN 0001-4966 (Print) 1520-8524 (Electronic).
Goss, S.A.; Johnston, R.L. & Dunn, F. (1978). Comprehensive compilation of empirical ultrasonic properties of mammalian tissues. *The Journal of the Acoustical Society of America,* Aug;64(2):423-457, ISSN 0001-4966 (Print) 1520-8524 (Electronic).
Griffin, X.L.; Costello, I. & Costa, M.L. (2008). The role of low intensity pulsed ultrasound therapy in the management of acute fractures: a systematic review. *The Journal of Trauma,* Dec;65(6):1446-1452, ISSN 0022-5282 (Print) 1529-8809 (Electronic).
Hadjiargyrou, M.; McLeod, K.; Ryaby, J.P. & Rubin, C. (1998). Enhancement of fracture healing by low intensity ultrasound. *Clinical Orthopaedics and Related Research,* Oct;(355 Suppl):S216-229, ISSN 0009-921X (Print) 1528-1132 (Electronic).
Handolin, L.; Kiljunen, V.; Arnala, I.; Kiuru, M.J.; Pajarinen, J.; Partio, E.K. & Rokkanen, P. (2005). No long-term effects of ultrasound therapy on bioabsorbable screw-fixed lateral malleolar fracture. *Scandinavian Journal of Surgery* 94(3):239-242, ISSN 1457-4969 (Print).
Harris, G.R. (2005). Progress in medical ultrasound exposimetry. *IEEE Transactions on Ultrasonics, Ferroelectrics, and Frequency Control,* May;52(5):717-736, ISSN 0885-3010 (Print) 1525-8955 (Electronic).
Hartig, M.; Joos, U. & Wiesmann, H.P. (2000). Capacitively coupled electric fields accelerate proliferation of osteoblast-like primary cells and increase bone extracellular matrix formation in vitro. *European Biophysics Journal,* 29(7):499-506, ISSN 0175-7571 (Print) 1432-1017 (Electronic).
Hata, T.; Aoki, S.; Manabe, A.; Hata, K. & Miyazaki, K. (1997). Three dimensional ultrasonography in the first trimester of human pregnancy. *Human Reproduction (Oxford, England),* 12(8):1800-1804, ISSN 0268-1161 (Print) 1460-2350 (Electronic).
Heckman, J.D.; Ryaby, J.P.; McCabe, J.; Frey, J.J. & Kilcoyne, R.F. (1994). Acceleration of tibial fracture-healing by non-invasive, low-intensity pulsed ultrasound. *The Journal of Bone and Joint Surgery. American volume,* Jan;76(1):26-34, ISSN 0021-9355 (Print) 1535-1386 (Electronic).
Hill, G.E.; Fenwick, S.; Matthews, B.J.; Chivers, R.A. & Southgate, J. (2005). The effect of low-intensity pulsed ultrasound on repair of epithelial cell monolayers in vitro. *Ultrasound in Medical & Biology,* Dec;31(12):1701-1706, ISSN 0301-5629 (Print) 1879-291X (Electronic).
Holy, C.E.; Shoichet, M.S. & Davies, J.E. (2000). Engineering three-dimensional bone tissue in vitro using biodegradable scaffolds: investigating initial cell-seeding density and culture period. *Journal of Biomedical Materials Research,* Sep 5;51(3):376-382, ISSN 0021-9304 (Print) 1097-4636 (Electronic).
Horwitz, E.M. (2004). Dkk-1-mediated expansion of adult stem cells. *Trends in Biotechnology,* Aug;22(8):386-388, ISSN 0167-7799 (Print) 1879-3096 (Electronic).
Hsieh-Bonassera, N.D.; Wu, I.; Lin, J.K.; Schumacher, B.L.; Chen, A.C.; Masuda, K.; Bugbee, W.D. & Sah, R.L. (2009). Expansion and redifferentiation of chondrocytes from osteoarthritic cartilage: cells for human cartilage tissue engineering. *Tissue*

engineering, Part A, May 20 [Epub ahead of print], ISSN 1076-3279 (Print) 1557-8690 (Electronic).

Hsu, S.H.; Kuo, C.C.; Whu, S.W.; Lin, C.H. & Tsai, C.L. (2006). The effect of ultrasound stimulation versus bioreactors on neocartilage formation in tissue engineering scaffolds seeded with human chondrocytes in vitro. *Biomolecular Engineering,* Oct;23(5):259-264, ISSN 1389-0344 (Print).

Ikeda, K.; Takayama, T.; Suzuki, N.; Shimada, K.; Otsuka, K. & Ito, K. (2006). Effects of low-intensity pulsed ultrasound on the differentiation of C2C12 cells. *Life Sciences,* Oct 12;79(20):1936-1943, ISSN 0024-3205 (Print) 1879-0631 (Electronic).

Inagaki, H.; Suzuki, J.; Ogawa, M.; Taniyama, Y.; Morishita, R. & Isobe, M. (2006). Ultrasound-microbubble-mediated NF-&kappaB decoy transfection attenuates neointimal formation after arterial injury in mice. *Journal of Vascular Research,* Feb;43(1):12-18, ISSN 1018-1172 (Print)1423-0135 (Electronic).

Inubushi, T.; Tanaka, E.; Rego, E.B.; Kitagawa, M.; Kawazoe, A.; Ohta, A,; Okada, H.; Koolstra, J.H.; Miyauchi, M.; Takata, T. &, Tanne, K. (2008). Effects of ultrasound on the proliferation and differentiation of cementoblast lineage cells. *Journal of Periodontology,* Oct;79(10):1984-1990, ISSN 0022-3492 (Print).

Iwabuchi, S.; Ito, M.; Hata, J.; Chikanishi, T.; Azuma, Y. & Haro, H. (2005). In vitro evaluation of low-intensity pulsed ultrasound in herniated disc resorption. *Biomaterials,* Dec;26(34):7104-7114, ISSN 0142-9612 (Print) 1878-5905 (Electronic).

Iwashina, T.; Mochida, J.; Miyazaki, T.; Watanabe, T.; Iwabuchi, S.; Ando, K.; Hotta, T. & Sakai, D. (2006). Low-intensity pulsed ultrasound stimulates cell proliferation and proteoglycan production in rabbit intervertebral disc cells cultured in alginate. *Biomaterials,* Jan;27(3):354-361, ISSN 0142-9612 (Print) 1878-5905 (Electronic).

Jacobsen, S.E.; Okkenhaug, C.; Myklebust, J.; Veiby, O.P. & Lyman, S.D. (1995). The FLT3 ligand potently and directly stimulates the growth and expansion of primitive murine bone marrow progenitor cells in vitro: synergistic interactions with interleukin (IL) 11, IL-12, and other hematopoietic growth factors. *The Journal of Experimental Medicine,* Apr 1;181(4):1357-1363, ISSN 0022-1007 (Print)1540-9538 (Electronic).

Jingushi, S.; Mizuno, K.; Matsushita, T. & Itoman, M. (2007). Low-intensity pulsed ultrasound treatment for postoperative delayed union or nonunion of long bone fractures. *Journal of Orthopaedic Science,* Jan;12(1):35-41, ISSN 0949-2658 (Print) 1436-2023 (Electronic).

Kaigler, D.; Cirelli, J.A. & Giannobile, W.V. (2006). Growth factor delivery for oral and periodontal tissue engineering. *Expert Opinion on Drug Delivery.* Sep;3(5):647-662, ISSN 1742-5247 (Print)1744-7593 (Electronic).

Kamimura, K. & Liu, D. (2009). Physical approaches for nucleic acid delivery to liver. *The AAPS Journal,* Jan;10(4):589-595, ISSN 1550-7416 (Electronic).

Kimelman, N.; Pelled, G.; Gazit, G.P.Z. & Gazit, D. (2006). Applications of gene therapy and adult stem cells in bone bioengineering. *Regenerative Medicine,* Aug;1(4):549-561, ISSN 1746-0751 (Print)1746-076X (Electronic).

Kobayashi, Y.; Sakai, D.; Iwashina, T.; Iwabuchi, S. & Mochida, J. (2009). Low-intensity pulsed ultrasound stimulates cell proliferation, proteoglycan synthesis and expression of growth factor-related genes in human nucleus pulposus cell line. *European Cells & Materials,* Jun 30;17:15-22, ISSN 1473-2262 (Electronic).

Korstjens, C.M.; van der Rijt, R.H.; Albers, G.H.; Semeins, C.M. & Klein-Nulend. J. (2008). Low-intensity pulsed ultrasound affects human articular chondrocytes in vitro. *Medical & Biological Engineering & Computing,* Dec;46(12):1263-1270, ISSN 0140-0118 (Print)1741-0444 (Electronic).

Kristiansen, T.K.; Ryaby, J.P.; McCabe, J.; Frey, J.J. & Roe, L.R. (1997). Accelerated healing of distal radial fractures with the use of specific, low-intensity ultrasound. A multicenter, prospective, randomized, double-blind, placebo-controlled study. *The Journal of Bone and Joint Surgery. American Volume,* Jul;79(7):961-973, ISSN 0021-9355 (Print) 1535-1386 (Electronic).

Leddy, H.A.; Awad, H.A. & Guilak, F. (2004). Molecular diffusion in tissue-engineered cartilage constructs: effects of scaffold material, time, and culture conditions. *Journal of Biomedical Materials Research. Part B, Applied Biomaterials,* Aug 15;70(2):397-406, ISSN 0142-9612 (Print) 1878-5905 (Electronic).

Lefebvre. P.; Lin, J.; Linzer, D.I. & Cohen, I. (2001). Murine prolactin-like protein E synergizes with human thrombopoietin to stimulate expansion of human megakaryocytes and their precursors. *Experimental Hematology,* Jan;29(1):51-58, ISSN 0301-472X (Print) 1873-2399 (Electronic).

Leung, K.S.; Cheung, W.H.; Zhang, C.; Lee, K.M. & Lo, H.K. (2004). Low intensity pulsed ultrasound stimulates osteogenic activity of human periosteal cells. *Clinical Orthopaedics and Related Research,* Jan;(418):253-259, ISSN 0009-921X (Print)1528-1132 (Electronic).

Li,Y.; Ma, T.; Kniss, D.A.; Lasky, L.C. & Yang, S.T. (2001). Effects of filtration seeding on cell density, spatial distribution, and proliferation in nonwoven fibrous matrixes, *Biotechnology Progress,* Sep-Oct;17(5):935:944, ISSN 8756-7938 (Print) 1520-6033 (Electronic).

Li, Y.P.; Zhou, S.X.; Schmelz, A. & Bachem, M.G. (2007a). [Molecular mechanisms of low intensity pulsed ultrasound-mediated cellular behavior in human primary macrophages]. *Zhongguo Shi Yan Xue Ye Xue Za Zhi/Zhongguo Bing Li Sheng Li Xue Hui,* Dec;15(6):1253-1256, ISSN 1009-2137 (Print).

Li, Y.P.; Zhou, S.X.; Andreas, S. & Max, G.B. (2007b). Effect of LIPUS on the cellular behavior of human primary macrophages. *Xi Bao Yu Fen Zi Mian Yi Xue Za Zhi,* Dec;23(12):1113-1116, ISSN 1007-8738 (Print).

Lu, H.; Qin, L.; Cheung, W.; Lee, K.; Wong, W. & Leung, K. (2008). Low-intensity pulsed ultrasound accelerated bone-tendon junction healing through regulation of vascular endothelial growth factor expression and cartilage formation. *Ultrasound in Medicine & Biology,* Aug;34(8):1248-1260, ISSN 0301-5629 (Print) 1879-291X (Electronic).

Lubbert, P.H.; van der Rijt, R.H.; Hoorntje, L.E. & van der Werken, C. (2008). Low-intensity pulsed ultrasound (LIPUS) in fresh clavicle fractures: a multi-centre double blind randomised controlled trial. *Injury,* Dec;39(12):1444-1452, ISSN 0020-1383 (Print) 1879-0267 (Electronic).

Maylia, E. & Nokes, L.D. (1999). The use of ultrasonics in orthopaedics. a review. *Technology and Health Care,* 7(1): 1–28, ISSN 0928-7329 (Print).

Mayr, E.; Rudzki, M.M.; Rudzki, M.; Borchardt, B.; Häusser, H. & Rüter, A. (2000). [Does low intensity, pulsed ultrasound speed healing of scaphoid fractures?].

Handchirurgie, Mikrochirurgie, Plastische Chirurgie, Mar;32(2):115-122, ISSN 0722-1819 (Print) 1439-3980 (Electronic).

Mende, U.; Zoller, J.; Dietz, A.; Wannenmacher, M.; Born, I.A. & Maier, H. (1996). [Ultrasound diagnosis in primary staging of head-neck tumors]. *Der Radiologe,* Mar;36(3):207-216, ISSN 0033-832X (Print)1432-2102 (Electronic).

Mostafa, N.Z.; Uludağ, H.; Dederich, D.N.; Doschak, M.R. & El-Bialy, T.H. (2009). Anabolic effects of low-intensity pulsed ultrasound on human gingival fibroblasts. *Archives of Oral Biology,* Aug;54(8):743-8, ISSN 0003-9969 (Print) 1879-1506 (Electronic).

Mukai, S.; Ito, H.; Nakagawa, Y.; Akiyama, H.; Miyamoto, M. & Nakamura, T. (2005). Transforming growth factor-beta1 mediates the effects of low-intensity pulsed ultrasound in chondrocytes. *Ultrasound in Medicine & Biology,* Dec;31(12):1713-1721, ISSN 0301-5629 (Print) 1879-291X (Electronic).

Nolte, P.A.; van der Krans, A.; Patka, P.; Janssen, I.M.; Ryaby, J.P. & Albers, G.H. (2001). Low-intensity pulsed ultrasound in the treatment of nonunions. *The Journal of Trauma,* Oct;51(4):693-702; discussion 702-3, ISSN 0022-5282 (Print) 1529-8809 (Electronic).

Nozaki, T.; Ogawa, R.; Watanabe, A.; Nishio, R.; Fuse, H. & Kondo, T. (2006). [Ultrasound-mediated gene transfection: problems to be solved and future possibilities]. *Journal of Medical Ultrasonics.* Nov;33(3):135-142, ISSN 1346-4523 (Print) 1613-2254 (Electronic).

Osawa, K.; Okubo, Y.; Nakao, K.; Koyama, N. & Bessho, K. (2009). Osteoinduction by microbubble-enhanced transcutaneous sonoporation of human bone morphogenetic protein-2. *The Journal of Gene Medicine,* Jul;11(7):633-641, ISSN 1099-498X (Print) 1521-2254 (Electronic).

Panossian, A.; Ashiku, S.; Kirchhoff, C.H.; Randolph, M.A. & Yaremchuk, M.J. (2001). Effects of cell concentration and growth period on articular and ear chondrocyte transplants for tissue engineering. *Plastic and Reconstructive Surgery,* Aug;108:392-402, ISSN 0032-1052 (Print) 1529-4242 (Electronic).

Puelacher, W.C.; Kim, S.W.; Vacanti, J.P.; Schloo, B.; Mooney, D. & Vacanti, C.A. (1994). Tissue-engineered growth of cartilage: the effect of varying the concentration of chondrocytes seeded onto synthetic polymer matrices. *International Journal of Oral and Maxillofacial Surgery,* Feb;23(1):49-53, ISSN 0901-5027 (Print) 1399-0020 (Electronic).

Purpura, K,A,; Aubin, J.E. & Zandstra, P.W. (2004). Sustained in vitro expansion of bone progenitors is cell density dependent. *Stem Cells.* 2004;22(1):39-50, ISSN 0250-6793.

Rabie, A.B.; Shum, L. & Chayanupatkul, A. (2002). VEGF and bone formation in the glenoid fossa during forward mandibular positioning. *American Journal of Orthodontics and Dentofacial Orthopedics,* Aug;122(2):202–209, ISSN 0889-5406 (Print) 1097-6752 (Electronic).

Rabie, A.B.; Zhao, Z.; Shen. G.; Hagg, E.U. & Robinson, W. (2001). Osteogenesis in the glenoid fossa in response to mandibular advancement. *American Journal of Orthodontics and Dentofacial Orthopedics,* Apr;119(4):390–400, ISSN 0889-5406 (Print) 1097-6752 (Electronic).

Reher, P.; Doan, N.; Bradnock, B.; Meghji, S. & Harris, M. (1999). Effect of ultrasound on the production of IL-8, basic FGF and VEGF. *Cytokine,* Jun;11(6):416-423, ISSN 1043-4666 (Print) 1096-0023 (Electronic).

Rue, J.P.; Armstrong, D.W. 3rd; Frassica, F.J.; Deafenbaugh, M. & Wilckens, J.H. (2004). The effect of pulsed ultrasound in the treatment of tibial stress fractures. *Orthopedics.* Nov;27(11):1192-1195, ISSN 0147-7447 (Print) 1938-2367 (Electronic).

Saini, S., & Wick, T.M. (2003). Concentric cylinder bioreactor for production of tissue engineered cartilage: effect of seeding density and hydrodynamic loading on construct development. *Biotechnology Progress,* Mar-Apr;19(2):510-521, ISSN 8756-7938 (Print)1520-6033 (Electronic).

Saito, M.; Soshi, S.; Tanaka, T. & Fujii, K. (2004). Intensity-related differences in collagen post-translational modification in MC3T3-E1 osteoblasts after exposure to low- and high-intensity pulsed ultrasound. *Bone.* Sep;35(3):644-655, ISSN 8756-3282 (Print) 1873-2763 (Electronic).

Saito, M.; Mazda, O.; Takahashi, K.A.; Arai Y.; Kishida, T.; Shin-Ya, M.; Inoue, A.; Tonomura, H.; Sakao, K.; Morihara, T.; Imanishi, J.; Kawata, M. & Kubo, T. (2007). Sonoporation mediated transduction of pDNA/siRNA into joint synovium in vivo. *Journal of Orthopaedic Research,* Nov;25(10):1308-1316, ISSN 0736-0266 (Print) 1554-527X (Electronic).

Sakai, T.; Kawaguchi, M. & Kosuge, Y. (2009). siRNA-mediated gene silencing in the salivary gland using in vivo microbubble-enhanced sonoporation. *Oral Diseases,* Jun 10 (Epub ahead of print), ISSN 1354-523X (Print) 1601-0825 (Electronic).

Scheven, B.A.; Man, J.; Millard, J.L.; Cooper, P.R.; Lea, S.C.; Walmsley, A.D. & Smith, A.J. (2009a). VEGF and odontoblast-like cells: stimulation by low frequency ultrasound. *Archives of Oral Biology,* Feb;54(2):185-191, ISSN 0003-9969 (Print) 1879-1506 (Electronic).

Scheven, B.A.; Millard, J.L.; Cooper, P.R.; Lea, S.C.; Walmsley, A.D. & Smith, A.J. (2007). Short-term in vitro effects of low frequency ultrasound on odontoblast-like cells. *Ultrasound in Medicine & Biology,* Sep;33(9):1475-1482, ISSN 0301-5629 (Print) 1879-291X (Electronic).

Scheven, B.A.; Shelton, R.M.; Cooper, P.R.; Walmsley, A.D. & Smith, A.J. (2009b). Therapeutic ultrasound for dental tissue repair. *Medical Hypotheses,* Oct;73(4):591-593, ISSN 0306-9877 (Print) 1532-2777 (Electronic).

Schumann, D.; Kujat, R.; Zellner, J.; Angele, M.K.; Nerlich, M.; Mayr, E. & Angele, P. (2006a). Treatment of human mesenchymal stem cells with pulsed low intensity ultrasound enhances the chondrogenic phenotype in vitro. *Biorheology.* 43(3-4):431-43, ISSN 0006-355X (Print) 1878-5034 (Electronic).

Schumann, D.; Kujat, R.; Nerlich, M. & Angele, P. (2006b). Mechanobiological conditioning of stem cells for cartilage tissue engineering. *Bio-medical Materials and Engineering,* 16(4 Suppl):S37-52, ISSN 0959-2989 (Print) 1878-3619 (Electronic).

Sena, K.; Leven, R.M.; Mazhar, K.; Sumner, D.R. & Virdi, A.S. (2005). Early gene response to low-intensity pulsed ultrasound in rat osteoblastic cells. *Ultrasound in Medicine & Biology,* May;31(5):703-708, ISSN 0301-5629 (Print) 1879-291X (Electronic).

Shen, Z.P.; Brayman, A.A.; Chen, L. & Miao, C.H. (2008). Ultrasound with microbubbles enhances gene expression of plasmid DNA in the liver via intraportal delivery. *Gene Therapy,* Aug;15(16):1147-1155, ISSN 0969-7128 (Print) 1476-5462 (Electronic).

Sheyn, D.; Kimelman-Bleich, N.; Pelled, G.; Zilberman, Y.; Gazit, D. & Gazit, Z. (2008). Ultrasound-based nonviral gene delivery induces bone formation in vivo. *Gene Therapy*. Febr;15(4):257-266, ISSN 0969-7128 (Print) 1476-5462 (Electronic).

Suzuki, A.; Takayama, T.; Suzuki, N.; Kojima, T.; Ota, N.; Asano, S. & Ito, K. (2009). Daily low-intensity pulsed ultrasound stimulates production of bone morphogenetic protein in ROS 17/2.8 cells. *Journal of Oral Science,* Mar;51(1):29-36, ISSN 1343-4934 (Print) 1880-4926 (Electronic).

Tachibana, K.; Irie, Y. & Ogawa, K. (2006). [Emerging technologies in therapeutic ultrasound]. *Journal of Medical Ultrasonics,* Feb;33(6):631-639, ISSN 1346-4523 (Print) 1613-2254 (Electronic).

Takagi, M. (2005). Cell processing engineering for ex-vivo expansion of hematopoietic cells. *Journal of Bioscience and Bioengineering,* Mar;99(3):189-196, ISSN 1389-1723 (Print) 1347-4421 (Electronic).

Takeuchi, R.; Ryo, A.; Komitsu, N.; Mikuni-Takagaki, Y.; Fukui, A.; Takagi, Y.; Shiraishi, T.; Morishita, S.; Yamazaki, Y.; Kumagai, K.; Aoki, I. & Saito, T. (2008). Low-intensity pulsed ultrasound activates the phosphatidylinositol 3 kinase/Akt pathway and stimulates the growth of chondrocytes in three-dimensional cultures: a basic science study. *Arthritis Research & Therapy,* 10(4):R77, ISSN 1478-6354 (Print) 1478-6362 (Electronic).

Tam, K.F.; Cheung, W.H.; Lee, K.M.; Qin, L. & Leung, K.S. (2008). Osteogenic effects of low-intensity pulsed ultrasound, extracorporeal shockwaves and their combination - an in vitro comparative study on human periosteal cells. *Ultrasound in Medicine & Biology,* Dec;34(12):1957-1965, ISSN 0301-5629 (Print) 1879-291X (Electronic).

Tamama, K.; Fan, V.H.; Griffith, L.G.; Blair, H.C. & Wells, A. (2005). Epidermal growth factor as candidate for ex vivo expansion of bone marrow-derived mesenchymal stem cells. *Stem Cells,* Mar;24(3):686-695, ISSN 0250-6793.

Tien, Y.C.; Lin, S.D.; Chen, C.H.; Lu, C.C.; Su, S.J. & Chih, T.T. (2008). Effects of pulsed low-intensity ultrasound on human child chondrocytes. *Ultrasound in Medicine & Biology,* Jul;34(7):1174-1181, ISSN 0301-5629 (Print) 1879-291X (Electronic).

Tsai, C.L.; Chang, W.H. & Liu, T.K. (1992). Preliminary studies of duration and intensity of ultrasonic treatments on fracture repair. *The Chinese Journal Of Physiology,* 35(1):21-26, ISSN 0304-4920 (Print). Erratum in: *The Chinese journal of physiology,* 35:168, 1992.

Turnbull, D.H. & Foster, F.S. (2002). In vivo ultrasound biomicroscopy in developmental biology. *Trends in Biotechnology,* 20(8):S29-33, ISSN 0167-7799 (Print) 1879-3096 (Electronic).

Walker, N.A.; Denegar, C.R. & Preische, J. (2007). Low-intensity pulsed ultrasound and pulsed electromagnetic field in the treatment of tibial fractures: a systematic review. *Journal of Athletic Training,* Oct-Dec;42(4):530-535, ISSN 1062-6050 (Print) 1938-162X (Electronic).

Wang, Y.; Singh, A.; Xu, P.; Pindrus, M.A.; Blasioli, D.J. & Kaplan, D.L. (2006). Expansion and osteogenic differentiation of bone marrow-derived mesenchymal stem cells on a vitamin C functionalized polymer. *Biomaterials,* Jun;27(17):3265-3273, ISSN 0142-9612 (Print) 1878-5905 (Electronic).

Warden, S.J.; Bennell, K.L.; McMeeken, J.M. & Wark, J.D. (2000). Acceleration of fresh fracture repair using the sonic accelerated fracture healing system (SAFHS): a review. *Calcified Tissue International,* Feb;66(2):157-163, ISSN 0171-967X (Print) 1432-0827 (Electronic).

Wu, C.C.; Lewallen, D.G.; Bolander, M.E.; Bronk, M.E.; Kinnick, R. & Greenlead, J.F. (1996). Exposure to low-intensity ultrasound stimulates aggrecan gene expression by cultured chondrocytes. *Transactions - Orthopaedic Research Society,* 1996;21:622.

Young, S.R. & Dyson, M. (1990a). Macrophage responsiveness to therapeutic ultrasound. *Ultrasound in Medicine & Biology,* 16(8):809-816, ISSN 0301-5629 (Print) 1879-291X (Electronic).

Young, S.R. & Dyson, M. (1990b). Effect of therapeutic ultrasound on the healing of full-thickness excised skin lesions. *Ultrasonics,* May;28(3):175-180, ISSN 0041-624X (Print) 1874-9968 (Electronic).

Young, S.R.; & Dyson, M. (1990c). The effect of therapeutic ultrasound on angiogenesis. *Ultrasound in Medicine & Biology,* 16(3):261-269, ISSN 0301-5629 (Print) 1879-291X (Electronic).

Zhang, C.C.; Kaba, M.; Ge, G.; Xie, K.; Tong, W.; Hug, C. & Lodish, H.F. (2006). Angiopoietin-like proteins stimulate ex vivo expansion of hematopoietic stem cells. *Nature Medicine,* Feb;12(2):240-245, ISSN 1078-8956 (Print) 1546-170X (Electronic).

Zhou, S.; Schmetz, A.; Seufferlein, T.; Li, Y.; Zhao, J. & Bachem, M.G. (2004). Molecular mechanisms of low intensity pulsed ultrasound in human skin fibroblasts. *The* Journal of Biological Chemistry, Dec 24;279(52):54463-9, ISSN 0021-9258 (Print) 1083-351X (Electronic).

Ziskin, M.C. (1987). Applications of ultrasound in medicine: comparison with other modalities, Ultrasound: medical applications, biological effects, and hazard potential, Repacholi, M.H., Grandolfo, M., Rindi, A., (Eds.), 49-59, Plenum Press, ISBN 0306424118, New York, NY.

Bioreactors in Tissue Engineering

S. Partap[1,2], N. A. Plunkett[1,2] and F. J. O'Brien[1,2]
[1]*Department of Anatomy, Royal College of Surgeons in Ireland, 123 St. Stephen's Green, Dublin 2*
[2]*Trinity Centre for Bioengineering, Department of Mechanical Engineering, Trinity College Dublin, Dublin 2*
Ireland

1. Introduction

1.1 Tissue Engineering

The field of tissue engineering (regenerative medicine) aims to repair and regenerate damaged tissues by developing biological substitutes that mimic the natural extracellular matrix to help guide the growth of new functional tissue *in vitro* or *in vivo* to restore, maintain or improve tissue function. Tissue engineering technologies are based on a biological triad and involve the successful interaction between three components: (1) the scaffold that holds the cells together to create the tissue's physical form, (2) the cells that synthesise the tissue and (3) signalling mechanisms (*i.e.* mechanical and chemical signals) that direct the cells to express the desired tissue phenotype (Langer & Vacanti, 1993). Bioreactors can be used to provide the signals in this latter area in order to influence biological processes by the application of a mechanical stimulus, and may also be used as an alternative to, or in conjunction with growth factors.

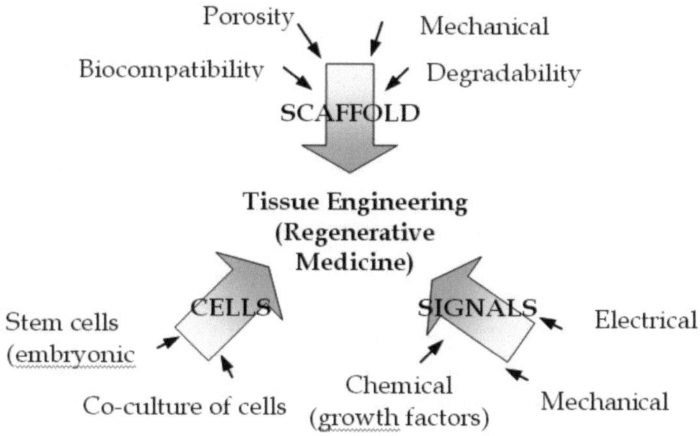

Fig. 1. The tissue engineering triad; factors that need to be considered when designing a suitable structure for tissue engineering applications (Lyons, *et al.*, 2008).

1.2 Use of Bioreactors in Tissue Engineering

A tissue engineering bioreactor can be defined as a device that uses mechanical means to influence biological processes (Darling & Athanasiou, 2003). Bioreactors can be used to aid in the *in vitro* development of new tissue by providing biochemical and physical regulatory signals to cells and encouraging them to undergo differentiation and/or to produce extracellular matrix prior to *in vivo* implantation. Bioreactors are devices in which biological or biochemical processes develop under a closely monitored and tightly controlled environment.

Cells respond to mechanical stimulation and bioreactors can be used to apply mechanical stimulation to cells. This can encourage cells to produce extracellular matrix (ECM) in a shorter time period and in a more homogeneous manner than would be the case with static culture. For example, in comparisons between ECM protein levels of equine articluar chondrocytes cultured on polyglycolic acid scaffolds after 5 weeks in culture, constructs cultured under hydrostatic pressure showed significant improvements over constructs cultured in static medium (Carver & Heath, 1999). A benefit of ECM production is the increase in mechanical stiffness that it provides to the construct. A six-fold increase in equilibrium aggregate modulus (an intrinsic property of cartilage which is a measure of stiffness) was found after 28 days of culture in a compression bioreactor compared to free swelling controls (Mauck, *et al.*, 2000). Another important application of bioreactors is in cellular differentiation. Mechanical stimulation can be used to encourage stem cells down a particular path and hence provide the cell phenotype required. Bioreactors can provide biochemical and physical regulatory signals that guide differentiation (Altman, *et al.*, 2002). There is great potential for using mesenchymal stem cells and other multipotent cells to generate different cell types and bioreactors can play an important role in this process.

As well as providing mechanical stimulation, bioreactors can also be used to improve cellular spatial distribution. A heterogeneous cell distribution is a major obstacle to developing any three-dimensional tissue or organ *in vitro*. Defects requiring tissue engineering solutions are typically many millimetres in size (Goldstein, *et al.*, 2001). Scaffolds in such a size range are easily fabricated, however, problems arise when culturing cells on these scaffolds. As the size of the scaffold increases, diffusion of nutrients to the centre of the construct becomes more difficult. Static culture conditions result in scaffolds with few cells in the centre of the construct (Cartmell, *et al.*, 2003). It is hypothesised that this is due to limited cell penetration during seeding, cell migration to the scaffold periphery during culture, or cell death in the centre of the scaffold (Glowacki, *et al.*, 1998, Ishaug-Riley, *et al.*, 1998, Goldstein, *et al.*, 2001, Yu, *et al.*, 2004, Gomes, *et al.*, 2005). The only mechanism by which nutrients and waste can move when a scaffold is in static culture is by diffusion. It has been shown that despite homogeneous cell seeding, after long periods in culture, more cells are found on the periphery of constructs (Cartmell, *et al.*, 2003) leading to peripheral encapsulation which hinders nutrient and waste exchange from the centre, resulting in core degradation of tissue engineered constructs. This is of major concern in the field of tissue engineering, and is a major obstacle in the formation of a viable tissue *in vitro*. For this reason, for a number of tissue types, the move towards clinical trials has been slow and progress to date in engineering significant quantities of functional tissue *in vitro* for implantation in humans *in vivo* has been somewhat disappointing.

Thus, bioreactors can be used in tissue engineering applications to overcome problems associated with traditional static culture conditions, improve cellular distribution and accelerate construct maturation (Freed, *et al.*, 2006) whilst applying biophysical signals to constructs to improve tissue formation *in vitro* prior to *in vivo* implantation. In general, bioreactors are designed to perform at least one of the following five functions, to (1) provide a spatially uniform cell distribution, (2) maintain the desired concentration of gases and nutrients in culture medium, (3) facilitate mass transport to the tissue, (4) expose the construct to physical stimuli and/or (5) provide information about the formation of 3D tissue (Vunjak-Novakovic, *et al.*, 1998, Bancroft, *et al.*, 2003)

1.3 Bioreactor Design Requirements

The design of the bioreactor should be as simple as possible *e.g.* avoiding the introduction of machined recesses which could become breeding grounds for micro-organisms. Simplicity in design should also mean that the bioreactor is quick to assemble and disassemble. Apart from being more efficient, this ensures that cell-seeded constructs inserted into the bioreactor are out of the incubator for the minimum amount of time possible. This minimises the risk to the cells and the experiment being undertaken.

The detailed requirements for bioreactor design are tissue- and/or application- specific, however, there are a few general principles which have to be adhered to when developing a bioreactor. The material selection is very important as it is vital to ensure that the materials used to create the bioreactor do not elicit any adverse reaction from the cultured tissue. Any material which is in contact with media must be biocompatible or bioinert. This eliminates the use of most metals, although stainless steel can be used if it is treated so that chromium ions do not leach out into the medium. Numerous plastics comply with this constraint but there are further limitations on material selection that must also be kept in mind. Materials must be usable at 37°C in a humid atmosphere. They must be able to be sterilised if they are to be re-used. Bioreactor parts can be sterilised by autoclaving or disinfected by submersion in alcohol. If they are to be autoclaved, materials that can withstand numerous cycles of high temperature and pressure must be used in bioreactor manufacture. Alternatively, some non-sterilisable disposable bioreactor parts may be used which can be replaced after each use of the bioreactor. Other material choices are between transparent or opaque and flexible or inflexible materials. Materials with different properties are needed for various components in the bioreactor. For example, transparent materials can be of benefit in allowing the construct to be monitored in the bioreactor during culture while flexible tubing can help with assembly of the bioreactor.

The specific application of the bioreactor must be kept in mind during the design process to ensure that all the design constraints are met. If various parameters such as pH, nutrient concentration or oxygen levels are to be monitored, these sensors should be incorporated into the design. If a pump or motor is to be used, it must be small enough to fit into an incubator and also be usable at 37°C and in a humid environment. The forces needed for cellular stimulation are very small so it is important to ensure that the pump/motor has the capability to apply small forces accurately. In any design involving fluids, problems can arise with leaking fluid seals and, if possible, the need for seals should be reduced. However, in most cases, fluid seals are necessary and good design should decrease the problems with them. If a prototype bioreactor is being designed, it is worthwhile thinking about scale up opportunities for the bioreactor from the outset. This may mean designing a device that is relatively easy to enlarge without changing its characteristics or designing a simple device of which many more can be made so that numerous scaffolds can be cultured at one time.

2. Types of Bioreactors in Tissue Engineering

Bioreactors are laboratory tissue culture devices that provide a controllable, mechanically active environment that can be used to study and potentially improve engineered tissue structure, properties and integration. Many tissue engineering studies employ conventional methods (static seeding) that result in constructs comprising a thin tissue like layer at the base of the scaffold due to gravitiational settling of the cells. In contrast convective mixing (spinner flasks) and convective flow (flow perfusion) can improve initial cell seeding and homogeneity, and thereby improve tissue architecture (Freed, *et al.*, 2006). Numerous types of bioreactors including spinner flasks, rotating wall, compression, strain and flow perfusion bioreactors are used in various tissue engineering applications, which will all be discussed below. All of these bioreactors rely on forced media flow through and/or around the scaffold to provide nutrient and waste exchange within the scaffold.

2.1 Spinner Flask Bioreactors

One of the most basic bioreactors, a spinner flask induces mixing of oxygen and nutrients throughout the medium and reduces the concentration boundary layer at the construct surface. In a spinner flask, scaffolds are suspended at the end of needles in a flask of culture media. A magnetic stirrer mixes the media and the scaffolds are fixed in place with respect to the moving fluid (figure 2). Flow across the surface of the scaffolds results in eddies which are turbulent instabilities consisting of clumps of fluid particles that have a rotational structure superimposed on the mean linear motion of the fluid particles. They are associated with transitional and turbulent flow. It is *via* these eddies that fluid transport to the centre of the scaffold is thought to be enhanced (Goldstein, *et al.*, 2001). Typically, spinner flasks are around 120 mL in volume (although much larger flasks of up to 8 litres have been used), are run at 50-80 rpm and 50% of the medium used in them is changed every two days (Freshney, 2000). Cartilage constructs have been grown in spinner flasks to thicknesses of 0.5 mm (Freed & Vunjak-Novakovic, 2000). While this is an improvement on cartilage grown in static culture, it is still too thin for clinical use. Mass transfer in the flasks is not good enough to deliver homogeneous cell distribution throughout scaffolds and cells predominantly reside on the construct periphery.

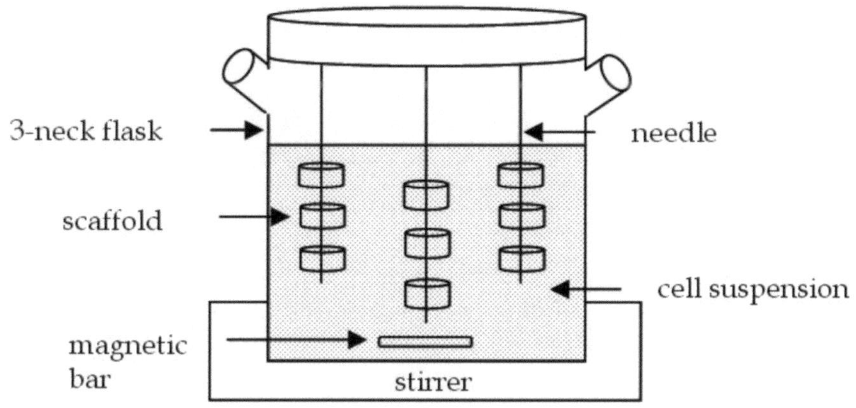

Fig. 2. A spinner flask bioreactor. Scaffolds are suspended in medium and the medium is stirred using a magnetic stirrer to improve nutrient delivery to the scaffold

2.2 Rotating Wall Bioreactors

The rotating wall bioreactor was developed by NASA (Schwarz, *et al.*, 1992). It was originally designed with a view to protecting cell culture experiments from high forces during space shuttle take off and landing. However, the device has proved useful in tissue engineering here on Earth. In a rotating wall bioreactor, scaffolds are free to move in media in a vessel. A rotating wall vessel bioreactor consists of a cylindrical chamber in which the outer wall, inner wall, or both are capable of rotating at a constant angular speed (figure 3). The vessel wall is then rotated at a speed such that a balance is reached between the downward gravitational force and the upward hydrodynamic drag force acting on each scaffold. The wall of the vessel rotates, providing an upward hydrodynamic drag force that balances with the downward gravitational force, resulting in the scaffold remaining suspended in the media. Dynamic laminar flow generated by a rotating fluid environment is an alternative and efficient way to reduce diffusional limitations of nutrients and wastes while producing low levels of shear. Media can be exchanged by stopping the rotation temporarily or by adding a fluid pump whereby media is constantly pumped through the vessel. Fluid transport is enhanced in a similar fashion to the mechanism in spinner flasks and the devices also provide a more homogeneous cell distribution than static culture. Gas exchange occurs through a gas exchange membrane and the bioreactor is rotated at speeds of 15-30 rpm. Cartilage tissue of 5 mm thickness has been grown in this type of bioreactor after seven months of culture (Freed, *et al.*, 1997). As tissue grows in the bioreactor, the rotational speed must be increased in order to balance the gravitational force and ensure the scaffold remains in suspension.

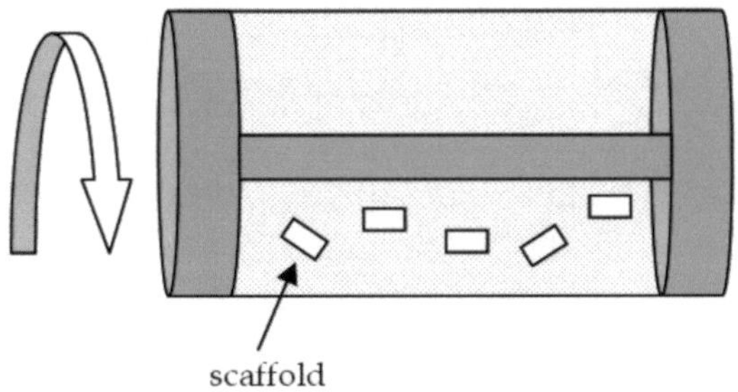

Fig. 3. A rotating wall vessel bioreactor. Scaffolds are suspended in medium due to opposing gravitational and drag forces

2.3 Compression Bioreactors

Another widely used type of bioreactor is the compression bioreactor. This class of bioreactor is generally used in cartilage engineering and can be designed so that both static and dynamic loading can be applied. This is because static loading has been found to have a negative effect on cartilage formation in comparison to dynamic loading, which is more representative of physiological loading (Darling & Athanasiou, 2003). In general, compression bioreactors consist of a motor, a system providing linear motion and a controlling mechanism used to provide displacements of different magnitudes and frequencies. A signal generator can be used to control the system including load cells while transformers can be used to measure the load response and imposed displacement (Huang, *et al.*, 2004). The load can be transferred to the cell-seeded constructs *via* flat platens which distribute the load evenly (Thorpe, *et al.*, 2008), however in a device for stimulating multiple scaffolds simultaneously, care must be taken that the constructs are of similar height or the compressive strain applied will vary as the scaffold height does. Mass transfer is improved in dynamic compression bioreactors over static culture (as compression causes fluid flow in the scaffold) which results in the improvement of the aggregate modulus of the resulting cartilage tissue to levels approaching those of native articular cartilage (Mauck, *et al.*, 2000).

2.4 Strain Bioreactors

Tensile strain bioreactors have been used in an attempt to engineer a number of different types of tissue including tendon, ligament, bone, cartilage and cardiovascular tissue. Some designs are very similar to compression bioreactors, only differing in the way the force is transferred to the construct. Instead of flat platens as in a compression bioreactor, a way of clamping the scaffold into the device is needed so that a tensile force can be applied. Tensile strain has been used to differentiate mesechymal stem cells along the chondrogenic lineage. A multistation bioreactor was used in which cell-seeded collagen-glycosaminoglycan scaffolds were clamped and loaded in uniaxial tension (McMahon, *et al.*, 2008). Alternatively, tensile strain can also be applied to a construct by attaching the construct to anchors on a rubber membrane and then deforming the membrane. This system has been used in the culture of bioartificial tendons with a resulting increase in Young's modulus over non-loaded controls (Garvin, *et al.*, 2003).

2.5 Hydrostatic Pressure Bioreactors

In cartilage tissue engineering, hydrostatic pressure bioreactors can be used to apply mechanical stimulus to cell-seeded constructs. Scaffolds are usually cultured statically and then moved to a hydrostatic chamber for a specified time for loading. Hydrostatic pressure bioreactors consist of a chamber which can withstand the pressures applied and a means of applying that pressure. For example, a media-filled pressure chamber can be pressurised using a piston controlled by an actuator (Darling & Athanasiou, 2003). For sterility, the piston can apply pressure *via* an impermeable membrane so that the piston itself does not come into contact with the culture media. Variations on this design include a water-filled pressure chamber which pressurises a media-filled chamber *via* an impermeable film and is controlled using a variable backpressure valve and an actuator (Watanabe, *et al.*, 2005).

2.6 Flow Perfusion Bioreactors

Culture using flow perfusion bioreactors has been shown to provide more homogeneous cell distribution throughout scaffolds. Collagen sponges have been seeded with bone marrow stromal cells and perfused with flow. This has resulted in greater cellularity throughout the scaffold in comparison to static controls, implying that better nutrient exchange occurs due to flow (Glowacki, *et al.*, 1998). Using a calcium phosphate scaffold, abundant extracellular matrix (ECM) with nodules of calcium phosphate was noted after 19 days in steady flow culture (Janssen, *et al.*, 2006). In comparisons between flow perfusion, spinner flask and rotating wall bioreactors, flow perfusion bioreactors have proved to be the best for fluid transport. Using the same flow rate and the same scaffold type, while cell densities remained the same using all three bioreactors, the distribution of the cells changed dramatically depending on which bioreactor was used. Histological analysis showed that spinner flask and static culture resulted in the majority of viable cells being on the periphery of the scaffold. In contrast, the rotating wall vessel and flow perfusion bioreactor culture resulted in uniform cell distribution throughout the scaffolds (Goldstein, *et al.*, 2001, Yu, *et al.*, 2004).

Flow perfusion bioreactors generally consist of a pump and a scaffold chamber joined together by tubing. A fluid pump is used to force media flow through the cell-seeded scaffold. The scaffold is placed in a chamber that is designed to direct flow through the interior of the scaffold. The scaffold is kept in position across the flow path of the device and media is perfused through the scaffold, thus enhancing fluid transport. Media can easily be replaced in the media reservoir (figure 4). However, the effects of direct perfusion can be highly dependent on the medium flow rate. Therefore optimising a perfusion bioreactor for the engineering of a 3D tissue must address the careful balance between the mass transfer of nutrients and waste products to and from cells, the retention of newly synthesised ECM components within the construct and the fluid induced shear stresses within the scaffold pores.

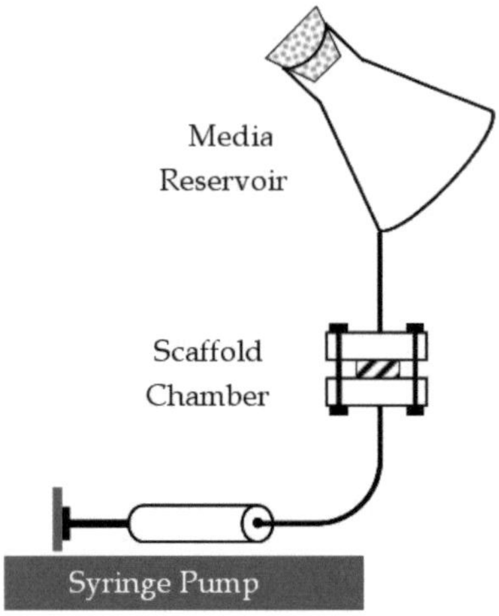

Fig. 4. A flow perfusion bioreactor, media is forced through the scaffold in the scaffold chamber by the syringe pump (Jaasma, *et al.*, 2008)

3. Tissue Formation in Bioreactor Systems

3.1 Flow Perfusion Bioreactors for Bone Tissue Engineering

Bone grafts are required to aid bone defect and non-union healing, and restore function to the damaged area as quickly and completely as possible (Perry, 1999). They are required in a number of procedures *e.g.* replacing diseased bone, filling bone voids, reconstructive surgery and in spinal fusion operations. The use of bone grafts substitutes (autografts, allografts and xenografts) are associated with donor site morbidity, small volume of donor tissue harvested, disease transmission, infection and chronic pain (Perry, 1999, Langer, 2000). Therefore, attention has turned to bone tissue engineering.

It has been shown that fluid flow can stimulate bone cells to increase levels of bone formation markers (Reich & Frangos, 1991, Klein-Nulend, *et al.*, 1997, You, *et al.*, 2001, Cartmell, *et al.*, 2003, Li, *et al.*, 2004, Batra, *et al.*, 2005, Kreke, *et al.*, 2005, Jaasma & O'Brien, 2008) and its use could improve mineralisation of the scaffold on which cells are seeded. Flow perfusion bioreactors increased alkaline phosphatase (ALP) expression after 7 and 14 days of culture compared to constructs cultured in spinner flasks or rotating wall vessels (Goldstein, *et al.*, 2001), and are more commonly used than any other bioreactor for use in 3-D stimulation studies. In one study, pre-osteoblastic MC3T3-E1 cells were seeded on decalcified human trabecular bone, the flow rate of perfusion altered and the mRNA expression of Runx2, Osteocalcin (OC) and ALP measured (Cartmell, *et al.*, 2003). It was found that using a steady flow rate of 1 mL/min killed a majority of the cells on the scaffold after 7 days in culture. However, a flow rate of 0.01 mL/min led to a high proportion of viable cells both on the surface and inside the scaffold. This compared favourably to static culture, where cells were predominantly on the periphery (Cartmell, *et al.*, 2003).

During locomotion, bone cells are primarily subjected to oscillatory and pulsatile flow *in vivo* (Jacobs, *et al.*, 1998). Results from 2D cultures show that osteoblasts are more responsive to pulsatile than steady and oscillatory flow (Jacobs, *et al.*, 1998). In our laboratory, a flow perfusion bioreactor has been developed to examine the effects of different flow profiles on cell-seeded collagen-glycosaminoglycan scaffolds (Jaasma, *et al.*, 2008). The scaffold chamber was specifically designed to ensure that the compliant scaffold was under perfusive flow. This involved using a scaffold of larger diameter than the flow path and using spacers to ensure the scaffold was under 10% compression during culture. A programmable syringe pump was used in order to stimulate the cell-seeded constructs using different flow profiles. This device demonstrated that intermittent flow perfusion is advantageous for mechanically stimulating osteoblasts while maintaining cell viability. It was found that intermittent dynamic flow caused greater stimulation than a continuous low flow rate (Jaasma & O'Brien, 2008). Cyclyoxygenase-2 (COX-2) and osteopontin (OPN) expression increased due to culture in the bioreactor, as did prostaglandin E_2 (PGE_2) production. Whilst, in a more recent study in our lab, it was found that the insertion of short term rest periods (5, 10 and 15 seconds) combined with long term rests (7 hours) provided an enhanced osteogenic response compared to constructs cultured in static culture, whilst simultaneously improving cellular spatial distribution. The use of short term rests upregulated COX-2, PGE_2 and OPN (Partap, *et al.*, 2009, Plunkett, *et al.*, 2009). PGE_2 levels were also found to increase over static controls when a calcium phosphate scaffold was cultured in a flow perfusion bioreactor with a flow rate of 0.025 mL/min. A stimulus of 30 minutes of oscillatory flow at 1 Hz with a 40 mL/min peak superimposed on steady flow increased levels of PGE_2. The number of cells left residing on the scaffolds decreased due to this large dynamic stimulus but this decrease was not found to be statistically significant to static culture (Vance, *et al.*, 2005).

Results suggest that the increased cell proliferation and matrix mineralisation in the centre of the scaffold resulting from culture in a flow perfusion bioreactor are primarily due to better nutrient and waste exchange. However, the increased mineral deposition with an increase in flow rate and numerous 2D experiments indicate that cells are also stimulated by the flow-induced shear stress. Sikavitsas *et al.* (2003) added various amounts of dextran to the media to increase its viscosity. This allowed for the shear stress magnitude experienced by the cells on the scaffold to be varied independently while keeping the flow rate constant. Mineral deposition on the scaffold was found to be proportional to the magnitude of the shear stress imposed on the cells. Thus, while flow perfusion does appear to increase the ability of cells to remain viable at the centre of the scaffold, it also serves as a means of increasing matrix production and mineralisation on the scaffold (Sikavitsas, *et al.*, 2003).

3.2 Compression Bioreactor for Cartilage Tissue Engineering

Cartilage is a supporting tissue containing chondroitin sulphates, collagen and elastic fibres and cells. The cells present in cartilage are known as chondrocytes and they are situated in lacunae in the cartilage matrix. Nutrient and waste product exchange occurs purely by diffusion through the cartilage matrix. It is an avascular, aneural and alymphatic tissue. The healing response of damaged cartilage is dependent upon the depth of the lesion. When minor damage occurs to cartilage, it can be repaired by appositional growth, but when severe damage occurs, it is repaired without complete restoration. This is thought to be due to the low metabolic activity of chondrocytes and its avascular nature (Martini, 2002). Currently, autologous chonrdocyte implantation (ACI), osteochondral grafting and bone marrow stimulation techniques are used to stimulate the regeneration of native cartilage. However, these techniques generally result in the formation of fibrocartilage which has poor mechanical properties and does not perform as well as native cartilage. Cartilage tissue engineering may offer a solution to this problem. There are three types of cartilage: hyaline, elastic and fibrocartilage. Joints can contain both hyaline cartilage and fibrocartilage, with the more flexible hyaline cartilage covering the bone and the more durable fibrocartilage acting as a shock-absorber between bones. As a joint moves, there is motion between two articulating layers of cartilage. This deforms the cartilage, causes fluid flow within it and induces a hydrostatic pressure load on it. These mechanical forces affect the chondrocytes in the cartilage. The force applied, along with the length of time it is applied for and the frequency of application modifies the response of chondrocytes (Mauck, *et al.*, 2007). This is useful in bioreactor design and for use in cartilage tissue engineering; if the correct stimulation pattern is used, chondrocytes can be induced to produce more extracellular matrix and this can result in more cartilage-like tissue being formed.

After twenty weeks in static culture, the aggregate modulus of tissue-engineered cartilage was 179±9 kPa. This is 40% of the value reported for native cartilage (Ma & Langer, 1999). At twenty five weeks, the modulus had not increased further so this may be the closest approximation to cartilage that can be cultured without the aid of a bioreactor. The most commonly used bioreactors in cartilage tissue engineering are compression bioreactors. When free swelling controls were compared to dynamically loaded agarose gels, after 28 days in culture, there was a six-fold increase in the equilibrium aggregate modulus for the loaded gels (Mauck, *et al.*, 2000). A sinusoidal strain of 10% at 1 Hz was applied to the gels for five days per week for a total of three hours per day with a rest period of one hour between each hour of loading. This complex loading pattern was deemed to be physiological and it resulted in increased glycosaminoglycan content over free swelling controls after 21 days in culture. The combination of increased modulus and increased glycosaminoglycan formation over free swelling controls after only four weeks in culture demonstrates the benefits of bioreactor culture in cartilage tissue engineering. Compression bioreactors have also been used to examine the effect of loading on the differentiation of bone marrow mesenchymal stem cells down the chondrocytic lineage. Growth factors such as transforming growth factor (TGF-β) can also be used to encourage differentiation. In a study comparing the use of compressive loading, the use of TGF-β and the use of a combination of loading and TGF-β, it was found that compressive loading alone was just as effective at inducing chondrogenic differentiation as TGF-β or TGF-β plus loading (Huang, *et al.*, 2004).

4. Bioreactors: State of the Art and Future Directions

The use of bioreactors has brought us a step closer to engineering numerous tissue types. Recently, a bioreactor was used to culture a tissue engineered trachea that was successfully implanted into a patient. A donor trachea was decellularized and subsequently seeded with the patients own cells. The bioreactor was specifically designed to seed and culture different cell types on either side of the donor tissue to allow for nutrient supply and waste removal, to provide biomechanical cues in the form of a hydrodynamic shear stress, as well as being designed to be autoclaved and handled in a sterile manner. The bioreactor rotated the trachea around its longitudinal axis so that a shear stress was applied to the cells to stimulate them and to ensure the even distribution of nutrients and waste. In addition to the rotation of the graft, the culture medium was continuously mixed to increase oxygenation and the exchange of waste and nutrients in the bioreactor chamber (Macchiarini, *et al.*, 2008).

However, most bioreactors at present are specialised devices with a low volume output. Their assembly is often time consuming and labour intensive. Many also exhibit operator dependent variability. Ways of minimizing the time and effort needed to form tissue must be sought if costs are to be reduced so that engineered tissues can be routinely used in clinical environments. In the future, scaled up versions of some devices (which may be automated) could potentially supply larger amounts of tissue. The ideal bioreactor would generate the required amount of tissue after a defined culture period. In addition, development of tissue could be monitored throughout the culture period through the incorporation of for example video microscopy and microcomputed tomography (μCT) for observing the structural properties of the growing tissue. A better understanding of the different effects of mechanical stimulation on cell signalling and mechanotransduction is also needed. This can be achieved through the use of existing simple bioreactors in conjunction with numerical simulation of culture conditions to minimise the number of experiments needed. This may be the future for bioreactors in tissue engineering.

5. Acknowledgements

Funding for the salaries of Sonia Partap and Niamh Plunkett was provided by Science Foundation Ireland (President of Ireland Young Researcher Award, 04/Yl1/B531) and the Irish Research Council for Science, Engineering and Technology (RS/2005/173) respectively.

6. References

Altman, G.H.; Horan, R.L.; Martin, I.; Farhadi, J.; Stark, P.R.; Volloch, V.; Richmond, J.C.; Vunjak-Novakovic, G. & Kaplan, D.L. (2002). Cell Differentiation by Mechanical Stress. *Faseb J,* 16, 270-272.

Bancroft, G.N.; Sikavitsas, V.I. & Mikos, A.G. (2003). Design of a Flow Perfusion Bioreactor System for Bone Tissue-Engineering Applications. *Tissue Eng,* 9, 549-554.

Batra, N.N.; Li, Y.J.; Yellowley, C.E.; You, L.; Malone, A.M.; Kim, C.H. & Jacobs, C.R. (2005). Effects of Short-Term Recovery Periods on Fluid-Induced Signaling in Osteoblastic Cells. *J Biomech,* 38, 1909-1917.

Cartmell, S.H.; Porter, B.D.; Garcia, A.J. & Guldberg, R.E. (2003). Effects of Medium Perfusion Rate on Cell-Seeded Three-Dimensional Bone Constructs in Vitro. *Tissue Eng,* 9, 1197-1203.

Carver, S.E. & Heath, C.A. (1999). Semi-Continuous Perfusion System for Delivering Intermittent Physiological Pressure to Regenerating Cartilage. *Tissue Eng,* 5, 1-11.

Darling, E.M. & Athanasiou, K.A. (2003). Biomechanical Strategies for Articular Cartilage Regeneration. *Ann Biomed Eng,* 31, 1114-1124.

Darling, E.M. & Athanasiou, K.A. (2003). Articular Cartilage Bioreactors and Bioprocesses. *Tissue Eng,* 9, 9-26.

Freed, L.E.; Guilak, F.; Guo, X.E.; Gray, M.L.; Tranquillo, R.; Holmes, J.W.; Radisic, M.; Sefton, M.V.; Kaplan, D. & Vunjak-Novakovic, G. (2006). Advanced Tools for Tissue Engineering: Scaffolds, Bioreactors, and Signaling. *Tissue Eng,* 12, 3285-3305.

Freed, L.E.; Langer, R.; Martin, I.; Pellis, N.R. & Vunjak-Novakovic, G. (1997). Tissue Engineering of Cartilage in Space. *Proceedings of the National Academy of Sciences of the United States of America,* 94, 13885-13890.

Freed, L.E. & Vunjak-Novakovic, G. Tissue Engineering Bioreactors. In: Lanza, R.P., Langer, R., and Vacanti, J., eds. Principles of Tissue Engineering. San Diego, CA: Academic Press, 2000, pp. 143-156.

Freshney, R.I. Culture of Animal Cells. New York: Wiley-Liss, 2000.

Garvin, J.; Qi, J.; Maloney, M. & Banes, A.J. (2003). Novel System for Engineering Bioartificial Tendons and Application of Mechanical Load. *Tissue Eng,* 9, 967-979.

Glowacki, J.; Mizuno, S. & Greenberger, J.S. (1998). Perfusion Enhances Functions of Bone Marrow Stromal Cells in Three-Dimensional Culture. *Cell Transplant,* 7, 319-326.

Goldstein, A.S.; Juarez, T.M.; Helmke, C.D.; Gustin, M.C. & Mikos, A.G. (2001). Effect of Convection on Osteoblastic Cell Growth and Function in Biodegradable Polymer Foam Scaffolds. *Biomaterials,* 22, 1279-1288.

Gomes, M.E.; Reis, R.L. & Mikos, A.G. (2005). Bone Marrow Stromal Cells Cultured on Starch Based Three-Dimensional Scaffolds in a Flow Perfusion Bioreactor: A Promising in-Vitro Approach for Obtaining Bone Tissue Substitutes. *Trans. Orthop. Res. Soc.,* 1743.

Huang, C.Y.; Hagar, K.L.; Frost, L.E.; Sun, Y. & Cheung, H.S. (2004). Effects of Cyclic Compressive Loading on Chondrogenesis of Rabbit Bone-Marrow Derived Mesenchymal Stem Cells. *Stem Cells,* 22, 313-323.

Ishaug-Riley, S.L.; Crane-Kruger, G.M.; Yaszemski, M.J. & Mikos, A.G. (1998). Three-Dimensional Culture of Rat Calvarial Osteoblasts in Porous Biodegradable Polymers. *Biomaterials,* 19, 1405-1412.

Jaasma, M.J. & O'Brien, F.J. (2008). Mechanical Stimulation of Osteoblasts Using Steady and Dynamic Fluid Flow. *Tissue Eng Part A,* 14, 1213-1223.

Jaasma, M.J.; Plunkett, N.A. & O'Brien, F.J. (2008). Design and Validation of a Dynamic Flow Perfusion Bioreactor for Use with Compliant Tissue Engineering Scaffolds. *J Biotechnol.,* 133, 490.

Jacobs, C.R.; Yellowley, C.E.; Davis, B.R.; Zhou, Z.; Cimbala, J.M. & Donahue, H.J. (1998). Differential Effect of Steady Versus Oscillatory Flow on Bone Cells. *Journal of Biomechanics,* 31, 969-976.

Janssen, F.W.; Oostra, J.; Oorschot, A. & van Blitterswijk, C.A. (2006). A Perfusion Bioreactor System Capable of Producing Clinically Relevant Volumes of Tissue-Engineered Bone: In Vivo Bone Formation Showing Proof of Concept. *Biomaterials,* 27, 315-323.

Klein-Nulend, J.; Burger, E.H.; Semeins, C.M.; Raisz, L.G. & Pilbeam, C.C. (1997). Pulsating Fluid Flow Stimulates Prostaglandin Release and Inducible Prostaglandin G/H Synthase Mrna Expression in Primary Mouse Bone Cells. *Journal of Bone and Mineral Research,* 12, 45-51.

Kreke, M.R.; Huckle, W.R. & Goldstein, A.S. (2005). Fluid Flow Stimulates Expression of Osteopontin and Bone Sialoprotein by Bone Marrow Stromal Cells in a Temporally Dependent Manner. *Bone,* 36, 1047-1055.

Langer, R. (2000). Biomaterials in Drug Delivery and Tissue Engineering: One Laboratory's Experience. *Acc Chem Res,* 33, 94-101.

Langer, R. & Vacanti, J.P. (1993). Tissue Engineering. *Science,* 260, 920-926.

Li, Y.J.; Batra, N.N.; You, L.; Meier, S.C.; Coe, I.A.; Yellowley, C.E. & Jacobs, C.R. (2004). Oscillatory Fluid Flow Affects Human Marrow Stromal Cell Proliferation and Differentiation. *J Orthop Res,* 22, 1283-1289.

Lyons, F.; Partap, S. & O'Brien, F.J. (2008). Part 1: Scaffolds and Surfaces. *Technol Health Care,* 16, 305-317.

Ma, P.X. & Langer, R. (1999). Morphology and Mechanical Function of Long-Term in Vitro Engineered Cartilage. *J Biomed Mater Res,* 44, 217-221.

Macchiarini, P.; Jungebluth, P.; Go, T.; Asnaghi, M.A.; Rees, L.E.; Cogan, T.A.; Dodson, A.; Martorell, J.; Bellini, S.; Parnigotto, P.P.; Dickinson, S.C.; Hollander, A.P.; Mantero, S.; Conconi, M.T. & Birchall, M.A. (2008). Clinical Transplantation of a Tissue-Engineered Airway. *Lancet,* 372, 2023-2030.

Martini, F. Fundamentals of Anatomy and Physiology. Prentice Hall, 2002.

Mauck, R.L.; Byers, B.A.; Yuan, X. & Tuan, R.S. (2007). Regulation of Cartilaginous Ecm Gene Transcription by Chondrocytes and Mscs in 3d Culture in Response to Dynamic Loading. *Biomech Model Mechanobiol,* 6, 113-125.

Mauck, R.L.; Soltz, M.A.; Wang, C.C.; Wong, D.D.; Chao, P.H.; Valhmu, W.B.; Hung, C.T. & Ateshian, G.A. (2000). Functional Tissue Engineering of Articular Cartilage through Dynamic Loading of Chondrocyte-Seeded Agarose Gels. *J Biomech Eng,* 122, 252-260.

McMahon, L.A.; Reid, A.J.; Campbell, V.A. & Prendergast, P.J. (2008). Regulatory Effects of Mechanical Strain on the Chondrogenic Differentiation of Mscs in a Collagen-Gag Scaffold: Experimental and Computational Analysis. *Ann Biomed Eng,* 36, 185-194.

Partap, S.; Plunkett, N.A. & O'Brien, F.J. (2009). Stimulation of Osteoblasts Using Rest Periods During Bioreactor Culture on Collagen-Glycosaminoglycan Scaffolds. *Journal of Materials Science-Materials in Medicine,* Submitted.

Perry, C.R. (1999). Bone Repair Techniques, Bone Graft, and Bone Graft Substitutes. *Clin Orthop Relat Res,* 71-86.

Plunkett, N.A.; Partap, S. & O'Brien, F.J. (2009). Osteoblast Response to Rest Periods During Bioreactor Culture of Collagen-Glycosaminoglycan Scaffolds, Tissue Eng. Part A, Accepted for Publication.

Reich, K.M. & Frangos, J.A. (1991). Effect of Flow on Prostaglandin E2 and Inositol Trisphosphate Levels in Osteoblasts. *Am J Physiol,* 261, C428-432.

Schwarz, R.P.; Goodwin, T.J. & Wolf, D.A. (1992). Cell Culture for Three-Dimensional Modeling in Rotating-Wall Vessels: An Application of Simulated Microgravity. *J Tissue Cult Methods,* 14, 51-57.

Sikavitsas, V.I.; Bancroft, G.N.; Holtorf, H.L.; Jansen, J.A. & Mikos, A.G. (2003). Mineralized Matrix Deposition by Marrow Stromal Osteoblasts in 3d Perfusion Culture Increases with Increasing Fluid Shear Forces. *Proc Natl Acad Sci U S A,* 100, 14683-14688.

Thorpe, S.D.; Buckley, C.T.; Vinardell, T.; O'Brien, F.J.; Campbell, V.A. & Kelly, D.J. (2008). Dynamic Compression Can Inhibit Chondrogenesis of Mesenchymal Stem Cells. *Biochem Biophys Res Commun,* 377, 458-462.

Vance, J.; Galley, S.; Liu, D.F. & Donahue, S.W. (2005). Mechanical Stimulation of Mc3t3 Osteoblastic Cells in a Bone Tissue-Engineering Bioreactor Enhances Prostaglandin E2 Release. *Tissue Eng,* 11, 1832-1839.

Vunjak-Novakovic, G.; Obradovic, B.; Martin, I.; Bursac, P.M.; Langer, R. & Freed, L.E. (1998). Dynamic Cell Seeding of Polymer Scaffolds for Cartilage Tissue Engineering. *Biotechnology Progress,* 14, 193-202.

Watanabe, S.; Inagaki, S.; Kinouchi, I.; Takai, H.; Masuda, Y. & Mizuno, S. (2005). Hydrostatic Pressure/Perfusion Culture System Designed and Validated for Engineering Tissue. *J Biosci Bioeng,* 100, 105-111.

You, J.; Reilly, G.C.; Zhen, X.; Yellowley, C.E.; Chen, Q.; Donahue, H.J. & Jacobs, C.R. (2001). Osteopontin Gene Regulation by Oscillatory Fluid Flow Via Intracellular Calcium Mobilization and Activation of Mitogen-Activated Protein Kinase in Mc3t3-E1 Osteoblasts. *J Biol Chem,* 276, 13365-13371.

Yu, X.; Botchwey, E.A.; Levine, E.M.; Pollack, S.R. & Laurencin, C.T. (2004). Bioreactor-Based Bone Tissue Engineering: The Influence of Dynamic Flow on Osteoblast Phenotypic Expression and Matrix Mineralization. *Proc Natl Acad Sci U S A,* 101, 11203-11208.

High Resolution X-Ray Tomography - 3D Imaging for Tissue Engineering Applications

Zehbe Rolf *, Haibel Astrid, Schmidt Franziska, Riesemeier Heinrich, Kirkpatrick C. James, Schubert Helmut and Brochhausen Christoph
Technische Universität Berlin, Institute of Materials Science and Technologies
Germany

1. Introduction

Synchrotron radiation-based micro computed tomography (SR-μCT) has become a valuable tool for the structural analyzes of different types of biomaterials in tissue engineering research. This methodology allows to investigate millimetre sized specimens in their three dimensional context - even on a cellular level without destroying the material.

In this chapter we elucidate the use of SR-μCT to analyze several biomaterials for tissue engineering research, which are targeted at different applications in the life sciences and which feature completely different properties. Among the investigated materials are biopolymeric scaffolds for cartilage tissue engineering, a commercial fibre based scaffold, biodegradable microspheres as drug release system and two multi-layered materials for possible applications in guided nerve growth. Further, as the used SR-μCT beam line allows the generation of coherent and monochromatic X-rays, limitations due to low differences in absorption contrast can be solved by using the enhanced phase contrast imaging capabilities of the beam line station. The three-dimensional data acquired for each investigated sample allowed for qualitative and quantitative results unprecedented by light microscopy or electron microscopy. While the main focus has been set on imaging aspects, the results obtained are discussed in view of the respective literature indicating trends in future research.

---

*Author for correspondence: Rolf Zehbe (rolf.zehbe@tu-berlin.de), other authors affiliations as follows:
Zehbe, Schmidt & Schubert: TU Berlin, Institute of Materials Science & Technologies, Englische Strasse 20, 10587 Berlin, Germany
Haibel: GKSS Research Centre at DESY, Petra III, Notkestrasse 85, 22607 Hamburg, Germany
Riesemeier: BAM, Federal Institute for Materials Research & Testing, Division I.3, Richard Willstätter Strasse 11, 12489 Berlin, Germany
Brochhausen & Kirkpatrick: Johannes Gutenberg University, Institute of Pathology, Langenbeckstrasse 1, 55101 Mainz, Germany

2. Fundamentals of X-Ray Imaging and Tomography

In 1895 Wilhelm Conrad Röntgen discovered a new type of radiation being emitted from a vacuum tube while investigating cathode rays (Röntgen, 1895). Röntgen named this type of radiation X-rays and produced the first medical X-ray image showing his wife's hand. X-rays are part of the electromagnetic spectrum, although with a much shorter wavelength than visible light and with the ability to penetrate matter depending on atomic mass (μ) and density (respectively the thickness d) according to the Lambert-Beer's law (1).
Especially biological matter, being of low density and being mostly composed of elements with a low atomic mass, is penetrated rather well. Consequently, the application of X-rays to medical sciences was established from the very beginning, but soon extended to applications in chemistry, physics and materials sciences.

$$I = I_0 e^{-\mu d} \tag{1}$$

Just over twenty years later in 1917 Johann Radon (Radon, 1917) developed the mathematical base to calculate cross-sectional views from series of rotational projection images - an approach which ultimately laid the basis for X-ray tomography and which is known as the projection-slice theorem. Simplified, the projection image of an object at a certain rotation angle $\boldsymbol{\theta}$ can be described as a set of line integrals, which themselves represent the attenuation of the X-ray after traversal through the object. This intensity information is, like the first X-ray image of Röntgen's wife, a two-dimensional image. Following, the mathematical fundamentals to derive three-dimensional data from these sets of projection images in parallel beam geometry are briefly summarized. Although, other beam geometries, like conical or fan beams exist, the simplest mathematical approach is given considering a parallel beam of X-rays. This setup was furthermore exclusively used in all experiments described in this chapter. For a more in depth description, the reader is referred to the work of Kak and Slaney (Kak & Slaney, 1988). Briefly, the total attenuation $\boldsymbol{p}$ of a ray at an angle $\boldsymbol{\theta}$ and a position $\boldsymbol{r}$ relative to the rotation centre as displayed for a phantom object in Figure 1a is given by equation (2), where $\boldsymbol{I}$ is the intensity and $\boldsymbol{\mu}$ is the attenuation coefficient corresponding to the Lambert-Beer's law.

$$p(r,\theta) = \ln(I / I_0) = -\int \mu(x,y) ds \tag{2}$$

Consequently, a point (x,y) is projected onto $r(\boldsymbol{\theta})$ according to equation (3).

$$r = x\cos\theta + y\sin\theta \tag{3}$$

Using above formula, equation (2) can be rewritten in its transformed form (4), which is called the radon transform or sinogram (Figure 1b).

$$p(r,\theta) = \int_{-\infty}^{\infty}\int_{-\infty}^{\infty} f(x,y)\delta(x\cos\theta + y\sin\theta - r)dxdy \tag{4}$$

In tomography, the sinogram is used to reconstruct the sliced data from the projection data via the inverse of equation (4). The exact reconstruction of a given object in its sliced form requires an infinite amount of projections due to the projection-slice theorem.
Practically, a finite amount of projections in the range of 10^3 is used to optimize between data size/ processing time and reconstructed details.
Figure 1c displays the back projection obtained from the sinogram, which appears blurred and features a distinct halo. Applying a filter significantly reduces this effect and is consequently termed filtered back projection (Figure 1d, some artefacts can be observed as displayed in the cut-out). A parallel beam geometry has another favourable impact on the reconstruction of tomographic data as projections from 0°-180° are identical to projections from 180°-360°, which is, for example, not the case for a conical or fan beam geometry. This reduced amount of needed measurements directly impacts computing time and data size.

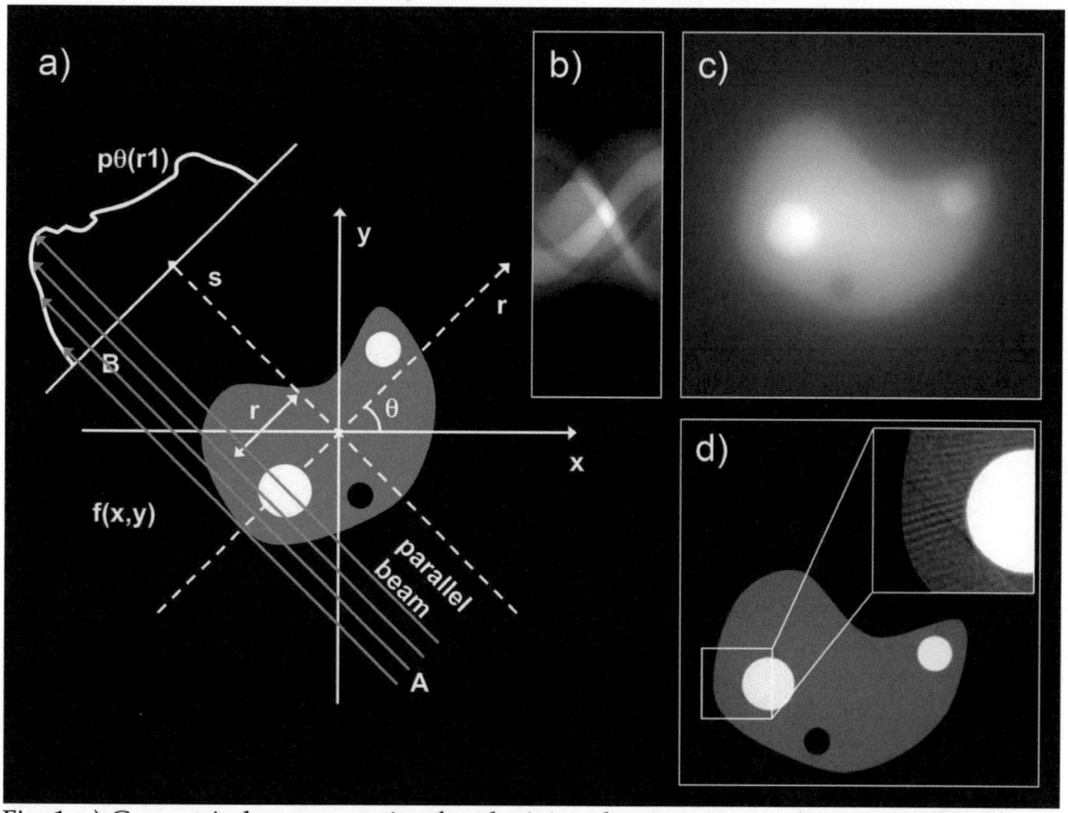

Fig. 1. a) Geometrical representation for obtaining the sinogram as demonstrated in b) and which is back projected as shown in c) without applying filtering and d) applying a ramp filter (the cut-out has been enhanced in contrast to display artefacts due to reconstruction).

3. Synchrotrons as Source for Superior X-Rays

X-rays can be produced in two different ways. The first approach dates back to Röntgen and uses X-ray tubes, essentially evacuated glass tubes with an electrically heated cathode, which emits electrons by thermionic emission from e.g. a tungsten filament. These electrons are accelerated towards the anode (e.g. copper) by applying a high voltage in the range of some kilovolts to hundred kilovolts. Furthermore, the anode's plain surface is positioned at an angle relative to the incident electrons at around 5° to 20° and it is made of a heat resistant material like tungsten or molybdenum as most of the energy is dissipated as heat upon deceleration of the electrons in the anode material. Only a few percent of the energy is converted into X-rays which are emitted as a fan or cone beam.

Although, current X-ray tube-based experimental setups can achieve high quality scientific data, some beam related limitations cannot be resolved. Here, synchrotrons offer a more advanced source for X-rays by accelerating charged particles (e.g. electrons) in storage rings up to relativistic speeds. Although electromagnetic radiation (or synchrotron radiation) is already emitted tangentially solely due to the particles being accelerated on a circular pathway, the efficacy can be raised significantly using undulators or wigglers. These so called insertion devices produce forward directed synchrotron radiation by using a series of alternating magnets, forcing the particles into oscillations and to loose energy in the form of synchrotron radiation. This radiation offers a high photon flux over a large range of energies, a high brilliance, a high level of polarization and coherency. Due to the high photon flux, wavelength monochromatization is a feasible approach (Dilmanian, 1992; Bonse & Bush, 1996). In this context, methods based on synchrotron radiation, allow for faster measurements and a broader spectrum of applications. For a more detailed comparison between synchrotron generated and tube-generated X-rays, the reader is referred to Bernhard et al. (Bernhard et al., 2004).

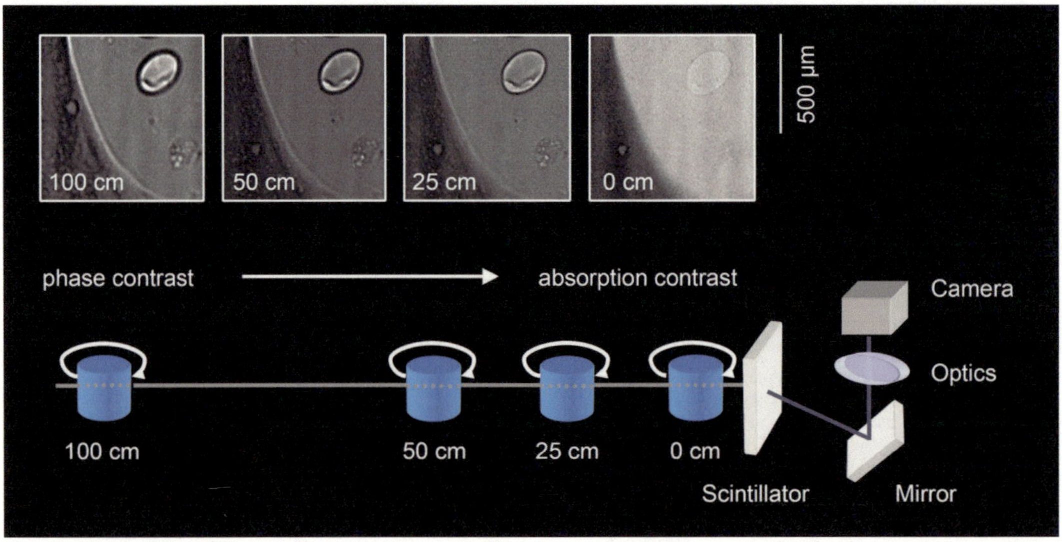

Fig. 2. Basic SR-µCT experimental setup, allowing measurement in both absorption contrast and phase contrast. Top: real world projection images of a thin sectioned articular cartilage specimen showing the impact of the distance between scintillator and sample on absorption contrast/ phase contrast (adapted from Zehbe[1] et al., 2009).

A promising methodology for certain imaging applications, where it is impossible to rely solely on absorption contrast, was introduced at third generation synchrotron sources (e.g.

ESRF, BESSY II or in the future PETRA III). This methodology requires a partially coherent beam and allows applying phase contrast to X-ray imaging techniques. While imaging in absorption contrast is well adapted to samples featuring a prominent difference in density or atomic mass, phase contrast imaging allows detection of structural details which are not well visible in absorption. Here, the contrast is achieved by retardation or refraction of coherent X-rays at phase boundaries inside the sample (Cloetens et al., 2006; Rack et al., 2008; Zehbe[1] et al., 2009). Fig. 2 illustrates the basic experimental setup which was used throughout this chapter and details the impact of the sample position relative to the scintillator on the resulting image contrast. The top left of Fig. 2 displays the projection images of a sliced bovine cartilage specimen mounted on a microscope slide being imaged at the same X-ray energy of 10 keV but different distances to the scintillator. Well visible is the increasing edge retrieval for the elliptical object with greater distance to the scintillator.

4. X-Ray Tomography and the Life Sciences

X-ray based tomography was first established by Hounsfield and Cormack (Hounsfield, 1973, Cormack, 1973) using an X-ray tube based setup. This work cumulated in the introduction of the first clinical X-ray tomography system in 1975. With the advancement of superior experimental setups new scientific applications became possible, allowing adapting tomographic experiments to much smaller samples while obtaining an unprecedented resolution on the micrometer scale. This methodology was consequently termed micro computed tomography (μCT).

One of the first medical related research fields in μCT was the analysis of trabecular bone samples by Feldkamp et al. (Feldkamp et al., 1989) but was soon extended to other areas including tissue engineering (Ho & Hutmacher, 2006). The superior properties of synchrotron generated X-rays further benefitted the tomographic analyses in the life sciences and became known as SR-μCT (Bonse & Bush, 1996). Finally, μCT (regardless of the origin of the X-ray) is heavily dependent on advanced computer technology benefiting significantly from the improvements in this field over the last 20 years. In this context, it should be noted that current tomographic data sets can easily amount to several hundred gigabytes in uncompressed size (e.g. approximately 280 Gigabytes for a data set of 4096 pixel x 4096 pixel x 4096 pixel in 32 bit grey values). Even nowadays the processing of such data can take a very long time or might even be impossible for single workstation computer systems. Concerning the experimental conditions, enhancements in scintillators (which are needed for the conversion of X-rays into visible light) and of course the high mechanical precision of modern rotational stages are prerequisites for acquiring high resolution tomographic data. In experiments which rely mainly on absorption contrast to image biological specimen, it is usually mandatory to significantly enhance this contrast. In this regard, metal staining commonly used in electron microscopy is a well adapted methodology especially if a discrimination of soft tissue structures is to be considered. One of the principal metal stains is osmiumtetroxide (OsO_4) which is a potent lipid stain and allows detection of cells or other lipid tissue structures. Using SR-μCT, OsO_4 has been used among many others by Ginty et al. to investigate cell seeded scaffolds (Ginty et al., 2006), by Lareida et al. to investigate the microstructure of the inner ear (Lareida et al., 2009) and by ourselves to detect cells inside a multilayered sheet scaffold as will be described later in section 7.4. A more specific staining can be achieved using immunhistochemical methods or

probes. In our own research (see section 7.1) we have established Au-lysine as a potent marker for anionic sites at the cell surface (Zehbe[2] et al., 2009).
To circumvent the usage of staining agents, which possibly introduce artefacts or which might occlude certain structural details from view, the application of phase contrast imaging is a valuable alternative. While a measurement in absorption contrast relies on the adaptation of the beam energy to the sample, phase contrast imaging relies solely on a coherent beam and therefore can be adjusted by varying the distance between sample and scintillator. In this measurement mode it is further possible to significantly reduce the dose applied to the sample by increasing the beam energy. In principle, this mode of operation allows imaging both strongly and weakly absorbing structures simultaneously. The reader is referred to the work of Cloetens et al. (Cloetens et al., 2006) and our own work (Zehbe[1] et al., 2009) for a more detailed discussion.

5. Tissue Engineering

In regenerative medicine, tissue engineering can be described as a tool to re-establish lost functionality and morphology to a previously damaged tissue. Over the years, several strategies were developed with combinations of cells and biomaterials playing a crucial role. These constructs are designed to be transplanted into the defect site and are already in clinical application. In orthopaedic surgery this methodology is usually termed matrix based autologeous chondrocyte transplantation. In this context, artificial scaffolds were introduced, allowing predefining a structure, into which cells can be expanded and immobilized *in vitro* or which serve as filler for migrating cells *in vivo*. Current tissue engineering strategies further adapt biochemical targeting to specifically stimulate proliferation, matrix synthesis or to control cell differentiation or cell migration. This last approach usually involves drug release concepts.
The archetype for all these tissue engineering strategies is the native and healthy tissue which should be regenerated and remodelled. From an investigational viewpoint, the gold standard defining this native structure is given by the analysis of histolochemically stained slices of the specimen. Unfortunately, this approach results in the complete loss of all volumetric morphological information, and requires complex procedures if it is to be recovered, usually involving alignment and registration of subsequent serial sections (Braverman et al., 1986). Due to its non destructive nature and its ability to resolve the true volumetric representation of a given specimen, μCT and especially SR-μCT have become important methods in the life sciences.
The following research details some possible applications in tissue engineering, putting a focus on cartilage tissue engineering. Later on, this chapter deals with approaches using stacks of polymer sheets for possible applications in neuronal tissue engineering.

6. Experimental Setup

Computed tomography (CT) requires a setup where either the sample or the beam source/ detector system (e.g. clinical CTs) are rotated relative to each other. For synchrotron radiation based experimental setups, a rotation of the beam source/ detector system is impossible. In this case, the sample is positioned on a multiaxial high precision rotational stage, allowing adjustment of translation, rotation and tilting. The experimental setup as displayed in Fig.2 has been used predominantly in this chapter and is described in several studies (Zehbe[1] et al., 2007; Zehbe[1] et al., 2009; Rack et al., 2008). Although the photon flux density is several orders of magnitude higher than in X-ray tubes, it is not constant over the X-ray energy. Fig. 3 displays the dependency of the photon flux density from the X-ray energy as was calculated according to Schäffers et al. for the BAM*line* experimental station at BESSY II (Schäffers et al., 1996). Here, at around 10 keV, the photon flux density is dramatically reduced due to the used monochromator (DMM, see inset) being coated with tungsten (L3ab absorption edge at 10.198 keV). Accordingly, measurements in this energy range usually require longer exposure times due to the reduced photon flux when compared to for example measurements at around 22 keV. In the research detailed in this chapter, the energy was set to different values ranging from 7 keV to 30 keV, while further using different sample to scintillator distances as to achieving the desired contrast mode. The scintillator consisted of a thin, single $CdWO_4$ crystal and was coupled to light optics (up to 10x magnification) and a CCD camera (2048x2048, peltier cooled to -30 °C). The acquired projection images were recorded in multi-image file formats.

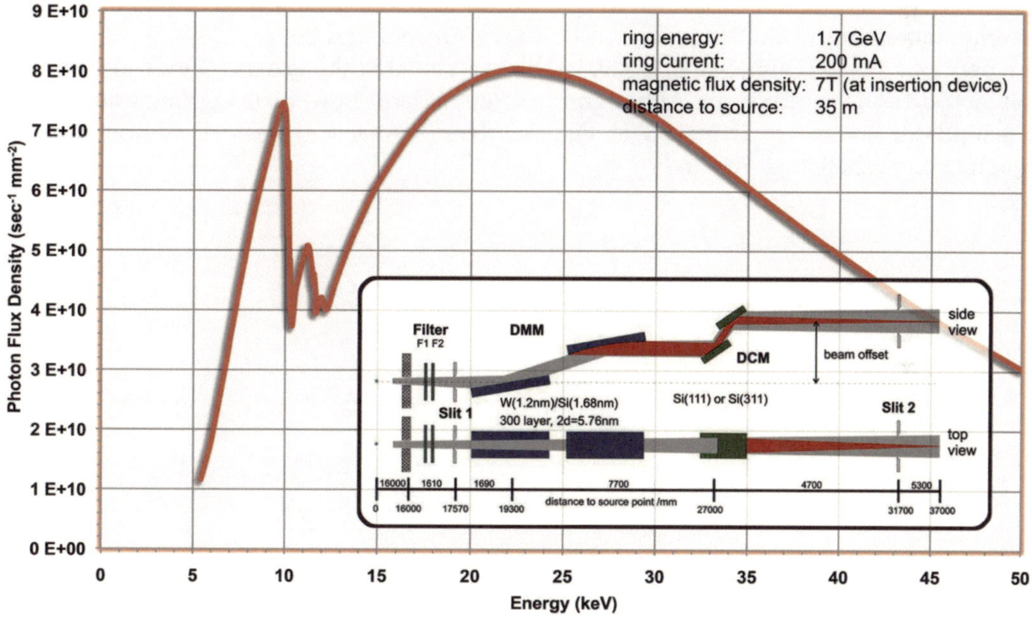

Fig. 3. Photon flux density behind the double multilayer monochromator (DMM) at the BAM*line* according to Schäffers et al. and Rack et al. (Schäffers et al., 1996, Rack et al., 2008). Around 10 keV the flux is reduced; data from BAM*line* description (Riesemeier et al. 2007).

7. Research

For an in-depth description of the experimental methodology and the results obtained, the reader is kindly referred to Table 2 at the end of this chapter and the cited references. Rendered data has been re-coloured for visualization purposes.

7.1 Cartilage Tissue Engineering

Degenerative, rheumatic or traumatic processes are the predominant causes for articular cartilage damage. The tissue itself has only a very limited regenerative potential and consequently, cartilage tissue engineering aims at either repopulating a defect with an excess amount of autologeous cells, stabilizing the defect area by supporting cell-scaffold constructs or enhancing the regeneration by biochemical targeting mechanisms. To better understand these different tissue engineering approaches it is mandatory to comprehend structure and function of the native tissue. Articular cartilage covers the ends of the long bones and is well known for its mechanical and tribological properties featuring a specific zonal structure of the extracellular matrix (primarily collagen II and proteoglycans) and the integrated cartilage cells, the chondrocytes. The following regions can be identified as follows: the surface zone, the middle zone, the deep zone and the subchondral bone (Buckwalter & Mankin, 1997). Depending on the zonal localisation, the chondrocytes appear in different morphologies. They are stretched along the surface as are the collagen fibres, while being mostly elliptical or rounded in the middle zone, where some cells form isogeneous cell groups. Towards the deep zone, the collagen fibres become more perpendicularly oriented to the surface resulting in a rise in stiffness. This zone further becomes increasingly calcified and connects to the subchondral bone.

In a previous study (Zehbe[1] et al., 2009) we have elucidated the use of SR-µCT to image the volumetric tissue morphology. Although the original data showed strong ring artefacts due to scintillator errors, as displayed in Fig. 4a, these artefacts were well compensated by sinogram correction as shown in Fig. 4b.

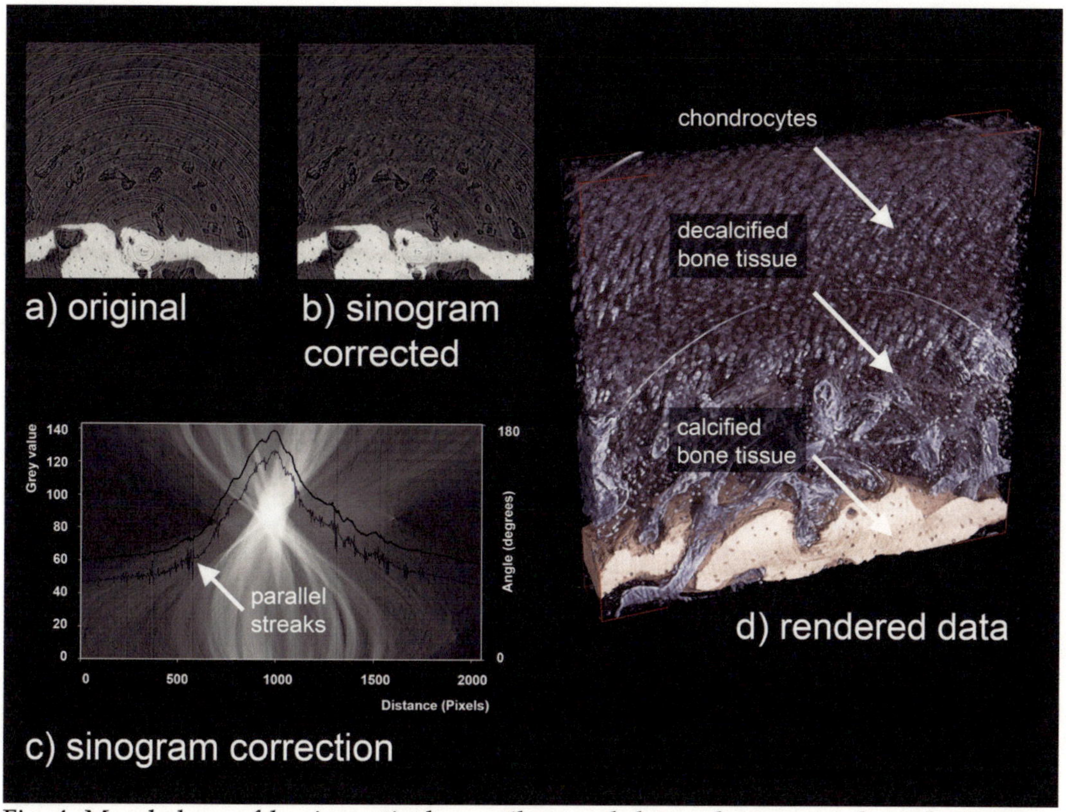

Fig. 4. Morphology of bovine articular cartilage and data enhancement through sinogram correction (a-c) resulting in the rendered data (d).

In the sinogram data (Fig. 4c), ring artefacts appear as parallel streaks or lines and can be easily filtered with a median filter. The data was rendered in a user defined colour scheme as displayed in Fig. 4d, showing the three-dimensional distribution and orientation of the chondrocytes depending on their zonal localisation.

Consequently, a tissue engineering approach should adapt as many of these properties as possible. As pure cell based tissue engineering can only partly re-establish the cellularity, usage of scaffold systems offers a methodology not only to restore the cellularity but also structure and function. These scaffolds are usually made of synthetic or natural polymers by techniques like fibre bonding, solvent casting, gas foaming or phase separation (Mikos & Temenoff, 2000). In our own research on scaffolds for cartilage tissue engineering, we developed a technique, which uses a directional freezing process to structure water based solutions of gelatine. This process was further designed to allow for the electrolysis of water prior to freezing and consequently results in the introduction of gas bubbles inside the gelatine solution before it is directionally frozen. Addition of other components like salts, acids, ceramics or polymer particles allows synthesising composite scaffolds or achieving different pore morphologies (Zehbe et al., 2005; Zehbe[1] et al., 2007; Zehbe[2] et al., 2009).

Fig. 5 displays some rendered representations of scaffolds made by ourselves and which are designed to degrade biologically.

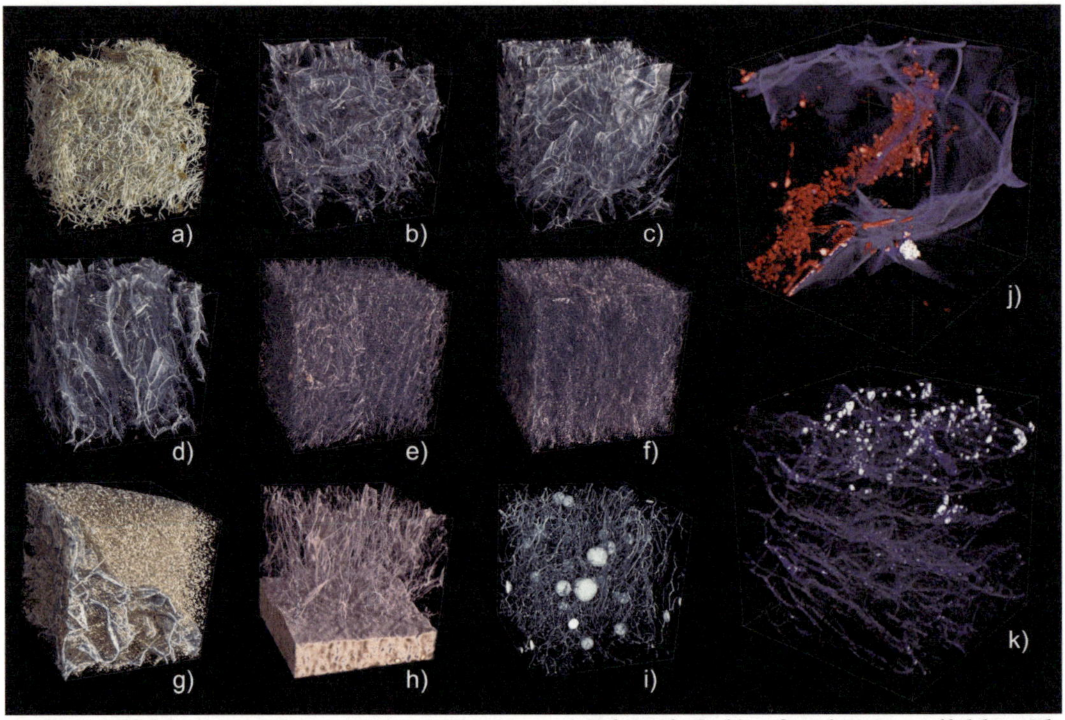

Fig. 5. Scaffolds for cartilage tissue engineering. a) Ethisorb ®, b) – f) gelatine scaffolds with increasing pore diameter, g) gelatine scaffolds with cells in fibrin matrix, h) gelatine-hydroxyapatite-composite scaffolds, i) PLGA-microsphere modified gelatine scaffold, j) scaffold with manual separated cells (red) and Au/ Ag stained cells (white) and k) scaffold with Au/ Ag stained cells (white) at the surface only.

Fig. 5a shows a competing commercial type scaffold (Ethisorb ®) which is made of PLGA-fibres being connected by poly-p-dioxanone and which was cultivated with porcine chondrocytes ($2 * 10^6$ cells) for 1 week, although the SR-μCT data did not reveal cells, the scaffold morphology was well imaged.

Figs. 5b to 5f demonstrate the effects of different freezing temperatures and differing ion contents (here, addition of hydrochloric acid) on the pore morphology as outlined in Table 1. The Feret's diameter of the pores was calculated as mean value from each slice of the respective tomographic data (Zehbe[2] et al., 2009). All these gelatine scaffolds were cultivated with porcine chondrocytes similar to the commercial Ethisorb® scaffold as described above. Further, the cultivated cells were labelled with a cell surface specific metal stain (Au-lysine in combination with a silver enhancer) to better visualize individual cells in the tomographic data.

Scaffold (as labelled in Fig. 5)	b)	c)	d)	e)	f)	j)	k)
Freezing temperature (°C)	-10	-10	-10	-20	-30	*-10*	*-20*
HCl content (1 molar) (ml/ml)	0.08	0.04	0.00	0.00	0.00	*0.08*	*0.00*
Pore diameter (Feret) (μm)	606	374	240	131	67	*606*	*131*

Table 1. Parameters influencing the pore morphology of gelatine scaffolds (scaffold b) and j) and scaffold e) and k) are identical).

Interestingly, this staining did not label all cells as intensely as desired (Fig. 5j). Consequently, cells were individually separated using a software-based volumetric mark-up tool. As a result, it was only possible to demonstrate a homogeneous and dense population of cells inside the larger pored scaffolds (Fig. 5b/ 5j), while the denser scaffolds appeared to show intensely stained cell-clusters at the surface only (Fig. 5e/ 5k).
In a different approach we infiltrated scaffolds with a cell-suspension containing fibrin to immobilize the cells immediately after seeding inside the scaffold. This approach was partly successful, as it allowed for a better seeding efficacy but resulted in a layer of fibrin and cells at the surface of the scaffold only (Fig. 5g). Due to the increased phase contrast in this SR-μCT experiment, it was possible to identify individual cells rather easily (cells can be seen as grainy structure on top of the scaffold).
Another two completely different gelatine scaffolds were made by incorporating hydroxyapatite (HAp) as a bottom layer being designed to seal against the subchondral bone *in vivo* (Fig. 5h) and by distributing polymeric poly-lactide-co-glycolide (PLGA) microspheres inside the gelatine suspension (Fig. 5i). These microspheres were intended as possible carriers for targeting molecules (here: prostaglandine E_2, PGE_2) in a drug-release setup. The morphology and the design of the drug-loaded PLGA-microspheres are further detailed in section 7.3. From a tomographic viewpoint these different components were easily separated from the gelatine network, revealing their respective structures rather well. Due to all components having differing elemental compositions or densities, imaging was performed in absorption contrast (Zehbe[1] et al., 2007).
On the other hand absorption contrast cannot be well utilized for scaffolds in a watery or swollen state. Consequently, further experiments were conducted using phase contrast imaging to elucidate the resistance of the scaffolds against compression in a wet environment as detailed in the following section.

7.2 Mechanics and Morphology of Scaffolds for Cartilage Tissue Engineering

Articular cartilage demonstrates mechanical and tribological properties allowing lifelong relative movement of adjacent bones. Tissue engineering concepts, especially those dealing with artificial scaffolds, should, if possible, adapt these properties to some extend.
In this regard, an early study of us (Zehbe et al., 2004) proved near native mechanical properties for gelatine scaffolds with a highly directional pore structure as discussed in the previous section. Mechanical testing of the water soaked scaffolds in confined and unconfined conditions showed a strain increase caused by fluid efflux, typical for biphasic

materials. Young's moduli were determined to E_s = 0.67 ± 0.02 MPa in unconfined compression, aggregate moduli to H_A = 0.84 ± 0.03 MPa in confined compression and the Poisson's ratio to ε_s = 0.27. The obtained values correspond well with bovine cartilage tissue measured by Korhonen et al. (Korhonen et al., 2002).

Until now, we were not able to determine the three-dimensional structural changes of these scaffolds under mechanical load. Therefore, we adapted SR-μCT in phase contrast imaging mode to determine structural changes, while completely immersing the scaffolds in water.

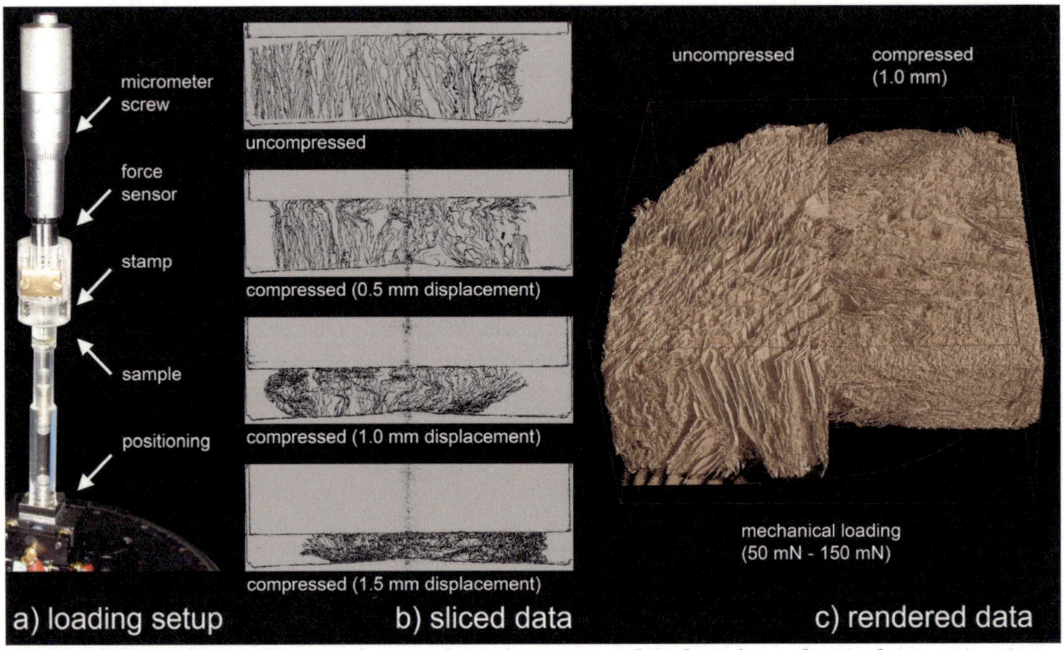

Fig. 6. a) Experimental setup for combined tomographical and mechanical investigations, b) acquired sliced data of subsequent mechanical displacements and c) rendered data comparing two different compression levels.

Although the absorption coefficients for water and gelatine are almost similar and would make it extremely difficult to achieve any tomographic details in absorption, phase contrast imaging in this case allows for an excellent contrast. Consequently, we have established an experimental setup (Fig. 6a) allowing simultaneous mechanical loading and tomographic measurement of water soaked gelatine scaffolds. The loading setup consisting of a micrometer screw, a plexiglas rod and a simple force sensor is described in more detail in Thiem et al. (Thiem et al., 2009). Briefly, a water soaked and air bubble free scaffold was placed inside the plexiglas rod while carefully adjusting the on top stamp and calibrating the force sensor. This whole setup was placed on the sample stage and was adjusted in the beam. The maximum possible distance of 109 cm between sample and scintillator and a beam energy of 30 keV ensured sufficient phase contrast to detect the scaffold structure in the watery environment as seen in the sliced data (Fig. 6b) and in the rendered data (Fig. 6c). The parallel pore network appears to bulge at specific sites (Fig. 6b – 1.0 mm displacement) upon increasing the load. As a result most of the mechanical energy is dissipated at these sites, while leaving the overall scaffold structure mostly intact. Upon further loading, the pore network cannot withstand and the pore channels collapse (Fig. 6b – 1.5 mm displacement). Summarizing the previously measured biphasic material properties and the newly acquired three-dimensional morphological data, we expect good mechanical properties *in vivo* as well.

7.3 Polymeric Drug Release System for Cartilage Tissue Engineering

Current tissue engineering concepts consider biomolecular targeting mechanisms aimed at specific cells the most feasible approach towards successful tissue regeneration. Consequently, the integration of targeting molecules into scaffold materials might significantly improve the quality and long-term performance of the regenerating tissue by maintaining cellular differentiation and optimal tissue functions. In this context, numerous substances are known to promote tissue regeneration by stimulation of specific cellular pathways. Well established in bone and cartilage tissue engineering are the members of the TGF-β superfamily, including the bone morphogenic proteins (BMP) and other related substances (Nicoll et al., 1998).

From the viewpoint of cartilage tissue engineering another promising class of targeting molecules has been found in the metabolites of the arachidonic acid – namely prostaglandine E_2, which has been demonstrated being important in both developmental processes and in tissue regeneration (Sandulache et al., 2006) including bone repair and cartilage regeneration. One limiting factor which applies to many targeting molecules is their short half-life *in vivo*. One possibility to prolong the availability *in vitro* or *in vivo* is to encapsulate the targeting molecule directly inside the scaffold material or indirectly by using a more defined carrier system like polymeric microspheres.

In previous studies we have adapted this later approach and were able to successfully encapsulate PGE_2 into polymeric polylactide-co-glycolide microspheres (Watzer et al., 2009). For more detailed information, the reader is referred to the chapter by Brochhausen et al. in this same book. The integration of these microspheres into a gelatine scaffold was briefly described in section 7.1. Unfortunately, the data obtained by SR-μCT did not reveal ultrastructural details, probably due to a too low difference in absorption contrast, which was the dominant image contrast. A reinvestigation (Zehbe[3] et al., 2009) of single microspheres at a slightly higher resolution but in phase contrast imaging mode revealed the data presented in Fig. 7. Apparently, the PGE_2-loaded PLGA-microspheres feature a porous shell with a compact core (Fig. 7a and Fig. 7c). It is currently unknown if this porosity is due to local solvent-entrapment and accumulation, while the PLGA continuously hardens, forming hollow structures or, some other mechanism.

This structural data might explain the release profile (Brochhausen et al., 2008), which is shown in Fig. 7b, showing a burst release early on with a plateau after 24 h and decreasing release afterwards. The more porous shell can probably release the PGE_2 much faster than the compact core which functions as diffusion barrier limiting the progress of the target molecule from the inside to the outside.

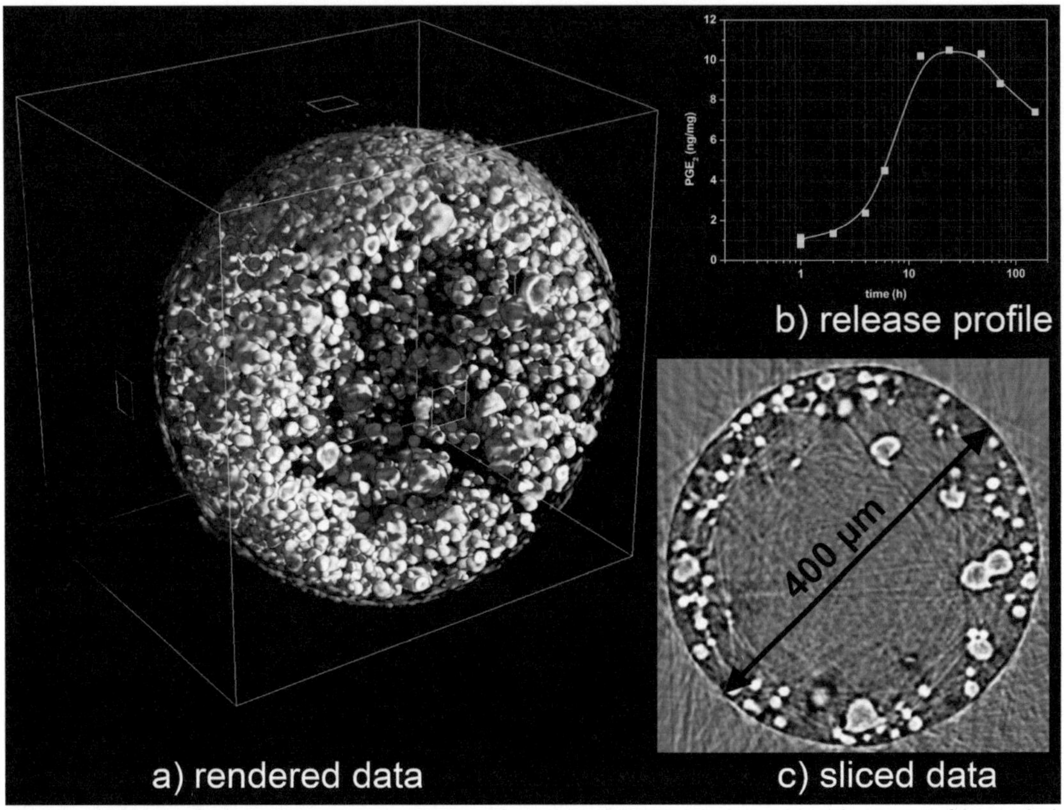

Fig. 7. PGE_2-loaded PLGA microsphere: a) rendered SR-µCT data, b) release profile and c) sliced SR-µCT data.

7.4 A Different Type of Cell Sheet Engineering investigated by SR-µCT

A few years ago a novel cell culture technology was developed and termed *cell sheet engineering* (Shimizu et al., 2003). This technology uses a thermoresponsive polymer (poly(N-isopropylacrylamide, PIPAAm) to detach cultivated monolayer cells from tissue culture ware by changing the PIPAAm from a hydrophilic state to a hydrophobic state by decreasing the temperature below 32 °C. The main advantage of this technology is that a pure (autologeous) cell sheet can be harvested without any supporting biomaterial which can be beneficial in applications, where an artificial material is not desired. Further, these cell sheets can be stacked. Due to the monolayer nature of this tissue engineering technique, only relatively thin cell layer aggregates can be achieved. To achieve a thicker more scaffold like structure, we have developed a different approach by cultivating cells on thin composite coatings of PLGA and gelatine on cell culture ware. As the gelatine gradually swells in cell culture and partly dissolves, the resulting porous PLGA-gelatine sheet and the cells on top can be easily detached from the cell culture ware either for further *in vitro* experimentation or by stacking these sheets obtaining a layered structure of cells and supporting biomaterial.

Fig. 8 displays a stack of 13 PLGA-gelatine sheets which were cultivated with a chondrocyte cell line (CHON-001) and which were afterwards fixated using glutaraldehyde and stained with OsO_4. Due to the enhanced absorption contrast resulting from using OsO_4 a morphological differentiation between cells and supporting PLGA-gelatine-sheets can be achieved via SR-μCT.

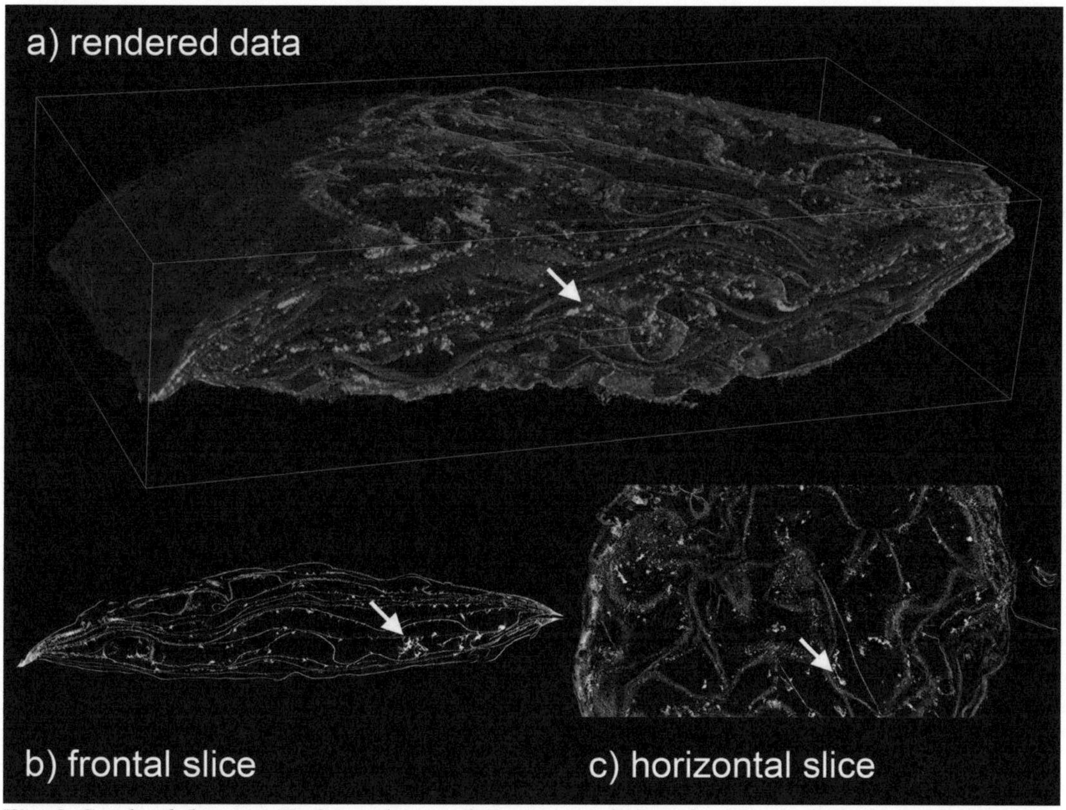

Fig. 8. Stack of thirteen PLGA-gelatine sheets pre-cultivated with a cell line, glutaraldehyde fixated and OsO_4 stained. a) rendered data, b) and c) sliced data (frontal and horizontal).

While Fig. 8a displays the rendered SR-μCT data, Figure 8b and 8c display the corresponding frontal and horizontal sliced data. Arrows indicate exemplarily locations of OsO_4 stained cells, which exhibit a strong absorption contrast sometimes resulting in artefacts due to shading effects. Overall, the structural data obtained indicates possible applications in the area of the tissue engineering of articular cartilage, skin or nerve tissue. Further, nerve tissue engineering is the topic of the following section describing yet another approach using stacked polymer sheets.

7.5 Neuronal Tissue Engineering

Restoring peripheral nerve damage is an important research field in tissue engineering sciences. One promising approach uses so called nerve conduits consisting of biodegradable polymers. The intention for using these conduits is to establish a preformed structure serving as guidance for axonal growth *in vivo*.

In this context, it can be beneficial to prearrange autologeous neuronal cells inside these conduits to achieve a better control over axonal growth e.g. by structural, biochemical or physical means. Biochemical stimulation can be achieved by integrating certain targeting molecules into the conduits similar to the methodology described in section 7.3. Here, the nerve growth factor protein (NGF) is probably the most relevant biochemical substance in neuronal tissue engineering (Bhang et al., 2007). Its usage is one of several other biochemical and physical mechanisms that can be applied to control neuronal regeneration (Schmidt[1] et al., 2009). Concerning physical stimulation, it is well known, that neuronal cells are especially susceptible to electrical potentials. Therefore, in the following research, we present a rather simple methodology to control cell growth by applying an electrical potential and we demonstrate that SR-µCT can be used to confirm the efficacy from a three-dimensional morphological viewpoint (Schmidt[2] et al., 2009).
One major problem occurring in tissue engineering can be attributed to the fact that cells usually grow in a non ordered fashion *in vitro*. Consequently, we have developed a technique to print microelectrodes on polymer sheets via a method named *inverse ink-jet printing* (using a sputter coater to establish a gold metallization) allowing for vital co-deposition of cells and the protein fibrin on the anode part of the electrodes (Zehbe[2] et al. 2007). While the structural resolution of the electrodes is limited by the used printer, a minimal line width of 35 µm was achievable with a HP deskjet 3520. At first, this method was designed for two-dimensional cell cultivation purposes, but was later extended into the third dimension by stacking - similar to section 7.4.
The electrodes being designed for an intended use in neuronal tissue engineering feature parallel aligned rows of thin gold made by the sputter coating process. All rows were contacted anodically allowing for deposition of cells and fibrin. Afterwards, several microelectrodes with the on top cells were stacked, fixated using glutaraldehyde, rinsed in distilled water and freeze-dried for further SR-µCT investigations. One microelectrode was used for separate vital staining using fluoresceindiacetate/ ethidiumbromide and displays nearly 100% vitality (see Fig. 9c). Moreover, the cells were found to adhere solely on the anode parts of the electrodes - an effect which we think results from the co-deposited fibrin network, serving as artificial extracellular matrix immobilizing the deposited cells.
Concerning the following SR-µCT imaging experiments, the final imaging results were rather interesting and unexpected. As the experiments were conducted in absorption contrast only, the expected result would have been to observe strongly absorbing deposited gold structures and a weekly absorbing polymer substrate. Further, as the deposited cells were not stained with any metal stain (like OsO_4), we did not expect to observe the cells themselves, as they have a low density and are thin compared to the polymer substrate.
But interestingly, cells were imaged well (Fig. 9a and 9b) and in good accordance with the fluorescence microscopic images acquired previously. This effect is even more astonishingly, as the cells appear as even stronger absorbing structures than the deposited gold layer. Currently, we do not have a valid explanation for this imaging result.

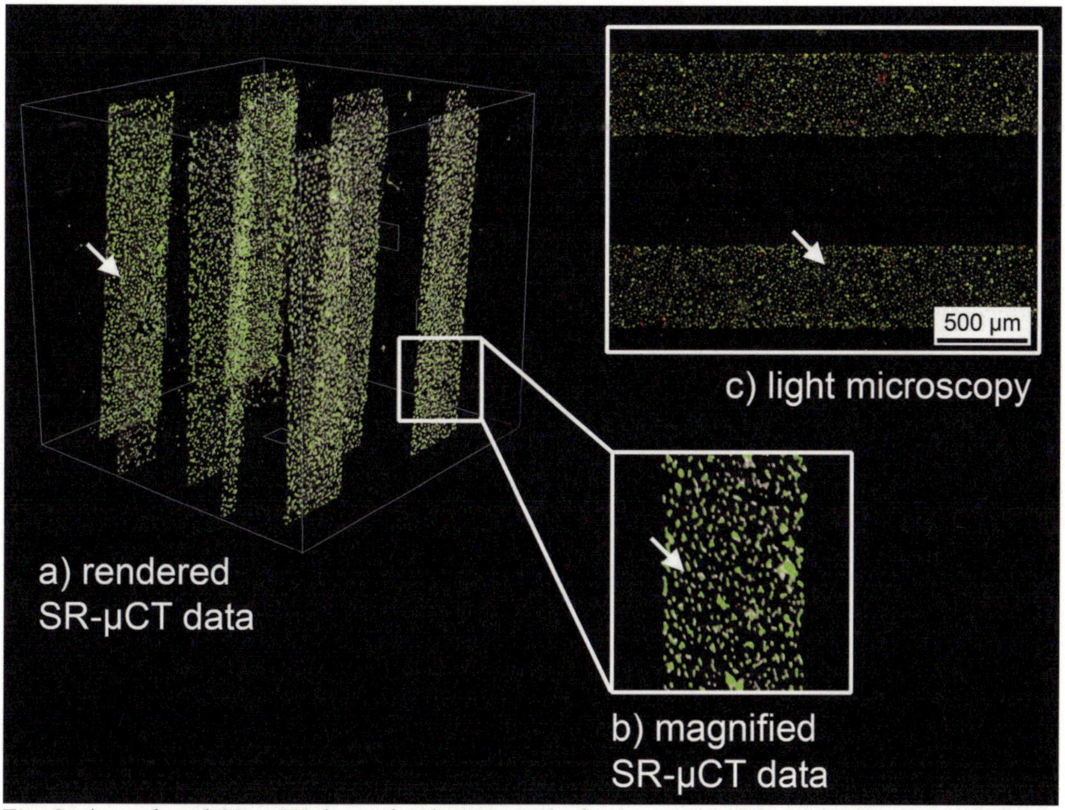

Fig. 9. a) rendered SR-µCT data of a design study for a possible 3D-nerve conduit consisting of stacked micro electrodes, b) magnified SR-µCT data showing single cells and c) corresponding fluorescence microscopic image of a vital stain showing 100 % vital cells.

8. Discussion and Conclusion

In this chapter we have given a brief overview of some tissue engineering strategies and biomaterials concepts while focussing on high resolution SR-µCT imaging.

As was stated previously the presented research should be viewed in the light of the respective references, which give more detailed information. Among the described tissue engineering solutions, the most extensive research has been conducted on gelatine scaffolds with an adjustable pore network for articular cartilage tissue engineering. These scaffolds were designed to allow for further optimizations by introducing other substances like hydroxyapatite as filler for subchondral bone defects or by addition of biochemical targeting molecules to stimulate tissue regeneration. Here, prostaglandine E_2 was immobilized exemplarily in polylactide-co-glycolide microspheres for extended release, while distributing the microspheres in the pore structure of the gelatine scaffolds.

The presented tomographic data for this tissue engineering solution detailed both cells and scaffold materials down to a voxel resolution of 1.6 µm.

One major advantage of synchrotron radiation based µCT in combination with a highly coherent beam was found in the enhanced phase contrast imaging capabilities of the used

beamline in some experiments. Here, the demonstration of the pore channel collapse under mechanical loading resolved structures otherwise not visible in absorption contrast.
A resolution up to single biological cells was demonstrated in different imaging conditions. While OsO_4 and a combination of Au-lysine with a silver enhancer were used to increase the imaging results in absorption contrast, phase contrast imaging was adapted to image individual cells in both native cartilage tissue and inside gelatine scaffolds as well. A rather interesting result was found for a stack of microelectrodes, where exceptionally well resolved single cells were found in absorption contrast.
In conclusion, SR-μCT is a promising methodology to investigate tissues or tissue engineered biomaterial-cell constructs. Due to the superior photon quality generated in modern synchrotron facilities a far more adaptable setup is established to solve specific scientific problems, otherwise not possible in X-ray tube μCT experimental setups. For example, this technique opens up the possibilities for *in situ* measurements of living tissues e.g. mechanical loading experiments, acquiring time-lapse tomographic data.

9. Acknowledgments

This study was supported partly by the Deutsche Forschungsgemeinschaft (DFG) by grants SCHU 679/27-1 & SCHU 679/27-2.
The authors would like to thank U. Gross, A. Thiem, B. Watzer, R. Grupp and M. Dilger for discussions and technical assistance.

10. References

Bernhardt, R.; Scharnweber, D.; Müller, B.; Thurner, P.; Schliephake, H.; Wyss, P.; Beckmann, F.; Goebbels, J. & Worch, H. (2004). Comparison of microfocus- and synchrotron X-ray tomography for the analysis of osteointegration around Ti6Al4V implants, *European cells & materials*, 7, 42-51

Bhang, S.H.; Jeon, O.; Choi, C.Y.; Kwon, Y.H. & Kim, B.S. (2007). Controlled release of nerve growth factor from fibrin gel, *J Biomed Mater Res A*, 80, 998-1002

Bonse, U. & Busch, F. (1996). X-ray computed microtomography (μCT) using synchrotron radiation (SR), *Prog. Biophys. molec. Biol.*, 65, 1-2, 133-169

Braverman, M.S. & Braverman, I.M. (1986). Three-dimensional reconstructions of objects from serial sections using a microcomputer graphics system, *J. Invest. Dermatol.*, 86, 290–294

Brochhausen, C.; Zehbe, R.; Watzer, B.; Halstenberg, S.; Gabler, F.; Schubert, H. & Kirkpatrick, C.J. (2008), Immobilization and controlled release of prostaglandin E_2 from poly-L-lactide-co-glycolide microspheres, *J Biomed Mat Res A*, published online, DOI: 10.1002/jbm.a.32215

Buckwalter, J.A. & Mankin, H.J. (1997). Articular Cartilage, Part I: Tissue Design and Chondrocyte-Matrix Interactions, *Journal of Bone and Joint Surgery*, 79A, 600-611

Cloetens, P.; Mache, R.; Schlenker, M. & Lerbs-Mache, S. (2006). Quantitative phase tomography of Arabidopsis seeds reveals intercellular void network, *PNAS*, 103, 39, 14626-14630

Cormack, A.M. (1073). Reconstruction of densities from their projections, with applications in radiological physics, *Physics in Medicine and Biology*, 18, 195-207

Dilmanian, F.A. (1992). Computed tomography with monochromatic X-rays, *American journal of physiologic imaging*, 7, 3-4, 175-93

Feldkamp, L.A.; Goldstein, S.A.; Parfitt, A.M.; Jesion, G. & Kleerekope, M. (1989). The direct examination of three-dimensional bone architecture *in vitro* by computed tomography, *J Bone Min Res*, 4, 3-11

Ginty, P.J.; Howard, D.; Rose, F.R.A.J.; Whitaker, M.J.; Barry, J.J.A.; Tighe, P.; Mutch, S.R.; Serhatkulu, G.; Oreffo, R.O.C.; Howdle, S.M. & Shakesheff, K.M. (2006). Mammalian cell survival and processing in supercritical CO_2, *PNAS*, 103, 19, 7426-7431

Ho, S.T. & Hutmacher D.W. (2006). A comparison of micro CT with other techniques used in the characterization of scaffolds, *Biomaterials*, 27, 1362–1376

Hounsfield, G.N. (1973). Computerized transverse axial scanning (tomography). Part I: Description of system. Part II: Clinical applications, *British Journal of Radiology*, 46, 1016-1022

Kak, A. & Slaney M. (1988). *Principles of Computerized Tomographic Imaging*, IEEE Press, New York

Korhonen, R.K.; Laasanen, M.S.; Töyräs, J.; Rieppo, J.; Hirvonen, J.; Helminen, H.J. & Jurvelin, J.S. (2002). Comparison of the equilibrium response of articular cartilage in unconfined compression, confined compression and indentation, *Journal of Biomechanics*, 35, 903-909

Mikos, A. & Temenoff, J. (2000). Formation of highly porous biodegradable scaffolds for tissue engineering, *Electronic Journal of Biotechnology*, 3, 2, 1-6

Nicoll, S.B.; Denker, A.E. & Tuan, R.S. (1998). Mesenchymal cell-based repair of connective tissue defects: Application of transforming growth factor-b superfamily members and biodegradable polymer scaffolds, *Cell Mater*, 8, 99–122

Rack, A.; Zabler, S.; Müller, B.R.; Riesemeier, H.; Weidemann, G.; Lange, A.; Goebbels, J.; Hentschel, M. & Görner, W. (2008). High resolution synchrotron-based radiography and tomography using hard X-rays at the BAM*line* (BESSY II). *Nuclear Instruments and Methods in Physics Research A*, 586, 327-344

Riesemeier, H.; Müller, B.R.; Radtke, M. (2007). BAM*line* Description (ID-02-2, 7T-WLS-BAMline, October 2007), http://www.bessy.de/upload/bitpdfs/ID_02_2.pdf

Röntgen, W.C. (1895). On a New Kind of Rays, *Nature*, 53, 274-276

Sandulache, V.C.; Parekh, A.; Li-Korotky, H.S.; Dohar, J.E. & Hebda, P.A. (2006). PGE_2 differentially modulates human fetal and adult dermal fibroblast migration and contraction: implication for wound healing, *Wound Rep Reg*, 14, 633-643

Schäfers, F. & Krumrey, M. (1996). REFLEC, Program to calculate VUV/ X-ray optical elements and synchrotron radiation beamlines, *BESSY, TB 201/96*

Schmidt[1], F.; Ghalanbor, Z.; Thormann, F.; Kühbacher, M.; Gross, U.; Bodmeier, R.; Schubert, H. & Zehbe, R. (2009). Drug Loaded, Biodegradable Nerve Conduits For The Simultaneous Chemical And Electrical Stimulation Of Neural Cells As A Therapeutic Approach For Peripheral Nerve Regeneration, *Materials Science Forum/ Advanced Materials Research*, Trans Tech Publications, accepted

Schmidt[2], F.; Kühbacher, M.; Gross, U.; Kyriakopoulos, A.; Schubert, H. & Zehbe, R. (2009). From 2D Slices to 3D Volumes: Characterization of Neuronal Cells Deposited on Anodic, Cationic and Uncharged Surfaces Using Dual Beam SEM/ FIB Serial Sectioning, submitted to *Ultramicroscopy*

Shimizu, T.; Yamato, M.; Kikuchi, A. & Okano, T. (2003). Cell sheet engineering for myocardial tissue reconstruction, *Biomaterials*, 24, 13, 2309-2316

Thiem, A.; Lum, V.; Grupp, R.; Riesemeier, H.; Bordia, R.; Schubert, H. & Zehbe, R. (2009). Synchrotron μCT Investigation of the Collapsing Pore-Network of Gelatin Scaffolds under Compression, *Materials Science Forum/ Advanced Materials Research*, Trans Tech Publications, accepted

Watzer, B.; Zehbe, R.; Halstenberg, S.; Kirkpatrick, C.J. & Brochhausen, C. (2009). Stability of prostaglandin E_2 (PGE_2) embedded in poly-D,L-lactide-co-glycolide microspheres - a pre-conditioning approach for tissue engineering applications, *J Mater Sci: Mater Med*, 20, 1357-1365

Zehbe, R.; Gross, U. & Schubert, H. (2004). Oriented collagenbased/hydroxyapatite matrices for articular cartilage replacement, *Key Engineering Materials*, 254-256, 1083-1086

Zehbe, R.; Libera, J.; Gross, U. & Schubert, H. (2005). Short-term human chondrocyte oriented collagen coated gelatine for cartilage replacement, *Bio-Medical Materials and Engineering*, 15, 445–454

Zehbe[1], R.; Haibel, A.; Brochhausen, C.; Gross, U.; Kirkpatrick, C.J. & Schubert, H. (2007). Characterization of oriented protein-ceramic and protein-polymer-composites for cartilage tissue engineering using synchrotron μ-CT, *Int J Mat Res*, 98, 562-568

Zehbe[2], R.; Gross, U.; Knabe, C.; Radlanski, R.J. & Schubert, H. (2007). Anodic cell-protein deposition on inverse inkjet printed micro structured gold surfaces, *Biosensors & Bioelectronics*, 22, 1493-1500

Zehbe[1], R.; Haibel, A.; Riesemeier, H.; Gross, U.; Kirkpatrick, C.J.; Schubert, H.; Brochhausen, C. (2009). Going beyond Histology: Synchrotron μCT as a Complementary Methodology for Biological Tissue Characterization, *Journal of the Royal Society: Interface*, published online: DOI: 10.1098/rsif.2008.0539

Zehbe[2], R.; Goebbels, J.; Ibold, Y.; Gross, U. & Schubert, H. (2009). Investigation of the Spatial Distribution of *in vitro* Cultivated Chondrocytes in Gelatine Scaffolds with varying Porosities – a Tomographic Approach, submitted to *Acta Biomaterialia*

Zehbe[3], R.; Watzer, B.; Grupp, R.; Halstenberg, S.; Riesemeier, H.; Kirkpatrick, C.J.; Schubert, H. & Brochhausen, C. (2009). Tomographic and Topographic Investigation of Poly-D,L-Lactide-Co-Glycolide Microspheres Loaded with Prostaglandin E_2 for Extended Drugrelease Applications, *Materials Science Forum/ Advanced Materials Research*, Trans Tech Publications, accepted

11. Experimental data

Section	Sample	Sample preparation	SR-μCT parameters *)					Reference
			Contrast mode	D_{ss} (cm)	E (keV)	t_e (s)	s_v (μm)	
7.1	bovine joint cartilage	aldehyde fixation, EDTA treatment, and ethanol dehydration	Phase	15	14	2.0	~1.6	Zehbe[1] et al., 2009
7.1	commercial scaffold (Ethisorb)	Seeded and cultivated with chondrocytes for 1 week. Following, aldehyde fixation and Au/Ag-staining, followed by rapid freezing and freeze-drying.	Absorption	<1.5	10	0.5	~3.7	-
7.1	gelatine scaffolds (according to Figs. 5b – 5f)	The pore morphology was modified during synthesis by varying the freezing temperature (d, e, f) and the ion content (b, c, d). Further sample treatment identical to Ethisorb scaffold	Absorption	<1.5	10	0.5 - 1.0	~3.7	Zehbe[2] et al., 2009
7.1	gelatine-fibrin composite scaffold	Similar to scaffold d) but infiltrated with cells suspended in fibrin glue.	Phase	60	20	1.5	~2.1	-
7.1	gelatine-HAp-composite scaffold	Deposition of hydroxyapatite (HAp) as bottom layer by sedimentation.	Absorption	<1.5	15	0.7	~3.6	Zehbe[1] et al., 2007
7.1	PLGA-microsphere gelatine scaffold	Inclusion of PLGA-microspheres into scaffold by suspension/ gelling	Absorption	<1.5	10	1.2	~3.6	Zehbe[1] et al., 2007
7.2	mech. loaded gelatine scaffolds	Samples measured in wet condition inside a loading apparatus.	Phase	109	30	0.7 - 0.8	~4.3	Thiem 2009
7.3	PGE_2 loaded PLGA-microsphere	Drug loaded microsphere, synthesized via emulsion.	Phase	30	20	1.8	~2.1	Zehbe[3] et al., 2009
7.4	PLGA-gelatine multilayer	Layer of 13 PLGA-Gelatine sheets (40 % PLGA, 60 % Gelatine) cultivated with CHON-001 cells and OsO_4 stained.	Absorption	1.5	7	5.5	~2.1	-
7.5	microelectrode multilayer	Layer of 4 microelectrodes cultivated with fibroblasts	Absorption	<1.5	10	0.6	~3.6	Schmidt[1] et al., 2009

Table 2. Condensed experimental data for all samples in this chapter (for a detailed description, the reader is referred to the corresponding section and the cited references).

*) D_{ss}: Distance between sample and scintillator, E: X-ray energy, t_e: exposure time, s_v: voxel-size

18

In vivo and In vitro Models of Psoriasis

Jessica Jean and Roxane Pouliot
Laboratoire d'Organogenèse Expérimentale
Centre de recherche FRSQ du CHA universitaire de Québec, Hôpital du St-Sacrement
Faculté de Pharmacie, Université Laval
Québec, Canada

1. Introduction

1.1 Skin

The evolution of life in the terrestrial environment required the development of a waterproof integument: the skin (Loden and Maibach, 2006). Skin is an extensive organ covering the entire exterior of the body (Stevens and Lowe, 2005). It provides the primary barrier against chemical and biological external agents and water loss (Hadgraft, 2001). The skin also plays an important role in thermoregulation, sensory perception and vitamin D metabolism (McKay and Leigh, 1995).

The skin is composed of three main layers: the epidermis, the dermis and the hypodermis. The epidermis is the protective skin layer in contact with the external environment (Stevens and Lowe, 2005). This skin layer consists mainly of a stratified squamous keratinized epithelium (Junqueira and Carneiro, 2005). The epidermis cells, the keratinocytes, divide in the basal layer and differentiate throughout their migration to the surface. The epidermis is divided in 5 different layers (stratum basale, stratum spinosum, stratum granulosum, stratum lucidum and stratum corneum). The dermis is the feeder layer of the epidermis and provides most of the skin's mechanical resistance and elasticity. It is mainly composed of fibroblasts, epidermal appendages, blood vessels, nerves and nerve endings (Stevens and Lowe, 2005). Finally, the hypodermis is the deepest layer of the skin. It varies in size and content, but is usually composed of adipocytes which form the adipose tissue (Stevens and Lowe, 2005). Many severe skin diseases can be observed in human beings such as psoriasis.

1.2 Psoriasis

Epidemiology

Psoriasis is an ancient chronic skin disease (Nickoloff and Nestle, 2004). Uncommon under the age of 10 years, it appears between the ages of 15 and 30 years affecting men and women (Fitzpatrick and Wolff, 2008). Psoriasis is universal in occurrence, but its prevalence in different populations varies from 0.1% to 11.8% (Raychaudhuri and Farber, 2001). In fact, this pathology shows a significant geographical variability with the lowest incidence seen at the equator and increasing frequency towards the poles (Kormeili et al., 2004). Psoriasis affects about 25 million people in North America and Europe (Lowes et al., 2007).

Characteristics

Psoriasis is characterized by red and scaly plaques on the skin with a predilection for certain areas (elbows, knees and scalp) (Chapman et al., 1990; Lowes, et al., 2007). There are five clinical variants of psoriasis: guttate, erythrodermic, pustular, inverse and psoriasis vulgaris, the last of wich is the most common type seen in approximately 90% of patients (Fitzpatrick and Wolff, 2008). The severity can be divided into benign, moderate and severe psoriasis. Histological appearance of lesions shows elongation of the rete ridges, disappearance of the granular layer and persistence of the keratinocytes nuclei in the stratum corneum of the epidermis (Danilenko, 2008). In psoriasis, epidermal hyperproliferation, abnormal keratinocyte differentiation, angiogenesis with blood vessel dilatation and excess Th-1 and Th-17 inflammation can be observed (Azfar and Gelfand, 2008). However, although genetic, immunological and environmental factors seem implied, the exact cause is not yet known and even today, psoriasis is not well understood (Bowcock, 2005).

Treatments

A broad spectrum of anti-psoriatic treatments, both topical and systemic, is available for the management of psoriasis (Fitzpatrick and Wolff, 2008). However, it is often resistant to treatment or, else, frequently relapses upon cessation of medication after partial or acceptable clearance is obtained (Kormeili, et al., 2004). The severity of the disease usually determines the therapeutic approach. Among the treatments, there are topical, phototherapy, biologic and systemic treatments (Table 1). Approximately, 70 to 80% of all patients with psoriasis can be treated adequately with topical therapy (Schon and Boehncke, 2005). For others, phototherapy and systemic treatments are effective; however, the duration of a treatment is restricted because of the cumulative toxicity potential of an individual therapy (Kormeili, et al., 2004). For example, some treatments may increase the risk of cancer (phototherapy) while others can induce disorders in the liver (methotrexate) (Dubertret, 2004). Sometimes, treatment efficacy may diminish with time and it must be replaced by another therapy (Fitzpatrick and Wolff, 2008). At the present time, there is still no curative treatment for psoriasis.

Topical	Phototherapy	Biologic	Systemic
Corticosteroids	Narrowband UVB	Alefacept	Cyclosporine A
Vitamin D analogues	Broadband UVB	Efalizumab	Methotrexate
Tazarotene	Psoralen-UVA	Etanercept	Acitretin
Calcineurin inhibitors	Excimer laser	Infliximab	Fumaric acid esters
		Adalimumab	Sulfasalazine
			Mycophenolate mofetil
			6-Thioguanine
			Hydroxyurea

Table 1. Anti-psoriatic treatments

2. Tissue engineering

Although conventional approaches for organ replacement, such as transplantation, autografts and implantation of engineered prostheses, are extensively used, the process by which a patient can regenerate a lost organ is more attractive (Yannas, 2004). Recent biotechnological progress in the tissue engineering field allows us to conceive, develop and produce biomaterials which can replace tissues or organs (Bernard et al., 2007). It has been demonstrated that tissue reconstruction can be applied to several different tissues such as skin, blood vessels, cartilage, bones and corneas (Arosarena, 2005). Tissue engineering can be used in experimental and clinical applications (Auger et al., 2004).

3. In vivo and in vitro models versus pharmaceutical researches

3.1 Development of drugs

The development of a new drug takes approximately 20 years. During this period of time, pharmaceutical industries spend millions of dollars on the research for new, more effective drugs. The cost of bringing new drugs to the market has recently increased and today, it costs approximately 1 billion US dollars to bring a new medicine to the market. A large amount of research is devoted every year to the discovery of new pharmacologically effective substances and many molecules reach the level of pre-clinical and clinical phase trials. However, attrition rates in clinical development are still very high and up to 90 % of new compounds fail in clinical phases I-III (Zollner et al., 2004). Late-stage clinical failure can be, to a great extent, attributed to a lack of clinical efficacy, indicating a strong need for highly predictive in vivo and in vitro models. Consequently, the pharmaceutical industry stays alert for the development of more relevant pathological in vivo and in vitro models to improve the success rate of new drugs (Zollner, et al., 2004). Highly predictive models can easily translate into significant savings of time and money for the pharmacological industry.

3.2 Experiments on psoriatic models

Some models have already been used to evaluate the effect of various molecules on psoriasis. For example, a team tested an antibody directed against interleukin 15 that was known to inhibit the production of T lymphocytes as well as the liberation of TNF-α in vitro (Krueger and Bowcock, 2005). Application of this antibody on severe combined immunodeficient mice (SCID) led to the disappearance of psoriatic characteristics. This suggested that the molecule could be targeted for future studies to develop a new anti-psoriatic treatment.

4. In vivo models

Animal models are very popular for the study of psoriasis. Many approaches are currently followed in order to obtain a representative animal model of the disease with all the characteristics of the human pathology. Many immunological and genetic models have been developed to date. However none of these models show all the characteristics of psoriasis. Researchers look for a perfect animal model which would have many genetic, histological and morphological similitudes with human beings and react to treatments in a similar way (Zollner, et al., 2004). Furthermore, the models must be easily reproducible, inexpensive and ethical.

4.1 Spontaneous mutations

Over the years, many mutations were described as being responsible for abnormal changes in the skin or hair of mice (Sundberg et al., 1990). Among these mutations, some were studied for their psoriasis-like characteristics such as thickening of the skin and the formation of scales (Mizutani et al., 2003). However, the resulting mutants do not closely mimic the disease enough to be considered as good models of psoriasis. They must rather be used to compare local pathogenic events such hyperkeratosis, regulation of neutrophil infiltration and microabscess formation or dermal angiogenesis (Schon, 2008). Nearly a hundred mouse mutations that lead to psoriasiform phenotypes have been documented (Sundberg, et al., 1990).

Homozygous asebia ($Scd1^{ab}/Scd1^{ab}$)

The asebia mouse mutation was one of the first in vivo models used to study therapies directed at hyperkeratotic disorders (Schon, 2008). These homozygous asebia mutant mice are characterized by a hypoplasia of the sebaceous glands resulting from a defect in the stearoyl coenzyme A desaturase-1 (Scd1) gene (Zheng et al., 1999). This model exhibits moderate epidermal acanthosis, increased dermal vascularization, a dermal infiltrate composed of macrophages and mast cells, but neither T cells nor neutrophils (Schon, 2008). This lack of T cells and neutrophils do not mirror psoriatic lesions. Moreover, alterations of the cutaneous lipid metabolism seem different from psoriasis.

Flaky skin mice (Ttc^{fsn}/Ttc^{fsn})

The spontaneous chronic proliferative dermatitis mutation ($Sharpin^{cpdm}/Sharpin^{cpdm}$) (HogenEsch et al., 1993) and the flaky skin (Sundberg, et al., 1990) mice show a more interesting psoriatic phenotype than homozygous asebia mice. The flaky skin mice are probably the best spontaneous model of psoriasis described (Danilenko, 2008). Its spontaneous mutation induces proliferation and hyperkeratosis of stratified squamous epithelia including the nonglandular forestomach (Stratis et al., 2006). Previous studies have shown a positive Koebner reaction after tape-stripping, which resolves after 6 weeks of treatment with oral, but not topical, cyclosporine A, topical EGF, or UVB exposure (Sundberg et al., 1994). The flaky skin phenotype is very complex and comprises aspects not present in psoriasis. To verify if plaques had an inflammatory origin, the mice were treated with cyclosporine. The results showed that immunosuppressive treatment had no effect on the psoriatic lesions of those mice. This observation demonstrated that such models are not complete because they lack the immunological side of the pathology (Schon, 1999).

Spontaneous chronic proliferative dermatitis mutation

The spontaneous chronic proliferative dermatitis mutation shows various characteristics found in psoriasis such as hyperproliferative skin, infiltration of inflammatory cells in the skin and dilation of blood vessels in the dermis (Schon, 1999). However, as with the flaky skin mice, the value of the spontaneous chronic proliferative dermatitis mutation for psoriasis research is limited by the lack of a T cell based immunopathogenesis. Moreover, as immunosuppressive therapeutic regimens used to treat psoriasis fail to improve skin lesions with these mutations, it appears uncertain whether they can be used to test potential therapeutic compounds (Schon, 1999).

4.2 Genetically engineered models

Genetically engineered mice represent the largest category of psoriasis models. These include the transgenic and knockout models. In this section, some of these models will be discussed. The complete list of in vivo psoriatic skin models can be seen in table 2.

HLA-B27 rat

The human leukocyte antigen B27 (HLA-B27) transgenic rats are normal at birth but develop chronic inflammation of multiple organ systems as they age (Keith et al., 2005). In this model, the human HLA-B27 and β_2-microglobulin proteins are overexpressed and an epidermal acanthosis with epidermal infiltration of both CD4+ and CD8+ T cells, as well as immune-mediated arthritis and inflammatory bowel disease, which seem similar to the spondyloarthropathies in humans that have been associated with the HLA-B27 and β_2-microglobulin genes, such as acanthosis can be observed (Breban et al., 1996; Keith, et al., 2005; Taurog et al., 1993; Yanagisawa et al., 1995). The HLA-B27 transgenic rat model has been used for several years to evaluate the activity and mechanisms of action of anti-inflammatory molecules (Chadwick et al., 2005; Harnish et al., 2004; Peterson et al., 2002). Results show that broad-spectrum antibiotic therapy can produce significant remissions of inflammatory lesions, but relapse occurs when antibiotic therapy stops (Keith, et al., 2005). In this model, the occurrence of psoriatic skin lesions is less consistent than is the occurrence of the other immune-mediated disorders and there are no published reports of therapeutic efficacy testing for the psoriatic lesions (Danilenko, 2008; Taurog, et al., 1993; Yanagisawa, et al., 1995).

CD18 hypomorphic

The CD18 hypomorphic mice model targets leukocytes (Bullard et al., 1996). This model shows a decreased expression of the common β_2 chain of the leukointegrin adhesion molecule complex (Danilenko, 2008). Beta2 integrins are leukocyte adhesion molecules exclusively expressed on hematopoietic cells and are responsible for cell-cell contacts in a variety of inflammatory interactions (Kess et al., 2003). When the CD18 mutation was crossed onto the PL/J strain of mice, it developed a psoriasiform inflammatory skin condition with a predominantly lymphocyte infiltration (Bullard, et al., 1996). However, these mice exhibit nonpsoriasiform characteristics such as the absence of hyperproliferation markers.

K14/VEGF and Tie2

Mice overexpressing VEGF in epidermis via a keratin 14 promoter develop a phenotype very similar to psoriasis with an epidermal acanthosis, abnormal differentiation, inflammatory infiltrate and inflamed dermal blood vessels (Kunstfeld et al., 2004; Xia et al., 2003). Even if this model has many immunological, vascular and epidermal resemblances with psoriasis, the lesions appear to be vascular-based and there is the presence of a major dermal infiltration of mast cells (Danilenko, 2008). Similar elements were observed in the Tie2 transgenic mice. With the activation of the transgene, the mice develop erythema with silver-white scaling, acanthosis, elongation of the rete ridges, hyperkeratosis, parakeratosis, increase in blood capillaries, infiltration of inflammatory cells and neutrophilic microabscess (Gudjonsson et al., 2007; Voskas et al., 2005). This model also reacts to cyclosporine A when used as a psoriatic therapy.

K14/TGF-α, K5/TGF-β_1, K14/KGF and K14/IL-20

Many transgenic mice models were produced to target, via the keratin 5 or keratin 14 promoter, the expression of epithelial growth factors such as TGF-α (Vassar and Fuchs, 1991), TGF-β_1 (Li et al., 2004) and KGF (Guo et al., 1993) into the basal layer. In each model, psoriasiform phenotype can be observed such as acanthosis, but no cutaneous inflammation is displayed into any of them. In the K14/IL-20 model, the expression of IL-20 was targeted to the basal layer under the control of a keratin 14 promoter (Blumberg et al., 2001). In this model, some psoriatic characteristics can be observed such as abnormal keratinocyte differentiation and an expression of the keratin 6, a hyperproliferation marker (Blumberg, et al., 2001). However, the lack of inflammatory components is a severe drawback of this model, as it is for K14/TGF-α, K5/TGF-β_1, K14/KGF models.

IKK2 and JunB/c-Jun

The immunological theory of psoriasis was challenged by more recent studies using animal models, including the IKK2 (Pasparakis et al., 2002) and JunB/c-Jun transgenic mice (Zenz et al., 2005). Results obtained with these two models have suggested revisiting the primary pathogenic role for epidermal keratinocytes. In the IKK2 knockout mice model, the deletion of the IKK2 catalytic subunit of the IκB kinase complex caused mice to develop a psoriasiform cutaneous inflammation (Pasparakis, et al., 2002). This model displays many features of psoriasis, including dependence on the dermal expression of TNF-α, acanthosis and abnormal differentiation. However, some characteristics of this model, such as T-cell-independent inflammation, keratinocyte apoptosis and early death are not found in psoriasis (Pasparakis, et al., 2002; Stratis, et al., 2006). In the JunB/c-Jun mice model, JunB and c-Jun are knocked out in the epidermis of postnatal mice with $Junb^{tm3Wag}$ or Jun^{tm4Wag}, or both (Zenz, et al., 2005). In this model, affected skin areas showed infiltration of neutrophils and lymphocytes with upregulation of several cytokines and chemokines typical of psoriasis. However, other factors such as INF-γ were only slightly upregulated while IL-12 or IL-18 were absent (Gudjonsson, et al., 2007). In this model, cutanous inflammation is not dependant on T cells nor is it independent of TNF signalling (Zenz, et al., 2005).

K5.Stat3C

The K5.Stat3C mice model is characterized by the activation of the signal transducer and activator of transcription 3 (Stat3) in basal keratinocytes under the control of the keratin 5 promoter (Sano et al., 2005). Stat3 plays an important role in various biological activities including cell proliferation, survival and migration (Hirano et al., 2000). These mice developed psoriatic-like skin lesions characterized by a keratinocyte hyperplasia, a loss of the granular layer and parakeratosis. The presence of many dilated blood vessels and a leukocytic infiltration of lymphocytes and neutrophils were observed (Sano, et al., 2005). To generate a psoriatic phenotype in transplanted SCID mice, the injection of activated lymphocytes and the Stat3 transgenic skin are necessary. These results allow to establish a link between keratinocytes and CD4+ T lymphocytes in psoriasis (Danilenko, 2008).

K14/IL-6 and K14/IL-1α

A human keratin 14 promoter was used to express IL-6 in the basal cells of epidermal mice (Turksen et al., 1992). IL-6 expression did not lead to enhanced epidermal proliferation, but it did result in a thicker stratum corneum with an otherwise seemingly normal differentiation program. However, IL-6 expression did not lead to leukocytic infiltration (Turksen, et al., 1992). In 1995, Groves et al., observed characteristics of psoriasis in a transgenic mouse model that expresses high levels of interleukin 1α in basal epidermis. This model displayed many characteristics of psoriasis. Uninvolved skin of these animals was characterized by hyperkeratosis and dermal mononuclear cell infiltrate of macrophage/monocyte lineage (Groves et al., 1995). Inflammatory lesions were marked by a mixed cellular infiltrate, acanthosis and parakeratosis in some cases. These results strongly indicate that IL-1 is a cytokine which plays an important role in psoriasis and that it is capable of inducing an inflammatory reaction (Groves, et al., 1995). Murine transgenic models can provide representative models of the disease, but they can also target or confirm if a cytokine such as interleukin 1 or 6 has a role to play in psoriasis.

Chymotryptic enzyme

In 2002, Hansson et al., developed a transgenic mouse strain that overexpresses the chymotryptic enzyme, which is also overexpressed in the stratum corneum of psoriatic skin. This model allows to observe pathological characteristics of psoriasis such as an increase of epidermal thickness, hyperkeratosis, severe pruritus and an inflammation of the dermis (Hansson et al., 2002).

4.3 Xenotransplantation

Animal models based on transgenic technology have been used extensively to study the pathogenesis of various diseases, including psoriasis (Raychaudhuri et al., 2001). Xenotransplantation experiments were performed in which a skin biopsy from a patient or a skin equivalent produced in vitro was transplanted on mice from spontaneously mutated or genetically modified strains.

Athymic nude mouse

Athymic nude mouse have no thymus and therefore no T cells (Raychaudhuri, et al., 2001). This model has been used in laboratory to gain insights into the immune system and autoimmune diseases. The absence of a functional humoral immune system allows transplantation from another species without graft rejection. However, psoriatic skin transplanted on athymic mice develops certain histological changes that are not typical of psoriasis, such as the absence of parakeratosis and the presence of a granular layer (Raychaudhuri, et al., 2001). In past years, athymic nude mice were mainly used to verify if there was a difference between involved and uninvolved psoriatic skin (Fraki et al., 1983; Krueger et al., 1981). The results showed that involved psoriatic skin maintains its psoriasiform histology when transplanted onto nude athymic mice such as epidermal thickness and papillomatosis (Fraki, et al., 1983). However, not all the characteristics of the pathology were preserved. In fact, psoriatic epidermis did not contain polymorphonuclear leucocytes after grafting on athymic nude mice. In this study, uninvolved psoriatic epidermis from psoriatic patients seemed to be able to display markers of involved psoriatic epidermis independently from the psoriatic host. The authors suggested that the skin itself could be the primary cause of psoriasis. However, some years later, other results obtained in a new mouse model, the severe combined immunodeficient mice, suggested that psoriasis is a T-cell-mediated disease (Fraki, et al., 1983).

Severe combined immunodeficient mice (SCID)

The severe combined immunodeficient mouse model shows a lack of T and B cells, but it contains functional neutrophil and mature natural killer (NK) cells with normal cell activity (Gudjonsson, et al., 2007). This model has a mutation in the DNA-dependant protein kinase gene that is required for successful T-cell and B-cell development (Gudjonsson, et al., 2007). It is probably the most widely used relevant model for psoriasis, but the presence of NK which are involved in rejection of xenogeneic tissue is a severe inconvenience (Gourlay et al., 1998). In fact, single-cell suspensions are rapidly recognized and lysed by mice NK cells (Meyerrose et al., 2003). However, in SCID mice, grafts of solid tissues are well tolerated and psoriatic characteristics are maintained for several months in the transplantations involving psoriatic skin (Raychaudhuri, et al., 2001; Takizawa et al., 1995). Raychaudhuri et al., noticed that clinical, histological and immunological features of psoriasis could be maintained for durations of 12-16 weeks, while Sugai et al. conserved the human psoriatic skin transplant for up to 22 weeks (Sugai et al., 1998). In this latter study, the psoriatic phenotype was well maintained but the histological and immunohistochemical characteristics gradually disappeared as lymphocytic infiltration of the psoriatic lesion declined. However, Ghilhar et al. showed that injection of T cells from psoriatic plaques in SCID mouse could maintain the pathological characteristics longer than in the absence of T cells (Gilhar et al., 1997). These results show that the presence of inflammatory cells is necessary to maintain the psoriatic phenotype in the SCID model and suggest a role of inflammatory cells in the appearance of psoriasis (Gilhar, et al., 1997).

Spontaneous AGR129 model

A new model, the AGR129 mice, shows a lack of T and B cells but, contrary to the severe combined immunodeficient mice, it has immature NK (Boyman et al., 2004). While the natural killer cells are recognized to have a role in transplant rejection (Gourlay, et al., 1998), the AGR129 model tolerates xenogenic grafts better than the SCID model. The AGR129 mice are deficient in type I(A) and type II(G) IFN receptors in addition to being RAG-2$^{-/-}$. Disruption of both IFN receptors has been previously shown to lead to decreased NK cytotoxic activity in vitro and in vivo (Lee et al., 2000). Such deficiencies imply that these mice possess immature NK cells which are much less cytotoxic than mature ones (Boyman, et al., 2004). Thus, transplant rejection is reduced in this type of mice. Boyman et al. demonstrated that human uninvolved psoriatic skin grafted onto AGR129 mice spontaneously developed psoriatic plaques without injection of activated immune cells or any other exogenous factor. In fact, skin grafts developed a psoriatic phenotype in 28 of 31 (90 %) grafted mice (Gudjonsson, et al., 2007). Histology of developed plaques was comparable to psoriatic lesion biopsies from the same patient. Boyman's team also noticed that when they injected an inhibitor of T cells (monoclonal anti-CD3), the psoriatic phenotype disappeared. The same phenomenon occurred when they injected an inhibitor of TNF-α. Furthermore, in this model, the normal control skin did not develop a psoriasis-like phenotype. These observations indicate that activation and proliferation of resident T cells are necessary and sufficient to drive psoriasis formation, underlining an essential role for immune cells residing within symptomless pre-psoriatic skin (Conrad and Nestle, 2006).

Model	Epidermal thickness	Abnormal differentiation	Increased vascularization	Epidermal T-cell infiltration	References
Spontaneous mutation					
Homozygous asebia	+	-	+	-	(Brown and Hardy, 1988; Brown and Hardy, 1989)
Flaky skin	+	+	+	-	(Sundberg, et al., 1994; Sundberg et al., 1997)
Chronic proliferative dermatitis	+	+	+	-	(HogenEsch, et al., 1993)
Genetically engineered models					

Targeting the immune system					
HLA-B27/β2 microglobulin rat	+	+	+	+	(Breban, et al., 1996; Keith, et al., 2005)
Hypomorphic CD18	+	+	+	+	(Bullard, et al., 1996; Kess, et al., 2003)
αE (CD103)	+	+	?	+	(Schon et al., 2000)
K14/p40	+	?	?	+	(Kopp et al., 2001)
Targeting vascular endothelium					
pTek-*tTA*/Tie2	+	+	+	+	(Voskas, et al., 2005)
K14/VEGF	+	+	+	+	(Xia, et al., 2003)
Targeting epidermal proteins					
K5/Stat3C	+	+	+	+	(Sano, et al., 2005)
IKK2	+	+	?	-	(Pasparakis, et al., 2002)
c-Jun/JunB	+	+	+	+	(Zenz, et al., 2005)
K14/KGF	+	+	+	-	(Guo, et al., 1993)
K14/TGF-α	+	+	?	Some animals	(Vassar and Fuchs, 1991)
K14/IL-20	+	+	-	-	(Blumberg, et al., 2001)
K14/amphiregulin	+	+	+	+	(Cook et al., 1997)
K14/IL-1α	+	+	-	?	(Groves, et al., 1995)
K14/IL-6	+	-	-	-	(Turksen, et al., 1992)
K10/BMP-6	+	+	+	+	(Blessing et al., 1996)
Involucrin/integrins	+	+	+	+	(Carroll et al., 1995)
Involucrin/MEK1	+	+	?	+	(Hobbs et al., 2004)
Involucrin/amphiregulin	+	+	+	+	(Cook et al., 2004)
Involucrin/IFN-γ	+	+	+	-	(Carroll et al., 1997)
Chymotryptic enzyme	+	+	?	+	(Hansson, et al., 2002)
Xenotransplantation					
Athymic nude mouse	+	+	?	-	(Fraki, et al., 1983; Krueger, et al., 1981)
SCID	+	+	+	+	(Boehncke et al., 1994; Gilhar, et al., 1997; Nickoloff et al., 1995; Raychaudhuri, et al., 2001; Sugai, et al., 1998; Takizawa, et al., 1995)
AGR129	+	+	+	+	(Boyman, et al., 2004)

Table 2. In vivo models of psoriasis

5. In vitro Models

Contrary to in vivo models, in vitro models are generally characterized by a lack of inflammatory cells. Since psoriasis has been described to be an autoimmune disease (Bowcock, 2005; Krueger and Bowcock, 2005; Lowes, et al., 2007; Schon and Boehncke, 2005), the pertinence of these models has been discussed. It is important to note that (1) inflammatory cells can be included in most of the in vitro models, (2) these models allow to dissect step by step the mechanisms of psoriasis by isolating or combining the cell types and (3) recent studies using animal models, including the IKK2 (Pasparakis, et al., 2002) and JunB/c-Jun transgenic mice (Zenz, et al., 2005) are challenging the immunological theory of psoriasis.

5.1 Monolayer

Monolayer models allow, from a small biopsy of pathological skin, the generation of a large number of cells sufficient for multiple experiments. In monolayer models, only one cell type is considered at a time. Thus, either keratinocytes or fibroblasts will be used to test different conditions or observe various characteristics of the disease, such as cell proliferation or differentiation. These models allow the isolation of a normal or pathological cell type to better understand its specific role. However, they preclude the study of interactions among many cellular types (e.g. interactions between the dermis and the epidermis).

Despite the absence of interaction between the different cell types, monolayer models have helped in the discovery of several interesting facts about psoriasis and a better understanding of the disease. For example, monolayer cell cultures have helped to realize that TGF-α regulates VEGF expression in psoriasis through an autocrine mechanism, leading to vascular hyperpermeability and angiogenesis (Detmar et al., 1994). In other experiments, van Ruissen et al. found that psoriatic keratinocytes display a lower number of cells in S-phase and a shorter duration of G1 compared to normal keratinocytes (van Ruissen et al., 1996). Monolayer cultures are widely used and have led to many critical observations (Table 3).

5.2 De-epidermized dermis

Organ culture on de-epidermized dermis

Mils et al. developed a reconstructed epidermal model with a de-epidermized dermis. A psoriatic or normal biopsy was placed dermal side down on the epidermal surface of non viable de-epidermized dermis, thus in contact with the remnant basement membrane component (Mils et al., 1994). The dead de-epidermized dermis was maintained at the air-liquid interface on a metallic support. In approximately 15 days, the epidermal layer grew out from the punch biopsy to cover the entire de-epidermized dermis. In this model, the psoriatic substitutes did not differ significantly from normal substitutes. The only observed difference was in LH8 labelling between normal and psoriatic substitutes (Mils, et al., 1994). No distinguishing histological or biochemical criteria could be established between normal and psoriatic equivalents.

Keratinocytes on de-epidermized dermis

Another reconstructed epidermal model was developed with normal adult human keratinocytes seeded on de-epidermized dermis (Tjabringa et al., 2008). This model allowed controlled induction of psoriasis-associated features and gene expression by the addition of relevant pro-inflammatory cytokines (TNF-α, IL-1α, IL-6 and IL-22) and primary keratinocytes obtained from donors without history of psoriasis. To produce this model, a hollow metal ring was placed on de-epidermized dermis and keratinocytes were seeded within the ring (Tjabringa, et al., 2008). Results showed that after the addition of a pro-inflammatory cytokine mixture, the expression of psoriasis-associated proteins hBD-2 and SKALP/elafin for the pro-inflammatory cytokines IL-8 and of TNF-α were increased in the skin equivalent, as well as keratinocyte hyperproliferation cells. Tjabringa et al. also showed that the addition of all-trans retinoic acid inhibits the expression of psoriasis-associated proteins hBD-2 and SKALP/elafin in cytokine-stimulated skin equivalents and reduces expression of the normal differentiation marker: keratin 10, as such as acanthosis can be observed in psoriatic skin in vivo (Tjabringa, et al., 2008).

5.3 Collagen gels

Organ culture

To observe cell proliferation, some studies were done by making a small full-thickness punch biopsy into a dermal equivalent, which itself originated from the contraction of a collagen gel by dermal fibroblasts (Saiag et al., 1985). The total surface area covered by the keratinocytes was used to calculate the cell proliferation percentage. Higher keratinocyte proliferation values were obtained in the presence of psoriatic fibroblasts (Saiag, et al., 1985). Furthermore, this model led to the conclusion that normal fibroblasts are unable to suppress the hyperproliferative growth of psoriatic keratinocytes and that hyperproliferation of normal epidermis can be induced both by uninvolved and involved psoriatic fibroblasts (Saiag, et al., 1985).

Models using many cellular types

Other teams developed skin equivalents composed of two or more cell types to better understand cell interactions such as those between keratinocytes and fibroblasts. In general, these models imply the isolation of cells from biopsies of normal or affected persons. Fibroblasts are extracted from the dermis, expanded and embedded in a collagen gel. Keratinocytes are extracted from the epidermis, expanded and plated on top of a collagen gel containing fibroblasts.

Konstantinova et al. have presented a psoriatic collagen gel model which showed a psoriasiform phenotype. In fact, in this model, the psoriatic skin equivalent showed a thicker epidermis and a loss of filaggrin (Konstantinova et al., 1996). These models enabled to examine the effects of normal and psoriatic (involved and uninvolved) fibroblasts on epidermal differentiation and cytokine expression. This model led to a number of interesting conclusions, such as the confirmation that involved and uninvolved fibroblasts induce higher amounts of IL-8 than normal fibroblasts (Konstantinova, et al., 1996). This model allowed the observation of interactions between fibroblasts and keratinocytes, but contained exogenous material. As previously described, the collagen gel can be very useful for rapid production of skin substitutes (Bell et al., 1981a; Bell et al., 1981b; Bell et al., 1979) but, the presence of an exogenous scaffold can be disadvantageous for mechanical studies of the extracellular matrix and a severe surface reduction of dermal substitutes can be observed (Auger et al., 1998; Germain and Auger, 1995). Contraction can be prevented using anchorage methods in vitro (Eckes et al., 1995; Germain and Auger, 1995; Grinnell and Lamke, 1984; Parenteau et al., 1991; Xu et al., 1996). However, these anchoraged models are often fragile and difficult to handle.

Other studies have used similar in vitro models. In 2004, Barker et al. developed and characterized a new psoriatic skin model in vitro using collagen gels. In this model, psoriatic skin substitutes maintained many characteristics of psoriasis such as hyperproliferation, overexpression of chemokine receptor CXCR2 and pro-inflammatory genes (TNF-α, INF-γ and IL-8) and also increased levels of pro-inflammatory cytokines IL-6 and IL-8 (Barker et al., 2004). The substitutes derived from uninvolved psoriatic skin showed the same gene expression profile as those derived from involved skin substitutes with an increased proliferation rate when compared to normal substitutes. This model suggests that psoriatic individuals possess an inherent predisposition to develop the disease phenotype even in the absence of T cells (Barker, et al., 2004). As with the model proposed by Konstantinova et al., this model contains exogenous material.

Models	Epidermal thickness	Abnormal differentiation	Increased vascularization	Epidermal T-cell infiltration	References
Monolayer					
IL-15	?	?	?	?	(Krueger and Bowcock, 2005)
TGF-α regulates VEGF	?	?	?	?	(Detmar, et al., 1994)
S-phase	?	?	?	?	(van Ruissen, et al., 1996)
De-epidermized dermis					
Organ culture	-	-	?	-	(Mils, et al., 1994)
Keratinocytes	+	+	?	-	(Tjabringa, et al., 2008)
Collagen gel					
Organ culture	?	+	?	?	(Saiag, et al., 1985)
Skin substitutes	+	+	?	-	(Barker, et al., 2004; Konstantinova, et al., 1996)
Self-assembly					
Skin substitutes	+	+	-	-	(Bernard, et al., 2007; Jean et al., 2009)

Table 3. In vitro models of psoriasis

6. A new in vitro psoriatic skin model

In pharmaceutical sciences, pathological skin substitutes can be very useful for toxicity or for cutaneous cellular and molecular biology studies. To be effective, a pathological substitute must mimic as closely as possible, morphological and ultrastructural characteristics of the pathology. At the present time, although some in vitro and in vivo models of psoriasis have been reported to replicate some aspects of the disease, research into

psoriasis and the subsequent development of therapeutic strategies have been hindered by the absence of more relevant models (Jean, et al., 2009).
As previously shown, a number of teams have carried out in vitro monolayer and stratified keratinocyte culture studies to better understand the behaviour of cell types both in an individual way, but also to study the interactions between cells with co-culture and conditioned media techniques (Mils, et al., 1994; Saiag, et al., 1985; van Ruissen, et al., 1996). Some of these studies have shown the importance of the interaction between fibroblasts and keratinocytes in psoriatic models. Amongst in vivo models (Gudjonsson et al., 2004), there are animal models primarily consisting of grafting a psoriatic skin athymic (Fraki, et al., 1983) or on severe combined immunodeficient mice (Raychaudhuri, et al., 2001). While the clinical and histological characteristics of psoriasis can be maintained for a sufficient period of time to test pharmacological drugs, mouse skin is not like human skin. The graft of human psoriatic plaques on mice might cause in those animal models an immune response that differs from that in humans. Mice used for xenografts are immunocompromised to prevent rejection or graft shrinkage; there is no organized lymphatic system to drain the grafted skin and a lack of lymphocyte recirculation.

7. Self-assembly method

7.1 Introduction

Our team has developed a new psoriatic skin model produced by the self-assembly method. This tissue engineered approach is based on the capacity of fibroblasts to create their own extracellular matrix in vitro, which makes it possible to obtain cell sheets which are easy to handle. A cohesive dermal tissue is obtained by stacking two of these sheets together upon which keratinocytes are seeded. This leads to a complete bilayered skin substitute devoid of exogenous extracellular matrix proteins and synthetic material. This substitute has many histological characteristics close to those of normal human skin (Michel et al., 1999; Pouliot et al., 2002). While previous studies using the self-assembly model only used healthy cells, the present study creates pathological substitutes with cells isolated from psoriatic biopsies. This work strives to develop and characterize a novel in vitro psoriatic skin model produced by tissue engineering, which could be used to investigate the mechanisms of abnormal keratinocyted growth and to study cell-cell interactions. In this section, previously published results will be discussed (Jean, et al., 2009).

7.2 Methods

The psoriatic substitutes were produced using our modified version of the self-assembly method (Fig. 1). The method is, briefly, as follows: human fibroblasts were cultured in the presence of ascorbic acid, thus forming manipulatable sheets, which were superimposed and incubated for 7 days to form the dermal component. After 7 days of culture, keratinocytes were seeded upon this tissue to form the epidermal layer. After another 7 days of culture, the substitutes were raised to the air–liquid interface (Michel, et al., 1999; Pouliot, et al., 2002). Biopsies were taken after 7, 14 and 21 days of culture at the air–liquid interface and analyzed using histological and immunohistochemical techniques. Four different combinations of healthy and/or psoriatic cells were used to produce skin substitutes:
A. Healthy fibroblast/healthy keratinocyte (controls)

B. Psoriatic fibroblast/psoriatic keratinocyte
C. Psoriatic fibroblast/healthy keratinocyte
D. Healthy fibroblast/psoriatic keratinocyte

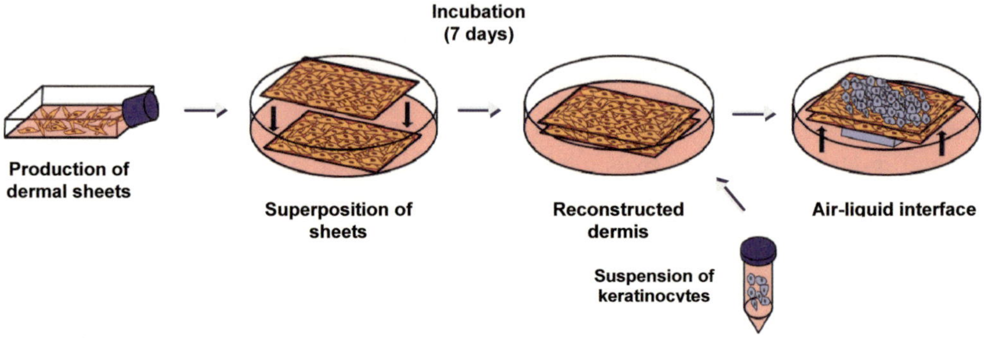

Fig. 1. Self-assembly method

7.3 Results

In the presence of healthy fibroblasts and healthy keratinocytes (controls), the substitutes showed a uniform, smooth and white surface (Fig. 2A). When psoriatic fibroblasts and keratinocytes were used to produce the substitutes, the aspect of the substitutes was less uniform showing a surface that was thick and whitish in some regions and thinner in others (Fig. 2B). The same characteristics were observed with a combination of healthy fibroblasts and psoriatic keratinocytes (Fig. 2D). Skin substitutes produced with psoriatic fibroblasts and healthy keratinocytes showed the presence of some protuberances (Fig. 2C).

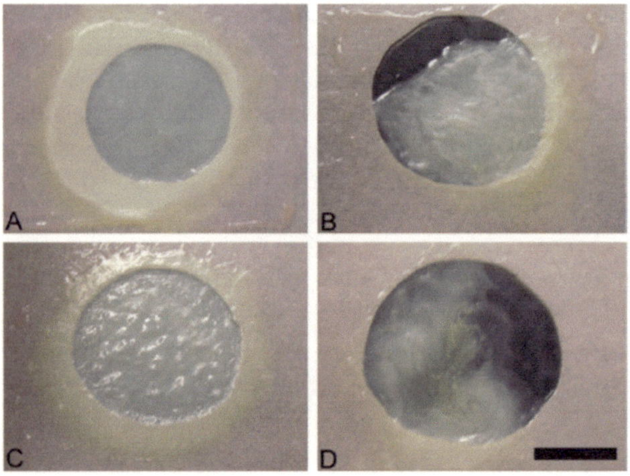

Fig. 2. Macroscopic analysis of skin substitutes produced with healthy fibroblasts and healthy keratinocytes (A), psoriatic fibroblasts and psoriatic keratinocytes (B), psoriatic fibroblasts and healthy keratinocytes (C), and healthy fibroblasts and psoriatic keratinocytes (D). Pictures were taken after 21 days of culture at the air-liquid interface (scale bar = 2.2 cm). Reproduced from Jean et al., 2009 © Journal of Dermatological Science.

Masson's trichrome staining of 5-µm thick biopsies from skin substitutes showed that those produced with psoriatic keratinocytes (Fig. 3B and D) had a significantly thicker epidermis than the controls (Fig. 3A). The basal layer of the epidermis was less organized in psoriatic substitutes (Fig. 3B–D). A compact cornified layer, partly removed during the sectioning process of some constructs (ex; Fig. 3D), was observed in combinations containing psoriatic keratinocytes. No significant histological difference was noticed between the control and the psoriatic fibroblast/healthy keratinocyte combination (Fig. 3A and C), even if the macroscopic appearance of this combination showed some protuberances (Fig. 2C), which could suggest differences in the thickness of the epidermis.

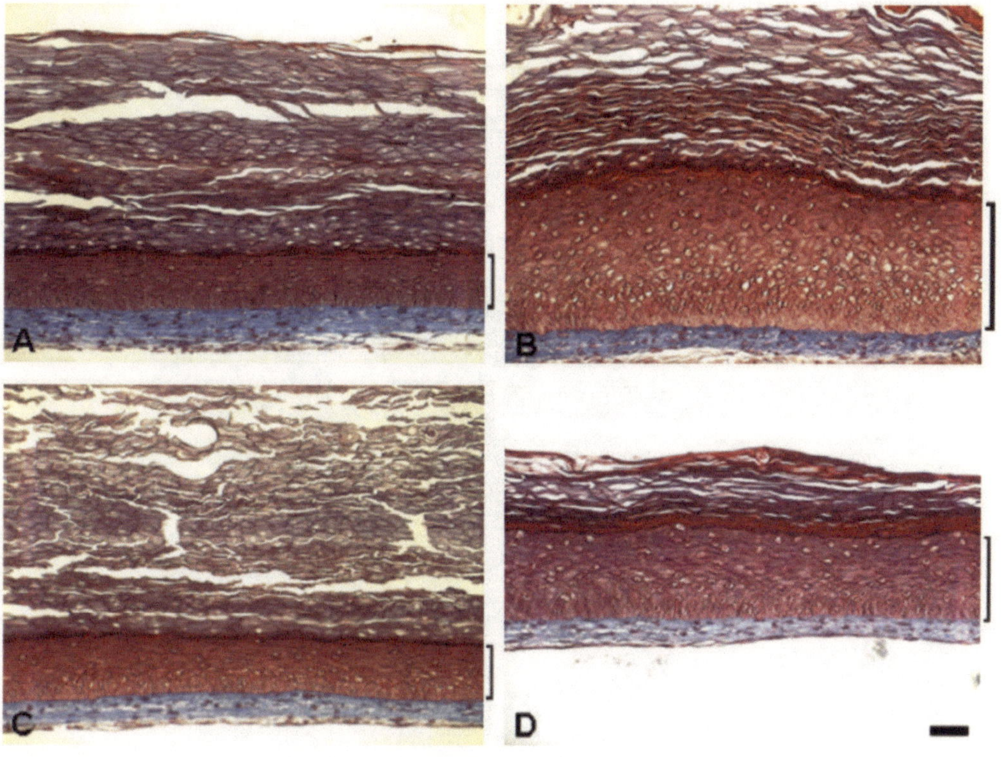

Fig. 3. Histological analysis of skin substitutes produced with healthy fibroblasts and healthy keratinocytes (A), psoriatic fibroblasts and psoriatic keratinocytes (B), psoriatic fibroblasts and healthy keratinocytes (C), and healthy fibroblasts and psoriatic keratinocytes (D) (scale bar = 5 µm). Note: the brackets indicate the measured epidermis. Reproduced from Jean et al., 2009 © Journal of Dermatological Science.

Epidermal differentiation markers, such as involucrin, appeared in deeper epidermal layers of substitutes produced with psoriatic fibroblasts and psoriatic keratinocytes, compared to those produced with healthy fibroblasts and healthy keratinocytes (controls) (Fig. 4). An overexpression of involucrin, relative to the controls was observed in substitutes produced with psoriatic keratinocytes (both combinations). Filaggrin staining showed a diminution or absence of filaggrin expression in substitutes with psoriatic cells compared to those produced with healthy fibroblasts and healthy keratinocytes (controls), in which filaggrin expression was present from the granular layer (Fig. 4). In the combination of psoriatic fibroblasts and healthy keratinocytes, the expression of loricrin was similar to levels seen in controls but in the presence of psoriatic keratinocytes (both combinations) its expression was partially or completely absent (Fig. 4). In substitutes produced with psoriatic keratinocytes, keratin 10 expression was less intense and appeared in the upper layers of the epidermis compared to those produced with healthy keratinocytes (both combinations) (Fig. 4). Some of our immunohistological results, such as loricrin and filaggrin stainings, showed the presence of blue nuclei stained with Hoechst in the upper layers of the stratum corneum, which is characteristic of parakeratosis.

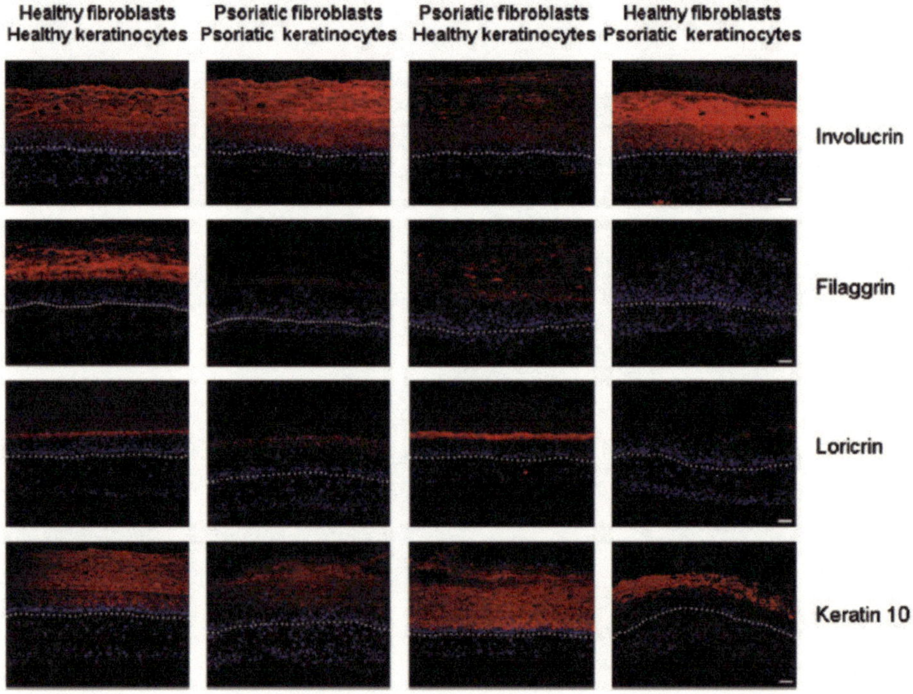

Fig. 4. Markers of keratinocyte differentiation in skin substitutes. Samples were prepared after 21 days of culture at the air–liquid interface (scale bar = 5 µm). The dotted line indicates the dermo-epidermal junction. Reproduced from Jean et al., 2009 © Journal of Dermatological Science.

In previous studies, the organization of the stratum corneum lipids was analyzed by attenuated total reflectance Fourier transform infrared spectroscopy (ATR-FITR) (Bernard, et al., 2007). Results suggested that the stratum corneum of involved psoriatic skin substitutes is less organized than that of normal skin substitutes. Bernard et al., also showed that the properties of uninvolved psoriatic skin may vary with the severity of the disease. In fact, in substitutes made with cells from a patient with 20% of affected body area, the ATR-FITR results are similar to controls and uninvolved psoriatic substitutes. However, in substitutes made with cells from a patient with an affected body area 40%, the ATR-FITR results of uninvolved psoriatic substitutes are different from those of the control (Bernard, et al., 2007). To summarize, this study showed that psoriatic skin substitutes produced by the self-assembly method have a higher level of lipid disorder and a modification of protein conformation in involved psoriatic stratum corneum layer compared to the stratum corneum of healthy substitutes (Bernard, et al., 2007).

7.4 Discussion

The psoriatic substitutes produced by the self-assembly method maintained many psoriasis-like characteristics such as thicker epidermis, abnormal differentiation, increased proliferation and higher levels of lipid disorganization of the stratum corneum. The four different keratinocyte/fibroblast combinations show different intensities in the expression of the psoriatic phenotype. These combinations allow the observation of interactions between healthy/psoriatic cells and fibroblast/keratinocyte cells. This model could become the basis of a robust technique to better understand the mechanisms involved in psoriasis and an excellent tool for the study of accelerated cellular differentiation of psoriatic keratinocytes. In an effort to resolve this complex skin disease, our research first endeavoured to develop a psoriatic model devoid of complicating elements such as immunocytes, in order to dissect step by step the mechanisms of this pathology. However, this model will be further refined to include elements of the immune system which may help establish the roles of different T cell subsets and cytokines in psoriasis.

8. Future research

A large number of psoriatic models have been developed. However, although many characteristics of the disease are preserved, no model shows all of them. Experts agreed that there was no available model that served all their needs (Zollner, et al., 2004). The choice of the most suitable model must always depend on the particular scientific question. The ultimate purpose of future research is to find the exact causes of psoriasis in order to find specific and effective treatments that will provide a definitive cure for psoriasis with a minimum of side effects. In vitro or in vivo pathological models that faithfully mimic this disease would be of significant help in the research and development of new treatments. For that reason, the development of a more relevant model of psoriasis is a priority for the various research efforts that are targeting this disease and would be an important step towards its cure.

9. Acknowledgements

Special thanks to Dr Dan Lacroix and Todd Galbraith for the revision of the manuscript.

10. References

Arosarena, O. (2005). Tissue engineering. *Curr Opin Otolaryngol Head Neck Surg,* 13, 4, 233-41.

Auger, F. A., Berthod, F., Moulin, V., Pouliot, R., and Germain, L. (2004). Tissue-engineered skin substitutes: from in vitro constructs to in vivo applications. *Biotechnol Appl Biochem,* 39, Pt 3, 263-75.

Auger, F. A., Rouabhia, M., Goulet, F., Berthod, F., Moulin, V., and Germain, L. (1998). Tissue-engineered human skin substitutes developed from collagen-populated hydrated gels: clinical and fundamental applications. *Med Biol Eng Comput,* 36, 6, 801-12.

Azfar, R. S., and Gelfand, J. M. (2008). Psoriasis and metabolic disease: epidemiology and pathophysiology. *Curr Opin Rheumatol,* 20, 4, 416-22.

Barker, C. L., McHale, M. T., Gillies, A. K., Waller, J., Pearce, D. M., Osborne, J., Hutchinson, P. E., Smith, G. M., and Pringle, J. H. (2004). The development and characterization of an in vitro model of psoriasis. *J Invest Dermatol,* 123, 5, 892-901.

Bell, E., Ehrlich, H. P., Buttle, D. J., and Nakatsuji, T. (1981a). Living tissue formed in vitro and accepted as skin-equivalent tissue of full thickness. *Science,* 211, 4486, 1052-4.

Bell, E., Ehrlich, H. P., Sher, S., Merrill, C., Sarber, R., Hull, B., Nakatsuji, T., Church, D., and Buttle, D. J. (1981b). Development and use of a living skin equivalent. *Plast Reconstr Surg,* 67, 3, 386-92.

Bell, E., Ivarsson, B., and Merrill, C. (1979). Production of a tissue-like structure by contraction of collagen lattices by human fibroblasts of different proliferative potential in vitro. *Proc Natl Acad Sci U S A,* 76, 3, 1274-8.

Bernard, G., Auger, M., Soucy, J., and Pouliot, R. (2007). Physical characterization of the stratum corneum of an in vitro psoriatic skin model by ATR-FTIR and Raman spectroscopies. *Biochim Biophys Acta,* 1770, 9, 1317-23.

Blessing, M., Schirmacher, P., and Kaiser, S. (1996). Overexpression of bone morphogenetic protein-6 (BMP-6) in the epidermis of transgenic mice: inhibition or stimulation of proliferation depending on the pattern of transgene expression and formation of psoriatic lesions. *J Cell Biol,* 135, 1, 227-39.

Blumberg, H., Conklin, D., Xu, W. F., Grossmann, A., Brender, T., Carollo, S., Eagan, M., Foster, D., Haldeman, B. A., Hammond, A., Haugen, H., Jelinek, L., Kelly, J. D., Madden, K., Maurer, M. F., Parrish-Novak, J., Prunkard, D., Sexson, S., Sprecher, C., Waggie, K., West, J., Whitmore, T. E., Yao, L., Kuechle, M. K., Dale, B. A., and Chandrasekher, Y. A. (2001). Interleukin 20: discovery, receptor identification, and role in epidermal function. *Cell,* 104, 1, 9-19.

Boehncke, W. H., Sterry, W., Hainzl, A., Scheffold, W., and Kaufmann, R. (1994). Psoriasiform architecture of murine epidermis overlying human psoriatic dermis transplanted onto SCID mice. *Arch Dermatol Res,* 286, 6, 325-30.

Bowcock, A. M. (2005). The genetics of psoriasis and autoimmunity. *Annu Rev Genomics Hum Genet,* 6, 93-122.

Boyman, O., Hefti, H. P., Conrad, C., Nickoloff, B. J., Suter, M., and Nestle, F. O. (2004). Spontaneous development of psoriasis in a new animal model shows an essential role for resident T cells and tumor necrosis factor-alpha. *J Exp Med,* 199, 5, 731-6.

Breban, M., Fernandez-Sueiro, J. L., Richardson, J. A., Hadavand, R. R., Maika, S. D., Hammer, R. E., and Taurog, J. D. (1996). T cells, but not thymic exposure to HLA-

B27, are required for the inflammatory disease of HLA-B27 transgenic rats. *J Immunol,* 156, 2, 794-803.

Brown, W. R., and Hardy, M. H. (1988). A hypothesis on the cause of chronic epidermal hyperproliferation in asebia mice. *Clin Exp Dermatol,* 13, 2, 74-7.

Brown, W. R., and Hardy, M. H. (1989). Mast cells in asebia mouse skin. *J Invest Dermatol,* 93, 5, 708.

Bullard, D. C., Scharffetter-Kochanek, K., McArthur, M. J., Chosay, J. G., McBride, M. E., Montgomery, C. A., and Beaudet, A. L. (1996). A polygenic mouse model of psoriasiform skin disease in CD18-deficient mice. *Proc Natl Acad Sci U S A,* 93, 5, 2116-21.

Carroll, J. M., Crompton, T., Seery, J. P., and Watt, F. M. (1997). Transgenic mice expressing IFN-gamma in the epidermis have eczema, hair hypopigmentation, and hair loss. *J Invest Dermatol,* 108, 4, 412-22.

Carroll, J. M., Romero, M. R., and Watt, F. M. (1995). Suprabasal integrin expression in the epidermis of transgenic mice results in developmental defects and a phenotype resembling psoriasis. *Cell,* 83, 6, 957-68.

Chadwick, C. C., Chippari, S., Matelan, E., Borges-Marcucci, L., Eckert, A. M., Keith, J. C., Jr., Albert, L. M., Leathurby, Y., Harris, H. A., Bhat, R. A., Ashwell, M., Trybulski, E., Winneker, R. C., Adelman, S. J., Steffan, R. J., and Harnish, D. C. (2005). Identification of pathway-selective estrogen receptor ligands that inhibit NF-kappaB transcriptional activity. *Proc Natl Acad Sci U S A,* 102, 7, 2543-8.

Chapman, M. L., Dimitrijevich, S. D., Hevelone, J. C., Goetz, D., Cohen, J., Wise, G. E., and Gracy, R. W. (1990). Inhibition of psoriatic cell proliferation in in vitro skin models by amiprilose hydrochloride. *In Vitro Cell Dev Biol,* 26, 10, 991-6.

Conrad, C., and Nestle, F. O. (2006). Animal models of psoriasis and psoriatic arthritis: an update. *Curr Rheumatol Rep,* 8, 5, 342-7.

Cook, P. W., Brown, J. R., Cornell, K. A., and Pittelkow, M. R. (2004). Suprabasal expression of human amphiregulin in the epidermis of transgenic mice induces a severe, early-onset, psoriasis-like skin pathology: expression of amphiregulin in the basal epidermis is also associated with synovitis. *Exp Dermatol,* 13, 6, 347-56.

Cook, P. W., Piepkorn, M., Clegg, C. H., Plowman, G. D., DeMay, J. M., Brown, J. R., and Pittelkow, M. R. (1997). Transgenic expression of the human amphiregulin gene induces a psoriasis-like phenotype. *J Clin Invest,* 100, 9, 2286-94.

Danilenko, D. M. (2008). Review paper: preclinical models of psoriasis. *Vet Pathol,* 45, 4, 563-75.

Detmar, M., Brown, L. F., Claffey, K. P., Yeo, K. T., Kocher, O., Jackman, R. W., Berse, B., and Dvorak, H. F. (1994). Overexpression of vascular permeability factor/vascular endothelial growth factor and its receptors in psoriasis. *J Exp Med,* 180, 3, 1141-6.

Dubertret, L. (2004). "Le psoriasis : de la clinique au traitement." Éditions MED'COM.

Eckes, B., Krieg, T., Nusgens, B. V., and Lapiere, C. M. (1995). In vitro reconstituted skin as a tool for biology, pharmacology and therapy: a review. *Wound Repair Regen,* 3, 3, 248-57.

Fitzpatrick, T. B., and Wolff, K. (2008). "Fitzpatrick's dermatology in general medicine." New York.

Fraki, J. E., Briggaman, R. A., and Lazarus, G. S. (1983). Transplantation of psoriatic skin onto nude mice. *J Invest Dermatol,* 80 Suppl, 31s-35s.

Germain, L., and Auger, F. A. (1995). *In* "Encyclopedic handbook of biomaterials and bioengineering Part B: Applications" (E. R. Schwartz, ed.), Vol. 1, pp. 699-734. Marcel Dekker Inc., New York.

Gilhar, A., David, M., Ullmann, Y., Berkutski, T., and Kalish, R. S. (1997). T-lymphocyte dependence of psoriatic pathology in human psoriatic skin grafted to SCID mice. *J Invest Dermatol,* 109, 3, 283-8.

Gourlay, W. A., Chambers, W. H., Monaco, A. P., and Maki, T. (1998). Importance of natural killer cells in the rejection of hamster skin xenografts. *Transplantation,* 65, 5, 727-34.

Grinnell, F., and Lamke, C. R. (1984). Reorganization of hydrated collagen lattices by human skin fibroblasts. *J Cell Sci,* 66, 51-63.

Groves, R. W., Mizutani, H., Kieffer, J. D., and Kupper, T. S. (1995). Inflammatory skin disease in transgenic mice that express high levels of interleukin 1 alpha in basal epidermis. *Proc Natl Acad Sci U S A,* 92, 25, 11874-8.

Gudjonsson, J. E., Johnston, A., Dyson, M., Valdimarsson, H., and Elder, J. T. (2007). Mouse models of psoriasis. *J Invest Dermatol,* 127, 6, 1292-308.

Gudjonsson, J. E., Johnston, A., Sigmundsdottir, H., and Valdimarsson, H. (2004). Immunopathogenic mechanisms in psoriasis. *Clin Exp Immunol,* 135, 1, 1-8.

Guo, L., Yu, Q. C., and Fuchs, E. (1993). Targeting expression of keratinocyte growth factor to keratinocytes elicits striking changes in epithelial differentiation in transgenic mice. *Embo J,* 12, 3, 973-86.

Hadgraft, J. (2001). Skin, the final frontier. *Int J Pharm,* 224, 1-2, 1-18.

Hansson, L., Backman, A., Ny, A., Edlund, M., Ekholm, E., Ekstrand Hammarstrom, B., Tornell, J., Wallbrandt, P., Wennbo, H., and Egelrud, T. (2002). Epidermal overexpression of stratum corneum chymotryptic enzyme in mice: a model for chronic itchy dermatitis. *J Invest Dermatol,* 118, 3, 444-9.

Harnish, D. C., Albert, L. M., Leathurby, Y., Eckert, A. M., Ciarletta, A., Kasaian, M., and Keith, J. C., Jr. (2004). Beneficial effects of estrogen treatment in the HLA-B27 transgenic rat model of inflammatory bowel disease. *Am J Physiol Gastrointest Liver Physiol,* 286, 1, G118-25.

Hirano, T., Ishihara, K., and Hibi, M. (2000). Roles of STAT3 in mediating the cell growth, differentiation and survival signals relayed through the IL-6 family of cytokine receptors. *Oncogene,* 19, 21, 2548-56.

Hobbs, R. M., Silva-Vargas, V., Groves, R., and Watt, F. M. (2004). Expression of activated MEK1 in differentiating epidermal cells is sufficient to generate hyperproliferative and inflammatory skin lesions. *J Invest Dermatol,* 123, 3, 503-15.

HogenEsch, H., Gijbels, M. J., Offerman, E., van Hooft, J., van Bekkum, D. W., and Zurcher, C. (1993). A spontaneous mutation characterized by chronic proliferative dermatitis in C57BL mice. *Am J Pathol,* 143, 3, 972-82.

Jean, J., Lapointe, M., Soucy, J., and Pouliot, R. (2009). Development of an in vitro psoriatic skin model by tissue engineering. *J Dermatol Sci,* 53, 1, 19-25.

Junqueira, L. C., and Carneiro, J. (2005). "Basic Histology: Text and Atlas." New York.

Keith, J. C., Jr., Sainz, I. M., Isordia-Salas, I., Pixley, R. A., Leathurby, Y., Albert, L. M., and Colman, R. W. (2005). A monoclonal antibody against kininogen reduces inflammation in the HLA-B27 transgenic rat. *Arthritis Res Ther,* 7, 4, R769-76.

Kess, D., Peters, T., Zamek, J., Wickenhauser, C., Tawadros, S., Loser, K., Varga, G., Grabbe, S., Nischt, R., Sunderkotter, C., Muller, W., Krieg, T., and Scharffetter-Kochanek, K.

(2003). CD4+ T cell-associated pathophysiology critically depends on CD18 gene dose effects in a murine model of psoriasis. *J Immunol,* 171, 11, 5697-706.

Konstantinova, N. V., Duong, D. M., Remenyik, E., Hazarika, P., Chuang, A., and Duvic, M. (1996). Interleukin-8 is induced in skin equivalents and is highest in those derived from psoriatic fibroblasts. *J Invest Dermatol,* 107, 4, 615-21.

Kopp, T., Kieffer, J. D., Rot, A., Strommer, S., Stingl, G., and Kupper, T. S. (2001). Inflammatory skin disease in K14/p40 transgenic mice: evidence for interleukin-12-like activities of p40. *J Invest Dermatol,* 117, 3, 618-26.

Kormeili, T., Lowe, N. J., and Yamauchi, P. S. (2004). Psoriasis: immunopathogenesis and evolving immunomodulators and systemic therapies; U.S. experiences. *Br J Dermatol,* 151, 1, 3-15.

Krueger, G. G., Chambers, D. A., and Shelby, J. (1981). Involved and uninvolved skin from psoriatic subjects: are they equally diseased? Assessment by skin transplanted to congenitally athymic (nude) mice. *J Clin Invest,* 68, 6, 1548-57.

Krueger, J. G., and Bowcock, A. (2005). Psoriasis pathophysiology: current concepts of pathogenesis. *Ann Rheum Dis,* 64 Suppl 2, ii30-ii36.

Kunstfeld, R., Hirakawa, S., Hong, Y. K., Schacht, V., Lange-Asschenfeldt, B., Velasco, P., Lin, C., Fiebiger, E., Wei, X., Wu, Y., Hicklin, D., Bohlen, P., and Detmar, M. (2004). Induction of cutaneous delayed-type hypersensitivity reactions in VEGF-A transgenic mice results in chronic skin inflammation associated with persistent lymphatic hyperplasia. *Blood,* 104, 4, 1048-57.

Lee, C. K., Rao, D. T., Gertner, R., Gimeno, R., Frey, A. B., and Levy, D. E. (2000). Distinct requirements for IFNs and STAT1 in NK cell function. *J Immunol,* 165, 7, 3571-7.

Li, A. G., Wang, D., Feng, X. H., and Wang, X. J. (2004). Latent TGFbeta1 overexpression in keratinocytes results in a severe psoriasis-like skin disorder. *Embo J,* 23, 8, 1770-81.

Loden, M., and Maibach, H. I. (2006). "Dry skin and moisturizers : chemistry and function."

Lowes, M. A., Bowcock, A. M., and Krueger, J. G. (2007). Pathogenesis and therapy of psoriasis. *Nature,* 445, 7130, 866-73.

McKay, I. A., and Leigh, I. M. (1995). Altered keratinocyte growth and differentiation in psoriasis. *Clin Dermatol,* 13, 2, 105-14.

Meyerrose, T. E., Herrbrich, P., Hess, D. A., and Nolta, J. A. (2003). Immune-deficient mouse models for analysis of human stem cells. *Biotechniques,* 35, 6, 1262-72.

Michel, M., L'Heureux, N., Pouliot, R., Xu, W., Auger, F. A., and Germain, L. (1999). Characterization of a new tissue-engineered human skin equivalent with hair. *In Vitro Cell Dev Biol Anim,* 35, 6, 318-26.

Mils, V., Basset-Seguin, N., Moles, J. P., Tesniere, A., Leigh, I., and Guilhou, J. J. (1994). Comparative analysis of normal and psoriatic skin both in vivo and in vitro. *Differentiation,* 58, 1, 77-86.

Mizutani, H., Yamanaka, K., Konishi, H., and Murakami, T. (2003). Animal models of psoriasis and pustular psoriasis. *Arch Dermatol Res,* 295 Suppl 1, S67-8.

Nickoloff, B. J., Kunkel, S. L., Burdick, M., and Strieter, R. M. (1995). Severe combined immunodeficiency mouse and human psoriatic skin chimeras. Validation of a new animal model. *Am J Pathol,* 146, 3, 580-8.

Nickoloff, B. J., and Nestle, F. O. (2004). Recent insights into the immunopathogenesis of psoriasis provide new therapeutic opportunities. *J Clin Invest,* 113, 12, 1664-75.

Parenteau, N. L., Nolte, C. M., Bilbo, P., Rosenberg, M., Wilkins, L. M., Johnson, E. W., Watson, S., Mason, V. S., and Bell, E. (1991). Epidermis generated in vitro: practical considerations and applications. *J Cell Biochem,* 45, 3, 245-51.

Pasparakis, M., Courtois, G., Hafner, M., Schmidt-Supprian, M., Nenci, A., Toksoy, A., Krampert, M., Goebeler, M., Gillitzer, R., Israel, A., Krieg, T., Rajewsky, K., and Haase, I. (2002). TNF-mediated inflammatory skin disease in mice with epidermis-specific deletion of IKK2. *Nature,* 417, 6891, 861-6.

Peterson, R. L., Wang, L., Albert, L., Marchese, E., Erickson, J., Wong, A., Mounts, W. M., Hayes, L., Bouchard, P., Keith, J., and Dorner, A. J. (2002). Pharmacogenomic analysis of rhIL-11 treatment in the HLA-B27 rat model of inflammatory bowel disease. *Pharmacogenomics J,* 2, 6, 383-99.

Pouliot, R., Larouche, D., Auger, F. A., Juhasz, J., Xu, W., Li, H., and Germain, L. (2002). Reconstructed human skin produced in vitro and grafted on athymic mice. *Transplantation,* 73, 11, 1751-7.

Raychaudhuri, S. P., Dutt, S., Raychaudhuri, S. K., Sanyal, M., and Farber, E. M. (2001). Severe combined immunodeficiency mouse-human skin chimeras: a unique animal model for the study of psoriasis and cutaneous inflammation. *Br J Dermatol,* 144, 5, 931-9.

Raychaudhuri, S. P., and Farber, E. M. (2001). The prevalence of psoriasis in the world. *J Eur Acad Dermatol Venereol,* 15, 1, 16-7.

Saiag, P., Coulomb, B., Lebreton, C., Bell, E., and Dubertret, L. (1985). Psoriatic fibroblasts induce hyperproliferation of normal keratinocytes in a skin equivalent model in vitro. *Science,* 230, 4726, 669-72.

Sano, S., Chan, K. S., Carbajal, S., Clifford, J., Peavey, M., Kiguchi, K., Itami, S., Nickoloff, B. J., and DiGiovanni, J. (2005). Stat3 links activated keratinocytes and immunocytes required for development of psoriasis in a novel transgenic mouse model. *Nat Med,* 11, 1, 43-9.

Schon, M. P. (1999). Animal models of psoriasis - what can we learn from them? *J Invest Dermatol,* 112, 4, 405-10.

Schon, M. P. (2008). Animal models of psoriasis: a critical appraisal. *Exp Dermatol,* 17, 8, 703-12.

Schon, M. P., and Boehncke, W. H. (2005). Psoriasis. *N Engl J Med,* 352, 18, 1899-912.

Schon, M. P., Schon, M., Warren, H. B., Donohue, J. P., and Parker, C. M. (2000). Cutaneous inflammatory disorder in integrin alphaE (CD103)-deficient mice. *J Immunol,* 165, 11, 6583-9.

Stevens, A., and Lowe, J. S. (2005). "Human histology." Toronto.

Stratis, A., Pasparakis, M., Rupec, R. A., Markur, D., Hartmann, K., Scharffetter-Kochanek, K., Peters, T., van Rooijen, N., Krieg, T., and Haase, I. (2006). Pathogenic role for skin macrophages in a mouse model of keratinocyte-induced psoriasis-like skin inflammation. *J Clin Invest,* 116, 8, 2094-104.

Sugai, J., Iizuka, M., Kawakubo, Y., Ozawa, A., Ohkido, M., Ueyama, Y., Tamaoki, N., Inokuchi, S., and Shimamura, K. (1998). Histological and immunocytochemical studies of human psoriatic lesions transplanted onto SCID mice. *J Dermatol Sci,* 17, 2, 85-92.

Sundberg, J. P., Beamer, W. G., Shultz, L. D., and Dunstan, R. W. (1990). Inherited mouse mutations as models of human adnexal, cornification, and papulosquamous dermatoses. *J Invest Dermatol,* 95, 5, 62S-63S.

Sundberg, J. P., Dunstan, R. W., Roop, D. R., and Beamer, W. G. (1994). Full-thickness skin grafts from flaky skin mice to nude mice: maintenance of the psoriasiform phenotype. *J Invest Dermatol,* 102, 5, 781-8.

Sundberg, J. P., France, M., Boggess, D., Sundberg, B. A., Jenson, A. B., Beamer, W. G., and Shultz, L. D. (1997). Development and progression of psoriasiform dermatitis and systemic lesions in the flaky skin (fsn) mouse mutant. *Pathobiology,* 65, 5, 271-86.

Takizawa, Y., Saida, T., Tokuda, Y., Dohi, S., Ikegawa, S., and Ueyama, Y. (1995). Engraftment of precursor lesions of human cutaneous neoplasms onto C.B-17 SCID mice: a useful in vivo experimental model of carcinogenesis in human skin. *Arch Dermatol Res,* 287, 3-4, 237-41.

Taurog, J. D., Maika, S. D., Simmons, W. A., Breban, M., and Hammer, R. E. (1993). Susceptibility to inflammatory disease in HLA-B27 transgenic rat lines correlates with the level of B27 expression. *J Immunol,* 150, 9, 4168-78.

Tjabringa, G., Bergers, M., van Rens, D., de Boer, R., Lamme, E., and Schalkwijk, J. (2008). Development and validation of human psoriatic skin equivalents. *Am J Pathol,* 173, 3, 815-23.

Turksen, K., Kupper, T., Degenstein, L., Williams, I., and Fuchs, E. (1992). Interleukin 6: insights to its function in skin by overexpression in transgenic mice. *Proc Natl Acad Sci U S A,* 89, 11, 5068-72.

van Ruissen, F., de Jongh, G. J., van Erp, P. E., Boezeman, J. B., and Schalkwijk, J. (1996). Cell kinetic characterization of cultured human keratinocytes from normal and psoriatic individuals. *J Cell Physiol,* 168, 3, 684-94.

Vassar, R., and Fuchs, E. (1991). Transgenic mice provide new insights into the role of TGF-alpha during epidermal development and differentiation. *Genes Dev,* 5, 5, 714-27.

Voskas, D., Jones, N., Van Slyke, P., Sturk, C., Chang, W., Haninec, A., Babichev, Y. O., Tran, J., Master, Z., Chen, S., Ward, N., Cruz, M., Jones, J., Kerbel, R. S., Jothy, S., Dagnino, L., Arbiser, J., Klement, G., and Dumont, D. J. (2005). A cyclosporine-sensitive psoriasis-like disease produced in Tie2 transgenic mice. *Am J Pathol,* 166, 3, 843-55.

Xia, Y. P., Li, B., Hylton, D., Detmar, M., Yancopoulos, G. D., and Rudge, J. S. (2003). Transgenic delivery of VEGF to mouse skin leads to an inflammatory condition resembling human psoriasis. *Blood,* 102, 1, 161-8.

Xu, W., Germain, L., Goulet, F., and Auger, F. A. (1996). Permanent grafting of living skin substitutes: surgical parameters to control for successful results. *J Burn Care Rehabil,* 17, 1, 7-13.

Yanagisawa, H., Richardson, J. A., Taurog, J. D., and Hammer, R. E. (1995). Characterization of psoriasiform and alopecic skin lesions in HLA-B27 transgenic rats. *Am J Pathol,* 147, 4, 955-64.

Yannas, I. V. (2004). Synthesis of tissues and organs. *Chembiochem,* 5, 1, 26-39.

Zenz, R., Eferl, R., Kenner, L., Florin, L., Hummerich, L., Mehic, D., Scheuch, H., Angel, P., Tschachler, E., and Wagner, E. F. (2005). Psoriasis-like skin disease and arthritis caused by inducible epidermal deletion of Jun proteins. *Nature,* 437, 7057, 369-75.

Zheng, Y., Eilertsen, K. J., Ge, L., Zhang, L., Sundberg, J. P., Prouty, S. M., Stenn, K. S., and Parimoo, S. (1999). Scd1 is expressed in sebaceous glands and is disrupted in the asebia mouse. *Nat Genet,* 23, 3, 268-70.

Zollner, T. M., Renz, H., Igney, F. H., and Asadullah, K. (2004). Animal models of T-cell-mediated skin diseases. Bioessays, 26, 6, 693-6.

Growth Factors and Signalling Molecules for Cartilage Tissue Engineering – from Embryology to innovative release strategies for Guided Tissue Engineering

Christoph Brochhausen*, Rolf Zehbe*, Bernhard Watzer, Sven Halstenberg, Helmut Schubert and C. James Kirkpatrick
REPAIR-lab, Institute of Pathology, Johannes Gutenberg-University Mainz Germany

1. Introduction

Modern Tissue Engineering represents a highly complex interdisciplinary branch of science which brings the life sciences and engineering together to design innovative solutions for the treatment of critical tissue defects after traumatic or degenerative damage. In this area an interesting evolution can be observed in the approach to achieve this goal: Originally understood as a principally biomaterial-based strategy using natural and synthetic materials as implants to be integrated into the neighbouring tissue, during the last two decades a change of paradigm has taken place which involves a movement away from the structural replacement of damaged tissue towards strategies to regenerate functional tissue (Kirkpatrick et al., 2006). In this approach cells play an important role, for example in seeding scaffold materials, which give a preformed structure for the regenerating tissue. A major aim is the synthesis of so-called tissue-like structures *in vitro* which can be implanted in the damaged tissue site and should be integrated into the body's own tissue. Relevant obstacles to this strategy are the proper morphological integration of the cell-material construct in the structural context of the damaged tissue site as well as the maintenance of its function. The use of cells in combination with different biomaterials gives an important

---

*Both authors contributed equally to this work.
Author for correspondence: Dr. Christoph Brochhausen (brochhausen@pathologie.klinik.uni-mainz.de)
Authors affiliations:
R. Zehbe & H. Schubert: TU Berlin, Institute of Materials Science & Technologies, Englische Strasse 20, 10587 Berlin, Germany
B. Watzer: Philipps-University, Mother-Child Medical Center, Department of Pediatric Science, Baldinger Straße 1, 35033 Marburg, Germany
C. Brochhausen, S. Halstenberg & C.J. Kirkpatrick: Johannes Gutenberg University, Institute of Pathology, Langenbeckstrasse 1, 55101 Mainz, Germany

input for the further development of innovative materials with the important aims of preserving cell viability, improving the cell seeding efficiency of a scaffold material and by this to enhance their functionality.

Due to the limited regenerative potential of cartilage, traumatic damage to articular cartilage is a unique example for the clinical relevance of tissue engineering strategies to treat damaged tissue. In this context, the establishment of the autologous chondrocyte transplantation by Brittberg et al. (1994) represents the first innovative step to treat articular cartilage damage via tissue engineering. With the help of a syringe a suspension of *ex vivo* expanded chondrocytes was injected into the defect which had been covered with a periosteal flap. The current approach involves the expansion and transplantation of the autologous chondrocytes with the help of a matrix. For this pupose different material-based matrices are already on the market and have been developed to provide the seeded cells with a suitable biomimetic microenvironment.

Modern tissue engineering approaches integrate the combined use of cells, scaffolds, and growth factors or signalling molecules respectively. There is the hope that such molecules will stimulate the controled proliferation and proper differentiation of the seeded cells in the scaffold. Furthermore, growth factors should also enhance the migration, proliferation, and differentiation of cells from the edges of the treated defect. Thus, such factors can serve first to optimize tissue engineered cell-scaffold constructs *in vitro*, and second to improve their performance and integration *in vivo*. For the targeting of potential signalling molecules and growth factors it is important to realize that regeneration in part recapitulates developmental processes. Therefore, it can be postulated that for the improvement of regeneration in the context of tissue engineering it is essential to understand the regulation of developmental processes (Reddi, 2003). With this porpose in mind our group has demonstrad that the growth plate of the long bones represents a suitable developmental model to target interesting molecules for tissue engineering application of cartilage and bone (Brochhausen et al., 2009).

In this contribution we firstly focused on the structural organization and regulation of the growth plate. Based on this we present the development of an integrated release-system in an oriented gelatine-based scaffold material for cartilage tissue engineering.

2. Long bone development and endochondral ossification

Two different steps in bone development are distinguishable, both being important for potential tissue engineering strategies: first, the condensation of mesenchymal cells as a morphological template and its development to a cartilaginous template during fetal development, and second, the longitudinal growth of the long bones within the area of endochondral ossification (fig. 1). In the present report we very briefly present the process of foetal bone development and then concentrate on endochondral ossification and its regulation.

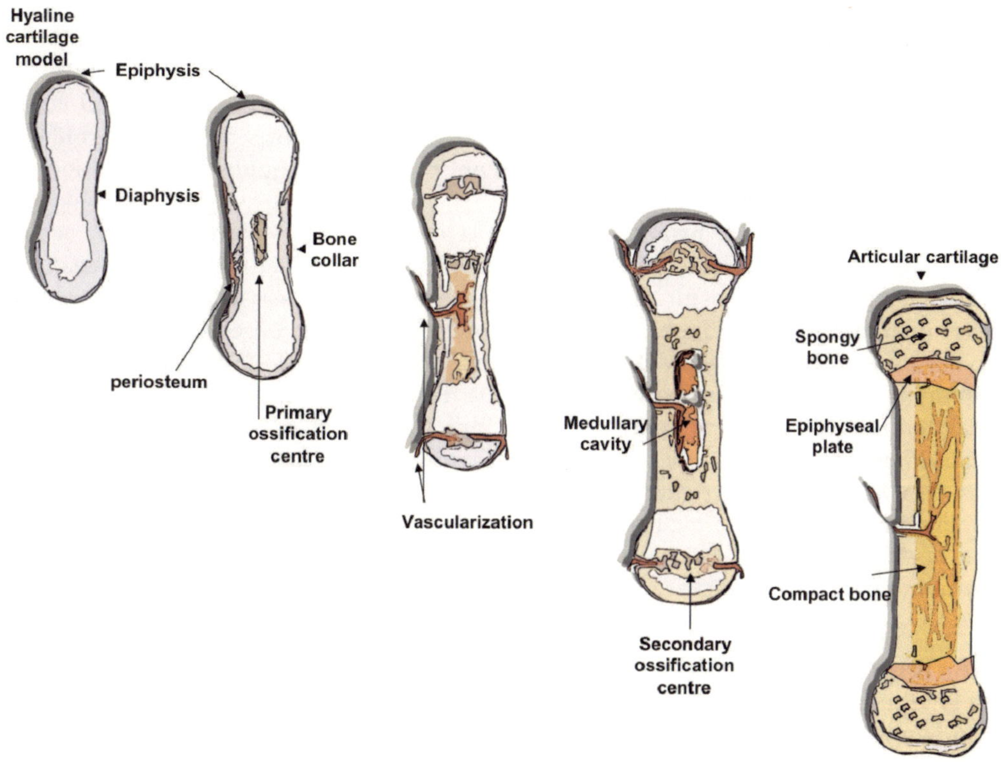

Fig. 1. Schematic depiction of bone development from the hyaline cartilage model to the ossified bone with the growth plate as the epicentre of the long bone growth.

The growth plate of long bones represents the epicentre of the long bone longitudinal growth. In this structurally highly organized region between the epiphysis and the metaphysis of the bone a cartilaginous template is responsible for longitudinal growth, which results in terminal ossification and thus leads to the elongation of the long bone (fig.2). The characteristic features of this region are chondrocytes in different proliferation and differentiation stages integrated in a scaffold of oriented extracellular matrix components, mainly consisting of different types of collagen. Since proliferation and differentiation in this region is a well regulated process the growth plate can be divided into characteristic horizontal zones. In these zones not only the chondrocytes showed a specific phenotype, but also specific types of collagen can be found within the different zones.

2.1 The resting zone

The zone closest to the epiphyseal cartilage is the smallest zone, called the resting zone. The latter consists of monomorphic, small chondrocytes with large central nuclei and a narrow cytoplasmatic rim. Nowadays, the resting zone cells are seen as being the cell pool for the adjacent proliferative zone, since they contain so called "stem-like" cells to supply new clones of proliferative zone chondrocytes. This hypothesis could be confirmed by findings of Abad et al. (2002), who demonstrated that, after removing the proliferative and the hypertrophic zones of the growth plate, a complete proliferative and hypertrophic zone was regenerated, based on cells from the residual resting zone alone. Furthermore, the chondrocytes of the resting zone might produce a morphogen, so-called Growth Plate-Orienting Factor (GPOF) that directs the alignment of proliferative clones into columns parallel to the long axis of the developing bone. The GPOF diffuses into the proliferative zone along a concentration gradient. This gradient is thought to guide the orientation of the proliferative chondrocytes.

The stem-like cells in the resting zone have a defined proliferative capacity, which is gradually exhausted and leads to fusion of the growth plate. Thus, the epiphyseal fusion process seems to be the result and not the cause of growth cessation at the moment of skeletal maturation. Accordingly, the term "growth plate senescence" has been introduced by Nilsson and Baron (2004). In this context, Stevens et al. (1999) demonstrated in transplantation experiments the dependence of the growth rate on the age of the donor animal, not that of the recipient. This finding indicates that the programmed senescence is not caused by hormonal or other systemic mechanisms but is intrinsic to growth plate cells themselves. In further investigations Gafni et al. (2001) showed that senescence is a function of proliferation: After systemic treatment with dexamethasone, a glucocorticoid that slows chondrocyte proliferation, the treated growth plates showed a delay in senescent decline of the growth rate, and the epiphyseal fusion occurred later.

2.2 The proliferative zone

After entering the cell cycle the cells begin to proliferate. During mitosis the cells divide in the horizontal axis followed by a rotation and finally the arrangement in the vertical axis forming the characteristic columns parallel to the longitudinal axis of the bone (Aszodi et al., 2003). These columns are separated from each other by matrices with large amounts of collagen Type II. Each column contains 10 to 20 cells, depending on the mitotic activity of the chondrocytes. In the proximal portion of each column, multiple mitotic figures can be observed. It is interesting that the achromatic figure spindles are arranged parallel to the long axis of the bone and that the chondrocytes migrate after division so that eventually the daughter cells are lined in the longitudinal axis (Trueta & Morgan, 1960; Stevenson et al., 1990). The proliferating zone is the only region in the growth plate where chondrocytes have relevant mitotic activity. Biochemically, in this zone the high oxygen content is noteworthy, as well as the high glycogen content and increased mitochondrial ATP production in chondrocytes (Brighton & Heppenstall, 1971).

2.3 The hypertrophic zone

After 5-10 mitotic cycles the chondrocytes lose their proliferative capacity and enter the so-called G0 phase for further differentiation. After passing a prehypertrophic stage chondrocyes enter a hypertrophic differentiation stage, which in the growth plate unidirectionally leads to terminal differentiation. During this process the chondrocytes increase their volume, especially in their longitudinal diameter, about 5- to 10-fold. In this way, hypertrophic differentiation makes an important contribution to longitudinal growth. Furthermore, the hypertrophic cells synthesize high amounts of extracellular matrix components, especially collagen Type X (Hunziker et al., 1987; Breur et al., 1991). In comparison to the resting zone and the proliferative zone, the hypertrophic zone is characterized by a distinctly different extracellular matrix composition: The synthesis of the collagens Type II, IX and XI is dramatically reduced, whereas the production of collagen Type X increases. In fact, collagen Type X is only expressed in this zone of the growth plate and hence, it is considered to be a specific marker for chondrocytic hypertrophy. Furthermore, hypertrophic chondrocytes show a significant increase of intracellular calcium concentration. Intracellular calcium is essential for the production of release vesicles (Wang & Kirsch, 2002), which are rich in several types of Ca^{2+}-binding annexins which promote Ca^{2+} uptake by these vesicles (Kirsch et al., 2000; Wang et al., 2003). Calcium phosphate, hydroxyapatite, phosphatases such as alkaline phophatase, and metalloproteinases are secreted via exocytosis, thus driving proteolytic remodelling and mineralization of the surrounding matrix (Lewinson & Silbermann, 1992; Vu et al., 1998). Moreover, the hypertrophic chondrocyte is a crucial master regulatory cell with an active metabolism. It prepares the matrix for calcification and vascularization (Forriol & Shapiro, 2005). In this context the expression of Vascular Endothelial Growth Factor (VEGF) by hypertrophic chondrocytes is of special interest for the initiation of the terminal ossification process.

2.4 The invasive zone

The expression of VEGF attracts invading blood vessels and, concomitantly, osteoblastic cells (Gerber et al., 1999). After vascularization of the mineralized cartilaginous matrix, the matrix is degraded by immigrant chondroclasts/ osteoclasts (Lewinson & Silbermann, 1992; Vu et al., 1998). The main part of the longitudinal septa (up to 80%) within the hypertrophic cartilage is rapidly resorbed behind the invading blood vessels, whereas the remaining septa serve as scaffolds on which osteoblasts deposit bone matrix (Salle et al., 2002). Finally, triggered by caspases, the hypertrophic chondrocytes undergo apoptosis or are resorbed (Gibson, 1998; Salvesen & Dixit, 1997).

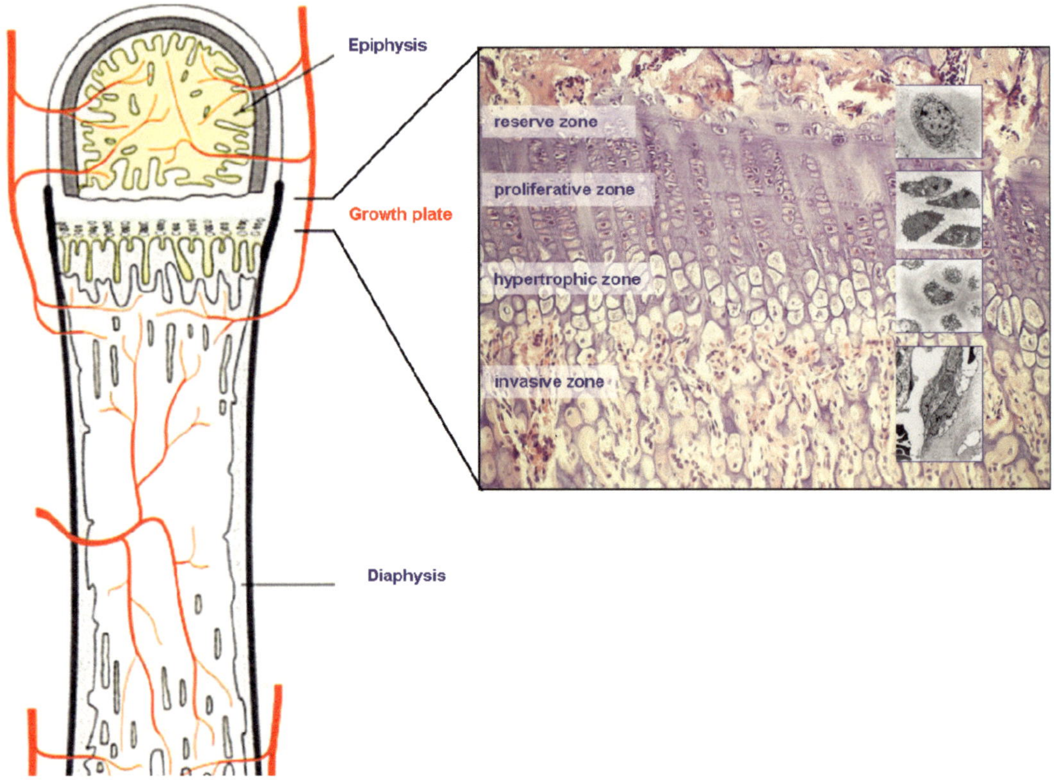

Fig. 2. The growth palte is a well organized structure between the epihysis and the diaphysis of the long bones. The chondrocytes are arranged according to there proliferation and differentiation stage with small round shaped cells in the resting zone, flat, proliferating cells in the proliferatve zoe and great cells in the hypertrophic zone. In the invasive zone vessels grow in and give rout for the final ossification (HE and transmission electron microscopical images).

3. Molecular Regulation of Endochondral Ossification

The complex interplay between chondrocyte proliferation and differentiation, which is responsible for the characteristic morphological features of the growth plate, is highly orchestrated by the interaction of multiple systemic hormones and paracrine factors, which finally lead to changes in gene expression of the chondrocytes in the growth plate. The fundamental importance of endochondral ossification in long bone growth and development as well as in repair mechanisms of the skeleton is one reason for the increasing interest in the molecular factors that regulate this critical pathway. This section describes the latest insights into the regulation of chondrocyte proliferation and differentiation within the growth plate and focusses on molecules that play key roles in embryology and development. The description represents a choice of some important molecules without any claim to be exhaustive. A more prescise review of the different pathways involved in growth plate with respect to its potential relevance for tissue engineering is given elsewhere

(Brochhausen et al. 2009). Table 1 summarizes the most important characteristics and effects of the signalling molecules and growth factors, which are described in detail below.

3.1 Bone Morphogenetic Proteins

Bone Morphogenetic Proteins (BMPs) represent a group of well known growth factors of the Transforming Growth Factor superfamily (TGF-beta). They were first identified and described by their capacity to stimulate ectopic cartilage and bone formation (Wozney, 1989). On this basis, BMP's represent a prime example of the high potential of growth factors for tissue engineing strategies. Today their important functional role in osteogenesis and chondrogenesis is well established. In early fetal development BMPs play important roles in the aggregation of mesenchymal cells during limb bud formation (Pizette & Niswander, 2000). In addition, BMPs influence the complex mechanisms in developing bone at later stages of skeletal development. The cellular effects of BMPs are mediated by two type I receptors, namely BMPRIA and BMPRIB, which heterodimerize with a type II receptor (BMPRII). The expression of BMPRIB was described by Zou et al. (1997) in early embryological development during mesenchymal cell condensations, which gives evidence that this receptor is involved in early bone and cartilage formation. The receptor BMPRIA could be detected in prehypertrophic chondrocytes (Zou et al., 1997). The differential expression of these receptors indicates the involvement of BMP-related pathways in different steps of bone and cartilage development. In this context it is of interest that BMPs have different expression patterns within bone and cartilage: BMP2, -3, -4, -5, and -7 are expressed in the perichondrium. BMP2 and -6 are expressed in hypertrophic chondrocytes and BMP7 is expressed in proliferating chondrocytes.

BMPs are integrated in the Ihh-PTHrP (Indian hedgehog-parathyroid hormone-related peptide) signalling loop. They are capable of inducing the expression of Ihh and, hence, increase chondrocyte proliferation (Grimsrud et al., 2001). Furthermore, BMP pathways inhibit FGF signalling by constraining the expression of FGFR1 (fibroblast growth factor receptor 1) (Yoon et al., 2006). Within the growth plate a concentration gradient of BMP may be a key feature responsible for spatial regulation of chondrocyte proliferation and differentiation (Nilsson et al., 2007).

Regarding tissue engineering strategies it is not surprising that recombinant human BMP's have already found their way into clinical practice for the treatment of critical sized bone defects. To further optimize the use of these factors in tissue engineering the identifcation and synthesis of well-defined carriers with suitable BMP release profiles is necessary as described by Bessa et al. (2008). A detailed description of the physiological function of the different BMP's during bone development is given by Kronenberg (2003), Forriol and Shapiro (2005).

3.2 Fibroblast Growth Factor-signalling

Fibroblast Growth Factors (FGFs) represent a superfamily of peptide growth factors synthesized by a variety of human and animal cells and act in a paracrine manner on their target cells. The cellular effects are mediated by plasma membrane tyrosine kinase receptors. The 22 distinct human FGFs are involved in diverse biological processes. In this context, a temporal selectivity in their expression is involved in the regulation of these pathways: the expression of FGF3, FGF4, FGF8 and FGF17 is restricted to embryonic tissue, whereas other

FGFs play significant roles not only in developmental processes but also in the biology of mature tissues (Jones, 2008). A further dicrimination of the effects in the FGF-related pathways is given by the mediation via four different types of FGF receptors (FGFR).
Regarding skeletal development FGF1, 2, 17, and 19 are the predominant FGF members present in developing human cartilage (Krejci et al., 2007). In the early stage of endochondral bone formation multiple FGF's as well as FGFR2 are expressed in condensing mesenchymal cells, which is thought to stimulate SOX9 expression (Murakami et al., 2000). FGFR3 is expressed by proliferating chondrocytes in the human growth plate, whereas FGFR1 is located in prehypertrophic and hypertrophic chondrocytes. Additionally, FGFR2 can be detected on perichondral cells. Each of these receptors has distinct roles in bone development. At the moment FGFR3 is best understood: FGFR3 inhibits the proliferation of growth plate chondrocytes, especially by down-regulating Ihh expression in hypertrophic chondrocytes. Hence, FGF signalling shortens the proliferative columns. Furthermore, FGFs decelerate chondrocytic terminal differentiation. Thus, they decrease matrix synthesis and consequently delay hypertrophy of the growth plate chondrocytes (Kato & Iwamoto, 1990). FGF18 is expressed in the perichondrium and by an activation of FGFR3 this factor is thought to regulate skeletal vascularization (Liu et al., 2007). The crucial role of FGFR3-mediated pathways in skeletal development is illustrated by the fact that a specific mutation in FGFR3 causes achondroplasia, the most common dwarfism in humans (Bellus et al., 1995; Shiang et al., 1994). A comprehensive overview of FGF and its functions during skeletal development is given by Olsen et al. (2000), Kronenberg (2003) as well as by Forriol and Shapiro (2005).

3.3 Parathyroid Hormone-related Peptide (PTHrP) and Indian hedgehog (Ihh)

PTHrP is a peptide which shares a receptor with the parathyroid hormone and has at its amino-terminus the identical 13 aminoacids as in the hormone itself. However, in contrast to the hormone PTHrP is detectable in many adult and developing tissues (Ingleton and Danks, 1996). Functionally, PTHrP is much more related to developmental growth factors such as FGF or TGF-beta than to a hormone. The identification of the PTHrP/Ihh feedback loop led to a major advancement in our understanding of growth plate regulation since it became clear that this pathway regulates the entry of chondrocytes into the hypertrophic phase of growth (Alvarez et al. 2003). The interplay between PTHrP and Ihh determines the moment when the chondrocytes leave the proliferative zone and enter the transition to hypertrophic chondrocytes. Perichondral cells and chondrocytes in the periarticular perichondrium synthesize PTHrP. This protein diffuses toward the prehypertrophic zone where cells have a high expression level of PTH/PTHrP receptors. After binding, PTHrP inhibits further differentiation to hypertrophic chondrocytes and keeps the chondrocytes in the proliferative stage (Lee et al., 1996). In the moment when the diffusion distances from the PTHrP-producing cells become too long, the effector cells are no longer sufficiently stimulated by PTHrP, which results in a delay of proliferation and the synthesis of Ihh, a member of the hedgehog family. Indian hedgehog is not only important in embryological development but also in the regulation of the growth plate. In this context Ihh is synthesized by prehypertrophic chondrocytes. As a consequence, the rate of chondrocyte proliferation increases and, by mechanisms that are still not fully understood, Ihh stimulates the production of PTHrP at the bone ends. Furthermore, by action of Ihh perichondral cells convert into osteoblasts and thus, enhance the formation of the bone collar. The essential role of Ihh in the postnatal integrity of the growth plate and endochondral ossification was clearly demonstrated by Maeda et al. (2007).

3.4 Vascular Endothelial Growth Factor

In the final ossification of the growth plate which leads to the formation of new bone the vascularisation of hypertrophic cartilage represents a crucial step. Thus, it is important that hypertrophic chondrocytes of the growth plate produce and secrete Vascular Endothelial Growth Factor (VEGF), one of the most potent angiogenetic factors (Gerber et al., 1999). VEGF in turn stimulates the proliferation and migration of endothelial cells. It enhances the formation of blood vessels in the hypertrophic zone of the growth plate and with this paves the way for ingrowth of osteoblasts, thus leading to ossification. Furthermore, the vascularization of the calcified cartilage within the invasive zone is essential for the resorption of chondrocytes by chondroclasts. Experimental data were able to demonstrate that the absence of VEGF causes profound disturbances in the architecture of the growth plate and affects significantly the longitudinal growth. In the absence of VEGF calcified cartilage persists as a result of a decrease in the recruitment and differentiation of osteoblasts and /or chondroclasts. As a morphological consequence a widening of the hypertrophic region and diminished trabecular bone formation could be observed as a result of the reduced elimination of terminally differentiated chondrocytes (Gerber et al., 1999). Thus, VEGF-mediated capillary invasion is an essential signal that triggers cartilage remodelling and hence influences growth plate morphogenesis. Interestingly, hypertrophic chondrocytes also express receptors for VEGF, which leads to the hypothesis that this factor could have an autocrine function in these cells (Carlevaro et al., 2000). In addition to VEGF, ligands for the tyrosine kinase receptor Tie2 are further key regulators of angiogenesis and post-angiogenic blood vessel maturation (Asahara T, 1998, Suri C et al., 1996). The investigations of Horner et al. (2001) revealed the expression and spatial distribution of angiopoietin-1 and -2 (Ang-1 and Ang-2), the ligands for the endothelial receptor Tie-2 in the human growth plate. This group described Ang-1 and Ang-2 as an important regulator in the orchestration of neovascularization in the growth plate. In this context, Ang-1 and Ang-2, which were expressed by hypertrophic chondrocytes of the growth plate, seemed to modulate the effects of VEGF (Horner et al., 2001). The fundamental role of VEGF during long bone development is reviewed in detail by Ballock and O'Keefe (2003).

3.5 Sox transcription factors

Sox proteins are a member of the High-Mobility-Group (HMG) superfamily, which represents DNA-binding proteins, whose sequence is at least 50% identical with the sequence of the HMG domain of the sex-determining region located on the Y chromosome of human and other mammals (Sinclair et al., 1990). This relation is responsible for the name of the Sox protein family (Wegner, 1999). During long bone development Sox9 is critically involved in several stages during the developmental processes. In early embryological development Sox9 is required for the aggregation of mesenchymal cells (Bi et al., 1999). Sox5 and Sox6, two other members of the Sox family, are coexpressed with Sox9 in chondroprogenitor cells and there is strong evidence that these genes are clearly needed for the differentiation of the chondrocytic lineage (Lefebvre et al., 1998; Akiyama et al., 2002). Furthermore, Sox9 is necessary for the expression of various extracellular matrix components, especially the collagen Types II, IX, XI as well as aggrecan (Bi et al., 1999; Lefebvre and de Crombrugghe, 1998). At later stages of development, the presence of Sox9, partly mediated by Sox5 and Sox6, is essential for chondrocyte proliferation and the alignment of proliferative clones into columns parallel to the longitudinal axis of the developing bone, indicating their important role in the proliferative zone of the growth plate (Akiyama et al., 2002). This fact is further ilustrated by the fact that the expression of the Sox5, Sox6, and Sox9 genes is completely turned off in hypertrophic chondrocytes (Lefebvre et al., 1998). Functionally, in the proliferative zone of the growth plate Sox9 prevents the transition of proliferating chondrocytes into hypertrophic chondrocytes, and is thus involved in the control of subsequent endochondral ossification (Akiyama et al., 2002). Furthermore, Zehentner et al. (1999) provided evidence that BMPs are also modulated by the expression of Sox9. With the help of transducing experiments Hardingham et al. (2006) demonstrated the critical role of Sox9 in chondrogenesis of isolated human articular chondrocytes. Regarding possibile tissue engineering strategies for cartilage, Khan et al. (2009) concluded that chondroprogenitors maintain SOX9 expression during extended monolayer culture and further retain chondrogenic potential. These findings demonstrate that the sox transcription factors not only play a crucial role in endochondral ossification but also in establishing the phenotype of articular cartilage both *in vivo* and *in vitro*, indicating that the regulation of the growth plate represents a suitable model to target potential signalling molecules and growth factors for cartilage tissue engineering.

3.6 Runx2

Runx2 (also known as Cbfa1) is another important transcription factor of the Runt-domain family (Ito, 1999). Runx2 is expressed during condensation of mesenchymal cells in early bone development and later in osteoblasts, but also in hypertrophic chondrocytes (Ducy et al., 1997; Inada et al., 1999; Kim et al., 1999; Otto et al., 1997). The analyses of Runx2 deficient mice gave evidence for the crucial functions of Runx2 during skeletal development: This transcription factor is firstly required for the differentiation of mesenchymal progenitor cells into osteoblasts. Later on in the growth plate of the long bones Runx2 enhances chondrocyte terminal differentiation (Inada et al., 1999; Kim et al., 1999) opening up the way for subsequent ossification. In Runx2-deficient mice, no endochondral and no intramembranous bone formation is present due to an arrest in osteoblast differentiation (Otto et al., 1997; Komori et al., 1997). In such mice, most bones either lack, or have reduced numbers of hypertrophic chondrocytes. In addition, hypertrophic chondrocytes fail to mineralize their matrix and have reduced or undetectable expression levels of osteotypical genes, such as osteopontin and MMP-13 (Takeda et al., 2001; Ueta et al., 2001). Thus, Runx2 is required for the differentiation of mesenchymal progenitor cells into osteoblasts. Furthermore, in the growth plate of long bones, this transcription factor enhances chondrocyte terminal differentiation (Inada et al., 1999; Kim et al., 1999) and is therefore a crucial factor for the definitive formation of trabecular bone.

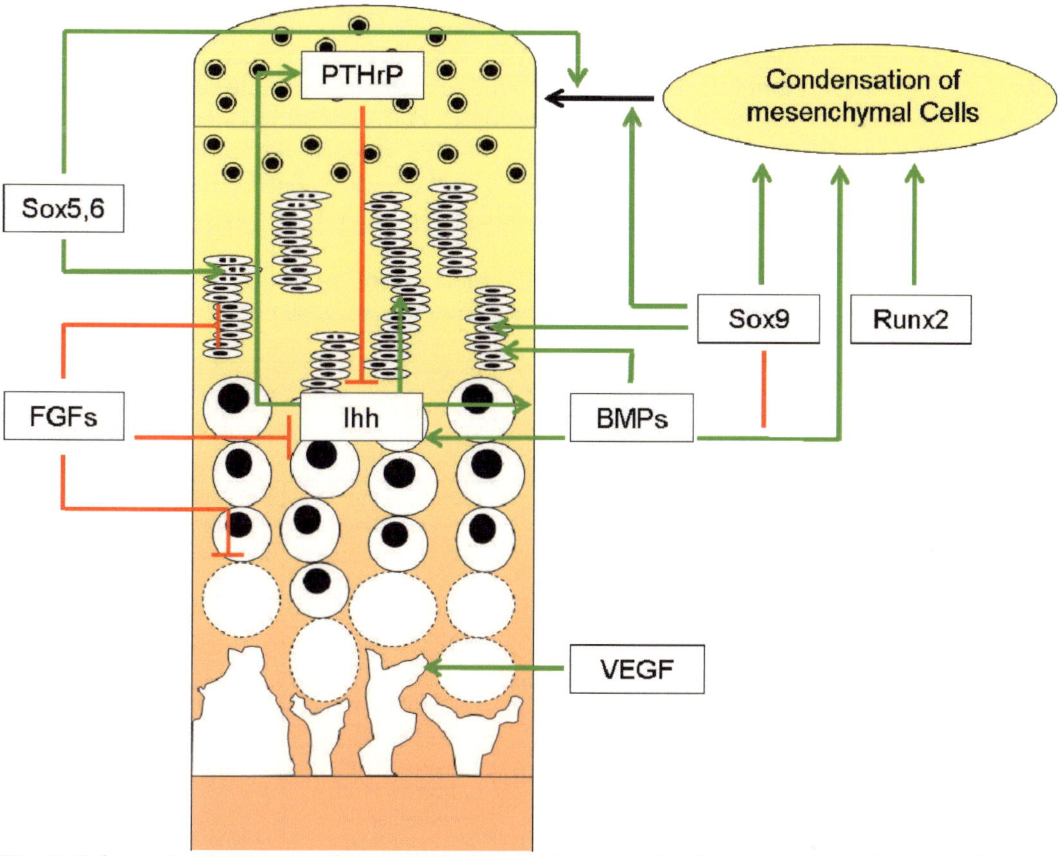

Fig. 3. Schematic depiction of the regulation of the growth plate with stimulating molecules (green arrows) and inhibitory molecules (red lines).

3.7 Prostaglandin E_2

Prostaglandin E_2 (PGE_2) is a metabolite of the arachidonic acid cascade, an evolutionary highly conserved familiy of lipid mediators. Prostaglandins are involved in numerous physiological and pathophysiological mechanisms, as well as in the regulation of developmental processes. In the latter context, PGE_2 also plays a significant role in skeletal development. In osteoblast-like cells, endogenous PGE_2 was shown to affect proliferation and differentiation by stimulation of DNA synthesis and alkaline phosphatase activity (Igarashi et al., 1994). Furthermore, PGE_2 influences the differentiation of chondrocytes as previously shown for the chondrocyte cell line RCJ3.1C5.18 (Lowe et al. 1996) and for rat growth plate chondrocytes (Schwartz et al. 1998, O'Keefe et al., 1992, Brochhausen et al., 2006). In addition to such findings in cell culture, the physiological role of prostaglandins was clarified by its stimulating effect on bone formation and increase in bone mass after systemic administration of PGE_2 to infants (Ueda et al., 1980) and animals (Suponitzki & Weinreb, 1998). Interestingly, also local administration of PGE_2 resulted in osteogenesis in situ (Yang et al., 1993, Marks & Miller, 1988). The cellular effects of PGE_2 are mediated by the binding to transmembranal G protein-coupled receptors. Four different EP receptors are known, which are linked to different intracellular signal transduction pathways (Narumiya

et al., 1999). The EP1 receptor is coupled to intracellular Ca^{2+} mobilization, while the EP2 and EP4 receptors increase intracellular cAMP accumulation. By contrast, EP3 inhibits intracellular cAMP accumulation. Regarding bone formation and bone resorption the EP4 receptor has been shown to be essential in terms of PGE_2 action in bone (Yoshida et al., 2002). In this respect the critical involvement of EP4 receptor in fracture healing was demonstrated very recently (Yoshida et al., 2002, Xie et al., 2009). Furthermore, EP2 and EP4 receptors were shown to be required for PGE_2-dependent chondrocyte differentiation (Miyamato et al., 2003).
Despite their obvious role in bone and cartilage metabolism, prostaglandins are not in the main focus regarding potential tissue engineering strategies of bone and cartilage.

Name	Substance class	Receptors	Effect	Reference
BMP	Protein growth factor	BMPRI-A, BMPRI-B	Cartilage & bone formation	Grimsrud et al. (2001), Pizette & Niswander (2000)
FGF	peptide growth factor	FGFR1, FGFR2, FGFR3	Mesenchymal condensation, Osteogenic differentiation	Kronenberg (2003), Hassan et al. (2007) Krejci et al. (2007)
PTHrP	Signalling peptide	G-protein G_S and G_q receptors	Chondrocyte proliferation	Lee et al. (1996)
Ihh	Signalling protein	Patched-1, Smo	Chondrocyte proliferation	Kronenberg (2003)
VEGF	Protein growth factor	VEGF-R	Vascularization, ingrowth of osteobalsts & chondroblasts	Gerber et al. (1999)
Sox	Transcription factor	DNA-binding	Chondrocyte differentiation & proliferation	Lefebvre et al. (1998) Akiyama et al. (2002)
Runx 2	Transcription factor	DNA-binding	Matrix mineralization, chondrocyte differentiation	Inada et al. (1999) Kim et al. (1999), Takeda et al. (2001), Ueta et al. (2001)
PGE_2	Lipid mediator	G-protein receptors, EP1, EP2, EP3, EP4	Chondrocyte proliferation, bone formation	Brochhausen et al. (2006) Xie et al. (2009)

Table 1. Summary of important growth factors and signalling molecules with relevant impact for the regulation of proliferation and differenzation of growth plate chondrocytes.

4. Experimental Setup

A principal aim of our research efforts was to develop an integrated release-system in an oriented scaffold for cartilage tissue engineering. Regarding the scaffold, one goal was to imitate the physiological fibre orientation of articular cartilage to achieve a topographically beneficial microenvironment. Articular cartilage represents a highly organized tissue not only at the cellular level, but also at the level of extracellular matrix components. Based on the spatial distribution, the morphological characterization and the biochemical activity of cells three zones are distinguishable in the articular cartilage (fig. 4): the superficial zone with flat aligned chondrocytes and perpendicularly arranged fibres, the middle zone with clusters of round shaped chondrocytes in a scaffold of parallel aligned collagen fibres and the deep zone with calcified cartilage.

Fig. 4. Histomorphology of articular cartilage of the rat (HE, x200) with schematic depiction of the fibre arrangement.

Reviewing the previously described signalling molecules, there are several reasons for the use of PGE_2 in cartilage tissue engineering: First of all several *in vitro* and *in situ* data demonstrate an important role for PGE_2 in cartilage development. Furthermore, this molecule, its metabolism, biological effects and possible side effects are already well understood and investigated. Finally, this substance is already on the market since decades, and is thus available in pharmaceutical quality and for a price less expensive than many of the modern developments. Since PGE_2 has a short half-life, which ranges under physiological conditions between 30 seconds in the cardiovascular system and 35 minutes in amniotic fluid (Bygdeman 2003, Ghodgaonkar et al. 1979) we established a release system for this molecule as an important prerequisite for its potential use in tissue engineering strategies. As release sytem we developed poly-D,L-lactide-co-glycolide (PLGA) microspheres, which were originally designed for the expansion of chondrocytes (Gabler et

a., 2007). Polylactide and its polymers are well established as drug release systems and are already in clinical use. Some examples for polylactide based release systems which are already in clinical application or hich are used in anmimal models are presented in Table 2, indicating that various substances can be integrated in corresponding systems. Thus, the polymer used in our experiments has also been shown to be could also be suitable for the integration and controlled release of further potential signalling moecules or growth factors.
For the analyses of the PGE_2 levels a novel sample preparation procedure was developed based on a previously published gas chromatography tandem mass spectrometry (GC-MS/MS) method for prostaglandin detection (Schweer et al., 1994). The modification of this method wich is demonstrated in detail elswere (Watzer et al., 2009, Brochhausen et al., 2009) was necessary to determine the incorporated PGE_2 levels.
The combination of two different material approaches for scaffold and release sytems was given by the idea to create a veritable toolbox which permits the use of different factors in the same three-dimensional scaffold material combined with a systematic analysis of their effects.

Active Agent	Release systeme	Indication	Reference
TGF-beta 1	PLGA (scaffold)	Cartilage/bone grwoth	Gombotz et al. (1994)
BMP	PLGA (scaffold)	Bone growth	Ferguson et al. (1987), Johnson et al. (1988)
VEGF	PLGA-PEG (microspheres)	Angiogenesis	King & Patrick (2000)
PGE_1	Lipid-microspheres	Vasodilatation	Kanamaru et al. (1998)

Table 2. Summary of active agents integrated in a PLGA related release-system, which are either used in animal models or already in clinical application.

5. Research

5.1 Scaffold design

From the biomaterial point of view the basis for the scaffold system was the development of a processing route for a material with similar structural, biochemical and mechanical properties as native articular cartilage. The basic approach and methodology for the synthesis route is based on a directional freezing technique which can be further combined with an electrolytic process (Zehbe et al., 2005). For this purpose, a gelatin solution was exposed to an electric field which leads to an arrangement of the fibres along the gas bubble formation. The container with the gelatin solution was positioned on the cold side of a peletier element. Thus, variation of the applied temperature was possible. A detailed description of the combination process which adapts an electrolytic decomposition process of the gelatin solution to introduce another pore fraction prior to freezing is precisely decribed in Zehbe et al. (2004, 2005). Freezing led to a stabilization of the fibre arrangement along the direction of ice crystal formation. After complete freezing, the sample was freeze dried to remove the ice crystals. Since a non-cross linked gelatin-based sample would rapidly dissolve under the fluid conditions *in vitro* or *in vivo*, the material was finally cross-linked to guarantee chemical stability of the material. As a result, a porous network was achieved after complete removal of the formed ice crystals.

Interestingly, the degree of directionality of the pores as well as the pore size is dependent on the freezing temperature, the ion content and the gelatin concentration. Thus, scaffolds with suitable directional pore channels could be synthesized at lower freezing temperatures without additional ions, while scaffolds with larger, non-directional pores could be made using higher freezing temperatures and elevated ion contents (fig 5a). In addition, based on this original scaffold processing route, we have introduced a hydroxyapatite layer to mimick the subchondral bone and address the potential application in tissue engineered-based treatment of deep cartilage defects (fig 5b).

Furthermore, we integrated in this scaffold drug release components in the form of PLGA-microspheres, which themselves were further enhanced by the integration of an active agent to meet so-called 3rd generation tissue engineering requirements.

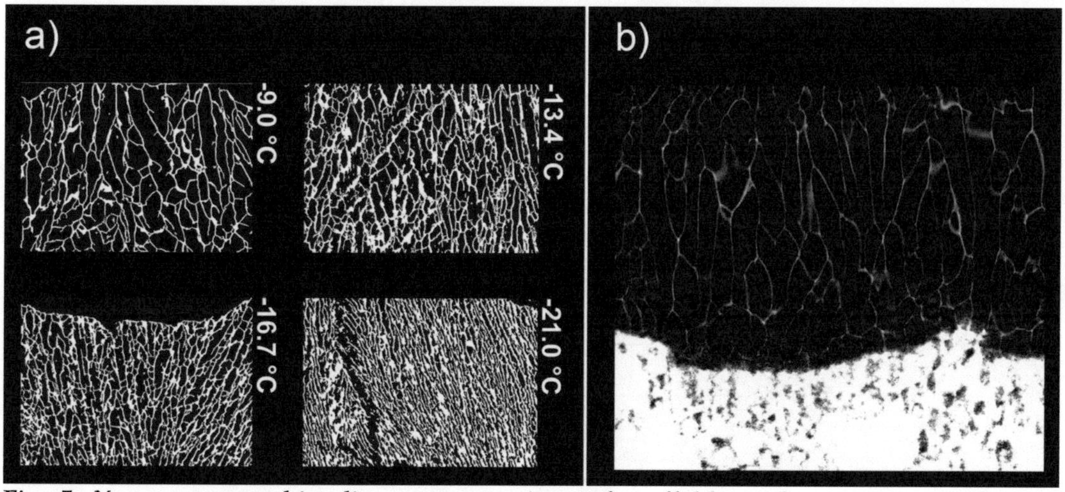

Fig. 5. X-ray tomographic slice representations of scaffolds with varying pore-channel networks (a) and sliced representation of a scaffold featuring hydroxyapatite as bone-like layer (b) (adapted from Zehbe et al., 2007).

5.2 PLGA- release system for prolonged PGE_2 release

As already mentioned above, the possible use of PGE_2 is highly limited due to its short half-life under physiological conditions. Therefore, for tissue engineering strategies this molecule should be used in the form of a release-system to guarantee a prolonged bioavailability. With respect to the microsphere scaffold modification we adapted an emulsion-based synthesis pathway which is described in detail by Brochhausen et al. (2007, 2009). In these protocols the microspheres were synthesized using a co-solvent emulsification method. The PLGA (Resomer RG859SR, Boehringer Ingelheim) used in our experiments had a lactide-glycolide-ratio of 85:15, a residual monomer content of < 2.0 % and an inherent viscosity of 2.5 - 3.5 dl/g. In the first step, 0.1 g PLGA granules were dissolved in 1.0 $CHCl_3$. Then, PGE_2 was dissolved in hexafluoroisopropanol (HFIP) in a concentration of 1.0 mg/ml. The PLGA solution was mixed with the PGE_2 solution giving the following composition: 0.1 g PLGA, 3.5 µg PGE_2, 1.0 ml CHCl3 and 3.5 µl HFIP. To allow the complete removal of the organic solvents by evaporation, the resulting solution was stirred over night in 0.5 % polyvinyl alcohol at 300 rpm with the help of a multi-position magnetic stirrer (RO 15 power Ikamag, IKA, Germany) according to the adaption of a previously described method (Gabler et al., 2007). Finally, the resulting microspheres were washed in sterile deionized water.

The characterization of the microspheres especially with respect to their size distribution was performed according to a protocol developed in our research group: In brief, SEM-images showing approximately 50 to 100 well dispersed microspheres were imaged top down and were then analyzed with the software ImageJ according to Gabler et al. (2007). In addition, the degradation changes regarding mass loss, pH-shift and the morphological changes of the microspheres in cell culture medium were documented over a timeframe of up to 100 days. Regarding the release profile we were able to show the kinetics of PGE_2 release over a period of 154 hours as measured by gas chromatography-mass spectroscopy (GC-MS/MS) and mass spectroscopy (Brochhausen et al.,2009). For this purpose the method of Schweer et al. (1994) was adapted. An important advantage of this method is that the structurally and thus functionally intact molecule could be detected even at very low concentrations. As a result of our analyses, we were able to determine reliably the amount of incorporated PGE_2 in the PLGA microspheres. Interestingly, the stability of PGE_2 was pH dependent: Strong acidic or basic environments reduced the half-life from 300 hours (pH 2.6 - 4.0) to below 50 hours at pH 2.0 or pH 8.8. Furthermore, we showed that the half-life of incorporated PGE_2 is highly temperature dependent (fig. 6). Thus, the half life could be increased from 70 days at 37 °C to 300 days at 8 °C. In our kinetic study we demonstrated a gradual concentration increase of PGE_2 in DMEM over the first 13 h, followed by a concentration plateau for 35 h and a slight decline of the PGE_2 concentration after the 48 h time point.

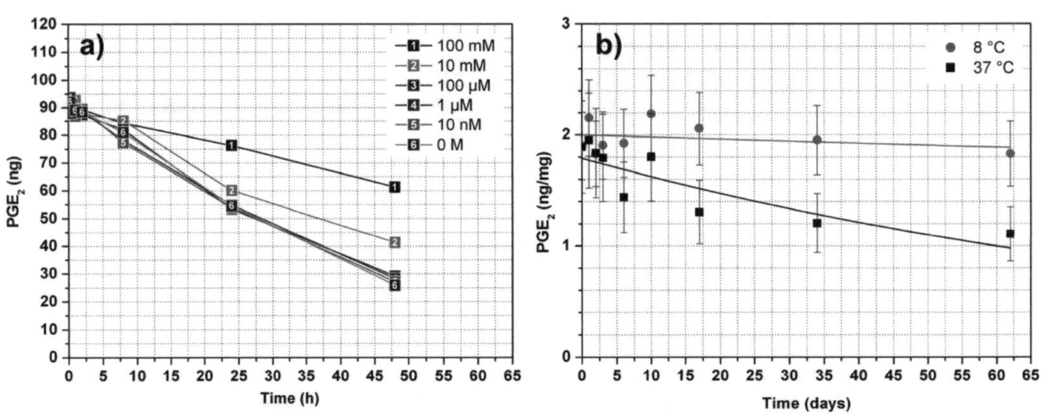

Fig. 6. pH dependent (a) and temperature dependent stability of PGE_2 (b), graphs adapted from Watzer et al., 2009.

For the used PLGA-type with a lactide-glycolide-ratio of 85:15 we hypothesized that the PGE_2 is primarily released from the outer shell of the microsphere as depicted in figure 7. Our image shows a microsphere after degradation for 80 days, as well as a superimposed graph of the kinetic experiments. After 80 days in cell culture medium the interior of the sphere is highly degraded, while the shell is almost intact. Further analysis with SR-µCT revealed that the microspheres consist of a porous shell and a compact core. These findings could be confirmed by scanning electron microscopy (SEM) and atomic force microscopy (AFM) which clearly showed the internal and external structure of the microspheres with a distinct nanostructure of the polymeric phase and both nano- and micro-porosity (Watzer et al., 2009). Thus, the rapid increase of PGE_2 could be the result of the release out of the porous shell, whereas the later PGE_2 values could be the result of the release out of the core. Regarding the degradation of the microspheres we hypothesized that the slow erosion of the surfaces results in a semi-permeable membrane which delays the release of degradation products and thus results in autocatalytic processes, amplifying the degradation inside the material (Heidemann et al., 2002, Maspero et al., 2002, Gabler et al., 2007).

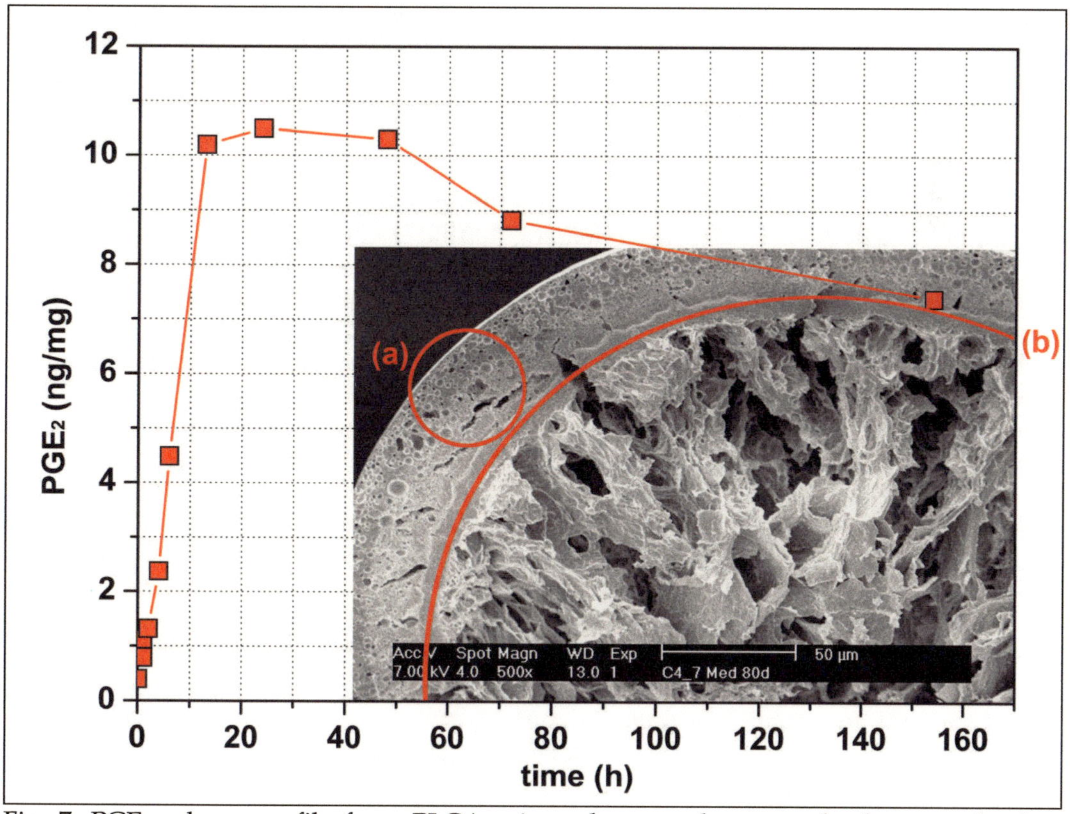

Fig. 7. PGE_2 release profile from PLGA microspheres and proposed release mechanism derived from SEM investigations. Initial release from the outer shell (a) and sustained release from the core after more than 160 h (b).

For the combined drug-release-scaffold system PGE_2-loaded microspheres were added to the gelatin solution and stirred until the gelatin gelled. By this procedure sedimentation of the spheres could be prevented so that a homogeneous distribution of microspheres could be reached without negatively affecting the casting of the gelatin in the freezing apparatus. After the freezing process, the microsphere-modified frozen gelatin samples were freeze dried and chemically cross linked as described above. We analysed the distribution of the microspheres by high resolution X-Ray tomography as described in detail in the chapter by Zehbe et al. in this book.

5.2 Chondocyte isolation and Cell Seeding of Scaffolds

For the cell culture experiments with the release-scaffold system bovine chondrocytes were used. For this purpose, fetlock joints of 3-4 months old calves were obtained directly after slaughter and processed under aseptic conditions. Articular cartilage chondrocytes were harvested aseptically from dissected flakes of the full thickness cartilage. After harvesting the cartilage flakes were washed twice for 15 min with Tyrode's balanced salt solution (TBSS) containing penicillin, streptomycin and amphotericin B. The cartilage flakes were then predigested with 1 mg/mL of pronase (Roche Diagnostics GmbH, Mannheim, G) on a magnetic stirrer at 37°C for 2 h, followed by washing with TBSS. Afterwards the specimens were digested with 600 units/mL of collagenase type II (Worthington Biochemical Corp, Lakewood, NJ) in DMEM by stirring at 37°C for a maximum of 15 h. The suspension was then filtered, and the isolated chondrocytes were washed 3 times with DMEM containing 10% fetal bovine serum (FBS). The viability of the isolated cells was determined using the trypan blue exclusion method. The cells were cultured in 175 cm^2 culture flasks at 5% CO2, 95% humidity and 37°C in DMEM containing 10% FBS. Chondrocytes were cultured in monolayers until confluence and then seeded onto the scaffolds. Cell-seeded scaffolds were cultured in 12-well tissue culture plates in DMEM supplemented with 10% FCS, nonessential amino acids, and 40 µg/mL of l-proline. After 2 days, 50 µg/mL of ascorbic acid was added. The medium was changed three times per week. Cell scaffold constructs were analyzed after 8 days of culture.

In these experiments the cultured cells showed a round shaped chondrocytic phenotype as seen by histological examination. Regarding the distribution of cells the majority of cells was detectable in the upper half of the scaffold. In this area cell clusters consisting of two or more cells could be seen, which are typical for the middle zone of articular cartilage. With the help of alcian blue staining, the production of extracellular matrix components was demonstrated especially surrounding the cell clusters (fig 8).

In further experiments with cultured human chondrocytes from patients who underwent implantation of endoprothesis due to osteoarthrosis, preliminary data revealed that in the presence of low dose PGE_2 the seeded cells showed a chondrocytic, round shaped phenotype with arrangement of multiple clusters, whereas cells without PGE_2 showed a spindle shaped fibrocytic phenotype (Brochhausen et al., 2008).

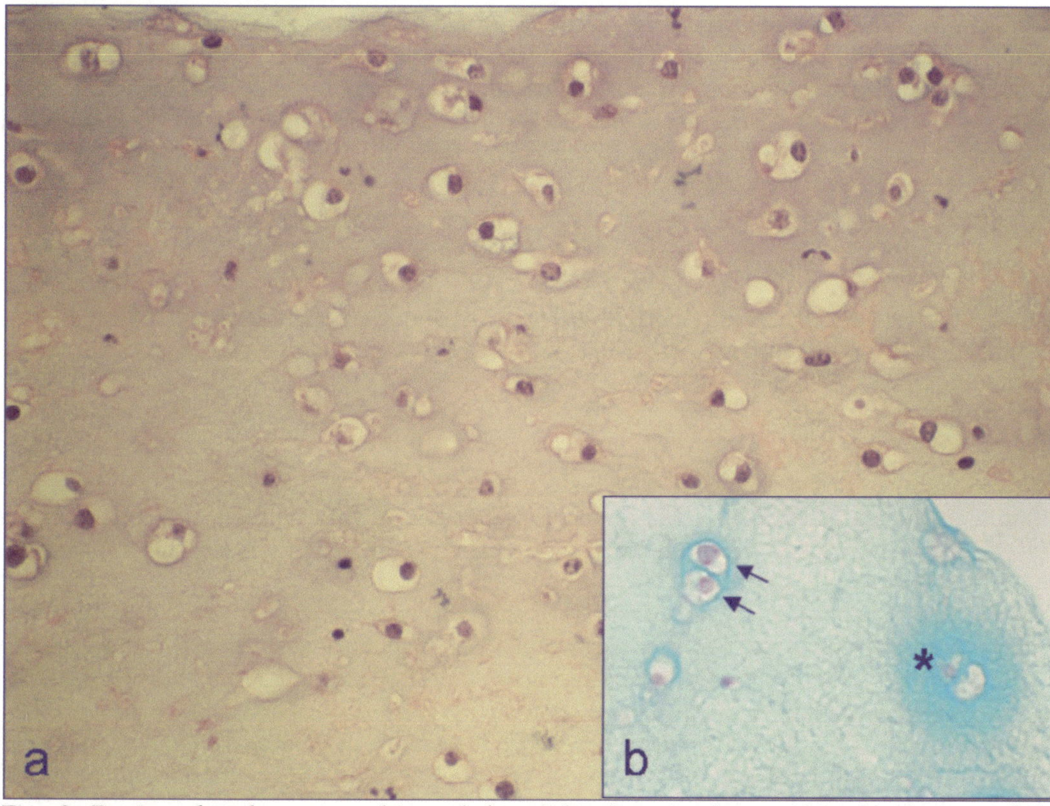

Fig. 8. Bovine chondrocytes cultivated for eight days in the combined release-scaffold system with round shaped chondroid phenotype (a, HE, x400), formation of chondrocyte clusters (b, arrow) and relevant extracellularmatrix production (b, alcian blue, x400)

6. Conclusion

In the present report we demonstrate that growth plate cartilage is a suitable model for strategies to target potential signalling molecules and growth factors for tissue engineering. Furthermore, we described the synthesis of a scaffold material with a structural design that gives ingrowing cells a pre-defined microenvironment similar to the cellular organization in native articular cartilage. Based on knowledge gained from the development of bone we used PGE_2 as a signalling molecule for cartilage tissue engineering strategies. In this context we firstly immobilized this molecule in polyester-based (PLGA) microspheres and then incorporated these in our gelatin-based orientated scaffold. The present experiments confirmed the prolonged release of PGE_2 from the PLGA microspheres and the beneficial effects of low dose PGE_2 for the phenotype of bovine and human articular chondrocytes seeded on the three-dimensional scaffold.

Since PLGA release systems are already used for various substances the present microsphere-scaffold system opens the perspective for the incorporation of further signalling molecules and growth factors. Furthermore, our construct facilitates the systematic analysis of different factors influencing the proliferation and differentiation of chondrocytes in an identical three-dimensional environment. For further tissue engineering

applications it is of interest to create gradients of signalling molecules and growth factors within the scaffold by establishing a graded microsphere distribution.
In conclusion, we demonstrate in this contribution an innovative scaffold material with properties of a biphasic composition and the incorporation of signalling molecules with the perspective to improve our understanding of the triad of cells, materials and signalling molecules as essential elements for tissue engineering strategies.

7. Acknowledgements

This work was performed under the umbrella of the European network of excellence EXPERTISSUES. The experiments were supported in part by the German Research Foundation (DFG) (SCHu 679/27-1, SCHU 679/27-2; Se 263/17-1).

8. References

Abad, V.; Meyers, J. L.; Weise, M.; Gafni, R. I.; Barnes, K. M.; Nilsson, O.; Bacher, J. D.; Baron, J. (2002). The role of the resting zone in growth plate chondrogenesis. *Endocrinology*, 143, 1851-1857

Akiyama, H.; Chaboissier, M. C.; Martin, J. F.; Schedl, A.; de Crombrugghe, B. (2002). The transcription factor Sox 9 has essential roles in successive steps of the chondrocyte differentiation pathway and is required for expression of Sox 5 and Sox 6. *Genes Dev*, 16, 2813-2828

Alvarez, J.; Sohn, P.; Zeng, X.; Doetschman, T.; Robbins, D. J.; Serra, R. (2002). TGFbeta2 mediates the effects of hedgehog on hypertrophic differentiation and PTHrP expression. *Development*, 129, 1913–1924

Asahara, T.; Chen, D. H.; Takahashi, T.; Fujikawa, K.; Kearney, M.; Magner, M.; Yancopoulos, G. D.; Isner, J. M. (1998). Tie2 receptor ligands, angiopoietin-1 and angiopoietin-2 modulate VEGF-induced postnasal neovascularisation. *Circ Res*, 83, 233–240

Aszodi, A.; Hunziger, E. B.; Brakrbusch, C.; Faessler, R. (2003). Beta 1 integrins regulate chondrocyte rotation, G1 progression, and cytokinesis. *Genes Dev*, 17, 2465-2479

Ballock, R. T.; O'Keefe, R. J. (2003). The biology of the growth plate. *J Bone Joint Surg*, 85, 715-726

Bellus, G. A.; Hefferon, T. W.; Ortiz de Luna, R. I.; Hecht, J. T.; Horton, W. A.; Machado, M.; Kaitila, I.; McIntosh, I.; Francomano, C. A. (1995). Achondroplasia is defined by recurrent G380R mutations of FGFR3. *Am J Hum Genet*, 56, 368-373

Bessa, P. C.; Casal, M.; Reis, R. L. (2008). Bone morphogenetic proteins in tissue engineering: the road from laboratory to clinic, part II (BMP delivery). *J Tiss Eng Reg Med*, 2, 81-96

Bi, W.; Deng, J. M.; Zhang, Z.; Behringer, R. R.; de Crombrugghe, B. (1999). Sox 9 is required for cartilage formation. *Nat Genet*, 22, 85-89

Breur, G. J.; Van Enkevort, B. A.; Farnum, C. E.; Wilsman, N. J. (1991). Linear relationship between the volume of hypertrophic chondrocytes and the rate of longitudinal bone growth in growth plates. *J Orthop Res*, 9, 348-359

Brighton, C. T.; Heppenstall, R. B. (1971). Oxygen tension in zones of the epiphyseal plate, the metaphysis and diaphysis: An in vitro and in vivo study in rats and rabbits. *J Bone Joint Surg*, 53, 719-728

Brittberg, M.; Lindahl, A.; Nilsson, A.; Ohlsson, C.; Isaksson, O.; Peterson, L. (1994). Treatment of deep cartilage defects in the knee with autologous chondrocyte transplantation. *N Engl J Med*, 331, 889-895

Brochhausen, C.; Halstenberg, S.; Sanchez, N.; Zehbe, R.; Watzer, B.; Meurer, A.; Schubert, H.; Kirkpatrick, C. J. (2008). Prostaglandin E_2 as an innovative signalling molecule for the tissue engineering of cartilage. *Tiss Eng*, 14, 771

Brochhausen, C.; Lehmann, M.; Halstenberg, S.; Meurer, A.; Klaus, G.; Kirkpatrick, C. J. (2009). Signalling molecules and growth factors for tissue engineering of cartilage - What can we learn from the growth plate? *J Tiss Eng Reg Med*, 3, 416-429

Brochhausen, C.; Neuland, P.; Kirkpatrick, C. J.; Nüsing, R. M.; Klaus, G. (2006). Cyclooxygenases and prostaglandin E_2 receptors in growth plate chondrocytes in vitro and in situ - PGE_2 dependent proliferation of growth plate chondrocytes. *Arthritis Research and Therapy*, 8, R78

Brochhausen, C.; Zehbe, R.; Watzer, B.; Halstenberg, S.; Kirkpatrick, C. J. (2009). Immobilization and Controlled Release of Prostaglandin E_2 (PGE_2) from Poly-l-lactide-co-glycolide Microspheres. *J Biomed Mat Res A*, 91, 454-62

Brochhausen, C.; Zehbe, R.; Watzer, B.; Halstenberg, S.; Schubert, H.; Kirkpatrick, C. J. (2007). PLGA-microspheres as a release system for signalling molecules in tissue engineering - first results of prostaglandin E_2 Release. *Eur Cell Mat*, 14 (supl. 1), 57

Bygdeman, M. (2003). Pharmacokinetics of prostaglandins. *Best Pract Res Clin Obstet Gynaecol*, 17, 707-716

Carlevaro, M. F.; Cermelli, S.; Cancedda, R.; Descalzi Cancedda, F. (2000). Vascular endothelial growth factor (VEGF) in cartilage neovascularization and chondrocyte differentiation: Auto-paracrine role during endochondral bone formation. *J Cell Sci*, 113, 59-69

Ducy, P.; Zhang, R.; Geoffroy, V.; Ridall, A. L.; Karsenty, G. (1997). Osf2/ Cbfa1: A transcriptional activator of osteoblast differentiation. *Cell*, 89, 747-754

Ferguson, D.; Davis, W.L.; Urist, M.R., Hurt, W.C., Allen, E.P. (1987). Bovine bone morphogenetic protein (bBMP) fraction-induced repair of craniotomy defects in the rhesus monkey (Macaca speciosa). *Clin Orthop Relat Res*, 219, 251-258

Forriol, F.; Shapiro, F. (2005). Bone development - Interaction of molecular components and biophysical forces. *Clin Orthop Relat Res*, 432, 14-33

Gabler, F.; Frauenschuh, S.; Ringe, J.; Brochhausen, C.; Schneider, C.; Götz, P.; Kirkpatrick, C. J.; Sittinger, M.; Schubert, H.; Zehbe, R. (2007). Emulsion-based Synthesis of PLGA-Microspheres for the in vitro expansion of Porcine Chondrocytes. *Biomol Eng*, 24, 515-520

Gafni, R. I.; Weise, M.; Robrecht, D. T.; Meyers, J. L.; Barnes, K. M.; De-Levi, S.; Baron, J. (2001). Catch- up growth is associated with delayed senescence of the growth plate in rabbits. *Pediatr Res*, 50, 618-623

Gerber, H. P.; Vu, T. H.; Ryan, A. M.; Kowalski, J.; Werb, Z.; Ferrera, N. (1999). VEGF couples hypertrophic cartilage remodeling, ossification and angiogenesis during endochondral bone formation. *Nat Med*, 5, 623-628

Ghodgaonkar, R. B.; Dubin, N. H.; Blake, D. A.; King, D. M. (1979). 13,14-dihydro-15-keto-prostaglandin F2alpha concentrations in human plasma and amniotic fluid. *Am J Obstet Gynecol*, 134, 265-269

Gibson G. (1998). Active role of chondrocyte apoptosis in endochondral ossification. *Microsc Res Tech*, 43, 191-204

Gombotz, W.R.; Pankey, S.C.; Bouchard, L.S.; Phan, D.H.; Puolakkainen, P.A. (1994). Stimulation of bone healing by transforming growth factor-beta 1 released from polymeric or ceramic implants. *J Appl Biomater*, 5, 141-150

Grimsrud, C. D.; Romano, P. R; D'Souza, M.; Puzas, J. E.; Schwarz, E. M.; Reynolds, P. R.; Roiser, R. N.; O'Keefe, R. J. (2001). BMP signalling stimulates chondocyte maturation and expression of indian hedgehog. *J Orthop Res*, 19, 18-25

Hardingham, T. E.; Oldershaw, R. A.; Tew, S. R. (2006). Cartilage, SOX9 and Notch signals in chondrogenesis. *J Anat*, 209, 469-480

Hassan, M.Q.; Tare, R.; Lee, S.H.; Mandeville, M.; Weiner, B.; Montecino, M.; van Wijnen. A.J.; Stein, J.L.; Stein, G.S.; Lian, J.B. (2007). HOXA10 controls osteoblastogenesis by directly activating bone regulatory and phenotypic genes. *Mol Cell Biol* 27, 3337-3352

Heidemann, W.; Jeschkeit-Schubbert, S.; Ruffieux, K.; Fischer, J. H.; Jung, H.; Krueger, G.; Wintermantel, E.; Gerlach, K. L. (2002). pH-stabilization of predegraded PDLLA by an admixture of water-soluble sodiumhydrogenphosphate--results of an in vitro- and in vivo-study. *Biomaterials*, 23, 3567-3574

Horner, A.; Bord, S.; Kelsall, A. W.; Coleman, N.; Compston, J. E. (2001). Tie2 ligands angiopoietin-1 and angiopoietin-2 are coexpressed with vascular endothelial cell growth factor in growing human bone. *Bone*, 28, 65-71

Hunziker, E. B.; Schenk, R. K.; Cruz-Orive, L. M. (1987). Quantitation of the chondrocyte performance in growth-plate cartilage during longitudinal bone growth. *J Bone Joint Surg*, 69, 162-173

Igarashi, K.; Hirafuji, M.; Adachi, H.; Shinoda, H.; Mitani, H. (1994). Role of endogenous PGE_2 in osteoblastic functions of a clonal osteoblast-like cell, MC3T3-E1. *Prostaglandins Leukot Essent Fatty Acids*, 50, 169-172

Inada, M.; Yasui, T.; Nomura, S.; Miyake, S.; Deguchi, K.; Himeno, M.; Sato, M.; Yamagiwa, H.; Kimura, T.; Yasui, N.; Ochi, T.; Endo, N.; Kitamura, Y.; Kishimoto, T.; Komori, T. (1999). Maturational disturbance of chondrocytes in Cbaf1- deficient mice. *Dev Dyn*, 214, 279-290

Ingleton, P.M.; Danks, J.A. (1996).Distribution and functions of parathyroid hormone-related protein in vertebrate cells. *Int Rev Cytol* ;166, 231-280

Ito, Y. (1999). Molecular basis of tissue-specific gene expression mediated by the runt domain transcription factor PEBP2/CBF. *Genes to Cells*, 4, 685-696

Johnson, E.E.; Urist, M.R.; Finerman, G.A. (1988). Bone morphogenetic protein augmentation grafting of resistant femoral nonunions. A preliminary report. *Clin Orthop Relat Res*, 1988, 230, 257-265

Jones, S. (2008).Mini-Review: Endocrine actions of fibroblast growth factor 19. *Mol Pharm*,5: 42-48

Kanamaru, K.; Waga, S.; Kuga, Y.; Nakamura, F.; Kamata, N. (1998). Transcranial Doppler pattern after intracarotid papaverine and prostaglandin E1 incorporated in lipid microsphere in patients with vasospasm. *Neurol Med Chir (Tokyo),* 1998, 38 Suppl, 152-155

Kato, Y.; Iwamoto, M. (1990). Fibroblast growth factor is an inhibitor of chondrocyte terminal differentiation. *J Biol Chem,* 265, 5903-5909

Khan, I. M.; Bishop, J. C.; Gilbert, S.; Archer, C. W. (2009). Clonal chondroprogenitors maintain telomerase activity and Sox9 expression during extended monolayer culture and retain chondrogenic potential. *Osteoarthritis Cartilage,* 17, 518-528

Kim, I. S.; Otto, F.; Zabel, B.; Mundlos, S. (1999). Regulation of chondrocyte differentiation by Cbfa1. *Mech Dev,* 80, 159-170

King, T.W.; Patrick, C.W.. Jr. (2000). Development and in vitro characterization of vascular endothelial growth factor (VEGF)-loaded poly(DL-lactic-co-glycolic acid)/poly(ethylene glycol) microspheres using a solid encapsulation/single emulsion/solvent extraction technique. *J Biomed Mater Res,* 2000, 51, 383-390

Kirkpatrick, C. J.; Fuchs, S.; Peters, K.; Brochhausen, C.; Hermanns, M. I.; Unger, R. E. (2006). Visions for regenerative medicine: interface between scientific fact and science fiction. *Artificial Organs,* 30, 822-827

Kirsch, T.; Harrison, G.; Golub, E. E.; Nah, H. D. (2000). The roles of annexins and types II and X collagen in matrix vesicle-mediated mineralization of growth plate cartilage. *J Biol Chem,* 275, 35577-35583

Komori, T.; Yagi, H.; Nomura, S.; Yamaguchi, A.; Sasaki, K.; Deguchi, K.; Shimizu, Y.; Bronson, R. T.; Gao, Y. H.; Inada, M.; Sato, M.; Okamoto, R.; Kitamura, Y.; Yoshiki, S.; Kishimoto, T. (1997). Targeted disruption of Cbfa1 results in a complete lack of bone formation owing to maturational arrest of osteoblasts. *Cell,* 89, 755-764

Krejci, P.; Krakow, D.; Mekikian, P. B.; Wilcox, W. R. (2007). Fibroblast growth factors 1, 2, 17, and 19 are the predominant FGF ligands expressed in human fetal growth plate cartilage. *Pediatr Res,* 61, 26-72

Kronenberg, H. M. (2003). Developmental regulation of the growth plate. *Nature,* 423, 332-336

Kronenberg, H. M. (2006). PTHrP and skeletal development. *Ann N Y Acad Sci,* 1068, 1-13

Lee, K.; Lanske, B.; Karaplis, A. C.; Deeds, J. D.; Kohno, H.; Nissenson, R. A.; Kronenberg, H. M.; Segre, G. V. (1996). Parathyroid hormone- related peptid delays terminal differentiation of chondrocytes during endochondral bone development. *Endocrinology,* 137, 5109-5118

Lefebvre, V.; de Crombrugghe, B. (1998). Toward understanding Sox9 function in chondrocyte differentiation. *Matrix Biol,* 16, 529-540

Lefebvre, V.; Li, P.; de Crombrugghe, B. (1998). A new long form of Sox5 (L-Sox5), Sox6 and Sox9 are coexpressed in chondrogenesis and cooperatively activate the type II collagen gene. *EMBO J,* 17, 5718-5733

Lewinson, D.; Silbermann, M. (1992). Chondroclasts and endothelial cells collaborate in the process of cartilage resorption. *Anat Rec,* 233, 504-514

Liu, Z.; Lavine, K. J.; Hung, I. H.; Ornitz, D. M. (2007). FGF18 is required for early chondrocyte proliferation, hypertrophy and vascular invasion of the growth plate. *Dev Biol,* 302, 80-91

Lowe, G. N.; Fu, Y. H.; McDougall, S.; Polendo, R.; Williams, A.; Benya, P. D.; Hahn, T. J. (1996). Effects of prostaglandins on deoxyribonucleic acid and aggrecan synthesis in the RCJ 3.1C5.18 chondrocyte cell line: role of second messengers. *Endocrinology*, 137, 2208-2216

Maeda, Y.; Nakamura, E.; Nguyen, M. T.; Suva, L. J.; Swain, F. L.; Razzaque, M. S.; Mackem, S.; Lanske, B. (2007). Indian hedgehog produced by postnatal chondrocytes is essential for maintaining a growth plate and trabecular bone. *Proc Natl Acad Sci U S A*, 104, 6382-6387

Marks, S.C., Jr; Miller, S. (1988). Local infusion of prostaglandin E1 stimulates mandibular bone formation in vivo. *J Oral Pathol*, 17, 500-505

Maspero, F.A.; Ruffieux, K.; Müller, B.; Wintermantel, E. (2002). Resorbable defect analog PLGA scaffolds using CO2 as solvent: structural characterization. *J Biomed Mater Res*, 62, 89-98

Miyamoto, M.; Ito, H.; Mukai, S.; Kobayashi, T.; Yamamoto, H.; Kobayashi, M.; Maruyama, T.; Akiyama, H.; Nakamura, T. (2003). Simultaneous stimulation of EP2 and EP4 is essential to the effect of prostaglandin E_2 in chondrocyte differentiation. *Osteoarthritis Cartilage*, 11, 644-652

Narumiya, S.; Sugimoto, Y.; Ushikubi, F. (1999). Prostanoid receptors: structures, properties, and functions. *Physiol Rev*, 79, 1193-1226

Nilsson, O.; Baron, J. (2004). Fundamental limits on longitudinal bone growth: Growth plate senescence and epiphyseal fusion. *Trends Endocrinol Metab*, 15, 370-374

Nilsson, O.; Parker, E.A.; Hegde, A.; Chau, M.; Barnes, K.M.; Baron, J. (2007) Gradients in bone morphogenetic protein-related gene expression across the growth plate. *J Endocrinol*, 193, 75-84

O'Keefe, R. J.; Crabb, I. D.; Puzas, J. E.; Rosier, R. N. (1992). Influence of prostaglandins on DNA and matrix synthesis in growth plate chondrocytes. *J Bone Miner Res*, 7, 397-404

Olsen, B. R.; Reginato, A. M.; Wang, W. (2000). Bone development. *Annu Rev Cell Dev Biol*, 16, 191-220

Otto, F.; Thornell, A. P.; Crompton, T.; Denzel, A.; Gilmour, K. C.; Rosewell, I. R.; Stamp, G. W. H.; Beddington, R. S. P.; Mundlos, S.; Olsen, B. R.; Selby, P. B.; Owen, M. J. (1997). Cbfa1, a candidate gene for cleidocranial dysplasia syndrome, is essential for osteoblast differentiation and bone development. *Cell*, 89, 765-771

Pizette, S.; Niswander, L. (2000). BMPs are required at two steps of limb chondrogenesis: Formation of prechondrogenic condensations and their differentiation into chondrocytes. *Dev Biol*, 219, 237-249

Reddi, A. H. (2003). Morphogenesis and tissue engineering of bone and cartilage: inductive signals, stem cells, and biomimetic biomaterials. *Tissue Eng*, 6, 351-359

Salle, B. L.; Rauch, F.; Travers, R.; Bouvier, R.; Glorieux, F. H. (2002). Human fetal bone development: Histomorphometric evaluation of the proximal femoral metaphysic. *Bone*, 30, 823-828

Salvesen, G. S.; Dixit, V. M. (1997). Caspases: Intracellular signalling by proteolysis. *Cell*, 91, 443-446

Schwartz, Z.; Gilley, R. M.; Sylvia, V. L.; Dean, D. D.; Boyan, B. D. (1998). The effect of prostaglandin E_2 on costochondral chondrocyte differentiation is mediated by cyclic adenosine 3',5'-monophosphate and protein kinase C. *Endocrinology*, 139, 1825-1834

Schweer, H.; Watzer, B.; Seyberth, H. W. (1994). Determination of seven prostanoids in 1 ml of urine by gas chromatography-negative ion chemical ionization triple stage quadrupole mass spectrometry. *J Chromatogr*, 11, 221-217

Shiang, R.; Thompson, L. M.; Zhu, Y. Z.; Church, D. M.; Fielder, T. J.; Bocian, M.; Winokur, S. T.; Wasmuth, J. J. (1994). Mutations in the transmembrane domain of FGFR3 cause the most common genetic form of dwarfism, achondroplasia. *Cell*, 78, 335-342

Sinclair, A. H.; Berta, P.; Palmer, M. S.; Hawkins, J. R.; Griffiths, B. L.; Smith, M. J.; Foster, J. W.; Frischauf, A. M.; Lovell-Badge, R.; Goodfellow, P. N. (1990). A gene from the human sex-determining region encodes a protein with homology to a conserved DNA-binding motif. *Nature*, 346, 240-244

Stevens, D. G.; Boyer, M. I.; Bowen, C. V. (1999). Transplantation of epiphyseal plate allografts between animals of different ages. *J Pediatr Orthop*, 19, 398-403

Stevenson, S.; Hunziker, E. B.; Herrmann, W.; Schenk, R. K. (1990). Is longitudinal bone growth influenced by diurnal variation in the mitotic activity of chondrocytes of the growth plate? *J Orthop Res*, 8, 132-135

Suponitzky, I.; Weinreb, M. (1998) Differential effects of systemic prostaglandin E_2 on bone mass in rat long bones and calvariae. *J Endocrinol*, 156, 51-57

Suri, C.; Jones, P. F.; Patan, S.; Bartunkova, S.; Maisonpierre, P. C.; Davis, S.; Sato, T. N.; Yancopoulos, G. D. (1996). Requisite role of angiopoietin-1, a ligand for Tie2 receptor, during embryonic angiogenesis. *Cell*, 87, 1171–1180

Takeda, S.; Bonnamy, J. P.; Owen, M. J.; Ducy, P.; Karsenty, G. (2001). Continuous expression of Cbfa1 in nonhypertrophic chondrocytes uncovers its ability to induce hypertrophic chondrocyte differentiation and partially rescues Cbfa1-deficient mice. *Genes Dev*, 15, 467-481

Trueta, J.; Morgan, J. D. (1960). The vascular contribution to osteogenesis. I. Studies by the injection method. *J Bone Joint Surg*, 42, 97-109

Ueda, K.; Saito, A.; Nakano, H.; Aoshima, M.; Yokota, M.; Muraoka, R.; Iwaya, T. (1980). Cortical hyperostosis following long-term administration of prostaglandin E1 in infants with cyanotic congenital heart disease. *J Pediatr*, 97, 834-836

Ueta, C.; Iwamoto, M.; Kanatani, N.; Yoshida, C.; Liu, Y.; Enomoto-Iwamoto, M.; Ohmori, T.; Enomoto, H.; Nakata, K.; Takada, K.; Kurisu, K.; Komori, T. (2001). Skeletal malformations caused by overexpression of Cbfa1 or its dominant negative form in chondrocytes. *J Cell Biol*, 153, 87-100

Vu, T. H.; Shipley, J. M.; Bergers, G.; Berger, J. E.; Helms, J. A.; Hanahan, D.; Shapiro, S. D.; Senior, R. M.; Werb, Z. (1998). MMP-9/ gelatinase B is a key regulator of growth plate angiogenesis and apoptosis of hypertrophic chondrocytes. *Cell*, 93, 411-422

Wang, W.; Kirsch, T. (2002). Retinoic acid stimulates annexin-mediated growth plate chondrocyte mineralization. *J Cell Biol*, 157, 1061-1069

Wang, W.; Xu, J.; Kirsch, T. (2003). Annexin-mediated Ca2+ influx regulates growth plate chondrocyte maturation and apoptosis. *J Biol Chem*, 278, 3762-3769

Watzer, B.; Zehbe, R.; Halstenberg, S.; Kirkpatrick C. J.; Brochhausen, C. (2009). Stability of prostaglandin E_2 (PGE_2) embedded in poly-D,L-lactide-co-glycolide microspheres - a pre-conditioning approach for tissue engineering applications. *J Mater Sci: Mater Med*, 20, 1357-1365

Wegner M. (1999). From head to toes: The multiple facets of Sox proteins. *Nucleic Acids Res*, 27, 1409-1420

Wozney, J. M. (1989). Bone morphogenetic proteins. *Prog Growth Factor Res*, 1, 267-280

Xie, C.; Liang, B.; Xue, M.; Lin, A.S.; Loiselle, A.; Schwarz, E.M.; Guldberg, R.E.; O'Keefe, R.J.; Zhang, X. (2009) Rescue of impaired fracture healing in COX-2-/- mice via activation of prostaglandin E_2 receptor subtype 4. *Am J Pathol*, 175, 772-785

Yang, R. S.; Liu, T. K.; Lin-Shiau, S. Y. (1993). Increased bone growth by local prostaglandin E_2 in rats. *Calcif Tissue Int*, 52, 57-61

Yoon, B. S.; Pogue, R.; Ovchinnikov, D. A.; Yoshii, I.; Mishina, Y.; Behringer, R. R.; Lyons, K. M. (2006). BMPs regulate multiple aspects of growth-plate chondrogenesis through opposing actions on FGF pathways. *Development*, 133, 4667-4678

Yoshida, K.; Oida, H.; Kobayashi, T.; Maruyama, T.; Tanaka, M.; Katayama, T.; Yamaguchi, K.; Segi, E.; Tsuboyama, T.; Matsushita, M. et al. (2002). Stimulation of bone formation and prevention of bone loss by prostaglandin E EP4 receptor activation. *Proc Natl Acad Sci U S A*, 99, 4580-4585

Zehbe, R.; Gross, U.; Schubert, H. (2004). Oriented collagenbased/hydroxyapatite matrices for articular cartilage replacement. *Key Engineering Materials*, 254-256, 1083-1086

Zehbe, R.; Haibel, A.; Brochhausen, C.; Gross, U.; Kirkpatrick, C. J.; Schubert, H. (2007). Characterization of oriented protein-ceramic and protein-polymer-composites for cartilage tissue engineering using synchrotron μ-CT. *Int J Mat Res*, 98, 562-568

Zehbe, R.; Libera, J.; Gross, U.; Schubert, H. (2005). Short-term human chondrocyte oriented collagen coated gelatine for cartilage replacement. *Bio-Medical Materials and Engineering*, 15, 445–454

Zehentner, B. K.; Dony, C.; Burtscher, H. (1999). The transcription factor Sox9 is involved in BMP-2 signalling. *J Bone Miner Res*, 14, 1734-1741

Zou, H.; Wieser, R.; Massague, J.; Niswander, L. (1997). Distinct roles of type I bone morphogenetic protein receptors in the formation and differentiation of cartilage. *Genes Dev*, 11, 2191-2203

20

Scaffold-free cartilage tissue by mechanical stress loading for tissue engineering

Katsuko S Furukawa, Masato Sato, Toshihiro Nagai
Stephanie Ting, Joji Mochida and Takashi Ushida
University of Tokyo, Tokai University
Japan

1. Introduction

Articular cartilage has a limited repair capacity due to its low mitotic activity, avascular nature, and sparse cell population. Osteoarthritis, a painful condition of the joints, has affected over 200 million people worldwide. Cell-based tissue engineering is emerging as a promising clinical option to address tissue and organ failure by implanting biological substitutes for the compromised tissue.

Various dynamic stresses such as shear flow, hydrostatic pressure, and direct compression exist in living bodies and are thought to contribute to cartilage formation. A rotational culture can provide such a mechanical load in a simple manner. In this chapter, we introduce our recent scaffoldf-free cartilage tissue engineering approaches using such a rotational culture.

2. Articular cartilage

Articular cartilage is a form of hyaline cartilage that envelopes the surface of diarthrodial joints in the human body to distribute the force and allow smooth movement with minimal friction between two bones. Diarthroidial joints are joints where threre is a lot of motion between opposing bones; for example, the ankle, knee, hip, shoulder, or elbow joints. The joints can bear very large compressive loads and has a very low coefficient of friction (0.005 to 0.05). Unlike other connective tissues, cartilage does not contain anyblood vessels, nerves, or lymphatics. Therefore, it receives nutrients though convection from the surrounding synovial fluid, or diffusion, which is aided by the pumping action created during compression of the cartilage. Cartilage is made up of chondrocytes embedded in a rich matrix of collagen and proteoglicans (Fig. 1).

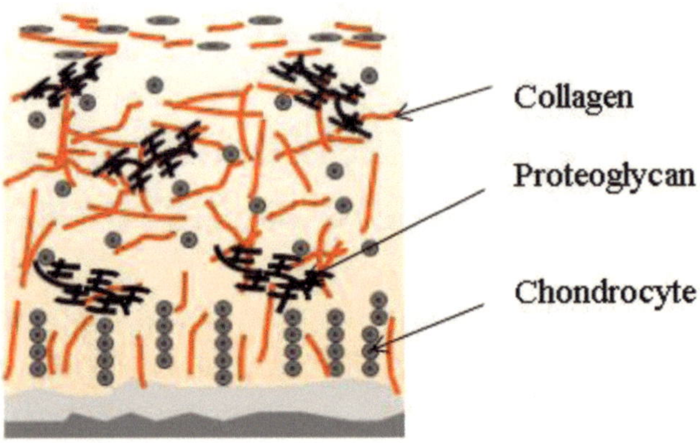

Fig. 1. Cartilage composition

Cells in articular cartilage , which are known as chondrocytes, account for only 5% of the volume of the cartilage. Approximately 70-80% of the articular cartilage is composed of water, whereas the solid phases in composed primarily of type II collagen and aggrecan, a chondroitin and keratan sulohate proteoglycan (Tablel).

Water	66-79%
Solid	21-34%
Minerals	
Residual ash	5-6%
Organic materials	
Collagen	48-62%
Glycosaminoglycan	8-15%
Protein	14-23%
Hyaluronic acid	<1%
Sialic acid	<1%
Lipid	<1%
Glycoprotein	<1%

Table 1. Biochemical composition of articular cartilage

Collagen forms a network of fibrils, which resist the swelling pressure generated by the proteoglycans. Aggrecan, because of its tendency to non-covalemtly interact with hyaluronic acid, forms huge aggregates that become trapped in the collagen network. Because of their numerous negatively charged sulfate groups, these proteoglycan aggregates attract cations, which in turn bring in water to minimize differences in osmotic pressure. Thus, thpe II collagen and proteoglycans create a swollen, hydrated tissue that resists compression. Therefore, the articular cartilage functions as load bearing, shock absorbing, and friction reducing material in diarthrodial joints.

Cartilage tissue has only a small capacity for self-repair. The avascular nature, the sparse cell population, and the low mitotic activity of the chondrocytes severely limit the natural healing of these cartilage defects(Campbell et al., 1969; Mankin et al., 1982; Ushida et al., 2002). Many patients with damaged cartilage due to trauma or diseases such as osteoarthritis and osteochondrosis dissecans are existing in the world. The degradation of cartilage eventually leads to osteoarthritis, which is the thinning of the articular cartilage resulting in a bone-to-bone joint, causing pain and a reduction in notion. In this disease, there is often the loss of proteoglycans, which leads to decreased compressive stiffness of the tissue. When osteoarthritis reaches the subcondral bone, an ingress of cells from the bone marrow will generate fibrocartilagegious repair. This leads to a tissue that has lower proteoglycan and high levels of type I collagen. Fibrocartilage is mechanically insufficient as it cannot withstand the normal physiological loads and wear stresses on the knee during work or sports. In very severe cases, both bones of the joint may grow into one antoher, causing them to fuse together. The joints most commonly affected by osteoarthritis include the hips, knees, wrists, between the fingers, and between the vertebrae.

3. Current treatment for osteoarthritis

Current therapies include transplantation of healthy host cartilage or implantation of artificial prosthetic devices. However, problems remained in these treatments. The amount of donor tissue for transplantation is limited, and durability of the prosthetic devices (Fig. 2) is not good by wear debris or adhesive breakdown at the host/prosthesis interface. For these reasons, there is considerable interest in developing cell therapies and tissue-engineered cartilages to treat damaged cartilage(Brittberg et al., 1994; Freed et al., 1994a, Freed et al., 1994b, Freed et al., 1994c).

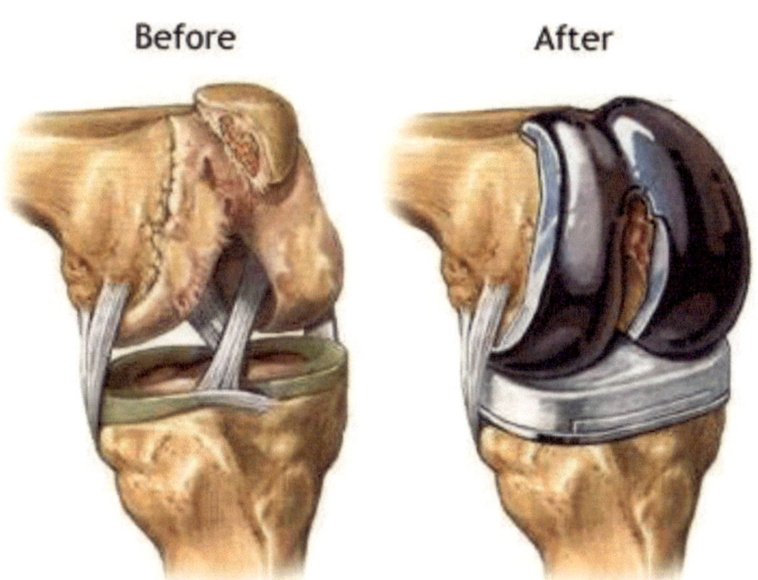

Before After

Fig. 2. Knee joint replacement prosthesis
(http:// assets.aarp.org/external_sites/adam/html/2/9494.html)

Still, many problems remain in cartilage tissue engineering using cells. When grown in two-dimensional culture, adult articular chondrocytes are capable of proliferation, promoting researchers to use cultured autologous chondrocytes to accelerate cartilage regeneration. However, the proliferating chondrocytes gradually lose their differentiated phenotype(von der Mark et al., 1977; Benya et al., 1982, Fig.3), as indicated by the loss of synthesis of aggrecan and type II collagen and the increase of synthesis of type I collagen (Fig. 4, Chen et al., 2003, Ushida et al., 2004). Therefore, in order to form a tissue-engineered cartilage using proliferated chondrocytes with a dedifferentiated phenotype, it is thought that reexpression of the chondrogenic phenotype should be induced before implantation.

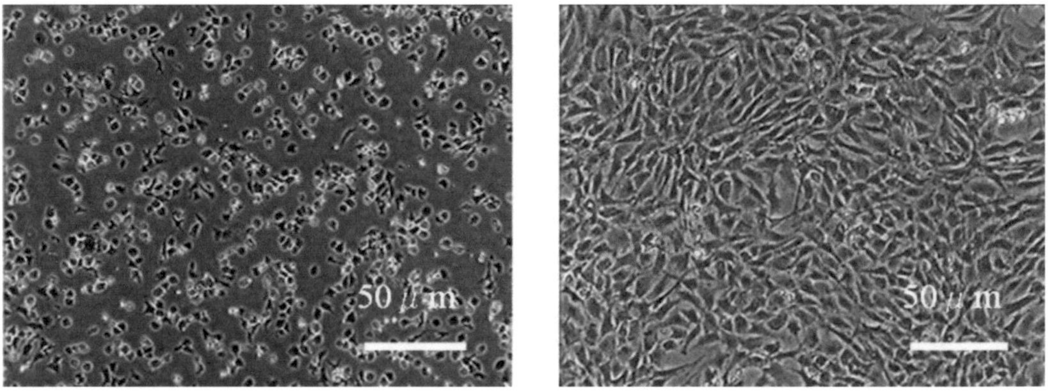

Fig. 3. Chondrocytes in 2-dimensional culture. Left: Primary chondrocytes, right: dedifferentiated chondrocytes

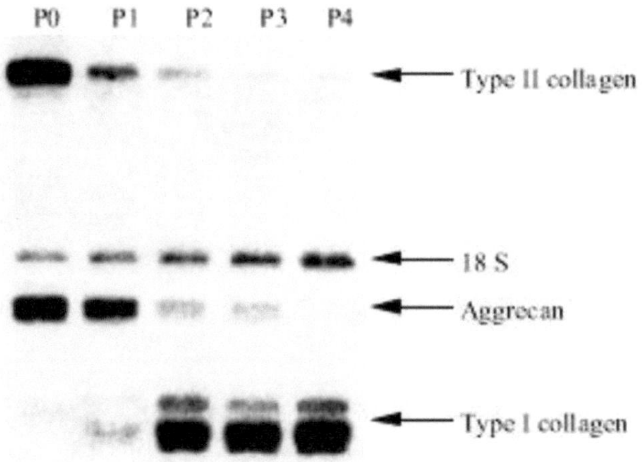

Fig. 4. Northern blot analysis of the mRNA encoding type I collagen. Type II collagen, and aggrecan of bovine chnodrocytes cultured in monolayer for 0, 1, 2, 3 and 4 passages (Chen et al., 2003, Ushida et al., 2004).

To date, "pellet" culture systems (Fig. 5) under static conditions have been reported as a method for preventing and reversing the phenotypic modulation of chondrogenesis *in vitro(Hay* et al., 2000; Johnstone et al., 1998; Lunstrum et al., 1999; Mackay et al., 1998; Oberlender et al., 1994; Pittenger et al., 1999; Tracy et al., 1994; Yoo et al., 1998).

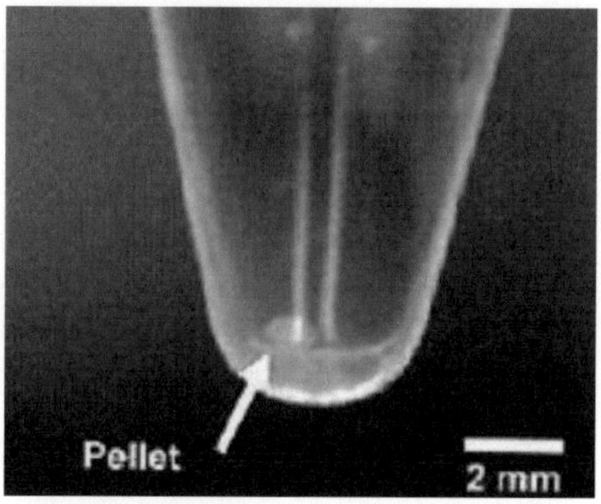

Fig. 5. Pellet culture by dedifferentiated chondrocytes (Tsuchiya et al., 2005).

This culture system allows cell-cell interactions analogous to those that occur in condensation during embryonic development. Although the system facilitates chondrogenic differentiation from dedifferentiated chondrocytes or mesenchymal stem cells with high reproducibility, its clinical application to tissue engineering has not been successful thus far. Because the "pellet" culture system forms only one small cell-aggregate per tube by a centrifugator, it is difficult to yield sufficient numbers of pellets with the differentiated phenotype. Therefore, a cartilage tissue with a large size in which cell-cell interactions exist, as in a pellet culture, without using a scaffold was tried to form for tissue engineering.

4. Scaffold-free cartilage model by dynamic rotational culture

By using a high porous simple mold, scaffold-free cartilage tissue of arbitrary shapes was tried to fabricate (Furukawa et al., 2008). Various dynamic stresses such as shear flow, hydrostatic pressure, and direct compression exist in living bodies and are though to contribute a scaffold-free rotational culture provided such a mechanical load in a simple manner. Thus, it was demonstrated that a scaffold-free cartilage tissue with a large size and arbitrary shape for tissue engineering could be regenerated without any scaffolds, by making cell-cell interaction and mechanical stress loading for tissue engineering.
When a cylindrical glass mold with a diameter of 1 cm was used, dedifferentiated chondrocytes of 1.5×10^7 cells were inoculated into the mold using a commercially available culture-insertion film with pores of 0.4 |_im. Oxygen and nutrition can diffuse from the lower and upper sides of cells. After 8 hours of cell inoculation, the mold was removed from the membranous surface, and cell plates were left on there with the same shape as the mold. After 24 hours' static culture for shape stabilization, scaffold-free tissue was moved to a 6-well culture plate without any shape changing adding 5 ml of differentiation medium, and was begun on rotational culture. In this point, the scaffold-free plate was very fragile; therefore, displacement of the tissue to the rotational culture plate was done with great care. A disk-like cartilage tissue was formed by 3 weeks' culture as shown in Fig. 6.

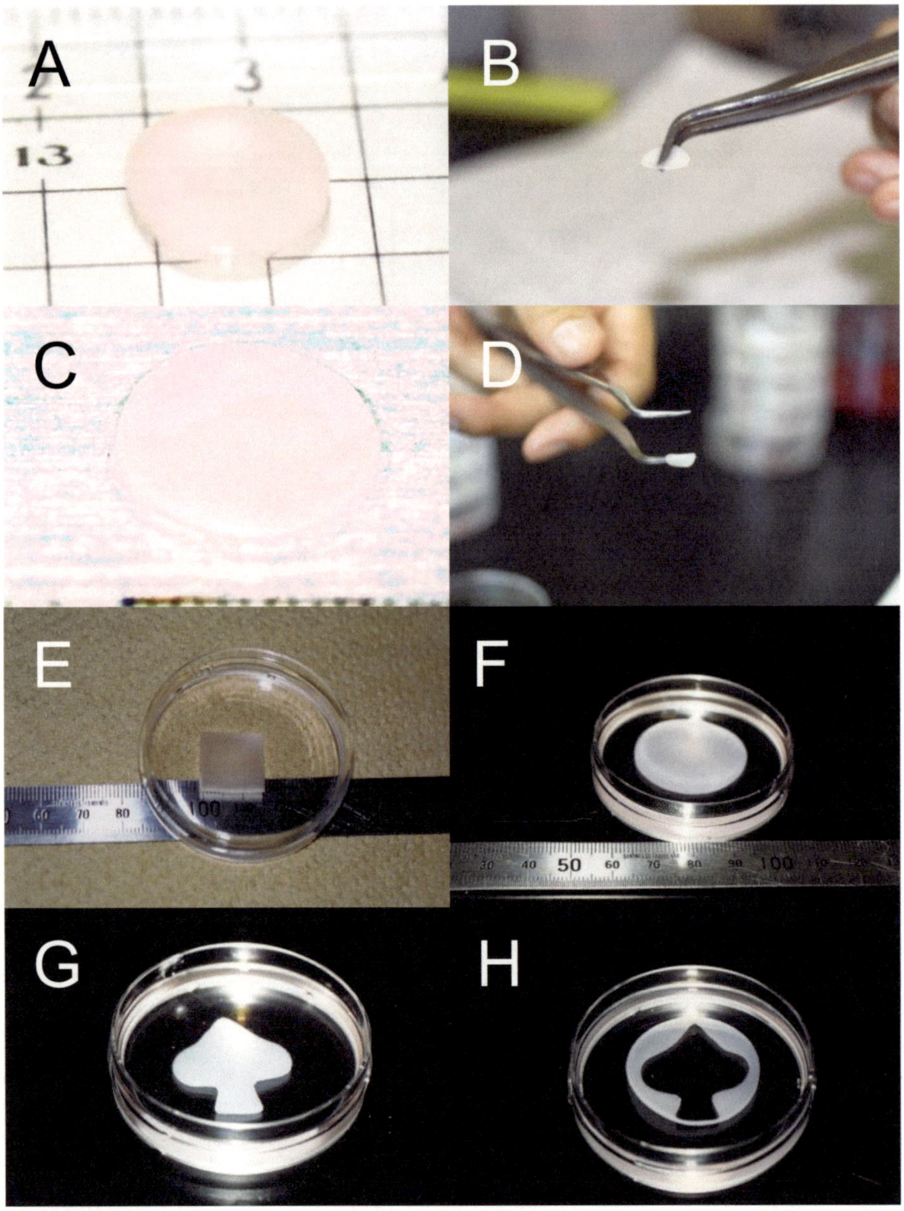

Fig. 6. Appearance of the scaffold-free cartilage tissue after 3-week culture. A, B, E-H: Rotationcal culture, C, D: static culture. A-D, F: mold shapes were circle. E: Mold shape was square. G: Spade shape formed by stamping out with a metal mold. H: F-C.

When the cartilage plate without any scaffold cultured under shear flow conditions (rotational culture condition) was grasped, the shape of the plate with a disk-like form was maintained (Fig. 6B). In contrast, the cartilage plates cultured under static conditions

changed to a dogleg shape when grasped (Fig. 6D); often they could not keep their shape for 3 weeks if we changed the medium. Therefore, it was obvious that mechanical conditions played an important role in the mechanical properties of a scaffold-free cartilage tissue. When a square mold was used, a square-shaped cartilage plate was formed under shear flow conditions (Fig. 6E). The formation of a comparatively large cartilage tissue was also possible as shown in Fig. 6F. Furthermore, it was possible to form and cut it into spade shapes, i.e., arbitrary shapes, by stamping out with a metal mold (Fig. 6G, 6H).
The thickness, diameter, and volume of cartilage tissue formed by a glass mold with a diameter of 1 cm at the start point of culture were measured. The thickness of the cartilage tissue continued to increase for the 3-week rotation culture (Fig. 7A). The diameter increased slightly for 1 week, then remained the same (Fig. 7B). The volume continued to increase for the 3-weeks culture (Fig. 7C). Similar results were obtained in control static culture. To confirm the viability of the chondrocytes in the cartilage tissue, the scaffold-free cartilage with a diameter of 1 cm cultured under shear flow conditions was stained with calcein-AM and propium iodide. Necrosis in the center of the tissue was not observed; the majority of the cartilage cells in the central part of the cartilage tissue were understood to be living. Therefore, it was suggested that the chondrocytes included in the cartilage tissue were alive and maintained their growth for 3 week *in vitro.*

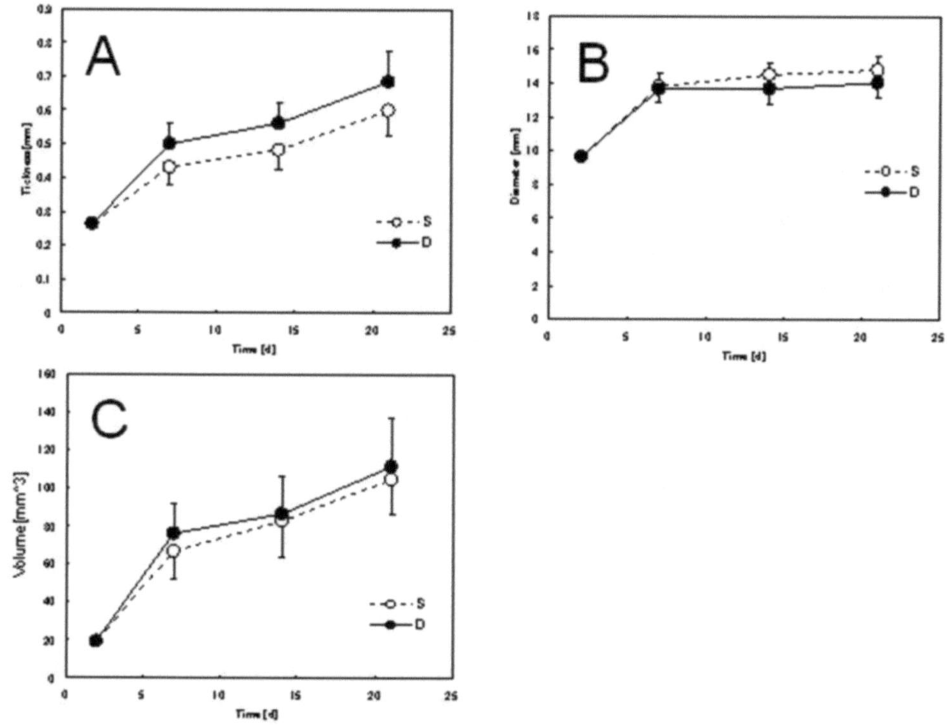

Fig. 7. Time changes in thickness, diameter, and volume of the cartilage tissue for 3-week rotational culture. D: Dynamic (rotational) culture of the cartilage tissue, S: Static culture. A) Thickness of the cartilage plate was measured by a mechanical spectrometer (AGS-G, Shimazu, Kyoto, Japan), mean±SE (n=4). B) Diameter was measured by a caliper square. C)

Volumes of the cartilage plate were calculated from the data of the thickness and the diameter of the cartilage plates, independently.

To investigate the existence of proteoglycans, the cartilage tissue with a diameter of 1 cm was stained with safranin-O and toluidine blue. The cartilage tissue cultured under dynamic conditions (rotational culture) for 3 weeks showed strong staining properties (Fig. 8B and 8D) compared with that of the static culture (Fig. 8A and 8C). In rotational culture, cells aligned horizontally to the surface were observed (Fig. 9B). In contrast, cartilage tissue cultured under static conditions did not have such alignment; instead, cells near the surface had a round shape without any orientation (Fig. 9A).

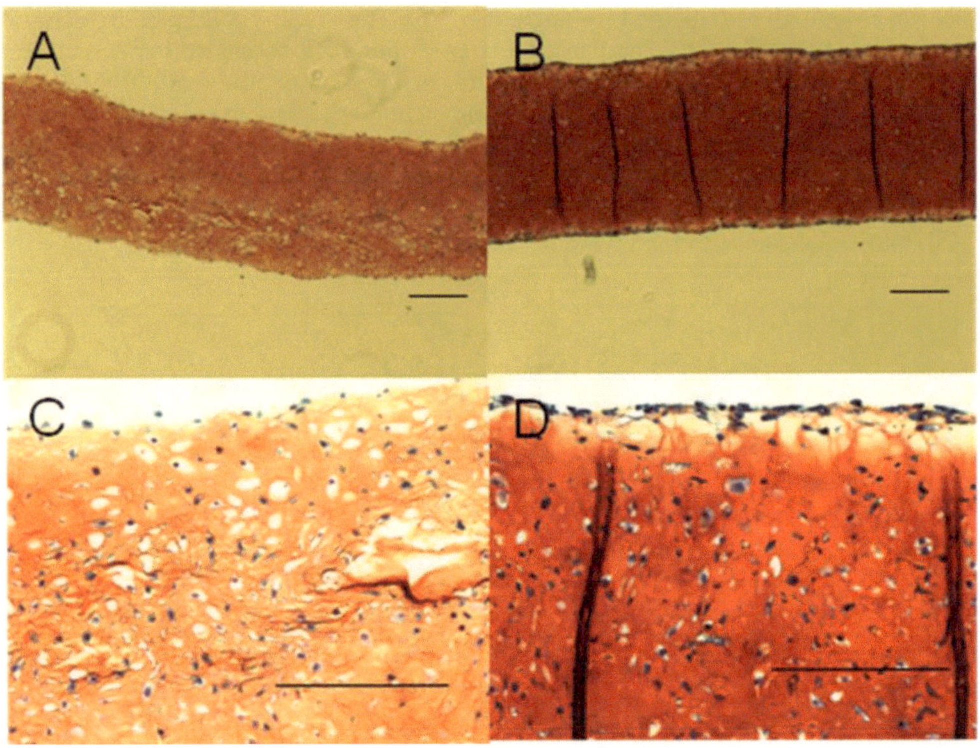

Fig. 8. Distribution of proteoglycan was investigated. A and C) Scaffold-free cartilage tissue cultured under a static condition without flow was stained with safranin-O after cultivation of 3 weeks. B and D) Scaffold-free cartilage tissue cultured by rotational culture was stained with safranin-O after 3 weeks. Scales, 200|im. Magnifications of objective lens, A, B; x4, C, D; x20.

The edge of the cartilage tissue cultured under static conditions was not smooth while that of rotational culture was extremely smooth. In cartilage tissue by rotational culture, there were parts, near the surface, not stained with safranin-O (Fig. 3B and 3D). The results by toluidine blue were similar to the results of safranin-O (data was not shown). The cartilage tissue formed by rotational culture presented metachromasia, and the color was stronger than those of the static cultures. The tendencies of the cell shape and tissue formation were also similar to those of safranin-O.

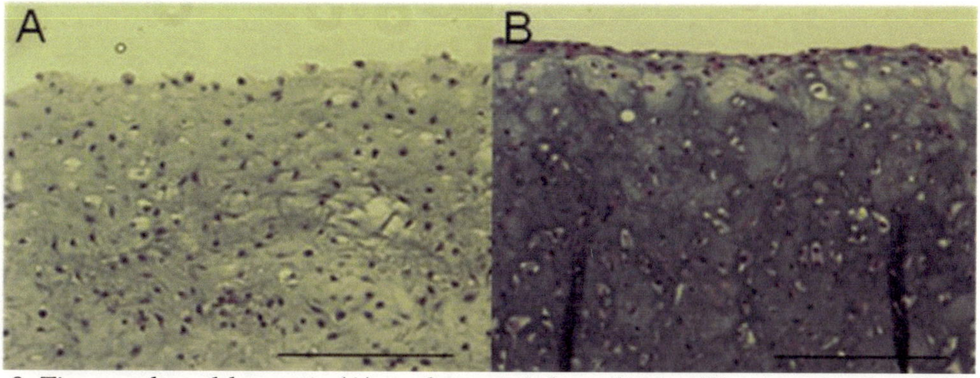

Fig. 9. Tissue cultured by static (A), and rotational conditions (B) was stained with HE in order to observe cell distributions. A: Magnification of objective lens=x5, B: Objective lens=x20. Scales, 200 | im.

Irnmunohistochernical analysis showed that type II collagen protein was abundantly distributed in scaffold-free cartilage tissue cultured by rotational culture (Fig.lOA). On the other hand, cartilage tissue formed by static culture showed inhomogeneous distribution of type II collagen protein (Fig. 10B).

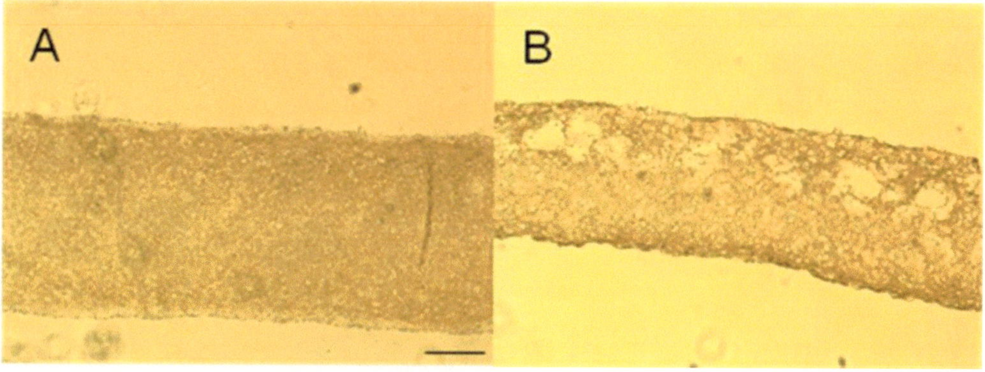

Fig. 10. Cartilage tissue by 3-week culture was stained with type II collagen antibody. A) Cartilage tissue cultured by rotational conditions. B) Cartilage tissue cultured by static conditions. Objective lens: x5, Scales; 200 | im.

The cartilage tissue formed by 3 weeks' rotational culture had scarcely any type I collagen protein (Fig. 11A). In static culture, type I collage protein appeared at the edge of the cartilage tissue as shown in brown color (Fig 11B).
The amounts of proteoglycan and total collagen protein secreted were quantitatively investigated. As shown in Fig. 12, rotational culture significantly increased the protein production. The rotational culture was effective in terms of matrix production per DNA contents in the cartilage tissue with a diameter of 1 cm when compared to static culture. There were statistical differences between dynamic and static cultures (PG/DNA; $p<0.05$, CN/DNA; $p<0.01$).

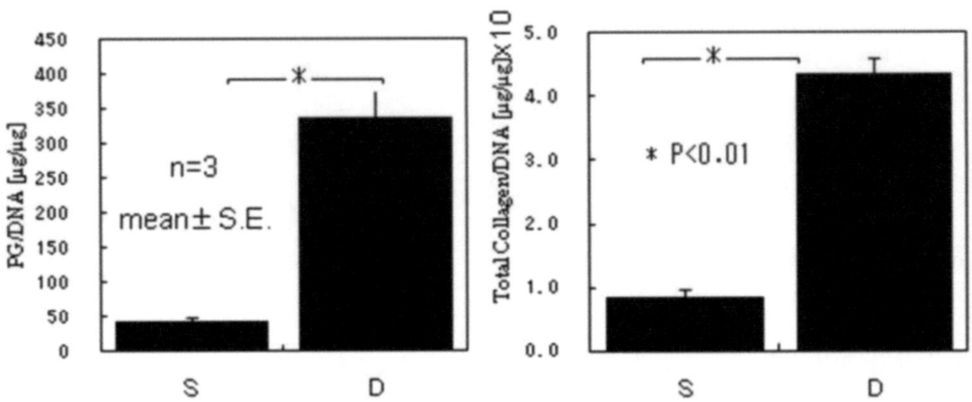

Fig. 12. Proteoglycan and total collagen protein expressed in the cartilage tissue waere measured by DMMB method and hydroxyproline assay. S) Static culture of the cartilage tissue, D) Dynamic (rotational) culture of the cartilage tissue.

The expression levels of mRNA of type II collagen and proteoglycan in cartilage tissue cultured under shear flow conditions were higher than that of static, culture (Fig. 13). On the other hand, type I collagen mRNA was expressed at a lower level in the cartilage tissue under rotational culture than in static cultures.

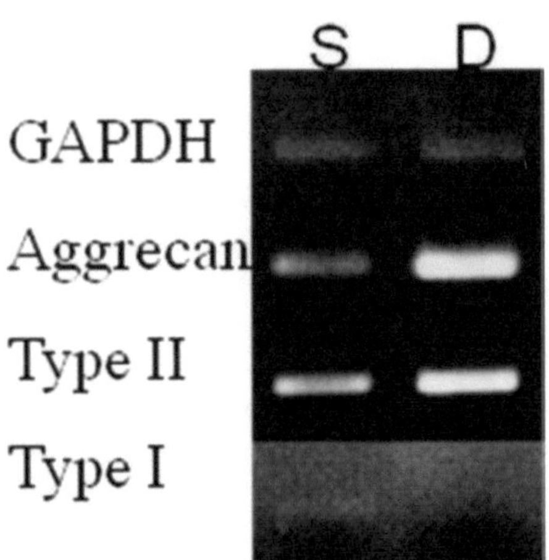

Fig. 13. Gene expression of GAPDH, aggrecan, type II collagen, and type I collagen of the tissue was investigated using RT-PCR methods. S) Static culture of the cartilage tissue, D) Rotational culture of the cartilage tissue.

To obtain quantitative data of the mechanical properties of the cartilage tissue, a stress-strain curve (Fig. 14B) was obtained using dumbbell-shaped culturing samples (Fig. 14C) prepared by stamping out with a metal mold (Fig. 14A). The values of calculated rupture strength and Young's modulus from the data of Fig. 13C significantly increased in cartilage tissue cultured under shear flow conditions. There were statistically significant differences in the values between dynamic and static cultures. The tactile impression of the cartilage tissue by rotational culture was like a plastic coaster.

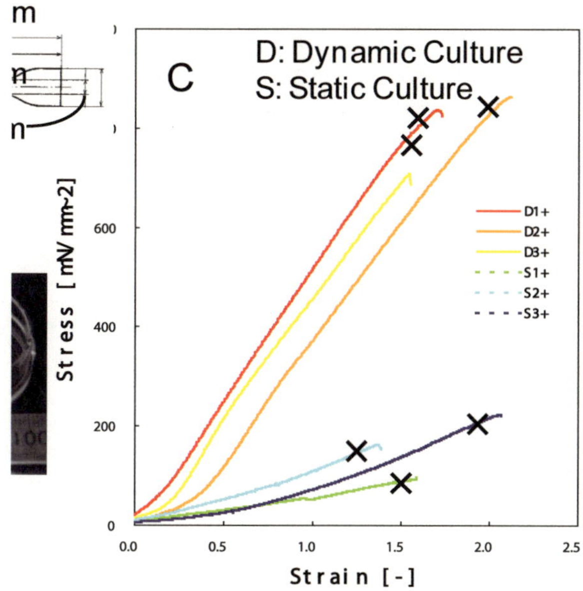

Fig. 14. Mechanical properties of the cartilage tissue. A) Shape of the mold for evaluating the mechanical properties of cartilage tissue. B) Stress-strain curve was investigated. D1, D2 and D3 are data of the samples cultured by rotational culture. S1, S2 and S3 are those of samples cultured under static conditions. C) Dumbbell-shaped samples were prepared for the mechanical evaluations of the cartilage tissue.

5. *In vivo* implantaion experiments of scaffold-free cartilage tissues cultured under dynamic rotational conditions

For animal transplantations of a scaffold-free cartilage tissue cultured under shear flow conditions, the tissue by dedifferentiated chondrocytes derived from Japanese white rabbits was tried to form (Nagai, Furukawa et al, 2008; Sato et al., 2008). After 7 days weeks of rotational culture, the plate was cultured under dynamnic conditions, using rotational culture. After 2-3 weeks of rotational culture, the chondrocyte plate maintained a constant form and was considered stable enough to be handled with surgical pincers (Fig. 15). Conversely, after 3 weeks of static culture, the plate gradually changed into an arch over that time (data was not shown). Histological and immunohistochemical evaluations indicated that the plate had cartilaginous qualities in terms of cell distribution and organization and the production of glycosaminoglycans and type II collagen in rotational cultures. The chondrocyte-plates measuring about 1.1 mm in thickness after 2 weeks and about 1.4 mm after 3 weeks of cultivation had a regular cylindrical shape and the macroscopic appearance of cartilage. The chondrocytes-plate was thus able to be made without losing its regular cylindrical shape regardless of the cellular passage number (Fig.15).

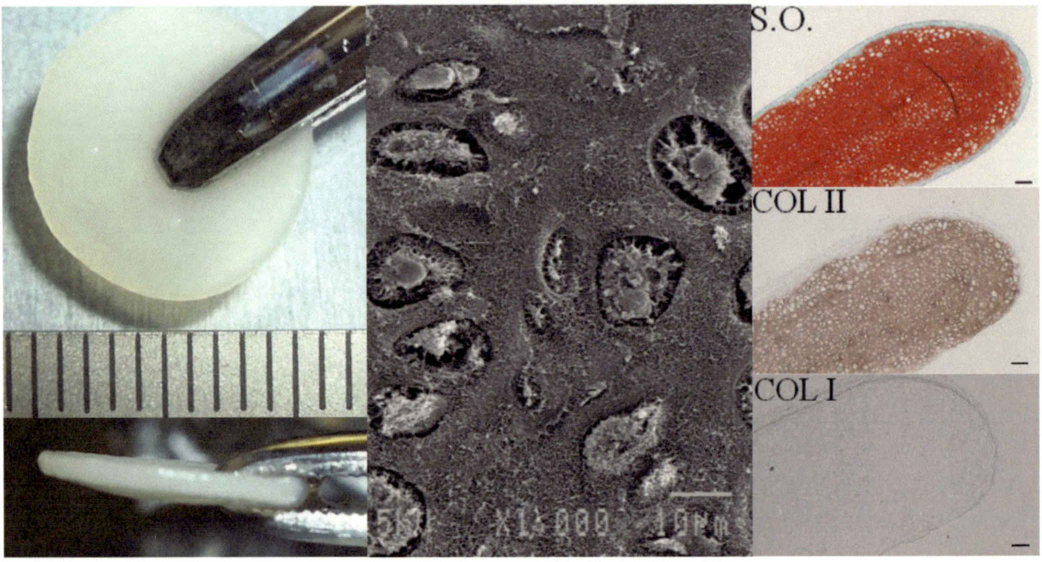

Fig. 15. Scaffold-free cartilage tissue by dedifferentiated rabbit chondrocytes culutured under dynamic rotational culture conditions. Left columns were macroscopic appearances. Center was the scanning electron microscopic appearance. Right columns show the cros-sections stained with safranin-O (S.O.), collagen type II (COL II) and collagen type I (COL I) antibodies.

As a preliminary in vivo implantation experiment of the scaffold-free cartilage tissue cultured under dynamic rotational culture conditions, one-week and 2-week plates were transplated to the total thickness-defect model in patellar grooves of rabbits. At 4 weeks after implantation, fibrous tissue ad replaced most of te 1-week plates, and many inflammatory cells were seen in the subchondral bone. Alternatively, 2-week plates maintained their shape and integrated with the host cartilage(Fig. 16).

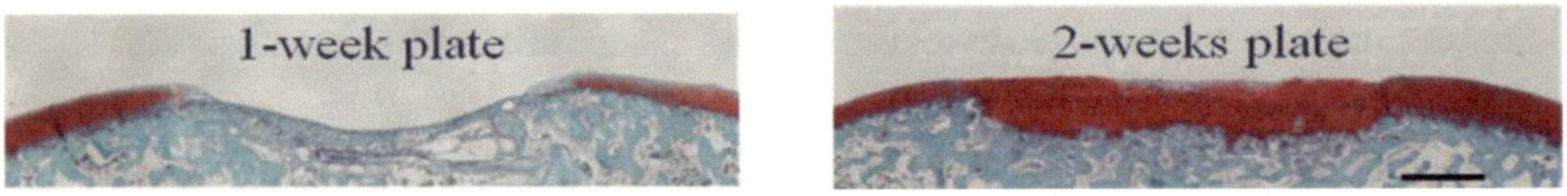

Fig. 16. Preliminary in vivo implantation experiments. Histological analysis with safranin O and Masson's trichrome at 4 weeks. Left picture in the 1 week chondrocyte plate insertion group and right in the 2 week chondrocyte plate insertion group (scale bars=1mm).

At 4 weeks after implantation, according to macroscopic observation, the chondrocyte plate insertion group seemed to show better results in terms of the integration of host cartilage, and the defects repaired by the plate were shoother than in the noninsertion group (Fig. 17, Nagai et al., 2008). In the chondrocyte plate insertion group (CP), the repair site appeared to be filled with cartilaginous tissues, which were strongly stained with safranin O (Fig. 17), positive for type II collagen(Fig. 17), and negative for tupe I collagen (data was not shown). Lateral integration of the cnohdrocyte plate was well bounded at both sides of host cartilage. Basal integration of the plate was also good. Each implanted chondrocytes plate was clearly visible but was thinner than before implantation (650-700|_im thick). The interfacial adhesion between the plate and the lower portion of the repair tissue contained hypertrophic chondrocytes that were remodeling the subchondral bone. Structural integrity of the plate was shown by the begining of columnar organization of rounded chondrocytes. No infiltration of inflammatory cells within the subchondral bone was seen. Alternatively, in the noninsertion group, the defects were filled with mainly fibrous tissue concealing the lower portion of the repair tissue. The lower portions of the regions stained with safranin O were positive for type II collagen but not for type I.

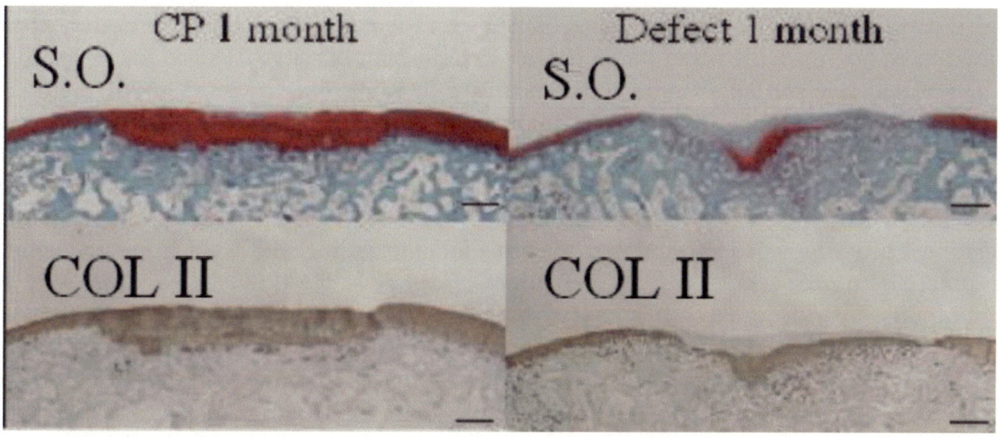

Fig. 17. Macroscopic and histological analysis of the in vivo study for safranin O and for type II collagen. Macroscopic observations on femoral chondyles at 4 weeks after surgery. In the chondrocyte plate group (CP), the defect seemed to be better regarding the integration of host cartilage than in the noninsertion control group (Defect). Scale bars=250 |_im

6. Perspectives for the scaffold-free cartilage formation cultured under dynamic rotational conditions

The scaffold-free cartilage tissue by rotational culture was proven to have better mechanical, biochemical, and histochemical properties than that of scaffold-free cartilage tissue under static conditions. During a 3-week culture, cartilage tissue composed of dedifferentiated chondrocytes and extracellular matrices was formed, without use of a scaffold. Based on our analysis of tissue growth after 3 weeks, the scaffold-free cartilage tissue by rotational culture was able to assume arbitrary shapes according to the mold. If shear force is never loaded onto the cartilage tissue, the mechanical properties become extremely week; hence, the rotational culture may be an essential factor in terms of shape control of scaffold-free cartilage tissue with an arbitrary shape. From the view point of clinical medicine, although we must obtain detailed data on differences for scaffold-free cartilage formation of factors such as cellular species, culture period, passage number, number of cells inoculated to a mold and so on, this work should suggest an evocative concept for future research.

Many papers have reported on high-cell-density cultures in which cells contact, creating cell-cell interactions, to promote differentiation from dedifferentiated chondrocytes(Koch, 2002) and/or stem cells such as mesencymal stem cells(Denker, 1999). Moreover, it is also known that cartilage differentiation is promoted when cells condense in the stage of generating(Summerbell, 1972; Thorogood, 1975). Furthermore, it is also reported that the mechanical stresses existing under physiological conditions such as a walking(Hodge, 1986) regulate the differentiation of chondrocytes(Smith, 2000, Matsuda, 2003). Therefore, these backgrounds may support the reason why dramatic facilitatory effects in differentiation occurred by loading of shear flow (rotational culture) to scaffold-free cartilage tissue.
As a mechanism of the effect of rotational culture on the differentiation of dedifferentiated chondrocytes and the increase in the mechanical properties, involvements of shear stress, oxygen and nutrition diffusion were considered. Comparing the scaffold-free cartilage tissue cultured under static conditions with that under dynamic conditions, the static-culture cartilage tissue had surface irregularities and was not smooth; the exterior surface seemed to be open. On the other hand, a smooth tight membrane existed on the surface area of the scaffold-free cartilage tissue cultured under dynamic conditions (rotational culture). The cells in the membrane aligned horizontal to the surface. The appearance of this smooth layer was considered to have enabled accumulation of an extracellular matrix such as proteoglycan and collagen in the cartilage tissue. As a result, the method may have succeeded at tissue formation with abundant matrices like cartilage and improvement of mechanical properties. It is speculated that this smooth membrane layer appeared only in the dynamic culture and is due to the direct effect of shear flow. The reason why the membranous layer appeared only in the dynamic culture system was not clear, but it is obvious that cartilage cells can detect mechanical stresses such as shear stress, hydrostatic pressure, direct compression, etc., and that the stresses promote the production and deposition of a matrix similar to that in cartilage. There is no doubt that a low shear force exists in the rotational culture. This shear force may stimulate scaffold-free cartilage tissue to differentiate without cell damage. Therefore, it was suggested that a combination of mechanical stress and scaffold-free cartilage tissue composed solely of chondrocytes at the start point of culture may well become a suitable model for tissue-engineered cartilage.
As a cartilage tissue engineering, autologous chondrocytes transplantation was tried (Brittberg et al., 1994). But problems in this method due to loss of viability in the transplanted cells and the difficulty of fixing chondrocytes into a defect remained. To improve the cell loss, wakitani et al. embedded allograft articular chondrocytes in collagen gel (Wakitani et al., 1989; 1996). However, it was difficulty that the collagen gel keeps the shape with cells. Currently biodegradable scaffolds either synthetic or naturally derived ones, are applying to control the shapes. But the phenotype control of chondrocytes in the scaffold was not enough. On the other hand, scaffold-free cartilage tissues formed by methods using alginate gel(Matsuda, 2003; Stoddart et al., 2006a), agarose gel (Kelm et al., 2004), rotational wall vessel(Marlovits, 2003), special system(Maini-Varlet, 2001; Grogan, 2003; Wang, 2004; Park, 2006), and a suspension culture of aggregating chondrocytes(Furukawa, 2003; Tsuchiya, 2005) are reported, and the phenotype of chondrocytes in the scaffold-free cartilage was possible to control , but these methods could not freely form cartilage tissues into arbitrary shapes. Furukawa et al (2008) succeeded in the formation of cartilage tissue into arbitrary shapes using two kinds of techniques. Rotational culture realized cartilage tissue formation with arbitrary shape. It was thought

that the induction of mechanical properties by rotational culture contributed to the shape retention. In addition, the next point was based on a simple mold system. For oxygen and nutrition supply from the lower and upper sides of the cartilage tissue, a culture insertion with a highly porous film (pores 0.4 | m in diameter) was used as a basement support at the only start point of tissue formation. Chondrocytes were not able to pass through the porous film. When a mold was set on the porous film, cartilage tissues of arbitrary shapes were easily induced. Because to adhere cells each other, a large amount of oxygen and nutrition may be needed. If we use normal culture dish or plastic rigid plate without pores at this point, all cells died and did not form tissue. In order to design complicated three-dimensional (3-D) molds for shapes such as a total knee cartilage, design of 3-D molds will be needed in the future, but this is a realizable technology.

To compare the mechanical properties of tissue engineered cartilages cultured by static and dynamic conditions, tensile tests were carried out. It was proven that the loading of mechanical stress increased the mechanical properties of the tissue by using the method. Therefore, it was understood that improvement of the mechanical properties realized formation of tissue engineered cartilage tissue with arbitrary shapes from the analysis. When the tensile testing data of scaffold-free cartilage tissue cultured under dynamic shear flow conditions to bovine native cartilage of 5 month old (Williams, 2001) were compared, the value of rapture strength was 58%, and Young's modulus was 6.4%. On the other hand, the values of scaffold-free tissue cultured under static conditions compared to the native cartilage tissue were 11, and 1.2%, respectively. These data suggested that the scaffold-free cartilages formed under dynamic and static culture conditions had inferior tensile properties to native cartilage tissue. Shear stress, oxygen and nutrition diffusion may reinforce the mechanical properties of scaffold-free cartilage by rotational culture. From the study, it was not clear which one has most prominent effect on mechanical properties of them. On the other hand, many researchers had already reported that the positive effects of mechanical stresses such as shear flow (Freed et al., 1994a, Freed et al., 1994b, Freed et al., 1994c, Vunjak-Novakovic et al., 1996), hydrostatic pressure with flow (Allemann et al., 2000, Mizuno et al., 2002), hydrostatic pressure without flow (Smith et al., 1996, Smith et al., 2000, Angle et al., 2003, Toyoda, et al., 2003, Kawanishi et al., 2007), dynamic compression(Elder et al., 2001), microgravity (Freed et al., 1997) on extracellular matrix production of chondrocytes. It was obvious that extracellular matrix production increase the mechanical properties of tissue engineered cartilage. Therefore, probably shear stresses by rotational culture also reinforce the mechanical properties of our scaffold-free cartilage tissue.

Recently, Elder et al reported an effect of hydrostatic pressures, and Stoddart et al (2006b) did an effect of of compressive force, on the formation of scaffold-free cartilage tissue from immature primary bovine chondrocytes. According to their studies, constant hydrostatic pressures of 5 and 10 MPa were more effective than oscilatory ones. The cyclical load of compressive force increased the chondrogenic protein and mRNA productions. However we can not directly compare the effects of the dynamic rotation, hydrostatic pressure and compression on matrix production of scaffold-free cartilage tissues, because the cell population was different; Elder et al and Stoddart et al. used immature chondrocytes (differentiated chondrocytes), and Furukawa et al used dedifferentiated bovine chondrocytes.The hydrocstatic pressure and compressive force loading machines have complicated structures. Therefore, they have a higher risk of bacteria contaminations into their systems. In addition,the complicated structures take a great deal of time and effort to load the mechanical stresses on tissue-engineered consructs. In contrast, in dynamic rotational culture, a simple and straightforward procedure is enough to load a mechanical stress on scaffold-free cartilage tissues. For clinical treatment of cartilage diseases, usages of a simple stractured machine and protocol are important point for the success, therefore, the dynamic rotational culture will facilitate future clinical trials.

According to conventional tissue engineering, the mass balance between the disappearance (hydrolysis) of a biodegradable scaffold and the production/deposition of extracellular matrix from cells inoculated to the scaffold has been thought to be important for regulation of shape(Freed, 1997; Furukawa, 2002). The cartilage tissue which constituted of chondrocytes and ECM produced by inoculated dedifferentiated chondrocytes has arbitrary shapes, and did not need any scaffold to control the shape.In scaffold-free cartilage tissue examined by this study, although the thickness of the cartilage tissue increased linearly with time, a change in a transverse direction was not observed (constant). As a result, it turned out that the volume of the cartilage tissue also increased linearly with time. Therefore, a quantitative prediction of the tissue growth and shape change by a calculation may be possible. So, the model may point out the possibility for a new tissue design by shape control not only by using a biodegradable scaffold, but also by using a mold with an arbitrary shape and calculation of the cartilage growth.

From the view point of the clinical medicine, our scaffold-free cartilage tissue cultured by shear flow conditions was thought as a suitable model, because the scaffold-free cartilage tissue had promising biochemical, mechanical properties as a tissue-engineered cartilage tissue. In addition, it was cultured by medium without special expensive factors such as transforming growth factor and bone morphologic growth factor etc. The useless of special growth factors not only will prevent the tissue-engineered cartilage tissue from a risk of carcinogenesis, but also realize cost-cutting of medical expenses. Then, transplantation of biomaterial , a kind of foreign body, to our body was not needed in our approach. Therefore, it was though that the barrier to the promotion of clinical trial is lower.

7. Conclusion

In conclusion, a scaffold-free cartilage tissue with arbitrary shapes and a large size with promising biological, mechanical properties was formed. The scaffold-free cartilage loaded mechanical stress based on a high porous simple mold system may become tools of not only therapies for diseases such as osteoarthritis and osteochondrosis dissecans as a tissue-engineered cartilage, but also clarification of the mechanisms of cartilage formation and therapeutic treatment instead of conventional pellet culture system.

8. References

Allemann, F.; Mizuno, S.; Eid, K.; Yates, K.E.; Zaleske, D. & Growacki, J. (2000). Effects of hyaluronan on engineered articular cartilage extracellular gene expression in 3-dimensional collage scaffolds. *Journal of Biomedical Material Research.* 55,3, 13-19.

Angele, P.; Yoo, J.U.; Smith, C.; Mansour, J.; Jepsen, K.J.; Nerlich, M. & Johnstone, B. (2003).Cyclic hydrostatic pressure enhances the chondrogenic phenotype of human mesenchymal progenitor cells differentiated in vitro. *J Orthoaedic Res* 21, 451-457.

Benya, P.D. & Shaffer, J.D. (1982). Dedifferentiated chondrocytes reexpress the differentiated collagen phenotype when cultured in agarose gels. *Cell* 30, 215-224.

Brittberg, M.; Lindahl, A.; Nilsson, A.; Ohlsson, C.; Isaksson, O. & Peterson, L. (1994). Treatment of deep cartilage defects in the knee with autologous chondrocyte transplantation. *N. Engl. J. Med.* 331, 889-895.

Campbell, C.J. (1969). The healing of cartilage defects. Clin. Orthop. Rel. Res. 64, 45-63. Mankin, H.J. (1982).The response of articular cartilage to mechanical injury. *J. Bone Joint Surg.* 64-A, 460-466.

Chen, G.; Sato, T.; Ushida, T.; Hirochika, R. & Tateishi, T. (2003). Redifferentiation of dedifferentiated bovine chondrocytes when cultured in vitro in a PLGA-collagen hybrid mesh. FEBS Lett 8, 542, 1-3, 95-99.

Denker, A.E.; Haas, A.R.; Nicoll, S.B. & Tuan, R.S. (1999). Chondrogenic differentiation of murine C3H10T1/2 multipotential mesenchymal cells: I. Stimulation by bone morphogenetic protein-2 in high-density micromass cultures. *Diffenrentiation* 64, 67-76.

Elder, S.H.; Goldstein, S.A.; Kimura, J.; Soslowsky, L.J. & Spemgler, D.M. (2001). Chondrocyte differentiation is modulated by frequenvy and duration of cyclic compressive loading. *Annals of Biomedical Engineering* 29, 476-482.

Freed, L.E.; Grande, D.A.; Lingbin, Z.; Emmanual, J.; Marquis, J.C. & Langer, R. (1994a). Joint resurfacing using allograft chondrocytes a synthetic biodegradable polymerscaffolds. *Journal of Biomedical Material Research* 28, 891-899.

Freed, L.E.; Marquis, J.C.; Vunjak-Novakovic, G.; Emmanual, J. & Langer, R. (1994b). Composition of cell-polymer cartilage implants. *Biotechnology and Bioengineering* 43, 605-614.

Freed, L.; Vunjak-Novakovic, G.; Biron, R.J.; Eagles, D.B.; Lesnoy, D.C.; Barlow, S.K. & Langer, R. (1994c). Biodegradable polymer scaffolds for tissue engineering. *Bio/Technology* 12, 689-693.

Freed, L.E.; Langer, R.; Martin, I.; Pellis, N. & Vunjak-Novakovic, G. (1997). Tissue engineering of cartilage in space. *Proc Natl Acad Sci USA* 94, 13885-13890.

Furukawa,K.S.; Ushida, T.; Sakai, Y.; Suzuki, M.; Tanaka, J. & Tateishi, T. (2001). Formation of human fibroblasts-aggregates (spheroids) by rotational culture for tissue-engineered skin. *Cell Transplantation* 10, 441-445.

Furukawa, K.S.; Ushida, T.; Toita, K.; Sakai, Y. & Tateishi, T. (2002). Hybrid of Gel-cultured smooth muscle cells with PLLA sponge as a scaffold towards blood vessel regeneration. *Cell Transplantation,* 11, 5, 475-480.

Furukawa, K.S.; Suenaga, H.; Toita, K.; Numata, A.; Tanaka, J.; Ushida, T.; Sakai, Y. & Tateishi, T. (2003). Rapid and large-scale formation of chondrocytes aggregates by rotational culture. *Cell Transplantation* 12,5, 475-479.

Furukawa, K.S.; Imura, K., Tateishi, T. & T Ushida. T. (2008). Scaffold-free cartilage by rotational culture for tissue engineering. *Journal of Biotechnology* 133, 134-145.

Grogan, S.P.; Reiser, F.; Winkelmann, V.; Berardi, S.; Mainil-Varlet, V. (2003). A static, closed and scaffold-free bioreactor system that permits chondrogenesis in vitro. *OsteoArthritis and Cartilage* 11, 403-411.

Hay, W.; Lemonnier, J.; Modrowski, D.; Lomri, A.; Lasmoles, F. & Marie, P.J. (2000). N- and E- cadherin mediate early human calvaria osteoblast differentiation promoted by bone morphogenetic protein-2. *Journal of Cellular physiology* 183, 117-128.

Hodge, W.A.; Fijan, R.S.; Carlson, K.L.; Burgess, R.G.; Harris, W.H. & Mann, R.W. (1986). Contact pressures in the human hip joint measured in vivo. *Proc Natl Acad sci USA*.83, 2879- 883.

Johnstone, B.; Hering, TM.; Caplan, AI.; Goldberg, VM. & Yoo, JU. (1998). In vitro chondrogenesis of bone marrow-derived mesenchymal progenitor cells. Experimental *Cell Research* 238, 265-272.

Kawanishi, M.; Ohura, A.; Furukawa, K.; Tateishi, T.; Fukubayashi, T. & Ushida, T. (1997). Redifferentiation of dedifferentiated bovine articular chondrocytes enhanced by cyclic hydrostatic pressure under a gas-controled system. *Tissue Engineering* 13, 5, 957-964.

Kelm, J. & Fussenegger, M. (2004). Microscale tissue engineering using gravity-enforced cell assembly. *Trends Biotechnol* 22, 4, 195-202.

Koch, R.J. & Gorti, G..K. (2002). Tissue engineering with chondrocytes. *Facial Plast. Surg.* 18, 59-68.

Lunstrum, G.P.; Douglas, R.; Keene, D.R.; Weksler, N.B.; Cho, Y.J.; Cornwall, M. & Horton, W.A. (1999). Chondrocyte differentiation in a rat mesenchymal cell line. *The Journal of Histochemistry & Cytochemistry* 47, 1, 1-6.

Mackay, A.M.; Beck, S.C.; Murphy, J.M.; Barry, F.; Chichester, C.O. & Pittenger, M.F. (1998). Chondrogenic differentiation of cultured human mesenchymal stem cells from marrow. *Tissue Engineering* 4, 4, 415-428.

Maini-Varlet, P.; Rieser, F.; Grogan, S.; Mueller, W.; Saager, C. & Jakob, R.P. (2001). Articular cartilage repair using a tissue-engineered cartilage-like implant: an animal study. *Osteroarthritis and Cartilage* 9 Supplement A, S6-S15.

Marlovits, S.; Brigitte, T.; Michaela, T.; Gruber, D. & Vecsei, V. (2003). Chondrogenesis of aged human articular cartilage in a scaffold-free bioreactor. *Tissue Engineering* 9, 6, 1215-1226.

Matsuda, K.; Sah, R.L.; Hejna, M.J. & Thonar, E.J.-M.A. (2003). A novel two - step method for the formation of tissue-engineered cartilage by mature bovine chondrocytes: the alginate-recovered-chondrocyte (ARC) method. *J Orthopaedic Research* 21, 139-148.

Nagai, T.; Furukawa, K.S.; Sato, M.; Ushida, X. & Mochida, J. (2008). Characteristics of a scaffold-free articular chondrocyte plate grown in rotational culture. *Tissue Eng* 14(7):1183-93.

Nagai, T.; Sato, M.; Furukawa, K.S.;, Kutsuna, T.; Ohta, N.; Ushida, T. & Mochida, J. (2008).Optimization of allograft implantation using scaffold-free chondrocyte plates. Tissue Eng Part A. 14, 7,1225-35.

Sato, M.; Ishihara, M.; Furukawa, K.; Kaneshiro, N.; Nagai, T.; Mitani, G.; Kutsuna, T.; Ohta, N.; Kokubo, M.; Kikuchi, T.; Sakai, H.; Ushida, T.; Kikuchi, M. & Mochida, J. (2008). HRecent technological advancements related to articular cartilage regeneration. Med Biol Eng Comput. 46, 8, 735-43.

Mizuno, S.; Tateishi, T.; Ushida, T. & Glowachi, J. (2002). Hydrostatic fluid pressure enhances matrix synthesis and accumulation by bovine chondrocytes in three-dimensional culture. *J Cell Physiol* 193, 319-327.

Oberlender, S.A. & Tuan, R.S. (1994). Spatiotemporal profile of N-cadherin expression in the developing limb mesenchyme. Cell Adhesion and Communication 2, 521-537.

Park, K.; Huang, J.; Azar, F.; Jin, R.L.; Min, B.H.; Han, D.K. & Hasty, K. (2006). Scaffold-free, engineered porcine cartilage construct for cartilage defect repair-in vitro and in vivo study. *Artificial Organs* 30, 8, 586-596.

Pittenger, M.R.; Mackay, A.M.; Beck, S.C.; Jaiswal, R.K.; Douglas, R.; Mosca, J.D.; Moorman, M.A.; Simonetti, D.W.; Craig, S. & Marshal, D.R. (1999). Multilinease potential of adulf human mesenchymal stem cells. *Science* 284, 143-147.

Smith, R.L.; Rusk, S.F.; Ellison, B.E.; Wessells, P.; Tsuchiya, K.; Carter, D.R.; Caler, W.E.; Sandell, L.J. & Schurman D.J. (1996). In vitro stimulation of articular chondrocytes mRNA and extracellular matrix synthesis by hydrostatic pressure. Journal of *Orthopaedic Research* 14, 53-60.

Smith, R.L.; Lin, J.; Trindade, M.C.D.; Shida, J.; Kajiyama, G.; Vu, T.; Hoffman, A.R.; van der Meulen, M.C.H.; Goodman, S.B.; Schurman, D.J. & Carter, D.R. (2000). Time-dependent effects hydrostatic pressure on articular chondrocytes type II collagen and aggrecan mRNA expression. *J of Rehabilitation Res and Dev* 37, 153-161.

Stoddart, M.J.; Ettinger, L. & Hauselmann, H.J. (2006a). Generation of a scaffold free cartilage-like implant from a small amount of starting material. *Journal of Cell Molecular Medicine* 10, 2, 480-492

Stoddart, M.J.; Ettinger, L. & Hauselmann, H.J. (2006b). Enhanced matrix syhthesis in de novo, scaffold free cartilage-like tissue subjected to compression and shear. Biotech Bioeng 95, 6, 1043-1051.

Summerbell, D. & Wolpert, L. (1972). Cell density and cell division in the early morphogenesis of the chick wing. *Nature New Biol* 239, 24-26.

Thorogood, P.V. & Hinchliffe, J.R. (1975). An analysis of the condensation process during chondrogenesis in the embryonic chick hind limb. *J Embryol Exp Morphol* 33, 581606.

Toyoda, T.; Seedhon, B.B.; Yao, J.Q.; Kirkham, J.; Brookes, W. & Bonass, W. (2003). Hydrostatic pressure modulates proteoglycan metabolism in chondrocytes seeded in agarose. *Arthritis & Rheumatism* 48, 10, 2865-2872. Tracy, R. & Reddi, A.H. (1994). Thyroxine is the serum factor that regulates morphogenesis of columnar cartilage from isolated chondrocytes in chemically defined medium. The *Journal of Cell Biology* 126, 5, 1311-1318.

Tsuchiya, K.; Furukawa, K. & Ushida, T. (2006). Microelements for cartilage tissue engineering. *Biomechanics at micro-and nanoscale levels. Vol. II,* Wada, H., (ed.), 87-95, World Scientific Publishing Co. Pte. Ltd., 981-256-746-1. Shingapore

Ushida, T.; Furukawa, K.; Toita, K. & Tateishi, T. (2002). Three-dimensional seeding of chondrocytes encapsulated in collagen gel into PLLA scaffold. *Cell Transplantation* 11, 489-494.

Ushida, T.; Furukawa, K.S.; Chen, G.; Tateishi, T. (2004). Engineering approaches to regulate cell differentiation and tissue regeneration. *Biomechanics at micro-and nanoscale levels Vol. I,* Wada, H., (Ed.), 91-99, World Scientific Publishing Co. Pte. Ltd., 981-256-098-X, Shingapore

von der Mark, K.; Gauss, V. & Muller, P. (1977). Relationship between cell shape and type of collagen synthesized as chondrocytes lose their cartilage phenotype in culture. *Nature* 265, 531-532.

Vunjak-Novakovic, G.; Freed, L.E.; Biron, R.J. & Langer, R. (1996). Effects of mixing on the composition and morphology of tissue-engineered cartilage. *AIChE Journal* 42, 3, 850-860.

Wakitani, S.; Kimura, T.; Hirooka, A.; Ochi, T.; Yoneda, M.; Yasui, N.; Owaki, H. & Ono, K. (1989). Repair of rabbit articular surfaces with allograft chondrocytes embedded in collagen gel. *J bone Joint Surg.* 71-B, 74-80.

Wakitani, S.; Goto, T.; Pineda, S.J.; et al., (1995). Mesenchymal cell-based repair of large, full-thickness defects of articular cartilage. *J. Bone Joint Surg. Am.* 76A, 579-592.

Wang, X.; Grogan, S.P.; Riser, F.; Winkelmann, V.; Maquet, V.; Berge, M.L. & ainil-Varlet, P. (2004). Tissue engineering of biphasic cartilage constructs using various biodegradable scaffolds: an in vitro study. *Biomaterials* 25, 3681-3688.

Williams, J.L.; Do, P.D.; Eick, J.D. & Schmidt, T.L. (2001). Tensile properties of the physis vary with anatomic location, thickness, strain rate and age. *Journal of Orthopaedic Research* 19, 1043-1048.

Yoo, J.U.; Barthel, T.S.; Nishimura, K.; Solchaga, L.; Caplan, A.I.; Goldberg, V.M. & Johnstone, B. (1998). The chondrogenic potential of human bone-marrow-derived mesencymal progenitor cells. The Journal of Bone and Joint Surgery 80, 12, 1745-1757.

Tissue engineering for meniscus regeneration

Elizaveta Kon*, Giuseppe Filardo*, Marco Delcogliano**, Giuseppe Peretti°, Alessandro Di Martino* and Maurilio Marcacci*
*IX Division, Biomechanics Lab, Rizzoli Orthopaedic Institute, Bologna, Italy
°Orthopaedic Department, Ospedale San Raffaele, Università Vita Salute, Milan, Italy
**Orthopaedics and Traumatology Department, Ospedale San Carlo di Nancy, Rome, Italy

1. Introduction

Menisci represent fundamental structures for the maintenance of knee homeostasis, playing a key role in knee biomechanics. Lesions of the meniscus are frequently observed in orthopedic practice. Injury to the meniscus is one of the most common problems in the knee joint, with a mean annual incidence of 60 to 70 per 100 000 knee injuries (Erkman et al. 1975; Hede et al 1990; Neilson et al 1991; Renstrom P 1990; Weinand et al 2006). Injury or loss of meniscal tissue leads to pain, knee dysfunction and osteoarthritis at long term (Cook et al 2005; Wyland 2002). Studies have demonstrated that knee degeneration is inversely related to the amount of meniscal tissue resected (McDermott and Amis 2006; Rijk 2004). Unfortunately, their intrinsic regenerative potential is poor. Healing is usually limited to the vascularized areas in the outer one third of the meniscus (Arnoczky and Warren 1983; Miller 1994). When a lesion involves the avascular portion of the meniscus, the reparative process cannot occur, since it emanates from bleeding and subsequent inflammation. As a result, the standard biological healing process produces limited results. Thus, a large proportion of meniscal tears observed at arthroscopy remain irreparable, and partial, subtotal or even total meniscectomy is often necessary, regardless of the recognized consequence. Even after an only partial meniscectomy, knee mechanics are subject to dramatic changes, leading often to the development of early osteoarthritis. Therefore, menisci should be repaired whenever possible.

In the last decades, tissue engineering approaches have been advocated to improve the reparative processes of joint tissues. The possibility to entirely reproduce the meniscus structure and function is highly attractive, and some new biomaterials have been studied and applied in preclinical and clinical studies.

2. Preclinical studies

In the last decade, replacement strategies have been applied to the menisci tissue in order to enhance their intrinsic reparative properties (Peretti et al. 1999; Peretti et al. 2004; Weinand et al. 2006) or even to replace them with an engineered tissue. Several materials have been tested as partial meniscus substitutes in preclinical studies. Veth et al (Veth et al. 1986) used carbon fiber for meniscus repair in dogs, with poor results. Small intestine submucosa (SIS) was successfully used to repair posterior vascular meniscal defects (Cook et al. 2006), but not for complete substitution. Total meniscus substitution remains difficult and has been poorly described in the literature. A polyvinyl alcohol-hydrogel meniscus in rabbits showed interesting results in terms of chondroprotection, but certain unresolved problems persisted: durability of the polymer, fixation method, complete tissue regeneration in a material that does not adhere to tissue (De Groot et al. 1997; Kobayashi et al.2003). Van Tienen et al (Van Tienen et al 2006) studied two different porous polyester urethane polymers as meniscus substitutes in dogs. Despite the promising results in tissue formation, one of the materials (aromatic 4,4,-diphenylmethanediisocynate) is thought to degrade into toxic products. Other polymer implants did not show to prevent cartilage degeneration or were not suitable as meniscal substitute because of poor tissue ingrowth related to polymer degradation rate and poor mechanical properties (Van Tienen et al. 2002; Van Tienen et al 2003). Tissue engineering has recently been proposed as a possible solution for meniscal regeneration. A few animal studies investigated the possibility of using cells in combination with different scaffold biomaterials for a partial or total meniscal substitution (Van Tienen et al. 2002; Van Tienen et al 2003; Van Tienen et al 2006; Weinand et al 2006; Chiari et al. 2006; Kon et al 2008).

Several polymers, both natural and synthetic, have been tested for engineering meniscal and cartilage tissue in vitro or in vivo. Among the synthetic scaffolds, favorable polymers are open lattice structures with large pores into which cartilage matrix is permitted to form and new synthetic hydrogels are also good candidate scaffolds for generating meniscal tissue.

Collagen sponges have many desirable properties as a biological scaffold, including porosity, biodegradability, and biocompatibility. Multiple methods for preparing collagen and collagen-GAG scaffolds have been reported. Generally, collagen scaffolds are made from animal tissues such as type I collagen from bovine tendon. It is also possible to chemically modify the biomechanical and biological properties of the collagen scaffolds to enhance certain characteristics that promote tissue formation (Speer et al. 1979). Open lattice collagen scaffolds, some of which also include glycosaminoglycans, have been synthesized and used for generating new matrix. Following the list of the potential biological scaffolds for meniscus repair, hyaluronan could play an important role. A resorbable biomaterial consisting of hyaluronic acid and polycaprolactone was also tested for total meniscal substitution (Kon et al; 2008).

The other critical element for engineering meniscal tissue is finding the most suitable cell source. Various cell-sources have been evaluated in vitro to find the most suitable source for cell augmentation of tissue engineered meniscus. Meniscal cells can be isolated by enzymatic digestion of the tissue. Histological evidence suggests that the meniscal cells are capable of generating fibrocartilagenous tissue resembling meniscus. It is unlikely, however, that autologous meniscal fibrochondrocytes will be used to engineer one's own meniscus because of the lack of expendable donor sites. It is conceivable that allogeneic meniscal cells could be used as an option for overcoming the donor site related problems. The central issue involving the use of allogeneic chondrocytes is their potential for eliciting an immune response once the extracellular matrix is removed and MHC antigens are exposed (Tiku et al. 1985).

Different cell lines have been proposed as alternative solution for meniscal tissue engineering. Chondrocytes could be used to generate meniscal tissue or used to induce meniscal repair. Articular chondrocytes showed the highest tissue regeneration capacity, with cellular and matrix phenotypes similar to those of both the inner and outer meniscus regions (Ibarra et al. 2000; Marsano et al. 2007).

Moreover, cartilage harvesting from a less loaded joint area would be less damaging to the knee biomechanics as compared to healthy meniscal tissue harvesting. Peretti et al (Peretti et al; 2004) have shown that articular chondrocytes seeded onto a devitalized allogenic meniscal slices or on a Vicryl mesh scaffold are capable of inducing a healing process in swine meniscus. Other alternative cartilage sources for obtaining cells include the cartilage portion of ribs or ear cartilage (Johonson et al 2004). MSCs are another valid candidate for meniscal bioengineering. Multiple experiments using human, chicken, dog, and rabbit MSCs have shown that, under controlled in vitro conditions, these cells can differentiate into bone, fat, tendon, muscle and cartilage-like tissues (Caplan et al. 2001; Pittenger et al. 1999; Conget et al. 1999).

All these cell sources could be used to engineer reparative meniscal tissue.

Walsh et al. (Wlash et al. 1999) used a collagenous sponge loaded with mesenchymal stem cells to treat a partial meniscus defect in rabbits. They reported that the presence of cells augmented the repair process but did not prevent knee degeneration. Martinek et al. (Martinek et al. 2006) documented better macroscopic and histological results in CMI implants seeded with meniscal fibrochondrocytes in comparison to cell-free implants in sheep. However, the tissue-engineered meniscus was biomechanically unstable and the implant size reduced during the three-month observation period. Therefore, the authors suggested that an improvement in scaffold and cell seeding procedure is required before human application.

Ibarra et al. also reported on a pilot study in sheep where they used autologous meniscal fibrochondrocytes and PGA polymer (Ibarra et al. 2000). The authors report that the constructs produced a new tissue with fibroblasts and chondrocytes. The presence of collagen fibers was observed histologically and the cells produced proteoglycans. The number of animals was small, but the authors demonstrated proof-of-principle for the technique. Polyurethane was initially developed and tested in animal studies for total meniscal replacement but is now being assessed as a scaffold for partial meniscal replacement as an alternative for the CMI (Van Tienen et al. 2003). Brophy and colleagues (Brophy et al. 2008) showed in sheep cadavers that the contact pressures after partial meniscectomy and replacement with a polyurethane scaffold were less than the contact pressures after partial meniscectomy only.

We have investigated the feasibility of using a new resorbable biomaterial consisting of hyaluronic acid and polycaprolactone for total meniscal substitution in a sheep model (Chiari et al. 2006; Kon et al. 2008). Twenty-four skeletally mature sheep were treated with total medial meniscus replacements while 2 meniscectomies served as empty controls. The animals were divided in two groups: cell free scaffold or scaffold seeded with autologous chondrocytes. Autologous chondrocytes were used as we have previously demonstrated that, both in vitro and following ectopic implantation, expanded articular chondrocytes were superior to meniscal cells, synovial or fat pad cells in their capacity to reach phenotypes typical of the inner and outer meniscus regions (Marsano et al. 2007). Two different surgical techniques were compared: in 12 animals the implant was sutured to the capsule and to the meniscal ligament and in the other 12 animals also a transtibial fixation of the horns was used. The animals were euthanized after 4 months. The specimens were assessed by gross inspection and histology. All implants showed excellent capsular ingrowth at the periphery. Marcoscopically, no difference was observed between Cell Seeded and Cell Free group. Better implant appearance and integrity was observed in the group without trans-osseous horn fixation. Using the latter implantation technique, lower joint degeneration was observed in the cell-seeded group with respect to cell-free implants. The histological analysis indicated cellular infiltration and vascularisation throughout the implanted constructs. Cartilaginous tissue formation was significantly more frequent in the cell-seeded constructs. Our study supports the potential of the Hyaff/PCL scaffold for total meniscal substitution and seeding of the scaffolds with autologous chondrocytes provides some benefit in the extent of fibrocartilaginous tissue repair.

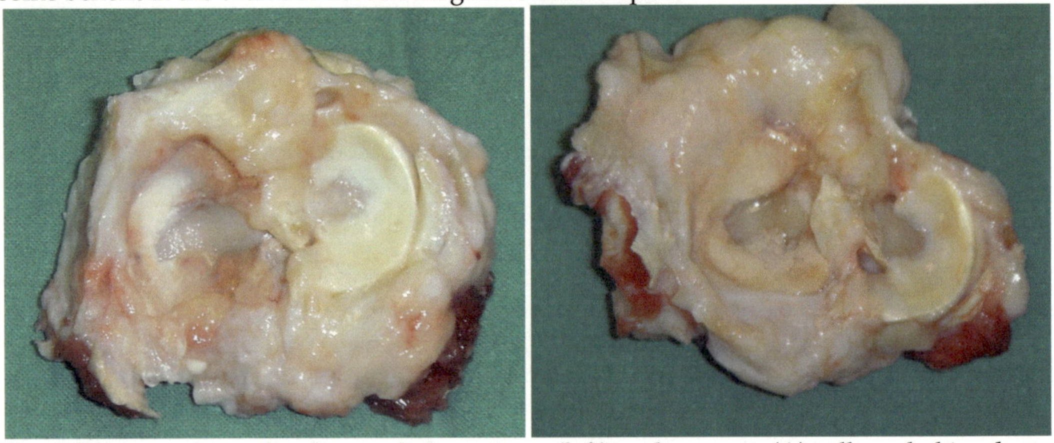

Fig. 1. Four month results after medial meniscus (left) replacement, (A) cell-seeded implants, (B) cell-free implants (left).

3. Clinical studies

Although the literature addresses merely the total replacement of the meniscus in animal studies, with or without cells, the current clinical studies only concern partial replacement of the resected meniscus with cell-free scaffolds. To the authors' knowledge, no clinical studies have been published in which the meniscus was replaced by a bioengineered meniscus.

Stone et al. (Stone et al. 1997) have developed a bioreadsorbable collagen matrix (CMI) which acts as a scaffold to restore the original medial meniscal. We studied (Zaffagnini et al. 2007) and prospectively evaluated the results of CMI implantation at a follow-up from a minimum of 6 to a maximum of 8 years. Eight patients (mean age 25) were evaluated at a final observation point from 6 to 8 years after CMI implantation. Inclusion criteria were an irreparable meniscal tear or a previous meniscectomy involving the medial meniscus. There were no complications related to the device. All patients were able to return to daily activities without limitations 3 months after surgery. Both subjective CKRS score and objective IKDC score showed improvement in all cases except one patient with an ACL re-injury. In two cases scores were slightly worse from 2 years after surgery to the final observation point. The other five cases obtained maximum score at final follow-up. In four cases the absence of pain remained until the final observation point, while in four cases a low entity of pain was described at long term follow-up. MRI showed in five cases mixoid degeneration signal, two had normal signal with reduced size, while one patient had no recognizable implant. Six patients had preserved cartilage and articular space, with no changes compared to pre-op control. Arthroscopic second look evaluation has been performed in three cases, revealing in two cases the presence of the implant, although with a reduced size compared to the original one, while in one case the CMI was almost disappeared. Our small series of eight patients prospectively followed from 6 to 8 years of follow-up has shown highly satisfactory results. Although the aspect of the implant was mostly abnormal, the implant may have helped reduce the deterioration of the knee joint at final observation time.

Recently, a new polyurethane scaffold, presented by Brophy and colleagues (Brophy et al. 2008), has been introduced in clinical practice. The polyurethanes are believed to have better material properties For fixation to the remaining native meniscus tissue and to resist the extreme forces within the knee joint. A prospective clinical trial, which is being performed at this moment, should reveal if this hypothesis would hold. The preliminary results seem promising.

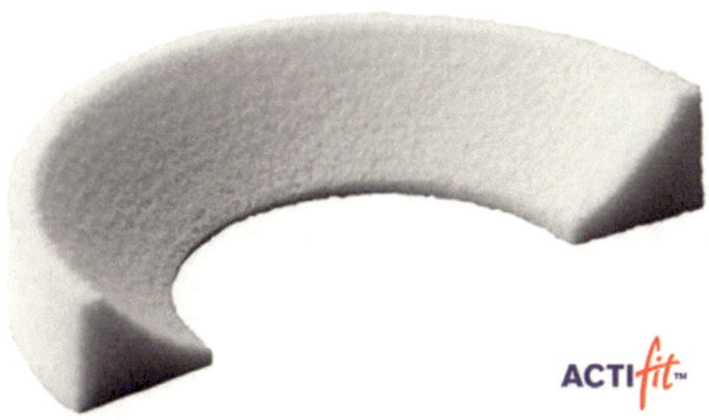

Fig. 2. New polyurethane scaffold. (Courtesy of Orteq Bioengineering)

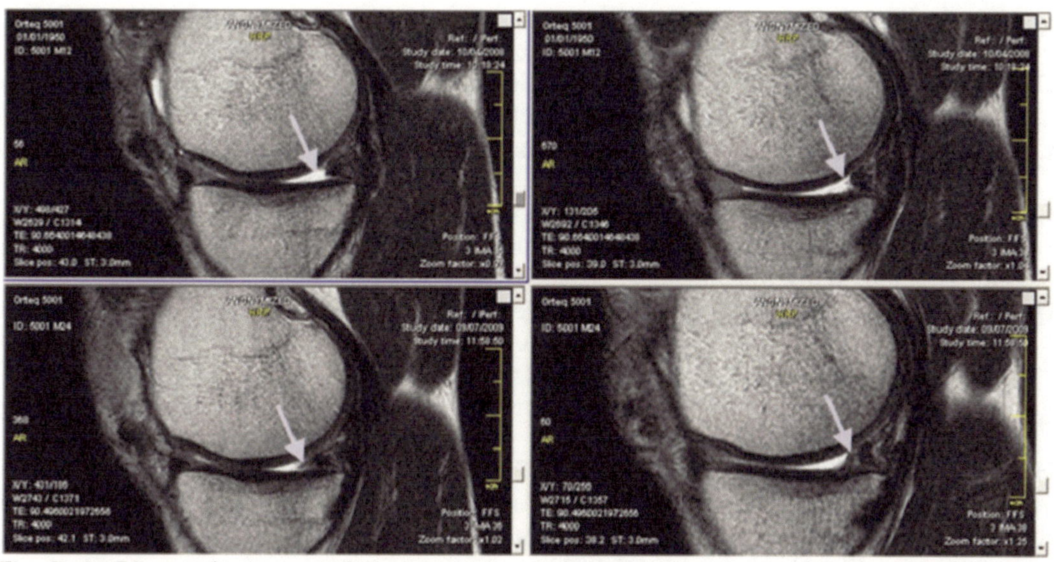

Fig. 3. At 24 months increased tissue can be seen anterior to the residual rim of the native meniscus when compared to the month 12 findings.

4. Conclusion

Subtotal or total meniscectomy is no longer considered a valid option for the treatment of meniscal tears, while preservation or regeneration of this tissue is now recommended, when possible, for the maintenance of a healthy knee. Menisci have poor intrinsic healing potential so that techniques based on fixation of the avascular regions of torn menisci often fail to allow for biological repair of the tissue. Until now, the properties of this tissue seemed hard to mimic. Tissue engineering for meniscal replacement represents an innovative and promising solution. Many materials have been proposed and tested in preclinical studies, but for the time being meniscus regeneration is still in its infancy, and further studies need to confirm the potential of the tissue engineering approach.

5. Acknowledgements

Angela Montaperto, Giulio Altadonna, Federica Balboni, Silvia Bassini: Biomechanics Lab, Rizzoli Orthopaedic Institute, Bologna, Italy

6. References

Arnoczky, S.P., and Warren, R.F. The microvasculature of the meniscus and its response to injury. An experimental study in the dog. Am J Sports Med 11, 131-41, 1983.

Brophy R, Cottrell J, Deng XH, et al. Dynamic contact mechanics of a scaffold for partial meniscal replacement [abstract 125]. In: Proceedings of the 75th Annual Meeting of the AAOS. San Francisco: American Academy of Orthopaedic Sur- geons; 2008.

Caplan AI, Bruder SP. Mesenchymal stem cells: building blocks for molecular medicine in the 21st century. Review. Trends Mol Med 7(6):259-64, 2001.

Chiari, C., Koller, U., Dorotka, R., Eder, C., Plasenzotti, R., Lang, S., Ambrosio, L., Tognana, E., Kon, E., Salter, D., and Nehrer, S. A tissue engineering approach to meniscus regeneration in a sheep model. Osteoarthritis Cartilage 14, 1056-65, 2006.

Conget PA, Minguell JJ. Phenotypical and functional properties of human bone marrow mesenchymal progenitor cells. J Cell Physiol 181(1):67-73, 1999.

Cook, J.L. The current status of treatment for large meniscal defects. Clin Orthop Relat Res , 88-95, 2005.

Cook, J.L., Fox, D.B., Malaviya, P., Tomlinson, J.L., Kuroki, K., Cook, C.R., and Kladakis, S. Long-term outcome for large meniscal defects treated with small intestinal submucosa in a dog model. Am J Sports Med 34, 32-42, 2006.

De Groot, J.H., Zijlstra, F.M., Kuipers, H.W., Pennings, A.J., Klompmaker, J., Veth, R.P., and Jansen, H.W. Meniscal tissue regeneration in porous 50/50 copoly(L-lactide/epsilon-caprolactone) implants. Biomaterials 18, 613-22, 1997.

Erkman MJ. The role of the menisci in force transmission across the knee. Clin Orthop Relat Res. 1975;109:184-192.

Hede A, Jensen DB, Blyme P, Sonne-Holm S. Epidemiology of meniscal lesions in the knee: 1,215 open operations in Copenhagen 1982-84. Acta Orthop Scand. 1990;61:435-437.

Ibarra, C., Koski, J.A., and Warren, R.F. Tissue engineering meniscus: cells and matrix. Orthop Clin North Am 31, 411-8, 2000.

Johnson TS, Xu JW, Zaporojan VV, Mesa JM, Weinand C, Randolph MA, Bonassar LJ, Winograd JM, Yaremchuk MJ. Integrative repair of cartilage with articular and non-articular chondrocytes. Tissue Eng 10(9-10):1308-15, 2004

Kobayashi, M., Toguchida, J., and Oka, M. Preliminary study of polyvinyl alcohol-hydrogel (PVA-H) artificial meniscus. Biomaterials 24, 639-47, 2003.

Kon E, Chiari C, Marcacci M, Delcogliano M, Salter DM, Martin I, Ambrosio L, Fini M, Tschon M, Tognana E, Plasenzotti R, Nehrer S. Tissue engineering for total meniscal substitution: animal study in sheep model. Tissue Eng Part A. 2008 Jun;14(6):1067-80

Martinek, V., Ueblacker, P., Bräun, K., Nitschke, S., Mannhardt, R., Specht, K., Gansbacher, B., and Imhoff, A.B. Second generation of meniscus transplantation: in-vivo study with tissue engineered meniscus replacement. Arch Orthop Trauma Surg 126, 228-34, 2006.

Marsano, A., Millward-Sadler, S.J., Salter, D.M., Adesida, A., Hardingham, T., Tognana, E., Kon, E., Chiari-Grisar, C., Nehrer, S., Jakob, M., and Martin, I. Differential cartilaginous tissue formation by human synovial membrane, fat pad, meniscus cells and articular chondrocytes. Osteoarthritis Cartilage 15, 48-58, 2007.

McDermott, I.D., and Amis, A.A. The consequences of meniscectomy. J Bone Joint Surg Br 88, 1549-56, 2006.

Miller MD, Warner JJ P, Harner CD . In: Fu FH, Harner CD, Vince KG Eds. Knee Surgery. Williams and Wilkins, 1994, .

Neilson AB, Yde J. Epidemiology of acute knee injuries: a prospective hospital investigation. J Trauma. 1991;31:1644-1648.

Peretti GM, Bonassar LJ, Caruso EM, Randolph MA, Zaleske DJ Biomechanical analysis of a cell-based model for articular cartilage repair. Tissue Eng 1999; 5(4):317–326

Peretti GM, Gill TJ, Xu JW, Randolph MA, Morse KR, Zaleske DJ Cell-based therapy for meniscal repair: a large animal study. Am J Sports Med 2004; 32(1):146–158

Pittenger MF, Mackay AM, Beck SC, Jaiswal RK, Douglas R, Mosca JD, Moorman MA, Simonetti DW, Craig S, Marshak DR. Multilineage potential of adult human mesenchymal stem cells. Science 284(5411):143-7, 1999.

Renstrom P, Johnson RJ. Anatomy and biomechanics of the menisci. Clin Sports Med. 1990;9:523-538.

Rijk, P.C. Meniscal allograft transplantation--part II: alternative treatments, effects on articular cartilage, and future directions. Arthroscopy 20, 851-9, 2004.

Speer DP, Chvapil M, Volz RG, Holmes MD. Enhancement of healing in osteochondral defects by collagen sponge implants. Clin Orthop 144: 326-335, 1979.

Stone KR, Steadman JR, Rodkey WG, et al. Regeneration of meniscal cartilage with use of a collagen scaffold. Analysis of preliminary data. J Bone Joint Surg Am 1997;79:1770–7.

Tiku ML, Liu S, Weaver CW, Teodorescu M, Skosey JL. Class II histocompatibility antigen-mediated immunologic function of normal articular chondrocytes. J Immunol 135(5):2923-2928, 1985.

Van Tienen, T.G., Heijkants, R.G.J.C., Buma, P., De Groot, J.H., Pennings, A.J., and Veth, R.P.H. Tissue ingrowth and degradation of two biodegradable porous polymers with different porosities and pore sizes. Biomaterials 23, 1731-8, 2002.

Van Tienen, T.G., Heijkants, R.G.J.C., Buma, P., De Groot, J.H., Pennings, A.J., and Veth, R.P.H. A porous polymer scaffold for meniscal lesion repair--a study in dogs. Biomaterials 24, 2541-8, 2003.

Van Tienen, T.G., Heijkants, R.G.J.C., de Groot, J.H., Schouten, A.J., Pennings, A.J., Veth, R.P.H., and Buma, P. Meniscal replacement in dogs. Tissue regeneration in two different materials with similar properties. J Biomed Mater Res B Appl Biomater **76**, 389-96, 2006

Veth, R.P., Jansen, H.W., Leenslag, J.W., Pennings, A.J., Hartel, R.M., and Nielsen, H.K. Experimental meniscal lesions reconstructed with a carbon fiber-polyurethane-poly(L-lactide) graft. Clin Orthop Relat Res. 1986;286-93,

Weinand C, Peretti GM, Adams SB Jr, Bonassar LJ, Randolph MA, Gill TJ.An allogenic cell-based implant for meniscal lesions. Am J Sports Med. 2006 Nov;34(11):1779-89.

Wyland, D.J., Guilak, F., Elliott, D.M., Setton, L.A., and Vail, T.P. Chondropathy after meniscal tear or partial meniscectomy in a canine model. J Orthop Res **20**, 996-1002, 2002.

Walsh, C.J., Goodman, D., Caplan, A.I., and Goldberg, V.M. Meniscus regeneration in a rabbit partial meniscectomy model. Tissue Eng 5, 327-37, 1999.

Zaffagnini S, Giordano G, Vascellari A, et al. Arthroscopic collagen meniscus im- plant results at 6 to 8 years follow up. Knee Surg Sports Traumatol Arthrosc 2007; 15:175–83.

Tissue engineering of the anterior cruciate ligament and meniscus using acellularized scaffolds

Gunther H. Sandmann and Thomas Tischer
Department of Orthopaedic and Trauma Surgery
Technische Universität München
Germany

Abstract

Injuries to the anterior cruciate ligament (ACL) and the meniscus are very common. Both the ACL and the meniscus play a crucial role in the complex biomechanics of the knee and operative treatment is often necessary. For example, in the US approximately 200,000 ACL reconstructions are performed annually. Usually, autogenous tendons like the mid- third of the patellar tendon or the hamstring tendons (semitendinosus/gracilis) are used. Nevertheless, the availability of autogenous tendons for ACL reconstruction is limited as well as the possibilities of meniscus repair. Ligament and meniscus allografts often show problems regarding long-term stability, immunological reactions and possible transmission of infectious diseases, whereas synthetic materials can cause foreign body reactions and often lack good initial biomechanical stability or fail to maintain long-term stability.

The idea of acellularizing ligaments and meniscus cartilage has arisen to remove the immunogenic cells to reduce the adverse immunological reactions. By preserving the extracellular matrix (e. g. collagen, glycosaminoglycans, proteoglycans), the main biomechanical properties are preserved and the so obtained scaffolds provide a natural environment for the ingrowing cells. The acellular scaffolds can be used as a scaffold for *in vivo* repopulation or can be seeded with autologous cells *in vitro* before implantation using different cell sources. By this, further improvement of the biochemical and biomechanical properties as well as the remodelling of the graft could be improved.

In the meantime, a variety of different methods has been described to acellularize tissues reaching from repetitive freezing/ thawing over the use of hydrostatic high pressure to different chemical methods. Only by chemical means are the cells actually removed. In terms of chemical acellularization, SDS (sodium dodecyl sulphate), Triton-X or TnbP (tri-nitro-butyl-phosphate) are the most popular substances, with SDS being the strongest cell removal agent for tendon and meniscus tissue. As *in vitro* tests have shown, tissue processing using SDS does not influence the biomechanical properties, while all cells - being responsible for the immunological effects - are removed. However, more research using *in vivo* animal experiments has to be performed before application in humans is possible.

This tissue engineering strategy might be suitable to satisfy the increasing demand for tissue engineered tissue coming along with the increased patients´ demands in the future.

1. Introduction

Injuries to the ACL and the meniscus are very common and negatively influence the complex biomechanics of the knee joint. Due to the poor vascularity, both structures show limited healing capacity and surgical repair is often necessary (Arnozcky 1983, Arnozcky 1983). In the US approximately 200,000 ACL reconstructions are performed per year and the number is still growing. The golden standard is the reconstruction of the injured ACL by using autogenous tendons like the hamstring tendons (semitendinosus/gracilis) or the middle- third of the patellar or quadriceps tendon. These methods have been used for several years now and show reliable results in the long- term. However, the problem of development of osteoarthrosis in the long term is not solved yet (Lohmander 2007, Messmer 1999). So far no study exists proving a chondro-protective effect after ACL reconstruction. The reported outcome 32 years after ACL replacement is unacceptable in about 50% of cases. In accompanying meniscus lesions the outcome is even worse (Taylor 2009). Whether this is mainly due to cartilage injury at the time of accident or due to altered joint biomechanics is not known yet. So far, there exists no clear consensus, which patients benefit from ACL replacement in the long run. New techniques of ACL reconstruction like the anatomic double-bundle reconstruction (anteromedial and posterolateral bundle) and new methods of fixation might improve the outcome in the future (Kondo 2008, Zantop 2008, Brucker 2006).
Though the ACL reconstruction using autogenous tendons is currently the gold standard, some problems still remain. There is on the one hand the problem of donor site morbidity with persisting anterior knee pain and the inability to kneel, especially after the use of patellar tendon transplants (Kartus 2001). Muscle strength deficits have been found after ACL reconstruction with patellar and hamstring grafts (Aune et al. 2001; Beard et al. 2001), especially the harvesting of hamstring grafts has been found to cause muscular weakness at high knee flexion angles (Makihara et al., 2006; Nakamura et al., 2002; Tashiro et al., 2003). Furthermore the availability of autogenous tendons is limited. In multi-ligamentous knee injuries the use of allografts is often required. Different allograft ligaments may be used: bone-patella-tendon-bone, hamstring, achilles or tibialis anterior/posterior tendons. However, no differences between the used allograft tissue have been reported, whereas significant differences exist, which sterilization processes were used (Carey et al. 2009; Krych et al. 2007). In a recent systematic review comparing ACL replacement using autografts with allografts, similar medium term results were reported in non-randomized studies (Carey et al., 2009). Still, immunogenic reactions may cause degenerative alterations and the potential risk of disease transmission remains. Using standard sterilization procedures like lyophilisation (Crawford et. al, 2004) or gamma irradiation (Fideler et al. 1995; Gibbons et al. 1991) retroviruses (i.e. HIV) can hardly be destroyed.
Apart from the ligaments, the meniscus plays a crucial role in the complex biomechanics of the knee, as it works as a shock absorber and transforms joint pressure in circular meniscus tension. This is mediated by the complex structure (mainly its circular collagen network) of the meniscus. Unfortunately its healing capacities are very poor. As soon as a radial tear destroys the circular collagen network or the horns are ruptured, the meniscus looses all biomechanical function. This leads to joint overload and might end in an overuse of the

cartilage and an early onset of osteoarthritis (Lohmander et al. 2007; Allaire et al. 2008). Many improvements have been made to reconstruct the injured meniscus, but due to its biomechanical and anatomical specialities this is still impossible in many cases.

The transplantation of meniscus allografts is a viable option but has its own problems as well (Packer et al. 2009). On the one hand there is the problem that there are too little donor organs as the meniscus of elderly people has usually degenerative alterations and can thereby not be used. Apart from the lack of organs it is difficult to get a suitable transplant fitting in size. In a recent work Van Thiel et al. 2009 could show that the fitting size of the meniscus is mainly correlated with height, weight and gender of the donor and recipient. Apart from that, the fixation is technically demanding. Because of this, the transplantation of meniscus is reserved to specialized centers. Unfortunately, long-term results still show problems due to immunogenic and degenerative alterations. Verdonk et al. (2006) examined the long- term results of meniscus transplantation clinically and radiologically and they could show, that the meniscus transplantation lead to an improvement of function and relief of pain. Nevertheless, substantial disability and knee pain were present at the follow up after 10 years. On MRI the meniscus transplants had signs of graft extrusion and an increased signal intensity indicating degeneration. The failure rate was 18% with conversion to a total arthroplasty (Verdonk 2006). Wirth et al. (2002) found similar results but could show, that deep frozen meniscal grafts were superior compared to lyophilized meniscal grafts at a follow up of 3 and 14 years. The latter even showed similar results to the control group after meniscectomy. If those problems could be solved in the future, the meniscus transplantation might gain much popularity over time.

As an alternative to allografts the use of synthetic tendon constructs (e.g. silk, Gore Tex, LAD) or meniscus prostheses (CMI®, Actifit® etc.) have emerged (Martinek 2006; van Tienen et. al 2009; Zaffagnini et al., 2008). Most of the prostheses used for ACL reconstruction have failed as they showed an adverse effect on the cartilage and caused in some cases a marked synovialitis. Though their biomechanical Properties were excellent initially, most of the prostheses had the problem of degradation, tunnel widening and high failure rates after several years (Muren et al., 2005; Roolker et al., 2000). Compared to the biomechanical properties of ligaments, meniscus tissue is even more complex and not surprising, the middle to long term results of the artificial meniscus constructs show different results from one study to the other, but in general do not reach the anatomic, biomechanical results (Buma et al., 2007). So far, artificial constructs (neither ACL nor meniscus) do not reach the biomechanical properties of autologous tissue and are therefore not an ideal long- term alternative.

To improve the long-term outcome of ligament and meniscus allografts, new strategies are needed. Recently the use of extracellular matrix (ECM) scaffolds by means of acellularization of allograft tissue has become popular (Badylak et al., 2007, Badylak et al., 2009). In the following we will depict the different means of acellularization, describe our used method and give future prospects in the field of tissue engineering concerning ligament and meniscus tissue.

2. Acellularization methods

Recently, biologic scaffold materials composed of extracellular matrix are widely used in surgical procedures for the reconstruction of numerous tissues and organs and their use in orthopaedic research surgery has emerged over the last years. As scaffolds allograft or possibly xenograft tendons/meniscus are used, which are acellularized by different methods.

The basic requirements for optimal acellularization methods are:

1. removal of all cellular components as the cells (especially fibroblasts) and macrophages seem to play a crucial role in the cell- mediated immunogenic reactions, i.e. host versus graft reaction (Badylak 2008).

2. preserving the extracellular matrix which is responsible for biomechanical stability and preserving chemoatractants and providing cell attachment sites for ingrowing cells (Reing 2009).

The ECM has chemotactic factors that attract cells (Beattie et al., 2009) and supportive function for the diffusion of nutrients from the blood to the cells. Most important, the biomechanical properties are mediated by the extracellular matrix. For example, the tensile strength of tendons is mediated by its collagen I fibrils, whereas the hyaline cartilage properties are maintained by the interplay of a collagen II network with large proteoglycans (aggrecan) and negatively charged glycosaminoglycans (Tischer et al., 2002). Finally, the process of repopulation is dependent on an intact matrix and the used method should not be cytotoxic to allow cellular ingrowth. The repopulation can then happen *in vitro* before implantation or *in vivo*.
Therefore different methods have been described for the acellularization of tissues and every method has its own advantages and disadvantages. Generally, two methods can be distuingished: first methods which destroy all cells, but leave cellular components and debris and second methods which remove all the cellular components from the tissue by chemical means. In the following section, the most common methods are now discussed in more detail.

2.1. The Gamma irradiation

The gamma irradiation is widely used and one of the methods with the longest experience. It is very effective in destroying cells and most of the pathogenic germs. Still the gamma irradiation faces a dilemma, as doses of 3.5 Mrad (35kGy) are needed to destroy the DNA of the HIV. However, using this amount of energy weakens the transplant, leading to loss of biomechanical stability, making it unsuitable for implantation (Fideler BM et al., 1995). In most cases irradiation doses of 1 to 2.5 Mrad (10- 25kGy) are used with a slight remaining risk of disease transmission (Buck et al., 1989). In a study by Curran it could be shown that irradiation of allografts with a dose of 20kGy lead to minor biomechanical properties compared to the untreated control group. Elongation was more than 27% higher and the load to failure was diminished as well (Curran et al., 2009). In a prospective randomized trial by Sun (Sun et al., 2009) the clinical outcome after ACL replacement with autograft patellar-tendon-bone and irradiated as well as non-irradiated allograft (all used allografts had been frozen) was examined. Whereas the non-irradiated allografts showed similar results to the autografts, the clinical outcome of the irradiated allografts was inferior in the clinical scores and the subjective instability was much more common in this group of patients. Taking that into account the irradiation can no longer be recommended as a sterilization procedure as it weakens the tissue and thereby negatively influences the biomechanical properties without reliably destroying retroviruses like HIV.

2.2. Lyophilization/cryopreservation

Lyophilization means the preparation of tissue by repetitive freezing and thawing. This leads to the destruction of cells, but as with irradiation the cellular components mainly remain within the extracellular matrix. The lyophiliation process has been widely used and has been proven as a simple and cheap method and is therefore most commonly used (Sass et al., 2009; Busam et. al, 2007; Lewis et al., 2008). Much work has been performed on the effects of cryopreservation on tendon allografts, but the potential adverse effects on mechanical properties and histological changes by ice crystal formation (Mahirogullari 2007). In a work by Park 2008 the effects of cryopreservation or heating on the mechanical properties and histomorphology of rat bone-patellar tendon-bones (BTBs) were investigated. BTBs were processed by cryopreservation at -80°C for 3 weeks, or heating at +80°C for 10 min. Tensile testing and histomorphological examinations were performed. The cryopreservation of tendons showed less influences on their mechanical properties. When cryopreserved BTBs were fixed in frozen state by the freeze-substitution method, many widened interfibrillar spaces were observed. These results suggest that the collagen fibres of cryopreserved tendons were histomorphologically affected by ice crystals. The cryopreservation does not only affect the cellular components, but also the collagen matrix, which might have effects on the biomechanical properties as well.
However, a new work by Gelber et al. (2009) showed that the ultra structure of the meniscus is not affected by cryopreservation. Therefore, an allograft stored in that way would not alter its biomechanical properties, although its cellular viability is highly unpredictable. As the immunogenicity is cell-mediated, this imponderability makes the cryopreservation not the method of choice for tissue engineering applications, especially as the fibril diameters in frozen menisci show a thinner diameter and had a higher degree of disarray. These alterations of the collagen network can partially explain the pathological changes found in shrunken menisci after transplantation (Gelber et al., 2008).

2.3. High pressure treatment

The treatment of tissue with hydrostatic high pressure has emerged over the last years. By this method tissue is exposed to hydrostatic high pressure of up to 600MPa for 10 minutes. Diehl et al. (2006) could show that this treatment had no adverse effect on the biomechanical properties of the treated tissue. The cells were destroyed by this treatment and similar to the lyophilization the cellular components remain in the extracellular matrix and might thereby still be immunogenic. Apart from the use in tissues, high hydrostatic pressure (HHP) is widely used in the food processing industry, for example to inactivate vegetative microorganisms in meat products, milk and juice, thereby avoiding the addition of any chemical preservatives. Besides this HHP is also an attractive novel approach to effectively kill tumor cells in bone, cartilage or tendon *ex vivo* while leaving the tissues' mechanical properties unimpaired, thus allowing possible reimplantation of the resected tissue explants (Diehl et al., 2006, Diehl et al., 2008). However, human studies on this topic are still missing.

2.4. Enzymatic treatment

Acellularization using enzymatic treatment uses various enzymes like trypsin, collagenase and others to open up the ECM and remove cellular components. For example Maier et al. described an enzymatic way of acellularization of ovine meniscus tissue (Maier et al. 2007). Forty-one menisci were used to establish their own protocol. They used distilled water, trypsin and collagenase A to acellularize meniscus samples and could show that the biomechanical and the biochemical properties were left almost unaffected. In a recently published work by Vavken et al. 2009, they could demonstrate the effectiveness of enzymatic acellularization of the ACL. Therefore, they used a combination of Trypsin, EDTA and PBS and could show that the samples could be acellularized by this enzymatic treatment. The glycosaminoglycan content was negatively affected as well.

Taking everything into account, we believe that enzymatic treatment and especially the combination of collagenase and trypsin has adverse effects on the extracellular matrix and the collagen network. These enzymes can actually destroy the ECM and this could cause problems in reseeding the acellularized samples or even the biomechanical results. Future research would be needed to prove the safety of these methods.

2.5. Chemical acellularisation by solvents/detergents

A newer promising approach is the extraction of cells by chemical solvents/detergents. The most widely used chemicals are sodium dodecyl sulphate (SDS, an ionic detergent), Triton-X (a non-ionic detergent) and tri-nitro-butyl phosphate (TnBP). They have been described to acellularize tissue and successful application was performed on small intestine (Badylak et al., 1995), the urinary bladder (Chen et al., 1999), the liver (Brown et al., 2006), arterial vasculature (Walles et al., 2003), dermis (Armour et al., 2006), heart valves (Cebotari et al., 2006) and oesophagus transplants (Bhrany et al. 2006; Gilbert et al., 2006) among others. In orthopaedic surgery there have been experiments using Graft jacket® in irreparable rotator cuff defects for tendon augmentation (Barber et al., 2006; Coons et al., 2006). Acellularization of tendons (Tischer et al., 2007; Gilbert et al., 2006) and meniscal samples (Sandmann et al. 2008; Stapleton et al., 2008) has caused further interest in this technique.

Bolland et al. aimed to produce a natural, acellular matrix from porcine bladder tissue for use as a scaffold in developing a tissue-engineered bladder replacement. They used full-thickness, intact porcine bladders, which were acellularised by distention and immersion in hypotonic buffer containing 0.1% (w/v) SDS and nuclease enzymes. Histological analysis of the resultant matrices showed that they were completely acellular whereas the major structural proteins had been retained. Intracellularly, some poorly soluble proteins remained. The amount of DNA per mg dry weight of fresh porcine bladder was 2.8 (+/-0.1) μg/mg compared to 0.1 (+/-0.1) μg/mg in acellularised bladder. In biomechanical testing the tensile testing indicated that acellularisation did not significantly compromise the tensile strength of the tissue. Cytotoxicity assays using porcine smooth muscle cell cultures excluded the presence of soluble toxins in the biomaterial (Bolland et al., 2007). A newer work by Gilbert et al. has shown that the presence of remaining intracellular DNA might not have as adverse effects as at first suspected. They examined the DNA content of commercially available ECM scaffold materials and compared their DNA content with scaffolds prepared in the laboratory using different acellularisation protocols. Most of the examined scaffolds still contained some amounts of DNA (4 out of 6 from the commercially available scaffolds and 2 out of 3 from the produced ECM scaffolds). Interestingly, scaffolds

which retained higher amounts of DNA did not show worse results than the scaffolds with less DNA (Gilbert et al., 2008). In a work by Brown et al., it could be shown in an animal model that the presence of cellular components within the extracellular matrix scaffold modulates the phenotype of the macrophages participating in the host response following implantation. Scaffolds containing cellular components (even autologous cellular components), elicited a predominantly M1 type macrophage response, resulting in a more dense connective tissue and/or scarring. Acellular scaffolds, however, were observed to elicit a predominantly M2 type macrophage response, which results in a more constructive type of remodelling response (Brown et al., 2009).
In the same animal model (abdominal wall defect in rats covered with several ECM scaffolds, i.e. porcine small intestinal submucosa (SIS), noncrosslinked SIS, and autologous body wall) the effect of macrophages that derive from peripheral blood in the degradation of ECM scaffolds was examined and Valentine et al. could show, that peripheral blood monocytes are required for the early and rapid degradation of both SIS scaffolds and autologous body wall, and that carbodiimide crosslinked SIS's are resistant to macrophage-mediated degradation (Valentin et al., 2009). Apart from the macrophages the T lymphocytes play a crucial role in organ rejection, especially in cell-mediated immune responses to xenografts (Strom et al., 1996). Th1 lymphocytes produce cytokines such as interleukin (IL)-2, interferon (IFN)-γ, and tumor necrosis factor (TNF)-α, leading to macrophage activation, stimulation of complement and differentiation of $CD8^+$ cells to a cytotoxic phenotype, which leads to transplant rejection (Chen et al., 1996). In contrast, Th2 lymphocytes produce IL-4, IL-5, IL-6, and IL-10, cytokines that do not activate macrophages. Resulting from this, the activation of the Th2 pathway is associated with transplant acceptance (Chen et al., 1995). The only ECM scaffold material for which the Th1/Th2 response has been characterized so far is small intestine ECM (Allman et al., 2001). Allman implanted SIS-ECM subcutaneously in mice and could show that the expression of IL-4 (Th2) was increased, while the expression of IFN-γ (Th1) was markedly decreased by 100-fold compared to the response elicited by the xenogeneic muscle group. Although there was an immune response to the SIS-ECM after implantation, this response was dominated by the Th2 pathway mediators, lacking an acute rejection.
Since similar studies have not been conducted for other forms of ECM scaffold materials, it is not possible to determine whether they would elicit the same type of host response. Thinking of the diversity of different tissues and scaffolds, it seems likely that the host response to biologic scaffolds will vary to a large degree following implantation. Further research is needed to understand the immunogenic reactions to the implanted scaffolds and the effects of acellularisation on the repopulation of these scaffolds. One step in this direction are two recent publications by Beattie et al. and Reing et al. who could show that the acellularisation process and the produced degradation products are important modulators for the recruitment and proliferation of appropriate cell types during the remodelling process (Beattie et al., 2009; Reing et al., 2009).
The first study using acellularized tissue in orthopaedic research was performed by Cartmell et al. (Cartmell et al., 2000). Therein, they investigated the effectiveness of treatment of rat tail tendons with SDS, Tri (n-butyl) phosphate and Triton-X and found that both the treatment with SDS (1% for 24h) or TnBP (1%for 48h) resulted in an acellular matrix with retention of almost normal histological structure. The biomechanical properties were similar to native tendons. In contrast, Triton-X (1% for 24h) disrupted the collagen fibers and

was not as effective in acellularization as the two other methods, causing a disruption of the normal D-period of reconstituted rat tail tendon collagen fibers and an approximate 50% decrease in wet tensile strength and stiffness values. Following this, in 2004 Cartmell and Dunn could show that bone-patella-tendon-bone ligaments from rabbits could be successfully acellularized and reseeded with fibroblasts. Fibroblast proliferation rate was thereby slightly retarded on SDS- treated patellar tendons, whereas the repopulation in the TnBP- treated group was successful. Ingram et al. examined the additional effects of ultrasound on the repopulation process of natural tendon scaffolds acellularized with 0.1% SDS and found that repopulation with tenocytes was improved (Ingram et al., 2007). The work of Woods et al. compared the effectiveness of three extraction protocols in the development of an acellularized porcine bone-anterior cruciate ligament- bone graft (Woods et al. 2005). Therefore, they combined Triton-X (1% for 24 hours) either with SDS (1% for 24 hours) or TnBP (1% for 24 hours). The combination of Triton-X with SDS was most effective at removing cell nuclei and intracellular protein (vimentin) from the ACL but affected both the collagen and glycosaminoglycan (GAG) components of the extracellular matrix while increasing the tensile stiffness of the ligament. In contrast, the combination of Triton-X with Triton-X was least effective of the three treatments in terms of cellular extraction, anyway its use did not alter the mechanical (stiffness and failure load) and biochemical (collagen and GAG content) properties of the ACL compared to the control group significantly. Triton-TnBP matched the level of decellularization achieved by Triton-SDS in terms of visible cell nuclei. It had lower effects on the extraction of the intracellular vimentin. Though Triton-TnBP did change the collagen content of the ACL (significant reduction of collagen content), the mechanical properties (tensile stiffness, failure load) were left unaffected. Overall, all three decellularization treatments maintained adequate mechanical and biochemical properties of B-ACL-B grafts. In a study by Vavken et al. 2009 the effectiveness of the acellularization of porcine ACL with Triton-X (0.25% for 24 hours), SDS (0.1% for 24 hours) and trypsin (0.1% for 24 hours) was examined. All acellularization protocols reduced DNA content, with triton-X treatment having the greatest effect. Concurrently, acellularization did not affect tissue collagen or total protein content, but did decrease glycosaminoglycan content. In this work Triton-X had the lowest effect on glycosaminoglycan depletion, so that the author favoured the acellularization with Triton-X. These different results show that future research is needed to find the most effective and most gentle way of acellularization to prepare the scaffold for repopulation.

For the acellularization of meniscal samples there is even more research needed. We could show (Sandmann et al., 2009) that the treatment of human meniscal samples with SDS 2% for 2 weeks is highly effective in acellularization (see chapter 5). In another work published by Stapleton et al. 2008 a different protocol was used for the acellularization of porcine meniscal samples. The menisci were acellularized by exposing the tissue to repetitive freeze-thaw cycles, incubation in hypotonic tris buffer, 0.1% (w/v) sodium dodecyl sulfate in hypotonic buffer plus protease inhibitors, nucleases and hypertonic buffer followed by desinfection using 0.1% (v/v) peracetic acid and final washing in phosphate-buffered saline. Histological, immunohistochemical, and biochemical analyses of the acellularized tissue confirmed the retention of the major structural proteins. However, a 59.4% loss of glycosaminoglycan content was noticed but with no significant alteration of the biomechanical characteristics. Furthermore, acellularized tissue and extracts were not cytotoxic to cells.

Up to now, there are many unanswered questions regarding acelluarized scaffolds. The method of acellularization seems to be very promising of creating a collagen scaffold for tissue engineering of ligaments and meniscus, as the extracellular matrix and thereby the biomechanical properties are mainly preserved. However, *in vivo* animal studies are still missing. In the future acellularized tendons or meniscus samples might play a crucial role in orthopaedic surgery as do several extracellular matrix materials, which have been commercialized so far and are applied in a variety of indications reaching from the reconstruction of soft tissue defects (Ueno et al., 2004) to grafts in vascular surgery (Hiles et al., 1993) or neurosurgery (LeVisage et al., 2005).

3. Repopulation of acellularized scaffolds

The repopulation of acellularized tissue using detergents/solvents might improve and fasten incorporation and remodelling processes and thereby improve the long- term results. Cell proliferation on acellularized tissue is dependent on the used chemical treatment, with SDS showing the best acellularization rate but slightly diminished cell proliferation compared to Triton-X or TnbP (Gratzer et al., 2006, also see previous section).

In a study published by Cartmell in 2004 the seeding of acellularized tendons was most effective in TnBP treated tendons. They were seeded with fibroblasts and viable tissue-engineered grafts could be created. These modified allografts could be developed into mechanically functional delivery vehicles for cells, gene therapy vectors, or other biological agents. Whereas multiple in vitro studies have been performed to repopulate acellular tissue (Cartmell et al., 2004; Gratzer et al., 2006), *in vivo* animal studies are sparse. In a study by Vavken et al. 2009 it could be shown that acellularized tendons treated with Triton-X could best be repopulated as the extracellular matrix was mainly unaffected. SDS treated tissue seems more difficult to repopulate, whereas it shows the best acellularization properties. Harrison et al., 2005 compared the effect of different culture condition variables (i.e. (1) the number of cells used for seeding, (2) the addition of epidermal growth factor (EGF), and (3) culture duration) on the repopulation of porcine ACL with fibroblasts. They found that Triton-X and TnBP acellularized ACL's were suitable for repopulation. Cellular ingrowth in the SDS group was successful, but less effective. At first it was supposed that this might be due to cytotoxic effects of SDS, but Gratzer et al., 2006 showed that not the cytotoxicity but probably matrix alterations were responsible for that.

Nevertheless, little is known about the optimal cell source for tissue engineering of ligaments. Van Eijk et al. 2004 searched for the optimal cell source for tissue engineering of the anterior cruciate ligament. They compared the use of bone marrow stromal cells (BMSCs), ACL fibroblasts and skin fibroblasts and seeded them onto a resorbable suture material [poly(L-lactide/glycolide) multifilaments] at five different seeding densities, and cultured them for up to 12 days. All cell types tested attached to the suture material, proliferated, and synthesized extracellular matrix rich in collagen type I. On day 12 the scaffolds seeded with BMSCs showed the highest DNA content ($p < 0.01$) and the highest collagen production ($p < 0.05$ for the two highest seeding densities). Scaffolds seeded with ACL fibroblasts showed the lowest DNA content and collagen production. Another interesting study examined the proliferation and survival of different cell types (ACL fibroblasts vs. skin fibroblasts) on ligament analogues in the harsh environment of the rabbit knee joint (Belincampini 1998). Liu et al 2008 concluded in their study, that BMSC are a better cell source than ACL fibroblasts when using silk scaffolds for ACL tissue engineering. The same conclusion was reached in the study be Ge et al. in 2005.

4. Tissue engineering of the ACL - own work

The aim of our study (Tischer et al. 2007) was to investigate investigate whether the treatment of rabbit semitendinosus tendons with SDS leads to cell free constructs and whether these could serve as a scaffold for tissue engineering of the ACL. Therefore allograft semitendinosus tendons were treated with SDS using a special protocol (see table 1).

Day 1	Deionized water	24h
Day 2	1% SDS	24h
Day 3	Deionized water	24h
Day 4	70% ethanol	24h
Day 5	Phosphate buffered saline (PBS)	Several times

Table 1. Acellularization protocol for rabbit semitendinosus tendons

Following the acellularization process, the tendons were seeded with autologous dermal fibroblasts. Therefore dermal fibroblasts were taken from a small skin biopsy, extracted and cultured for approximately 10 days until the cells were confluent. The cells were split and after confluence solutions were made with a cell number of $1.0x10^6$ cells/ 0.5ml medium. This suspension was injected into the acellular tendons in line with the collagen fibres. Additionally, cell suspension was applied for 4 days to the tendon surface for outer settlement of the cells.

Following, the tendons were tested biomechanically and histologically. For biomechanical testing three groups were examined, being nine native tendons, nine acellular tendons and nine tendons seeded with autologous dermal fibroblasts for 4 days of culture. The load to failure test was performed on a universal material test machine (Modell Zwicki 1120, Fa. Zwick, Germany) with a preload of 2N. The load- to failure test was performed by increasing the tensile loading continuously with a speed of 10mm/min. Apart from the load to failure the biomechanical properties elongation of tendon until rupture and the stiffness were recorded.

For histological examination six tendons were used in each group. As standard all longitudinal sections were stained with hematoxylin-eosin. For analysis of changes in the extracellular matrix, tendons from each group were stained with a panel of monoclonal antibodies directed against collagen 1, 3, 4, 6, Pro-collagen 1, versican and vimentin.

Histological staining with a standard Hematoxylin/Eosin staining revealed that all cells could be removed from the tendons and the constructs were cell-free (see figure 1). No more cell nuclei could be seen. The acellularization process lead to more interfascicular spaces in the SDS group. The group with the seeded tendon constructs lead to increased irregularities in the tendons. The cells were viable, but most of them did not show the typical spindle shape appearance. Survival of the cells was proven by immunohistochemical staining for pro-collagen I, the precursor molecule of collagen I, which was absent after acellularization but was found around the seeded cells after seeding.

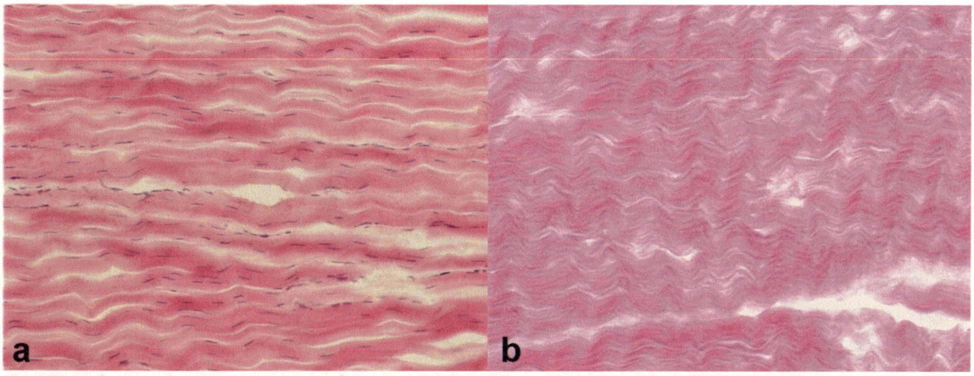

Fig. 1. Histological appearance of native (a) and acellular (b) rabbit semitendinosus tendon. No more cell nuclei can be found after acellularization (b). (HE staining, 100x magnification).

The treatment of the tendon constructs did not show any staining irregularities in the composition of the extracellular matrix. Staining for the collagens I, III, IV and VI did not show significant differences between the groups. Furthermore the distribution of macromolecules like versican was equally distributed in all three tested groups. Vimentin, an intermediary filament especially characteristic for fibroblasts was decreased, but clearly detectable.

Biomechanical results

For biomechanical testing nine native tendons, nine acellular tendons and nine cells seeded for 4 days were tested. The tests were done using a universal material test machine and with load- to failure testing. The maximum load to failure, the elongation of the tendon until rupture and the stiffness could be recorded (see table 2). The load to failure between the three groups was statistically not significant as is shown in the following table.

	Load to failure	Stiffness	Elongation
Native tendons	134.5 ± 12.9N	58.9 ± 6.57N/mm	2.47 ± 0.33%
Acellular tendons	118.5 ± 7.3N	48.5 ± 3.05N/mm	2.46 ± 0.06%
Seeded tendons	132.3 ± 5.6N	18.5 ± 1.73N/mm	7.50 ± 0.5%

Table 2. Results of load to failure testing of rabbit semitendinosus tendons

The elongation of the seeded tendons as well as the stiffness were significantly elevated compared to the acellular or the native control group. The most evident explanation might be that the seeded tendons underwent a swelling process during cell culture, which was not compensated for by our only short pretensioning before testing. However, for comparability between the other groups, pretensioning was not changed.

Animal model

While the *in vitro* results of repopulated acellular scaffold were very promising, in vivo results are missing. Therefore the above described scaffold was tested in an in vivo rabbit model of ACL replacement (see figure 2). Therein we could show that SDS acellularized, allogenic constructs colonized with autologous fibroblasts (Tischer et al. 2009) had significantly weaker biomechanical properties than autologous tendons. In histological examination inflammatory reactions and acellular regions could be noticed, which might be due to the used acellularization method. However, since this is the first study in the field, more research as to be performed.

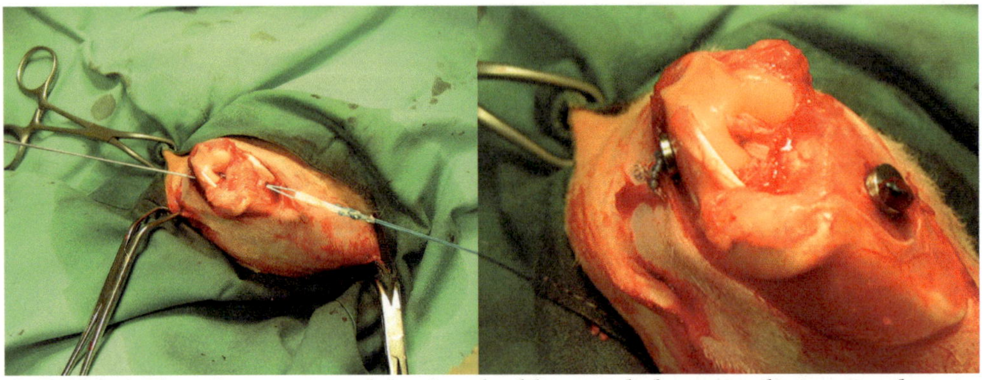

Fig. 2. Rabbit ACL replacement model using double stranded semitendinosus tendon at anatomical insertion site. Fixation was performed using custom made endobuttons.

A lot of ACL research has been done in the rabbit model and though it is a good animal model, future research in the field of double bundle research should take place in bigger animals, as their anatomy and the forces come closer to the forces in human knee joints. Questions, like the best fixation technique or the advantages of the double bundle reconstruction should be best addressed in bigger animals (Tischer et al., 2009).

Discussion

Solvents, like SDS or Triton- X have been shown to be effective agents in acellularizing tissues. In our work we focussed on SDS and our results show that with the solvent SDS it is possible to acellularize tendon tissue. The advantage compared to other methods is that by the acellularization with SDS the constructs are made cell- free and all the cell- detritus is washed out as well. This makes them less immunogenic as the cell-based reactions do not take place in the same way. The acellularization with SDS has been shown to be most effective.

Secondly, SDS acellularized constructs can be reseeded again. This might be a little more difficult than when using other methods due to an alteration of extracellular matrix hindering repopulation (Gratzer et al., 2006), but it has been shown to work (Ingram et al., 2007). Acellular and repopulated scaffolds might be very useful in the future as the integration of transplants can thereby be improved. A small dermal biopsy followed by cell culture might be enough to create a "custom-fit" transplant. Future research has to be done in this field to evaluate the possibilities.

5. Tissue engineering of the Meniscus - own work

Tissue engineering of the meniscus is even more complex than that of ligaments or tendons, as the meniscus has an even more complex collagen network and most of the methods described before have lead to a failure in acellularization or a deterioration of the biomechanical properties. Inspired by the work mentioned above, we tried to apply the same acellularization protocol and initially failed. After gradually improving the SDS concentration and time we finally succeeded in acellularization using SDS for meniscus samples (see table 3). Since the diffusion of chemicals is time dependent, larger tissue specimens are expected to take longer before acellularization is achieved. A higher concentration of SDS (2% compared to 1%) and a longer duration were needed to get a cell-free meniscus construct.

Day 1	Deionized water	24h
Day 2- 14	2% SDS	
Day 15	Deionized water	24h
Day 16	70% ethanol	24h
Day 17	Phosphate buffered saline (PBS)	Several times

Table 3. Extraction protocol for human meniscus constructs

Histological and immunhistochemical results

After treatment with SDS 2% for 2 weeks all meniscus samples (both medial and lateral) were identified as acellular by hematoxylin/eosin staining. Phase-contrast examinations revealed regular collagen bundle arrangement in the acellular specimens as seen in intact menisci. Immunohistochemically, no differences in the labelling patterns for collagen 1, 2 and 6 were observed when compared with intact menisci. Whereas for collagen I there was strong labelling in the whole meniscus, collagen 2 was labelled only in the fibrocartilaginous section of both groups. Collagen VI staining was evenly distributed throughout the meniscus tissue, both in acellularized and intact menisci.

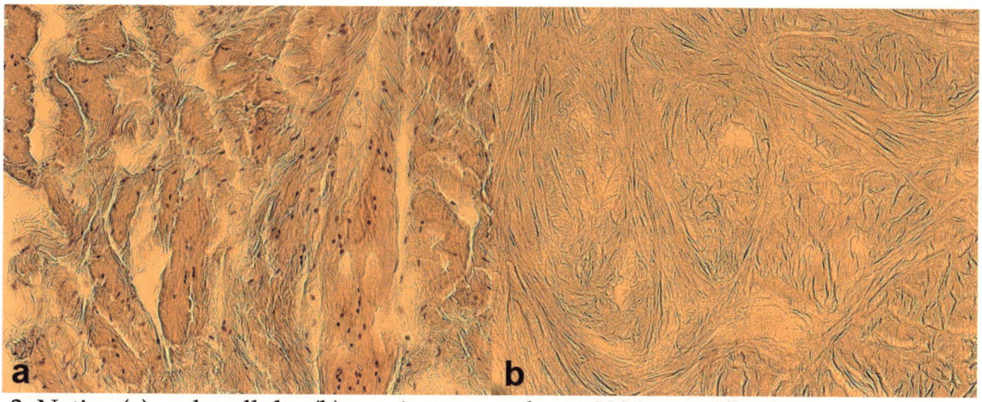

Fig. 3. Native (a) and acellular (b) meniscus samples at 200x magnification.

Biomechanical results

After the acellularization process a total of 5 medial and 5 lateral human menisci were examined using a ball indention test. The undersurface of the menisci was oriented perpendicular to the testing device. Three cylinders with a diameter of 5.0 mm and a height of 4.0 mm were punched out of each meniscus. These cylinders were put into a custom-made device and the upper part of the cylinder was shaped to create a surface parallel to the base. The sample was then put into a custom-designed metallic plate with a circular cavity (diameter 5.0 mm and depth 4.0 mm) to prevent the samples from dislocating during testing (see figure 4). The test was performed as a minimally constraint compression-relaxation test. During testing, the meniscus samples were kept moist using physiologic saline solution. A preload of 0.1 N was used and samples were checked intermittently during testing for displacement. The test cycle consisted of four phases: preloading of the sample with 0.1 N; dynamic compression with a constant load velocity of 5 mm/min until 7 N; static compression of the sample for 60 s with a load of 7 N; relaxation of the sample with a constant unload velocity of 1 mm/min until a load of 0.15 N. After an interval of 60 s, the new test cycle started until a total number of five test cycles were reached. Load, indentor position, and time were displayed by the test software TestXpert and the following three biomechanical values could be calculated: (1) *Relative sample compression* calculated by the indentor position in relation to absolute sample height at the end of the dynamic compression phase (Compression is an indicator for viscosity and characterizes the ability of the tested tissue to evade the indenter) (2)) *Stiffness* determined from the linear elastic slope of the loading curve between 2 and 5 N. High stiffness values mean high elasticity and vice versa. (3) *Residual force* (load measured at the end of the static compression phase). The residual force is influenced by the ability of tissue to evade the indenter in unconstraint compression (viscosity) as well as by the reset forces present in the tested tissue (elasticity). High residual forces show more elastic than viscous properties.

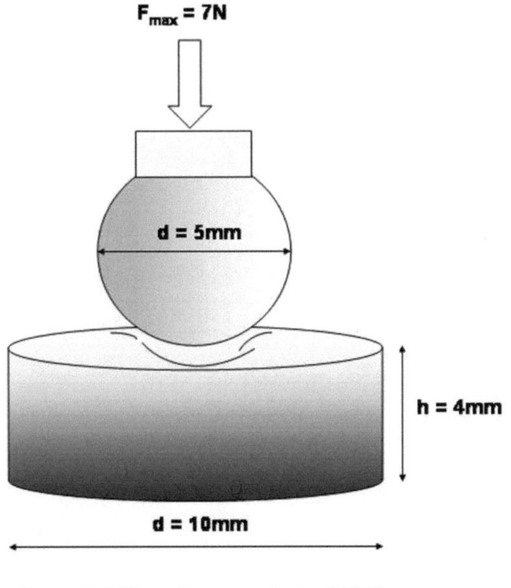

Fig. 4. Scheme of the ball indention trial (Sandmann et al., 2009).

The testing showed that the acellularization did not affect the biomechanical properties of the tested tissue. The test was performed with cyclic loading in 5 cycles with a maximum load of 7N. Between the cycles 60seconds of relaxation were used. Stiffness increased significantly throughout testing by 162% (native meniscus) and 143% (acellular meniscus). During each testing cycle the differences between the two groups—intact meniscus and acellular scaffold—were not statistically significant. Mean compression of sample height was 32% (±7.2%) for intact meniscus samples after the fifth cycle compared to 35% (±8.3%) in the scaffolds. Scaffold compression exceeded compression of intact meniscus by 9%, being not statistically significant ($p > 0.05$). The residual force of the two groups increased after each cycle, but no statistically significant difference could be noticed. Biomechanical results were not significantly different comparing processed medial or processed lateral meniscus samples with the control group.

	Stiffness (cycle 1)	Stiffness (cycle 5)	Mean compression height after 5 cycles	Residual force (cycle 1)	Resiudal force (cycle 5)
Native meniscus	11.6 ± 3.2 N/mm	31.1 ± 3.3 N/mm	32% (±7.2%)	1.0 ± 0.41 N	3.0± 0.36 N
Acellular meniscus	12.3 ± 2.9 N/mm	30 ± 3.2 N/mm	35% (±8.3%)	3.1 ± 0.4 N	3.2 ± 0.41 N

Table 4. Results of biomechanical testing

Discussion

Comparable to the tissue engineering of tendons the treatment of human meniscus samples with the potent solvent SDS generated cell-free meniscus scaffolds. Theses scaffolds have been shown to be biomechanically stable with characteristics similar to native meniscus samples. Special caution was given to the different biomechanical characteristics of meniscus tissue compared to tendon tissue. Seeding of the acellularized menisci has not been tested so far but future research might have to answer the question, whether repopulation of meniscus samples is possible and whether the repopulation rates can be improved by different stimulation methods like mechanical stimuli generated in bioreactors.

6. Outlook

The ACL and the meniscus are the both structures in the knee joint which are frequently injured and so far in most cases no restoration of the original anatomy is possible. Especially the development of osteoarthrosis in the long term is still an unsolved problem. Fortunately, many advances have been made in the surgical reconstruction of the ACL and the disastrous total meniscectomy has been abandoned long ago. The role of new surgical techniques like double bundle ACL surgery still has to be found.

Apart from the acellularization of tissues, the use of growth factors, additionally or alone, seems to be a very promising option in improving the long- term results. Especially in meniscus transplants those factors that stimulate matrix synthesis and inhibit degradation of the extracellular matrix could be very useful. Future research has to point on the repopulation process and its improvement. Angiogenic growth factors might contribute to even further improve the results. Maybe the use of bioreactors producing adequate mechanical stimuli, growth factors or synthetic scaffolds (e.g. silk) will help to solve the remaining questions.

7. Conclusion

Tissue engineering provides a lot of possibilities for improving tendon or meniscus scaffolds. Especially, the acellularization process with following cell seeding might be a very effective way of improving incorporation and remodelling processes as well as avoiding the cell-based immune reactions, which may be partially responsible for the deterioration of the allografts over the years. We have shown that the treatment of tendons or meniscus samples with the solvent SDS produces cell- free constructs. Despite acellularization, the extracellular matrix remains widely intact and the biomechanical properties remain unaltered. Whether the acellularization and the seeding will be successful in the long term has still to be shown in further animal models. Only then will human application be possible and improve operative treatment for our patients.

Big efforts have been made during the last years *in vitro* and *in vivo* and we believe that the tissue engineering of tendons and menisci samples might play a crucial role in the future of orthopaedic surgery.

8. References

Allman AJ, McPherson TB, Badylak SF, Merrill LC, Kallakury B, Sheehan C, Raeder RH, Metzger DW (2001). Xenogeneic extracellular matrix grafts elicit a TH2-restricted immune response, Transplantation 71, pp. 1631–1640

Allaire R; Muriuki M; Gilbertson L; Harner CD (2008). Biomechanical consequences of a tear of the posterior root of the medial meniscus. Similar to total meniscectomy. J Bone Joint Surg Am; Vol. 90; no. 9; 1922-31.

Armour AD; Fish JS; Woodhouse KA; Semple JL (2006). A comparison of human and porcine acellularized dermis: interactions with human fibroblasts in vivo. Plast Reconstr Surg; Vol. 117; No. 3, 845

Arnoczky SP; Warren RF (1983). The microvasculature of the meniscus and its response to injuries. An experimental study in the dog. American Journal of Sports Medicine. Vol. 11, No. 3, 131- 41.

Arnoczky SP (1983). Anatomy of the anterior cruciate ligament. Clin Orthop Relat Res.Vol 172,19-25

Aune A; Holm I; Risberg M; Jensen H; Steen H (2001). Four- strand hamstring tendon autograft compared with patellar tendon- bone autograft for anterior cruciate ligament reconstruction. American Journal of Sports Medicine; Vol. 32; 1639- 43

Badylak SF (2007). Review. The extracellular matrix as a biologic scaffold material. Biomaterials. Vol. 28, No. 25, 3587-93

Badylak SF, Freytes DO, Gilbert TW (2009). Extracellular matrix as a biological scaffold material: Structure and function. Acta Biomater. Vol. 5, No. 1, 1-13.

Badylak SF; Tullius R; Kokini K; Shelbourne KD; Klootwyk T; Voytik SL; Kraine MR; Simmons C (1995). The use of xenogeneic small intestinal submucosa as a biomaterial for Achilles tendon repair in the dog model. J Biomed Mater Res; Vol. 29; No 8; 977

Barber FA; Herbert MA; Coons DA (2006): Tendon augmentation grafts: biomechanical failure loads and failure patterns. Arthroscopy; Vol. 22; No. 5; 534-8

Beard D; Anderson J; Davies S; Price A; Dodd C (2001). Hamstring vs. patella tendon for anterior cruciate ligament reconstruction: a randomised controlled trial. Knee; Vol. 8; 45-50.

Bellincampi LD, Closkey RF, Prasad R, Zawadsky JP, Dunn MG (1998).Viability of fibroblast-seeded ligament analogs after autogenous implantation. J Orthop Res; Vol. 16; No. 4; 414-20.

Beattie AJ, Gilbert TW, Guyot JP, Yates AJ, Badylak SF. (2009). Chemoattraction of progenitor cells by remodeling extracellular matrix scaffolds. Tissue Eng Part A; Vol. 15; No. 5:1119-25.

Bhrany AD; Beckstead BL; Lang TC; Farwell DG; Giachelli CM; Ratner BD (2006). Development of an oesophagus acellular matrix tissue scaffold. Tissue Eng; Vol. 12; 319-30

Bolland F, Korossis S, Wilshaw SP, Ingham E, Fisher J, Kearney JN, Southgate J (2007). Development and characterisation of a full-thickness acellular porcine bladder matrix for tissue engineering. Biomaterials. Vol. 28; No 6:1061-70

Brown B; Lindberg K; Reing J; Stolz DB; Badylak SF (2006). The basement membrane component of biologic scaffolds derived from extracellular matrix. Tissue Eng; Vol. 12; No. 3; 519

Brown BN, Valentin JE, Stewart-Akers AM, McCabe GP, Badylak SF (2009). Macrophage phenotype and remodeling outcomes in response to biologic scaffolds with and without a cellular component. Biomaterials; Vol. 8:1482-91

Brucker PU; Lorenz S; Imhoff AB (2006). Aperture fixation in arthroscopic anterior cruciate ligament double-bundle reconstruction. Arthroscopy; Vol. 22; No 11; 1250.e1-6

Bryant AL; Pua YH; Clark RA (2009). Morphology of knee extension torque-time curves following anterior cruciate ligament injury and reconstruction. J Bone Joint Surg Am; Vol. 91; No. 6; 1424- 31

Buck RE; Malinin T; Brown MD (1989). Bone transplantation and human immunodeficiency virus: an estimate of risk of AIDS. Clin Orthop, Vol. 240; No. 129

Buma P; van Tienen T; Veth R (2007). The collagen meniscus implant. Expert Rev Med Devices; Vol. 4; No. 4; 507-16. Review

Busam ML, Rue JP, Bach BR Jr.(2007). Fresh-frozen allograft anterior cruciate ligament reconstruction.Clin Sports Med. Vol. 26; No.4:607-23

Carey JL; Dunn WR; Dahm DL; Zeger SL; Spindler KP (2009). A systematic review of anterior cruciate ligament reconstruction with autograft compared with allograft. J Bone Joint Surg Am; Vol. 91; No. 9; 2242-50

Cartmell JS; Dunn MG. (2000). Effect of chemical treatments on tendon cellularity and mechanical properties. J Biomed Mater Res. 2000 Jan;49(1):134-40.

Cartmell JS, Dunn MG. Development of cell-seeded patellar tendon allografts for anterior cruciate ligament reconstruction. Tissue Eng. 2004 Jul-Aug;10(7-8):1065-75

Cebotari S; Lichtenberg A; Tudorache I; Hilfiker A; Mertsching H; Leyh R; Breymann T; Kallenbach K; Maniuc L; Batrinac A; Repin O; Maliga O; Ciubotaru A; Haverich A (2006). Clinical application of tissue engineered human heart valves using autologous progenitor cells. Circulation; Vol. 14; No.1; 1132

Chen N, Field EH. (1995). Enhanced type 2 and diminished type 1 cytokines in neonatal tolerance, Transplantation, Vol. 59: 933–941

Chen N, Gao Q, Field EH. (1996). Prevention of Th1 response is critical for tolerance. Transplantation Vol.61:1076–1083.

Chen F; Yoo JJ; Atala A (1999). Acellular collagen matrix as possible „off the shelf" biomaterial for urethral repair. Urology; Vol. 54; No. 2; 407

Coons DA; Barber AF (2006): Tendons graft substitutes- rotator cuff patches. Sports Med Arthrosc; Vol. 14; No. 3; 185-90

Crawford MJ; Swenson CL; Arnoczky SP; O´Shea J; Ross H (2004). Lyophilzation does not inactivate infectious retrovirus in systemically infected bone and tendon allografts. American Journal of Sports Medicine, Vol. 32, No. 3.

Curran AR; Adams DJ; Gill JL; Steiner ME; Scheller AD (2009). The biomechanical effects of low- dose irradiation on bone- patellar- tendon- bone allograft. American Journal of Sports Medicine; Vol. 32; No. 5; 1131-1135.

Diehl P; Schauwecker J; Mittelmeier W; Schmitt M (2008). High hydrostatic pressure, a novel approach in orthopedic surgical oncology to disinfect bone, tendons and cartilage. Anticancer Res; Vol. 28; No. 6B; 3877- 83.

Diehl P; Steinhauser E; Gollwitzer H; Heister C; Schauwecker J; Milz S; Mittelmeier W; Schmitt M. (2006). Biomechanical and immunhistochemical analysis of high hydrostatic pressure-treated Achilles tendons. J Orthop Sci . Vol. 11: 380-385

Fideler BM; Vangsness CT; Lu B (1995). Gamma irradiation: effects on biomechanical properties of human bone- patellar- bone allografts. American Journal of Sports Medicine, Vol. 23: 643- 646.

Ge Z; Goh JC; Lee EH (2005). Selection of cell source for ligament tissue engineering. Cell Transplant. Vol. 14; No 8; 573-83.

Gelber PE; Gonzalez G; Lloreta JL; Reina F; Caceres E; Monllau JC (2008). Freezing causes changes in the meniscus collagen net: a new ultrastructural meniscus disarray scale. Knee Surg Sports Traumatol Arthrosc; Vol. 16; No. 4:353-9

Gelber PE; Gonzalez G; Torres R; Garcia Giralt N; Caceres E; Monllau JC (2009). Cryopreservation does not alter the ultrastructure of the meniscus. Knee Surg Sports Traumatol Arthrosc; Vol.17; No 6:639-44

Gibbons MJ, Butler DL, Grood ES, Bylski-Austrow DI, Levy MS, Noyes FR (1991). Effects of gamma irradiation on the initial mechanical and material properties of goat bone-patellar tendon-bone allografts. J Orthop Res; Vol. 9; No. 2; 209-18

Gilbert TW, Freundb S JM, Badylak SF (2008). Quantification of DNA in Biologic Scaffold Materials. J Surg Res. Vol. 152; No. 1; 135-39

Gilbert TW; Sellaro TL; Badylak SF (2006). Decellularization of tissues and organs. Biomaterials; Vol. 27; 3675-83

Gratzer PF, Harrison RD, Woods T (2006).Matrix alteration and not residual sodium dodecyl sulfate cytotoxicity affects the cellular repopulation of a decellularized matrix. Tissue Eng.; Vol. 12; No. 10; 2975-83.

Harrison RD, Gratzer PF.(2005). Effect of extraction protocols and epidermal growth factor on the cellular repopulation of decellularized anterior cruciate ligament allografts. J Biomed Mater Res AVol. 75; No.4:841-54.

Hiles MC; Badylak SF; Geddes LA; Kokini K; Morff RJ (1993). Porosity of porcine small – intestinal submucosa for use as a vascular graft. J Biomed Mater Res; Vol. 27; No. 2; 139-44

Ingram JH, Korossis S, Howling G, Fisher J, Ingham E. (2007). The use of ultrasonication to aid recellularization of acellular natural tissue scaffolds for use in anterior cruciate ligament reconstruction. Tissue Eng. Vol. 13; No. 7:1561-72.

Kartus J; Movin T; Karlsson J (2001). Donor-site morbidity and anterior knee problems after anterior cruciate ligament reconstruction using autografts. Arthroscopy; Vol 17; No. 9; 971- 80.

Kondo E, Yasuda K, Azuma H, Tanabe Y, Yagi T (2008). Prospective clinical comparisons of anatomic double-bundle versus single-bundle anterior cruciate ligament reconstruction procedures in 328 consecutive patients. American Journal of Sports Medicine; Vol. 36; No. 9, 1675-87

Krych AJ; Jackson JD; Hoskin TL; Dahm DL (2007). A meta-analysis of patellar tendon autograft versus patellar tendon allograft in anterior cruciate ligament reconstruction. Arthroscopy; Vol. 24; No. 3; 292-8

LeVisage C; Okawa A; Kadakia L; Yang S., Sieber AN; Kostuik JP (2005). Intervertebral disc regeneration using small intestinal submucosa as a bioscaffold. Comput Meth Biomech Biomed Eng; Suppl 1; No. 177

Lewis PB, Williams JM, Hallab N, Virdi A, Yanke A, Cole BJ (2008). Multiple freeze-thaw cycled meniscal allograft tissue: A biomechanical, biochemical, and histologic analysis. J Orthop Res. Vol. 26; No 1:49-55.

Liu H; Fan H; Toh SL; Goh JC (2008). A comparison of rabbit mesenchymal stem cells and anterior cruciate ligament fibroblasts responses on combined silk scaffolds. Biomaterials. Vol. 29; No. 10; 1443-53

Lohmander LS, Englund PM, Dahl LL, Roos EM (2007). The long-term consequence of anterior cruciate ligament and meniscus injuries: osteoarthritis. American Journal of Sports Medicine ; Vol. 35, No. 10, 1756-69.

Mahirogullari M; Ferguson CM; Whitlock PW; Stabile KJ; Poehling GG (2007). Freeze- dried allografts for anterior cruciate ligament reconstruction. Clin Sports Med; Vol 26; No. 4; 625- 37

Maier D; Braeun K; Steinhauser E; Ueblacker P; Oberst M; Kreuz PC; Roos N; Martinek V; Imhoff AB (2007). In vitro analysis of an allogenic scaffold for tissue-engineered meniscus replacement. J Orthop Res.; Vol. 25; No. 12; 1598-608

Makihara Y; Nishino A; Fukubayashi T; Kanamori A (2006). Decrease of knee flexion torque in patients with ACL reconstruction: combined analysis of the architecture and function of the knee flexor muscles. Knee Surg Sports Traumatol Arthrosc; Vol 14; 310-17

Martinek V, Ueblacker P, Bräun K, Nitschke S, Mannhardt R, Specht K, Gansbacher B, Imhoff AB (2006). Second generation of meniscus transplantation: in-vivo study with tissue engineered meniscus replacement. Arch Orthop Trauma Surg; Vol. 126; No 4; 228-34

Muren O, Dahlstedt L, Brosjö E, Dahlborn M, Dalén N. (2005). Gross osteolytic tibia tunnel widening with the use of Gore-Tex anterior cruciate ligament prosthesis: a radiological, arthrometric and clinical evaluation of 17 patients 13-15 years after surgery. Acta Orthop; Vol. 76; No. 2; 270-4

Nakamura N; Horibe S; Sasaki S; Kitaguchi T; Tagami M; Mitsuoka T; Toritsuka Y; Hamada M; Shino K (2002). Evaluation of active knee flexion and hamstring strength after anterior cruciate ligament reconstruction using hamstring tendons. Arthroscopy; Vol. 8; 337-42

Packer JD; Rodeo SA (2009). Meniscal allograft transplantation. Clin Sports Med; Vol. 28; No. 2; 259-83

Park HJ; Urabe K; Naruse K; Onuma K; Nemoto N; Itoman M (2009). The effect of cryopreservation or heating on the mechanical properties and histomorphology of rat bone-patellar tendon-bone. Cell Tissue Bank; Vol; 10; No. 1; 11-18

Reing JE; Zhang L; Myers- Irvine J; Cordero KE; Freytes DO; Heber- Katz E; Bedelbaeva K; McIntosh D; Dewilde A; Braunhut SJ; Badylak SF 2009. Degradation products of extracellular matrix affect cell migration and proliferation. Tissue Engineering ; Vol 15; No. 3; 605-13

Roolker W, Patt TW, van Dijk CN, Vegter M, Marti RK (2000). The Gore-Tex prosthetic ligament as a salvage procedure in deficient knees. Knee Surg Sports Traumatol Arthrosc; Vol. 8; No. 1; 20-5.

Sandmann GH; Eichhorn S; Vogt S; Adamczyk C; Aryee S; Hoberg M; Milz S; Imhoff AB; Tischer T (2009). Generation and characterization of a human acellular meniscus scaffold for tissue engineering. Journal of Biomedical Materials Research, Part A; Vol. 91A; No. 2; 567- 74.

Sass K, Corterier C, Brylla E, Löffler S, Steen M, Spanel-Borowski K(2009). Cryopreserved porcine tendons preserve cell viability after thawing. Transplant Proc. Vol. 41; No. 5:1911-3.

Stapleton TW, Ingram J, Katta J, Knight R, Korossis S, Fisher J, Ingham E (2008). Development and characterization of an acellular porcine medial meniscus for use in tissue engineering.Tissue Eng Part A; Vol. 14; No. 4:505-18.

Strom TB, Roy-Chaudhury P, Manfro R, Zheng XX, Nickerson PW, Wood K; Bushell A (1996) The Th1/Th2 paradigm and the allograft response. Curr Opin Immunol; Vol 8: 688–693.

Sun K, Tian S, Zhang J, Xia C, Zhang C, Yu T (2009). Anterior cruciate ligament reconstruction with BPTB autograft, irradiated versus non-irradiated allograft: a prospective randomized clinical study. Knee Surg Sports Traumatol Arthrosc. Vol. 17; No. 5; 464-74.

Tashiro T; Kurosawa H; Kawakami A; Hikita A; Fukui N (2003). The influence of medial tendon harvest on knee flexor strength after anterior cruciate ligament reconstruction: a detailed evaluation with comparison of single- and double tendon harvest; American Journal of Sports Medicine; Vol 31; 522- 29.

Taylor DC; Posner M; Curl WW; Feagin JA (2009).Isolated tears of the anterior cruciate ligament: over 30-year follow-up of patients treated with arthrotomy and primary repair. American Journal of Sports Medicine; Vol. 37; No. 1, 65-71

Tischer T, Milz S, Maier M, Schieker M, Benjamin M (2002). An immunohistochemical study of the rabbit suprapatella, a sesamoid fibrocartilage in the quadriceps tendon containing aggrecan. J Histochem Cytochem; Vol.50; No.7:955-60

Tischer T; Ronga M; Tsai A; Ingham SJ; Ekdahl M; Smolinski P; Fu FH (2009). Biomechanics of the goat three bundle anterior cruciate ligament. Knee Surg Sports Traumatol Arthrosc; Vol. 17; N0. 8; 935-40.

Tischer T; Vogt S; Aryee S; Steinhauser E; Adamczyk C; Milz S; Martinek V; Imhoff AB (2007). Tissue engineering of the anterior cruciate ligament: a new method using acellularized tendon allografts and autologous fibroblasts. Arch Orthop Trauma Surg, Vol. 127, 735- 741.

Tischer T; Aryee S; Wexel G; Steinhauser E; Adamczyk C; Eichhorn S; Milz S; Martinek V; Gänsbacher B; Imhoff AB; Vogt S. Tissue Engineering of the ACL - SDS Chemically Acellularized and Revitalized Tendons are Inferior to Native

Tendons. Tissue Engineering, accepted 2009

Ueno T; Pickett LC; de la Fuente SG; Lawson DC; Pappas TN (2004). Clinical application of porcine small intestinal submucosa in the management of infected or potentially contaminated abdominal defects. J Gastrointest Surg; Vol. 8; No. 1, 109- 112

Valentin JE, Stewart-Akers AM, Gilbert TW, Badylak SF (2009). Macrophage participation in the degradation and remodeling of extracellular matrix scaffolds. Tissue Eng Part A. Vol. 15; No. 7:1687-94

Van Eijk F, Saris DB, Riesle J, Willems WJ, Van Blitterswijk CA, Verbout AJ, Dhert WJ (2004). Tissue engineering of ligaments: a comparison of bone marrow stromal cells, anterior cruciate ligament, and skin fibroblasts as cell source. Tissue Eng.; Vol. 10; No. 5-6; 893-903.

Van Tienen TG; Hannink G; Buma P (2009). Meniscus replacement using synthetic materials. Clin Sports Med. Vol. 28; No. 1; 143-56.

Van Thiel GS; Verma N; Yanke A; Basu S; Farr J; Cole B (2009).Meniscal allograft size can be predicted by height, weight, and gender. Arthroscopy. Vol. 25; No. 7; 722-7.

Vavken P; Joshi S; Murray MM (2009). Triton-X is most effective among three decellularization agents for ACL tissue engineering. Journal of Orthopaedic Research.

Verdonk PC; Verstraete KL; Almqvist KF; DeCuyper K; Veys EM; Verbruggen G; Verdonk R (2006). Meniscal allograft transplantation: long-term clinical results with radiological and magnetic imaging correlations. Knee Surg Sports Traumatol Arthrosc; Vol. 14; 694-706

Walles T; Herden T; Haverich A; Mertsching H (2003). Influence of scaffold thickness and scaffold composition on bioartificial graft survival. Biomaterials; Vol. 24; No 7; 1233

Wirth CJ; Peters G; Milachowski KA; Weismeier KG; Kohn D (2002). Long-term results of meniscal allograft transplantation. American Journal of Sports Medicine; Vol. 30; No. 2; 174- 81

Woods T, Gratzer PF (2005). Effectiveness of three extraction techniques in the development of a decellularized bone-anterior cruciate ligament-bone graft; Biomaterials; Vol. 26; No. 35; 7339-49

Zaffagnini S, Marcheggiani Muccioli GM, Chatrath V, Bondi A, De Pasquale V, Martini D, Bacchelli B, Marcacci M. (2008). Histological and ultrastructural evaluation of Leeds-Keio ligament 20 years after implant: a case report. Knee Surg Sports Traumatol Arthrosc. Vol. 16; No. 11; 1026-9

Zantop T, Wellmann M, Fu FH, Petersen W (2008) Tunnel positioning of anteromedial and posterolateral bundles in anatomic anterior cruciate ligament reconstruction: anatomic and radiographic findings American Journal of Sports Medicine; Vol. 36; No 1; 65-72.

23

Bio-nanotechnology Approaches to Neural Tissue Engineering

Kun Zhou, David Nisbet, George Thouas,
Claude Bernard and John Forsythe
Monash University
Australia

1. Introduction

Injury to the peripheral nervous system (PNS) and central nervous system (CNS) may result in severe functional loss. Spontaneous regeneration is limited to small lesions within the injured PNS and is actively suppressed within the CNS. In this chapter we discuss the pathology and changes in the physiological environment following PNS/CNS injury. Several key factors such as the glial scar and inhibitory biomolecules are addressed. Neural tissue engineering approaches attempt to provide an alternative to nerve autografts, and have shown some promising regenerative outcomes. The scaffolds are designed to provide mechanical support for endogenous/transplanted cells at the lesion site, provide guidance cues to neurites and for the attachment or delivery of biomolecules that promote regeneration. Unfortunately many of the current scaffolds used for neural tissue engineering provide only limited regulation of cellular response due to insufficient control of physical support, topological stimulation, degradability and inflammatory/foreign body responses.

Nanostructured materials are currently being investigated for a new generation of neural tissue engineering scaffold. Bio-nanotechnology approaches extend current scaffold strategies for an enhanced manipulation of the physical, biochemical and biological cues that can significantly enhance cell survival and nerve regeneration. Two types of nanofabrication are discussed in this chapter:

(1) electrospinning, which produces nanofibers using a wide range of polymers with controlled fiber diameter, degree of alignment and pore size, and

(2) self-assembled peptide hydrogels which are particularly useful to fill cavities of irregular shape, especially in the case of brain and spinal cord injuries.

In both cases, the nanofibrillar scaffolds mimic some structural features of the extracellular matrix but importantly, the nano-size dimension allows for the high density presentation in three dimensions of peptides and growth factors important for neural repair. Biofunctionalisation strategies will therefore be discussed in some detail. Finally this chapter will provide some commentary on the short to long term view on neural tissue engineering and identify some key biological and engineering milestones required for successful nerve regeneration.

2. Physiological challenges for nerve repair in the PNS & CNS

In the United States, approximately 20,000 peripheral nerve injuries required clinical treatment per year (Neville, Huang et al. 2009). Most of the injuries are repaired using nerve autografts, however, autografts are limited to the source of donor nerves and can cause donor site morbidity. Injuries to peripheral nerves cause several deficits including loss of function of the innervated tissue and neuropathic pain. Peripheral nerve injuries are largely induced by trauma resulting from accidents. Peripheral nerves can spontaneously recover from injury when they are not completely severed however in most cases complete nerve transection occurs. After injury, the distal portion of the completely severed nerve begins to degenerate, while the proximal end of the nerve stump swells but undergoes less damage via retrograde degradation. Concurrently, Schwann cells surrounding the axons also degrade. The myelin and axonal debris are phagocytozed by macrophages and Schwann cells followed by regeneration at the proximal end toward the distal stump (Schmidt and Leach 2003). The regenerative outcome depends on the distance and the alignment between the proximal and distal ends. For large defects and non-aligned nerves, structural and functional recovery may not be achieved.

Unlike the PNS, CNS neurons are usually inhibited from regenerating across a lesion site. This inhibition results from a combination of biomechanical and biochemical factors, where healthy adult neurons are actively inhibited from regenerating (Tom, Steinmetz et al. 2004; Zhang, Uchimura et al. 2006; Busch and Silver 2007). The site-specific nature of this inhibition has been demonstrated in the 1980s by a series of experiments showing the failure of PNS axons to lengthen when encountered with CNS glia (David and Aguayo 1981). In the following years, many important inhibitory factors and the associated cell types have been identified at the site of injury following CNS damage. Of the biomechanical inhibitory factors, the major obstacle to regeneration after injury is the formation of a glial scar. Astrocytes undergo a morphological change, extending interwoven processes that lay down a rigid and relatively permanent collagenous scar that physically blocks axonal elongation and reconnection between proximal and distal ends (Fawcett 1997; Fitch and Silver 2008). The problem is compounded as astrocytes up-regulate glycoproteins like tenascin and a number of chondroitin sulphate proteoglycans which play an inhibitory role via growth cone repulsion and inhibition mechanisms to the regenerating neurites (Taylor, Pesheva et al. 1993; Fidler, Schuette et al. 1999; Chan, Roberts et al. 2008). Oligodendrocytes continue to express myelin proteins after injury. Both myelin proteins and myelin debris contain regeneration inhibitors including Nogo-A, myelin oligodendrocyte glycoprotein (MOG) and myelin-associated glycoprotein (MAG), which are the major contributors to the inhibition of axon regeneration post injury (GrandPre, Nakamura et al. 2000; Kottis, Thibault et al. 2002; Filbin 2003; He and Koprivica 2004; Karnezis, Mandemakers et al. 2004).

To date, much of the focus of neural tissue engineering in the CNS has been on curing spinal cord injuries (SCIs). This is largely due to the clearly defined role that scaffolds play in restoring neural pathways following SCI. There are about 12,000 new cases of SCI each year in the United States and the number of people with SCI has been estimated to be approximately 255,000. Statistical reports also show that the causes of traumatic SCI are primarily attributed to motor vehicle accidents (42%), followed by violence (15.3%), falls (27.1%), and recreational activities (7.4%) (National Spinal Cord Injury Statistical Center 2009). A recent report released in Australia on behalf of the Victorian Neurotrama Initiative (Access Economics 2009) states that the life time costs of disease and injury to the nervous system is in excess of $10.5AUD billion per annum, a significant figure for a population less than 25 million. Therefore such injuries are partly responsible for a major economic burden as well as the devastating personal consequences to the patient.

During primary injury of the spinal cord, vascular damage results in microhaemorrhages, which happen within minutes, radiating outward from the central grey matter over several hours. In the case of SCIs, the cord itself can swell significantly to occupy the whole spinal canal at the lesion site, resulting in secondary ischaemia when swelling exceeds venous blood pressure. Ischaemia leads to the release of cytokines and toxic levels of otherwise normally secreted factors from damaged cells, triggering an acute inflammatory cascade that attacks intact neighbouring cells (known as the by-stander effect). Secondary injuries include depolarization of neurons and abnormal release of amino acids. Following ischemia, hypersecretion of factors such as glutamate occurs, resulting in toxic levels in both the injured spinal cord and brain (Dirnagl, Iadecola et al. 1999; Ao, Wang et al. 2007). Excessive levels of this neurotransmitter, secreted from the injured spinal neurons, axons and astrocytes leads to a highly disruptive over-excitation of neighbouring neurons (excitotoxicity). Affected neurons undergo demyelination and increased calcium signalling, triggering apoptotic cascades and oxidative free radical damage. In addition, the humoral pro-inflammatory response of cytokine secretion causes neutrophil and macrophage infiltration into the injury site to ingest bacteria and debris, resulting in edematous cavities (Fitch, Doller et al. 1999).

3. Current scaffold strategies for new synaptic connections

3.1 Nerve guidance conduits

Gaps and cavities often appear in the injured PNS/CNS, resulting from traumaic injuries or macrophage infltration and post-injury inflammation (Blight 1994; Fitch, Doller et al. 1999). Scaffolds have been implanted or injected into cavities (lesion site) of the injured PNS/CNS normally in the form of conduits with impermeable/permeable walls (Heijke, Klopper et al. 2001; Lundborg, Rosen et al. 2004; Chew, Mi et al. 2007). These materials can be either biologically inert (Lundborg, Dahlin et al. 1991; Zhao, Lundborg et al. 1997; Heijke, Klopper et al. 2001) or biodegradable with a controlled degradation rate (Borkenhagen, Stoll et al. 1998; Xie, Li et al. 2008). Scaffolds are used to bridge or fill the gaps/cavities at the lesion site as well as provide support to surrounding nerve tissue and endogenous/implanted cells. Scaffolds provide adhesion sites to cells, allowing subsequent cell function including proliferation and differentation and even regulate specific cellular behavior by careful manipulation of the dimensional, topograpic and biochemical properties (Corey, Lin et al. 2007; Dodla and Bellamkonda 2008; Christopherson, Song et al. 2009).

The guidance conduit serves several important roles for nerve regeneration such as:

a) directing axonal sprouting from the regenerating nerve (Tsai, Dalton et al. 2004; Panseri, Cunha et al. 2008),
b) protecting the regenerating nerve by restricting the infiltration of fibrous tissue (Li and David 1996; Kitahara, Suzuki et al. 1998; Bundesen, Scheel et al. 2003; Yang, Xu et al. 2004; Lietz, Ullrich et al. 2006; Oh, Kim et al. 2008), and
c) providing a pathway for diffusion of neurotropic and neurotophic factors (Moore, Macsween et al. 2006; Chew, Mi et al. 2007).

Guidance conduites have several advantages over microsurgical autografting, the current gold standard for nerve repair. Suture line tension especially for long nerve gaps (>5 mm) is ineffective, however regeneration is improved using a guidance conduit (Terzis, Faibisoff et al. 1975; Schmidhammer, Zandieh et al. 2005). The use of donor nerve allografts is less feasible, due to the limited availability of tissue and morbidity following microsurgical procedures. Guidance conduits have therefore received considerable research focus over the past decade.

Early guidance conduits were primarliy made of silicone due to its stability under physiological conditions, biocompatibility, flexibility as well as ease of processing into tubular structures (Zhao, Lundborg et al. 1997; Chen, Hsieh et al. 2000). Although silicone conduits have proven reasonably successful as conduits for small gap lengths in animal models (<5 mm) (Lundborg, Dahlin et al. 1982), the non-biodegradability of silicone conduits has limited its application as a strategy for long-term repair and recovery. Tubes also eventually become encapsulated with fibrous tissue, which leads to nerve compression, requiring additional surgical intevention to remove the tube (Lundborg, Dahlin et al. 1991; Chamberlain, Yannas et al. 1998). Another limiting factor with inert guidance conduits is that they provide little or no nerve regeneration for gap lengths over 10 mm in the PNS (Lundborg, Dahlin et al. 1982) unless exogenous growth factors are used. In animal studies, biodegradable nerve guidance conduits have provided a feasible alternative, preventing neuroma formation and infiltration of fibrous tissue (Jansen, van der Werff et al. 2004; Oh, Kim et al. 2008). Biodegradable conduits have been fabricated from natural or synthetic materials such as collagen, chitosan and poly-L-lactic acid (Chamberlain, Yannas et al. 1998; Mingyu, Kai et al. 2004; Patel, Vandevord et al. 2008; Ichihara, Inada et al. 2009).

Guidance conduits can be used solely to form an open lumen (Panseri, Cunha et al. 2008), or in combination with molecular and/or biological cues. Open lumina are hypothesized to provide space for linear nerve outgrowth to enable axons to selectively reinnervate the appropriate target. However, a pitfall of an open lumen approach is that it provides limited physical support and can not adequately control the cellular microenvironment. This is due to inadequate sites for cell adhesion and attachment of biological molecules as well as insufficient topological stimulation, which limits the ability of the conduits to encourage cell migration. Therefore, scaffolds are more wildly used in combination with molecular cues to improve nerve regeneration (Labrador, But et al. 1998). Various types of peptides, proteins and polysaccharides, which enhance cell adhesion, proliferation and provide neutrophin support (Lee, Yu et al. 2003), can be used to fill the open lumen (Nakamura, Inada et al.

2004) or can be blended with the conduit forming material (Mingyu, Kai et al. 2004; Xie, Li et al. 2008). Other cellular cues can also be incorporated using such strategies (Sinis, Schaller et al. 2009). The reconnection of severed nerve axons in the PNS requires appropriate guidance of Schwann cells (Sinis, Schaller et al. 2009). In the case of large gaps between severed proximal and distal axons or severely damaged nerve bundles, guidance conduits with implanted stem/progenitor or supporting cells are essential for axon reconnection (Murakami, Fujimoto et al. 2003; Leaver, Harvey et al. 2006). For instance, conduits seeded with Schwann cells show a comparable effectiveness to autografts in sciatic nerve repair although control conduits with no Schwann cells contained significantly lower densities of nerve fibers, highlighting the importance of cellular biological cues (Neville, Huang et al. 2009).

Experimentally, cell-based therapies using stem and progenitor cell types have led to promising results for the treatment of both PNS and CNS injuries (Lu, Jones et al. 2005; Cummings, Uchida et al. 2006; Deshpande, Kim et al. 2006), with the intention of replacing lost and damaged cells in order to improve endogenous nerve regeneration. While functional recovery at the transplantation site requires cell survival in sufficient numbers, programmed cell death of the transplanted cells invariably occurs due a dearth of appropriate cell-matrix interactions. This has been a major limiting factor in the effectiveness of cell replacement therapies, with some reports of less than 10% of cells surviving after transplantation (Belle, Caldwell et al. 2004). Scaffolds which provide physical support, controlled degradability, controlled inflammatory/foreign body response and guidance cues must therefore also provide a suitable milieu for the survival and viability of transplanted cells. Unfortunately, current scaffolds for neural tissue engineering do not yet meet all these criteria. However recent progress in nanostructured and biomimetic scaffolds for neural tissue engineering has provided encouraging results, particularly when interfaced with stem cell therapies.

4. Nanostructured Scaffolds for Neural Tissue Engineering: Fabrication and Design

The neural architecture of the brain and spinal cord is more complex compared to the PNS, typically organized into discrete three-dimensional (3D) populations, depending on their roles in transmitting and processing signals according to the spatial body plan. At the micro- and nanoscale, cells of the CNS reside within functional microenvironments consisting of physical structures including pores, ridges, and fibers that make up the extracellular matrix (ECM) and plasma membrane cell surfaces of closely apposed neighboring cells (Desai 2000). Cell-cell and cell-matrix interactions contribute to the formation and function of this architecture, dictating signaling and maintenance roles in the adult tissue, based on a complex synergy between biophysical (e.g. contact-mediated signaling, synapse control), and biochemical factors (e.g. nutrient support and inflammatory protection). Neural tissue engineering scaffolds are aimed toward recapitulating some of the 3D biological signaling that is known to be involved in the maintenance of the PNS and CNS and to facilitate proliferation, migration and potentially differentiation during tissue repair. Nanostructured and biofunctional scaffolds are currently being investigated for the provision of microenvironments that are similar to those occurring *in vivo*, in the development of improved cellular function and successful regenerative outcomes (Stevens and George 2005). The scaffold fabrication techniques of electrospinning, self-assembly and phase separation are currently the mainstay of this approach, because they are capable of producing nanoscale (10 to 2000 nm) polymer fibers. This chapter will focus on two methods of fabricating nanostructured scaffolds; the "top-down" approach of electrospinning and the "bottom-up" of molecular self-assembly of peptides, both of which have recently received considerable attention in neural tissue engineering.

4.1 Electrospinning

Electrospinning was patented by Formhals in 1934 (Formhals 1934) and involves the production of a polymer filament using an electrostatic force. Electrospinning is a versatile technique that enables production of polymer fibers with diameters ranging from a few microns to tens of nanometers. A schematic for an experimental device for electrospinning is outlined in Fig. 1.

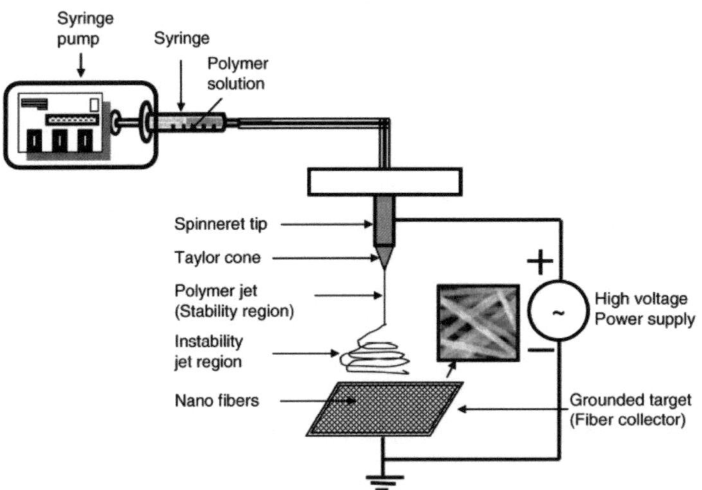

Fig. 1. An electrospinning device with a plate-shape metallic collector. The metallic plate can be either stationary or rotating to achieve different orientations of electrospun nanofibers (Reprinted from Tissue Engineering with permission of Mary Ann Liebert)

Although the electrospinning process is relatively simple in terms of its output, an understanding of the underlying mechanisms and several key processing conditions are essential for effective control of fiber properties. As the polymer solution is pumped to the spinneret tip, the high voltage induces charge accumulation to the solution. The droplet is then elongated into a conical shape, known as a Taylor cone (Taylor 1969) due to the electrostatic repulsive force. Electrospinning is initiated when the repulsive force of the solution exceeds the surface tension of the droplet. A finely charged jet is formed at the tip of the Taylor cone. The jet is then stretched and accelerated, accompanied by solvent evaporation, and eventually collected by the target. Adjustment of the applied DC voltage, feeding rate, polymer solution viscosity and working distance can be used to control the morphology of the collected fibers. Here, some key effects of these processing parameters are briefly introduced; for a more detailed review of the electrospinning process the authors direct you to the following reviews (see (Pham, Sharma et al. 2006) and (Boudriot and Wendorff 2006)). In most cases, increasing the applied DC voltage and resultant electric field will cause greater stretching of the polymer solution, which consequently reduces the diameter of the fibers. However, too high a voltage at a given feed rate will lead to a smaller and less stable Taylor cone, which can cause larger diameter bead formation along the fibers (Zong, Kim et al. 2002; Kim, Kim et al. 2005). The feed rate will also determine the volume of the polymer solution available for electrospinning at the spinneret tip. At a given voltage, higher feed rates generally yield fibers with larger diameters, although these rates are accompanied by slow solvent evaporation during flight time, leading to residual solvent and fusion of fibers. Working distance has less of an influence on fiber morphology, but smaller working distances results in an increased electric field strength and reduced flight time, which may also cause bead formation and fiber fusion.

Of physiological interest is the fabrication of scaffolds with variable fiber orientation and patterning, to create more sophisticated structures that can direct processes like cell migration and neurite outgrowth. A plate-shape metallic collector is often used to collect randomly orientated nanofibers, while aligned nanofibers can be collected by a rapidly rotating cylindrical target (mandrel) with controlled speed to give different degrees of fiber alignment. Alternatively, aligned nanofibers can be obtained between a parallel metallic collector gap, through the modified electrostatic field profile rather than mechanical rotation. The interfiber distance of the electrospun scaffolds can be controlled to some extent by means of adjusting the fiber diameter (Eichhorn and Sampson 2005), rotating speed of the collecting mandrel (Zhu, Cui et al. 2008), co-electrospinning a soluble or "sacrificial" fiber (Baker, Gee et al. 2008) or by combining with salt leaching (Nam, Huang et al. 2007).

4.2 Electrospun nanofibrous scaffolds for nerve generation

Fibers produced by thermal extrusion with diameters in the order of several hundred microns have been used in guidance tubes to provide structural support for cells (Wen and Tresco 2006). However cell orientation, directed process elongation and cell migration are very limited, especially for large diameter fibers of greater than 500 μm. There is also no appreciable induction of guidance effects on adhered cultured cells (Wen and Tresco 2006). Conversely, neurites follow the long axis of microfibers more obviously when the fiber diameters are in the order of tens of micrometers (Khan, Sayers et al. 1990; Smeal, Rabbitt et al. 2005).

Most cell types can actively sense scale, orientation, texture and stiffness of physical features in their immediate environment, at the microscale (cells and matrix) and nanoscale (macromolecular structures). As the electrospinning technique is capable of providing nanofibers of 10-1000 nm in diameter with some degree of control over the structural parameters, such as fiber alignment and interfiber distance, it has become a major research focus in neural tissue engineering. The "porosity" of electrospun scaffolds can be up to 70~90% (Yang, Xu et al. 2004; Gupta, Venugopal et al. 2009) which can facilitate ingrowth and migration of cells, as well as the transportation of nutrients (e.g. metabolites, oxygen) and signaling factors (e.g. growth factors, cytokines). The high surface-to-volume ratio of nanofibrous scaffolds, can provide many more binding sites to cell membrane receptors. This can lead to appropriate regulation of gene expression and amplify certain cellular processes. Although the mechanisms of how cells probe and respond to these nanofeatures is still not clear, different topographies of nanofibrous scaffolds have been demonstrated to induce diverse cellular behavior, including cell adhesion, migration (Yang, Xu et al. 2004), orientation (Schnell, Klinkhammer et al. 2007), infiltration and differentiation (Nisbet, Yu et al. 2008).

Nanofibrous scaffolds & adhesion:
Electrospun poly(L-lactic acid) (PLLA) nanofibrous scaffolds show significant improvement for neural stem cell adhesion *in vitro* compared with those cultured on solvent cast flat PLLA surfaces (Yang, Xu et al. 2004). PC12 cells cultured on either flat or nanostructured surfaces in close proximity, migrate from the flat region towards the nanostructured region of the surface (Cecchini, Bumma et al. 2007). A commonly held explanation for the enhanced adhesion is the increase in surface roughness and surface area, leading to an increase in non-specific protein adsorption and availability of focal adhesion sites (Fan, Cui et al. 2002; Manwaring, Walsh et al. 2004; Yao, O'Brien et al. 2009).

Nanofibrous scaffolds & stem/progenitor cell differentiation:
Embryonic stem (ES) cells provide a useful model for therapeutic investigation in tissue engineering due to their ability to give rise to all the main types of adult somatic cells (pluripotency). For the purpose of PNS/CNS regeneration, neural progenitor cells derived from ES cells are typically derived from 4-/4+ retinoic acid induction protocol (Bain, Kitchens et al. 1995). The 4-/4+ induction method avoids the inappropriate differentiation of ES cells within the adult PNS and CNS. During the time that ES cells are cultured as 3D aggregates called embryonic bodies (EBs) they are exposed to retinoic acid, giving rise to the three major types of nerve cells: neurons, astrocytes and oligodendrocytes.

Stem cells cultured *in vitro* on nanofibrous scaffolds can differentiate in response not only to the soluble differentiation factors, but to the structural features of their surrounding microenvironment. Nanofibrous environments often encourage the differentiation of stem cells and can bias their differentiation toward specific lineages providing enriched populations (Nisbet, Yu et al. 2008). Immunohistochemstry of EBs containing neural progenitor cells established by 4-/4+ induction showed preferential differentiation into neural lineages when cultured on electrospun PCL nanofibers with a mean fiber diameter of 250 nm (Fig. 2) (Xie, Willerth et al. 2009). About 60% of the cells within the EBs underwent differentiation when cultured on either aligned or random nanofibers (Xie, Willerth et al. 2009). Significantly lower numbers of glial fibrillary acidic protein (GFAP)-positive cells (astrocytes) were observed on aligned nanofibers compared with those on random nanofibers. Stem/progenitor cells normally show biased differentiation depending on substrate morphology. The majority of the human embryonic stem cells cultured on electrospun polyurethane scaffolds with a mean fiber diameter of 360±80 nm differentiate toward neuronal formation, while on 2D flat surfaces, a high proportion of GFAP positive cell types was evident (Carlberg, Axell et al. 2009). Fiber diameter also influences the differentiation of neural stem cells (NSCs) derived from mature CNS lineages (Christopherson, Song et al. 2009). Rat hippocampus-derived adult NSCs show enhanced differentiation on polyethersulfone nanofibrous substrates compared with that on tissue culture plastic (Fig. 3) (Christopherson, Song et al. 2009). However, an increase of oligodendrocytes was evident on smaller diameter fibers (283 nm), with the highest neuron cell number and lowest astrocyte numbers observed on the larger diameter (749 nm) fiber substrates. Cell therapy for spinal cord injuries in animal models using neural progenitors without a scaffold showed 10% of the transplanted cells becoming neurons, 60% oligodendrocytes, and about 30% astrocytes (McDonald, Becker et al. 2004). The intrinsic structural features of electrospun nanofibers, used as a "permanent" stimulus during nerve regeneration, may provide better cues to control cell differentiation compared with cell suspension injections.

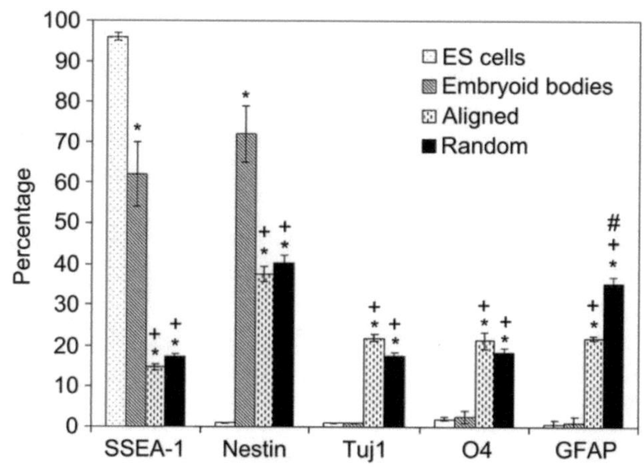

Fig. 2. Immunohistochemstry of ES cells, EBs (treated with by 4-/4+ induction) cultured on tissue culture plates without scaffolds and of EBs (treated with by 4-/4+ induction) cultured on aligned or random PCL nanofibrous scaffolds for 14 days. Cells were stained with SSEA-1 for undifferentiated mouse ES cells, Nestin for neural precursor cells, Tuj1 for neuronal cells, O4 for oligodendrocytes, and GFAP for astrocytes. * $p < 0.05$ compared with ES cells. + $p < 0.05$ compared with EBs. # $p < 0.05$ compared with random PCL nanofibers. (Reprinted from Biomaterials with permission of Elsevier) (Xie, Willerth et al. 2009).

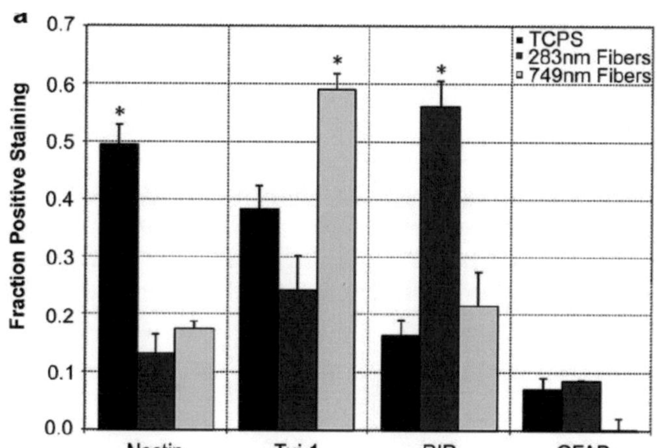

Fig. 3. Immunohistochemstry of rat NSCs cultured on tissue culture polystyrene (TCPS), electrospun polyethersulfone nanofibrous scaffolds with diameters of 283 and 749 nm respectively in differentiation medium. Cells were stained with Nestin for neural progenitor cells, Tuj-1 for neurons, RIP for oligodendrocytes, GFAP for astrocytes. * $p < 0.05$ compared

with cell staining on the other two culture substrates. (Reprinted from Biomaterials with permission of Elsevier) (Christopherson, Song et al. 2009)

Electrospun nanofibrous scaffolds and contact guidance:

In repairing intricate neural networks, different kinds of cell types need to be spatially distributed into discrete patterns, with axons precisely guided to their synaptic targets. The way spatial distribution mediates guidance of neurites depends on the following factors inherent to the nanostructured scaffold:

1) physical cues e.g. scaffold fiber dimension, scaffold topography, external stress and stiffness.
2) chemical cues e.g. surface-bound charged molecules and functional groups
3) biological cues e.g. signaling and growth factors and ECM molecules

The topography of the nanofibrous scaffold plays an important guidance role for neurite outgrowth. Primary rat dorsal root ganglia (DRG) cultured on aligned electrospun fibers of acrylonitrile-co-methylacrylate (PAN-MA) (diameter ranged between 400 and 600 nm) extend neurites parallel to the fiber direction (Fig. 4) (Kim, Valerie et al. 2008). Neurites extending out from EBs on electrospun poly-ε-caprolactone (PCL) also followed the aligned fiber direction (Xie, Willerth et al. 2009), indicating that cells at different stages can respond similarly to these topographies.

By comparing micro- with nanostructured scaffolds, submicron topography not only causes highly increased non-isotropic differentiation but also significantly influences the contact guidance cues for PC12 and substantia nigra cells (Manwaring, Walsh et al. 2004; Cecchini, Bumma et al. 2007). However, nanoscale features may only exert contact guidance cues within a certain window of nanodimensions (Manwaring, Walsh et al. 2004; Foley, Grunwald et al. 2005). Neurites on aligned fibers are significantly longer than that on randomly aligned fibers (Corey, Lin et al. 2007; Xie, Willerth et al. 2009; Yao, O'Brien et al. 2009), and this increase in length is also associated with changes in the polarity of individual cells due to re-organization of cytoskeletal components. Similar changes in cell structure are responsible for the elongated morphology of the cell body and neurite outgrowth for PC12 cells, human embryonic stem cells and neuroblastoma cells (Corey, Lin et al. 2007; Gerecht, Bettinger et al. 2007; Ferrari, Cecchini et al. 2009). Specifically, cytoskeletal filaments including vimentin, α-tubulin and calponin align in the same axis as the topographic nanoscale feature (Gerecht, Bettinger et al. 2007), with a reversal of filament alignment in response to treatment with actin polarization disrupters. However, the precise sequence of cytoskeletal events in association with the observed contact guidance remains unclear (Cecchini, Bumma et al. 2007).

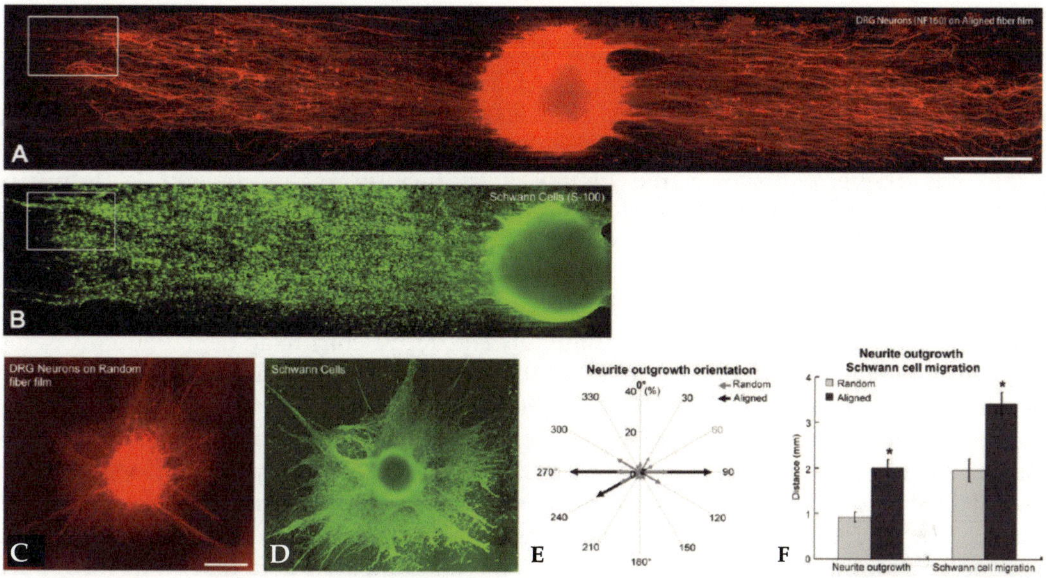

Fig. 4. Contact guidance role of aligned electrospun PAN-MA nanofibers of neurites extending from cultured DRG explants: (A) neurites extended parallel to aligned fibers, stained by NF160 for axons. (B) Schwann cell migration along aligned fibers, stained by S-100. (C, D) Random outgrowth of DRG neurites on random fibers, stained by NF160 and S-100 respectively. (E) Quantitative comparison of neurite outgrowth on aligned and random fibers, arrows indicate neurite extending direction, arrow length indicates the possibility of growth in that direction. (F) Quantitative comparison of neurite outgrowth length and Schwann cell migration on random and aligned fibers. *$p < 0.05$, error bar stands for standard error of mean (Reprinted from Biomaterials with permission of Elsevier) (Kim, Haftel et al. 2008).

In addition to topographical cues on neurite guidance, chemical cues, in many cases, are just as critical for nerve regeneration. After PNS/CNS injury, the secreted growth factors are normally at sub-optimal levels, and aligned nanofibrous scaffolds can potentially encourage neuritogenesis in cooperation with growth factors, especially for fibers with diameters less than 400 nm (Foley, Grunwald et al. 2005). Collectively, the alignment of nanofibrous scaffolds could potentially provide a better approach to PNS and CNS repair when bridging long gaps across a lesion site.

Building Biofunctionality into Electrospun Nanofibers:

Synthetic polymeric materials are widely used due to their superior mechanical properties and ease of electrospinning. At present, many of these biodegradable polymers have limited interaction with cells, relying largely on non-specific adsorption of proteins and the subsequent interaction with cell membrane processes and receptors which may give mixed and unwanted signaling (Cecchini, Bumma et al. 2007). The challenge remains to produce nanofibers with more bioactive surfaces, significantly improving specific targeting of cell-substrate interactions and consequently creating a more biomimetic microenvironment for implanted cells. There are several methods to improve bioactivity of nanofibrous scaffolds:

1) Polymer blending

Synthetic polymers can be blended with naturally derived biopolymers and ECM components such as hyaluronic acid (Ji, Ghosh et al. 2006) and collagen (Schnell, Klinkhammer et al. 2007; Jose, Thomas et al. 2009). Some of these polymers normally have specific biomolecular interactions at the cell surface to elicit specific responses and phenotypic changes, however the biomolecules themselves are limited in terms of practicality by their poor mechanical properties and formability. It is possible to fabricate electrospun scaffolds from blends of synthetic and natural polymers, which will then have improved cell-substrate interactions. Collagen for example, is the prevalent structural protein of the ECM, however it is relatively unstable due to the low fiber strength to withstand long-term support for cell adhesion and rapid enzymatic degradation (Okada, Hayashi et al. 1992). It should be noted that collagen denatures into gelatin when being electropun from fluoroalcohol solvents, therefore losing some of its ECM function (Zeugolis, Khew et al. 2008). However, using a mixture of phosphate-buffered saline and ethanol as solvent can avoid denaturing of collagen during the electrospinning process (Dong, Arnoult et al. 2009). Blending collagen with synthetic polymers has also assisted in overcoming these issues. The orientation of neurites from chick embryonic DRG is enhanced on aligned blended polycaprolactone/collagen (72:25) nanofibers compared with that on aligned, pure PCL (Fig. 5) (Schnell, Klinkhammer et al. 2007). Meanwhile, strong fasciculation of axons occurred on PCL nanofibers, which may suggest stronger axon-axon interactions in preference to axon-substrate interactions. The migration and proliferation of Schwann cells is also significantly improved on aligned PCL/collagen nanofibers, indicating more specific biomolecular interactions between cells and the collagen polymers on the nanofiber surface (Geiger, Bershadsky et al. 2001).

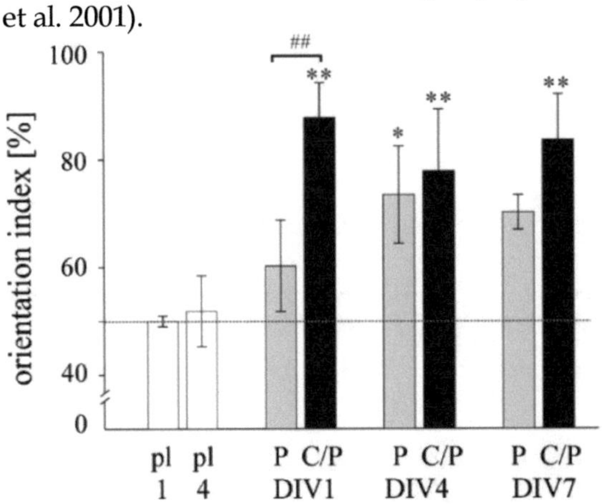

Fig. 5. Orientation of neurites extending from DRG explants on polylysine coated coverslips (pl), aligned PCL electrospun nanofibers (P), aligned collagen/PCL co-electrospun nanofibers (C/P), measured on 1, 4 and 7 days *in vitro* (DIV). The horizontal dotline (50% orientation index) indicates random alignment of neurites, 0% and 100% orientation index represent perpendicular and complete alignment respectively. *p<0.05, **p<0.01 compared with random alignment , ##p<0.01 between P and C/P at 1 day *in vitro*, error bars indicate standard error of the mean. (Reprinted from Biomaterials with permission of Elsevier) (Schnell, Klinkhammer et al. 2007).

2) Surface biofunctionalization
Although naturally derived polymers like collagen can be electrospun together with synthetic polymers to provide biomemitic nanofibrous scaffolds described previously, the concentration of the biological polymer in the blend is variable and limited in order to maintain desirable mechanical strength, diameter and orientated structure (Kwon and Matsuda 2005; Schnell, Klinkhammer et al. 2007). The bioactivity of the biological molecule in the blend largely depends on its degree of migration to the nanofiber surface where it is exposed to cells (Kwon and Matsuda 2005). Furthermore, low molecular weight molecules are difficult to be processed via electrospinning since a sufficient molecular weight is generally required to provide viscosities for stable fiber formation. However, specific peptide motifs derived from fibronectin and collagen VI have been identified to play an important role in nerve regeneration such as encouraging sprouting of hippocampal mouse neurons (Schense, Bloch et al. 2000), and these motifs are only short fragments of up to six amino acids. Therefore, immobilization of these molecules to the synthetic nanofiber surface provides an alternative method to render the fibers bioactive. Low molecular weight biological molecules can be covalently cross-linked to the surface, provided the surfaces are suitably functionalized for crosslinking, otherwise, pre-chemical treatment is required. The type and density of functional groups available for crosslinking will determine the degree of functionalization and subsequent bioactivity of the scaffold. Polyamine nanofibers with surface immobilized peptides derived from spliced fibronectin type III repeat D of human tenascin-C significantly enhance neuronal attachment, neurite generation and neurite extension in a range of cell types (Ahmed, Liu et al. 2006). Therefore, surface immobilization of these small molecules that are neuroactive can provide a great advantage for neural tissue engineering. In addition small molecules, growth factors such as brain-derived neurotrophic factor (Horne, Nisbet et al. 2009) and basic fibroblast growth factor (Patel, Kurpinski et al. 2007) that are immobilized can also promote cell survival and neurite outgrowth.

Although electrospun nanofibers have gained widespread use in neural tissue engineering and will undoubtedly continue to do so in the future, "bottom up" or self-assembly approaches to nanofabrication have recently been applied to produce nanofibrous scaffolds for nerve repair. Although the number of studies in this area is currently not as numerous as that for electrospinning, this will perhaps change in the future as advances in the self-assembly processes lead to improved features such as mechanical properties and controlled degradability. The remainder of the chapter will focus on self-assembling oligopeptides and polypeptides, and provides a constructive counterpoint for the previous discussion on electrospun nanofibers.

4.3 Lessons learned from nature: Self-assembling scaffolds

Self-assembly is ubiquitous in nature and presents another approach for producing nanofibrous scaffolds for neural tissue engineering. Molecular self-assembly is mediated by weak, non-covalent bonds, such as van der Waals forces, hydrogen bonds, ionic bonds, and hydrophobic interactions. Although these bonds are relatively weak, collectively they play a major role in the conformation of biological molecules found in nature. In particular self-assembling peptides, under physiological conditions can give hydrogels with nanofibrous substructures. The self assembling peptides can be injected, potentially with replacement cells, at the lesion site in a minimally invasive procedure.

Self-assembled amphiphilic polymers (oligopeptides & polypeptides)

Amphiphilic peptide molecules, possessing a hydrophilic head group and a hydrophobic hydrocarbon tail group, can undergo micellization in aqueous environments. During this process, the hydrophobic interaction is the main driving force bringing closely together the head groups on the surface of the self-assembled aggregate. The self-assembled aggregate transition results from the balance between the attractive hydrophobic interactions and repulsive forces experienced by the head groups. Therefore, the self-assembly process can be triggered by the addition of cell culture medium *in vitro* or physiological fluids *in vivo*. Here the presence of ions in the culture medium or physiological fluids screen electrostatic repulsion among the molecules and promotes self-assembly (Silva, Czeisler et al. 2004; Tysseling-Mattiace, Sahni et al. 2008). Upon the addition of cell culture medium or physiological fluids, some oligopeptides form β-sheet secondary structures, followed by the spontaneous formation of nanofibrous structures with chain directions perpendicular to the fibril cross section (Zhang, Holmes et al. 1995; Holmes, de Lacalle et al. 2000; Semino, Kasahara et al. 2004; Gelain, Bottai et al. 2006). The self-assembly process yields a product with gel-like properties, while the micro-architecture of the self-assembled aggregate depends on the molecular structure and concentration of the amphiphiles, solvent dielectric constant, pH, balance of counter ions, and type and concentration of salt. (Holmes, de Lacalle et al. 2000). The wormlike self-assembled aggregates with branched or randomly 3D network structures, normally have fiber cross-sectional diameters of 2-20 nm, and the apparent length can be as high as several micrometers (Cates and Candau 1990; Holmes, de Lacalle et al. 2000; Silva, Czeisler et al. 2004). In this chapter, the importance of nanofibrous structures via self-assembly that provide a 3D architecture for both endogenous and implanted cells will be highlighted.

The potential advantage of amphiphile molecular self-assembly is that the molecules can be injected into cavities within CNS/PNS lesion site which causes little tissue damage and minimal inflammation at the site of injection (Holmes, de Lacalle et al. 2000; Tysseling-Mattiace, Sahni et al. 2008; Yang, Song et al. 2009), while larger injuries caused by implantation of other tissue engineering scaffolds like electrospun nanofibers may be difficult to avoid. Amphiphilic oligopeptides can be prepared in isotonic glucose solutions (Yang, Song et al. 2009) and then injected into the lesion site, where the self-assembly process is triggered by ions in the aqueous physiological environment.

The use of self-assembled nanofibers for nerve regeneration has provided some promising results, partially due to their high surface-to-volume ratio, fiber diameter and 3D architecture that resemble the natural ECM, but also in the spatial and high density presentation of small bioactive molecules to cell receptors. Self-assembled oligopeptide nanofibers present a promising microenvironment to cells with improved regenerative outcomes, including extensive neurite outgrowth, formation of functional synaptic connections for both cultured neuronal cell lines and primary neuronal cells, with no significant difference compared with cells cultured *in vitro* on Matrigel (Holmes, de Lacalle et al. 2000). Despite the simple chemistry of many of these oligopeptides (Holmes, de Lacalle et al. 2000; Guo, Su et al. 2007), compared with Matrigel which contains a range of ECM components and growth factors that are well known to promote neuronal cell adhesion and functional development, the structural roles of self-assembled nanofibrous scaffolds have shown a complimentary effect to chemical cues. Specific amino acid sequences such as arginine-glycine-aspartic acid (RGD) and isolucine-lysine-valine-alanine-valine (IKVAV) derived from fibronectin and laminin respectively, are involved in receptor interactions which play an important role in neurite outgrowth and guidance of neurite extension (Tashiro, Sephel et al. 1991; Schense, Bloch et al. 2000). These amino acid sequences can also be appended with synthetic peptides for self-assembly to create more biomimetic microenvironments.

Potentially, low cost amphiphilic polymers can be used as a backbone appended with these various chemical and biological motifs, able to self-assemble into 3D nanofibrous scaffolds and trigger specific cellular responses (Fig. 6) (Silva, Czeisler et al. 2004; Gelain, Bottai et al. 2006) Moreover, nanofibrous scaffolds tailored with well defined, synthetic functional groups such as cell signalling motifs known to control specific gene expression or cell signalling processes provide research tools to investigate cellular response in a more *in vivo* like 3D environment where cells are more likely to receive more external stimuli rather than traditional 2D surfaces coated with these functional groups.

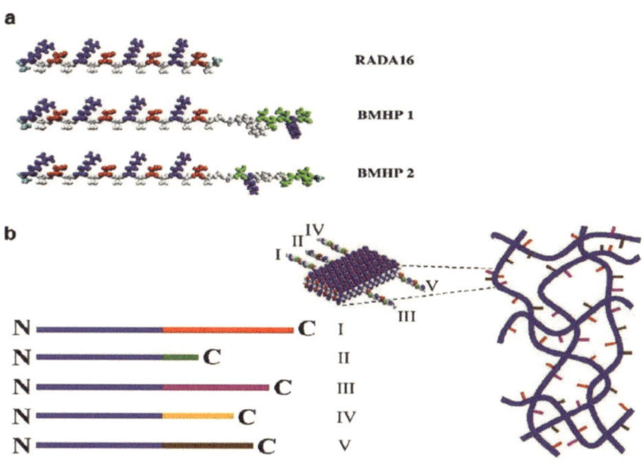

Fig. 6. (a) Schematic molecular structures of Arginine–Alanine–Aspartate (RAD) 16 peptide and Bone Marrow Homing Peptide (BMHP) 1, 2 appended RAD16. (b) Schematic structures of four different appended functional motifs RAD16 (I-V) and the structure of self-assembled peptide nanofibers. The enlarged insert shows the β-sheet with functional motifs extending out. (Reprinted from PLoS ONE with permission of Gelain et al.) (Gelain, Bottai et al. 2006).

Self-assembled oligopeptide nanofibers not only present these molecules to cells but arrange them in a spatial order that significantly changes the cellular response (Silva, Czeisler et al. 2004; Gelain, Bottai et al. 2006). For example, peptides coupled with IKVAV that self-assembled into nanofibers suppressed the differentiation of neural progenitor cells into astrocytes *in vitro*, however, non-bioactive IKVAV nanofibers mixed with soluble bioactive ones did not provide a biased differentiation effect (Silva, Czeisler et al. 2004). Self-assembled oligopeptides *in vivo* normally elicit a minimal inflammatory response (Guo, Su et al. 2007; Tysseling-Mattiace, Sahni et al. 2008). They can also suppress the hyperplastic response post CNS injury, but do not affect the early hypertrophic response, which may promote blood-brain barrier repair (Faulkner, Herrmann et al. 2004; Tysseling-Mattiace, Sahni et al. 2008). The reduced apoptosis of oligodendrocytes post injury within self-assembled oligopeptides can potentially contribute to the remyelination of regenerating axons at the injury site (Karimi-Abdolrezaee, Eftekharpour et al. 2006).

Diblock copolypeptides (about 200 amino acid residues) can also self assemble into nanofibrous hydrogels. The copolypeptides contain a hydrophilic polyelectrolyte block and a hydrophobic α-helical domain (Nowak, Breedveld et al. 2002; Breedveld, Nowak et al. 2004; Deming 2005) (Fig. 7). The polyelectrolyte block can be composed of a series of bioactive peptides like polylysine and polyglutamate with controlled degrees of polymerization which subsequently influences the gel point and strength of the resultant nanofibrous hydrogels *in vitro* and *in vivo* (Breedveld, Nowak et al. 2004; Yang, Song et al. 2009). The combination of hydrophobic α-helical block composed with a sufficient number of peptides residuals and the hydrophilic block helps the formation of a stiff gel and gelation at very low polypeptide concentrations (Nowak, Breedveld et al. 2002), potentially providing a wider range of hydrogel porosities. However, the properties of the hydrogel *in vitro* may not directly translate to hydrogels *in vivo,* due to the effect of different ions of body fluid on the self-assembly process (Beniash, Hartgerink et al. 2005). Unlike self-assembled oligopeptides, polypeptides can be assembled *in vitro* with or without cell suspensions, then injected to the lesion site by a micro-syringe due to shear thinning, followed by rapid structural recovery.

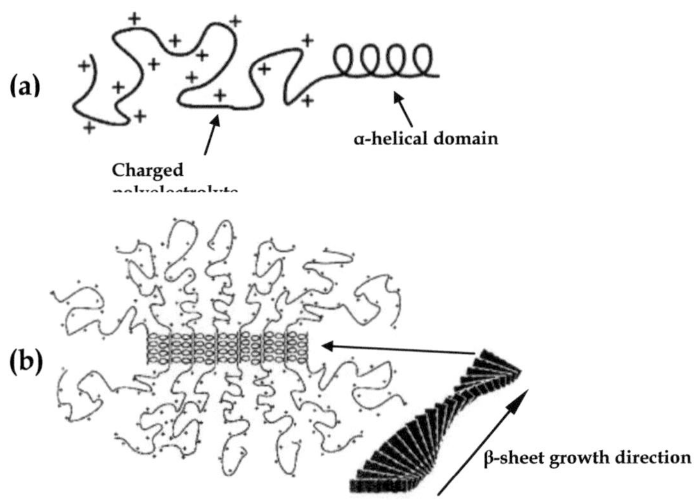

Fig. 7. Schematic structure of self-assembled polypeptides (a) Building block of self-assembled polypeptides, composed of charged polyelectrolyte block and oligopeptides that forms an α-helical domain. (b) Schematic structure of fibril-like nanostructure formed via β-sheet growth (no side chain shown). The enlarged insert shows the structure of β-sheet with polyelectrolyte side chains (Reprinted with permission from Breedveld et al. Copyright 2009 American Chemical Society) (Breedveld, Nowak et al. 2004).

Self-assembled peptide nanofibers can potentially reduce the formation of a glial scar by means of either suppressing the astrogliosis post injury within the self-assembled scaffold due to the nanofiber dimension and surface chemistry (Tysseling-Mattiace, Sahni et al. 2008) or by preventing infiltration of glial cells into the hydogel which can last for up to 8 weeks (Yang, Song et al. 2009). Fig. 8 shows an example of injecting self-assembling lysine-leucine block co-polymer ($K_{180}L_{20}$) into the mouse forebrain, which triggered moderate astrogliosis comparable to saline injection controls. The astrocyte activity reduced remarkably after 8 weeks within the host tissue with no evidence of glial cell ingrowth into the self-assembled scaffold. The minimal inflammatory response of the host tissue (Holmes, de Lacalle et al. 2000; Yang, Song et al. 2009) to self-assembled peptides reveals the potential to reduce inflammation-induced secondary injuries as well as the level of inhibitory molecules to promote functional recovery for the PNS/CNS.

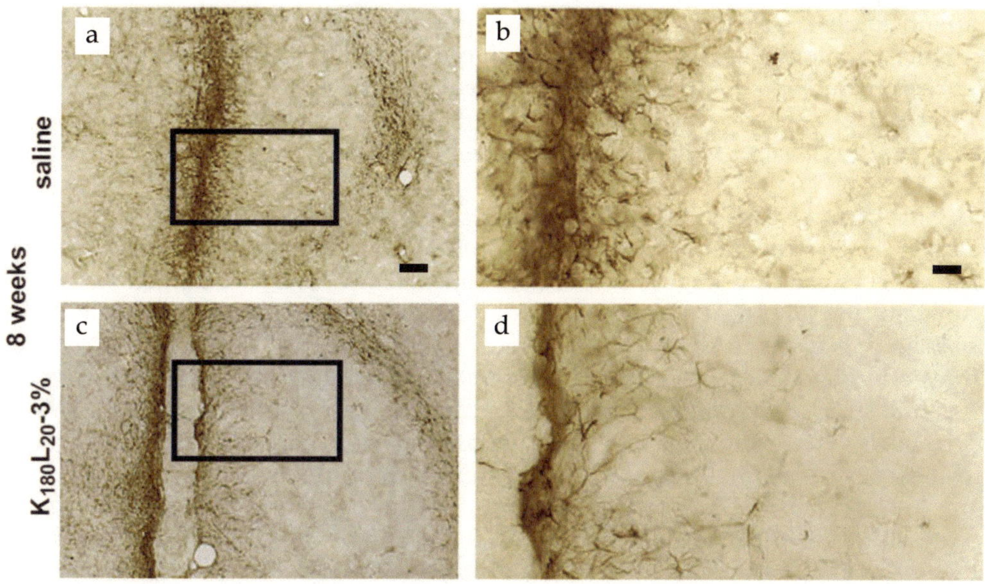

Fig. 8. Astrogliosis examined 8 weeks after physiological saline (a, b) and 3% $K_{180}L_{20}$ self-assembly polypeptide (c, d) injection into mouse forebrain. All tissue sections were immunochemically stained with GFAP for astrocytes. (b, d) are corresponding enlarged images of the box section in (a, c). Scale bar indicates 150 µm in (a, c) and 45 µm in (b, d) respectively. (Reprinted from Biomaterials with permission from Elsevier) (Yang, Song et al. 2009)

5. Long term view on neural tissue engineering

Although both electrospun and self-assembled peptide fibrous scaffolds show promising results in nerve regeneration after PNS/CNS injuries, there is still an essential requirement to further investigate the cellular response *in vivo* with scaffolds. The importance of transplantation of stem/progenitor cells arises because of the complicated dynamic *in vivo* microenvironment. The presence of proteins with non-specific adsorption properties on the scaffold surface can possibly cause unexpected outcomes *in vivo* (Yang, Song et al. 2009).

This may lead to differences in cellular responses such as adhesion and differentiation. Besides, various types of resident cells presented at the lesion site such as endogenous astrocytes, oligodendrocytes, and microglia together with their molecular products post injury may contribute to a more pernicious environment to neurons and implanted cells. In the case of SCI, the state of the axons as well as the extent and chronicity of the lesions are an important determinant of locomotor ability, e.g. the time point at which transplanted oligodendrocytes remyelinate regenerating axons most effectively (Keirstead, Nistor et al. 2005). There is an urgent need to define at which stage of differentiation neural-committed progenitor cells should be transplanted, and therefore optimize the regenerative functions of transplanted cells.

An understanding of the mechanisms of the inhibitory environment and barriers for regeneration post PNS/CNS injury is still limited. Although many inhibitory factors, like chondroitin sulphate proteoglycans and axon outgrowth inhibitors have been identified, a large number of molecules that are being produced by various cell types post injury, have not yet been characterized and their role in nerve regeneration remains to be defined (Fitch and Silver 2008). Moreover, the specific stimuli that trigger astrocyte and oligodendroctye activation remains unclear, which restricts the use of these as support cells.

The current "design rules" of tissue engineering scaffolds used for nerve repair only mimic a small part of the structural features of natural ECM and incorporate a limited number of biochemical cues that are known to promote functional recovery of injured nerves. The inflammation and foreign body response due to the introduction of the scaffold and transplanted cells needs to be controlled to reduce further activation of cells like microglia, astrocytes and oligodendrocytes which contribute to the inhibitory environment. Studies using scaffolds with/without transplanted cells for nerve regeneration are largely based on animal models. However, *in vivo* studies in patients will be essential to evaluate the effectiveness of such neural tissue engineering approaches as therapeutic strategies.

There are still numerous physical and biochemical features (and combinations thereof) which remain uncharacterized, and may be of great importance for cell survival and regulation of the cellular response. For instance *in vitro* studies have revealed some of the cellular responses to specific physical or biochemical cues presented by the scaffold. However a detailed understanding of the cellular response to the combination of these external cues within 3D architectures is limited at this stage.

Many challenges lay ahead to achieve successful nerve regeneration using scaffolds. The current regenerative outcome obtained both *in vitro* and *in vivo* in animal models using nanostructured scaffolds, reveals the exciting possibility of manipulating cell behaviour that can promote cell survival, neurite outgrowth, appropriate reinnervation and consequently the functional recovery post PNS/CNS injuries.

6. Acknowledgements

This work is supported by grants from the the Australian Research Council (DP0985433) the Baker Foundation, Bellberry Ltd, the National Health and Medical Research Council of Australia and the National Multiple Sclerosis Society of New York.

7. References

Access Economics (2009). Spinal Cord Injury Facts and Figures at a Glance. Birmingham, Alabama, National Spinal Cord Injury Statistical Center.

Ahmed, I., H.-Y. Liu, et al. (2006). "Three-dimensional nanofibrillar surfaces covalently modified with tenascin-C-derived peptides enhance neuronal growth in vitro." Journal of Biomedical Materials Research Part A 76A(4): 851-860.

Ao, Q., A. J. Wang, et al. (2007). "Combined transplantation of neural stem cells and olfactory ensheathing cells for the repair of spinal cord injuries." Medical Hypotheses 69(6): 1234-1237.

Bain, G., D. Kitchens, et al. (1995). "Embryonic stem cells express neuronal properties in vitro." Developmental Biology 168(2): 342-357.

Baker, B. M., A. O. Gee, et al. (2008). "The potential to improve cell infiltration in composite fiber-aligned electrospun scaffolds by the selective removal of sacrificial fibers." Biomaterials 29(15): 2348-2358.

Belle, J. E. L., M. A. Caldwell, et al. (2004). "Improving the survival of human CNS precursor-derived neurons after transplantation." Journal of Neuroscience Research 76(2): 174-183.

Beniash, E., J. D. Hartgerink, et al. (2005). "Self-assembling peptide amphiphile nanofiber matrices for cell entrapment." Acta Biomaterialia 1(4): 387-397.

Blight, A. R. (1994). "Effects of silica on the outcome from experimental spinal cord injury: Implication of macrophages in secondary tissue damage." Neuroscience 60(1): 263-273.

Borkenhagen, M., R. C. Stoll, et al. (1998). "In vivo performance of a new biodegradable polyester urethane system used as a nerve guidance channel." Biomaterials 19(23): 2155-2165.

Boudriot, U. and R. D. A. G. J. H. Wendorff (2006). "Electrospinning Approaches Toward Scaffold Engineering: A Brief Overview." Artificial Organs 30(10): 785-792.

Breedveld, V., A. P. Nowak, et al. (2004). "Rheology of block copolypeptide solutions: hydrogels with tunable properties." Macromolecules 37(10): 3943-3953.

Bundesen, L. Q., T. A. Scheel, et al. (2003). "Ephrin-B2 and EphB2 regulation of astrocyte-meningeal fibroblast interactions in response to spinal cord lesions in adult rats." Journal of neuroscience 23(21): 7789-7800.

Busch, S. A. and J. Silver (2007). "The role of extracellular matrix in CNS regeneration." Current Opinion in Neurobiology 17(1): 120-127.

Carlberg, B., M. Z. Axell, et al. (2009). "Electrospun polyurethane scaffolds for proliferation and neuronal differentiation of human embryonic stem cells." Biomedical Materials 4(4): 045004.

Cates, M. E. and S. J. Candau (1990). "Statics and dynamics of worm-like surfactant micelles." Journal of Physics: Condensed Matter 2(33): 6869-6892.

Cecchini, M., G. Bumma, et al. (2007). "PC12 differentiation on biopolymer nanostructures." Nanotechnology 18(50): 505103.

Chamberlain, L. J., I. V. Yannas, et al. (1998). "Early peripheral nerve healing in collagen and silicone tube implants: Myofibroblasts and the cellular response." Biomaterials 19(15): 1393-1403.

Chan, C. C. M., C. R. Roberts, et al. (2008). "Aggrecan components differentially modulate nerve growth factor-responsive and neurotrophin-3-responsive dorsal root ganglion neurite growth." Journal of Neuroscience Research 86(3): 581-592.

Chen, Y.-S., C.-L. Hsieh, et al. (2000). "Peripheral nerve regeneration using silicone rubber chambers filled with collagen, laminin and fibronectin." Biomaterials 21(15): 1541-1547.

Chew, S. Y., R. Mi, et al. (2007). "Aligned protein–polymer composite fibers enhance nerve regeneration: a potential tissue-engineering platform." Advanced Functional Materials 17(8): 1288-1296.

Christopherson, G. T., H. Song, et al. (2009). "The influence of fiber diameter of electrospun substrates on neural stem cell differentiation and proliferation." Biomaterials 30(4): 556-564.

Corey, J. M., D. Y. Lin, et al. (2007). "Aligned electrospun nanofibers specify the direction of dorsal root ganglia neurite growth." Journal of Biomedical Materials Research Part A 83A(3): 636-645.

Cummings, B. J., N. Uchida, et al. (2006). "Human neural stem cell differentiation following transplantation into spinal cord injured mice: association with recovery of locomotor function." Neurological Research 28(5): 474-481.

David, S. and A. J. Aguayo (1981). "Axonal elongation into peripheral nervous system "bridges" after central nervous system injury in adult rats." Science 214(4523): 931-933.

Deming, T. J. (2005). "Polypeptide hydrogels via a unique assembly mechanism." Soft Matter 1: 28-35.

Desai, T. A. (2000). "Micro- and nanoscale structures for tissue engineering constructs." Medical Engineering & Physics 22(9): 595-606.

Deshpande, D. M., Y.-S. Kim, et al. (2006). "Recovery from paralysis in adult rats using embryonic stem cells." Annals of Neurology 60(1): 32-44.

Dirnagl, U., C. Iadecola, et al. (1999). "Pathobiology of ischaemic stroke: an integrated view." Trends in Neurosciences 22(9): 391-397.

Dodla, M. C. and R. V. Bellamkonda (2008). "Differences between the effect of anisotropic and isotropic laminin and nerve growth factor presenting scaffolds on nerve regeneration across long peripheral nerve gaps." Biomaterials 29(1): 33-46.

Dong, B., O. Arnoult, et al. (2009). "Electrospinning of collagen nanofiber scaffolds from benign solvents." Macromolecular Rapid Communications 30(7): 539-542.

Economics, A. (2009). The economic cost of spinal cord injury and traumatic brain injury in Australia, Access Economics for the Victorian Neurotrauma Initiative.

Eichhorn, S. J. and W. W. Sampson (2005). "Statistical geometry of pores and statistics of porous nanofibrous assemblies." Journal of the royal society Interface 2: 309-318.

Fan, Y. W., F. Z. Cui, et al. (2002). "Culture of neural cells on silicon wafers with nano-scale surface topograph." Journal of Neuroscience Methods 120(1): 17-23.

Faulkner, J. R., J. E. Herrmann, et al. (2004). "Reactive astrocytes protect tissue and preserve function after spinal cord injury." Journal of Neuroscience 24(9): 2143-2155.

Fawcett, J. W. (1997). "Astrocytic and neuronal factors affecting axon regeneration in the damaged central nervous system." Cell and Tissue Research 290(2): 371-377.

Ferrari, A., M. Cecchini, et al. (2009). "Directional PC12 cell migration along plastic nanotracks." IEEE Transactions on Biomedical Engineering 56: 2692-2696.

Fidler, P. S., K. Schuette, et al. (1999). "Comparing astrocytic cell lines that are inhibitory or permissive for axon growth: the major axon-inhibitory proteoglycan is NG2." Journal of Neuroscience 19(20): 8778-8788.

Filbin, M. T. (2003). "Myelin-associated inhibitors of axonal regeneration in the adult mammalian CNS." Nature Reviews Neuroscience 4(9): 703-713.

Fitch, M. T., C. Doller, et al. (1999). "Cellular and molecular mechanisms of glial scarring and progressive cavitation: in vivo and in vitro analysis of inflammation-induced secondary injury after CNS trauma." Journal of Neuroscience 19(19): 8182-8198.

Fitch, M. T. and J. Silver (2008). "CNS injury, glial scars, and inflammation: Inhibitory extracellular matrices and regeneration failure." Experimental Neurology 209(2): 294-301.

Foley, J. D., E. W. Grunwald, et al. (2005). "Cooperative modulation of neuritogenesis by PC12 cells by topography and nerve growth factor." Biomaterials 26(17): 3639-3644.

Formhals, A. (1934). Process and apparatus for preparing artificial threads. 1975504.

Geiger, B., A. Bershadsky, et al. (2001). "Transmembrane crosstalk between the extracellular matrix and the cytoskeleton." Nature Reviews. Molecular Cell Biology 2(11): 793-805.

Gelain, F., D. Bottai, et al. (2006). "Designer self-assembling peptide nanofiber scaffolds for adult mouse neural stem cell 3-dimensional cultures." PLoS ONE 1(1): e119.

Gerecht, S., C. J. Bettinger, et al. (2007). "The effect of actin disrupting agents on contact guidance of human embryonic stem cells." Biomaterials 28(28): 4068-4077.

GrandPre, T., F. Nakamura, et al. (2000). "Identification of the Nogo inhibitor of axon regeneration as a Reticulon protein." Nature 403(6768): 439-444.

Guo, J., H. Su, et al. (2007). "Reknitting the injured spinal cord by self-assembling peptide nanofiber scaffold." Nanomedicine: Nanotechnology, Biology and Medicine 3(4): 311-321.

Gupta, D., J. Venugopal, et al. (2009). "Aligned and random nanofibrous substrate for the in vitro culture of Schwann cells for neural tissue engineering." Acta Biomaterialia 5(7): 2560-2569.

He, Z. and V. Koprivica (2004). "The Nogo signaling pathway for regeneration block." Annual Review of Neuroscience 27(1): 341-368.

Heijke, G. C. M., P. J. Klopper, et al. (2001). "Silicone rubber tubulization in peripheral sensory nerve reconstruction: An experimental study in rabbits." Microsurgery 21(7): 306-316.

Holmes, T. C., S. de Lacalle, et al. (2000). "Extensive neurite outgrowth and active synapse formation on self-assembling peptide scaffolds." Proceedings of the National Academy of Sciences of the United States of America 97(12): 6728-6733.

Horne, M. K., D. R. Nisbet, et al. (2009). "Three dimensional nanofibrous scaffolds incorporating immobilized BDNF promote proliferation and differentiation of cortical neural stem cells." Stem Cells and Development. accepted

Ichihara, S., Y. Inada, et al. (2009). "Development of new nerve guide tube for repair of long nerve defects." Tissue Engineering Part C: Methods 15(3): 387-402.

Jansen, K., J. F. A. van der Werff, et al. (2004). "A hyaluronan-based nerve guide: in vitro cytotoxicity, subcutaneous tissue reactions, and degradation in the rat." Biomaterials 25(3): 483-489.

Ji, Y., K. Ghosh, et al. (2006). "Electrospun three-dimensional hyaluronic acid nanofibrous scaffolds." Biomaterials 27(20): 3782-3792.

Jose, M. V., V. Thomas, et al. (2009). "Fabrication and characterization of aligned nanofibrous PLGA/Collagen blends as bone tissue scaffolds." Polymer 50(15): 3778-3785.

Karimi-Abdolrezaee, S., E. Eftekharpour, et al. (2006). "Delayed transplantation of adult neural precursor cells promotes remyelination and functional neurological recovery after spinal cord injury." Journal of Neuroscience 26(13): 3377-3389.

Karnezis, T., W. Mandemakers, et al. (2004). "The neurite outgrowth inhibitor Nogo A is involved in autoimmune-mediated demyelination." Nature Neuroscience 7: 736-744.

Keirstead, H. S., G. Nistor, et al. (2005). "Human embryonic stem cell-derived oligodendrocyte progenitor cell transplants remyelinate and restore locomotion after spinal cord injury." Journal of Neuroscience 25(19): 4694-4705.

Khan, T., S. Sayers, et al. (1990). "Carbon filaments provide support and directionality to growing rat fetal spinal cord explants." Neuroscience Letters 118(2): 172-176.

Kim, H. S., K. Kim, et al. (2005). "Morphological characterization of electrospun nano-fibrous membranes of biodegradable poly(L-lactide) and poly(lactide-co-glycolide)." Macromolecular Symposia 224(1): 145-154.

Kim, Y.-t., V. K. Haftel, et al. (2008). "The role of aligned polymer fiber-based constructs in the bridging of long peripheral nerve gaps." Biomaterials 29(21): 3117-3127.

Kitahara, A., Y. Suzuki, et al. (1998). "Evaluation of collagen nerve guide in facial nerve regeneration." Journal of Artificial Organs 1(1): 22-27.

Kottis, V., P. Thibault, et al. (2002). "Oligodendrocyte-myelin glycoprotein (OMgp) is an inhibitor of neurite outgrowth." Journal of Neurochemistry 82(6): 1566-1569.

Kwon, I. K. and T. Matsuda (2005). "Co-electrospun nanofiber fabrics of poly(l-lactide-co-e-caprolactone) with Type I collagen or heparin." Biomacromolecules 6(4): 2096-2105.

Labrador, R. O., M. But, et al. (1998). "Influence of collagen and laminin gels concentration on nerve regeneration after resection and tube repair." Experimental Neurology 149(1): 243-252.

Leaver, S. G., A. R. Harvey, et al. (2006). "Adult olfactory ensheathing glia promote the long-distance growth of adult retinal ganglion cell neurites in vitro." Glia 53(5): 467-476.

Lee, A. C., V. M. Yu, et al. (2003). "Controlled release of nerve growth factor enhances sciatic nerve regeneration." Experimental Neurology 184(1): 295-303.

Li, M. S. and S. David (1996). "Topical glucocorticoids modulate the lesion interface after cerebral cortical stab wounds in adult rats." Glia 18(4): 306-318.

Lietz, M., A. Ullrich, et al. (2006). "Physical and biological performance of a novel block copolymer nerve guide." Biotechnology and Bioengineering 93(1): 99-109.

Lu, P., L. L. Jones, et al. (2005). "BDNF-expressing marrow stromal cells support extensive axonal growth at sites of spinal cord injury." Experimental Neurology 191(2): 344-360.

Lundborg, G., L. B. Dahlin, et al. (1991). "Ulnar nerve repair by the silicone chamber technique." Scandinavian Journal of Plastic and Reconstructive Surgery and Hand Surgery 25(1): 79 - 82.

Lundborg, G., L. B. Dahlin, et al. (1982). "Nerve regeneration in silicone chambers: Influence of gap length and of distal stump components." Experimental Neurology 76(2): 361-375.

Lundborg, G., B. Rosen, et al. (2004). "Tubular repair of the median or ulnar nerve in the human forearm: A 5-year follow-up." Journal of hand surgery 29(2): 100-107.

Manwaring, M. E., J. F. Walsh, et al. (2004). "Contact guidance induced organization of extracellular matrix." Biomaterials 25(17): 3631-3638.

McDonald, J. W., D. Becker, et al. (2004). "Repair of the injured spinal cord and the potential of embryonic stem cell transplantation." Journal of Neurotrauma 21(4): 383-393.

Mingyu, C., G. Kai, et al. (2004). "Surface modification and characterization of chitosan film blended with poly-L-lysine." Journal of biomaterials applications 19(1): 59-75.

Moore, K., M. Macsween, et al. (2006). "Immobilized Concentration Gradients of Neurotrophic Factors Guide Neurite Outgrowth of Primary Neurons in Macroporous Scaffolds." Tissue engineering 12(2): 267-278.

Murakami, T., Y. Fujimoto, et al. (2003). "Transplanted neuronal progenitor cells in a peripheral nerve gap promote nerve repair." Brain Research 974(1-2): 17-24.

Nakamura, T., Y. Inada, et al. (2004). "Experimental study on the regeneration of peripheral nerve gaps through a polyglycolic acid-collagen (PGA-collagen) tube." Brain Research 1027(1-2): 18-29.

Nam, J., Y. Huang, et al. (2007). "Improved Cellular Infiltration in Electrospun Fiber via Engineered Porosity." Tissue engineering 13(9): 2249-2257.

National Spinal Cord Injury Statistical Center (2009). "Spinal Cord Injury Facts and Figures at a Glance". Birmingham, Alabama.

Neville, C. M., A. Y. Huang, et al. (2009). "Neural Precursor Cell Lines Promote Neurite Branching." International Journal of Neuroscience 119(1): 15-39.

Nisbet, D. R., L. M. Y. Yu, et al. (2008). "Characterization of neural stem cells on electrospun poly(ε-caprolactone) submicron scaffolds: evaluating their potential in neural tissue engineering." Journal of Biomaterials Science, Polymer Edition 19: 623-634.

Nowak, A. P., V. Breedveld, et al. (2002). "Rapidly recovering hydrogel scaffolds from self-assembling diblock copolypeptide amphiphiles." Nature 417(6887): 424-428.

Oh, S. H., J. H. Kim, et al. (2008). "Peripheral nerve regeneration within an asymmetrically porous PLGA/Pluronic F127 nerve guide conduit." Biomaterials 29(11): 1601-1609.

Okada, T., T. Hayashi, et al. (1992). "Degradation of collagen suture in vitro and in vivo." Biomaterials 13(7): 448-454.

Panseri, S., C. Cunha, et al. (2008). "Electrospun micro- and nanofiber tubes for functional nervous regeneration in sciatic nerve transections." BMC Biotechnology 8: 39.

Patel, M., P. J. Vandevord, et al. (2008). "Collagen--Chitosan Nerve Guides for Peripheral Nerve Repair: A Histomorphometric Study." Journal of Biomaterials Applications 23(2): 101-121.

Patel, S., K. Kurpinski, et al. (2007). "Bioactive Nanofibers: Synergistic Effects of Nanotopography and Chemical Signaling on Cell Guidance." Nano Letters 7(7): 2122-2128.

Pham, P. Q., U. Sharma, et al. (2006). "Electrospinning of Polymeric Nanofibers for Tissue Engineering Applications: A Review." Tissue engineering 12(5): 1197-1211.

Schense, J. C., J. Bloch, et al. (2000). "Enzymatic incorporation of bioactive peptides into fibrin matrices enhances neurite extension." Nature Biotechnology 18(4): 415-419.

Schmidhammer, R., S. Zandieh, et al. (2005). "Effects of alleviated tension at the nerve repair site using biodegradable tubular conduits: Histological, electrophysiological and functional results in a rat model." European Surgery 37(4): 213-219.

Schmidt, C. E. and J. B. Leach (2003). "Neural Tissue Engineering: Strategies for Repair and Regeneration." Annual Review of Biomedical Engineering 5(1): 293-347.

Schnell, E., K. Klinkhammer, et al. (2007). "Guidance of glial cell migration and axonal growth on electrospun nanofibers of poly-e-caprolactone and a collagen/poly-e-caprolactone blend." Biomaterials 28(19): 3012-3025.

Semino, C. E., J. Kasahara, et al. (2004). "Entrapment of Migrating Hippocampal Neural Cells in Three-Dimensional Peptide Nanofiber Scaffold." Tissue engineering 10(3-4): 643-655.

Silva, G. A., C. Czeisler, et al. (2004). "Selective Differentiation of Neural Progenitor Cells by High-Epitope Density Nanofibers." Science 303(5662): 1352-1355.

Sinis, N., H.-E. Schaller, et al. (2009). "Nerve regeneration across a 2-cm gap in the rat median nerve using a resorbable nerve conduit filled with Schwann cells." Journal of Neurosurgery 103(6): 1067-1076.

Smeal, R., R. Rabbitt, et al. (2005). "Substrate Curvature Influences the Direction of Nerve Outgrowth." Annals of Biomedical Engineering 33(3): 376-382.

Stevens, M. M. and J. H. George (2005). "Exploring and Engineering the Cell Surface Interface." Science 310(5751): 1135-1138.

Tashiro, K.-I., G. C. Sephel, et al. (1991). "The RGD containing site of the mouse laminin A chain is active for cell attachment, spreading, migration and neurite outgrowth." Journal of Cellular Physiology 146(3): 451-459.

Taylor, G. (1969). "Electrically Driven Jets." Proceedings of the Royal Society of London. Series A, Mathematical and Physical Sciences 313(1515): 453-475.

Taylor, J., P. Pesheva, et al. (1993). "Influence of janusin and tenascin on growth cone behavior in vitro." Journal of Neuroscience Research 35(4): 347-362.

Terzis, J., B. Faibisoff, et al. (1975). "The Nerve Gap: Suture Under Tension Vs. Graft." Plastic and Reconstructive Surgery 56(2): 166-170.

Tom, V. J., M. P. Steinmetz, et al. (2004). "Studies on the Development and Behavior of the Dystrophic Growth Cone, the Hallmark of Regeneration Failure, in an In Vitro Model of the Glial Scar and after Spinal Cord Injury." The Journal of Neuroscience 24(29): 6531-6539.

Tsai, E. C., P. D. Dalton, et al. (2004). "Synthetic Hydrogel Guidance Channels Facilitate Regeneration of Adult Rat Brainstem Motor Axons after Complete Spinal Cord Transection." Journal of Neurotrauma 21(6): 789-804.

Tysseling-Mattiace, V. M., V. Sahni, et al. (2008). "Self-Assembling Nanofibers Inhibit Glial Scar Formation and Promote Axon Elongation after Spinal Cord Injury." The Journal of Neuroscience 28(14): 3814-3823.

Wen, X. and P. A. Tresco (2006). "Effect of filament diameter and extracellular matrix molecule precoating on neurite outgrowth and Schwann cell behavior on multifilament entubulation bridging device in vitro." Journal of Biomedical Materials Research, Part A 76(3): 626-637.

Xie, F., Q. F. Li, et al. (2008). "In vitro and in vivo evaluation of a biodegradable chitosan-PLA composite peripheral nerve guide conduit material." Microsurgery 28(6): 471-479.

Xie, J., S. M. Willerth, et al. (2009). "The differentiation of embryonic stem cells seeded on electrospun nanofibers into neural lineages." Biomaterials 30(3): 354-362.

Yang, C.-Y., B. Song, et al. (2009). "Biocompatibility of amphiphilic diblock copolypeptide hydrogels in the central nervous system." Biomaterials 30(15): 2881-2898.

Yang, F., C. Y. Xu, et al. (2004). "Characterization of neural stem cells on electrospun poly(L-lactic acid) nanofibrous scaffold." Journal of Biomaterials Science, Polymer Edition 15(12): 1483-1497.

Yao, L., N. O'Brien, et al. (2009). "Orienting neurite growth in electrospun fibrous neural conduits." Journal of Biomedical Materials Research Part B: Applied Biomaterials 90B(2): 483-491.

Zeugolis, D. I., S. T. Khew, et al. (2008). "Electro-spinning of pure collagen nano-fibres - Just an expensive way to make gelatin?" Biomaterials 29(15): 2293-2305.

Zhang, H., K. Uchimura, et al. (2006). "Brain keratan sulfate and glial scar formation." Annals of the New York Academy of Sciences 1086(Integrated Molecular Medicine for Neuronal and Neoplastic Disorders): 81-90.

Zhang, S., T. C. Holmes, et al. (1995). "Self-complementary oligopeptide matrices support mammalian cell attachment." Biomaterials 16(18): 1385-1393.

Zhao, Q., G. Lundborg, et al. (1997). "Nerve regeneration in a 'pseudo-nerve' graft created in a silicone tube." Brain Research 769(1): 125-134.

Zhu, X., W. Cui, et al. (2008). "Electrospun Fibrous Mats with High Porosity as Potential Scaffolds for Skin Tissue Engineering." Biomacromolecules 9(7): 1795-1801.

Zong, X., K. Kim, et al. (2002). "Structure and process relationship of electrospun bioabsorbable nanofiber membranes." Polymer 43(16): 4403-4412.

24

Evolution in Tissue Engineering for the Lower Urinary Tract

Bouhout Sara[1]; Ouellet Gabrielle[1];
Perron Émilie[2] and Bolduc Stéphane[3]
Laboratoire d'Organogenèse Expérimentale/LOEX, Centre de recherche FRSQ du CHA universitaire de Québec, Faculté de Médecine, Université Laval, Québec Québec, Canada

1. Introduction

The organism is a complex machinery, which is composed of different systems acting in synergy to maintain the homeostasis and the integrity of the human body. When one of these systems cannot provide its regulating function, the quality of life and the survival of the organism are imperiled. The purpose of the urinary tract is to ensure equilibrium and consistency of the blood components. Thus, it maintains the osmotic tension, the acido-basic and hydro-electrolytic balance. The urine is elaborated from the blood medium and then evacuated out of the body. It consists of metabolic, nitrogenous and exogenous wastes, which makes it a toxic biological liquid for cells and structural proteins.

The human urinary system is composed of two sections; the upper urinary tract which includes the kidneys and the ureters; and the lower urinary tact that includes the bladder and the urethra. The bladder is a watertight and compliant reservoir in charge of urine storage before its evacuation via the urethra under sphincter control. In order to preserve the integrity of the upper urinary system, the maximum capacity of the bladder is approximately 500 mL. This eliminates possible reflux which could damage the renal function. There are two main differences between the female and male urinary tracts; the presence of the prostate and the length of the urethra. As showed in figure 1, the length of the female urethra is around 3 to 4 centimeters compared to 20 centimeters for the male.

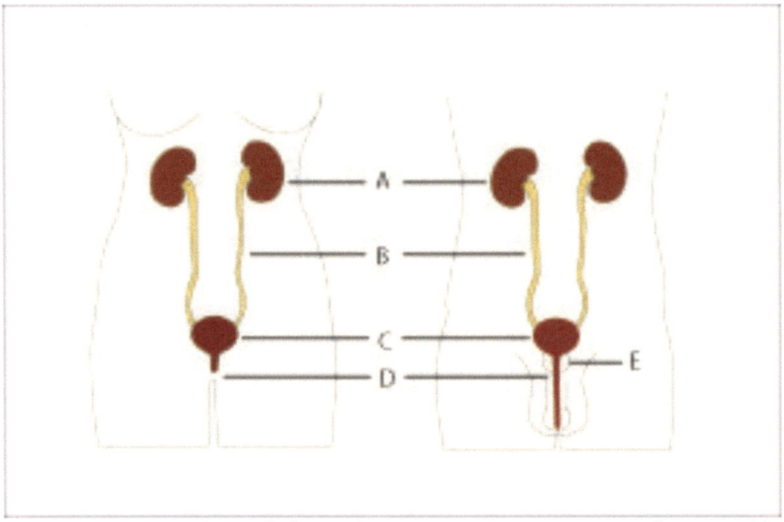

Fig. 1. Schematic representation of the urinary tract in woman and man. A) Kidneys. B) Ureters. C) Bladder. D) Urethra. E) Prostate.

As shown in figure 2, for the bladder and the urethra, the bladder mucosa is underlined with a unique and highly specialized epithelium, known as urothelium, which rests on a basal lamina. Underneath, the lamina propria is made of fibroblasts, microcapillaries and fibers of collagen mostly of type I and III. The detrusor muscle is formed by smooth muscle cells with interfascicular connective tissue containing fibers of collagen I and III. Finally, the presence of a serosa completes the bladder wall. The presence of collagen type I, III and elastin largely determine the mechanical properties of the bladder wall (McCarthy et al 2003). For the urethra, the mucosa is similar to the bladder but a smooth muscle cells layer replaces the detrusor and the serosa is not present.

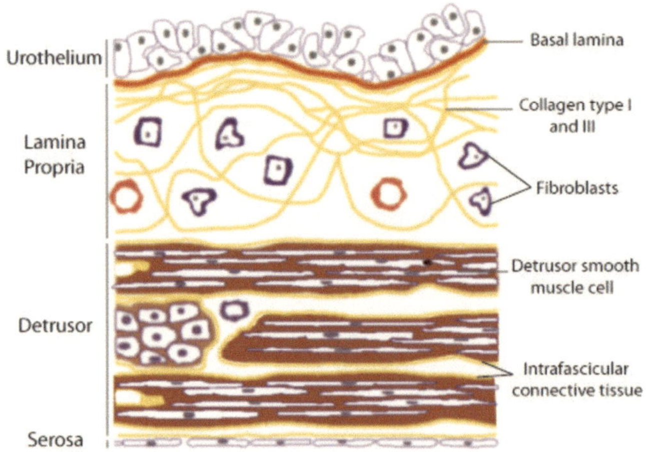

Fig. 2. Histological representation of the bladder structure of the mucosa.

All bladder layers have an important role, however we will focus our attention to the urothelium since it is the tissue which is directly in contact with the urine. In order to limit the passage of water and ions between urine and the bloodstream, the luminal superficial cell layer is composed of so called umbrella cells, which are derived from the basal urothelial cells precursors (Bolland & Southgate 2008). Umbrella cells are characterized by the presence of a specialized asymmetric apical membrane (AUM), which is an arrangement of uroplakin proteins by specific interactions, to form a protective barrier on the apical surface (Veranic et al. 2004). It is important to note that uroplakin organization is therefore a specific differentiation marker of the urothelium (Sun 2006).

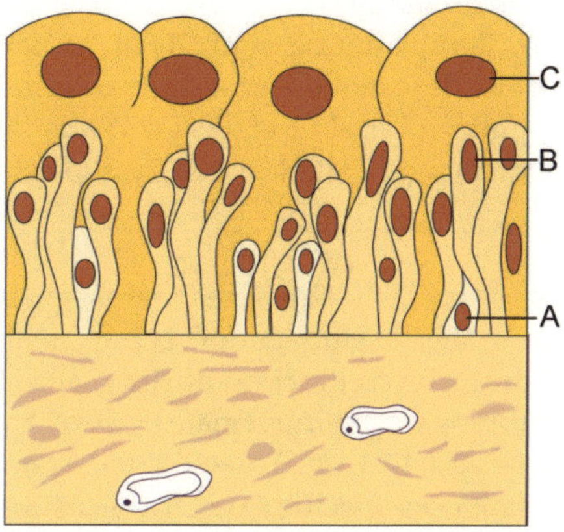

Fig. 3. Schematic representation of the pseudo-stratified urothelium. A) Basal cell. B) Intermediate cell. C) Superficial cell also known as umbrella cell.

Several conditions like congenital malformations, cancer, traumatisms and stricture can cause problems that will require surgical treatments to replace the injured section. To rectify those low urinary tract problems, partial or total reconstruction might be needed. In urology, the need for autologous tissues with properties similar to the native tissues is an important limitation. However, in organ regeneration, tissue engineering recently made great advances. It regroups biological, clinical and engineer disciplines that elaborate physiologically functional tissues similar to the native one found *in vivo*. Contrarily to the beginning of tissue engineering, cellular and molecular organizations are strongly considered nowadays. It is the reason why this discipline needs different matrices and host cells to reproduce a substitute conform to the original organ. Different types of substitutes, going from non-urologic tissues to different types of biomaterials, have been developed and used to guide *in vivo* regeneration on the injured site. The host cells migration from the extremity toward the center of the graft, consistently lead to post-surgical reorganization of the implanted substitute. This is why all current research studies head toward the fact that seeding cells *in vitro* on a matrix support will lead to better post-surgical results. This chapter will describe the tissue engineering position in urologic reconstruction and explore its evolution with the different types of strategies used.

2. Native Non-Urologic Tissues: the Beginning of Tissue Engineering

2.1 Bladder

In view of the necessity to maintain the compliance of pathological bladders, several types of tissues have been used unsuccessfully. The skin, the omentum, the stomach, the pericardium, representing native substitutes (Draper & Stark 1956, Goldstein et al. 1967, Andretto et al. 1981, Nguyen & Mitchell 1991), but also silicone, polyvinyl and Teflon, for the alloplastic materials (Kudish 1957, Bogash et al. 1960, Kelami et al. 1970), but they quickly encountered significant complications. The absence of biocompatibility, vascularization and the immunogenicity of some of these substitutes, as well as the direct contact with the toxic urine, lead mainly to the formation of fibrosis and contraction of the graft (Burbige & Hensle 1986, Vemulakonda et al. 2008). In parallel, the substitution of the bladder by the means of various intestinal segments has been practiced more successfully since the beginning of the eighties (Tariel et al. 2006). While taking into account the limits previously observed, this technique called enterocystoplasty is the clinical treatment currently used.

Unlike the first substitutes described, the intestinal segment allows improvement in compliance. Moreover, it has the advantage of being richly vascularized, preventing necrosis, inflammation, and thus, the contraction of the graft. However, like the first native substitutes, the mucous membrane of the intestinal segment has absorption properties and electrolytic secretion which are specific to it. Thus, the incorporation of these various native tissues involves metabolic imbalances specific to each one, as well as the formation of mucus and urolithiases, which are significant morbidity factors (Mundy & Nurse 1992, McDougal 1992). Its addition to the biochemical profiles of mucous membranes used, the bad contact with the toxic urine leads to risks of neoplasic transformation of the native substitutes (Woodhams et al. 2001, Ali-El-Dein et al. 2002), as well as structural deterioration of alloplastic substitutes (Cross et al. 2003). The incorporation of intestinal segments previously de-epithelialized was attempted to avoid mucus production and potential reabsorption within an animal model but it lead to fibrosis, and then, the contraction of the graft because of the urine being in direct contact with the underlying submucosa and extracellular matrix (Hafez et al. 2005).

2.2 Urethra

Several non-urologic tissues have also been tried for urethral reconstruction: ureter in 1909 (Schmieden 1909), saphenous vein in 1910 (Tuffier 1910) and appendix in 1911 (Lexer 1911). The skin was first used in 1914 (Nové-Josserand 1914) as a split thickness graft rolled into a tube. The results were not as expected because of the contracture of the graft, so it was abandoned until 1941 (Humby 1941) when a full thickness skin graft from the genital region was reported. In 1948 (Mc Indoe 1948), a split thickness graft technique was used. The patient had to wear a dilator for 6 months before the intervention as a tissue expander to obtain the harvest skin. The same year, the skin sample was used for urethral reconstruction (Vyas et al 1987, Young & Benjamin 1948).

The use of skin graft for urethroplasty is of two types, split or full thickness. For the full thickness skin graft, various donor sites like penile foreskin, supraclavicular skin, inner aspect of the upper arm, abdominal wall, buttock and thigh have been used for their different characteristics (Devine & Horton 1977, Hanna 1983, Hendren & Crooks 1980, Shapiro 1984). At the moment, the most commonly used remains the penile skin and foreskin as a flap, because they are more pliable and less thicker than the other types of sampling (Vyas et al 1987). Also the absence of hair in those two sites reduces the probabilities of complications. In general, good results have been obtained with the use of skin for urethroplasty. However, post-surgical complications still occurs. At the harvesting site, both type of skin graft can leads to poor cosmetic wound healing. Moreover, both types of skin grafts can lead to poor cosmetic wounds at the harvesting site. At the reconstruction site, complications like recurrent strictures or fistulas are found. Also, *Balanitis Xerotica Obliterans*, a genital skin inflammatory disease, can occur. Finally, hair growth from hair bearing harvest site, with or without concretion and graft contracture has been observed. The use of split thickness skin graft has shown a reduced tendency to contract, but it comes with a lack of adequate resiliency and strength (Markiewicz et al 2007). Actually, penile and foreskin skin flaps are considered the gold standard for urethroplasty, but when they are not available, due to use of that skin in a prior intervention, circumcision, presence of scar tissues or because of a long urethral defect, other tissues are needed for urethral reconstruction (Theodorescu et al 1998).

The first report using oral mucosa for urethroplasty was in 1941 (Humby 1941). It was not used again until 1992 (Burger et al 1992) where it was reintroduced as another free graft material for urethral reconstruction like penile skin. Two sites for oral mucosa are most commonly used for harvesting: the buccal and labial mucosa of the lower lip. If there is a lack of available genital skin, oral mucosa is a good replacement graft material because it is easily harvested, non-air bearing and in abundant supply even for adult cases. It has furthermore interesting properties like elasticity, abundant submucosal vascularity, thick mucosal layer and it is tolerant to air and liquid exposures (Caldamone et al 1998). Also, this avoids cosmetic disadvantages and consequences caused by the use of genital skin harvested from the penis (Barbagli & Lazzeri 2006). The morbidity associated with oral mucosa is pain, discomfort, limited range jaw opening, neurosensory defect and salivary flow modification. Those complications are temporary, going from 1 to several months depending on the harvesting site. The post-surgical complications found after urethral reconstruction are still fistula formation, stricture and meatal stenosis (Barbagli & Lazzeri 2006, Markiewicz et al 2008, Markiewicz et al 2007). A study compared oral mucosa with penile skin for urethroplasty and obtained a similar overall success rate with no significant difference and in conclusion, they were declared both excellent materials for urethral reconstruction (Alsikafi NF 2005).

Bladder mucosa for instance was first introduced in 1947 for urethroplasty, but significant complication rates delayed its clinical use until 1981 (Coleman et al 1981), where it was associated with good results in the case of a deficit of penile skin. It gained an immediate popularity with the clinicians because of its structural similarity with the urethra. However, due to the lower abdominal incision required to access the harvest site, the procedure is technically complex, it often results in poor esthetics and delay in wound healing occurs at the donor site. The principal complications found at the reconstructed site are mucosal prolapse and meatal stenosis (Markiewicz et al 2007, Vyas et al 1987). To overcome the complications found with the use of bladder mucosa, a combination of preputial or penile skin with bladder mucosa have been used (Duffy et al 1988). A reduction of the meatal stenosis was observed, but no other follow-up studies on this combination were found afterward.

For female urethral reconstruction, vaginal mucosa have been used (Brannan 1951). Urethral stricture or fistula in female are not as frequent and it is related to continence problems (McKinney 1979). The principal advantages of vaginal mucosa are the same as oral mucosa. It is easily available, hairless, elastic, naturally wet and it avoids any cosmetic wound problems. Good results have been obtained and limited complications have been reported post-surgically (Tsivian & Sidi 2006).

Other non urologic tissues have been studied for urethroplasty for specific situations when the gold standard is not available. In 1967, tunica vaginalis found around the testicule was first reported in an experimental study for urethral reconstruction (Ariyoshi 1967). It was recommended because it is a readily available source, simple to harvest and use. Other advantages such as a lack of hair follicles, a good tensile and physical property resulting in better sutures holding and a very good vascularization were reported. Over the years, it has been used in experimental studies for surgical techniques in urethroplasty (Calado et al 2005, Khoury et al 1989, Talja et al 1987) and in general, good results were observed. In 1998, a study compared an onlay flap with a tube flap performed on a New Zealand rabbit model with a control group. The New Zealand rabbit is the standard animal model used for urethral reconstruction. Their conclusion was that an onlay flap was more suitable for long term reconstruction, because it was further vascularized than the tube flap (Theodorescu et al 1998). In 2009, another study combined tunica vaginalis with preputial island flap, on a rabbit model for 12 weeks to overcome the complications found with the use of tunica vaginalis alone (Leslie et al 2009). Even with that combination, fistula and urethral diverticulum occurred. Clinically, the use of tunica vaginalis was evaluated in two studies with different conclusions. One found the use of tunica vaginalis useful mainly in difficult cases as long as there is no external exposure (Snow & Cartwright 1992) and the other found no advantage to its use as an onlay flap in proximal urethral repair due to its high stricture rate (Joseph & Perez 1999). The use of tunica vaginalis is now for coverage purposes, to protect the reconstructed area. Another tissue was used for an experimental study, the use of autologous vein graft on the rabbit model without strong conclusions (Kahveci et al 1995). More recently, colonic mucosa has been suggested for the urethral replacement in case of long and complex stricture. A dog model has been studied and it was also used for human urethral reconstruction with relatively good results. The principal complications at the harvest site were not described but at the reconstruction site: meatal stenosis, bulbar or bulbomembranous urethral stenosis and proximal anastomotic site stricture were observed with low complication rate (Xu et al 2004, Xu et al 2009, Xu et al 2003).

3. Acellular Tissue Matrices

The acellular matrix are obtained from native tissues, decellularized and sterilized. Thus, they are naturally cytocompatible and biodegradable, which represents a major asset as a substitution support. Various methods are used to carry out their preparation (Rosario et al.

2008), but the general outline presents a mechanical delamination of the tissue followed by an enzymatic treatment, an hypotonic solution to destroy cell residues and finally, the sterilization with ethylene oxide or peracetic acid. This kind of substitute has the advantage of presenting mechanical and biochemical properties, comparable to the native tissue. Indeed, the processing procedure allows preserving the three-dimensional environment, favorable to the cell infiltration, migration and proliferation within the substitute (Sutherland et al. 1996). The infiltrating cells then could have the suitable organization and differentiation to produce their own matrix during the resorption of the support, but this one is not in conformity with the physiological architecture. Indeed, these models cannot be identical since the mechanical and chemical treatment for the acellularization and sterilization, inevitably deteriorate the composition and organization of the matrix proteins (Brown et al. 2002). Let's remember that proteins are prone to unfolding which can be caused by temperature increase, other denaturing biophysical elements, like the pH, and biochemical agents, like urea. Thus, elasticity and mechanical resistance are decreased, which might not allow size variations during the bladder cycles of filling and emptying. Nevertheless, the aptitude to support the neovascularization is preserved (Lantz et al. 1993, Kajbafzadeh et al. 2007), at least the presence of endothelial cells precursors was observed post-graft, which is kinetically supported by the frequent use of a 100 μm thickness graft, instead of the full thickness.

3.1 Bladder

The small intestinal submucosa (SIS) was the first marketed model. After acellularization, a matrix mainly made up of collagen (I, II, VI), glycoaminoglycans, proteoglycans and glycoproteins like fibronectin was obtained. The combination of these various structural proteins allowed the storage of multiple growth factors, thus making them available to the surrounding cells. The heparan sulfate proteoglycans (HSPG), fibroblast growth factor 2 (FGF2), transforming growth factor beta (TGF-β), and the vascular endothelial growth factor (VEGF), essential to promote the graft neovascularization, were detected by immunofluorescence (Hurst & Bonner 2001), but in vivo regeneration showed only limited results in the rat and the dog models (Zhang et al. 2006). Indeed, the bladder capacity was improved partially even after 9 months post-operatively, however rapid deaths of one fourth (1/4) of the dogs due to perforation within the SIS had to be accounted for. To explain such results, an hypothesis was put forward by the same team at the sight of in vitro results, after urothelial and smooth muscle cells seeding (Zhang et al. 2000). Cells showed a good SIS adherence, but their organization and differentiation were not observed; which is however essential to preserve the graft size and to restore the functionality of the pathological bladder. A certain cytotoxicity was noticed concerning the SIS matrix, in addition to its immunogenic nature. Indeed, a series of postoperative inflammatory reactions (Ho et al. 2004) preceded the in vitro detection of porcine DNA residues on the commercially available SIS, by specific chromatin staining (Feil et al. 2006). These last reports made on this model, confirmed it as an unfavorable substitute for vesical reconstruction.

Recently, it was reported that the specific use of the organ to be regenerated is the most appropriate for the cellular organization and functionality (Sievert & Tanagho 2000). Thus, the bladder acellular matrix graft (BAMG) constitutes a logical substitute for the cell development. It was studied on various animal models, after being characterized in vitro. Under physiological conditions, the bladder matrix is made up of structure proteins, proteoglycans and glycoaminoglycans, able to trap and store the growth factors. Particularly, the matrix network constitutes a biochemical and spatial database, specific to the organ-source, thus informing the cells about the type of organization to be adopted. The bladder matrix therefore seems to be a substitute of choice, but after acellularization and sterilization treatments, the BAMG obtained displayed a diminished composition of elastin, laminin and fibronectin (Farhat et al. 2008). Collagen I and IV are preserved while the collagen III, one of the most abundant isoforms under the physiological conditions, seems to be degraded. The molecular architecture deterioration of the BAMG could potentially decrease its regenerating capacities. However, the dynamic culture of smooth muscle cells, within the BAMG, displayed encouraging results on the collagen I, III and IV synthesis (Farhat & Yeger 2008).

Beforehand, stretching forces were applied, which generated a growing interest to determine the impact of mechanical stress on the cell-cell and cell-matrix interactions (Baskin et al. 1993, Adam et al. 2004). However, this type of stress does not correspond to the physiological forces applied to the bladder. Dr. Farhat's team developed a bioreactor able to apply hydrostatic pressure waves, thus imitating the bladder filling and emptying (Wallis et al. 2008). Briefly, the seeded porcine BAMG, which is seeded with urothelial and smooth muscle cells, is placed between two chambers. One of which will be used to apply a pressure, gradually increasing on the presenting cell surface. At the end of four hours, the pressure reaches 10 cm H_2O, intravesical pressure at the fœtal stage, to fall abruptly in a few seconds to zero, to act like a voiding. The mRNA expression of matrix proteins increased significantly under mechanical stimulation and the mRNA expression of differentiation markers presented encouraging results. Indeed, mRNA coding for uroplakin II seemed to increase, which is promising for obtaining a watertight model and avoiding deterioration by the urine. This mode of dynamic preconditioning constitutes a promising alternative to counterbalance the deterioration attributable to the acellularized tissue preparation. Unfortunately in this study, only the gene expression was studied, which does not constitute sufficient information, since the protein functionality depends on its complete synthesis, its good folding and its appropriate localization, among other things. This bioreactor design is ideal to elaborate a bladder substitute under similar physiological conditions, but it still needs optimization. For example, the chamber design does not allow uniform distribution of the pressure waves on the tissue in place, but the evolution of this work is followed with attention. Meanwhile, studies in static mode continue and present relatively favorable results after implantation in a porcine model (Merguerian et al. 2000); in spite of the recurring complications observed, such as calcification formation and graft retraction. Urothelial and smooth muscle cells infiltration was noted but their localization remained limited to the graft periphery, and their organization was incomplete. The already raised hypothesis is that acellularization treatment affects the bladder matrix architecture. The BAMG is then obtained with a too high porosity to retain the growth factors and to promote the cell propagation (Farhat et al. 2003). In static culture conditions, porosity reduction was observed by hyaluronic acid incorporation during rehydratation phase of lyophilized BAMG (Cartwright et al. 2006). Hyaluronic acid is an essential glycoaminoglycan, known to support cell proliferation and migration in the extracellular matrix. The study's authors also claimed to make a barrier against urine with this method, but let us remember that urine pH and ionic forces have denaturing properties for nonspecific urothelium proteins.

BAMG and SIS are also limited by their xenogenic source. The rejection of this type of substitute is a serious possibility in sight of epitope presence, specific to the organ-source. The Gal oligosaccharide was found in the porcine SIS extracellular matrix, and that, even after the various preparation treatments (Badylak & Gilbert 2008). This antigen is present on the cell surface of many species, except human and a category of monkeys, which poses problems of immunological recognition post-graft. In the same way, it is not possible to eliminate any risk of viral contamination. To get around these undesirable characteristics, in addition to the differences inherent to the acellularized matrices preparation and their source (Kropp et al. 2004), the idea of constructing neobladders in a reproducible way was made possible by the use of synthetic polymers.

3.2 Urethra

The use of acellular matrix relies on the fact that it could have an effect with its environment because the extracellular matrix could influence the biological behavior of the cells (Farhat et al 2008) and lead to a better integration of the graft. In the literature, many acellular tissue matrices are called collagen matrix, for example, the bladder acellular matrix graft (BAMG). The principal advantages of using a BAMG matrix, as the other types of acellular matrices for urethral reconstruction is that it is an "off the shelf " material that eliminates all other surgery required for harvesting. It may also shorten the operative time, since it is easily prepared in large quantity in laboratory and can be stored until its use (Chen et al 1999, De Filippo et al 2002). Many studies have been done on the use of this type of material alone or with cells. A study was first done using xenogenic BAMG from a porcine source on the rabbit model for 6 months with an onlay fashion graft. Their goal was to evaluate if the use of BAMG would be suitable for a urethral reconstruction on an animal model (Chen et al 1999). Their results have demonstrated that it was appropriate since there was no evidence of infection, graft rejection, fistula or stone formation. In 2003, a study was done with an allogenic BAMG coming from cadaveric human bladder tissue on 28 adult patients (El-Kassaby et al 2003). A follow-up of 36 to 48 months was made and good results were observed, only 4 strictures occurred. Finally, another study was done with xenogenic BAMG in a tubular form from a porcine source, on a rabbit model for 1 month but reported opposite results (Dorin et al 2008). Their goal was to determine if there was a maximum limit for urethral defect in which an acellular matrix would be sufficient for replacement. The results showed that an unseeded tubularized BAMG may be successful when used for repair of a relatively small urethral defect of less than 0,5 centimeter. It is also well documented that re-anastomosis in those small urethral defects is very successful, so it would seem to limit the potential value of acellular tubularized matrix graft. The study suggested that for a urethral defect over 1 centimeter, the need to seed cells on acellular matrix may be necessary to allow the growth of healthy urethral tissue and avoid stricture formation and other complications. It also showed the discrepancy observed in the last decade from the literature regarding the success of acellular matrices for tissue regeneration on small animals and the failure on large animals and humans (Dorin et al 2008).

Due to these problems, studies began to report the importance of seeding cells on BAMG for urethral reconstruction. Those studies were done using xenogenic and allogenic BAMG seeded with cells and they were compared to unseeded BAMG as control. A first study using a porcine xenogenic BAMG in a tubular form was seeded with urothelial and smooth muscle cells extracted from New Zealand rabbit's bladders and implanted on a rabbit model for 6 months. The results showed that BAMG seeded with cells from normal bladder tissue can be used for a tubularized replacement, whereas BAMG without cells lead to poor tissue formation and stricture (De Filippo et al 2002). Another study, with an allogenic model consisting of BAMG seeded with foreskin epithelial cells from New Zealand rabbit's was performed on the same kind of animal for 6 months. The results were similar to the previous study, the BAMG seeded with epidermal cells obtained better outcomes than BAMG alone. Also, the research team claimed that the use of epidermal cells seemed adequate to replace urothelial cells (Fu et al 2007). The problem was that after 6 months, the epidermal cells kept their structural form rather than acquiring the transitional structure known to the urothelial cells. An additional study was done with rabbit allogenic BAMG. This time, the team seeded rabbit's oral keratinocytes and the study was done on rabbit model for a duration of 6 months. Their results showed that oral keratinocytes have a good biocompatibility with BAMG. It confirmed the prior results that seeded BAMG are superior to unseeded BAMG (Li et al 2008a, Li et al 2008b) and the oral keratinocytes kept their structure of stratified squamous epithelium rather than the transitional form of the urothelial cells, even after 6 months. In conclusion, cell seeded matrices seem to allow tissue regeneration across the defect and the maintenance of a non-obstructive outflow from the bladder; it confirms the need to incorporate cells.

For urethral reconstruction, other tissues were studied as acellular matrix graft. Small intestinal submucosa (SIS) was first suggested because it was shown to promote tissue specific regeneration in a variety of organs. SIS is composed of a collagen meshwork with various intrinsic growth factors, cytokines, structural proteins, glycoproteins and proteoglycans that may assist in cell migration as well as cell growth and differentiation during the regenerative process (Colvert et al 2002). The advantages of using this type of tissue are similar to BAMG matrices. A pilot study was done using xenogenic SIS, from a porcine source on a rabbit model for 3 months. Their goal was to evaluate the suitability of the SIS onlay patch in urethroplasty (Kropp et al 1998) and its feasibility, but it was a short-term study and not further mentioned. Also, the use of homologous acellular urethra was suggested for urethral reconstruction. The advantage of using an acellular urethra is that the extracellular matrix comes from the same place, so the various intrinsic growth factors; cytokines, structural proteins, glycoproteins and proteoglycans are identical. It allows a single-stage intervention with a tissue that has the same form and dimension. A research team studied the use of a homologous acellular urethra matrix for urethroplasty on a rabbit model for 8 months (Sievert et al 2000). Their goal was to evaluate urethral replacement by a free homologous graft of acellular urethral matrix on a rabbit model. They made a complete study of the different parameters on the take of the graft. Their results demonstrated a functional and histological regeneration. Another team did a similar study. They also used an allogenic acellular urethral matrix on a rabbit model. Their goal was the same but their implantation was for 6 months (Hu et al 2008). The results obtained were not convincing but they were similar. Also, an acellular aortic matrix was used for urethral reconstruction. The goal of this research team was to create an experimental model of urethral defect and to repair it using allogenic acellular aortic grafts as urethral substitutes (Parnigotto et al 2000). The advantages of using an aortic acellular matrix are the initial tubular structure with suitable dimensions and it is directly applicable. The surgeon can decide the length of the defect as well as the size of the urethra to be reconstructed. In general, they obtained good results, however, it was not further mentioned.

4. Synthetic polymers

Synthetic polymers are thermoplastic materials made of biodegradable composites that can be tailored to the patient's organ size. Mechanical and structural properties such as the size of the pores of these biomaterials can be controlled and their biological degradation is done by hydrolysis (Freed et al. 1994). It can also be produce in big quantity at a relatively low cost.

4.1 Bladder

The biomaterials synthesis, such as polyglycolic acid (PGA) and polylactic-co-glycolic acid (PLGA), offers a controlled composition after each production. They are authorized by the FDA and generally used for clinical drug encapsulation. This concept to isolate content from external space is particularly attractive for the bladder, since it can be used for the elaboration of a urinary storage bag. In vivo studies were performed where dogs were subjected to partial bladder replacement, using a PGA/PLGA scaffold, with cultured urothelial cells seeded on the luminal surface and smooth muscle cell on the exterior surface (Oberpenning et al. 1999). A functional evaluation was performed for 11 months which displayed an increased capacity to retain urine. This study demonstrated, most of all, that successful reconstitution of an organ was promoted when using cultured cells in vitro. The construct without cell seeding led to contraction and shrinkage already seen at the first post-graft month. Although histological results seemed to have a normal architecture, the study did not specify if the in vivo cellular organization was present just at the periphery or on the whole synthetic graft. Indeed, the migration of surrounding cells has not been frequently observed in the middle of the graft, where inosculation with host's vasculature takes more time in this area. Nanobiotechnology is well-positioned to influence synthetic scaffold optimization, but at cellular scale, the coating with a 3T3 mouse fibroblast as a feeder layer for urothelial cells already demonstrated an increased proliferation (Drewa et al. 2006).

In the same way as for the canine model, Atala's team continued clinical tests on candidate patients for cystoplasty suffering from high pressure bladders or pathological non-compliant bladder, by using the same scaffold wrapped with omentum but using a mixed of collagen and PGA (Atala et al. 2006). This study provided two key elements. First, the culture of urothelial and smooth muscle cells in a three-dimensional environment was feasible. Second, a nutritional support by the highly vascularized omentum wrapped around the synthetic scaffold provided positive effects on success. The young age of the patients theoretically promotes in vivo regeneration, but after 5 years, only one of the seven patients seemed to have significantly increased bladder compliance. In spite of everything, the biopsies obtained months after implantation showed a good organization of urothelial and smooth muscle cells, but the authors did not explain this paradox with the postoperative urodynamic results. Generally, the disadvantages of these polymers lie in the bad cicatrization with the host's tissue, as well as in the absence of biological recognition. However, on this last point, the nanobiotechnologies have the capacity to improve them by integrating cellular recognition domains in their structure, such as RGD peptide (Cook et al. 1997).

Harrington's review defines the discipline as the creation of objects or a surface whose unique functions are a direct result of their nanoscale dimension/organization (Harrington et al. 2008). The beginning of engineered synthetic matrices focused on the three-dimensional and porosity aspect, but to imitate the cellular environment with a more efficient approach, composites at a lower than 100nm were used to elaborate scaffolds since extracellular matrix proteins are nano-dimensional. To direct the cellular phenotype, the nanofibers composites are designed like collagen I and III fibrils, the most abundant isoforms of the native bladder. The conventional methods are based on a similar concept of adhesive protein interactions, such as non-covalent links, to form supramolecular polymer. Two synthesis techniques are commonly used. The « bottom-up » synthesis consists of triggered nanomolecules which self-assemble into a molecular object, like the micelles formation from individually charged lipids. In this case, the aggregate dimension is pre-determined through the incorporation of features with specific adhesive properties. The « top-down » approach is a technique in which untreated material is specifically degraded to result in a nanoscale patterned surface, as with lithographic techniques. Therefore, it appears evident that the new generation of biomaterials must manage the cells in their native language. Indeed, the incorporation of cytokines and the growth factor signaling could influence the cell behavior and promote controlled tissue regeneration. The challenge of this discipline is to precisely copy the extracellular matrix proteins architecture, to inform the cells of their physiological organization. In the bladder tissue engineering field, some studies of nanoscale materials were successfully completed (Baker et al. 2006, Han et al. 2006, McManus et al. 2007). Bladder smooth muscle cells adhesion and enhancement in elastin and collagen deposition were measured with the application of nano-structured PLGA/polyether-urethane scaffold (Pattison et al. 2007). The same study showed that synthetic composites are not yet spared from urine degradation, which led to post-graft bladder leaks in a rat model. Moreover, the clinical application of this cell-seeded construct still remains expensive.

4.2 Urethra

Many synthetic polymers have been studied over the years, alone or with cells to add more support to the material. The principal advantage using synthetic polymer is that it decreases the inflammatory reaction and other adverse effects, it is biodegradable and it eliminates the risk of hair growth or stone encrustation on hairs in the urethral lumen that can be found with the hair baring skin. Because it is an artificial material, it eliminates complications related to harvesting. Furthermore, it is possible to play on the different proprieties of the polymer like its porosity, the length and the diameter required for urethroplasty (Olsen et al 1992). At the end of the 70's, some alloplastics were studied for urethral reconstruction: Silicone, Dracon, Gore-Tex and Teflon were used without great success on dog and human (Court et al 1971, Court et al 1973, Gilbaugh et al 1969, Hakky 1977). They were unsuccessful and these materials could bring post-surgical complications like fistula and periurethral haematomas (Dreikorn et al 1979). With the improvement of technology, new synthetic polymers appeared and were studied for urethroplasty. In 1992, a tube formed of polyhydroxybutyric acid (PHB) coated with polyglycolic acid (PGA) on the inside was tested for urethroplasty on a dog model for 12 months. After 8 months, the polymer had completely disappeared and the lumen of the urethra was somewhat wider than normal and no complications were reported (Olsen et al 1992). These results were interesting but they were only based on macroscopic and histological examinations, no further studies were made after. In 1997, a group suggested to use guide implant for short urethral gaps. They made a pilot study on a new biodegradable and highly biocompatible polymer, the hyaluronan benzyl ester (Hyaff-11) on a rabbit model for a month. The implant guide was compared to two control groups; one using a silicone guide and a negative control. The principal advantages of the Hyaff-11 resides in its inertness, its possibility of producing devices with variable rates of biodegradation, its favorable profile of toxicity and biocompatibility, its properties of protecting delicate tissues and its potential for wound healing. In this pilot study, no complications were observed except at the suture site where microfistulas appeared. On histology, the use of Hyaff-11 seemed to promote the proliferation of urothelial cells in cuboidal form in opposition to the silicone guide, where the urothelial cells were in regression and in a squamous epithelium (Italiano et al 1997). No further studies were made with this material for urethroplasty.

More recently, a research team studied the use of a polyglycolic acid (PGA) stent coated with a polylactic-co-glycolic acid (PLGA) solution and seeded with bovine chondrocytes (Amiel et al 2001). They compared the model's reaction *in vitro*, in static condition, in a bioreactor and *in vivo*. It was implanted subcutaneously in a nude-mouse, for 4 and 10 weeks. The ability to implant chondrocytes in the genitourinary tract has been demonstrated in the literature for the treatment of vesicourethral reflux (Atala et al 1993, Atala et al 1994, Diamond & Caldamone 1999), urinary incontinence (Atala et al 1994) and penile prosthesis (Yoo et al 1998, Yoo et al 1999). Their final goal was to use autologous chondrocytes to obtain a biocompatible model. Generally, good results were obtained but further long-term studies have to be made on animal models. Also, the effect of toxic urine on the model has to be evaluated. Another team studied the use of synthetic polymer with cells on a rabbit model for 24 weeks (Fu et al 2009), a poly L-polylactic acid (PLLA) polymer stent seeded with autologous urothelial cells from the urethra. They used PLLA because it has a satisfactory mechanical strength in cross-section and moderate longitudinal flexibility, making it easy to adapt its expansion so that it can ensure support for the urethra and avoid collapse. The structure of the polymer also provides a good environment for cell adhesion. Good results were obtained. After 12 weeks, the polymer was degraded and after 24 weeks, the cells were implanted successfully, survived and regenerated well. However, it remains a short-term study. Like for other acellular tissue matrices, the use of cells in new synthetic polymers helped prevent complications and provided a better repair after urethroplasty. Even with these results, other materials are chosen before them and as one can see, not a lot of studies have been made to support their long-term use in urethral reconstruction.

5. Autologous approach: the self-assembly method

To this day, the integration of previously described substitutes has caused chronic inflammation or graft rejection. That is explain by the degradation of biosynthetic or xenogenous scaffold, that release non-recognized materials. The last few years, the organ reconstruction field knew the emergence of the self-assembly method (Auger et al. 2002), based on the cell capacity to produce their own matrix support. This approach consists of a cell culture in well defined conditions that will construct an autologous scaffold free of exogenous biomaterials. At the Laboratoire d'organogénèse expérimentale (LOEX), the self-assembly method proved its capacities in skin, blood vessel, and cornea regeneration (Pouliot et al. 2002, Tremblay et al. 2005, Germain et al. 1999), for graft or pharmacological studies. This is why our team works to elaborate autologous urological tissues. We investigate the feasibility to reproduce the physiological structure of bladder mucosa and genitourinary tubes. Dermal fibroblasts are interesting cells since, in presence of ascorbic acid (Chepda et al 2001, Davidson et al 1997), they have the ability to synthesize extracellular matrix in an increased way to form a collagen-layer. The autologous sheets could be then superposed or rolled to create a three-dimensional environment for the reorganization of additional cells specific to the reconstructed organ.

5.1 Bladder

The method starts by extracting urothelial, endothelial and smooth muscle cells, from a simple bladder biopsy (1cm by 5cm). This minimally invasive preoperative phase to harvest the three cell types of the bladder, was reported and confirmed good individual cell population purity (Magnan et al. 2006). The endothelialization of any substitute is necessary to promote the survival of the graft. Recently, LOEX laboratory obtained a rapid inosculation with a capillary-like network of reconstructed tissue and the host vasculature (Tremblay PL et al. 2005). Consequently, when cellular sheets were obtained, endothelial cells were added on, one day before they were superimposed, three by three. Urothelial cells were then seeded on top of the three-dimensional construction, and after a proliferation phase, the construction was elevated to the air/liquid interface as showed in figure 4. This technique was inspired from in vitro engineered epidermis, to induce cell maturation, because urothelial and epidermal cells are both epithelial in origin. However, epidermis and urothelium organization are not exactly the same. To generate a more physiological process, the replacement of the air/liquid phase by the dynamic bioreactor culture is currently being considered.

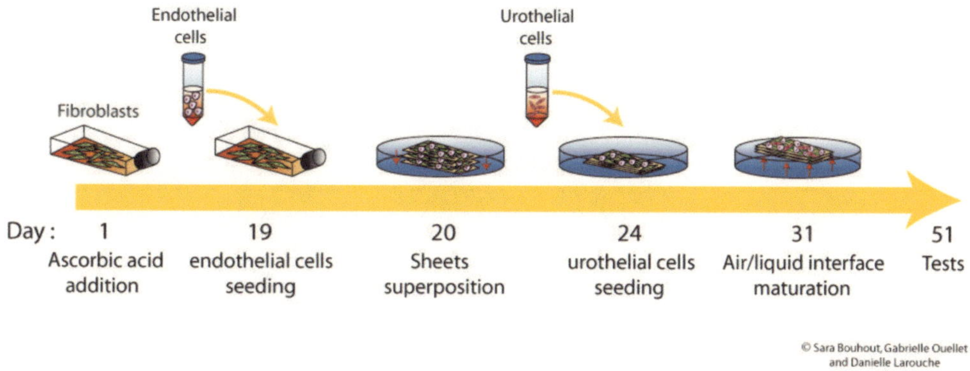

Fig. 4. Schematic representation of the bladder self-assembly approaches.

In the meantime, histological analysis showed a good distribution of the extracellular matrix and the cellular multi-layered morphology appears to be characteristic of normal bladder urothelium. Collagen I and basal membrane, essential to urothelial organization, were identified by immunofluorescence. More interestingly, cytokeratin 8/18 staining confirmed the good distribution of a stratified urothelium on the whole bioengineered support, which is important to avoid urine infiltration. Tissue permeability was evaluated by using Franz's cells, which consists of placing the vesical equivalent between the donor compartment containing radioactive urea and the receiver compartment, from which samples are obtained at different time points to determine permeation profiles. The endothelialized vesical equivalent displayed the same profile as the native bladder, contrarily to a construction of fibroblast only, which showed a fast diffusion of urea. This demonstration proved the necessity of having reconstructed urothelium covering substitutes, in order to avoid urine extravasation and secondarily, in vivo necrosis and fibrosis. Mechanical characteristics of the vesical equivalent seemed to have sufficient resistance to allow suturing. Even if the metalloproteinases released by endothelial cells reduced the vesical equivalent resistance, results remained satisfying because of their superiority compared to the minimal threshold measured from the native bladder (Dahms et al. 1998). Although this model still requires improvements, such as smooth muscle cells incorporation, its autologous origin and its efficiency as a barrier to urea, encourage further optimization in order to make this tissue-engineered equivalent, an ideal bladder replacement.

5.2 Urethra

Our tubular equivalent is build from the same cellular type. As showed in figure 5 and described previously (Magnan et al 2009), mature fibroblasts extracted from the skin are grown in culture medium for 28 days with ascorbic acid to stimulate the production of extracellular matrix. The presence of the extracellular matrix provides the formation of a fibroblasts sheet easy to handle, it enables us to roll it around a mandrel to obtain a tubular form. Then it is grown with another culture medium for 21 days with ascorbic acid, to increase the cohesion between the fibroblasts layers. When we obtain a uniform tube, urothelial cells are then seeded inside the tubular model and placed under perfusion at 15 ml per minute in a bioreactor with culture medium for 7 days with ascorbic acid to promote proliferation and maturation of the urothelial cells.

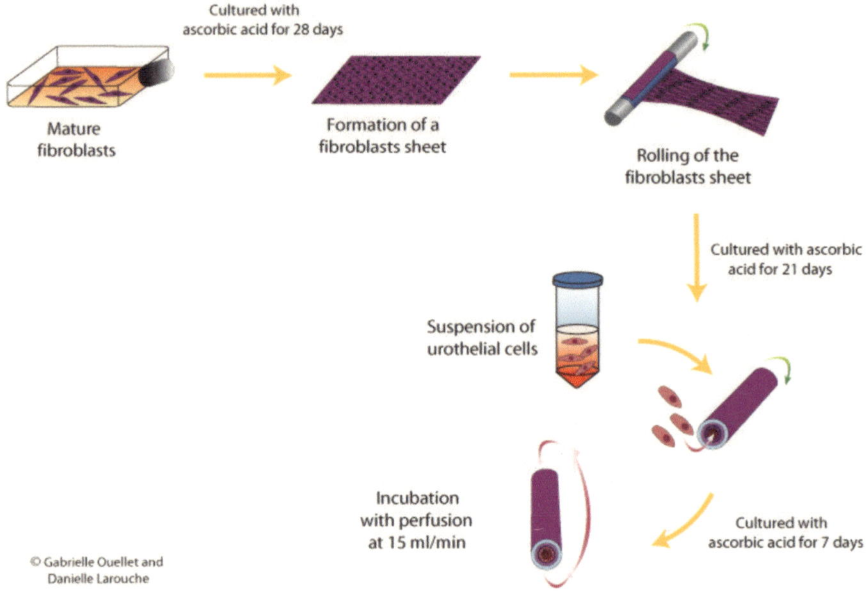

Fig. 5. Schematic representation of the urethral self-assembly approaches.

Macroscopically, the model was uniformly assembled, provided suture resistance and was easily handled. Histologically, we obtained a thick layer of fibroblasts in an abundant extracellular matrix and the urothelium was similar to a native urethra. The urethral model characterization with cytokeratin 8/18 by immunofluorescence and western blot confirmed the presence of a well differentiated and pseudostratified urothelium. The viability test demonstrated that the cells could be reextracted from the tube and reseeded with a 2% mortality rate, which is normally found in cell culture. Finally, the results of the burst pressure tests confirmed a higher resistance when compared to native porcine urethra.

Our autologous urethral equivalent is an innovative model that offers a promising alternative for urethral replacement or reconstruction. Like for alternative tissues, cell extraction will need harvesting from a skin biopsy, but in a smaller amount. To collect the urothelial cells, a bladder wash will be use. But the greatest advantage remains that our model will use the patient's cells with no xenogenic matrix, which should decrease the inflammatory reaction post-surgically.

6. Conclusion

Significant complications of alternative native tissue used for reconstruction raised the need for other types of materials. Tissue engineering had known successes and limits, but those limitations have contributed to move forward the knowledge in urologic tissue regeneration. Naturally derived materials and acellular tissue matrices have the potential advantages of familiar biological architecture. On the other hand, synthetic polymers can be obtained reproducibly on a large scale with control properties. But the challenge to produce a urologic tissue replacement with a scaffold that demonstrates functional compatibility with the native tissue still remains relevant. With the results obtained from the different studies, the protection against urine toxicity by a watertight construct and the quick vascularization of the urologic substitute are necessary to preserve the cellular growth and avoid graft contraction. The self-assembly approach seems a promising option to obtain an autologous engineered equivalent, endotheliazed and similar to the native tissue. For the future, all tissue engineering disciplines will need to explore new techniques, going from dynamic cell culturing to nanobiotechnologies, to obtain a unique structure and the full functionality of the urologic tissue to replace. We will have to consider the differences between a healthy animal model and the implantation of the substitute into a human patient with an active urologic pathology. This will be a major challenge.

7. References

Adam, RM. ; Eaton, SH. ; Estrada, C. ; Nimgaonkar, A. ; Shih, SC. ; Smith, LE. ; Kohane, IS. ; Bägli, D. & Freeman, MR. (2004). Mechanical stretch is a highly selective regulator of gene expression in human bladder smooth muscle cells. *Physiol Genomics.* 20(1):36-44.

Ali-El-Dein, B. ; El-Tabey, N. ; Abdel-Latif, M. ; Abdel-Rahim, M. & El-Bahnasawy, MS. (2002). Late uro-ileal cancer after incorporation of ileum into the urinary tract. *J Urol.* 167(1):84-8.

Alsikafi NF EM, McAninch JW. (2005). Long-term outcomes of penile skin graft versus buccal mucosal graft for substitution urethroplasty of the anterior urethra. *J. Urol.* 173: 87

Amiel GE, Yoo JJ, Kim BS, Atala A. (2001). Tissue engineered stents created from chondrocytes. *J Urol* 165: 2091-5

Andretto, R. ; Gonzales, J. ; Guidugli Netro, J. ; de Miranda, JF. & Antunes, AM. (1981). Experimental cystoplasty in dogs using preserved equine pericardium. *AMB Rev Assoc Med Bras.* 27(5):153-4.

Ariyoshi A. (1967). [Experimental studies of urethral reconstruction using tunica vaginalis graft]. *Nippon Hinyokika Gakkai Zasshi* 58: 417-32

Atala, A. ; Bauer, SB. ; Soker, S. ; Yoo, JJ. & Retik, AB. (2006). Tissue-engineered autologous bladders for patients needing cystoplasty. *Lancet.* 367(9518):1241-6.

Atala A, Cima LG, Kim W, Paige KT, Vacanti JP, et al. (1993). Injectable alginate seeded with chondrocytes as a potential treatment for vesicoureteral reflux. *J Urol* 150: 745-7

Atala A, Kim W, Paige KT, Vacanti CA, Retik AB. (1994). Endoscopic treatment of vesicoureteral reflux with a chondrocyte-alginate suspension. *J Urol* 152: 641-3; discussion 4

Auger, FA. ; Rémy-Zolghadri, M. ; Grenier, G. & Germain, L. (2002). A truly new approach for tissue engineering: the LOEX self-assembly technique. *Ernst Schering Res Found Workshop.* (35):73-88.

Badylak, SF. & Gilbert, TW. (2008). Immune response to biologic scaffold materials. *Semin Immunol.* 20(2):109-16.

Baker, SC. ; Atkin, N. ; Gunning, PA. ; Granville, N. ; Wilson, K. ; Wilson, D. & Southgate J. (2006). Characterisation of electrospun polystyrene scaffolds for three-dimensional in vitro biological studies. *Biomaterials.* 27(16):3136-46.

Barbagli G, Lazzeri M. (2006). Urethral reconstruction. *Curr Opin Urol* 16: 391-5

Baskin, L. ; Howard, PS. & Macarak, E. (1993). Effect of physical forces on bladder smooth muscle and urothelium. *J Urol.* 150(2 Pt 2):601-7.

Bogash, M. ; Kohler, FP. ; Scott, RH. & Murphy, JJ. (1960). Replacement of the urinary bladder by a plastic reservoir with mechanical valves. *Surg Forum.* 10:900-3.

Bolland, F. & Southgate, J. (2008). Bio-engineering urothelial cells for bladder tissue transplant. *Expert Opin Biol Ther.* (8):1039-49.

Brannan D. (1951). Stricture of the female urethra. *J Urol* 66: 242-53

Brown, AL. ; Farhat, W. ; Merguerian, PA. ; Wilson, GJ. ; Khoury, AE. & Woodhouse, KA. (2002). 22 week assessment of bladder acellular matrix as a bladder augmentation material in a porcine model. *Biomaterials.* 23(10):2179-90.

Burbige, KA. & Hensle, TW. (1986). The complications of urinary tract reconstruction. *J Urol.* 136(1 Pt 2):292-7.

Burger RA, Muller SC, el-Damanhoury H, Tschakaloff A, Riedmiller H, Hohenfellner R. (1992). The buccal mucosal graft for urethral reconstruction: a preliminary report. *J Urol* 147: 662-4

Calado AA, Macedo A, Jr., Delcelo R, de Figueiredo LF, Ortiz V, Srougi M. (2005). The tunica vaginalis dorsal graft urethroplasty: experimental study in rabbits. *J Urol* 174: 765-70

Caldamone AA, Edstrom LE, Koyle MA, Rabinowitz R, Hulbert WC. (1998). Buccal mucosal grafts for urethral reconstruction. *Urology* 51: 15-9

Cartwright, LM. ; Shou, Z. ; Yeger, H. & Farhat, WA. (2006). Porcine bladder acellular matrix porosity: impact of hyaluronic acid and lyophilization. *J Biomed Mater Res A.* 77(1):180-4.

Chen F, Yoo JJ, Atala A. (1999). Acellular collagen matrix as a possible "off the shelf" biomaterial for urethral repair. *Urology* 54: 407-10

Chepda T, Cadau M, Girin P, Frey J, Chamson A. (2001). Monitoring of ascorbate at a constant rate in cell culture: effect on cell growth. *In Vitro Cell Dev Biol Anim* 37: 26-30

Coleman JW, McGovern JH, Marshall VF. (1981). The bladder mucosal graft technique for hypospadias repair. *Urol Clin North Am* 8: 457-62

Colvert JR, 3rd, Kropp BP, Cheng EY, Pope JCt, Brock JW, 3rd, et al. (2002). The use of small intestinal submucosa as an off-the-shelf urethral sling material for pediatric urinary incontinence. *J Urol* 168: 1872-5; discussion 5-6

Cook, AD. ; Hrkach, JS. ; Gao, NN. ; Johnson, IM. ; Pajvani, UB. ; Cannizzaro, SM. & Langer, R. (1997). Characterization and development of RGD-peptide-modified poly(lactic acid-co-lysine) as an interactive, resorbable biomaterial. *J Biomed Mater Res.* 35(4):513-23.

Court B, Auvert J, Sausse A, Diep R. (1971). [Replacement of a urethral segment with a silicone elastomer tube in the male dog (preliminary note)]. *J Urol Nephrol (Paris)* 77: Suppl:562-4

Court B, Xerri A, Auvert J. (1973). [Replacement of the urethra in man using a silicone elastomer prosthesis. 3 cases. Preliminary results]. *J Urol Nephrol (Paris)* 79: 643-7

Cross, WR. ; Thomas, DF. & Southgate, J. (2003). Tissue engineering and stem cell research in urology. *BJU Int.* 92(2):165-71.

Dahms, SE. ; Piechota, HJ. ; Dahiya, R. ; Lue, TF. & Tanagho, EA. (1998). Composition and biomechanical properties of the bladder acellular matrix graft: comparative analysis in rat, pig and human. *Br J Urol.* 82(3):411-9.

Davidson JM, LuValle PA, Zoia O, Quaglino D, Jr., Giro M. (1997). Ascorbate differentially regulates elastin and collagen biosynthesis in vascular smooth muscle cells and skin fibroblasts by pretranslational mechanisms. *J Biol Chem* 272: 345-52

De Filippo RE, Yoo JJ, Atala A. (2002). Urethral replacement using cell seeded tubularized collagen matrices. *J Urol* 168: 1789-92; discussion 92-3

Devine CJ, Jr., Horton CE. (1977). Hypospadias repair. *J Urol* 118: 188-93

Diamond DA, Caldamone AA. (1999). Endoscopic correction of vesicoureteral reflux in children using autologous chondrocytes: preliminary results. *J Urol* 162: 1185-8

Dorin RP, Pohl HG, De Filippo RE, Yoo JJ, Atala A. (2008). Tubularized urethral replacement with unseeded matrices: what is the maximum distance for normal tissue regeneration? *World J Urol* 26: 323-6

Draper, JW. & Stark, RB. (1956). End results in the replacement of mucous membrane of the urinary bladder with thick-split grafts of skin. *Surgery.* 39(3):434-40.

Dreikorn K, Lobelenz J, Horsch R, Rohl L. (1979). Alloplastic replacement of the partially resected canine urethra by expanded polytetrafluoroethylene grafts. Preliminary results. *Urol Res* 7: 19-21

Drewa, T. ; Sir, J. ; Czajkowski, R. & Wozniak, A. (2006). Scaffold seeded with cells is essential in urothelium regeneration and tissue remodeling in vivo after bladder augmentation using in vitro engineered graft. *Transplant Proc.* 38(1):133-5.

Duffy PG, Ransley PG, Malone PS, Van Oyen P. (1988). Combined free autologous bladder mucosa/skin tube for urethral reconstruction: an update. *Br J Urol* 61: 505-6

El-Kassaby AW, Retik AB, Yoo JJ, Atala A. 2003. Urethral stricture repair with an off-the-shelf collagen matrix. *J Urol* 169: 170-3; discussion 3

Farhat, W. ; Chen, J. ; Erdeljan, P. ; Shemtov, O. ; Courtman, D. ; Khoury, A. & Yeger, H. (2003). Porosity of porcine bladder acellular matrix: impact of ACM thickness. *J Biomed Mater Res A.* 67(3):970-4.

Farhat WA, Chen J, Haig J, Antoon R, Litman J, et al. (2008). Porcine bladder acellular matrix (ACM): protein expression, mechanical properties. *Biomed Mater* 3: 25015

Farhat, WA. & Yeger, H. (2008). Does mechanical stimulation have any role in urinary bladder tissue engineering? *World J Urol.* 26(4):301-5.

Feil, G. ; Christ-Adler, M. ; Maurer, S. ; Corvin, S. ; Rennekampff, HO. ; Krug, J. ; Hennenlotter, J. ; Kuehs, U. ; Stenzl, A. & Sievert, KD. (2006). Investigations of urothelial cells seeded on commercially available small intestine submucosa. *Eur Urol.* 50(6):1330-7.

Freed, LE. ; Vunjak-Novakovic, G. ; Biron, RJ. ; Eagles, DB. ; Lesnoy, DC. ; Barlow, SK. & Langer, R. (1994). Biodegradable polymer scaffolds for tissue engineering. *Biotechnology (N Y).* 12(7):689-93.Fu Q, Deng CL, Liu W, Cao YL. 2007. Urethral replacement using epidermal cell-seeded tubular acellular bladder collagen matrix. *BJU Int* 99: 1162-5

Fu WJ, Zhang X, Zhang BH, Zhang P, Hong BF, et al. (2009). Biodegradable urethral stents seeded with autologous urethral epithelial cells in the treatment of post-traumatic urethral stricture: a feasibility study in a rabbit model. *BJU Int*

Germain, L. ; Auger, FA. ; Grandbois, E. ; Guignard, R. ; Giasson, M. ; Boisjoly, H. & Guérin, SL. (1999). Reconstructed human cornea produced in vitro by tissue engineering. *Pathobiology*. 67(3):140-7.

Gilbaugh JH, Jr., Utz DC, Wakim KG. 1969. Partial replacement of the canine urethra with a silicone prosthesis. *Invest Urol* 7: 41-51

Goldstein, MB. ; Dearden, LC. & Gualtieri, V. (1967). Regeneration of subtotally cystectomized bladder patched with omentum: an experimental study in rabbits. *J Urol*. 97(4):664-8.

Hafez, AT. ; Afshar, K. ; Bägli, DJ. ; Bahoric, A. ; Aitken, K. ; Smith, CR. & Khoury, AE. (2005). Aerosol transfer of bladder urothelial and smooth muscle cells onto demucosalized colonic segments for porcine bladder augmentation in vivo: a 6-week experimental study. *J Urol*. 174(4 Pt 2):1663-7.

Hakky SI. (1977). The use of fine double siliconised dacron in urethral replacement. *Br J Urol* 49: 167-71

Han, D. & Gouma, PI. (2006). Electrospun bioscaffolds that mimic the topology of extracellular matrix. *Nanomedicine*. 2(1):37-41.

Hanna MK. (1983). Single-stage hypospadias repair: techniques and results. *Urology* 21: 30-5

Harrington, DA. ; Sharma, AK. ; Erickson, BA. & Cheng, EY. (2008). Bladder tissue engineering through nanotechnology. *World J Urol*. 26(4):315-22.

Hendren WH, Crooks KK. (1980). Tubed free skin graft for construction of male urethra. *J Urol* 123: 858-61

Ho, KL. ; Witte, MN. & Bird, ET. (2004). 8-ply small intestinal submucosa tension-free sling: spectrum of postoperative inflammation. *J Urol*. 171(1):268-71.

Hu YF, Yang SX, Wang LL, Jin HM. (2008). Curative effect and histocompatibility evaluation of reconstruction of traumatic defect of rabbit urethra using extracellular matrix. *Chin J Traumatol* 11: 274-8

Humby G. (1941). A one-stage operation for hypospadias. *Brit. J. Surg*. 29: 84-92

Hurst, RE. & Bonner, RB. (2001). Mapping of the distribution of significant proteins and proteoglycans in small intestinal submucosa by fluorescence microscopy. *J Biomater Sci Polym Ed*. 12(11):1267-79.

Italiano G, Abatangelo G, Jr., Calabro A, Abatangelo G, Sr., Zanoni R, et al. (1997). Reconstructive surgery of the urethra: a pilot study in the rabbit on the use of hyaluronan benzyl ester (Hyaff-11) biodegradable grafts. *Urol Res* 25: 137-42

Joseph DB, Perez LM. (1999). Tunica vaginalis onlay urethroplasty as a salvage repair. *J Urol* 162: 1146-7

Kahveci R, Kahveci Z, Sirmali S, Ozcan M. (1995). Urethral reconstruction with autologous vein graft: an experimental study. *Br J Plast Surg* 48: 500-3

Kajbafzadeh, AM. ; Payabvash, S. ; Salmasi, AH. ; Sadeghi, Z. ; Elmi, A. ; Vejdani, K. ; Tavangar, SM. ; Tajik, P. & Mahjoub, F. (2007). Time-dependent neovasculogenesis and regeneration of different bladder wall components in the bladder acellular matrix graft in rats. *J Surg Res*. 139(2):189-202.

Kelâmi, A. ; Dustmann, HO. ; Lüdtke-Handjery, A. ; Cárcamo, V. & Herlld, G. (1970). Experimental investigations of bladder regeneration using teflon-felt as a bladder wall substitute. *J Urol.* 104(5):693-8.

Khoury AE, Olson ME, McLorie GA, Churchill BM. (1989). Urethral replacement with tunica vaginalis: a pilot study. *J Urol* 142: 628-30; discussion 31

Kropp BP, Ludlow JK, Spicer D, Rippy MK, Badylak SF, et al. (1998). Rabbit urethral regeneration using small intestinal submucosa onlay grafts. *Urology* 52: 138-42

Kropp, BP. ; Cheng, EY. ; Lin, HK. & Zhang, Y. (2004). Reliable and reproducible bladder regeneration using unseeded distal small intestinal submucosa. *J Urol.* 172(4 Pt 2):1710-3.

Kudish, HG. (1957). The use of polyvinyl sponge for experimental cystoplasty. *J Urol.* 78(3):232-5.

Lantz, GC. ; Badylak, SF. ; Hiles, MC. ; Coffey, AC. ; Geddes, LA. ; Kokini, K. ; Sandusky, GE. & Morff, RJ. (1993). Small intestinal submucosa as a vascular graft: a review. *J Invest Surg.* 6(3):297-310.

Leslie B, Barboza LL, Souza PO, Silva PS, Delcelo R, et al. (2009). Dorsal tunica vaginalis graft plus onlay preputial island flap urethroplasty: experimental study in rabbits. *J Pediatr Urol* 5: 93-9

Lexer E. (1911). On free transplantations. *Verh. Deutsch. Ges. Chir.* 40: 386

Li C, Xu Y, Song L, Fu Q, Cui L, Yin S. (2008a). Preliminary experimental study of tissue-engineered urethral reconstruction using oral keratinocytes seeded on BAMG. *Urol Int* 81: 290-5

Li C, Xu YM, Song LJ, Fu Q, Cui L, Yin S. (2008b). Urethral reconstruction using oral keratinocyte seeded bladder acellular matrix grafts. *J Urol* 180: 1538-42

Magnan, M. ; Berthod, F. ; Champigny, MF. ; Soucy, F. & Bolduc, S. (2006). In vitro reconstruction of a tissue-engineered endothelialized bladder from a single porcine biopsy. *J Pediatr Urol.* 2(4):261-70.

Magnan M, Levesque P, Gauvin R, Dube J, Barrieras D, et al. (2009). Tissue engineering of a genitourinary tubular tissue graft resistant to suturing and high internal pressures. *Tissue Eng Part A* 15: 197-202

Markiewicz MR, DeSantis JL, Margarone JE, 3rd, Pogrel MA, Chuang SK. (2008). Morbidity associated with oral mucosa harvest for urological reconstruction: an overview. *J Oral Maxillofac Surg* 66: 739-44

Markiewicz MR, Lukose MA, Margarone JE, 3rd, Barbagli G, Miller KS, Chuang SK. (2007). The oral mucosa graft: a systematic review. *J Urol* 178: 387-94

Mc Indoe A. (1948). Deformities of the male urethra. *Br J Plast Surg* 1: 29-47

McCarthy LS, Smeulders N, Wilcox DT. (2003). Cell biology of bladder development and the role of the extracellular matrix. *Nephron Exp Nephrol* 95: e129-33

McDougal, WS. (1992). Metabolic complications of urinary intestinal diversion. *J Urol.* 147(5):1199-208.

McKinney DE. (1979). Use of full thickness patch graft in urethrovaginal fistula. *J Urol* 122: 416

McManus, M. ; Boland, E. ; Sell, S. ; Bowen, W. ; Koo, H. ; Simpson, D. & Bowlin, G. (2007). Electrospun nanofibre fibrinogen for urinary tract tissue reconstruction. *Biomed Mater.* 2(4):257-62.

Merguerian, PA. ; Reddy, PP. ; Barrieras, DJ. ; Wilson, GJ. ; Woodhouse, K. ; Bagli, DJ. ; McLorie, GA. & Khoury, AE. (2000). Acellular bladder matrix allografts in the regeneration of functional bladders: evaluation of large-segment (> 24 cm) substitution in a porcine model. *BJU Int.* 85(7):894-8.

Mundy, AR. & Nurse, DE. (1992). Calcium balance, growth and skeletal mineralisation in patients with cystoplasties. *Br J Urol.* 69(3):257-9.

Nguyen, DH. & Mitchell, ME. (1991). Gastric bladder reconstruction. Urol Clin North Am. 18(4):649-57.

Nové-Josserand G. 1914. Late results of urethroplasty by tunnelization and dermoepidermal graft in severe forms of hypospadias and epispadias. *J. Urol.* 5: 393

Oberpenning, F. ; Meng, J. ; Yoo, JJ. & Atala, A. (1999). De novo reconstitution of a functional mammalian urinary bladder by tissue engineering. *Nat Biotechnol.* 17(2):149-55.

Olsen L, Bowald S, Busch C, Carlsten J, Eriksson I. (1992). Urethral reconstruction with a new synthetic absorbable device. An experimental study. *Scand J Urol Nephrol* 26: 323-6

Parnigotto PP, Gamba PG, Conconi MT, Midrio P. (2000). Experimental defect in rabbit urethra repaired with acellular aortic matrix. *Urol Res* 28: 46-51

Pattison, M. ; Webster, TJ. ; Leslie, J. ; Kaefer, M. & Haberstroh, KM. (2007). Evaluating the in vitro and in vivo efficacy of nano-structured polymers for bladder tissue replacement applications. *Macromol Biosci.* 7(5):690-700.

Pouliot, R. ; Larouche, D. ; Auger, FA. ; Juhasz, J. ; Xu, W. ; Li, H. ; Germain, L. (2002). Reconstructed human skin produced in vitro and grafted on athymic mice. *Transplantation.* 73(11):1751-7.

Rosario, DJ. ; Reilly, GC. ; Ali Salah, E. ; Glover, M. ; Bullock, AJ. & Macneil, S. (2008). Decellularization and sterilization of porcine urinary bladder matrix for tissue engineering in the lower urinary tract. *Regen Med.* 3(2):145-56.

Schmieden V. (1909). New method of operation for male hypospadias: free transplant of ureter to form urethra. *Arch. Klin. Chir.* 90: 748

Shapiro SR. (1984). Complications of hypospadias repair. *J Urol* 131: 518-22

Sievert KD, Bakircioglu ME, Nunes L, Tu R, Dahiya R, Tanagho EA. (2000). Homologous acellular matrix graft for urethral reconstruction in the rabbit: histological and functional evaluation. *J Urol* 163: 1958-65

Sievert, KD. & Tanagho, EA. (2000). Organ-specific acellular matrix for reconstruction of the urinary tract. *World J Urol.* 18(1):19-25.

Snow BW, Cartwright PC. (1992). Tunica vaginalis urethroplasty. *Urology* 40: 442-5

Sun, TT. (2006). Altered phenotype of cultured urothelial and other stratified epithelial cells: implications for wound healing. *Am J Physiol Renal Physiol.* 291(1):F9-21.

Sutherland, RS. ; Baskin, LS. ; Hayward, SW. & Cunha, GR. (1996). Regeneration of bladder urothelium, smooth muscle, blood vessels and nerves into an acellular tissue matrix. *J Urol.* 156(2 Pt 2):571-7.

Talja M, Kivisaari L, Makinen J, Lehtonen T. (1987). Free tunica vaginalis patch in urethroplasty. An experimental study. *Eur Urol* 13: 259-63

Tariel, E. ; Mongiat Artus, P. ; Meria, P. ; Cortesse, A. ; Desgrandchamps, F. & Teillac, P. (2006). Replacement enterocystoplasty in man (except Hautmann): principles and technical considerations. *Ann Urol. (Paris).* 40(6):368-94.

Theodorescu D, Balcom A, Smith CR, McLorie GA, Churchill BM, Khoury AE. (1998). Urethral replacement with vascularized tunica vaginalis: defining the optimal form of use. *J Urol* 159: 1708-11

Tremblay, PL. ; Hudon, V. ; Berthod, F. ; Germain, L. & Auger FA. (2005). Inosculation of tissue-engineered capillaries with the host's vasculature in a reconstructed skin transplanted on mice. *Am J Transplant.* 5(5):1002-10.

Tsivian A, Sidi AA. (2006). Dorsal graft urethroplasty for female urethral stricture. *J Urol* 176: 611-3; discussion 3

Tuffier T. (1910). À propos de greffes veineuses urétroplastiques. *Bull. et Mem. Soc. de Chir. de Paris* 36: 589

Vemulakonda, VM. ; Lendvay, TS. ; Shnorhavorian, M. ; Joyner, BD. ; Kaplan, H. ; Mitchell, ME. & Grady, RW. (2008). Metastatic adenocarcinoma after augmentation gastrocystoplasty. *J Urol.* 179(3):1094-6.

Veranic, P. ; Romih, R. & Jezernik, K. (2004). What determines differentiation of urothelial umbrella cells? *Eur J Cell Biol.* 83(1):27-34.

Vyas PR, Roth DR, Perlmutter AD. (1987). Experience with free grafts in urethral reconstruction. *J Urol* 137: 471-4

Wallis, MC. ; Yeger, H. ; Cartwright, L. ; Shou, Z. ; Radisic, M. ; Haig, J. ; Suoub, M. ; Antoon, R. & Farhat, WA. (2008). Feasibility study of a novel urinary bladder bioreactor. *Tissue Eng Part A.* 14(3):339-48.

Woodhams, SD. ; Greenwell, TJ. ; Smalley, T. & Mundy, AR. (2001). Factors causing variation in urinary N-nitrosamine levels in enterocystoplasties. *BJU Int.* 88(3):187-91.

Xu YM, Qiao Y, Sa YL, Wu DL, Zhang J, et al. (2004). 1-stage urethral reconstruction using colonic mucosa graft for the treatment of a long complex urethral stricture. *J Urol* 171: 220-3; discussion 3

Xu YM, Qiao Y, Sa YL, Zhang J, Fu Q, Song LJ. (2009). Urethral Reconstruction Using Colonic Mucosa Graft for Complex Strictures. *J Urol*

Xu YM, Qiao Y, Sa YL, Zhang J, Zhang HZ, et al. (2003). One-stage urethral reconstruction using colonic mucosa graft: an experimental and clinical study. *World J Gastroenterol* 9: 381-4

Yoo JJ, Lee I, Atala A. (1998). Cartilage rods as a potential material for penile reconstruction. *J Urol* 160: 1164-8; discussion 78

Yoo JJ, Park HJ, Lee I, Atala A. (1999). Autologous engineered cartilage rods for penile reconstruction. *J Urol* 162: 1119-21

Young F, Benjamin JA. (1948). Repair of hypospadias with free inlay skin graft. *Surg Gynecol Obstet* 86: 439-51

Zhang, Y. ; Frimberger, D. ; Cheng, EY. ; Lin, HK. & Kropp, BP. (2006). Challenges in a larger bladder replacement with cell-seeded and unseeded small intestinal submucosa grafts in a subtotal cystectomy model. *BJU Int.* 98(5):1100-5.

Zhang, Y. ; Kropp, BP. ; Moore, P. ; Cowan, R. ; Furness, PD. ; Kolligian, ME. ; Frey, P. & Cheng, EY. (2000). Coculture of bladder urothelial and smooth muscle cells on small intestinal submucosa: potential applications for tissue engineering technology. *J Urol.* 164(3 Pt 2):928-34.

25

Skin Substitutes

Barbara Zavan [1,*]; Vincenzo Vindigni [2],
Roberta Cortivo [1] and Giovanni Abatangelo[1]
1 *Dept of Histology, Microbiology and Biomedical Technologies, University of Padova Viale G. Colombo 3, 35131 Padova, Italy*
2 *Unit of Plastic and Reconstructive Surgery, University of Padova, Via Giustiniani 2, 35100 Padova, Italy*

On tissue engineering field the research about skin substitutes represents historically the most promising approach to heal acute and chronic skin wounds, reducing requirements for donor skin autografts. Aim of this chapter is to made the state of art of the most important skin substitutes and to resume the principal tissue engineering techniques for its in vitro reconstruction. In the first part we will pay the attention to the description of the evolutive course of skin substitutes, stressing the main steps happened thanks to the important inputs coming from biotechnologies progresses and clinical practice. Since then the research in bioengineering skin, moved by clinical pressing of reconstructive and burn surgery, made progresses trying to develop cutaneous substitutes very similar to native skin by a morpho-functional point of view. So it has been possible, starting from a 2X2 cm skin human biopsy, to realize completely autologous cutaneous substitutes not only composed of two many structures of skin, dermis and epidermis, but containing also other important components: microvascular network, micro nervous network, skin immonocompetent system and melanocyte system. The final goal is to develop effective and easy handling skin substitutes that could be reproduce human skin anatomy and physiology, introducing new advantages linked to successful grafting and then to satisfactory clinical results by functional and aesthetic point of view.

1. Introduction

Tissue Engineering (TE) is "an interdisciplinary field that applies the principles of engineering and life sciences toward the development of biological substitutes that restore, maintain, or improve tissue function or a whole organ". Tissue engineering has also been defined as "understanding the principles of tissue growth, and applying this to produce functional replacement tissue for clinical use."

Powerful recent developments in the multidisciplinary field of tissue engineering have yielded a novel set of tissue replacement parts and implementation strategies. Scientific advances in biomaterials, stem cells, growth and differentiation factors, and biomimetic environments have created unique opportunities to fabricate tissues in the laboratory from combinations of engineered extracellular matrices ("scaffolds"), cells, and biologically active

molecules. Among the major challenges now facing tissue engineering there is the need for more complex functionality with biomechanical stability in laboratory-grown tissues destined for transplantation.
Thus, the scientific and medical communities are working together to develop engineered tissues and regenerative approaches utilizing various combinations of stem and progenitor cells, biomaterials, growth factors, and gene therapy. However, after three decades, the potential to provide tissues and organs to millions of patients suffering from trauma, congenital defects, and chronic diseases has yet to be fully accomplished.
On the whole, tissue engineering appears to be the new frontier of medicine for its impact on regenerative and reconstructive procedures in humans. In fact, its ultimate goal is to develop powerful new therapies, namely "biological substitutes", for structural and functional disorders that have been proven to be difficult or impossible to tackle successfully with the current approaches of interventional medicine. In addition, providing "cell-to-tissue replacement parts" of the human body it can ultimately afford the irremediable shortage of transplantable organs.
Despite the rapid advancement of state of the art medically driven technologies, numerous challenges still exist for translating technology to the clinic including long and often undefined regulatory approval pathways, high translational costs, potential safety appropriate risk/benefit and cost/benefit ratios, and appropriate matching of technology and application. The latter may be one of the most difficult tasks faced by the biomedical engineer. Although development of general platform technologies can lead to major breakthroughs across multiple disciplines, conventional wisdom suggests rapid clinical translation and requires a systematic approach that involves first identifying and understanding a problem followed by engineering a focused solution. As technology rapidly advances, the bioengineer have more tools at their disposal than ever before, yet only if they are able to select the right tools and technologies, for society benefit.
Tissue engineered substitutes should have some essential characteristics which include: being easy to handle and apply to the injured site; provide vital barrier function with appropriate water flux; be readily adherent; have appropriate physical and mechanical properties; undergo controlled degradation; be sterile, non-toxic, non-antigenic; and evoke none or minimal inflammatory reactivity. Additionally, they should incorporate into the host with minimal scarring and pain and facilitate angiogenesis, while still being cost-effective.
The ability to create thick tissues is a major tissue engineering challenge, requiring the development of a suitable vascular supply. Current trends are seeing the utilization of cells seeded into hybrid matrix/scaffold systems to create in vitro vascular analogues. Approaches that aim to create vasculature in vitro include the use of biological extracellular matrices, co-culture of cells, incorporation of growth factors, culture in dynamic bioreactor environments, and combinations of these. Of particular interest are those approaches that aim to create bioengineered tissues in vitro that can be readily connected to the host's vasculature following implantation in order to maintain cell viability. A major consideration when developing a tissue engineering strategy for promoting repair and regeneration is to identify suitable sources of cells and mechanisms by which they can function and interact properly. These cells would also have to be abundant enough to be able to carry out the regeneration process completely.

Another obstacle to the large scale clinical application of cell therapy using conventional procedure is the extremely high cost of the treatment. The economic limit is due to the high cost of the actual manual operations for ex vivo cell expansion and differentiation in respect of the international regulations (Good laboratory and manufacture practices, GLP, GMP or Food and Drug Administration, FDA) that control the trading of clinical products.
Therefore there is a diffuse and pressing necessity to develop new methods and culture techniques that are able to control the cell micro-environment during both cell expansion and differentiation leading to repeatable clinical grade batches, operating in safety conditions and at low costs. This last point is often ignored but it is fundamental to make the clinical progresses available to everybody and not only to few people.

2. Anatomy of the skin

The skin is the largest organ of the body in vertebrates and is composed of the epidermis and dermis with a complex nerve and blood supply (Fig. 1). A third layer the hypodermis, is composed mainly of fat and a layer of loose connective tissue. These three layers play an important role in protecting the body from any mechanical damage such as wounding.

I. The epidermis is thin and highly cellular, composed mainly of keratinocytes and in the lower epidermal layers, melanocytes, for pigmentation. It does, however, have sufficient thickness to provide vital barrier function. Mammalian epidermis and its appendages (hair, nail, sweat and sebaceous glands) maintain homeostasis by constant recycling of the basal cell layer.

II. The dermis situated directly below the epidermis, constitutes the bulk of the skin and is composed of collagen with some elastin and glycosaminoglycans (GAG's). Hemidesmosomal structure are the main components of the basement membrane that mechanically stabilizes the interaction between epidermis and dermis. Basement membrane is an extremely complex and dynamic structure, which is responsible of the tight junction of the upper epidermis and underlying connective tissue. The dermis is a thicker tissue made essentially of a loose array of fibroblasts (mesenchymal cells) and vasculature (endothelial cells) forming an approximately 2-5 mm thick connective tissue sustaining the overlaid epidermis which is separated from the dermis by a 20 nm thick multilayered membrane: basement membrane. The dermal matrix, which provides considerable strength to skin by virtue of the arrangement of collagen fibers, has specialized components and structures. Collagenous mesh work is interwoven with varying contents of elastin fibers, proteoglycons (GAG being predominantly hyaluronic acid and dermatan sulfate with some chondroitin-6-sulphate and heparin sulphate), fibronectin and other components. (Metcalfe and Ferguson 2007)

III. The hypodermis, the final layer, contains adipose tissue that is well vascularised and contributes to both the thermoregulatory and mechanical properties of the skin.

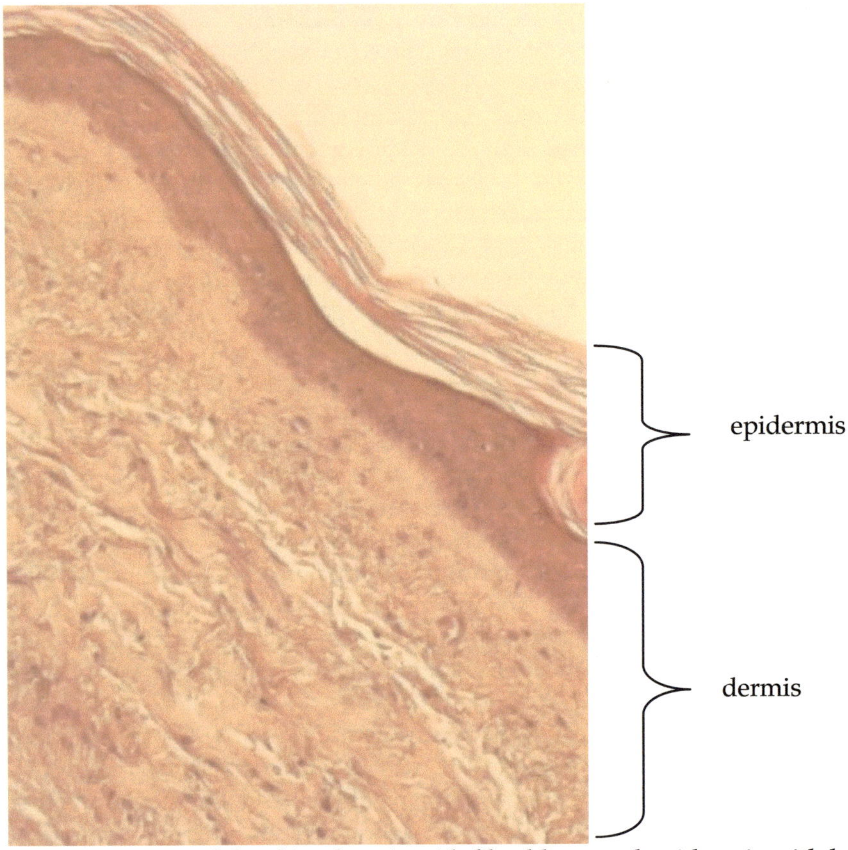

Fig. 1. Anatomy of the skin: dermis with fibroblasts and epidermis with keratinocytes

Skin is also composed of a series of appendages known as anexa, such as hair follicles, sweat glands and sebaceous glands generally embedded in the dermis. Because the skin serves as a protective barrier against the outside world, any break in it must be rapidly and efficiently amended. (Jennings RW and Hung TK 1982). The physiologic mechanism that coordinates the ordered repair of cutaneous tissues is generally referred to as wound healing and it consists in a well orchestrated series of temporary overlapping actions involving different cell and biochemical types. The process of wound healing involves interaction of a variety of different cell types and matrix components (R.A.F. Clark and P.M. Henson 1988). Different types of cells respond to environmental signals in a specific manner in order to carry out their genetic program of proliferation, differentiation and function. Cells synthesize different proteins necessary for their proliferation and migration, all of which are controlled in a phased manner (Zavan and Abatangelo 2004). Different phases are inflammation, granulation tissue formation, re-epithelialization, matrix production and remodelling Thus, each layer of skin plays its own pivotal role to play in the normal function of the skin, which otherwise can lead to several skin disorders. Healing of injured tissues involves both regeneration of the epidermis and repair of the dermis resulting in a scar tissue (M Calvin 1998).

3. Wound Healing

Wounds can arise either acutely from traumas such as burns, derive from surgical treatments such as skin graft harvesting or melanoma excision or also in a chronic fashion related to patho-physiological phenomena like insufficient vascularization, as in pressure wounds or diabetic foot ulcers. The underlying principles behind the use of skin substitutes to improve wound healing are valid for extensive wounds that cannot heal spontaneously because they are deep and affect a high percentage of the total body surface or for smaller wounds that cannot heal because of underlying factors (Auger and Lacroix 2009).

The normal acute wound healing process takes place in 4 steps (FIG.2):

1) hemostasis,
2) inflammation,
3) proliferation,
4) remodeling.

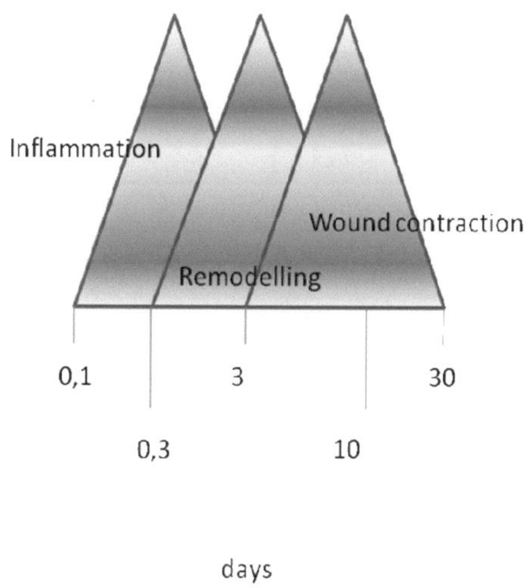

Fig. 2. Time table of wound healing process

Hemostasis

Unless there is severe arterial hemorrhage, hemostasis spontaneously begins immediately after the injury. It involves platelet aggregation, fibrin clot formation and activation of the coagulation pathways. Since the first aspects of wound treatment involve cleaning and debriding the wound, skin substitutes are rarely, if ever, used for hemostasis. However, it should be noted that chitosan dressings have been demonstrated to efficiently improve hemostasis compared to standard dressings, even with significant arterial hemorrhage [Garner and Brown R 2002]

Inflammation

Inflammation takes place 1–4 days after wounding. It involves the migration of neutrophils into the wound site, shortly followed by macrophages and later by lymphocytes. The fibrin clot matrix provides a three-dimensional initial scaffold through which immune cells will initially migrate to the wound and secrete a host of signaling molecules that will act as chemoattractants and growth factors. Platelets will produce factors such as platelet-derived growth factor (PDGF), epidermal growth factor (EGF), transforming growth factor-α (TGF-α), transforming growth factor- β (TGF- β) and vascular endothelial growth factor (VEGF). [Werner and Grose 2003; Spiekstra et al. 2007].

Proliferation

Factors secreted during inflammation will trigger and drive the proliferation stage, which normally takes place between days 4 and 21 following wounding. During that stage, fibroblasts are stimulated by fibroblast growth factor (FGF) and PDGF to invade the wound site, and will produce xtracellular matrix (ECM) components, such as collagen, elastin and glycosaminoglycans, to generate the granulation tissue. Fibroblasts secrete also FGF that can, along with the VEGF secreted by platelets and neutrophils, act as an angiogenic factor to stimulate endothelial cell proliferation and migration and thus promote vascularization at the healing site. While fibroblasts and endothelial cells mostly invade the wound site from its bed, keratinocytes at the margin of the wound undergo a transient burst of proliferation that will sustain epithelialization over the wound (FIG.3)[Laplante AF et al. 2001]

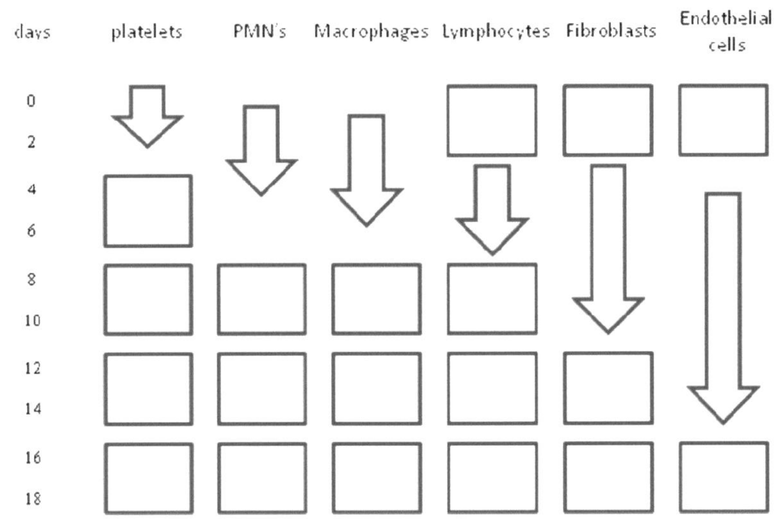

Fig. 3. Cells involved on wound healing

Remodeling

The last step of normal acute wound healing is remodeling, during which the ECM components will be modified by the balanced mechanisms of proteolysis and new matrix secretion, and during which the wound becomes gradually less vascularized. The initial granulation tissue is weak but will gain in strength over time because of remodeling effects such as the gradual replacement of immature type III collagen by mature type I collagen. The complete maturation of a wound may take months or even years to complete, depending on its initial extent and various factors that can affect the healing process [[Johnstone and Farley 2005]. Excessive contraction in large wounds can lead to contractures, while excess formation of scar tissue will form hypertrophic scars or even, if the scar tissue extends into the healthy tissue around the wound, a keloid scar. Wound contraction starts taking place around 6 days after injury and can last long after the epithelialization phase. The myofibroblast is considered to be the major cell responsible for contraction and is easily distinguished from normal fibroblasts by the presence of α-smooth muscle actin (α-SMA). α-SMA protein expression is undetectable until day 6 after wounding, after subsequently increases gradually for 15 days in the healing process and fades after 4 weeks [Brunicardi et al 2005].

3.1 Clinical application

Today, surgical grafting of split-thickness autologous skin is the mainstay for treatment of fullthickness burns and is often used to treat chronic wounds of all kinds. The depth of tissue destruction and the need to replace different functions will determine the type of skin substitute used:

a) Rapid replacement of the epidermal layer is important to restore, as fast as possible, control over fluid loss, body temperature and protection against bacteria. Epidermal or bilayer substitutes that do not provide autologous keratinocytes or immunocompatible living tissue, such as cadaveric skin, xenografts and inert membranes like silicone, can only provide temporary coverage. They are indicated if availability and expected healing speed make them the best or the only option at the beginning of the treatment. However, live keratinocytes have to be present to achieve permanent wound coverage and restore the epidermal barrier. With epidermal substitutes, the presence of autologous living cells is thus necessary for proper healing of extensive areas where spontaneous epithelialization is impossible or would take too long to complete. Autologous living cells can also be desirable to improve the speed and/or quality of the healing in wounds that otherwise would be small enough to epithelialize on their own [Macri et Clarke 2009].

b) In cases where the dermal layer is still at least partially present even after debridement, epidermal substitutes that provide living autologous keratinocytes can be applied directly on the wound bed and eventually achieve permanent wound closure. However, in cases where the dermal layer has been destroyed, the application of even the best epidermal substitute is not enough to ensure optimal healing.The dermal layer is essential for better elasticity and mechanical resistance, the absence of epithelial-mesenchymal communication increases the frequency of developing fibrotic conditions [Sawicki et al 2005] and evidence indicates its presence allows highly complex ex vivo function of epidermal cells. Thus dermal reconstruction should be an essential part of the treatment in cases where none of the original dermis is left on the wound bed prepared for grafting. The presence of living fibroblasts is not a necessity in dermal substitutes. Such acellular products can be sufficient to provide a scaffold that will be repopulated, revascularized and remodeled with the patient's own fibroblasts and endothelial cells. However, the presence of fibroblasts in dermal substitutes helps wound healing by providing factors of the FGF family, by the secretion of the ECM responsible for its mechanical properties and by the structuring effect they have on the formation of capillaries during angiogenesis [Mitrani et al 2005]. Dermal substitutes provide a crucial component to help repair full-thickness damage, but they do not create a barrier effect and need to be covered with an epidermal substitute to that end.

Basically, there are three methods of skin repair and replacement:
First: autografting.

This is a widely used method of treating organ loss. For instance, it is an effective procedure for the short-term treatment of full thickness skin loss, especially meshed skin in severely burned patients. Although autologous skin grafts remain the first option in the current standard of care for burn patients, there are clinical circumstances where using the donor's own skin might not be appropriate for reasons related to the quantity needed, as in the case of massive tissue injury such as extensive burns or because of healing deficiencies that lead to chronic wounds. The clinicians then need to turn towards alternative materials that will offer appropriate properties for the type and extent of the wound to be treated [Machens et al. 2000]

Second: cadaver skin

It has offered an alternate pathway to skin replacement for many years. Peter Medawar, in his early pioneer work on transplantation immunobiology during the 1930s and 1940s, used skin transplants in immunoincompatible animal models and thus outlined very early in the history of transplantation medicine that rejection and progressive loss of tissue is a major drawback for cadaver skin replacements. In recent decades the consequences of a promiscuous multiculture have taught us that additional problems are inherited with this type of surgery: hepatitis and the Aids virus are risks which cannot be excluded and make cadaver skin transplantation more and more obsolete. The potential of cadaveric allograft skin to transmit disease has been recognized as an increasing problem since the 1980s. Alternatives to cadaveric skin are represented by native biological substitutes, with or without living cells resumed in table 1.[Auger and Lacroix 2009]

Product	Commercial name
Fresh cadaveric skin	
Cryopreserved cadaveric skin	AllodermTM
Glycerolized cadaveric skin	
Fresh xenografts Usually fresh porcine disinfected skin	
Cryopreserved amniotic membrane	AmniograftTM
Porcine small intestine submucosa	OasisTM, FortaFlexTM
Porcine skin	PermacolTM

Table 1. commercial natural skin

Third: in vitro reconstructed skin

This procedure relies on the hypothesis that tissue can, in principle, be cultured in vitro. During the several preceding decades, there have been breathtaking achievements in medical science in the field of in vitro cell cultivation, tissue engineering and new biopharmaceutical as well as recombinant genetic therapies. The skin has been the first tissue-engineered organ from the lab bench to the patient [Rheinwald and Green, 1975]. Rheinwald and Green took 3T3 cells derived from mouse embryos (the 3T3 cell line established by Rheinwald and Green in particular is known as 3T3-J2), irradiated them to eliminate their capacity to divide, and co-cultured them with epidermal keratinocytes, sufficiently uppressing the differentiation of the latter to obtain cells with a high proliferative capacity (Hiroco, 2007). The greatest merit of cultured epidermis is that it enables the grafting of epidermal keratinocytes that have preserved sufficient proliferative capacity.

(Metcalfe) Engineering design specification in skin has relied upon the creation of both artificial dermal and epidermal components, which when combined produce a replacement of the skin, that can be grafted in place ([Supp and Boyce 2005]. Materials used as artificial ECM to date include those derived from naturally occurring materials and those manufactured synthetically [MacNeil 2007]. Examples of natural materials include polypeptides, GAG's, fibronectin, collagen, hydroxyapatites, hyaluronan, chitosan and alginates. These materials are advantageous in that they have low toxicity and a low chronic inflammatory response. Examples of synthetic materials include polyglycolide (PGA), polylactide (PLA) and polylactide-coglycolide (PLG), which have been used for sutures and meshes [[Metcalfe and Ferguson 2007]. Other examples include polytetrafluoroethylene (PTFE), polycaprolactone (PCL) and polyethylene terephthalate (PET). Matrices used routinely in therapeutic applications are made from polymers that are often resorbed by or degraded in the body. In the early days of tissue engineering, polymers were adapted from other surgical uses and over time have been shown to have deficiencies in terms of mechanical and degradation properties [Griffith 2002]. A greater challenge arises when functionalising these polymers into scaffolds that have defined shapes and a complex, porous internal architecture that can direct tissue growth [Griffith and Schwartz 2006].

Current skin constructs that are commercially available are generally based on the following techniques:

a) Epidermis: with only keratinocytes

The minimum requirement for a viable skin construct is to reestablish a barrier function to avoid infection and water loss. The horny layer of the epidermis (the product of terminal keratinocyte differentiation) plays this role. As mentioned above, after the breakthroughs of Rheinwald and Green in the mid '70s, the technology of cultured epithelial autograft (CEA) is now available to a large number of hospitals worldwide. In addition, cell culture parameters, harvest procedures and usage indications have been strictly defined and CEA production is also being performed on an industrial basis.

Epidermal substitutes

Commercial name	Cell source	biomaterial
EpicelTM	autologous keratinocytes from skin	petrolatum gauze backing
EpidexTM	autologous keratinocytes from hair follicles outer root sheath	silicone membrane
LaserskinTM,	autologous keratinocytes	esterified laser-perforated hyaluronic acid matrix
BioSeed-STM	Autologous oral mucosal cells	fibrin matrix
MyskinTM	autologous keratinocytes	specially treated silicone sheet

CellSprayTM	autologous keratinocytes	suspension for spray
CryoskinTM	Cryopreserved (viable cells) confluent allogeneic keratinocytes	gel-like proprietary chemical surface

The greatest merit of cultured epidermis is that it enables the grafting of epidermal keratinocytes that have preserved sufficient proliferative capacity. Whatever the circumstances, patients in whom epidermal damage has occurred all ultimately require an epidermis composed of epidermal keratinocytes. The only hope for patients with widespread epidermal damage, such as whole-body burns, is the localized capacity of surviving epidermal keratinocytes to divide and proliferate, and their treatment requires an extremely long period of time. Although autologous grafting of skin from other areas of the body is the surgical norm, the amount of skin that can be harvested is limited, and furthermore, new epidermal damage is created at the donor site. On the other hand, cultured epidermis is very thin, and its use is problematic in cases of deep skin damage such as for patients in whom the dermis has been completely destroyed. For this reason, other methods must be used in advance to create dermal tissue that can offer a favorable base for grafting. [Hichiro 2007]

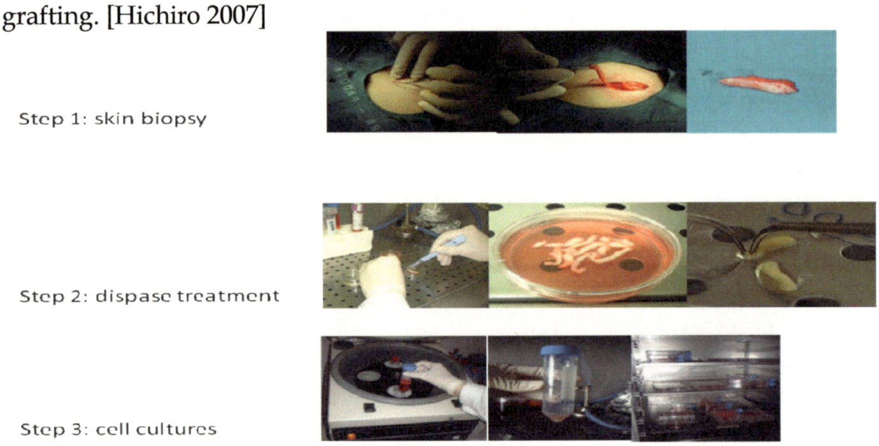

Fig. 4. Principal steps for in vitro epidermal skin reconstruction

Technique:
Keratinocytes are isolated from full-thickness skin biopsies (3×1 cm) (step1) obtained from healthy patients with ages ranging from 35 to 45 years. Keratinocytes are isolated and cultured according to a modified version of the Rheinwald and Green protocol. Briefly, dermal specimens are immersed in DMEM supplemented with 20% FBS and minced (step2).The small fragments obtained, are digested in 0.6 U/ml dispase II (Boehringer Mannheim) at 37° C for 1 h, at this time a manual separation between dermis and epidermis is performed (step2). The epidermis will be treated with trypsin ± EDTA (0.05%, 0.02 mM) solution (Seromed. Biochrom KG) for 10 min and the cells collected after a centrifugation and seeded in flask treated with a special keratinocytes attachment factors or in a feeder layer of 3T3 (step3).

For 3D coltures, keratinocytes are plated in secondary cultures at a density of 2 × 104 cells/cm^2 onto the biomaterials (e.g. hyaluronic acid: LaserSkin membrane) in the presence of a feeder layer of nonproliferating 3T3 mouse fibroblasts (step4).

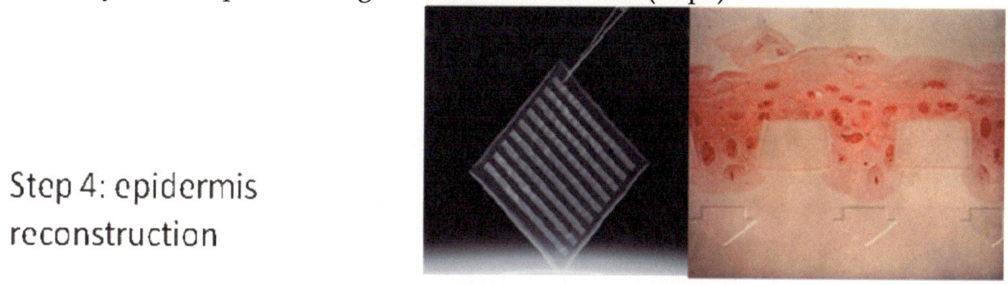

b) Dermis: presences of fibroblasts.

Fibroblasts are mesenchymal cells that can be readily cultured in the laboratory and play a significant role in epithelial–mesenchymal interactions, secreting various growth factors and cytokines that have a direct effect on epidermal proliferation, differentiation and formation of extracellular matrix. They have been incorporated into various tissue-engineered products such as Dermagraft_ (Advanced BioHealing, La Jolla, CA, U.S.A.) and Apligraf_ (Novartis, Basel, Switzerland) and used for a variety of clinical applications, including the treatment of burns, chronic venous ulcers and several other clinical applications in dermatology and plastic surgery. Fibroblasts are a heterogeneous population of cells found in numerous tissues and are of mesenchymal origin. Fibroblasts from different anatomical sites all have similar morphology but DNA-microarray studies have demonstrated that fibroblasts in different anatomical sites have their own gene-expression profile and characteristic phenotypes, synthesizing extracellular matrix (ECM) proteins and cytokines in a site-specific manner. Dermal fibroblasts have numerous functions, not only in synthesizing and depositing ECM components, but also proliferation and migration in response to chemotactic, mitogenic and modulatory cytokines, and also autocrine and paracrine interactions [Wong et al 2007].

Fibroblasts used in tissue engineering may be allogenic or autologous. In contrast to allogenic cells, autologous fibroblasts carry no risks of rejection or risk of cross-infection.

Cultured dermis using fibroblasts falls within this field, and has been produced with a number of materials. As the basic element, collagen or other biodegradable material has been used as a scaffold to create a tissue consisting of dermis-derived fibroblasts. Treatment using cultured dermis makes use of the wound-healing effects brought about by the physiologically active substances produced by fiobroblasts. Normally, the creation of granulation tissue mediated by localized inflammation is observed as a biological reaction at the site of skin loss. The reconstruction of connective tissue increases with time, mainly by the action of fibroblasts from the surrounding area, as does capillary regeneration. Cultured dermis, by enabling cultured fibroblasts to produce physiologically active substances that are important in the reconstruction of connective tissue, promotes favorable wound treatment. The intended effect of cultured dermis is the promotion of wound treatment by means of the physiologically active substances produced by fibroblasts contained within the culture. As results this treatment is characterized by the lack of any significant difference in outcome whether the fibroblasts used are autologous or allogeneic cells. In emergency use for burns, for example, allogeneic cultured dermis made in advance from allogeneic cells is of high clinical utility. It will be necessary to solve the variety of concerns related to the use of living cells, as well as to demonstrate this method's clear therapeutic advantages over the use of pharmaceuticals alone through numerous examples of clinical application. [Hichiro 2007]

Dermal substitutes

Commercial name	Cell source	biomaterial
DermagraftTM	Cryopreserved (viable cells) allogeneic fibroblast	derived dermal matrix
Hyalograft 3DTM	autologous fibroblasts	Esterified hyaluronic acid matrix
AllodermTM		Acellular (freeze-dried) allogeneic dermis
OasisTM		porcine small intestine submucosa
EZ-DermTM		Aldehyde-crosslinked porcine dermal collagen
RepliformTM		Acellular human dermal allograft
CymetraTM		Micronized particulate acellular cadaveric dermal matrix
FortaFlexTM		porcine small intestine submucosa
BiobraneTM		Porcine collagen chemically bound to silicone/nylon membrane
MatridermTM		Acellular scaffold made with bovine collagen types I, III, V and elastin

Technique:
Human dermal fibroblasts are prepared according to a modified version of the Rheinwald & Green protocol. After epithelial sheet dispase removal, dermis was cut into small pieces (2–3 mm^2) and fibroblasts were isolated by sequential trypsin (0,05%) and collagenase type I digestion. Cells isolated are then cultured with DMEM medium supplemented with 10% Foetal Bovine Serum (FBS). At confluence fibroblasts are harvested and seeded at a density of 3 × 105/cm2 on squares of biomaterials (i.e. hyaluronan non woven meshes) (1.5 × 1.5 cm) in the above-mentioned medium containing sodium ascorbate (50 mg/mL) (step 5). The nonwoven squares could be fixed on culture plates either by means of stainless steel rings or through a fibrin clot.

Step 5: dermis reconstruction

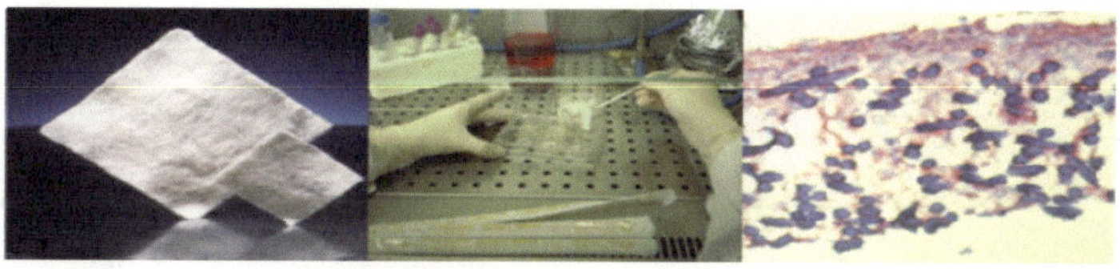

c) Co-coltures of fibroblasts and keratinocytes (Bilayer substitutes).

After these first attempts, given that fibroblasts are the main constituent of the dermis and are normally used to grow keratinocytes, the concept of a co-culture was introduced in order to produce in vitro a composite skin replacement composed of an epithelial layer overlaid onto a dermal substitute. The presence of fibroblasts within the grafted region is thus a major advantage. Several lines of research are being pursued to define the best conditions of use. These include studies of allogenic fibroblast persistence in humans. Such fibroblasts persist in animals, but in humans, even if allogenic fibroblasts give similar results to autologous fibroblasts, their ultimate fate is unknown. Definition of fibroblast populations within the dermis has been undertaken. There is clear evidence that, depending on the anatomical region or the depth within the dermis, fibroblasts can be distinguished on the basis of functional criteria. It would be useful to have markers to select those cells that would be most efficient within dermal substitutes. At present despite active research in this field there is still no commercially available fibroblast marker. Lastly, the possibility of using mesenchymal stem cells collected from bone marrow or peripheral blood, which seem to participate in wound healing, is being evaluated. In addition to skin substitutes for permanent skin replacement, other applications can help to rapidly cover patients' wounds (despite eventual rejection of allogenic cells) or to accelerate healing. It is already known that allogenic epidermal cell cultures promote external ulcer healing, [Zavan and Abatangelo 2004] probably by secreting factors (matrix metalloproteinases, growth factors, etc.) that promote wound cleansing and stimulate the activity of cells present at the wound site. At the end, the importance of the scaffold for a dermal-like tissue must be stressed too. In fact, the design of a dermal-like tissue critically depends on the use of an ideal scaffold which must allow the dermis to develop into a three-dimensional architecture. One of the most important features of the scaffold is its porosity, which has to fit the neo-vessel ingrowth from the host tissue (angiogenesis) during the wound healing process. In addition, the biomaterial should be fully biocompatible and totally degradable being substituted by a full functional ECM in the long term. Finally, the material should be resistant to mechanical forces and easy to apply on the wound area.

The following guidelines emerged from the clinical experience:

- The homogeneity of the cultured epidermal sheet is crucial for a good cosmetic result.
- One-step grafting of a cocultured dermis and epidermis is not yet satisfactory, and would be too lengthy for emergency use (e.g., burns).
- Preliminary grafting of a collagen matrix is hemostatic, reduces pain, and prepares the graft bed.
- Most importantly, the presence of living fibroblasts speeds up the reappearance of a functional neodermis (less than 1 year instead of 3 to 5years when the graft bed is prepared only with cryopreserved cadaver skin).

Bilayer substitutes

Commercial name	Cell source	biomaterial
OrCelTM	Allogeneic keratinocytes seeded over dermal scaffold containing allogeneic fibroblasts	bovine collagen sponge
ApligrafTM	Allogeneic keratinocytes seeded over dermal scaffold containing allogeneic fibroblasts	bovine collagen sponge
TissueTechTM autograft system	Combination of Hyalograft 3DTM and LaserskinTM	
PermaDermTM	Autologous keratinocytes seeded onto dermal substitute made with autologous fibroblasts in bovine collagen matrix	bovine collagen matrix
IntegraTM		Temporary silicone epidermal substitute over dermal scaffold made of collagen and chondroitin-6 sulfate

The success of a skin equivalent graft-take requires the viability of fibroblasts contained within and the keratinocytes adhering to the surface. When either keratinocytes or fibroblasts are directly introduced into a freshly made excisional wound, there is rapid necrosis of these transferred cells. When components are put together in vitro and transferred to an excisional wound site in vivo, the skin equivalent graft has optimal skinlike qualities.

Technique:

At confluence (10-12 days), pieces of membranes (1.5 × 1.5 cm) previously seed with keratinocytes (step 4) are laid down and fixed to the nonwoven meshes on which dermal fibroblasts had been grown for 15 days (step 5). The fibroblast-keratinocyte composite cultures were laid down carefully on sterile stainless steel grids and then cultivated for 15 days at the air-liquid interface (step 6).

Step 6: bilayer reconstruction

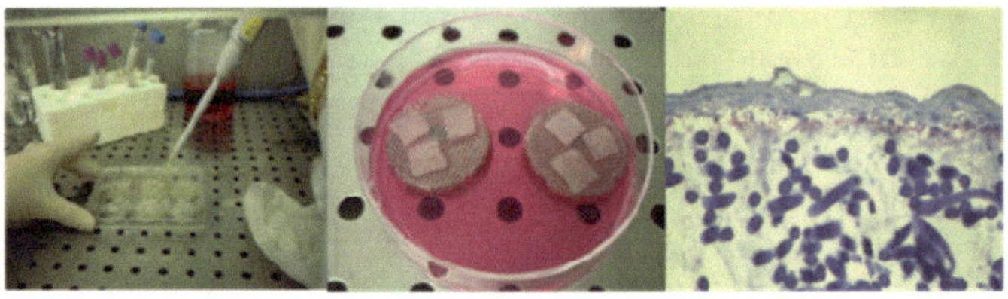

4. Conclusion

The many studies conducted so far reveal that Tissue Engineering of the skin is only at the beginning of its use in human applications. Burns patients were the first targets for such tissue substitutes, then chronic diseases, such as venous ulcers, have followed. The more experience is gained from the surgeon, the more feedback for the basic scientist to improve the product and to broaden clinical indications. Nowadays, progress in cell culture and biomedical material technologies have added two important spare parts: epidermis and dermis to the surgeon's toolbox which can be reconstituted in the laboratory from small biopsies of the same recipient. Other parts will follow in a few years, with the final aim to generate a full transplantable replica of the skin with adnexa and vasculature.

5. References

Anthony D. Metcalfe, Mark W.J. Ferguson. (2007) Bioengineering skin using mechanisms of regeneration and repair *Biomaterials* 28 5100–5113

Dunn DL, Hunter JG, Matthews JB, Pollock RE, Schwartz SI: Schwartz's (2005) Surgical grafting of skin substitutes. *Principles of Surgery,* ed 8. New York, McGraw-Hill Companies Inc.,

F.A. Auger D. Lacroix L. Germain (2009). Skin Substitutes and Wound Healing. *Skin Pharmacol Physiol*;22:94–102

Garner JP, Brown R (2002) Recent advances in topical agents for prehospital haemostasis. *Trauma*; 4: 203–209.

Griffith LG, Schwartz MA (2006). Capturing complex 3D tissue physiology in vitro. *Nat Rev Mol Cell Biol*;7(3):211–24.

Griffith LG.(2002) Emerging design principles in biomaterials and scaffolds for tissue engineering. *Ann NY Acad Sci*;961:83–95.

H.-G. Machensa,b A.C. Bergerb P. Mailaendera Bioartificial (2000). Skin. *Cells Tissues Organs* 167:88–94

Johnstone CC, Farley A: (2005). The physiological basics of wound healing. *Nurs Stand*; 19: 59–65; quiz 66. 15

Ken-Hichiro Hata, D.D.S. (2007) Current issues regarding skin substitutes using living cells as industrial materials *Artif Organs* 10:129–132

L. Macri a R.A.F. Clark (2009) Tissue Engineering for Cutaneous Wounds: Selecting the Proper Time and Space for Growth Factors, Cells and the Extracellular Matrix Skin *Pharmacol Physiol*;22:83–93

Laplante AF, Germain L, Auger FA, Moulin V (2001). Mechanisms of wound reepithelialization: hints from a tissue-engineered reconstructed skin to long-standing questions. *FASEB J.* Nov;15(13):2377-89

M Calvin (1998) Cutaneous wound repair Wounds. *A Compendium of Clinical Research and Practice* 12 p. 32.

MacNeil S.(2007) Progress and opportunities for tissue-engineered skin. *Nature* 445(7130):870–80.

Metcalfe AD, Ferguson MW. (2007) Tissue engineering of replacement skin: the crossroads of biomaterials, wound healing, embryonic development, stem cells and regeneration. *J R Soc Interface* 4(14):413–37

Mitrani E, Nadel G, Hasson E, Harari E, Shimoni Y (2005). Epithelial-mesenchymal interactions allow for epidermal cells to display an in vivo-like phenotype in vitro. *Differentiation* 73: 79–87.

R.A.F. Clark and P.M. (1988) Henson The molecular and cellular biology of wound repair., *Plenum Press*, New York.

Rheinwald and Green, (1975). Formation of a keratinizing epithelium in culture by a cloned cell line derived from a teratoma. *Cell.* Nov;6(3):317-30.

Sawicki G, Marcoux Y, Sarkhosh K, Tredget EE, Ghahary A (2005). Interaction of keratinocytes and fibroblasts modulates the expression of matrix metalloproteinases-2 and -9 and their inhibitors. *Mol Cell Biochem* 269: 209– 216.

Spiekstra SW, Breetveld M, Rustemeyer T, Scheper RJ, Gibbs S (2007). Wound-healing factors secreted by epidermal keratinocytes and dermal fibroblasts in skin substitutes. *Wound Repair Regen*; 15: 708–717.

Supp DM, Boyce ST. (2005) Engineered skin substitutes: practices and potentials. *Clin Dermatol* 23:403–12.

T. Wong, J.A. McGrath and H. Navsaria (2007) The role of fibroblasts in tissue engineering and regeneration. *British Journal of Dermatology* 156, pp1149–1155

Werner S, Grose R. (2003) Regulation of wound healing by growth factors and cytokines. *Physiol Rev*; 83: 835–870.

Zavan B, Abatangelo G (2004) . Bioartificial skin. *"Application of cell immobilization biotechnology"* (ed. Willaert) 55-68